The Rule Book

Measuring for the Trades

Jane Rees

Mark Rees

ISBN 13: 978-1-931626-26-2

Published by
THE ASTRAGAL PRESS
An Imprint of Finney Company
8075 215th Street West
Lakeville, Minnesota 55044
www.finneyco.com
www.astragalpress.com

Printed in the United States of America
1 3 5 7 9 10 8 6 4 2

Contents

Acknowledgements

Mark and I started work on this book ten years ago. Sadly, Mark did not live to see its completion but it is still very much a book authored by us both.

Inevitably, much of the text has been written since his death in 2001, but the inspiration and a considerable amount of the research and information is his. I have tried to follow our common vision and think he would have been pleased with the end result.

But this book could not have been written without the many people who have given generously of their time and knowledge, allowed me to study and photograph their collections, use their research and assisted and encouraged me in many other ways. They include Simon Barley, Robert Bousfield, David Bryden, John Collier, David Cox, Andrew Craner-Buckley, Clifford D. Fales, Richard Filmer, Roy Garnell, Ted Ingraham, the late Richard Knight, Don McConnell, J. Kenneth Major, David Millett, Louise Muse, Jim Packham, Jim Parker, Des Pawson, Arthur Peters, George Pocklington, Mike Rawlins, David M. Riches, Robert J. Sauer, Anthony Seo, Philip E. Stanley, Ian Simpson, Michael Taylor, Charles Tomlinson, Tony Turley, the late John Wilkins, Peter Williams and the many others who have shared information and drawn my attention to a variety of rules and other measuring devices.

I am very grateful to all the people at the Colonial Williamsburg Foundation in Virginia, and in particular those in the Historic Trades Department, who have most generously shared their knowledge and research with me.

I have read and worked in many libraries, including the Birmingham City Library, the Guildhall Library, the British Library and the Science Museum in London and the John D. Rockefeller Library in Williamsburg and many other local studies collections and thank all the librarians and other staff who have always been most helpful.

Don & Anne Wing have generously shared some of their research into the early rule makers and I am most grateful to them for their help.

I am most grateful to David and Ian Stanley for their support and help both in allowing me to use information from their illustrated catalogues and in generously supplying me with photographs.

P. A. (Tom) Martin shared his extensive knowledge of the Customs & Excise and kindly read the relevant chapter for me.

Ken Hawley, Joan Unwin and The Hawley Collection Trust Ltd. have been most generous in allowing me free access to study in the Collection and to photograph some of their material. I am particularly grateful to Ken Hawley for the time that he has spent explaining the intricacies of the rule making trade, for his insights into John Rabone & Sons, for giving freely and generously of his immense knowledge about rules and rule making and generally encouraging and helping me with this project.

I am also grateful to Roy and Julanne Arnold for their enduring belief in this book and for their constant support, help, advice and editing.

Jan Gilliam very kindly undertook the task of indexing the book for me.

A special "thank you" is due to Michael Taylor, who has supported this project throughout, not only allowing me to photograph and illustrate many of the rules in his magnificent collection but also giving generously of his time and expertise and sharing with me his thoughts, insights and extensive knowledge. Many of his rules have been used for the illustrations, too numerous to list individually, and without such generous access to his collection, this book would have been difficult to write.

There is one person who needs a very special "thank you" and without whom this book would never have been finished. Jay Gaynor, the Director of Historic Trades at the Colonial Williamsburg Foundation, gave generously of his help to Mark and me but it is his continued support and encouragement in the years since Mark died that has enabled me complete this work. He has given me a huge amount of time, reading and correcting the text, sharing his very extensive knowledge and his own research. His many ideas and suggestions have been invaluable, his patience has been inexhaustible and I give him my most grateful thanks.

Illustration credits:

The photographs are by Jane Rees except for the following: Richard Filmer: Figs. 10.1, 11.7, 11.34, 12.06, 12.18, 13.35, 13.40, 13.42, 13.43. David Stanley Auctions: Figs. 1.1, 2.29, 2.36, 8.27, 10.20, 11.20, 11.25, 12.03, 12.04, 12.07, 13.11, 13.14. Kenneth D. Roberts: Figs. 2.22, 2.23. Bonhams Auction House: Figs. 6.04.

Figs. 2.25, 2.26, 2.28, 2.31, 2.32, 2.33, 2.34, 2.38, 2.39, 4.15, 4.24, 4.26, 4.28, and 4.36 are reproduced by kind permission of the Hawley Collection Trust Ltd., Sheffield.

Figs. 2.22, 2.23 are reproduced by kind permission of Philip E. Stanley.

Figs. 3.10, 3.36 and 7.28 are reproduced by kind permission of the Colonial Williamsburg Foundation, Williamsburg, Virginia.

Figs. 6.20 and 13.28 are reproduced by kind permission of the British Library.

Mark Rees

An appreciation by David Millett

Mark Rees, who died in 2001, possessed that rare combination of qualities of mind and temperament which advance knowledge and understanding of the past in contexts where conventional approaches to research are not enough. In the history of technology, much that needs now to be rediscovered and understood was never recorded, and must be deduced without documentary evidence from artefacts and objects themselves. The task calls for breadth of knowledge, practical understanding of materials and processes, a disciplined approach to analysis, insatiable curiosity, and an indefinable spark of perceptiveness. Like Bill Goodman and Philip Walker before him, Mark had those qualities. And like them, he believed that knowledge should be shared, and was generous in passing his own discoveries and perceptions on to others.

Born in 1939 and trained as an architect, Mark practised in partnership with his wife Jane, specialising in the restoration and renovation of historic buildings. This work sharpened his interest in techniques and tools used by tradesmen in past centuries and he came to spend increasing amounts of his time on research, writing and lecturing on what might be termed the antiques of technology. He was a founder member of the Tools and Trades History Society, serving for many years on its committee and with Jane editing its *Newsletter* and its *Journal.* Mark was elected to the Board of Directors of the Early American Industries Association in 1999. He and Jane were awarded that Association's prestigious J. D. Hatch Award in recognition of their contribution to the advancement of knowledge of early American industries through their writing.

Mark's introduction to the reprint of the Preston 1909 Catalogue appeared in 1991. He and Jane produced a greatly expanded third edition of Goodman's *British Planemakers from 1700* in 1993 – an enormous undertaking. They edited and contributed to *The Tool Chest of Benjamin Seaton* (1994), wrote *Tools, A Guide for Collectors* (1986) and *Christopher Gabriel and the Tool Trade in 18th Century London* (1997) and a second edition of the *Guide* (1999).

Before the diagnosis of his final illness Mark and Jane had begun this new book on one of the subjects closest to his heart, *Measuring for the Trades.* They continued work on it after Mark knew that his remaining time was short and Jane has been able to complete the book for publication. It will be a fitting memorial to him, illustrating yet again the great breadth of his knowledge and enthusiasm.

Introduction

Let there be throughout our kingdom a single measure for wine and for ale and for corn, namely "the London quarter", and a single width of cloth (whether dyed, russet or halberjet) namely two ells within the selvedges; and let it be the same with weights as with measures. The Magna Carta of King John, 1215.

As long as there has been commerce, there has been a need to quantify and in order to quantify it is necessary to have equipment to measure. The growing complexity of commerce, the introduction of new products, the search for precision in manufacturing and the collection and processing of scientific data were the spur for ever more diverse and accurate means of measuring.

This book focuses on the measuring tools used by tradesmen and professionals in their everyday work, on the trades in which specific tools were used and on how these tools were made and by whom.

The intention is that the information contained herein will be of use to collectors and researchers by explaining how the measuring devices were used and by placing them in a trade and social context.

No book of this size could hope to describe in detail every type of measuring device so it is those measuring items likely to be found and collected today that have been included. Some are included because they are regularly encountered. Others because they reflect important historical trends or developments. Still others pose intriguing intellectual challenges. Finally, many are simply attractive objects. It is a truism that to measure accurately the measure itself must be finely made and in this can lie great appeal.

Linear measure

In order to limit further a task that might otherwise become overwhelming, this book has concentrated on *linear* measurement, although some other types of measurement have been included where it is thought that they would be of interest to the collector. The derivatives of linear measurement are area and cubic (volume) measurement so these must also form part of this study. But it is linear measurement and sizing that is the basis.

The boxwood rule

One of the most common of all measuring tools is the rule. This term should not be confused with the term ruler which is a tool used as a guide for drawing straight lines although a ruler may well be divided with measurements to enable it to fulfill both functions.

Look in any tool kit, into a desk drawer or on a market stall and it is very likely that you will find a 2 ft. 4-fold boxwood rule of some kind, for these were made in huge numbers over a period of 200 years. The boxwood rule in its many varieties is at the heart of this book.

The rule has been the symbol of carpenters for many years. Henry Mayhew, writing in 1850, says "The carpenters and joiners that work for the low speculating builders are, generally speaking, quite a different class of men to those who are in 'society'... They are generally to be seen on a Sunday lounging about the suburbs of London with their working clothes on, and their rules sticking from their side pockets..."[1] and a painting by William Hogarth, ca. 1725, of *The Carpenter's Yard* clearly shows folding rules in the workmen's pockets.

Rules achieved this symbolic status because standardisation and measurement were essential to the trade. The carpenter, when setting out, transfers the dimensions to the work piece by making marks opposite the appropriate dimensions on his rule. Alternatively he may use the rule to take the width of a board. In either case he is working by visual comparison with a standard – his rule – and his rule will have been made by reference to another known standard.

This book

Before it is possible to consider how a variety of trades measured, it is necessary to clarify the design, history and manufacture of rules and other measuring devices. In the first four chapters, these subjects are discussed and explained, with an overview of what is marked on rules, their historical development, how they were made and who made them.

The remaining ten chapters cover a variety of trades and professions, the measures that were used and the tools, often specialist, that were needed to take these measurements.

Appendices

After these chapters are four appendices containing transcripts of historical documents that elaborate on several topics. The first two are 1662 and 1866 descriptions of how rules were made and marked. The third is an 1851 inventory of tools and stock list of one of the most prolific rule makers, John Rabone & Sons of Birmingham. The fourth is the 1797-1818 ledger of a journeyman rule maker working for the London firm of Dring & Fage.

Checklist of makers

A checklist of rule makers follows. The majority of the earliest known makers of measuring devices were those involved in the trade that today collectors lump together as scientific instrument makers. However, this overall term only came into use in the last decades of the 19th century. Before that date the description of the makers' trades was more specific to the instruments being made, although the majority of makers made a variety of different categories.

The three main categories were mathematical instrument makers, which included all instruments that were used for calculating and measuring; philosophical instrument makers which included those instruments necessary for the study and demonstration of the natural and physical sciences; and optical instrument makers which covered instruments that were concerned with light, whether transmitted or absorbed.

It is only from the second half of the 17th century that a distinct trade of rule making can be even partially divided from that of mathematical instrument maker.

Metrication

The principal sources in this study are the rules and other measuring devices that are the concrete evidence of past practices. Mostly these measures are delineated in the feet and inches of the pre-metric era. But feet and inches are the easy part of the Imperial system – there were lines, yards, poles, perches, furlongs, ells and many other obscure units of measurement. The tendency for trades to use units of measurement convenient for their own purposes, even if these were obscure to outsiders, seems to have been universal and has occurred even with metric measurement.

The adoption in Britain of the metric system in the 1960s meant that large numbers of inch (Imperial) based measuring devices became obsolete and were thus available for the collector to gather up.

The younger generations are being taught only metric measurement – it cannot be long before most of our citizens are not conversant with feet and inches. This is undoubtedly a trend toward rationalisation and efficiency, but this book celebrates, and it is well that some of us preserve, the fascinating diversity of the past.

Illustrations

On a number of occasions, the same rule has been used to illustrate different points in different chapters. This has been done so that the reader does not have to go back and forth in the book looking for a particular rule.

References

The references for each chapter have been given at the end of the chapter.

References

1. Yeo, Eileen and Thompson, E. P., *The Unknown Mayhew*, New York, 1971. Originally published as letters in the *Morning Chronicle* between October 19th, 1849 and December 12th, 1850.

The Three Foot Rule

When I was bound apprentice, and learned to use my hands,
Folk never talked of measures that came from foreign lands:
Now I'm a British Workman, too old to go to school;
So whether the chisel or file I hold, I'll stick to my three-foot rule.

Some talk of millimetres, and some of kilogrammes,
And some of decilitres, to measure beer and drams;
But I'm a British Workman, too old to go to school,
So by pounds I'll eat, and by quarts I'll drink, and I'll work by my three-foot rule.

A party of astronomers went measuring the earth,
And forty million metres they took to be its girth;
Five hundred million inches, though, go through from pole to pole;
So let's stick to inches, feet and yards, and the good old three-foot rule.

William John Macquorn Rankine, 1820-1872

Chapter 1
What's on a rule

If Thou delightest in Timber-Measure, or hast Occasion for Much; Thou hast here a Line so Fit for thy Purpose, that thou canst not imagine, much less, wish a better. For what can be more ready and easie, than having set twelve to the length, to see the Content exactly against the Girt or Side of the Square. Henry Coggeshall, 1677. [1]

The formal name for the divisions on a rule is a line. Most rules have nothing other than a line of measure – the correct description of any line intended solely for the purpose of linear measurement – but there are many other lines.

This chapter is intended to be an overview of the various lines, tables and other information found on rules. More detailed descriptions and markings specific to various trades will be found in the relevant chapters later in the book.

Terminology

Until the 17th century the carpenter and the majority of other tradesmen used a measure of a simple lath type. Having no joint and being marked just with feet, inches, and parts of inches no smaller than eighths, such laths could be made by any reasonably skilled woodworker. Other simple lath type rules, or story rods, were used by various trades where frequent and repetitive measurements were required. With a suitable piece of wood and an accurate rule to copy, the result could be both durable and true.

Laths in various forms are still today used by some trades but a rule that folds is frequently more convenient when a long rule is desirable.

Sophisticated rules incorporating a developed form of rule joint were made in the Roman period, the surviving examples being made of bronze with square section legs. However the Roman carpenter, like his English counterpart in mediaeval or Renaissance Britain, was more likely to have used a wooden lath rule. The rule with a rule joint remained in the realm of gentlemen and science.

Since the 16th century, sectors have been made with a joint that is identical to the round joint commonly found in later 2-fold rules and this is the

Fig. 1.1a & b. A lath-type rule, dated 1552. Made of hardwood, it has been divided into inches and quarters, but not marked otherwise except for the date. It gives the appearance of being professionally made so the lack of numbers may indicate it was not completely finished or figures were not considered necessary. A more sophisticated lath type rule was recovered from the wreck of the Mary Rose. On one face it is marked with a line of measure, on the other a line of timber measure. This rule must be pre-1545, the date that the Mary Rose went down. A full description was written by Richard Knight.[2]

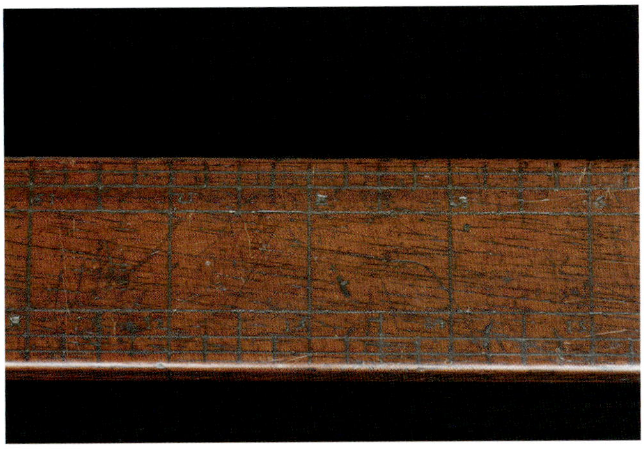

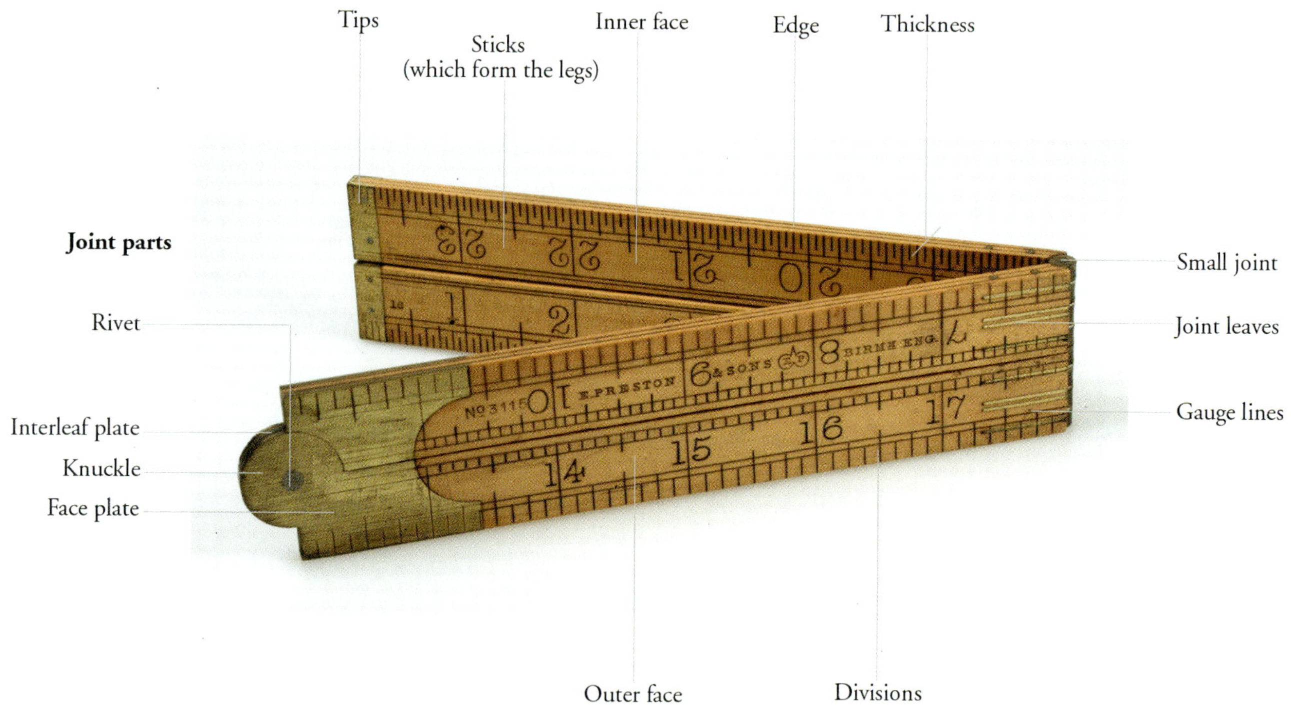

Fig. 1.2. The names of the parts of a rule. The shape of the face plate is used as the description of the joint. This rule would be described as a 2 ft. 4-fold arch joint rule.

heritage which the folding rule shared. By the 17th century ivory, boxwood and brass sectors with both round and arch type joints were being made, a development that is reflected in the history of the folding rule.

There is a rule bearing the date 1654 that is made with two 9 in. boxwood legs with a folding brass tongue to make the length 2 ft. when extended. This rule, clearly the work of a professional maker, provides evidence that by the latter part of the 17th century the folding rule was becoming available to the artisan.

The correct descriptions of folding rules can be somewhat misleading, so it is worth clarifying that:

A 2-fold rule has one joint and two legs.
A 4-fold rule has three joints and four legs.
A 6-fold rule has five joints and six legs.

Fig. 1.2 shows the terminology for the parts of a rule as used by the rule making trade and in this book.

Lines found on rules

Line of measure: A line of measure may be in imperial, metric or any other system. On almost all imperial rules the line of measure is marked in inches without acknowledgement of the foot. A complete description of a rule will state the subdivision of the inch and also any extra subdivision in some of the inches, usually the first inch. Many rules, particularly before the mid-19th century, were only subdivided into ⅛ in. Finer subdivision into 1⁄16 in. or less is often an indication of a rule of extra quality.

In addition to the line of measure marked on the faces of the rule, many

Fig. 1.3. A simple line of measure on a rule by John Rabone & Son (Birmingham, 1845–77).

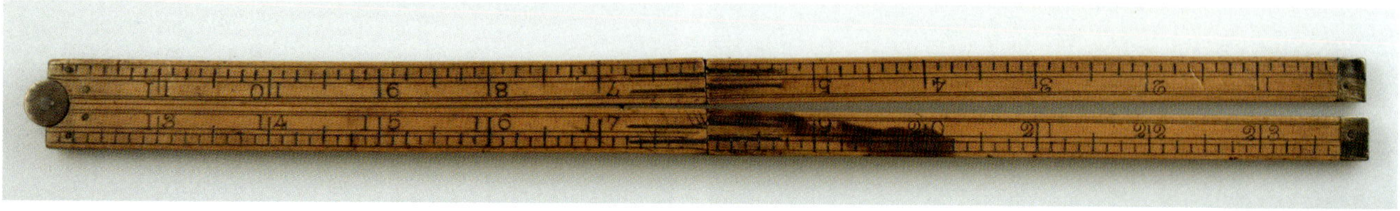

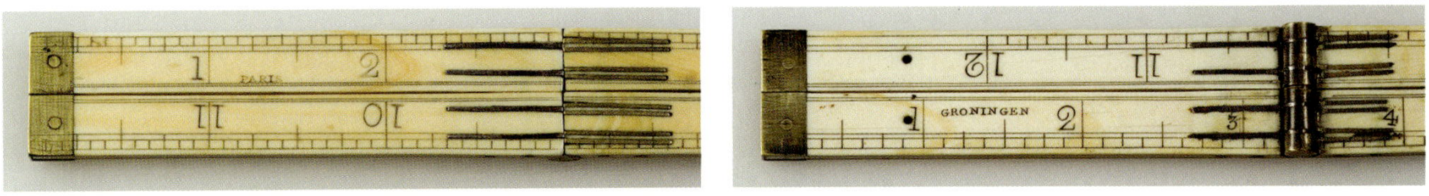

Fig. 1.4a & b. Two faces of an ivory rule marked with Paris inches and Groningen inches. Note that twelve Paris inches reach to the end of the rule, but the Groningen inches stop short of the end.

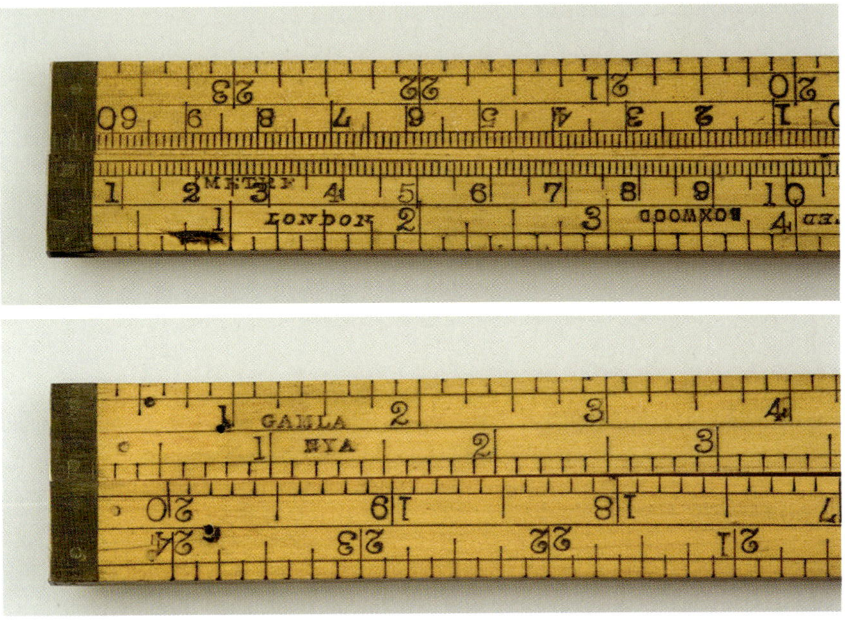

Fig. 1.5a & b. A 2 ft. 4-fold rule marked with London inches and metric on one face and Gamla and Nya inches on the other. Although unmarked, this rule was made by I. & D. Smallwood (Birmingham, 1858–1999).

rules dating from before 1800 also had a line of measure divided into 100 equal parts, which enabled decimal calculations if required.

Lines of European inches: Before the introduction of the metric system, many European countries, and often even individual towns, had their own sizes of inch, and rules marked with two, three or even four different European inches were made. A comprehensive list of these with their actual sizes can be found in Philip E. Stanley's *Source Book for Rule Collectors.*[3] The following are the more commonly found: French, Paris, Gamla (Swedish old inches), Nya (Swedish new inches), Spanish, Russian, Riga, Amsterdam, Groningen, Danish, German and Rheinland (also spelt Rhynland) and there are many others.

Scales: These are lines of measure, reduced by a known ratio, to allow an exact representation but at a reduced size. They are usually intended for architects and surveyors, but are also used for navigation and dialling (the design of sundials).

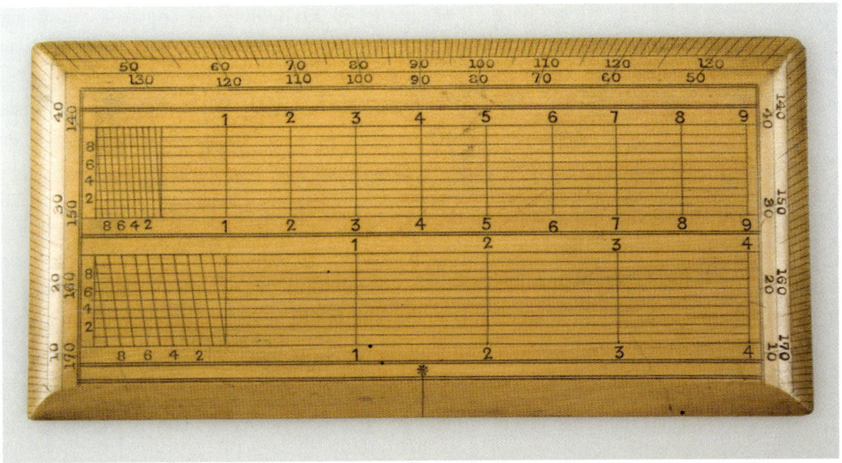

Fig. 1.6. A 6 in. boxwood rule marked with two different diagonal scales and a protractor scale (giving degrees of angle). The name, diagonal scale, derives from the subdivision of the first unit diagonally into 100ths. This allows more smaller divisions than could otherwise be placed on the rule.

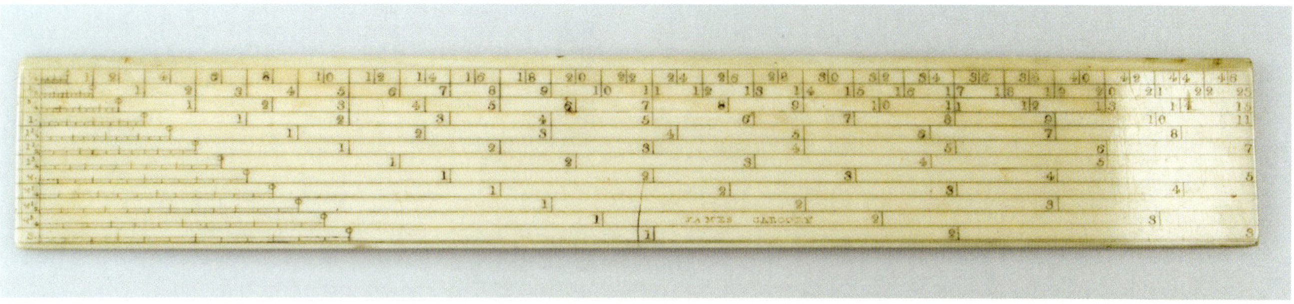

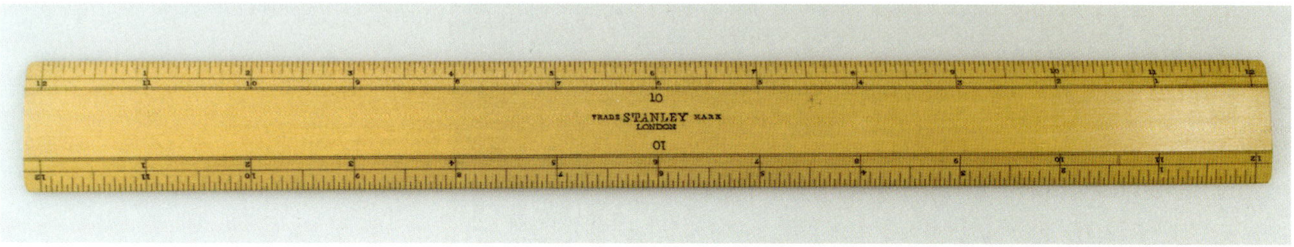

Top: Fig. 1.7. An ivory scale rule face marked with twelve different scales by James Gargory (London, 1835–62).

Bottom: Fig. 1.8. An edge marked scale by W. F. Stanley (London, 1853–1967). This rule is marked with surveying scales, but the same form is used for architects' scales.

Some confusion can arise with the term scale. Rules marked with scales are correctly known as scale rules. However, this is frequently abbreviated to just scales, so the same term can mean either what is marked on the rule or the rule itself.

Before the advent of edge marked scales around 1870, rules were marked on the face of the rule with either diagonal scales, or proportional scales. To use face marked scales, it was necessary to measure off the rule with a pair of dividers.

Line of board measure: A line which indicates the length of a board of a given width that has an area of 1 square foot. The line is used by finding the width on the line and then reading off the answer on the adjacent line of measure. For boards narrower than 6 in. the length exceeds 2 ft. so, on a rule of this length, a small table placed at the end of the line gives the lengths for boards from 1 in. to 6 in. wide.

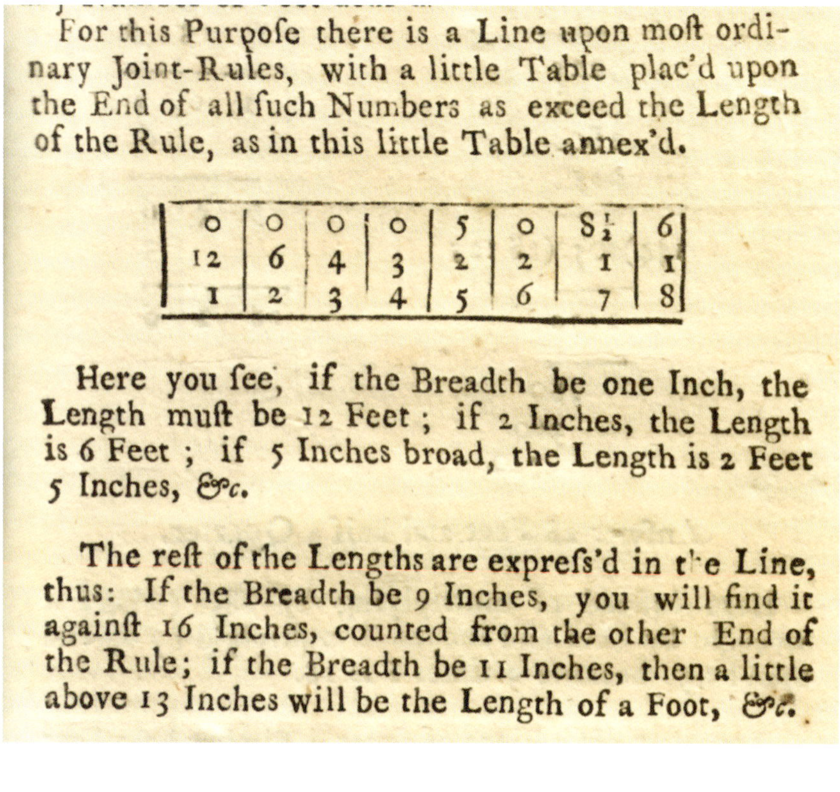

For this Purpoſe there is a Line upon moſt ordinary Joint-Rules, with a little Table plac'd upon the End of all ſuch Numbers as exceed the Length of the Rule, as in this little Table annex'd.

0	0	0	0	5	0	5½	6
12	6	4	3	2	2	1	1
1	2	3	4	5	6	7	8

Here you ſee, if the Breadth be one Inch, the Length muſt be 12 Feet; if 2 Inches, the Length is 6 Feet; if 5 Inches broad, the Length is 2 Feet 5 Inches, &c.

The reſt of the Lengths are expreſs'd in the Line, thus: If the Breadth be 9 Inches, you will find it againſt 16 Inches, counted from the other End of the Rule; if the Breadth be 11 Inches, then a little above 13 Inches will be the Length of a Foot, &c.

Fig. 1.9. Table of board measure from Hawney's Measurer.[4]

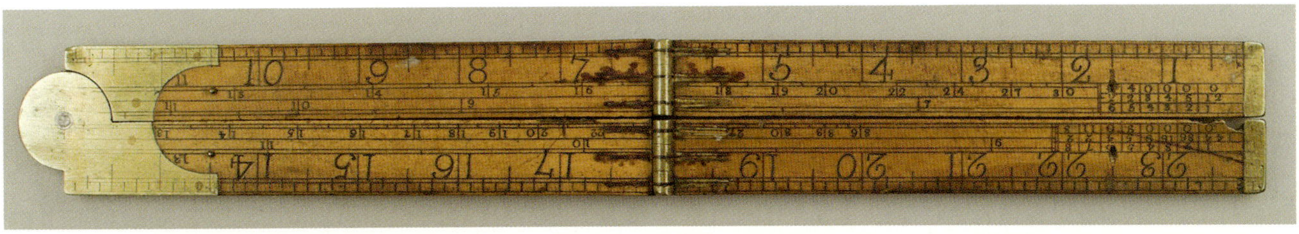

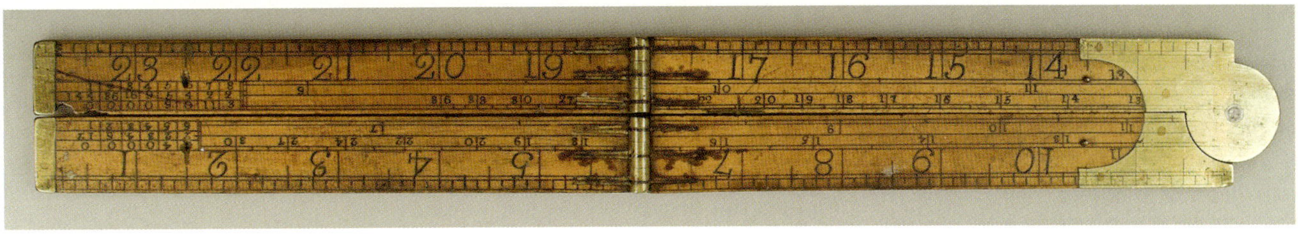

Top: *Fig. 1.10. Line of board measure on an anonymous late 18th or early 19th century rule.*

Bottom: *Fig. 1.11. The line of timber measure on the rule shown in Fig. 1.10.*

Line of timber measure: Often called simply a timber line, this indicates the length of a square timber of a given size that contains one cubic foot. Like the line of board measure the answer is found on the adjacent line of measure. For sizes of 8 in. square and under, where the length exceeds the length of a 2 ft. rule, the answer is indicated by a small table, placed at the end of the line.

1	2	3	4	5	6	7	8	*(size of square)*
144	36	16	9	5	4	2	2	*(feet)*
0	0	0	0	9	0	11	3	*(inches)*

Fig. 1.12. Table of timber measure giving lengths of timber containing one cubic foot.

Fig. 1.13. Table for line of timber measure on a walking stick, dated 1705, for timber less than 8 in. square.

E & M line: Actually two lines – the middle (M) line and the edge (E) line, these are also known as the 8-square line or the mast makers' scales. These give the dimensions needed to set out an octagon onto a square timber. Without these aids, trigonometry is required – or just trial and error! The

Fig. 1.14. 2 ft. 2-fold rule with 8-square line by Edward Preston & Sons (Birmingham, 1864–1932).

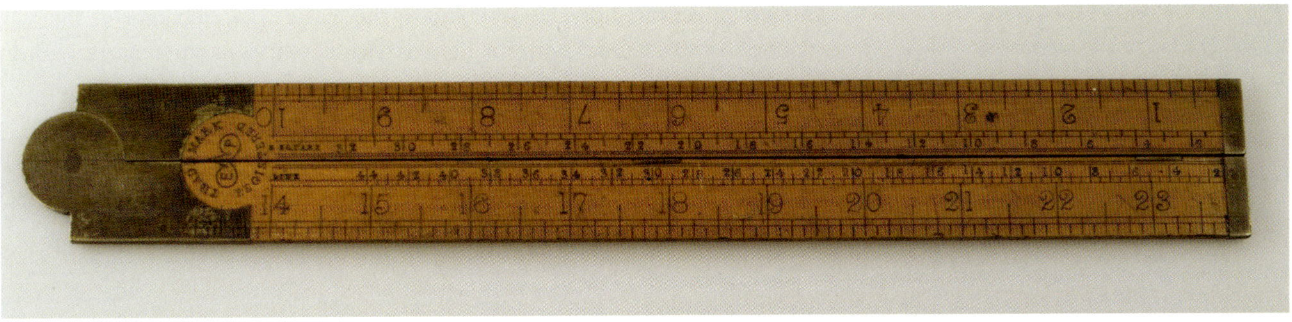

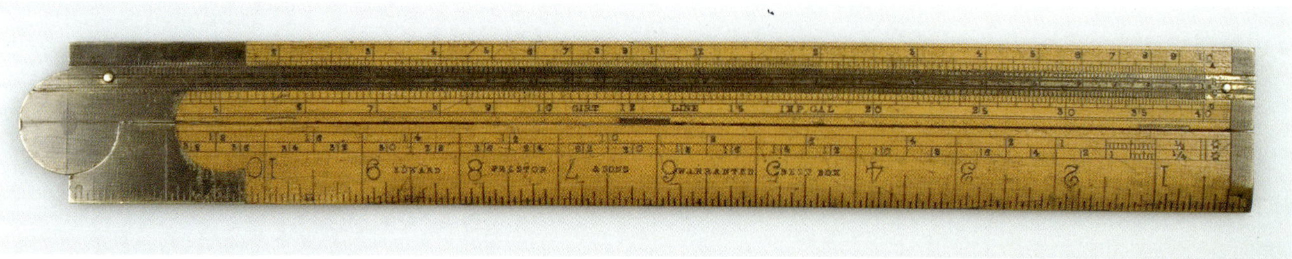

Fig. 1.15. A 2 ft. 2-fold rule with a brass slide by Edward Preston & Sons showing a line of logs (logarithms) – in this case, Gunter's lines – and a girt line. The "slide" is a thin strip, frequently of brass though it can be ivory or boxwood, let into a groove in the leg of a rule. The edges of the slide and the adjacent edges of the groove are gradutaed with logarithmic scales and, by moving the slide, these scales can be used for multiplication and division.

lines are not complete in themselves but must be placed on the rule to allow the answer to be read off an adjacent line of measure. They are direct reading, i.e. by looking across the rule from the size of square timber on the E or M line, the layout of the octagon, either in from the edge, or out from the middle, can be read off the line of measure.

Logarithmic line, also known as a line of numbers: Invented by Edmund Gunter (1581–1626) by 1623 and rapidly seized upon by the makers of rules as a feature that added usefulness, the line can be used for multiplication or division, either by using dividers or a slide.

On folding rules, Gunter's lines, as they are sometimes known, almost always have two cycles. That is, each scale starts at 1, reaches 10 in the centre of the rule and then repeats from 1 to 10 again to the end of the rule.

The design and disposition of logarithmic lines has exercised mathematicians for more than three centuries and has resulted in a wide range of lines which are to be found on both general and special purpose slide rules, but, essentially, none of these, except the two cycle line of numbers and Coggeshall's girt line, ever found a place on tradesmen's rules.

Girt line: On most rules where this line is present it is named in full. The line is a single line of numbers "folded" at 4. That is, the scale is offset from the normal line of logs, starting at 4 and continuing to 40. This arrangement was described by Henry Coggeshall in the late 17th century and named the girt line by him from its use in calculating the content of timber in the round by means of the quarter girth.[5] The presence of this line was the most long lasting of Coggeshall's many improvements – rules from which all other elements of his design had been lost were still being sold as "Coggeshall" rules until 1940.

Gauging lines: A large family of lines connected with the gauging (measuring) of casks and other cylindrical containers and their contents. These lines are common on the rods and other measuring tools used by officers of the Customs & Excise (who were known as gaugers) and are sometimes also to be found on more multi-purpose rules.

Sectoral lines: As the name indicates these are most commonly to be found on sectors but are also marked on some rules. They start at the hinge point and can be easily recognised as they are inscribed on the diagonal rather than parallel to the edges of the rule. Often referred to as the "double lines", the same line is repeated on the both legs of the sector. The calculations that can be done with these lines with the help of dividers (lines of equal parts, chords, sines, tangents and secants) use the properties of similar triangles.

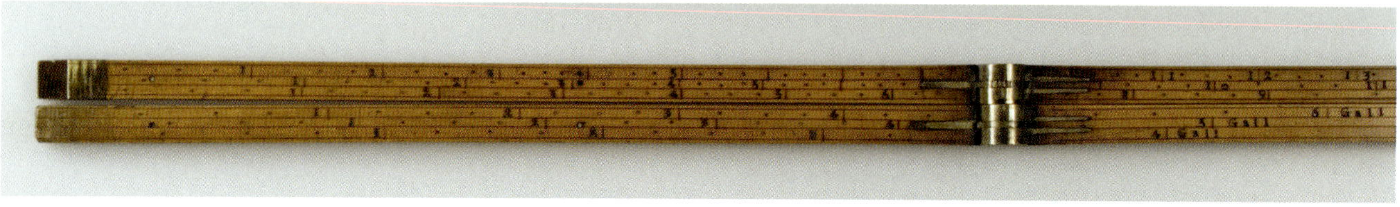

Fig. 1.16. A anonymous dip rod, dated 1788, marked to show the amount contained in a cask.

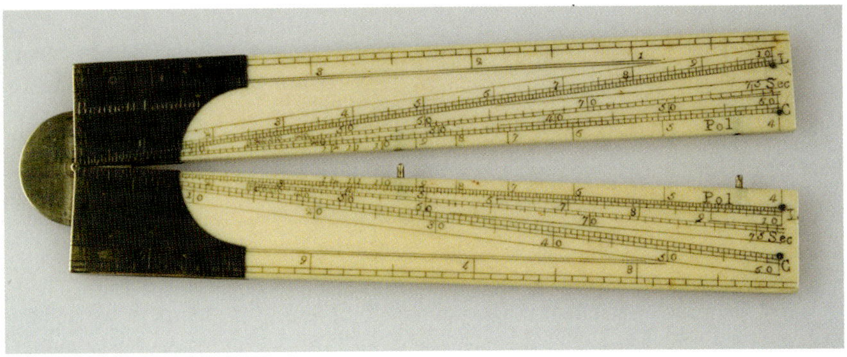

Fig. 1.17. Sectoral lines on a 6 in. ivory sector by John Bennett (London, 1735–70). Only those lines that start at the hinge are sectoral – the lines of measure along the outer edges are not.

Other lines on sectors which do not pass through the hinge point, such as the line of measure on the outer edges of the sector shown in Figure 1.17, are the "single lines" and are not sectoral.

Circumference line: A line shortened by a factor of 3.142 so that if the diameter of a circle is measured on the standard line of measure, by looking across to the circumference line at the same point, a direct reading of the circumference can be found. As rules won't bend round the circumference of a circle, this line cannot be used in the opposite direction – but some tape measures are marked with standard and circumferential measure and can thus be used either way.

Line of radii: Marked R^{D}, this line is used for calculations that require the subdivision of circles into equal parts.

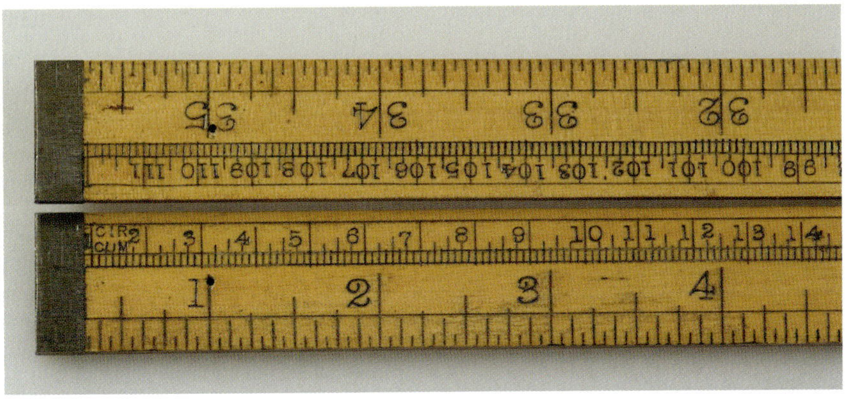

Fig. 1.18. Circumference line on Hays Improved Practical Mechanics' Rule.

Line of side of equal square: A line for converting between the dimensions of squares and circles of equal area, which is read in conjunction with a line of radii.

Line (of square) without: A line of numbers (a logarithmic scale) used for calculations that involve **Si** (0.707) which is the value of the side of a square inscribed within a circle with a diameter of 1.

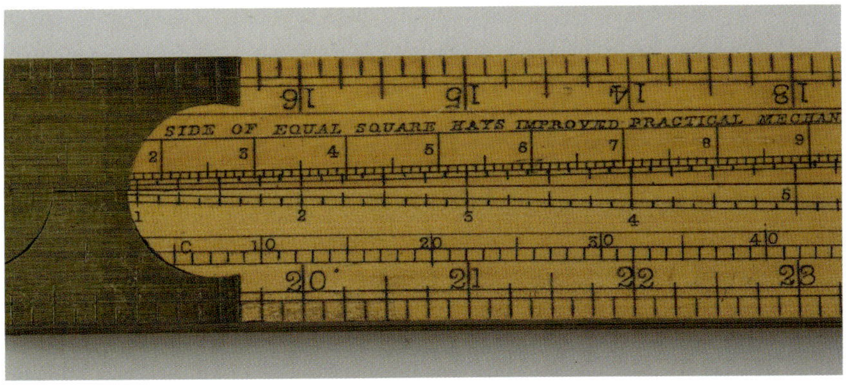

Fig. 1.19. The line of radii and the line of side of equal square on the same rule as shown in Fig. 1.18. A circle of 3 in. radius will have the same area as a square with sides 5.3 in. long.

Fig. 1.20. Gunter's scales on a 6 in. brass scale. This was probably originally part of a pocket drawing set.

Line (of circle) within: A line of numbers used for calculations that involve **Se** (0.886) being the side of a square with an area the same as a circle with the diameter of 1.

Lines on Gunter's scales: Based on another invention by Edmund Gunter and improved in the 17th century, these lines are used with dividers for navigation and dialling.

Plain or natural lines of numbers: These lines correspond to various trigonometrical functions and are used with dividers for calculations.

Lines of proportion: Used in architecture these lines were divided to give a draftsman the correct proportions for each of the five classical orders of architecture, Doric, Ionic, Corinthian, Tuscan and composite.

Conversion lines: These lines are intended for the conversion of one unit into another – for example, centimetres into inches or foreign inches to imperial inches.

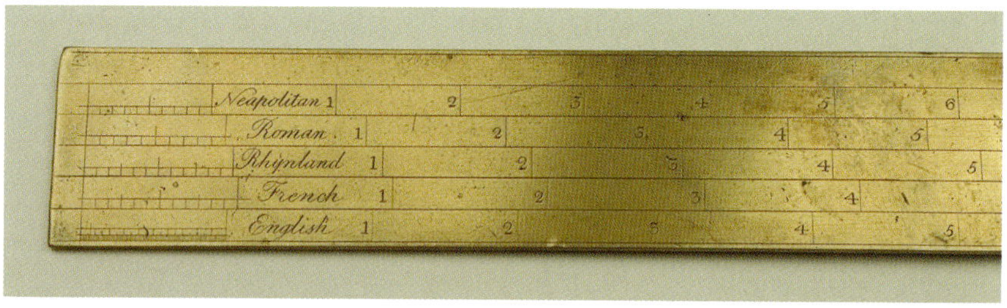

Fig. 1.21. An 18th century brass comparator rule marked with Neapolitan, Roman, Rhynland, French and English inches.

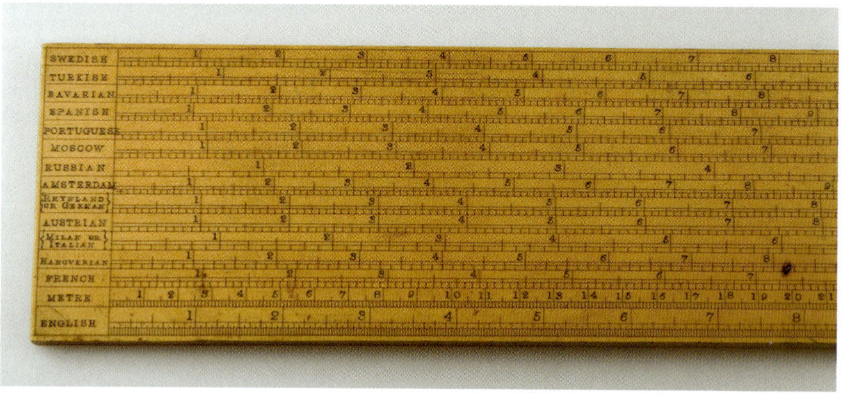

Fig. 1.22. Boxwood conversion rule by J. Dobson (London, 1850–52) with fourteen different inches and metric.

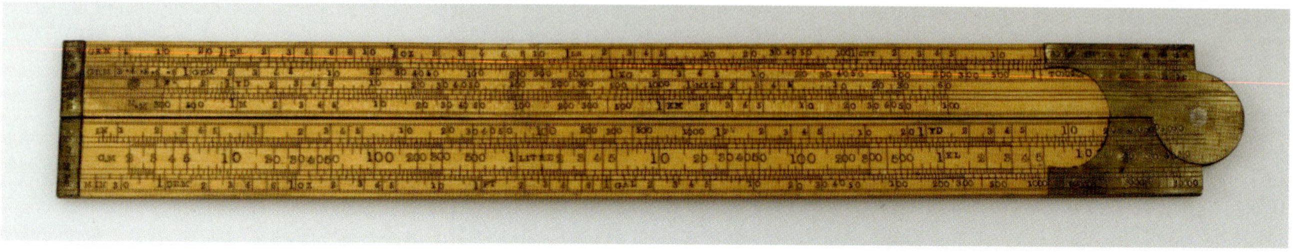

Fig. 1.23. An anonymous rule marked "Scotts English & metric equivalent rules". One face of this 2 ft. 2-fold rule is marked in inches and millimetres; the other is marked with four cycles of conversion lines, one for weights, two for linear measurement and one for liquid measure.

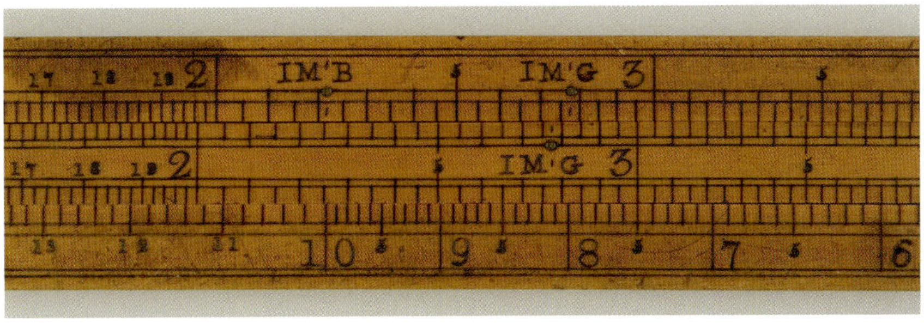

Fig. 1.24. Gauge points marked on a Customs & Excise slide rule.

Gauge points

These are points marked on a slide rule or other calculating rule, usually with a small brass pin, to assist in calculations. On rules without slides, the gauge points need to be used with dividers, so the brass pins may have been intended to act not only as markers but also a protection for the rule. However, on rules with slides the brass pins are purely accurate markers.

A gauge point may record a simple standard such as 2140, the number of cubic inches in a bushel. But in other cases they derived from the need to simplify complicated calculations such as those required to establish the duty to be paid to the Customs & Excise on the wide variety of goods taxed. The former type merely acts as an aide- mémoire but the latter example is a completed part of a calculation. A common principle of many 18th century calculations was to use gauge points to reduce the number of stages in a calculation, and if the calculation was "by sliding rule" preferably to one stage only.[6]

Most 18th century rules have the gauge points labeled with letters, although a few have been noted with the values written in figures, but it was not until around 1811, when Joshua Routledge produced the first engineers' rules, that numerous figured gauge points became a common feature on rules.

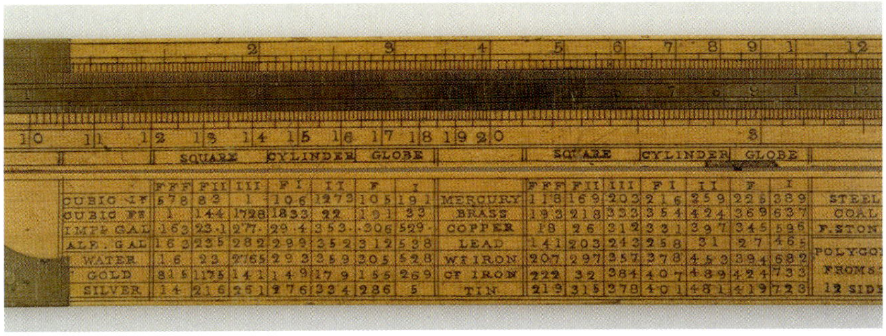

Fig. 1.25. Some of the gauge points found on a Routledge engineers' rule. These enabled calculations of a variety of geometric solids where the dimensions are in differing units.

One cannot help feeling that many of the users of gauge points understood little of the principles behind them and that they were viewed as magic numbers. In many instances, and this is particularly true of engineers' rules, the decimal point is not indicated; a further impediment to understanding.

By about 1870, with a growing body of scientifically trained engineers, calculation based on formulae became increasingly common as the old method using gauge points would no longer suffice. After this date some formulae start to appear on specialist slide rules and, very occasionally, on folding rules with log slides.

Tables

A convenient way of putting large amounts of information onto a rule, the earliest tables to appear on rules (certainly by 1660) were the two small tables for board measure and timber measure, previously described.

In many instances it was more convenient and accurate to make a table

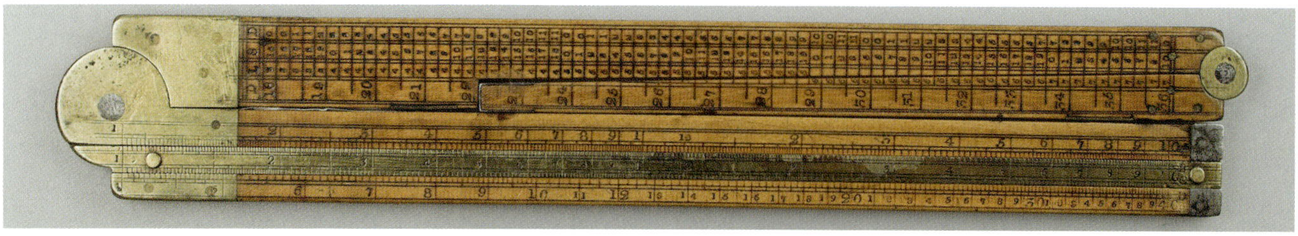

Fig. 1.26. An unmarked but almost certainly 18th century timber rule marked with a timber load table, the load in this instance being 50 cu. ft.

rather than inscribe a conversion line. However few tables are marked with anything other than the briefest indications of their meaning – perhaps it was all part of the mysteries of the particular trade to understand them.

Multiplication tables: The standard 2, 3, 4 etc. times multiplication tables are found on office type rules. Of less obvious use are such tables as times 27 and times 36, both of which have been recorded, but the use of which has not been established.

Multiplication tables involving a change of units: One of the most commonly found tables on rules and intended for many different trades, the "hundred weight table" is of this type. The conversion is from pence per pound (weight) to £. *s. d.* per hundredweight (cwt.).

As there are 112 pounds in a hundredweight, this is actually a 112 times multiplication table but the expression of the price per cwt. in pounds, shillings and pence tends to conceal this.

The other most commonly found multiplication table is the timber load table. This is another table that provides the cost in £. *s. d.* at a given price in pence per cubic foot of a load of timber (a load of timber was a specified amount, usually either 40 or 50 cu. ft.) and is therefore either a 40 or 50 times table.

Formulaic tables: These tables provided the solution to complex formulaic calculations, for example, the weight of round iron bar. The formula is πr^2 x Length x Density. As iron bar was available only in a limited number of standard sizes a table of maybe 20 sizes covered almost all the bar sold. In the past the sale of many commodities by weight was more common than today's size basis, so formulaic tables can be found on rules relating to a variety of trades and items.

Information tables: These tables give information that is not directly calculated by a formula, for example, the safe working load of different sizes of chain, which may be roughly proportional to the size but requires several factors to be considered.

False tables: These list information, usually gauge points, which are presented in the form of a table but are not a sequence – the table format is purely a convenience of layout.

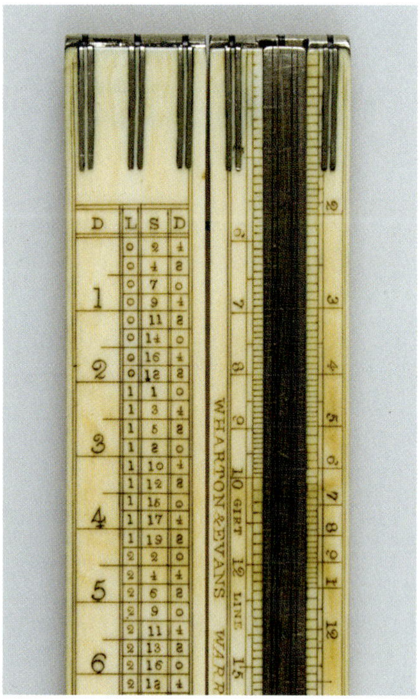

Fig. 1.27. The hundredweight table from an ivory 2 ft. 4-fold rule with nickel silver slide and fittings, a girt line and architects' scales, by Wharton & Evans (Birmingham 1858–77>). Wharton & Evans were listed as one of Rabone's customers in their 1870 "specials" order book and it is likely that this rule was made by them.[7]

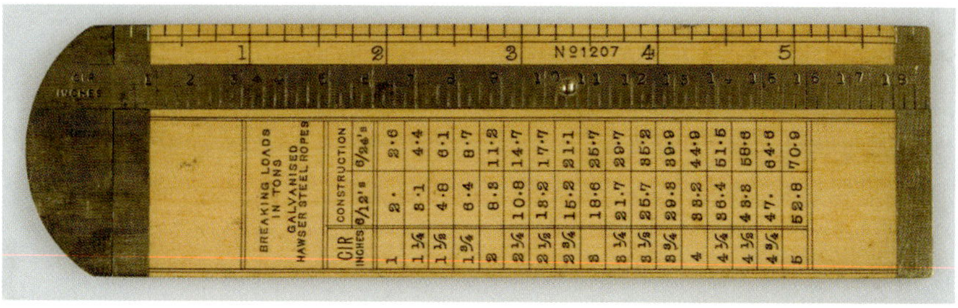

Fig. 1.28. An information table on a rope gauge giving the breaking strain of two different types of steel rope.

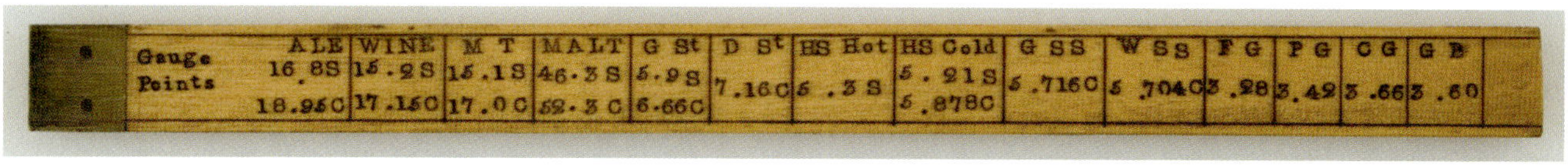

Fig. 1.29. Table of gauge points set out on the reverse of a slide from a 6 in. Customs & Excise slide rule

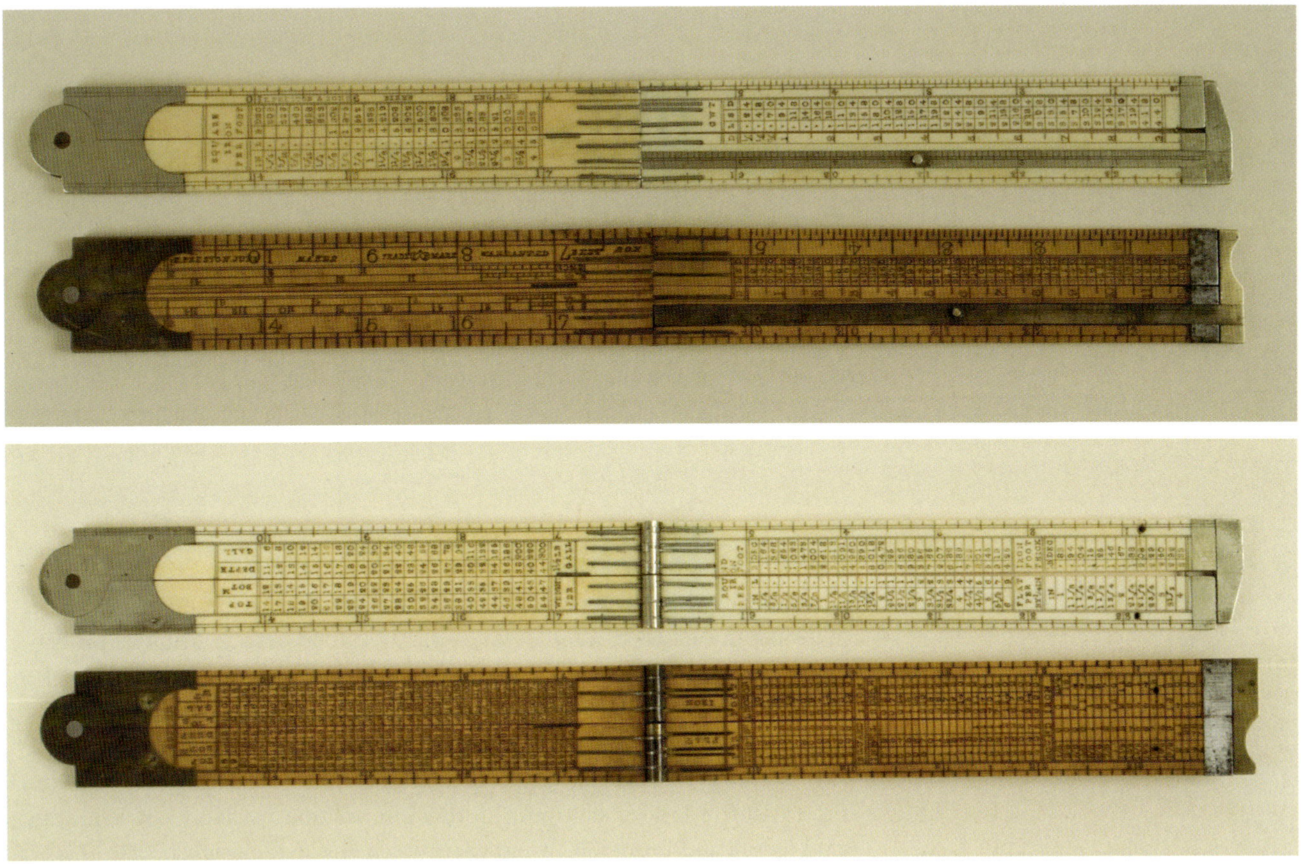

Fig. 1.30a & b. Two ironmongers' rules by Edward Preston (Birmingham, 1864–94). One is in boxwood and one in ivory with nickel silver fittings. Both rules include a caliper slide which is also a logarithmic slide.

The tables on the two rules are similar though not identical.

Both have a hundredweight table and both have a table for coppers (round bottomed vessels made of copper for boiling up liquids). This latter table is slightly different on each rule but both show the volume of liquid contained and the weight of coppers of various sizes. On both the measurements given for each size are: top diameter; bottom diameter; depth and capacity, given in gallons. In addition on the boxwood rule measurements are also given for "L to B" (lag to bottom – the lag is a wooden cover) and also the weight of each size is given – on the ivory rule only an average weight of 1½ lb. per gallon is given.

There are also tables for round iron, square iron and flat iron. These give the weight in pounds and parts of a pound, related to the dimensions of the section, for a one foot length.

Fig. 1.31a & b. 2 ft. 3-fold ironmongers' rule by Edward Preston & Sons (Birmingham, 1864-1932). This 3-fold form is rare and appears to have been used almost exclusively for ironmongers' rules.

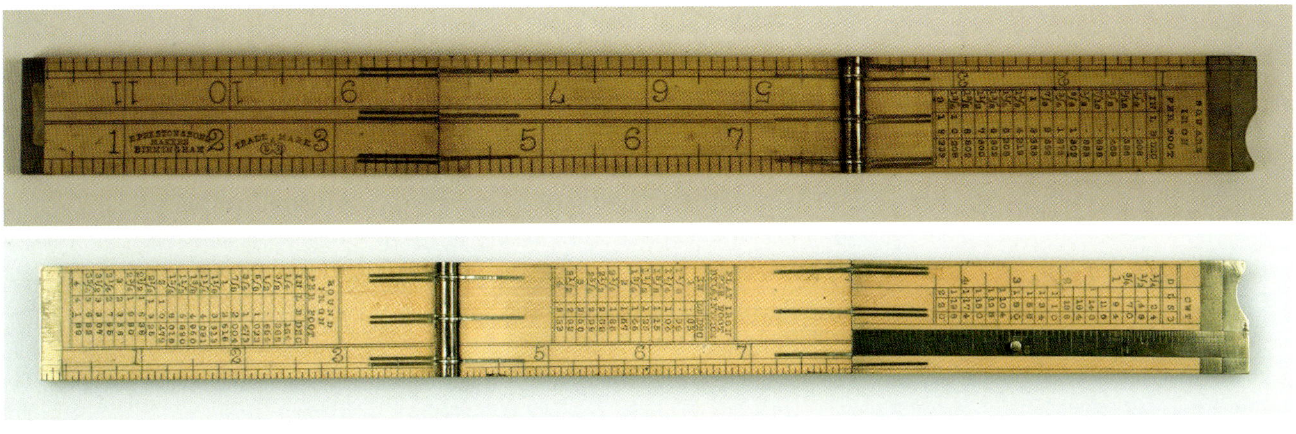

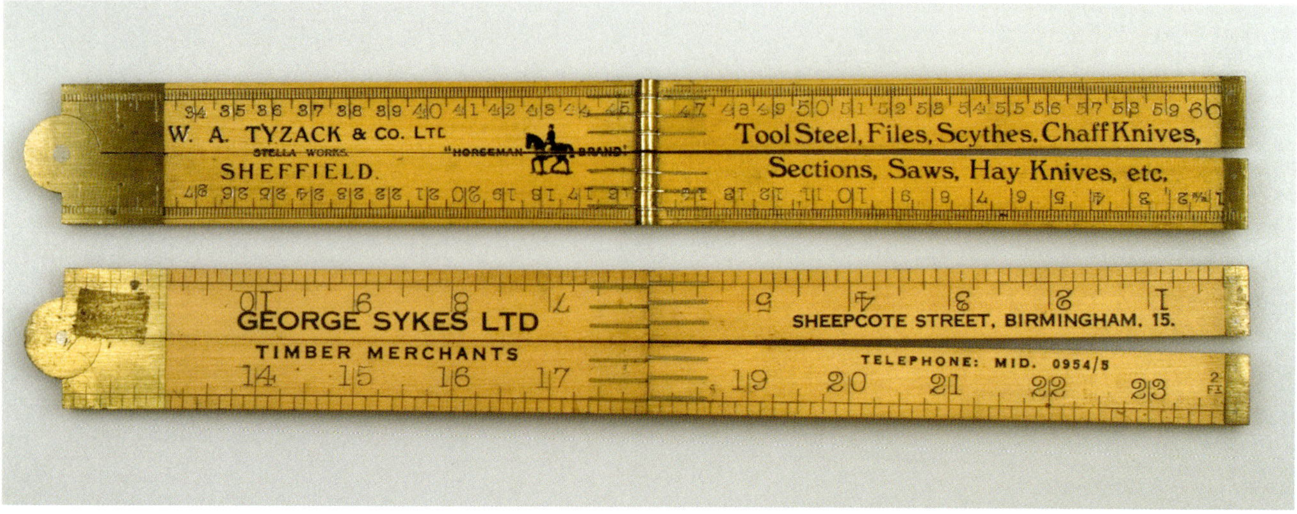

Fig. 1.32a & b. Two advertising rules. The lettering of the advertisers' name can appear on the outside face, the inside face, or frequently, on both. The top rule is anonymous, the lower by I. & D. Smallwood (Birmingham, 1858–1999).

A rule … or a book

Occasionally the desire of the designer to put information onto a rule overcame common sense. One 3 ft. 6-fold boxwood rule is known, probably intended for a civil engineer, where both faces of the rule are wholly taken up with fourteen different tables and the only measuring function is on one edge.[8]

It is extraordinary that any maker should have undertaken the laborious and expensive task of stamping all this information by hand – several thousand letters and numerals, when what was being made was a pocket book of useful information. Such a book could have been printed for a fraction of the cost and once checked would have been correct in all copies – something lacking in all too many hand stamped rules.

Advertising rules

In the 20th century rules became a popular medium for advertising and many can be found with the names of a wide variety of companies, being particularly favoured by timber merchants, builders' merchants, engineering firms, hatters and other clothing trades and rope makers. They are frequently marked with measurements relevant to the trade being advertised.

References

1. Coggeshall, Henry, *Timber Measure by a Line of more ease, dispatch and exactness, than any other way now in use, by a double scale …,* London, 1677.
2. Knight, Richard, "A Carpenters Rule from the Mary Rose", *Tools & Trades,* Vol. 6, The Tools & Trades History Society, 1990.
3. Stanley, Philip E., *A Source Book for Rule Collectors,* Mendham, NJ, 2003.
4. Hawney, William, *The Compleat Measurer…,* London, 1732.
5. Coggeshall, Henry, *The Art of Practical Measuring Easily Perform'd by a Two Foot Rule Which Slides to a Foot …,* 3rd ed., London, 1722.
6. Everard, Thomas, *Stereometry, or the Art of Gauging made easy…,* 7th ed., London, 1712.
7. Records of John Rabone & Sons, MS 1522, Archive Department, Birmingham Library Services.
8. Rees, Mark, "Not so much a rule – More a *vade mecum*", *Newsletter No. 45,* The Tools & Trades History Society, 1994.

Chapter 2
How rules were made

Scale rules: These are made from the finest central African tusks, sun bleached and seasoned from seven to fifteen years. They are the nicest scales to use, on account of their thinness and legibility, but we cannot guarantee that they will always keep perfectly accurate, as even after the long seasoning we give them there will sometimes be a slight further contraction when exposed to different atmospheric conditions. W. F. Stanley, 1919.[1]

Materials

There was a wide variety of materials from which rules were made at various times, but the most common throughout the history of rule making was undoubtedly boxwood.

Boxwood: This word applies to three different species of tree from different areas of the world. All three types have almost no discernible side grain though the growth rings are visible on the end grain. All boxwood is very light in colour when first cut but gradually darkens with age. This process however is slow and with a suitable finish boxwood will remain light coloured for a considerable length of time. Old boxwood, looked after with care, will take on a delicious golden colour.

Being tight grained, boxwood was used for many purposes such as small turnery, engraving blocks and decorative inlays, not just rule making. Both the tight grain and the light colour made it easy to mark and helped in retaining the marking, which was one of its stengths for rule making.

With this wide variety of uses, it is not surprising that by the 18th century supplies of English boxwood, never very plentiful, needed to be supplemented by imported wood.

English boxwood, *Buxus sempervirens*: The English native box tree has for centuries been widely planted in gardens for clipping to form dwarf hedges. Left untrimmed it will eventually develop into a small tree, perhaps up to 15 ft. or 18 ft. high. But it is very slow growing, the diameter of the stems increasing by less than 1 in. each year, and left untended it will develop a straggling habit with little wood of any thickness. The stems will eventually reach perhaps 8 in. in diameter.

English boxwood is notoriously difficult to season. It must be taken under cover as soon as it is cut and initially dried quickly or black fungal streaks

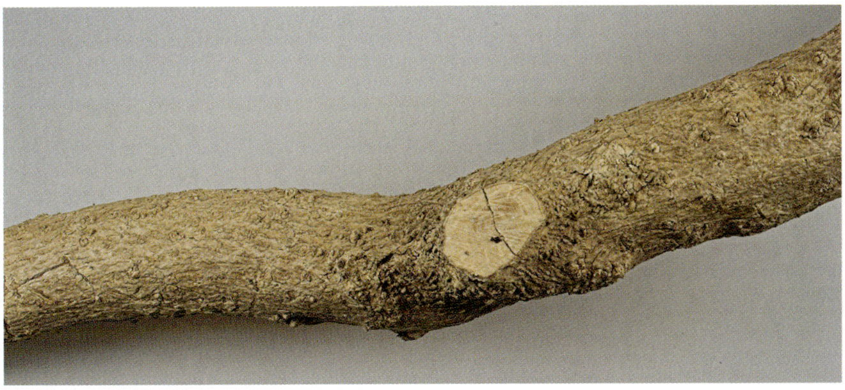

Fig. 2.1. A piece of English boxwood.

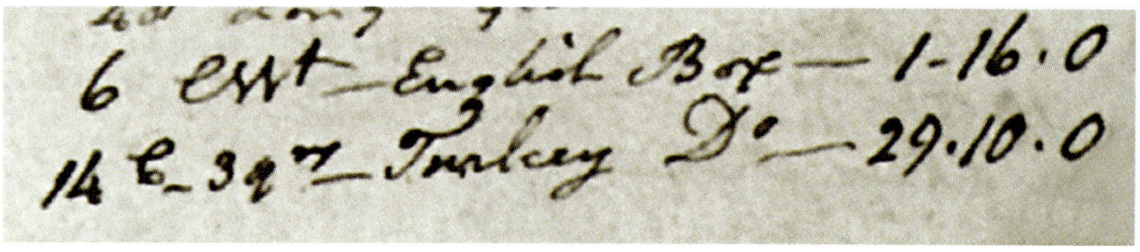
6 cwt English Box — 1-16.0
14 C 3 qrs Turkey Do — 29.10.0

Fig. 2.2. The entry in the 1800 Gabriel inventory listing English and Turkey boxwood. From this it can be seen that Turkey box was about seven times more expensive than English box.

will soon run through the wood. The drying then needs to be continued very slowly or the wood will develop numerous shakes. Even after many years' seasoning the wood will often distort when cut and it is prudent to rest the wood at each stage in the breaking down process. One may wonder why the craftsman, faced with such difficulties, used boxwood even after the advent of man-made materials, but it remained one of the best materials available to the rule maker. If it has a fault it is brittleness.

It is apparent that even by the late 18th century makers were having difficulty in getting suitable sized boxwood for items such as 12 in. gaugers' slide rules. Two such rules with small inset pieces to make good defects in the wood have been recorded.

Turkey boxwood, *Buxus balearica*: Named after the area from which it was shipped, the Turkey box tree grows across a wide area of the southern Balkans, Turkey and the Caucasus. It grows to about 30 ft. high and produces logs up to 12 in. in diameter. The wood is very similar to English boxwood and once made into a rule or other item it is almost impossible to distinguish between the two.

An inventory made in 1800 by Christopher Gabriel, a London plane and tool maker, lists "6 cwt of English box" and "14 cwt 3 quarters of Turkey Box".[2] Gabriel used this wood to reinforce the areas of greatest wear on planes and to make such items as tool handles.

At what date Turkey boxwood became readily available in London is uncertain but surviving artefacts would suggest that by the first quarter of the 18th century it was being used by the scientific instrument makers and that many rules made after 1770 are Turkey box and not the native wood.

Philip Stanley has stated that the American rule makers continued to use Turkey boxwood until around 1900 by which time supplies were no longer available.[3]

Maracaibo boxwood, *Gossypiospermum praecox*: Originally obtained from the area around Lake Maracaibo in Venezuela but later from other areas including Colombia, Haiti and some other West Indian islands, the tree will give logs up to 2 ft. 6 in. in diameter – a lot larger than either English or Turkey box. During a talk to the Slough Ironmonger's Assistants Association in 1936, a member of staff from John Rabone & Sons, one of the leading rule makers, referred to this wood as West Indian boxwood as it was imported via this route.[4] The wood whilst superficially similar to the other boxwoods is 20% lighter and somewhat softer – a thumb nail will often indent it. For the

Fig. 2.3. An unfinished rule from the I. & D. Smallwood factory in Birmingham. This rule is almost certainly Maracaibo boxwood as it is soft enough to dent with a finger nail, though the appearance is such that it is visually indistinguishable from other types of boxwood.

Fig. 2.4. A French rule in ebony and brass, marked on one side in French inches and on the other in metric.

rule maker, the advantages of larger sizes, a wood less prone to warping and checking and which is easier to work were obvious, though it did not make as good a rule. The size and stability of the Maracaibo wood allowed the makers to offer rules that would have been impossible before – John Rabone & Sons Catalogue No. 24 listed boxwood laths up to 120 inches long.[5]

As with English and Turkey box, the voracious appetite of the rule makers and other users eventually led to problems of supply, even for Maracaibo wood – indeed supply problems were cited by Rabone as a principal reason for the switch away from boxwood rules in the 1960s.

Ebony: The family of trees to which ebony belongs, *Diospyros,* grows in many parts of the world including India, Sri Lanka and Africa but the exact species is difficult to identify except by scientific means.

The ebonies are stable woods but their darkness is the problem for marked rules, requiring a white marking substance which seems to disappear far more readily than the black filling used on boxwood. Favoured by French rule makers, in Britain the ebonies were used only for such items as parallel rules that are not divided or marked.

One solution to the marking problem was to face the ebony with an ivory slip – parallel rules with this construction are quite common.

Rosewood (*Dalbergia* spp.): Like ebony this is often used for parallel rules but seldom for any marked items.

Other woods: Rule makers have occasionally used light-coloured woods other than boxwood but none of these have grain as fine as the boxwoods. They were generally used to make larger items such as tank dips where coarser markings were acceptable, or yardsticks where greater resistance to breakage was an advantage. Satinwood (*Chloroxylon swietenia*) is listed in the 1797 ledger of Dring & Fage (see Appendix 4) and in later ledgers, and lancewood (*Oxandra lanceolata*) is also named by them at various dates. Degame (a similar but less dense wood than lancewood) and beech also were occasionally used. North American makers sometimes made larger items in hard maple and other hardwoods. In recent years even less suitable woods, including the spongy and flecked ramin (*Gonystylus* spp.), have been used.

Metals

Brass: An alloy of copper and zinc, common brass, which consists of 63% copper and 37% zinc, is quite suitable for stamping and pressing. A higher

Fig. 2.5. A 12 in. brass rule by Edward Preston, also marked William Landell (Glasgow, 1846–1931), a plane, saw and edge-tool maker who also undoubtedly sold rules and other tools.

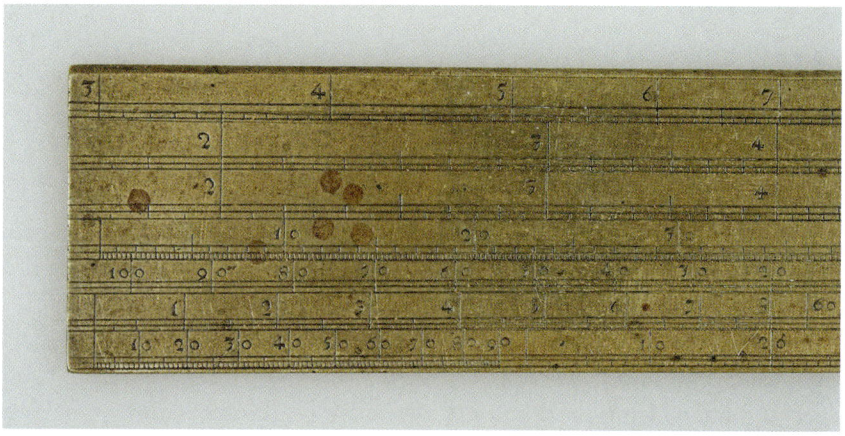

Fig. 2.6a & b. Front and reverse faces of a brass rule by Walter Hayes (London, 1648–84) showing blow holes that were plugged before the rule was divided.

percentage of copper makes a more malleable brass but for rule making common brass is suitable.

The early method of making brass was based on adding the zinc ore, calamine, to the copper. The composition of the brass made by this process is difficult to control and the regular production of brass suitable for rolling into sheets was only achieved in the mid-18th century. Before this date flat items such as rules were made from flat castings which were then filed up. This method of manufacture, which can be an indicator of age, is often shown by the presence of blow holes which have been drilled and plugged. The slight variation in the composition of the plugging material will invariably show up on an item which has not been polished for a long time.

Throughout the 18th century, brass became more readily available and cheaper, and casting methods improved, so rules with blow holes will not be found dating from after that period.

The least satisfactory aspect of brass is that it tarnishes speedily but the rule makers, like the scientific instrument makers, learned that a lacquer coating could keep brass a golden colour for years or even centuries. The problem is that on a rule which is subject to much handling this will wear through and the rule will tarnish.

Nickel silver (German silver, electrum): This alloy, known by a variety

Fig. 2.7. A 6 in. nickel silver rule marked by Jeremiah Leah (Manchester & Salford, <1875>). It is likely that this rule was intended for a pocket drawing set.

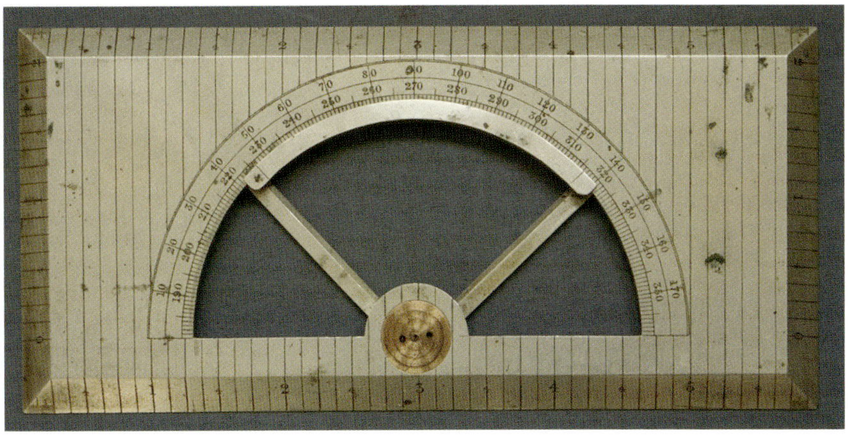

Fig. 2.8. A nickel silver protractor by William Cary, (London, 1789–1891). This is probably a military mapping and sketching protractor.

of names, is silver in colour and has the significant advantage over brass in that it is very resistant to tarnishing. Drawing instruments made of this alloy and kept in a box will remain almost as bright as the day they were made for 100 years. Exposed to the air, particularly in contaminated atmospheres, the material will blacken – but only slowly. The composition is copper 60%, zinc 20% and nickel 20% giving a metal with working characteristics very similar to common brass, although W. F. Stanley, writing in 1873 states that it should be composed of only pure nickel and copper.[6]

Although known in China for centuries, nickel silver only came into widespread use in Europe in the 1830s (a useful guide for dating). The makers of drawing instruments were one of the first to make extensive use of it – having made the compasses, etc. in a set from nickel silver it was only logical to use the same material for the sectors and rules.

Nickel silver looks well with ivory so once available it became the normal choice for fittings on ivory rules. Occasionally boxwood rules will be found with nickel silver fittings – an indication of special quality articles.

Silver and other precious metals: During the 18th century the best quality drawing instrument sets, or at least those that were intended to impress, were often made with silver instruments and scales. Sets of this type sometimes also included a small, all silver, 12 in. 2-fold rule. Small rules in silver, mother-of-pearl, gold plate or even gold continued to be included in fancier étuis, writing cases and sewing boxes right into the 20th century. Most are not hall-marked and when separated from the set are not easy to date.

Steel: Most of the major rule makers produced a range of steel rules and there were also a number of makers who produced only steel rules.

Steel rules are often quite flexible and are resistant to damage from oil, so were popular for engineers' rules. However, they are prone to rust. Chrome

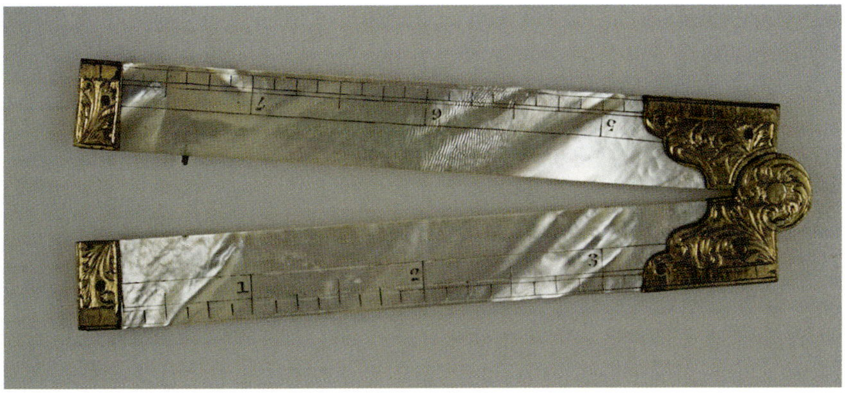

Fig. 2.9. A mother-of-pearl rule with gold fittings. This rule would have been intended for a ladies' workbox or sewing box.

Fig. 2.10. A 2 ft. 2-fold steel rule by I. & D. Smallwood (Birmingham, 1858-1999).

Fig. 2.11. A 12 in. 3-fold, pocket rule by Edward Preston & Sons (Birmingham, 1864–1932).

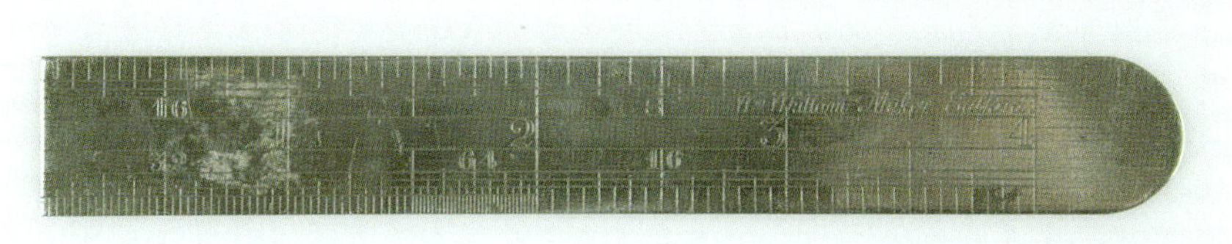

Fig. 2.12. A 4 in. chrome plated rule by William Whittam (Salford, 1880–1911). Whittam was a steel rule maker.

plating was sometimes used to combat this problem and rules made of stainless steel avoid the problem completely.

Steel was used also for spring measures and surveying tapes and bands. In the 1850s, James Chesterman & Co. of Sheffield invented a process for heat treating continuous lengths of steel strip, making steel tapes and spring measures easier and quicker to manufacture.

Iron: There are few commercially-made iron rules, though it was a popular material for early carpenters' squares. However, no doubt if a blacksmith was in need of a rule, he would mark a strip of iron.

Aluminium: Although not a very common material for rules, some can be found.

Invar: An alloy of iron that expands very little when heated. It contains 64% iron and 36% nickel. It was formerly used as a material for making standard length measures. Today it is used for measuring tapes and in instruments such as watches where sensitivity to temperature is of importance. The name was chosen in 1896 by its inventor, Swiss physicist C-E Guillaume, to indicate its invariability.

Ivory: Ivory is not a stable material however long it is seasoned. Leave a 12 in. scale on a drawing for an hour and the ends will curl up 3⁄16 in. off the

Opposite: Fig. 2.13. Standard Invar Bands from a James Chesterman & Co. Ltd. (Sheffield, 1821–1963) catalogue, n.d. but 20th century.

48

Standard Invar Bands

J C Marked with name and trade mark

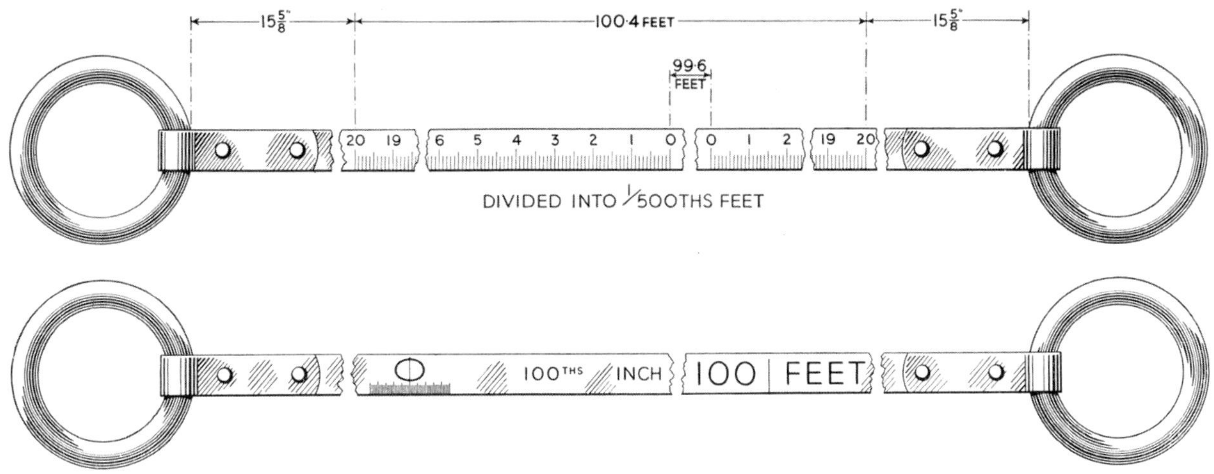

Invar Bands, made from a special alloy steel, are used as standards, and for determining measurements of great accuracy in base line work, etc.

They are generally supplied $\frac{1}{8}$ in. or $\frac{1}{4}$ in. wide, and are divided by fine lines direct on the band as specified below.

The bands are artificially aged and are supplied in one piece wound on a suitable reel.

They are usually made standard at 68° with a tension of 20 lb. when supported throughout, but they may also be made to meet special requirements.

If desired, these bands may be submitted to The National Physical Laboratory for tests in lineal accuracy, coefficient of expansion, etc. Expenses incurred charged extra at cost.

Graduated with a fine line at the terminals and at zero end only, has $\frac{1}{4}$ in. each side of the zero finely graduated into 100ths of an inch *or* decimals of a foot *or* 1 centimetre into millimetres.

P.S.—Bands can be graduated *each* end for a length of 0·4 ft. into 200 divisions at an extra price. See price list

$\frac{1}{8}$ in. and $\frac{1}{4}$ in. **INVAR BANDS** complete on Reels	
Type **1**	Graduated at zero and each 100 ft.
2	Graduated at every 25 ft.
3	Graduated at every 10 ft.
4	**Metre measurement** Graduated at zero and each 25 m.
5	Graduated at every 5 m.
6	Graduated at every 1 m.

100 feet	200 feet	300 feet

25 metres	50 metres	100 metres

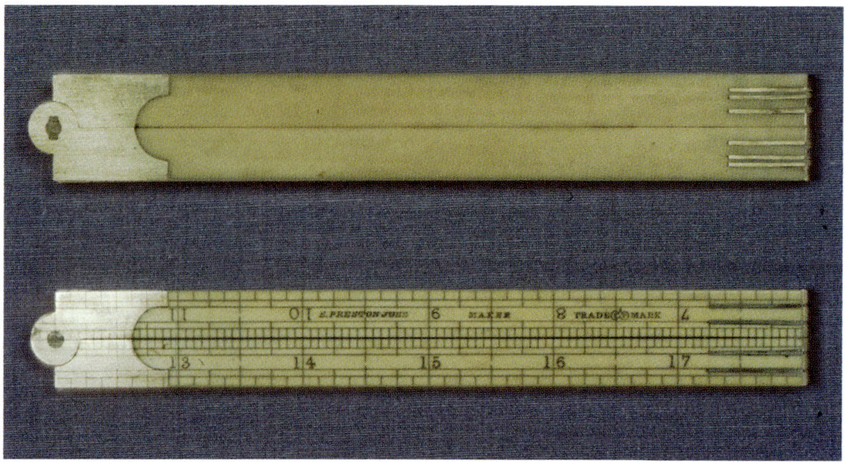

Fig. 2.14. A unfinished ivory rule and a finished ivory rule both by Edward Preston & Sons (Birmingham, 1864–1932).

Fig. 2.15. Green/grey iron staining from the iron interleaves in the hinges on an ivory rule by Patrick Adie (London, 1848–1931>).

paper. Turn it over and an hour later it will have bent in the opposite direction.

The problems of using ivory for drafting scales were well described in the W. F. Stanley & Co. 1919 Catalogue.[7] The firm's recommendation was that boxwood or the celluloid and boxwood composite scales were better.

Stability is not the only problem with ivory; it also discolours and stains. Discolouration from iron is easy to diagnose – the reason for many other pieces becoming discoloured is less obvious.

It is quite probable that the original seasoning has a bearing; it is not unusual to find folding rules with only one stick that has browned or yellowed. In these cases the quality of the original ivory is to blame, not later environmental factors. Ivory will also become stained by those inks that contain iron – but drawing inks, which are based on carbon black, do not usually penetrate and can be removed successfully.

Stanley describes a green semi-translucent ivory which originates in west Africa; this is stated to be a more transparent material.[8] The observant collector will today find some scales which are somewhat translucent.

Whatever its problems may be, until the advent of plastics, ivory was the only fine grained white material available that was suitable for making rules and was the choice for the best quality items.

Fig. 2.16. Whatever its problems in use, in the late 19th century, there was no problem obtaining ivory as shown in this photograph of the ivory warehouse in the London Docks from Round London, an album of photographs of the chief places of interest... *published by George Newnes Ltd. in 1896.*

Fig. 2.17. Four protractor scale rules in, from left to right, boxwood, ivory, brass and vulcanite.

Man-made materials

Occasionally rules are found that were made in man-made materials.

Vulcanite, also known as ebonite, is a hard black material made by the chemical reaction between sulphur and rubber. It was patented in both Britain and America in 1843. In their 1913 catalogue, W. F. Stanley lists vulcanite (hard rubber) scales but states that these are not recommended as they expand greatly with heat. These seem to have been dropped soon after.[9]

Bakelite: Based on a reaction between phenol and formaldehyde, Bakelite was developed in America between 1907 and 1909 by Dr. Leo Baekeland, who was born in Belgium and emigrated to America as a young man. Rules made in Bakelite are not common; those most frequently found are parallel or rolling rules.

Celluloid: Celluloid is a thermoplastic made from nitrocellulose and camphor. Generally regarded to be the first thermoplastic, it was created in 1856 in Birmingham by Alexander Parkes. In 1870 in America, two brothers, John and Isaiah Hyatt, patented a process that produced a similar material and the name "Celluloid" was registered in 1870 as a trademark of the Celluloid Manufacturing Company of Newark, New Jersey, which manufactured the Hyatts' material. There followed some extended legal battles to establish who was the true inventor, the eventual ruling being that it was Alexander Parkes. Translucent celluloid was used for rules and other measuring devices where it was helpful to see through the tool.

Ivorine, xylonite: Also known as zylonite in America, these are names for types of opaque celluloid that closely resemble ivory and which were used as ivory substitutes. It is often difficult to tell the difference between ivory and ivorine; the clue lies in the fact that ivorine has no grain.

Boxwood with celluloid edges: In order to solve the problems of ivory scale rules, from around 1890 scales started to be made with strips of plastic glued to boxwood cores. These were beautifully made and finely divided. Even so, W. F. Stanley, one of the leading makers of this type of scale, refused to guarantee them as "...celluloid was a chemically active substance..." that might shrink. By 1919, W. F. Stanley was producing scales with solid edges

Fig. 2.18. From the 1931 W. F. Stanley catalogue: An illustration showing the different sections of scales of composite construction available. Reverse bevel scales are rarely found.

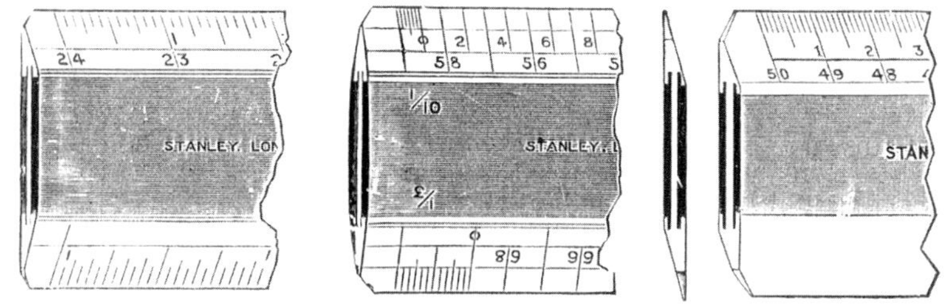

FLAT SECTION. OVAL SECTION. REVERSE BEVEL.

fixed with a tongue and groove joint.[10] By the 1930s the normal construction had changed to a celluloid core with the boxwood reduced to flat strips glued to each side.

Cardboard: Some manufacturers made cheap sets of drawing scales intended for school and student use from cardboard.

Linen (for tapes): One of the problems of linen tape used for making tape measures is its tendency to stretch. In 1842, James Chesterman & Co. of Sheffield patented a method of interweaving metal into linen tapes to address this problem. In 1895, W. F. Stanley wrote that the best tapes were made entirely from green unbleached flax, and made by hand. He also noted that they should only be used in fine weather.[11]

Rule making

To make a basic rule needs no particular expertise or equipment. A lath can be scribed with lines using a point, and then some divisions, which, if desired, can be indicated with a symbol such as a cross or circle.

Even today a bricklayer will sometimes mark out a softwood batten as a story rod – needed if the courses must work a little more or a little less than the usual four to the foot. The bricklayer makes the rod because he needs to repeatedly measure an awkward dimension and multiples thereof (for example 34 courses in 8 ft. 10 in. = $3\frac{2}{17}$ in. per course). If the rod is only needed for a short while, pencil marks will do; if something more durable is needed the divisions will be sawn in.

A carpenter, when setting out, transfers the dimensions to the work piece by making marks opposite the appropriate dimensions on his rule. Alternatively he may use the rule to take the width of a board. In either case he is working by visual comparison with a standard – his rule – and his rule will have been made by reference to a known standard.

Our imaginary carpenter, using his folding rule, at best measured and marked to an accuracy of about $\pm\frac{1}{16}$ in. on small work and $\frac{1}{8}$ in. on larger work. Even today $\pm\frac{1}{4}$ in. is the normal tolerance for building construction with many elements of a building being well outside this tolerance and yet still the finished building will be quite satisfactory.

The carpenter's boxwood rule will not have been made directly from the primary standard but will be at several removes. The accuracy of the rule will thus depend on the skill with which each of the copies has been made and finally the skill with which the final rule has been divided, which is usually very accurate. However the accuracy that the carpenter can achieve when marking out is likely to be much more dependent on his visual acuity and his manual dexterity than in any error in the rule.

The hand scribing of marks onto a wooden lath is the basic technique by which wooden rules were made until the introduction of machinery into the process. Early machinery mostly only duplicated the hand process, scribing lines singly or, in later machines, in multiples.

Rule making – an outline

The first process necessary for making a rule, whether wood or ivory, was to prepare and cut the sticks to size. The brass fittings were made and the rule assembled. Then the lines were scribed and the letters and numbers punched using stamps (also known as marks).

The lines, numbers, etc. were made clear by applying a black filling, the excess being cleaned off leaving only that caught in the scribed lines and the punched letters. For this part of the process, success depended on the surface being prepared to a very smooth finish. After blacking the rules were polished with shellac – this helped to keep the black in the divisions and the brass parts from oxidising, at least for a while. Finally the locating pins were fitted.

Boxwood rule making

Boxwood was imported in logs which had to be broken down into useable sizes. The Rabone firm had their own saw mills for processing from log to lath, but the smaller firms bought from timber merchants some of whom, such as John Hughes of Birmingham (1850–71), described themselves in the trades directories as rule makers' sawyers. These specialist merchants provided the boxwood laths sawn to the exact cross section required for the rule to be made.

As an example, the following steps were required to make a 2 ft. 4-fold rule:

- The sticks (woods) were sawn to both size and length.
- The sticks were milled out for tips and arch joints, if used, or cut out for a round joint. The slots for the mid-joint plates were sawn in.
- The tips, plates and pieces for the joints were made from brass (or in some cases, iron), either by casting and filing, or die stamping sheet in a fly press.
- The rule was assembled by riveting on the brass fittings. This was known as framing.
- The assembled rule was filed and possibly sanded, to provide an evenly

Fig. 2.19. Sticks for a 4-fold rule, milled out for the tips and arch joint and with the slots for the small joint plates sawn in. These rules came from the Smallwood factory in Birmingham when it closed down in 1999.

Fig. 2.20. Two unfinished rules. These rules have been framed but not yet divided. Also from the Smallwood factory.

Fig. 2.21. Two unfinished rules from the Smallwood factory. Top: This rule has been divided on the outside face but not yet on the inside face. Bottom: This rule is fully divided but has only been numbered on the outer face.

flat surface. (In the 20th century it became the practice to finish by machine sanding – however machines were unable to reach the inner surface near the mid joint and in many rules this is apparent).

- The gauge lines were marked.
- The rule was divided (marked).
- Numerals and other lettering were marked with punches and the maker's name applied and filled with blacking.
- A finish was applied over both the brass and the boxwood.
- The markings were filled and the excess blacking cleaned off.

A report in the *British Mercantile Gazette*, December 1874, describes a visit to the Rabone factory.[12] The reporter was shown round by Mr. Harry Rabone who described the various processes that were used to produce a rule. It is clear from this report that a considerable part of the production was by this date being carried out by machine. During the talk to the Slough Ironmonger's Assistants Association in 1936, previously mentioned, the member of Rabone's staff stated that "the ordinary 2-foot rule contains 52 separate parts and passed through 200 operations" during its manufacture.[13]

The sawing out of the sticks in the saw mill was by means of circular saws. One point of note is the report that the boxwood was imported from Turkey and Persia (modern day Iran).

In the sorting room, where the prepared strips were sorted according to quality and the kind of rule to be made and then stacked, the reporter estimated that there were more than 20,000 dozen woods stored.

It is clear from this report that Rabone were making their own brass parts with a description of long slips of brass being rolled to the required thickness and passed through a steam press to be cut. At this date, Rabone made their joints entirely out of brass (rather than using iron plates for the centre leaves of the joint) as then there was no danger of the joints rusting. The pins were also made in the factory from brass wire, using a specialised machine.

It was noted that once framed, the boxwood was finished by filing and smoothing, although the brass parts were finished before fixing.

Rule makers' planes

Boxwood can only be planed satisfactorily if the blade is set so near to vertical that it makes a scraping cut. When planes were used these were made in the rule makers' workshops to answer specific needs and had no general use and very few have survived. Their use was also limited in that much of the finishing work was done by filing and sanding.

Fig. 2.22. Rule makers' planes from the Corbishley family, rule makers working in London from 1917 to 1964. The irons in these planes are all set at 90°. Using these planes with the irons set at these angles would produce a scraping action rather than the usual cutting action, more suitable for use on boxwood.

The finishing planes intended for the faces and edges of the sticks have narrow blades, at most 1½ in. wide, and frequently narrower. The 1851 Rabone inventory (an extract of which is given in Appendix 3) mentions *edging* tables (i.e. sticking boards, a jig for holding a piece of wood for planing) and *planes.* Many of the rule makers' planes however were used for the making and fitting of moving slides. The *slide* and *size stick* planes which are also recorded in the 1851 Rabone inventory are examples of the specialised planes

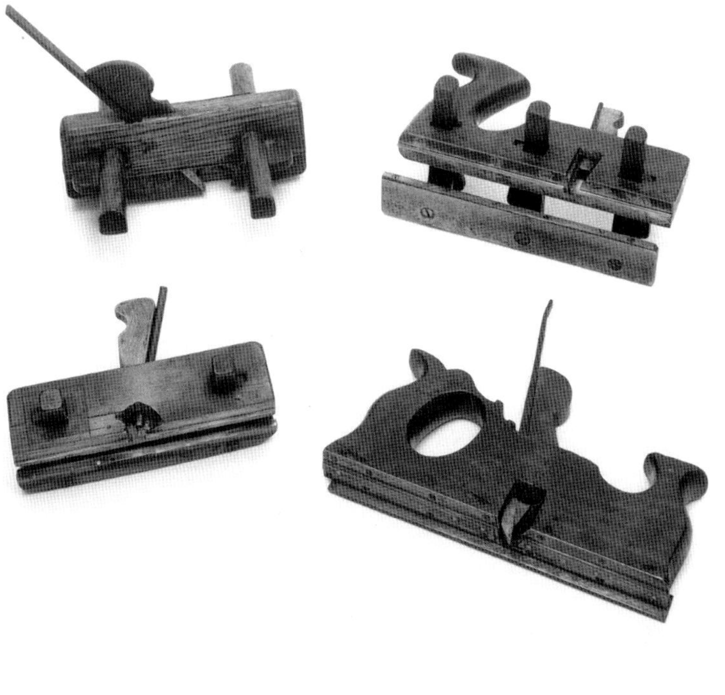

Fig. 2.23. Four different types of special plough planes used in the Corbishley workshop. These would have been the type of plane described in the Rabone inventory as slide and size stick planes. These are likely to have been used to cut grooves for slides. Three have irons set at 70°, the fourth at 45°.

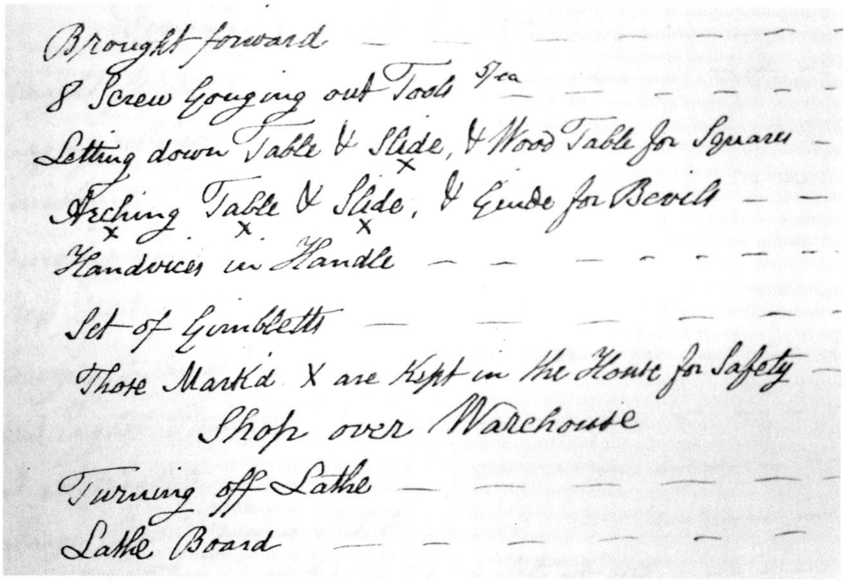

Brought forward
8 Screw Gouging out Tools 5/- ea
Setting down Table & Slide, & Wood Table for Squares
Arching Table & Slide, & Guide for Bevils
Handvices in Handle
Set of Gimbletts
Those Mark'd X are Kept in the House for Safety
Shop over Warehouse
Turning off Lathe
Lathe Board

Fig. 2.24. An extract from the list of tools in the Rabone inventory of 1851.

used in this work. The mention of sticking boards makes it clear that there were jigs and other devices used with the planes to ensure dimensional and locational accuracy.

A group of rule makers' planes used by the Corbishley family in London was found a few years ago. Abandoned by the users in a building that was subsequently sold, they found their way into the antique tool trade and are now in an American collection.

Simple machines

The 1851 Rabone inventory was made before the firm moved to steam-powered machinery and gives a good insight into rule making by hand.[14] Amongst the specialist tools listed are:

- Sawing off table and slide
- Slide rest and gouging out holder on top
- Sawing in table
- One spindle with six saws for tips
- Dividing patterns
- New dividing boards not made up
- Dividing knives hammers and squares
- Gunter punches
- Sector punches and boards

It is evident that jigs and small hand-powered machines were being used to provide speed and accuracy in the preparation of the wooden parts but the division and the marking of the rules was still entirely a hand process. Later, after the move to Hockley Abbey, Rabone did bring much more mechanisation to the manufacture of rules but the methods used by them in 1851 were in all probability similar to those being used by other makers of the time. A description of rule making in Birmingham and of the Rabone rule factory, written by John Rabone in 1865, is given in Appendix 2.

Brass fittings

Before the late 18th century, all the small brass parts were made by casting, clipping and sawing and then filing, a slow and laborious process. One of the major changes to affect the rule making trade – indeed it was what enabled Birmingham to dominate the trade in the 19th century – was the

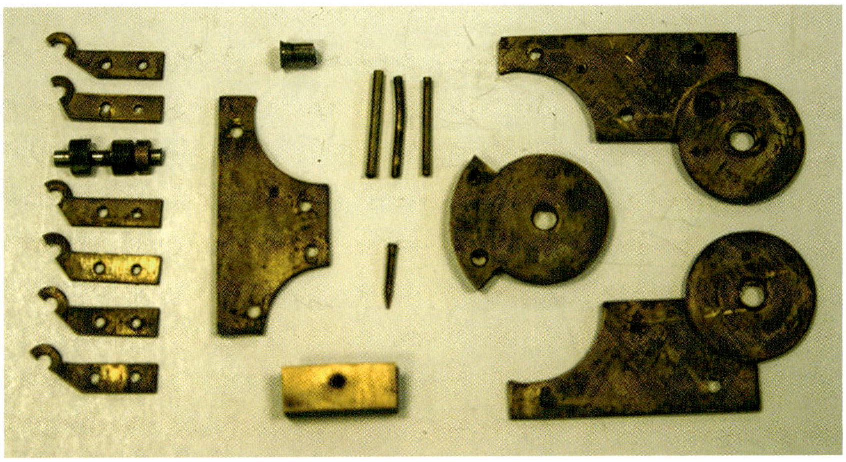

Fig. 2.25. Some of the small brass parts used in rule construction. From the Smallwood factory.

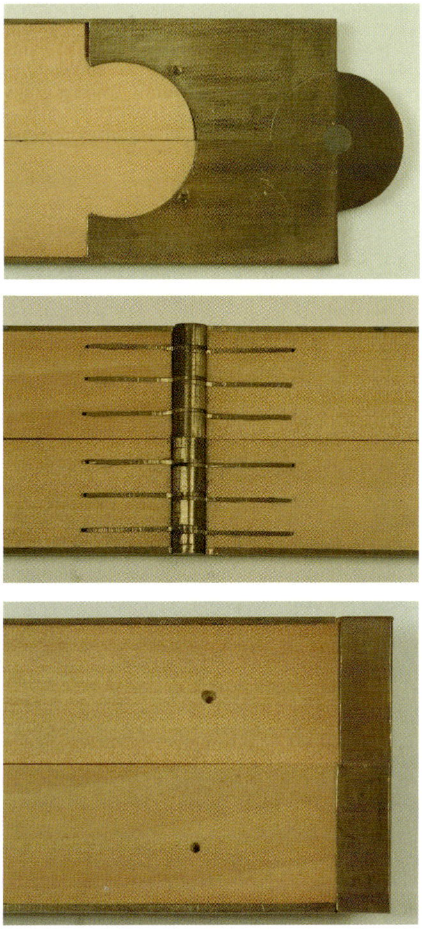

Fig. 2.26a, b & c. The brass parts fixed in position. Unfinished rule from the Smallwood factory.

introduction of the fly press. Invented in Birmingham in 1769 by Richard Ford, the fly press was a hand-operated stamping machine that allowed small metal items to be cut to a desired shape quickly and accurately. It worked by placing a small piece of thin metal between two dies and with one pull of a lever the shape was cut out. This technology was suitable for speeding up the production of many items traditionally made in Birmingham, including rule parts and was quickly accepted by the manufacturers, becoming a standard piece of equipment. There were few other places making sufficient quantity of this type of goods for the presses to be economic. This quick, easy and accurate method of manufacture of the small brass parts for rules was of huge significance.

Whilst some of the larger rule makers stamped out their own brass parts, it is likely that others purchased parts from specialist makers. Regardless, pressing produced parts so cheap as to immediately render hand production uneconomic.

By the last quarter of the 19th century, such specialist joint makers are documented. One such was Frederick Bampton, who issued a list of prices in which he offered a wide range of joints.

Fig. 2.27. Fly presses in a Birmingham factory in 1844. This was a button factory but the presses were the same as those used in the rule making trade.

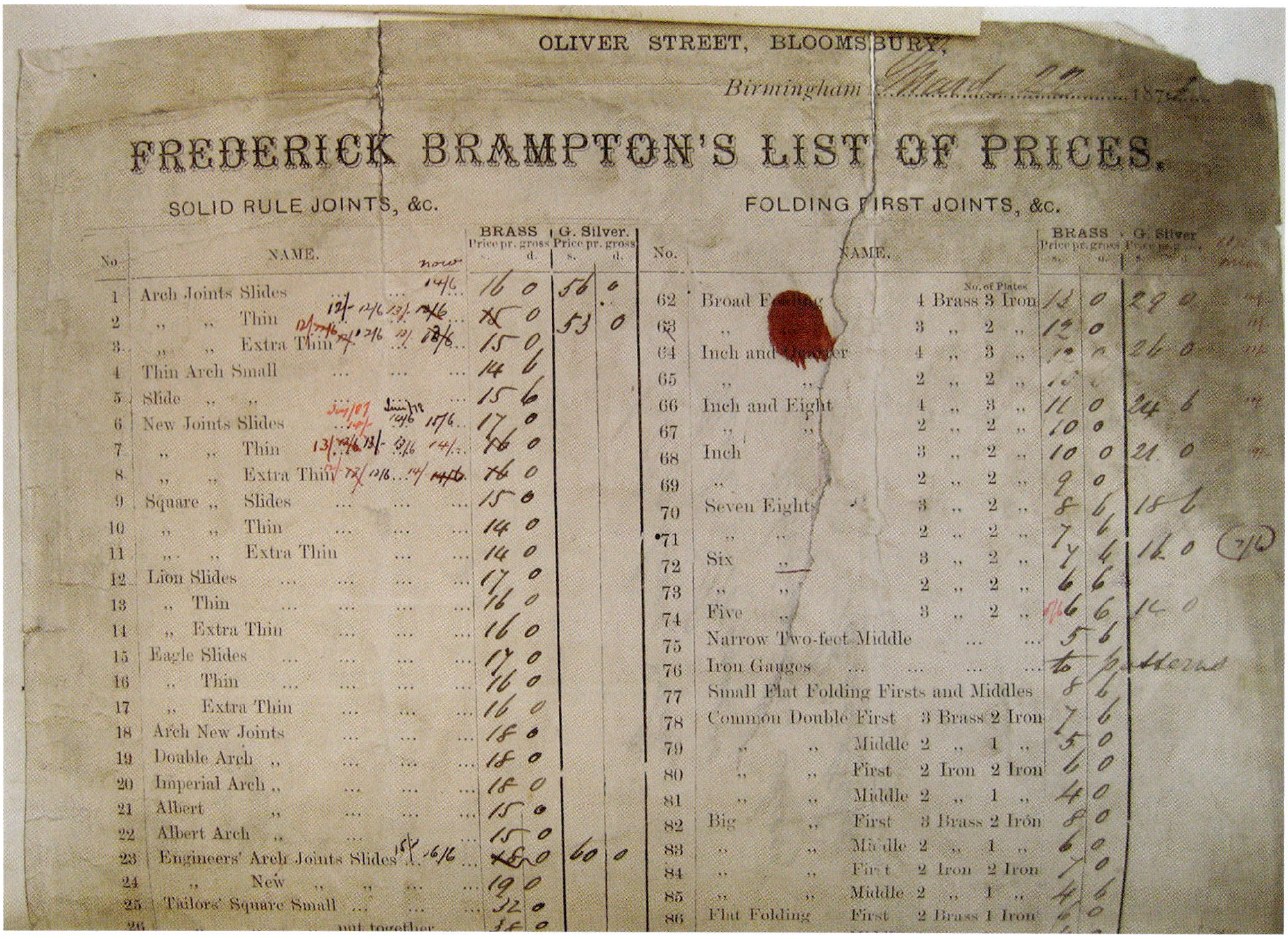

OLIVER STREET, BLOOMSBURY,

Birmingham 187

FREDERICK BRAMPTON'S LIST OF PRICES.

SOLID RULE JOINTS, &c.

No	NAME.	BRASS Price pr. gross s.	d.	G. Silver. Price pr. gross s.	d.
1	Arch Joints Slides	16	0	56	0
2	,, ,, Thin		0	53	0
3	,, ,, Extra Thin	15	0		
4	Thin Arch Small	14	6		
5	Slide ,, ,,	15	6		
6	New Joints Slides	17	0		
7	,, ,, Thin		0		
8	,, ,, Extra Thin		0		
9	Square ,, Slides	15	0		
10	,, ,, Thin	14	0		
11	,, ,, Extra Thin	14	0		
12	Lion Slides	17	0		
13	,, Thin	16	0		
14	,, Extra Thin	16	0		
15	Eagle Slides	17	0		
16	,, Thin	16	0		
17	,, Extra Thin	16	0		
18	Arch New Joints	18	0		
19	Double Arch ,,	18	0		
20	Imperial Arch ,,	18	0		
21	Albert ,,	15	0		
22	Albert Arch ,,	15	0		
23	Engineers' Arch Joints Slides		0	60	0
24	,, New ,, ,,	19	0		
25	Tailors' Square Small	32	0		

FOLDING FIRST JOINTS, &c.

No.	NAME.	No. of Plates	BRASS Price pr. gross s.	d.	G. Silver Price pr. gross s.	d.
62	Broad F[illegible]	4 Brass 3 Iron	13	0	29	0
63	,,	3 ,, 2 ,,	12	0		
64	Inch and [illegible]	4 ,, 3 ,,			26	0
65	,, ,,	2 ,, 2 ,,				
66	Inch and Eight	4 ,, 3 ,,	11	0	24	6
67	,, ,,	2 ,, 2 ,,	10	0		
68	Inch	3 ,, 2 ,,	10	0	21	0
69	,,	2 ,, 2 ,,	9	0		
70	Seven Eights	3 ,, 2 ,,	8	6	18	6
•71	,, ,,	2 ,, 2 ,,	7	6		
72	Six ,,	3 ,, 2 ,,	7	4	16	0
73	,, ,,	2 ,, 2 ,,	6	6		
74	Five ,,	3 ,, 2 ,,	6	6	14	0
75	Narrow Two-feet Middle		5	6		
76	Iron Gauges		to patterns			
77	Small Flat Folding Firsts and Middles		8	6		
78	Common Double First	3 Brass 2 Iron	7	6		
79	,, ,, Middle	2 ,, 1 ,,	5	0		
80	,, ,, First	2 Iron 2 Iron	6	0		
81	,, ,, Middle	2 ,, 1 ,,	4	0		
82	Big ,, First	3 Brass 2 Iron	8	0		
83	,, ,, Middle	2 ,, 1 ,,	6	0		
84	,, ,, First	2 Iron 2 Iron	7	0		
85	,, ,, Middle	2 ,, 1 ,,	4	6		
86	Flat Folding First	2 Brass 1 Iron				

Above: Fig. 2.28. Price list from Frederick Bampton, Birmingham, rule joint maker, dated 1873.

Below: Fig. 2.29. A set of salesman's sample rules from Sampson Aston showing the wide variety of small brass hinges used by this firm. Sampson Aston was a Birmingham rule maker who ceased trading in the 1870s.

Fig. 2.30. Assembling barrel measuring shoulder rods in the Dring & Fage factory.

Assembly

Once the parts had been made, the rule was assembled using pins. The pins were formed from soft brass wire which mushroomed out to create the smallest of heads when the pin was tapped with a hammer. Too much force would quickly deform the brass plates or bend the pin, splitting the boxwood. The 1865 description of this pinning process at the Rabone factory says that it was done by women "with a few taps of the hammer".[15] One of the secrets is to have a pin cut to exactly the right length but even so the operatives must have had considerable skill. They, like all production workers of the 18th and 19th centuries, became highly skilled and quick from constant repetition of the same minor tasks.

Dividing/marking

The process of dividing (marking) a simple rule by hand was as follows.

The longitudinal lines (gauge or limit lines) were scribed with a tool akin to a woodworkers' marking gauge.

The rule was then placed in a jig, usually described as a dividing board, alongside a pattern, i.e. a master rule, and using a scribing knife against a dividers' square, the graduations were cut. In its simplest form the square was aligned to the pattern by eye. In the more developed form the pattern was made with some form of register to help the operative quickly and accurately

Fig. 2.31. The tool used for marking the gauge lines. From the Smallwood factory.

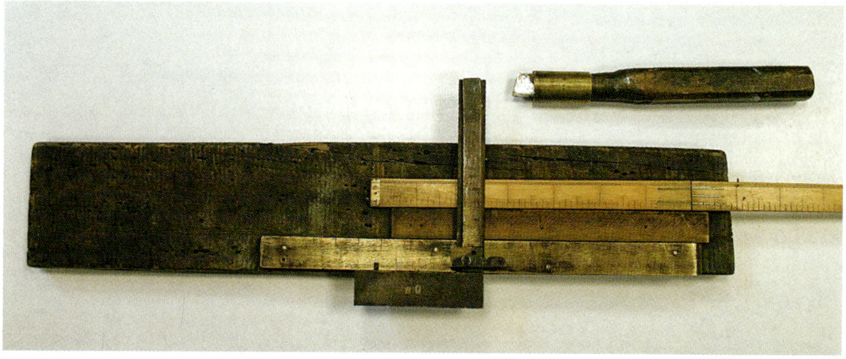

Fig. 2.32. A dividing board shown with the square and dividing knife. From the Smallwood factory.

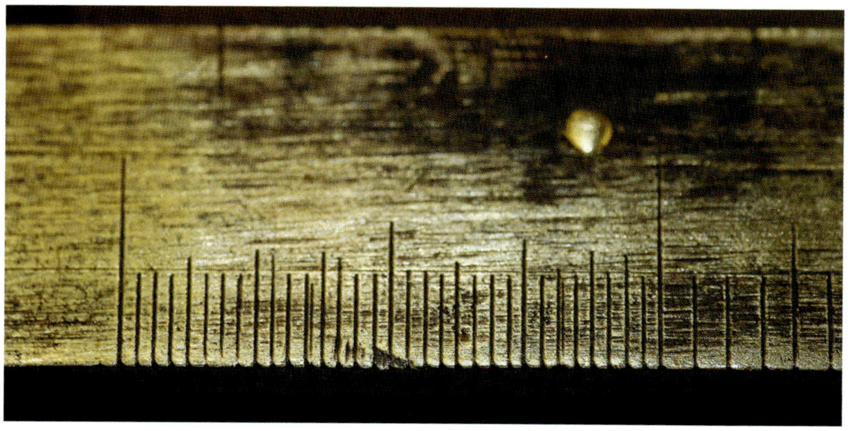

Fig. 2.33. The dividing pattern showing the inscribed lines into which the pin in the underside of the dividing square registers.

Fig. 2.34. The dividing square showing the pin on the underside that fits into the inscribed lines on the dividing pattern.

RULE MAKERS' KNIVES.—**HAND FORGED. SCALE TANG ROUND HANDLE.**

No. 5521

	$6\frac{1}{2}$	7 inches.
No. 5521 **Sycamore Handles, 2 Iron Pins, Scale Tang**	21/3	23/3 per doz.
No. 5602 **Yellow Ferruled Handles, Whittle Tang**	20/3	22/3 ,, ,,

Fig. 2.35. Rule makers' knives from the John Wilson (Sheffield) Ltd. Illustrated Price List of Butchers' Knives, Steels, Skinning, Stitching, Farriers' Shoe, Sheath, and Dagger Knives, etc. *n.d. but probably 1920s. This knife appears to be a different pattern than that used in either the Smallwood factory or the workshops of W. H. Harling.*

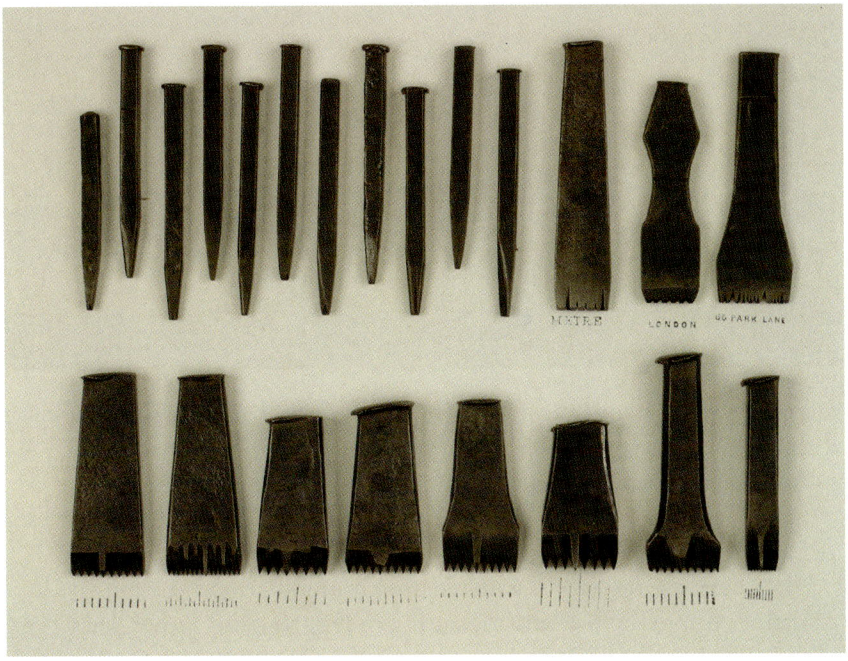

Fig. 2.36. A set of rule makers' marks (punches) from the workshop of James or Herbert Adams (Liverpool, 1873–1932) including number stamps, stamps for divisions, an address stamp "66 Park Lane" and stamps reading "Metre" and "London".

position the square for striking each line. The divisions were made from the gauge lines outwards; at the edge of the rule the knife ran out onto a piece of scrap.

Marking the divisions would seem to be a laborious task but a skilled person with the simple but effective equipment of a dividing board and square worked surprisingly quickly. It is recorded that a good worker could mark the 540 lines of a Gunter scale in 10 minutes – so a 2 ft. 4-fold rule with open divisions might only take a couple of minutes.

Evidence from a collection of marking punches from the workshop of James or Herbert Adams of Liverpool (1873–1932) suggests that some rule makers used punches to divide rules, though whether this was in addition to or instead of scribing is not known.

The accuracy of hand division of slide rules did lead to much adverse comment by later 19th century makers such as W. F. Stanley but he was seeking to promote the products of his dividing engine.[16] However for the tradesman's

Fig. 2.37. The dividing room in the Dring & Fage factory.

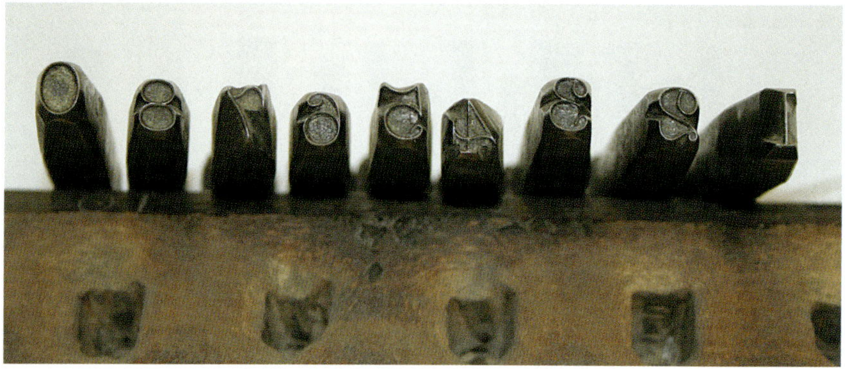

Fig. 2.38. A set of number stamps from the Rabone Chesterman factory in Sheffield which closed in the early 1990s.

folding rule the standard of accuracy obtained by hand division was more than adequate.

An interesting description of the method of common dividing can be found in Tomlinson's *Cyclopedia of Useful Arts and Manufactures* published in 1858, reprinted in Stanley's *Source Book for Rule Collectors.*[17]

After dividing the rule was marked with number punches and the lines named if required. On a rule measuring only in English inches, these were not usually named; however, on more complex rules with several measures of inches from different countries, the lines frequently were. In these instances, the English measure would usually be named "London". The maker's name might also be marked, sometimes with an address.

A very clear description of the hand dividing and marking of numerals on boxwood scale rules in 1963 has been written by David Goodson.[18] He visited the W. H. Harling workshops in Clapton, East London and recorded in some detail the making of scales both by hand division and on a dividing machine that made only one scale at a time. All the numerals were still marked by hand using the traditional method of hand-held punches.

This was the last days of hand working at the firm. The following year, in 1964, the company amalgamated with Blundell Rules Ltd. and the hand making of scales soon finished.

Mechanisation

By the middle of the 19th century, spurred by the great demand for the basic 2-fold and 4-fold rules, some manufacturers experimented with alternative

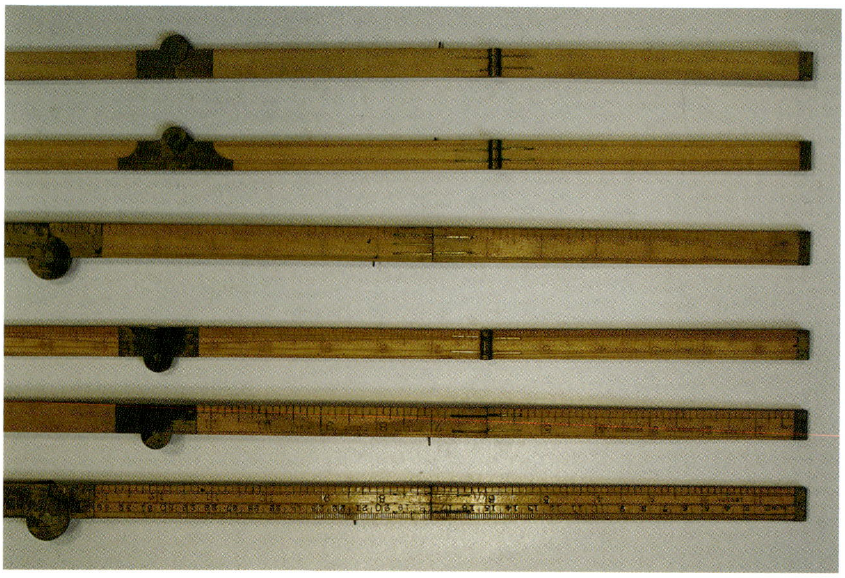

Fig. 2.39. Stages in dividing and marking rules. From the top: An unmarked rule; the gauge lines marked; the divisions marked; the numbers stamped; the markings blacked; the final rule, polished to complete. These rules all came from the Smallwood factory.

stamping or rolling techniques for dividing and applying the numerals, but these processes proved to be far more difficult to mechanise than might be supposed. The capital cost of doing so was considerable and the quality of the product was clearly poorer than the scribed rule.

The work that goes into the production of even an ordinary 2 ft. 4-fold rule is considerable. The rule has four wooden sticks, 26 metal parts in the joints, 18 rivets, and 3 pins. In addition to the longitudinal lines there are 384 lines if the rule is divided into ⅛ths and possibly around 500 if some part is divided into 1/16ths. About 68 numerals (some may be omitted near the joints) are required. These figures are for a simple rule; on a Routledge engineers' rule there are around 1400 inscribed lines and about 1400 letters or numerals. The earlier makers of all rules made each of these marks one at a time so that having the best possible aids to speed and accuracy was important. For this reason, later rule makers strived to apply machinery to all those parts of the work where it was possible.

Engine dividing

Although by 1777 Jesse Ramsden had made a practical dividing engine and others soon followed, these were machines for the accurate division of circles into precise increments, a requirement of many types of scientific instruments.

In 1861 W. F. Stanley constructed a straight line dividing engine intended to mark drafting scales more accurately than was attainable by hand. The machine was able to divide any space into an equal number of parts each of which could be as small as 1/200th of an inch. Mostly as the result of the fine work of which this machine was capable, Stanley received a medal for his mathematical instruments at the 1862 International Exhibition.

John Rabone, writing in 1865, is, perhaps deliberately, vague about the machinery used in the works for this purpose.[19] He mentions that machines had been made earlier by other Birmingham rule makers but these did not work (although there is an implication that some types of machine were in use in his factory). He states that the normal way of marking rules is with the dividing knife.

The rule makers' dividing machine

Towards the end of the 19th century, machines with multiple knives and capable of dividing 20 or more rules at one time became quite usual in the trade. This type of machine was sometimes called a pantograph but bears little similarity to the enlarging/reducing device for drawing normally associated with this name, other than that the dividing heads are controlled by a pointer that followed the pattern to be replicated. In some machines the master was followed by the operator whilst others were automatic – following the master until all the cuts were completed. With such machines the positioning and securing of the rules must have been the major manual task.

Evidence from the Rabone archive indicates, throughout the 19th century, a steadily increasing level of mechanisation in the John Rabone & Sons manufactory in Birmingham.[20] The other Birmingham makers in the mass market trade no doubt were also gradually mechanising but for the makers of short run items, and these were the majority, the methods of production changed but little over the three hundred years from 1660 to 1960. Oral evidence has indicated that in the Dring & Fage workshops the division of rules and the marking of numerals continued as a manual process until the firm closed in 1977.

Fig. 2.40. Three rules showing the effect of lines on the edge. Top: Without lines by John Rabone & Sons (Birmingham, 1877–1963). Centre: With a single line by I. & D. Smallwood (Birmingham, 1858–1999) Bottom: With two lines by Edward Preston & Sons (Birmingham, 1864–1932).

Inscribed lines on the edge

Before 1914 any quality rule was likely to have longitudinal lines inscribed on the edges, close to the faces and to line with the inset joint plates. The only purpose of these lines was to neaten the appearance of the inset plates. Possibly the lines are a residual feature from the time when the rebating was by hand and marking lines were needed but as the blanks were filed and finished after assembly, this is not an entirely convincing suggestion.

By the 1920s some rule makers, notably Rabone and Smallwood, were making rules with only one line on the edge and after 1945 many rules were made with no lines on the edge. Thus for the collector the presence or absence of lines is an indicator, though an imperfect one, of date and quality of the rule.

Filling and finishing

After dividing and numbering, the rules were finished by filling the lines and division with blacking.

In the Rabone factory, this operation was carried out by hand in the Rushing Shop. The blacking used was a mixture of charcoal and oil and the surplus cleaned off, originally with rushes (hence the name of the shop) but later by rubbing the rule in a box of sawdust. The rules were then finished with a coat of shellac and finally hand polished.[21]

Goodson records in the W. H. Harling workshops this material was a mixture of charcoal dust, whiting and linseed oil.[22]

The use of charcoal dust was perhaps because it was less fine than other black pigments and therefore less likely to become lodged where it was not required.

Impressed rules

The advantage in speed and cost to the rule maker of being able to print or impress all the marks on a rule from a master die seems obvious. However, the difficulties in obtaining a satisfactory result are greater than they appear. John Rabone relates an attempt by a Wolverhampton rule maker in about 1790 to mark rules by passing them between a pair of steel rolls, made at great expense.[23] The process was not a success.

The idea of pressing the markings onto a rule remained with the rule makers and did finally meet with some success in the early years of the 20th century but even then the rules made were of lower quality and the markings less permanent.

Fig. 2.41. An EesEseE rule by Edward Preston & Sons (Birmingham, 1864–1932) and below, an impresed rule by I. & D. Smallwood (Birmingham, 1858–1999).

In 1912 Edward Preston first listed the EesEseE rule which was marked with bold black figures applied in a press.[24] The divisions were also boldly marked but were still marked traditionally. The rule was described as "very useful to persons with defective eyesight" thus making a virtue out of what was the first step in the introduction in Britain of the impressed rule.

By the 1920s Preston, Smallwood and Rabone were all offering rules with extra heavy figures. It is surprising to note that these rules were offered at higher prices than the normal type of rule. In the Rabone Catalogue No. 24 (probably 1924), 2 ft. 4-fold arch joint boxwood rules 1⅜ in. wide were 22*s* 6*d* per doz. whilst the equivalent "Blindman's rule for use in mines and where light is not good" cost 29*s* 0*d* per doz.[25] A similar price differential is also apparent in the 1935 catalogue. Clearly the English manufacturers did not initially see the impress process as a means of making cheaper rules.

After 1945, with increasing labour costs the impress process was used more and more and after 1960 almost all rules were made this way.

References

1. Stanley, W. F., *Catalogue of Surveying and Drawing Instruments*, London, 1919.
2. Rees, Jane & Mark, *Christopher Gabriel and the Tool Trade in 18th Century London*, Needham Market, 1997.
3. Stanley, Philip E., *Boxwood & Ivory. Stanley Traditional Rules, 1855–1975*, Westborough, MA, 1984.
4. From the collection of scrapbooks kept by John Rabone & Sons from the late 19th century onwards, now in the Hawley Collection Trust, Sheffield.
5. John Rabone & Sons, *Catalogue No. 24,* Birmingham, n.d. but probably 1924.
6. Stanley, W. F., *A Descriptive Treatise on Mathematical Drawing Instruments,* 4th ed., London, 1873.
7. Stanley, W. F., 1919, *op. cit.*
8. *Ibid.*
9. Stanley, W. F., *Catalogue of Surveying and Drawing Instruments*, London, 1913.
10. Stanley, W. F., *Surveying and Levelling Instruments,* London, 1895.
11. *Ibid.*

12. Rabone scrapbooks, *op. cit.*
13. *Ibid.*
14. Records of John Rabone & Sons, MS 1522, Archive Department, Birmingham Library Services.
15. Timmins, Samuel ed., "Measuring Rules" by John Rabone, *Birmingham and the Midland Hardware District: The Resources, Products and Industrial History*, London, 1866.
16. Stanley, W. F., 1873, *op. cit.*
17. Tomlinson, Charles, "Graduation", *Cyclopaedia of Useful Arts & Manufactures,* ca. 1858. Reprinted in Stanley, Philip E., *A Source Book for Rule Collectors*, Mendham, NJ, 2003.
18. Goodson, David, "Crafting Scale Rules by Hand – An Historical Note" *Journal of the Oughtred Society*, Vol. 7, No 2, 1998.
19. Timmins, *op. cit.*
20. Records of John Rabone & Sons, *op. cit.*
21. Recorded conversation with a retired managing director of John Rabone & Sons. The Hawley Collection Trust, Sheffield, ref. AV37 & 38.
22. Goodson, *op. cit.*
23. Timmins, *op. cit.*
24. Edward Preston & Sons, *Catalogue No. 18. The "Preston" Catalogue,* 1909 with 1912 supplement, reprint, Mendham, NJ, 1991.
25. Rabone *Catalogue No. 24, op. cit.*

Chapter 3
Rule development

Seventy years ago the ordinary thickness of the boxwood carpenter's rule was not less than a quarter of an inch ... Many rules have been made of only the sixteenth of an inch in thickness, this small size of wood having to be further cut away to admit of the insertion of the joint, which is inlaid so as to lie level with the surface of the wood. Timmins, 1866.[1]

There are a number of features of rules that show a distinct chronology of development. Some are applicable to all rules, whether folding or not, square or flat section. Some are applicable only to folding rules and some only to flat stick rules.

They are all helpful in judging the approximate date of a rule, although care needs to be taken in assessing these as old-fashioned features often continued in use alongside more up-to-date ones.

Number stamps

It should be remembered that all number stamps were hand cut, so some variation in style is inevitable. The following are generalisations that usually apply to English rules of the given dates. The size and style of the numbers are one of the best indicators of the age of a rule, particularly with early ones, and can also be used to assist in dating both measuring and slide rules.

Up to late 17th century: The numbers are very small with some distinct characteristics. The number 1 has a flat head and foot, similar to an upper case I; the number 2 has a flat bottom; the number 3 has a flat top; the number 7 slants considerably; numbers 6 and 9 distinctly slant and are marked with the same stamp; number 0 is totally round.

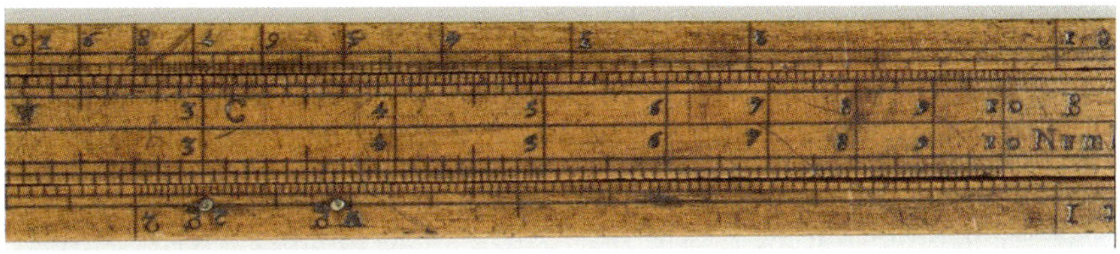

Fig. 3.1. A square section 2-slide slide rule, 12 in. long, by Isaac Carver (London, 1668–1713), dated 1690.

Early to mid-18th century: The style of the numbers has not changed but they are noticeably larger.

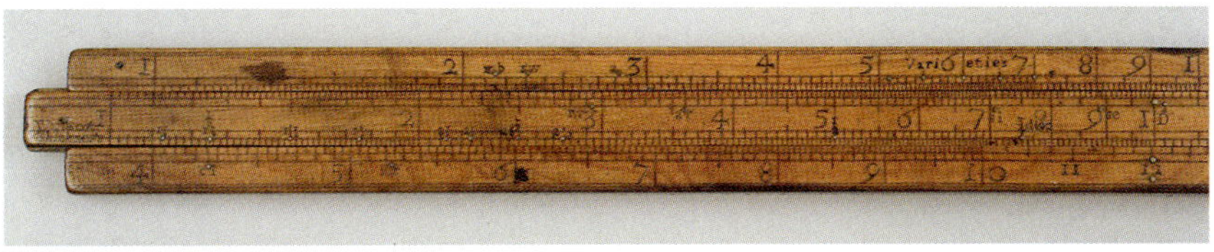

Fig. 3.2. A flat section one-slide slide rule, 18 in. long by Jacob Carver (London, 1717–42) dated 1717.

Mid-18th century: The numbers are yet larger still. The number 1 is still like an I; the number 2 has a more closed top; the number 3 is still flat

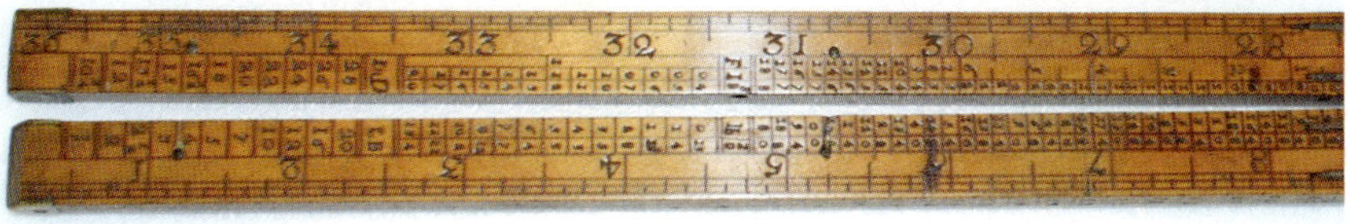

Fig. 3.3 A 2 ft. 4-fold rule by John Fry (London, <1747>).

Fig. 3.4. Slides from two square section 12 in. Customs & Excise slide rules, the top dating from the late 17th century, the lower from the early 18th century, showing the change in size of the numbers.

topped; the number 7 is less slanting; number 0 is becoming oval. All the numbers are more rounded than the earlier style.

Late 18th century: The numbers are still the same size as in the mid-century but the number 1 is now in a modern style (rather than like an upper case I); the number 3 has a rounded top; number 0 is more oval.

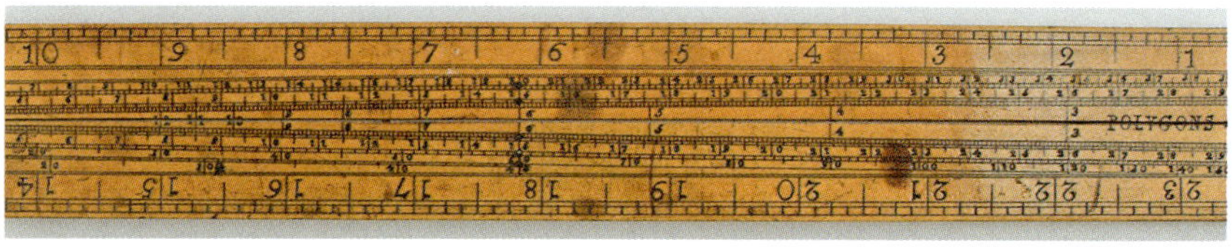

Fig. 3.5. A 2 ft. 2-fold rule by Robert Corson (Wolverhampton, 1767–70).

Ca. 1800: The numbers are larger and even more stylised. The number 2 has a top loop that is almost closed; the numbers 6 and 9 have curled round tails; the top bar of the number 7 is curved downwards at the end and the numbers 8 and 0 become increasingly more oval.

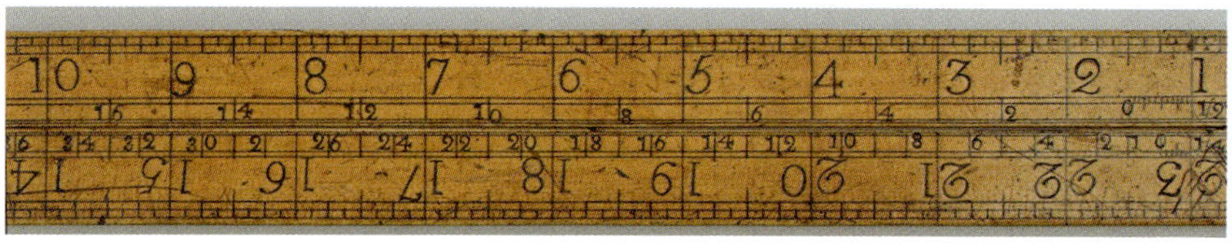

Fig. 3.6. A 2 ft. 2-fold rule by Thomas Onions (Birmingham, 1767–1801).

Early to mid-19th century onwards: The numbers have become much more regular and rounded, with tight circles. The number 7 has a distinct downward serif; 0 is completely oval.

Fig. 3.7. A 2 ft. 2-fold rule by Laban Cooke (London, 1826–35).

20th century: Although all makers continued to produce rules with stamped numbers, rules were also produced with bold black figures applied in a press, with stylised and bulbous numbers.

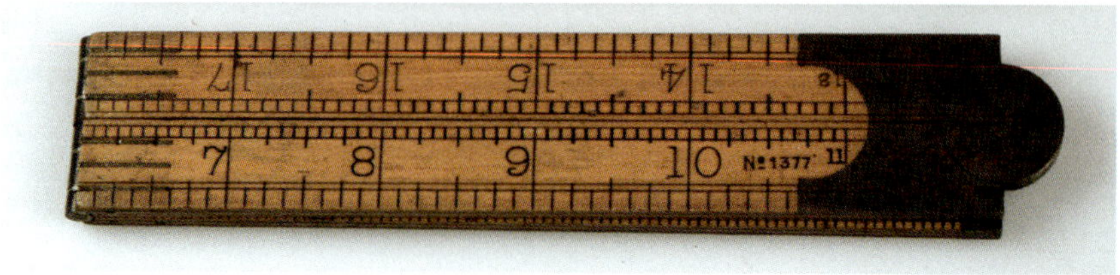

Fig. 3.8. A 20th century stamped rule by John Rabone & Sons (Birmingham, 1877-1963).

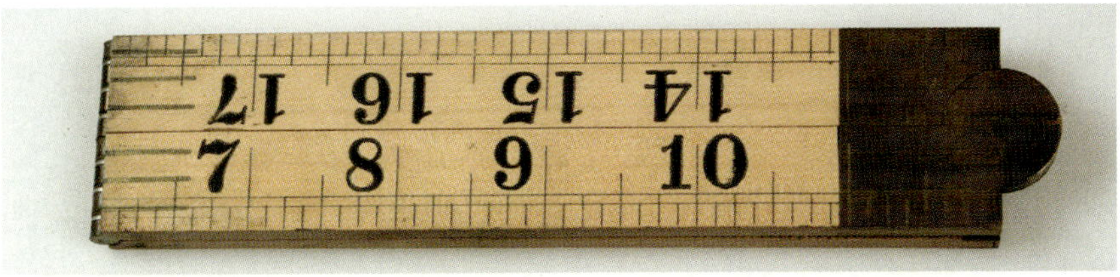

Fig. 3.9. A 20th century pressed rule by I. & D. Smallwood (Birmingham, 1858-1999).

Lines and divisions

The arrangement and number of both the gauge (limit) lines, marked horizontally on the rule, and the division lines, marked vertically, changed over time and can be significant in establishing the date of a rule.

17th century to circa 1775: Rules had three equally spaced gauge lines set in from the edge of the rule to which the ⅛ and ¼ divisions ran. The inch divisions terminated in a double gauge line, a feature which continued until the early 20th century.

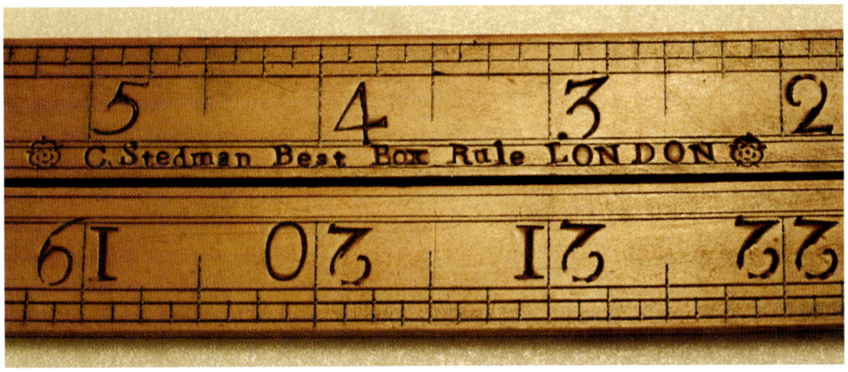

Fig. 3.10. A 2 ft. 2-fold rule, by Christopher Stedman (London, 1747–85).

Mid-18th century to late 19th century: For a long period rules were marked with three unequally spaced gauge lines with the larger space near the outer edge but set in from it.

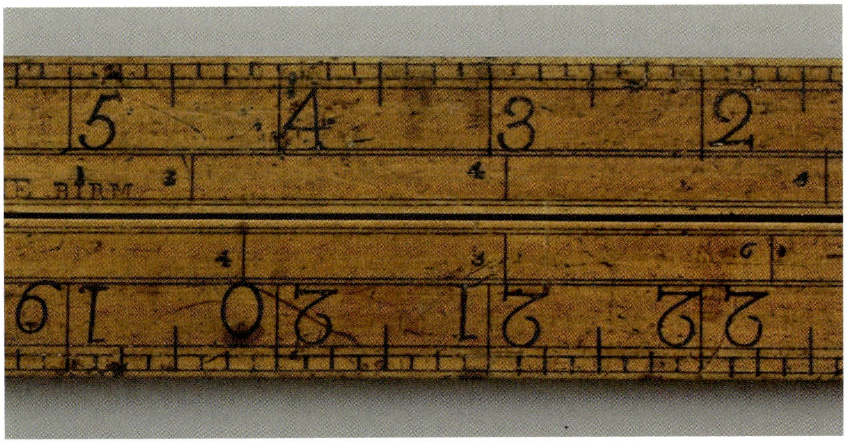

Fig. 3.11. A 2 ft. 2-fold rule by Hassall & Rabone (Birmingham, 1784–91).

Mid-19th to early 20th century: By the middle of the 19th century, the divisions on rules were taken right to the edge of the rule and there were two gauge lines marked that were used for designating the length of the ⅛ and ¼ divisions. There appears to be a time overlap in the use of this arrangement of lines with the earlier one.

20th century: On cheaply made rules, there was a single line from which the ⅛ divisions were taken to the edge of the rule.

Fig. 3.12. A 3 ft. 4-fold rule by Edward Preston & Sons (Birmingham, 1864–1932).

Direction of numbering – right to left/left to right

Rules made before the introduction, in the UK, of the imperial system by the Weights and Measures Act of 1824 will invariably measure from right to left. One of the requirements included in that act was that rules were to be marked from left to right and henceforth, therefore, all rules were marked in this direction.

However, rule makers in America continued in the old right to left arrangement. The direction of measurement is thus a useful tool in deciding whether a rule may be of British or American manufacture as well as a useful tool for determining the earliest date of a rule.

Fig. 3.13. A 2 ft. 2-fold rule by I. & D. Smallwood (Birmingham, 1858–1999).

Stick thickness

As manufacturing techniques improved, and in particular as the design of the rule joint changed to include proportionally more metal and less wood, rules became thinner.

Fig. 3.14. Top: A rule by Laban Cook (London, 1826–34) which reads from right to left. Bottom: A rule by Edward Preston & Sons, (Birmingham, 1864–1932) which reads from left to right. This rule is also marked T. C. Cooke & Son, 497 Harrow Road., probably the retailer.

17th century: The typical 17th century folding rule is a little less than ½ in. thick. George Atwel[l], in his instructions for making a carpenters' rule in *The faithfull surveyour …* published in 1662, states that the rule should be "a third part of an inch thick".[2]

Fig. 3.15. A 2 ft. 2-fold rule dated 1679. Thickness of sticks: 15⁄32 in.

18th century: Rules made during the 18th century generally measure a little over ¼ in. in thickness, a size that seems to have remained remarkably constant throughout the century.

Fig. 3.16. A 2 ft. 2-fold rule by George Wood (Wolverhampton, <1761>). Thickness of sticks: 5⁄16 in.

Toward the end of the century, rules became slimmer, and were nearer to ¼ in.

Fig. 3.17. A 2 ft. 2-fold rule by Thomas Onions (Birmingham, 1767–1801). Thickness of sticks: 9⁄32 in.

19th century: By about 1820 most boxwood rules were being made to a thickness of around 3⁄16 in. and this remained the usual dimension until boxwood rule making ceased.

The thickness of rules is a reflection partly of the skills and techniques that were available, but it may also have been a matter of aesthetics and the scarcity of materials. As the rule joints became more sophisticated, it was possible to make thinner rules which were considered more desirable. A shortage of boxwood may also have influenced the thickness.

The problems of making thinner rules are most acute for rules which include a log or other slide, and many rules did. There is a practical limit to how thin a boxwood slide can be made and the same is true of ivory, though ivory can be made a little thinner than boxwood. Only when hardened thin

Fig. 3.18. Two rules by John Rabone & Sons (Birmingham, 1877–1963) showing a regular thickness rule and an extra thin rule. Note that on the extra thin rule the joint, in this case, a Grecian joint, is not let into the rule.

brass became available (probably around 1800) was the challenge of making durable but thin slides solved. Even then it seems to have taken some years for the makers to understand that they could now fit slides into the thinner 3⁄16 in. rules.

The production of ever thinner rules seems to have been something of an ideal with rule makers. By the 1860s Rabone were making "extra thin rules" which were about half as thick as normal – only 3⁄32 in.! They are masterpieces of the rule maker's art. The supposed advantage was that they were flexible but very few were made and perhaps not surprisingly, even fewer have survived.

A word of caution, however, regarding the use of thickness to date a rule. One should not be misled by mid-19th to 20th century rules made "stout" for extra strength.

Joint design

The design of the joint in a folding rule can also be a useful indicator of the age of a rule. The earliest style of hinge that has survived in any quantity dates from the mid-17th century, or slightly before, and continued in use until the late 17th century.

Mid- to late 17th century: This early hinge takes the form of two wooden knuckles formed as an integral part of one leg with a central tongue of brass plate. The joint pivots on a large round brass pin, often peened over a large, usually square brass escutcheon.

Fig. 3.19a & b. A 2 ft. 2-fold rule, unnamed but dated 1679. Construction: Wood-brass-wood with a square washer and brass rivet.

Late 17th century to ca. 1775: Around the mid-17th century, a transitional type of joint was developed which continued to be used until around 1775. The construction was as earlier but the outsides of the joint were reinforced with brass plates that extended along the rule to form rectangular plates.

Fig. 3.20a & b. An anonymous 2 ft. 2-fold rule, probably late 17th century. Construction: Brass-wood-brass-wood-brass with a brass rivet, large knuckle and rectangular plate.

Mid- to late 18th century: By the mid-18th century, all brass joints appeared, formed by increasing the thickness of the plates at the knuckle so that the tongue pivoted directly between them. There were different qualities of joint, some with and some without plates, but all had a brass rivet.

A popular belief is that "early" (pre-1800) rules had small (7/16 in.) diameter knuckle joints and that later rules had large (1 in.) diameter joints. However, this statement requires qualification. The "small" diameter joints are found only on rules with slides. Using a small diameter joint avoided the complica-

Fig. 3.21. A 2 ft. 4-fold rule with an all brass joint dated 1756. This simple unplated joint is identical to that used on sectors.

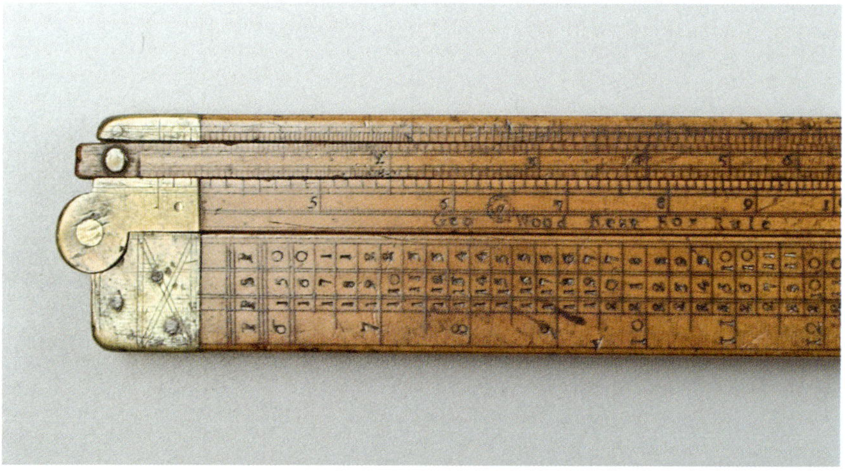

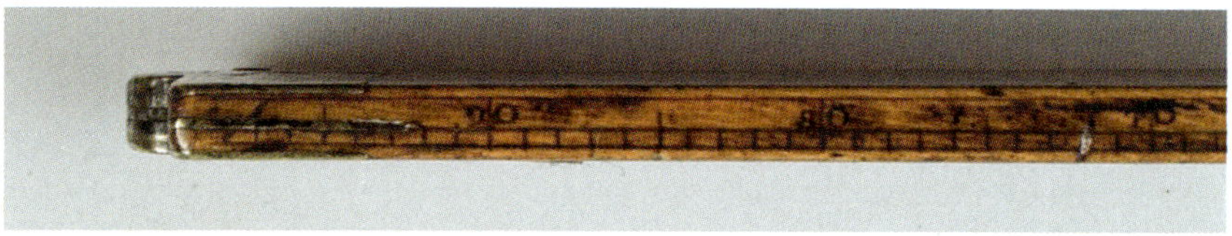

Fig. 3.22a & b. A 2 ft. 2-fold rule by George Wood (Wolverhampton, <1761>). Construction: All brass with brass plate and brass rivet, a small knuckle and a square plate.

tions of cutting the recess for the slide through it. This feature seems only to have lasted for a relatively short period of time, during the third quarter of the 18th century, though, like all features, there are a few examples that were made later. Early rules without slides may well have larger knuckles.

Towards the end of the 18th century, rule makers started to make rules with a groove cutting into the knuckle so the knuckle could be larger and therefore stronger, and most rules with slides were made this way.

Late 18th century onwards: By the late 18th and early 19th century, rule joints were still made completely of metal but with the brass tongue being replaced with an iron one. The pin was also made of iron. The reasons for this have not been established; it may have been an attempt to increase the strength of the hinge and/or to reduce the cost. There was also a belief at that time, quite erroneously, that two parts made of the same metal would destroy each other if moving against one another. The actual effect was quite the reverse. The iron plates frequently rusted and in so doing blew apart both the joint and the sticks. This construction continued into the 20th century.

Fig. 3.23a & b. A 2 ft. 2-fold mast makers' rule by Lorkin (London, 1789–1830). Construction: Brass-iron-brass with an iron rivet and tongue, a rectangular plate and large knuckle.

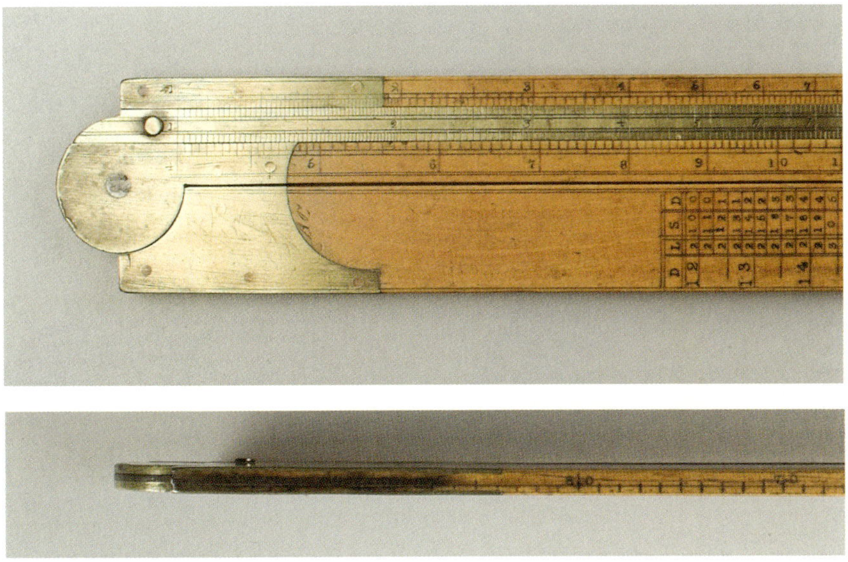

Fig. 3.24a & b. A 2 ft. 2-fold rule by Thomas Tustian (London, 1805–39). Construction: Brass-iron-brass; iron rivet; arched plate; large knuckle.

Early to mid-19th century: More elaborate designs of plates, such as arch joints, began to be introduced in the early to mid-19th century for better quality rules.

Mid-19th century onwards: Around the mid-19th century another style of all brass joint slightly different from the previous joint was introduced. The outer plates were thin pieces of brass whilst the inner knuckle of the joint was simply a thick disk of brass, which was not slotted into the sticks of the rule. The pin was made of either brass or steel. This type of joint did not supersede joints with iron interleaves. Both types were available.

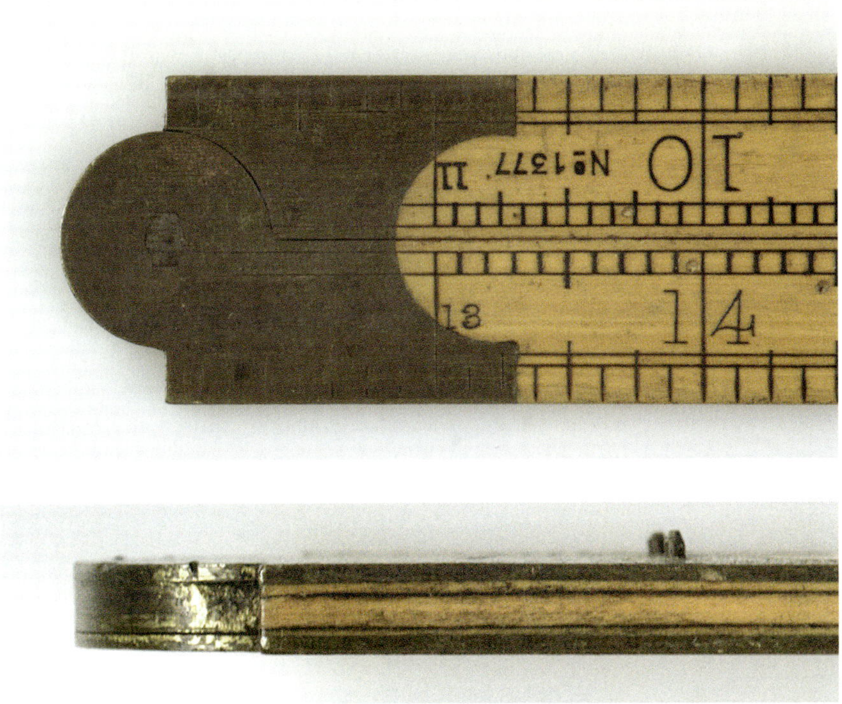

Fig. 3.25a & b. An all brass plated joint on a rule by John Rabone & Sons (Birmingham, 1877–1963) with a brass inner knuckle and steel pin. With this style of hinge both edges of the rule looked the same.

Continued development of the rule joint

Development and improvement of the rule joint continued into the 20th century. By the 1890s, both John Rabone & Sons' and Edward Preston & Sons' catalogues offered a variety of rule joints.[3]

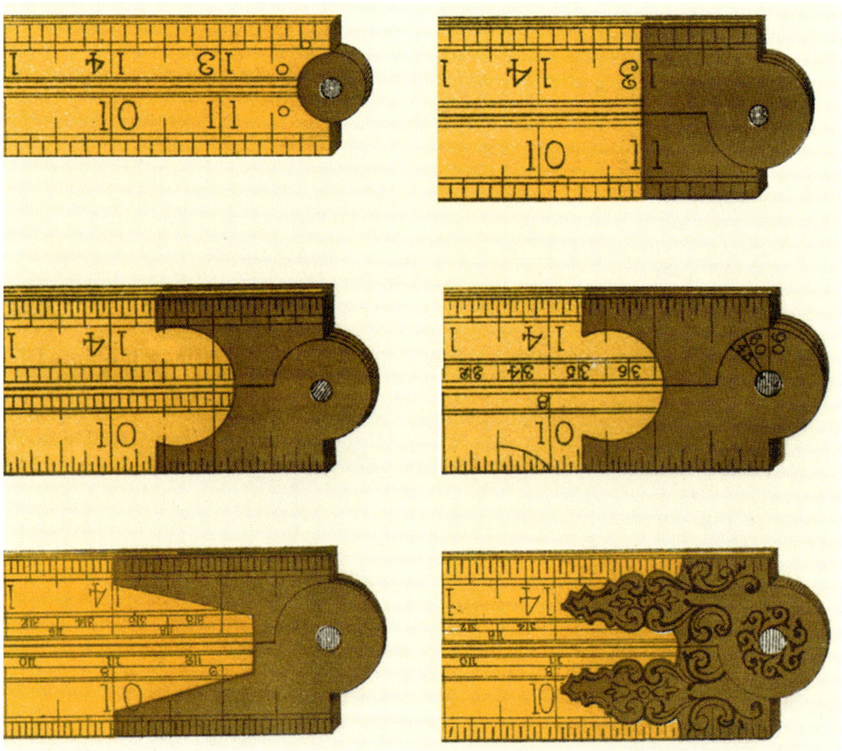

Fig. 3.26. The five joints offered by Preston in 1891 together with the arch joint rule with degrees. Top row: round joint and square joint. Centre row: arch joint and arch joint with degrees. Bottom row: Wellington joint and engraved Nonpareil joint.

Preston offered five different styles of joint for 2 ft. 4-fold rules:

- Round joint
- Square joint
- Arch joint
- Wellington joint
- Engraved Nonpareil joint

The round joint did not have plates; all other styles were purely alternative shapes for the plates. In addition arch joints were also offered with degrees marked on the joint.

Rabone offered a round joint, square joint and plain arch joint, an arch joint with degrees and an engraved arch joint.

Bolted joint

In 1876 Preston took over another Birmingham rule maker, Thomas Bradburn & Son, who held a patent for a "bolted joint". This was a small slide (the bolt) that engaged in a slot in the joint, making it possible to fix the joint fully open at 180° or partially open at 90°, 60° and 45°. This was a neat and effective addition that held the legs accurately at the required angle (though it was prone to inadvertent damage). This acquisition added yet another joint type to those available.

Below: Fig. 3.27. Left: Rabone's engraved arch joint. Right: Rabone offered a joint on their 2 ft. 4-fold rule with level that incorporated an additional plate attached to the joint marked with a full 180°.

Opposite: Fig. 3.28. An advertisement from Edward Preston & Sons 1914 catalogue for Bradburn's patent bolted joint.

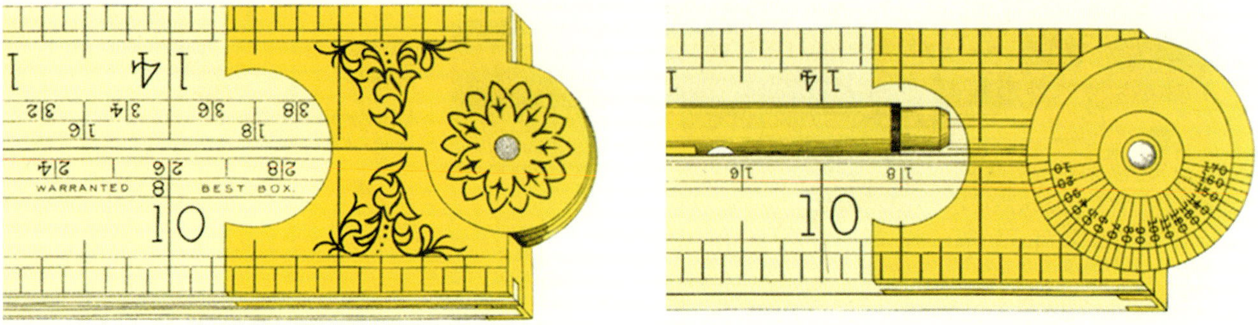

EDWARD PRESTON & SONS, LTD.

TRADE MARK
REGISTERED.

T. BRADBURN & SONS.

TRADE (EP) MARK

PROPRIETORS AND SOLE MANUFACTURERS OF

BRADBURN'S

PATENT BOLTED JOINT RULES.

(LION BRAND.)

THE ILLUSTRATIONS below will show the advantages these Bolted Joint Rules possess. They enable the user to bolt the Joint at either 45°, 60°, 90°, or straight open. The 1⅛in. and 1⅜in. Rules bolt at 45° and 90° and straight open. The 1½in. Rules bolt at 45°, 60°, and 90° and straight open.

They will be found most useful to Gentlemen, Architects, Draughtsmen, Mechanics, Carpenters, &c., being applicable to nearly every kind of *Best* Quality Rule at a small additional cost.

DIRECTIONS FOR USE.—Take one Leg of the Rule in each hand, fix the Rule by the mark on the Joint to the Angle required, and with Thumb of the left hand push the bolt firmly in.

When Ordering Patent Rule, place P before the List number.

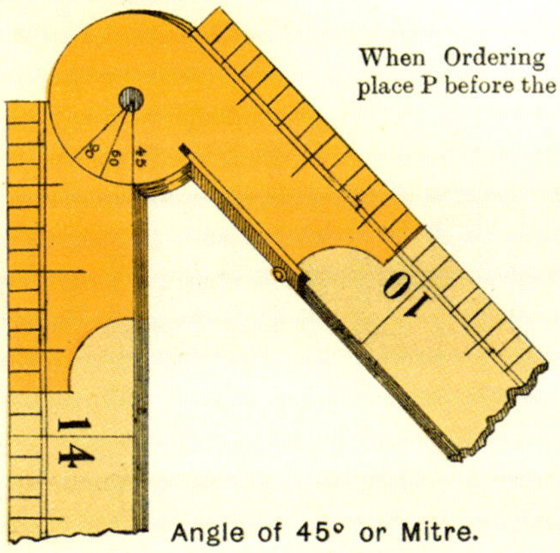

Angle of 45° or Mitre.

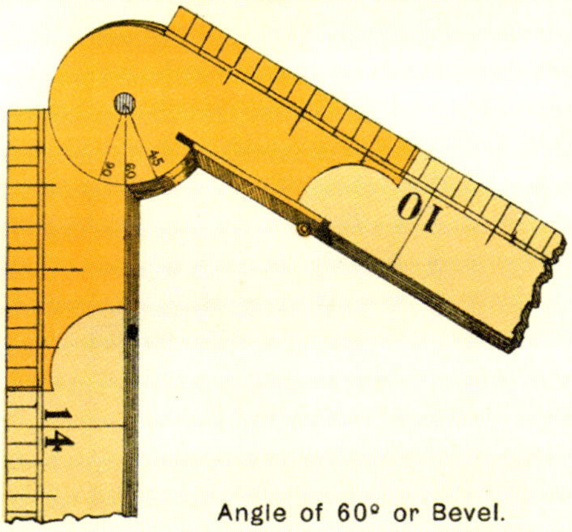

Angle of 60° or Bevel.

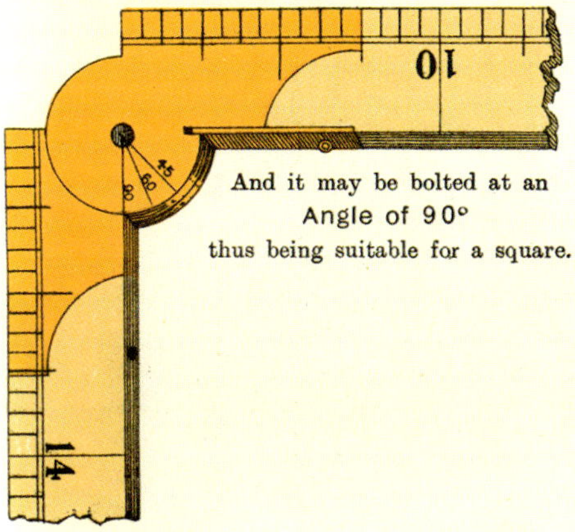

And it may be bolted at an Angle of 90° thus being suitable for a square.

Also it may be bolted straight open, making a firm, straight ruler for ruling.

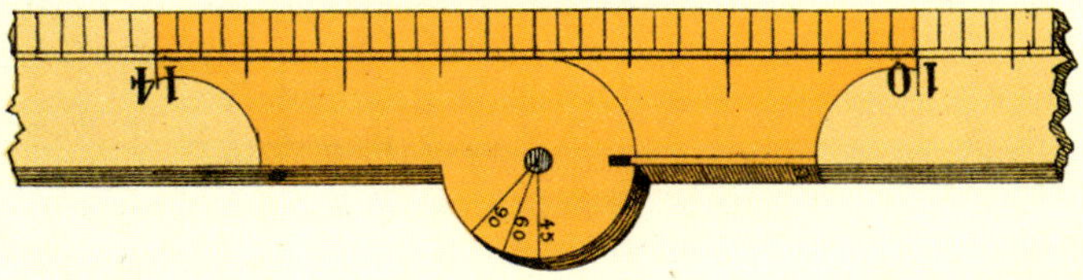

Fig. 3.29. Small joints of the simplest design on a rule by Edward Preston & Sons (Birmingham, 1864–1932).

Small joint

On a 4-fold rule, in addition to the main joint, there are joints, known as the small joints, that allow the four folds. These are of a different style as the sticks fold face to face rather than edge to edge. Again there were variations available, all of which were intended to strengthen the small joint.

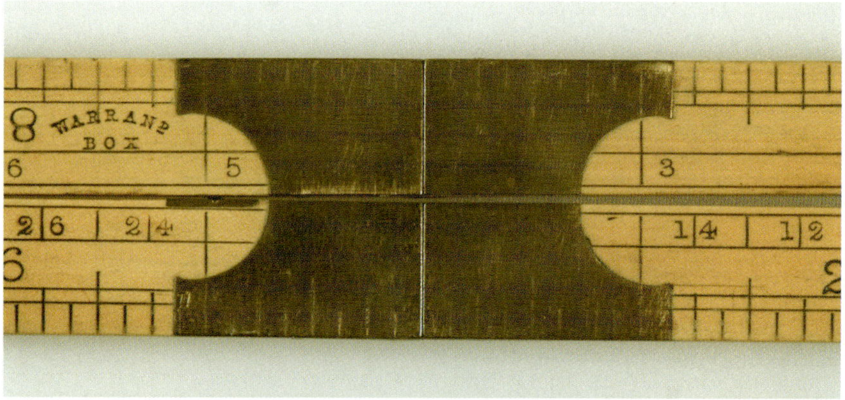

Fig. 3.30. Small joint with an arch joint on a rule by Edward Preston & Sons (Birmingham, 1864–1932).

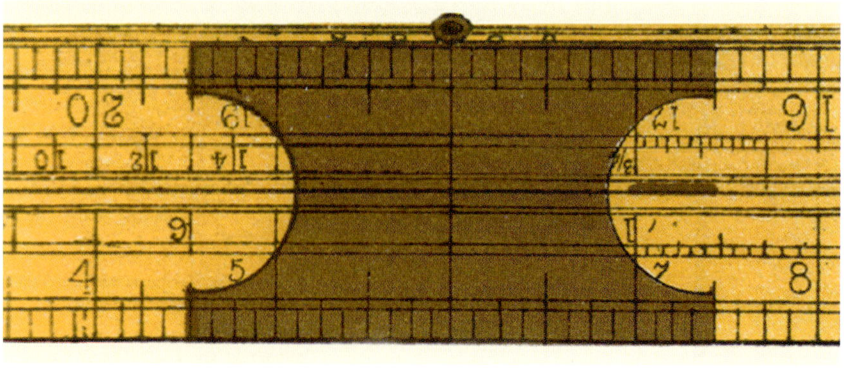

Fig. 3.31. The same joint as Fig. 3.30 as shown in the 1909 Preston catalogue.

Rules were also offered with three arch joints, the two additional ones being at the small joint. Small joints could be had with extra strong joints which had an extra interleaf plate in the joint.

Fig. 3.32. Top: The standard layout of plates in a small joint on a rule by I. & D. Smallwood (Birmingham, 1858–1999). Bottom: Extra strong small joint on a rule by Edward Preston & Sons (Birmingham, 1864–1932).

Both Preston and Rabone also offered rules with edge plates to the small joint.

Fig. 3.33. A rule by John Rabone and Sons (Birmingham, 1877–1963) with a small joint with edge plates.

Fig. 3.34. The same rule as Fig. 3.33 as illustrated in the John Rabone & Sons 1892 catalogue.

Tips

As with other features of rules, the style of the tips on the stick ends changed over time and is another helpful feature in assessing the date of a rule.

17th century: Rules frequently had no tips, though this possibly only applies to less expensive rules. Unfortunately not enough examples have come to light to be sure.

Fig. 3.35. A 2 ft. 2-fold rule, unnamed but dated 1679, without tips.

17th century to ca. 1775: There appear to have been two styles of tips in use during this period. One consisted of two separate flat plates pinned to the face of the sticks which did not extend over the end or the sides of the sticks.

In the other style, the tips were around the end of the sticks in an edge-end-edge fashion, formed with a "shoe" of brass that went from the edge of the stick across the end and back up the other edge.

Fig. 3.36. Tips formed by two separate plates one on each face of a rule by Christopher Stedman (London, 1747–85).

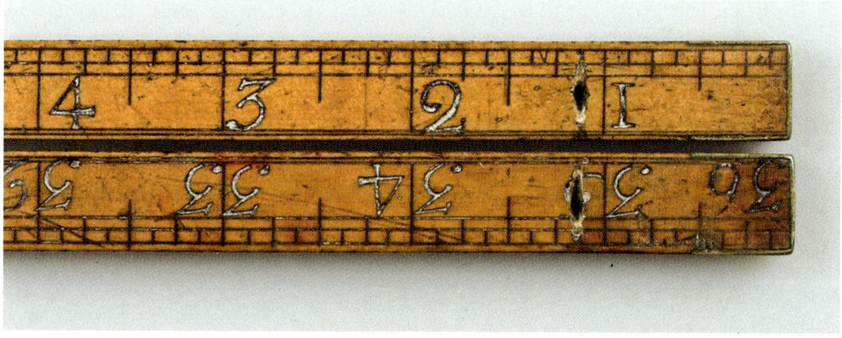

Above right: Fig. 3.37. A 3 ft. 4-fold rule also by Christopher Stedman (London, 1747–85) showing the edge-end-edge style of tips.

This style continued on those 2 ft. 3-fold rules that were formed with 9 in. sticks and a 6 in. fold-out brass tongue (usually mast makers' or plumbers' rules) until the mid-19th century, presumably to accommodate the brass tongue.

Fig. 3.38. A 2 ft. 3-fold mast makers' rule. Although anonymous, it was probably made towards the end of the 17th or early in the 18th century.

Mid- to late 18th century: There was a gradual change during the latter half of the 18th century from tips that wrapped the sticks edge-end-edge to those that wrapped from face, across the end, to the other face, i.e. face-end-face tips.

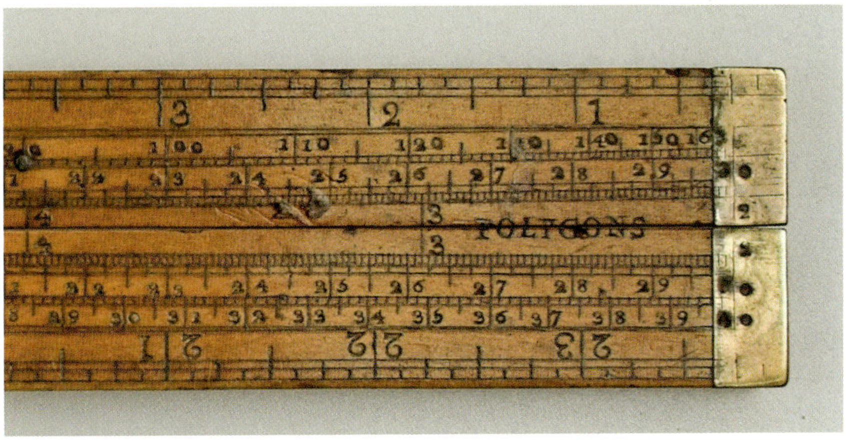

Fig. 3.39. A 2 ft. 2-fold rule by Samuel Davis (Wolverhampton, 1770–92) showing face-end-face tips.

Late 18th century onwards: Face-end-face tips in other than brass were introduced during the same period that iron interleaved joints appeared. However, brass face-end-face tips also continued to be used.

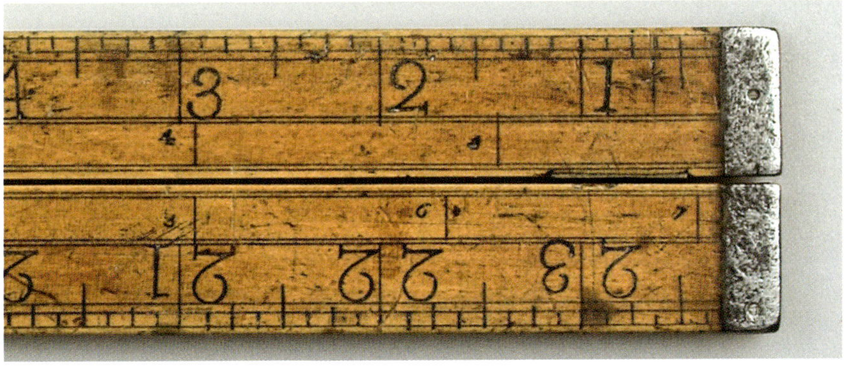

Fig. 3.40. A 2 ft 2-fold rule by Hassall & Rabone (Birmingham, 1784–91) with iron face-end-face tips.

Rules continued to be offered with both brass and iron or steel tips into the 20th century, in some cases, these materials being offered as alternatives.

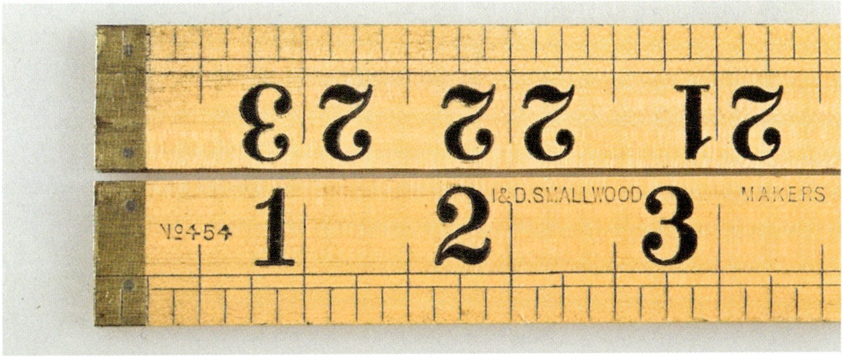

Fig. 3.41. A 2 ft. 4-fold rule by I. & D. Smallwood (Birmingham, 1858–1999) with brass face-end-face tips. This rule dates from the later years of this firm.

Variety and quality

From at least the latter years of the 18th century, and probably earlier, the variety of rules offered by the rule makers was phenomenal. In 1805, John Shaw and Sons of Birmingham offered over 90 different types of wood and ivory rules.[4] In 1798 Dring & Fage of London listed 56 different types of rules – all of which were specialist rules and all of which were supplied in a range of sizes.[5] And as can be seen in Appendix 3, the quantity of stock held in 1851 by John Rabone & Sons of Birmingham, was immense.

By the early years of the 20th century, all the rule makers were offering an immense range of rules, although the variations in some were very minor.

As an example, the Preston catalogue of 1914[6] contained seventy different types of 2 ft. 4-fold boxwood rules intended for general use. Also listed were 1 ft. rules, 3 ft. rules and many special purpose rules, rods and laths. In total one hundred and sixty different boxwood measuring devices were offered.

The seventy 2 ft. general purpose rules were all slightly different in construction – the cheapest were round joint rules; the most expensive had the outer edges bound in brass. In ascending quality and price these were:

- Round joint.
- Round joint extra thick (the actual thickness was not specified).
- Square joint.
- Arch joint offered in six widths from ¾ in. to 1½ in.
- Arch joint with extra strong small joints, offered in two widths.
- Arch joint with angles marked on joint, offered in three widths.
- Arch joint with the small joint with outside edge plates, offered in five widths.
- Arch joint with three arch joints, offered in four widths.
- Arch joint with brass bound outer edges, offered in four widths.
- Arch joint with brass bound edges, both inside and outside, offered in three widths.

The difference in price was considerable. The cheapest round joint rule cost 4*s* 6*d* per dozen whilst the widest arch joint, offered with both edges brass bound, cost 35*s* 0*d* per dozen – almost 8 times as much!

Opposite: Fig. 3.42. One of twenty-four pages of 2 ft. 4-fold rules from the 1914 Edward Preston & Sons catalogue.

Fig. 3.43. A selection of rules by Edward Preston & Sons. From the top: A brass bound rule 1 in. wide; an arch joint rule, 1⅜ in. wide with extra strong joint; an arch joint rule, 1¼ in. wide with extra strong joint and edge plates to the centre joint; a square joint rule, 1¼ in. wide.

17

Edward Preston & Sons, L^{TD.}

TRADE (E)(P) MARK

TWO-FEET TWO-FOLD BOXWOOD RULES.

French Polished. Inches marked in 8ths and 16ths. Packed in Dozens in Cardboard Boxes.

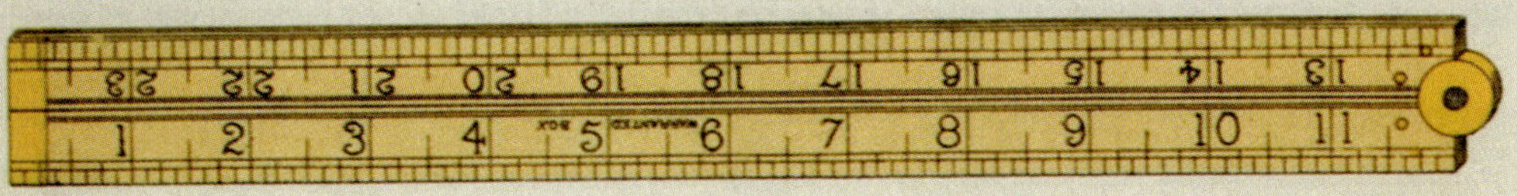

5543 — **4/6** per dozen.

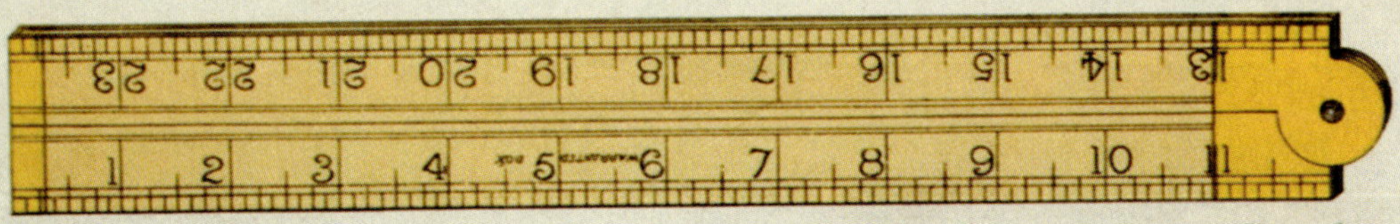

5533 — **6/-** per dozen.

5003 — **6/-** per dozen.

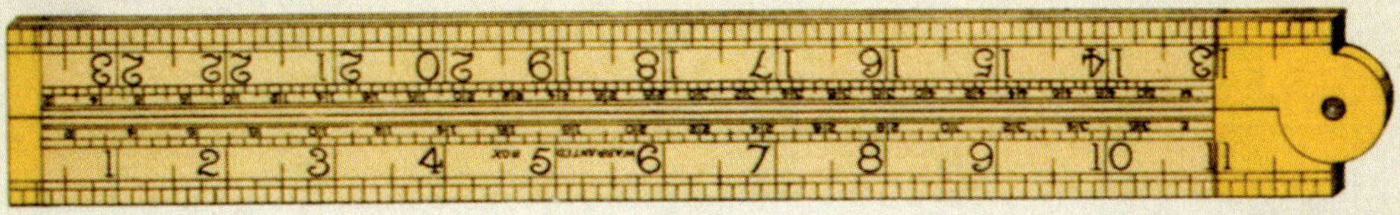

5535 — **6/9** per dozen.

5012 — **7/6** per dozen.

No. 5543.	Round Joint, marked 8ths only	1⅛-inch wide	4/6 per dozen.
,, 5533.	Square ,, ,, ,, ...	1⅝ ,,	6/- ,,
,, 5003.	,, ,, ,, 8ths, 16ths, 10ths and 12ths, Bevelled Edge	1⅜ ,,	6/- ,,
,, 5535.	,, ,, ,, 8-square Line, 8ths and 16ths	1⅝ ,,	6/9 ,,
,, 5012.	Arch ,, ,, ,, ,, ,, 10ths and 12ths, Bevelled Edge	1⅝ ,,	7/6 ,,

Drawings are Half Size.

Fig. 3.44. Four different styles of architects' rules. From the top: One bevel on the outside edge of one leg of the inner faces; one bevel on one leg of the outside edge of the outer face; four bevels on the inner edges of all legs of the inside face; eight bevels, one on every edge of every leg of the rule.

However, the alternatives did not end with construction and quality; a note in the same catalogue states that any of these rules could be had marked with a second measure (one of the European inches or metric) in addition to the London inches. The extra cost of this varied between 1*d* and 3*d* per rule.

In addition, almost all 2 ft. 4-fold rules were also offered as 3 ft. 4-fold rules.

Such options meant that the total number of alternatives available was almost infinite.

Specialised rules

In addition to the general-purpose rules, almost all the rule makers offered ranges for specialised use, some of which are noted below.

- Architects' rules with bevelled edges marked with architectural scales for use directly on drawings. Preston offered seventeen different 2 ft. 4-fold rules and eleven which were also available in a 3 ft. 4-fold format. These variations included brass mounts, German silver mounts,

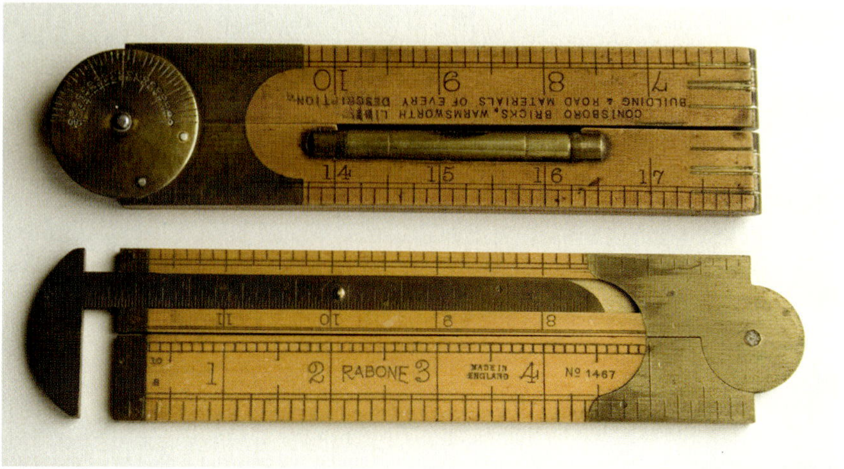

Fig. 3.45. A rule with a spirit level and a rule with a caliper, both by John Rabone & Sons (Birmingham, 1877–1963).

inside bevels, inside and outside bevels. and from four to sixteen different scales.

- Rules with spirit levels – these usually also had large diameter disc joints marked with degrees. These allowed the user to determine the angle to which the legs had been opened.
- Rules with a brass slide marked for measuring which was useful for establishing the depth of holes, etc.
- Rules with a caliper, useful for measuring diameters.
- Rules with a log slide and a timber load table for timber measuring.
- Rules for mechanics with circumference scales.

All in all, by the early years of the 20th century a vast array of general purpose and specialist purpose folding rules were readily available.

References

1. Timmins, Samuel ed., "Measuring Rules" by John Rabone, *Birmingham and the Midland Hardware District: The Resources, Products and Industrial History*, London, 1866.
2. Atwel[l], George, *The faithfull surveyour: teaching how to measure all manner of ground ... Shewing likewise the making and use of a new instrument, called a Pandoron ... with an appendix unfolding errours in board and timber-measure, with directions for making a carpenters-rule,* Cambridge, 1662.
3. Edward Preston & Sons Ltd., *Illustrated Price List of Measuring Rules, Planes, Spirit Levels ...*, Birmingham, 1891 and John Rabone & Sons, *Catalogue of Rules & Tapes,* Birmingham, 1892.
4. Stock List of John Shaw & Sons, ca. 1805, MSS DB/24/401, pp. 152-157, Records of John Shaw & Sons, Wolverhampton Borough Archives.
5. Records of Dring & Fage, Science Museum Library, London.
6. Edward Preston & Sons Ltd., *The Preston Catalogue No. 26 of rules, measures, levels ...*, Birmingham, 1914.

Chapter 4
The rule making trade

... abroad to do several small businesses, among others to find out one to engrave my tables upon my new sliding rule, it being so small that Browne that made it cannot get one to do it. So I found out Cocker, the famous writing master and got him to do it. Samuel Pepys, 1664.[1]

Rule making, as a distinct trade, developed from the trade of making mathematical instruments and only slowly became differentiated from these roots; indeed in the case of many firms the separation was never complete.

It is not surprising therefore that the majority of early rules are not just rules to measure but also have calculating features. Some of the 17th century instrument makers seem to have had some specialisation in making rules but none can be regarded solely as rule makers, with the possible exception of William Elmes (1668–75) who, in the records of the Clockmakers' Company of London, specifically described himself as a rule maker, listing no other trade. There are very few rules known that date to the 16th century; one that has survived is a brass rule marked with a line of inches and timber and board measures by Humfrey Cole (London, 1568-90). A few more have survived from the mid-17th century by makers such as Walter Hayes (London, 1648–84), Henry Sutton (London, 1649–65) and Christopher Coles (unknown location, <1664>), and there is evidence that by this date, some of the mathematical instrument makers were advertising rules as part of their production.

Walter Henshaw (London, 1659–1711) advertised in Coggeshall's *Timber Measure by a Line...*[2] that "he made the rules required, together with all other mathematical instruments". Thomas King (London, 1667-77) described his trade as gauging instrument maker, which almost certainly included rules. Other 17th century makers who were known to have supplied rules for sale, and can be presumed to have been the makers, were John Nash (London, 1668–1707) and Isaac Webb (London, 1668–1704). An early maker of boxwood rules was Isaac Carver (1668–1713) who seems to have specialised in gaugers' slide rules, advertising in the 7th edition (1712) of Thomas Everard's *Stereometry, or the Art of Gauging made easy...*[3], that he made "the *Sliding-Instruments*" very exactly. "Gauger" was the common name given to the officers of the Customs & Excise, their job being to "gauge", that is, to measure, the containers of alcoholic beverages for the purpose of levying duty.

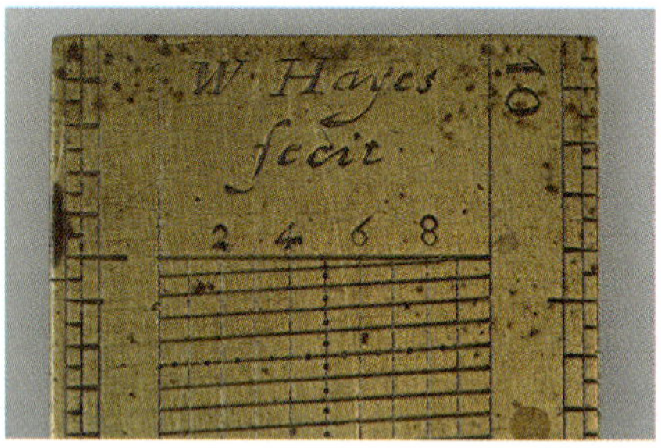

Fig. 4.1. The mark of Walter Hayes (London, 1648-84) on a brass Gunter's scale.

Fig. 4.2. A Customs & Excise slide rule dated 1690 by Isaac Carver (London, 1668-1713).

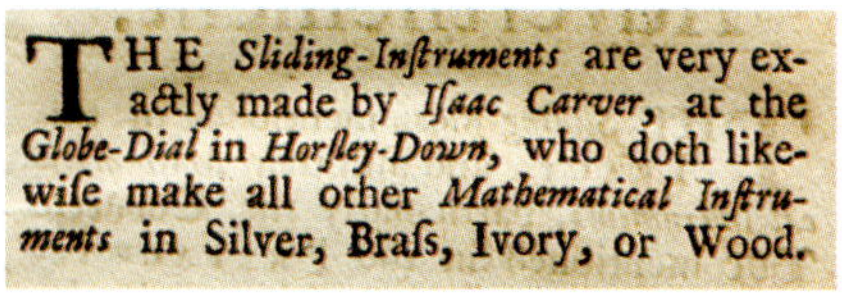

THE *Sliding-Instruments* are very exactly made by *Isaac Carver*, at the *Globe-Dial* in *Horsley-Down*, who doth likewise make all other *Mathematical Instruments* in Silver, Brass, Ivory, or Wood.

Fig. 4.3. Isaac Carver's advertisement from Everard's Stereometry, *1712.*

Fig. 4.4. The mark of Edward Roberts (London, 1759-95 in Dove Court) on a Customs & Excise slide rule.

Fig. 4.5. The mark of J. Rix (London, 1750-60) on a Customs & Excise slide rule

By the middle years of the 18th century rule making had become a trade in its own right when London makers such as J. Rix (1750–60) and Edward Roberts (1749–95), again both principally makers of slide rules for gauging purposes, became established.

Early rule making in London

The most prolific of the 18th century makers were the aforementioned Edward Roberts and his son of the same name, trading from 1759 to 1795, in Dove Court off Old Jewry. The elder Edward Roberts had been an apprentice of Thomas Cooke who had worked in Old Jewry from 1704. Cooke also advertised in Everard's *Stereometry, or the Art of Gauging made easy...*[4], although rules were not specifically mentioned. It is almost certain that Roberts continued this business. Cooke is clearly of some importance in the rule making

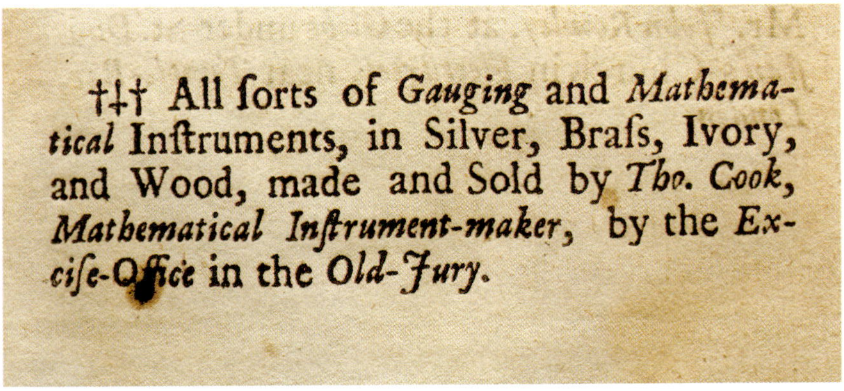

†‡† All ſorts of *Gauging* and *Mathematical* Inſtruments, in Silver, Braſs, Ivory, and Wood, made and Sold by *Tho. Cook, Mathematical Inſtrument-maker*, by the *Exciſe-Office* in the *Old-Jury*.

Fig. 4.6. Thomas Cooke's advertisement in Everard's Stereometry, *1712.*

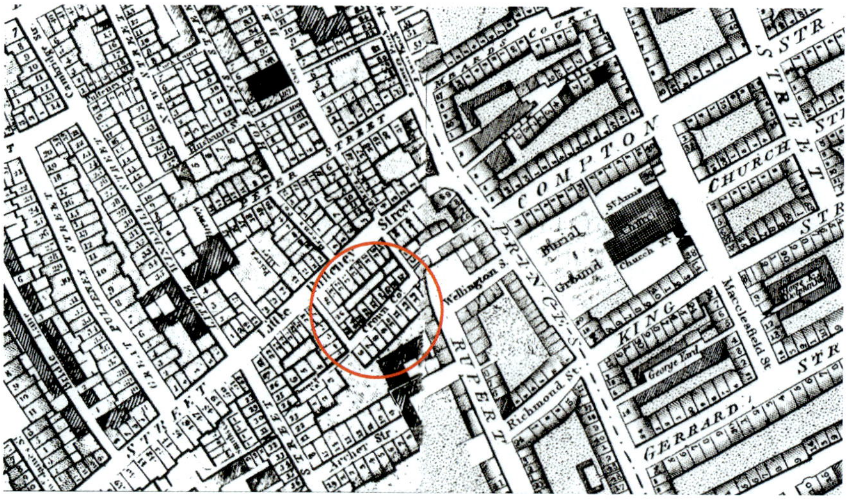

Fig. 4.7. A map of London in 1813, showing Crown Court, Soho.

trade as he was also the apprentice master of two other makers of gauging rules and equipment, William Mole and William Rix.

All these makers, and some others, were situated in or near Old Jewry in the City, conveniently close to the offices of the Commissioners of Excise whose officers were the principal buyers of gauging equipment. The Robertses became prolific makers of boxwood gaugers' slide rules. Almost nothing else has been found bearing their name.

Crown Court, Soho

Another centre of rule making in London was in Soho; this group seems to have developed from one maker though in this case there seems to be less logic for the choice of location. John Bennett (1735–70) was the first of a line of makers of rules to work in Crown Court, Soho. Crown Court is shown on Horwood's map of 1813, situated near the west end of Compton Street, a typical back land alley, small and doubtless dingy. It has now vanished.

Bennett made a range of instruments but may have had some specialisation in rules as he is mentioned by W. Flower in 1768 as the maker of his improved design of rule.[5] Bennett's business was continued by James Search (1771–81).

In 1784, Alexander Wellington, who had been an apprentice of James Search, is recorded at No. 20 Crown Court. The Wellington business was almost certainly a continuation of Search's. After Alexander's death in 1812, the business was carried on by his widow and son until 1827. For a short

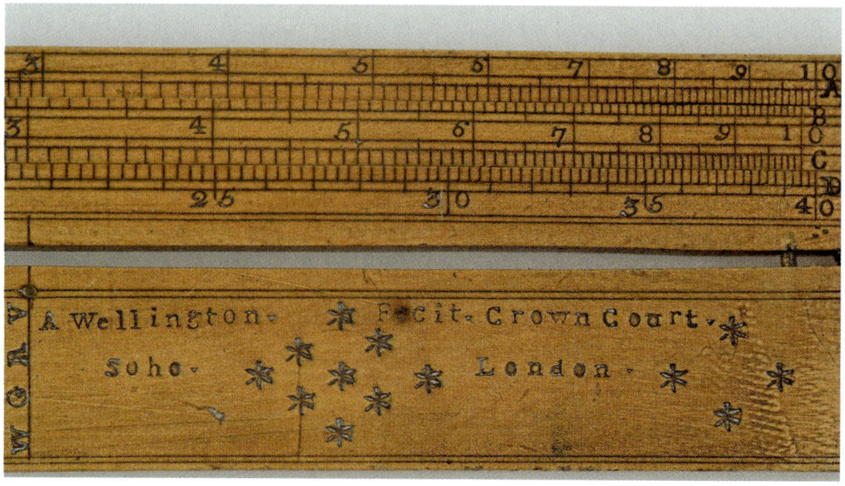

Fig. 4.8. The mark of Alexander Wellington (London, 1784-1812). The business was carried on by his family until 1827 and it is likely that they continued to use the same mark.

Fig. 4.9. *The mark of Laban Cook (London, 1826-34). Note that not only does he state that he is "Maker to the Hon. Board of Excise" but also "Late Wellington, Crown Court, Soho".*

while during this period in 1818/19, Fitch & Jones – a partnership between James Fitch, whose father had been apprenticed to John Bennett and who himself was apprenticed to James Search, and John Jones – also traded from these premises, with Jones continuing on by himself until 1827. In 1826 Laban Cook took over Alexander Wellington's business and in 1827 John Jones disappears from the records.

Within a few paces of the court, Isaac Aston and his successors traded at 25 Compton St. from 1814 and George Mander worked in the Court from 1830 to 1851. With his departure the long sequence of rule making in the Court finished, though Mander continued to trade in the renamed Old Compton St. until 1870.

Fig. 4.10. *The mark of Aston & Mander (London, 1870-1930) on a specialist slide rule designed for the timber trade.*

Wolverhampton

Another centre of rule making in the mid-18th century was the Wolverhampton/Birmingham area. At this period, Wolverhampton and Birmingham had approximately the same number of rule makers.

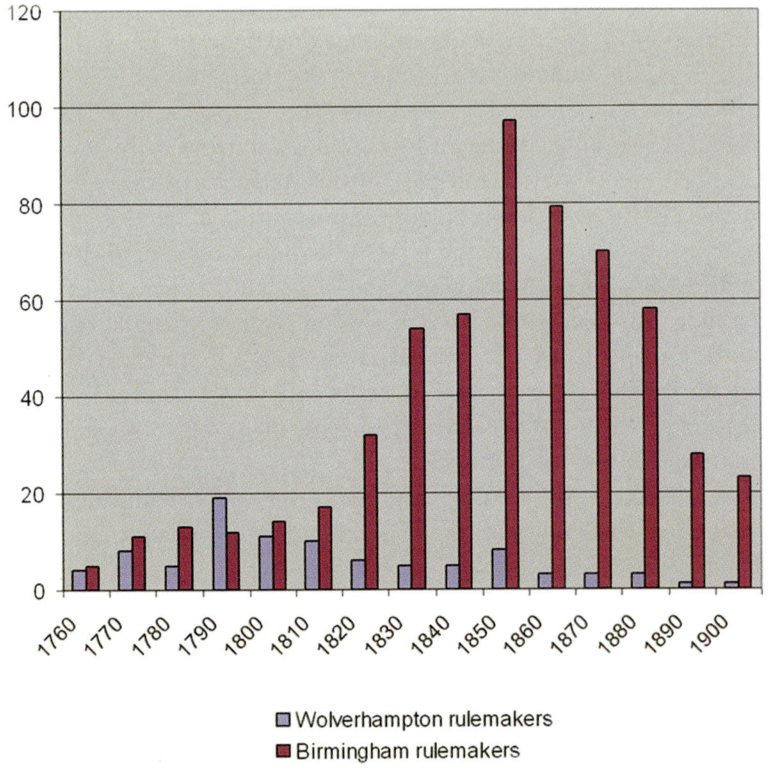

Fig. 4.11. *A graph comparing the number of rule makers in Wolverhampton to the number in Birmingham.*

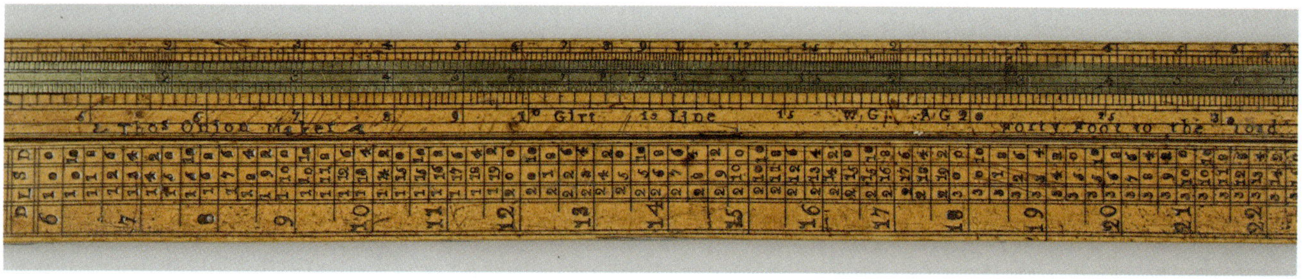

Fig. 4.12. A 2 ft. 2-fold rule by Thomas Onions with a brass log slide, girt line and a hundredweight table. Although Thomas Onions first established his business in Wolverhampton in 1767, by 1784 he had relocated to Birmingham where he remained in business until 1801, the business being continued by Benjamin Onions until 1852.

Fig. 4.13. The mark of Robert Corson (Wolverhampton, 1767-70). This type of mark, in the form of a die-struck disk, was only used by makers in Wolverhampton and Birmingham

The earliest recorded maker in Wolverhampton was George Wood working as a rule maker in 1761; in 1763, Christopher Jacob was in Bilstone Street, although he also had an address in Birmingham; and in 1767, Robert Corson was in Berry Street and Thomas Onions in Brick Kiln Lane. By 1770 there were eight rule makers established in Wolverhampton and by the 1790s nineteen makers were working there.

An indication of the size of the rule making trade in Wolverhampton around 1805 can be obtained from a study of the stock book of John Shaw & Sons.[6] Shaw was a hardware factor and merchant; the stock recorded in this book included over ninety different types of boxwood and ivory rules. Rules were described variously by their length, by their material, by the type of joint and whether they were best or common. Amongst the different joints listed in Shaw's stock book were common joint, solid joint, steel joint, outside solid joint, outside steel joint, double joint and large joint. Where the rules included a slide, this was specified as either wood or brass – there do not seem to be any ivory slides listed. The firm carried a variety of more specialist rules including shoe sticks and gauging rods, scales and sectors. There are Gunter's scales listed and where the rules carried tables or lines, this was noted. There were also mast makers' rules in stock.

However, as the 19th century progressed, the rule making business in Wolverhampton was soon overtaken by the rapidly expanding city of Birmingham.

Birmingham

Not far to the south of Wolverhampton, in the 1760s Birmingham had four rule makers, Joseph Edwards, Frost & Withnoll, Christopher Jacob and Wood & Lort; by the 1770s, this number had grown to eight; and by the 1790s there were twelve established.

Fig, 4.14. The mark of Hassall & Rabone (Birmingham, 1784-91).

In the 1810s, this number increased dramatically and it is probable that by then production had already surpassed that of London. By 1830, there were 54 makers, which number increased to 97 in 1850.

Amongst the Birmingham rule makers established in the late 18th and early 19th centuries were Michael Rabone, who worked in partnership with Joseph Hassall from 1784 to 1800 and then set up on his own in 1801; Samuel Aston who commenced in business in 1785; William Smallwood in 1823; and Thomas Bradburn in 1835, all of whom were to become major rule makers.

In Birmingham during the latter half of the 18th century the manufacture of small items such as jewellery, small silverware, plate, belt buckles, pressed brassware for making sword sheaths, buttons, etc. had expanded exponentially. The fly press was the core process of the manufacture of the items listed above and, as has been explained in Chapter 2, it was the application of this machine to the production of the many small brass parts needed to complete a rule that gave the Birmingham manufacturers an advantage over makers in London.

It changed the whole nature of the rule making trade. What had arrived in Birmingham as a hand-working trade, using only the simplest of jigs and fixtures, was quickly developed into a more productive business using sophisticated jigs and hand powered machines.

Appendix 3 contains a transcription of the tools that were being used by John Rabone & Son in 1851. At this time Rabone had not achieved the dominance of the trade that they were later to have and it is likely that other makers were similarly equipped. The items listed in the inventory, although from the descriptions appearing very simple, were capable of performing a large number of the preparatory stages needed to make a stick ready to receive the brass fitments.

It is interesting to note that at this date Rabone had but two fly presses; however such was the productive capacity of these that, with suitable tooling, they would have been more than capable of turning out all the small metal components needed.

Due to the mechanisation of the trade that they had developed, the Birmingham rule makers always concentrated on the production of mass market rules although it is apparent that they were also perfectly willing to make anything if required, however esoteric or special.

Birmingham also concentrated on an export trade. Even in the 18th century it is clear that the Wolverhampton and Birmingham manufacturers were exporting a proportion – possibly a considerable proportion – of their output.

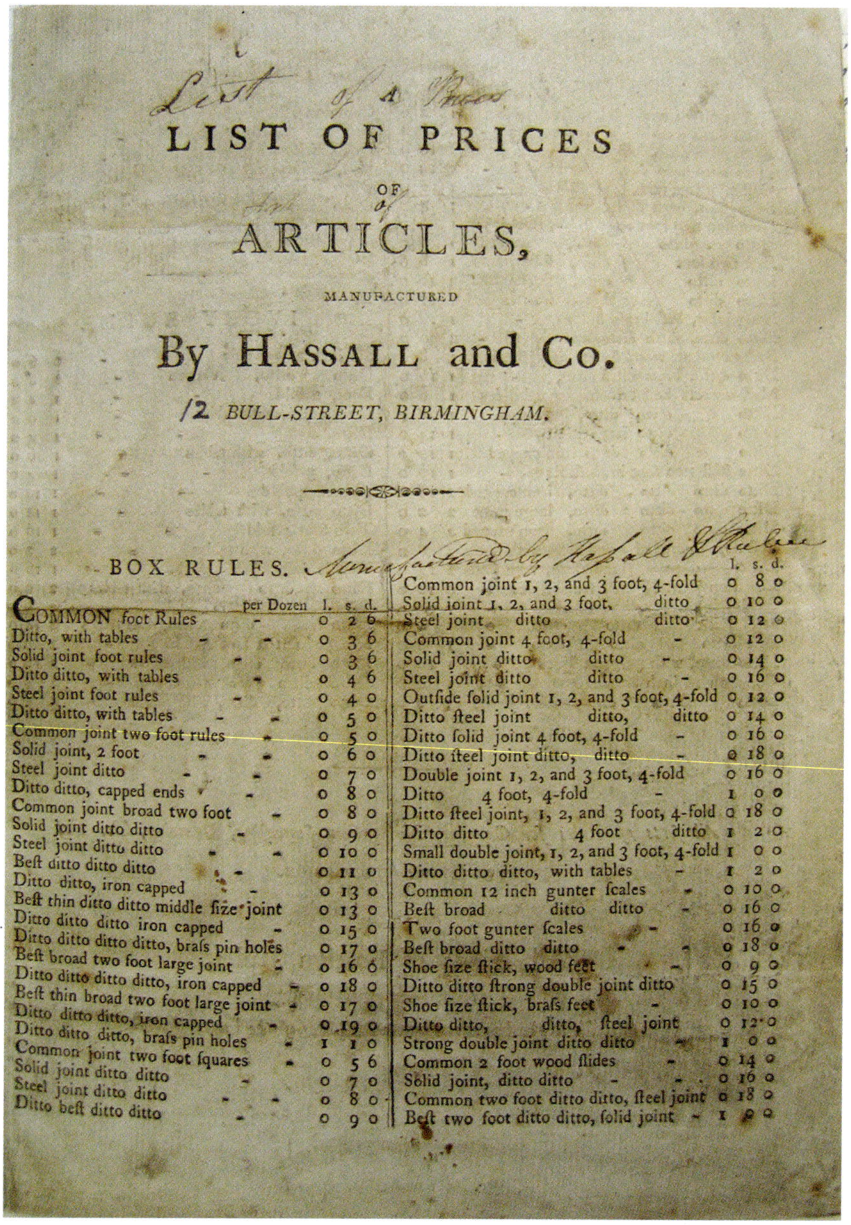

LIST OF PRICES

OF

ARTICLES,

MANUFACTURED

By HASSALL and Co.

12 BULL-STREET, BIRMINGHAM.

BOX RULES.

per Dozen	l.	s.	d.
COMMON foot Rules -	0	2	6
Ditto, with tables - -	0	3	6
Solid joint foot rules -	0	3	6
Ditto ditto, with tables -	0	4	6
Steel joint foot rules -	0	4	0
Ditto ditto, with tables - -	0	5	0
Common joint two foot rules -	0	5	0
Solid joint, 2 foot - -	0	6	0
Steel joint ditto - -	0	7	0
Ditto ditto, capped ends -	0	8	0
Common joint broad two foot -	0	8	0
Solid joint ditto ditto -	0	9	0
Steel joint ditto ditto - -	0	10	0
Beſt ditto ditto ditto -	0	11	0
Ditto ditto, iron capped -	0	13	0
Beſt thin ditto ditto middle ſize joint	0	13	0
Ditto ditto ditto iron capped -	0	15	0
Ditto ditto ditto ditto, braſs pin holes	0	17	0
Beſt broad two foot large joint -	0	16	6
Ditto ditto ditto ditto, iron capped -	0	18	0
Beſt thin broad two foot large joint -	0	17	0
Ditto ditto ditto, iron capped -	0	19	0
Ditto ditto ditto, braſs pin holes -	1	1	0
Common joint two foot ſquares -	0	5	6
Solid joint ditto ditto -	0	7	0
Steel joint ditto ditto - -	0	8	0
Ditto beſt ditto ditto -	0	9	0

	l.	s.	d.
Common joint 1, 2, and 3 foot, 4-fold	0	8	0
Solid joint 1, 2, and 3 foot, ditto	0	10	0
Steel joint ditto ditto	0	12	6
Common joint 4 foot, 4-fold -	0	12	0
Solid joint ditto ditto -	0	14	0
Steel joint ditto ditto -	0	16	0
Outſide ſolid joint 1, 2, and 3 foot, 4-fold	0	12	0
Ditto ſteel joint ditto, ditto	0	14	0
Ditto ſolid joint 4 foot, 4-fold -	0	16	0
Ditto ſteel joint ditto, ditto -	0	18	0
Double joint 1, 2, and 3 foot, 4-fold	0	16	0
Ditto 4 foot, 4-fold -	1	0	0
Ditto ſteel joint, 1, 2, and 3 foot, 4-fold	0	18	0
Ditto ditto 4 foot ditto	1	2	0
Small double joint, 1, 2, and 3 foot, 4-fold	1	0	0
Ditto ditto ditto, with tables -	1	2	0
Common 12 inch gunter ſcales -	0	10	0
Beſt broad ditto ditto -	0	16	0
Two foot gunter ſcales -	0	16	0
Beſt broad ditto ditto - -	0	18	0
Shoe ſize ſtick, wood feet -	0	9	0
Ditto ditto ſtrong double joint ditto	0	15	0
Shoe ſize ſtick, braſs feet -	0	10	0
Ditto ditto, ditto, ſteel joint	0	12	0
Strong double joint ditto ditto -	1	0	0
Common 2 foot wood ſlides -	0	14	0
Solid joint, ditto ditto - -	0	16	0
Common two foot ditto ditto, ſteel joint	0	18	0
Beſt two foot ditto ditto, ſolid joint -	1	[illegible]	0

Fig. 4.15. Price list from Hassall & Co. Birmingham. Although this company is only listed in the trades directory for the year 1797 it is in fact an alternative name for the partnership between Joseph Hassall and Michael Rabone, who were working as rule makers at 12 Bull Street. In addition Joseph Hassall had a wood turning business and sold ironmongery and household furniture. This price list covers the rules and drawing instruments etc. that the partnership made as well as the turned wooden items made by Joseph Hassall.

This included the whole range of small metal wares that went under the umbrella title of "Birmingham Goods", amongst which were "toys", small trade or domestic steel items. Illustrated pattern books and catalogues showing a wide range of both toys and other goods had been published before 1800.

These goods were being exported in quantity, particularly to North America, the trade being handled by factors with representatives or connections on both sides of the Atlantic, and rules were part of this trade.

London in the 19th and 20th centuries

In comparison to Birmingham, London never broke into the market for mass produced rules. By the 1820s the Midlands makers, as a result of their investment in more mechanised methods of production, could make rules in quantity much more cheaply than their London counterparts.

This did not mean that the rule trade in London ceased but thereafter, with the exception of rules for the Customs & Excise, the London makers concentrated on the more specialist short-run section of the trade for which the demand was modest.

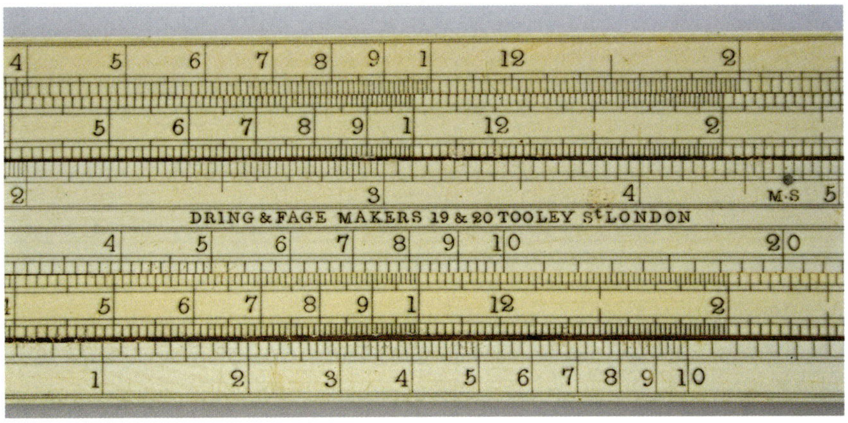

Fig. 4.16. The mark of Dring & Fage on an ivory proof rule. The firm was at 19 and 20 Tooley Street, London, from 1804 to 1882.

It was not economic for the London makers to invest in machines such as fly presses as there was no manufacture of the other types of goods that required this technology and which made the machines commercially viable. By the early years of the 19th century the London makers were buying stamped parts from Birmingham and as the century went by, they were also not averse to purchasing framed rules which they then divided and finished to their own requirements.

Indeed, later in the century, there is evidence that the London makers were purchasing fully finished rules from the Birmingham makers and passing them off as being of their own manufacture.

As previously mentioned, the one prominent area in which London makers did specialise was in gauging equipment. The manufacture of rules and other equipment for gauging, from early on, had been a significant part of their output and this remained a London speciality until the need for such items largely disappeared in the 1960s. A high proportion was purchased by the Customs & Excise, but with hundreds of small breweries operating in and around London, there was also a significant commercial demand.

The most prominent firm in this trade in the 19th and 20th centuries was Dring & Fage, who made the widest conceivable range of gauging equipment, and today's collector will find many surviving items bearing their name. Some indication of the quantity they produced can be seen in Appendix 4. Eventually the only survivor in this trade, the firm finally closed in 1977. Even in its heyday, they employed no more than a handful of men.

There were, as well as Dring & Fage, some other fine rule makers operating in London during the 19th century. Of particular note was the Tree family at Blackfriars. Their output was considerable and they appear to have specialised in rules with unusual features. One particular speciality was the London Pattern timber rule, a 9 in. 2-fold rule with slide and timber table and with a 6 in. extension folding out of one leg, making it a 2 ft. rule.

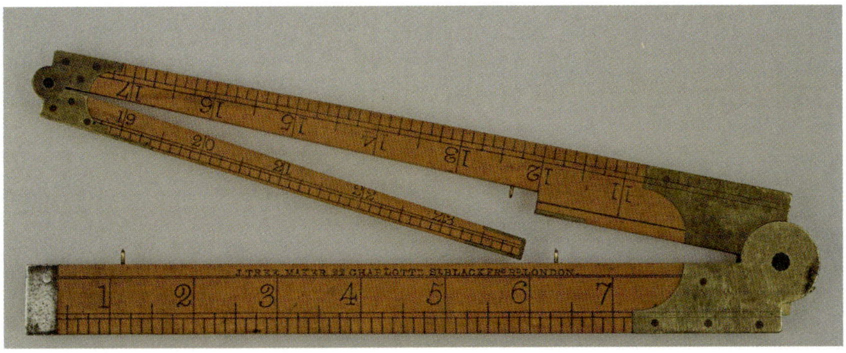

Fig. 4.17. A London Pattern timber rule by J. Tree. Tree occupied 22 Charlotte Street, Blackfriars from 1832 to 1895.

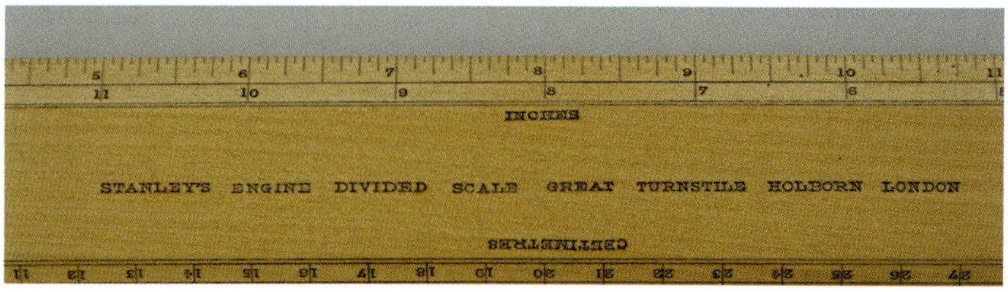

Fig. 4.18. The mark of W. F. Stanley (London, 1853-1967). This scale rule is unusual as it is divided in both imperial and metric. It is likely that this dates from the latter period of the firm's time in business.

It is likely that quite a large part of Tree's output was produced to order, for example the several versions of Ewart's cattle gauge bearing their name, which were clearly produced to the order of the inventor/promoter. Few of the items made by the firm are found in any quantity but the quality always seems to be excellent.

Another example of a specialised maker operating in London was W. F. Stanley in High Holborn, who made and sold almost everything that the surveyor or draughtsman might need.

Other locations

Although Birmingham and London were the main centres of rule making in the 19th century, there were many other locations where rule makers worked. This might be just one or two makers, catering to the local market, but there were also makers in larger towns and cities who must have supplied an area beyond their own immediate vicinity.

Towns in the latter category include Manchester, Liverpool, Sheffield and the Bristol / Bath area. However, with the exception of Bristol and Bath, none of the places mentioned had more than one maker established before 1800 and indeed there were only a handful overall before 1820.

After that date, in both Liverpool and Manchester, a reasonably extensive rule making trade developed, with 21 makers established in Manchester and 13 in Liverpool by the mid-19th century.

In Manchester, in the early years of the 19th century, Stephen Cooper and Thomas Underhill had substantial businesses. Two of the leading manufacturers of drawing instruments and scale rules, J. Halden & Co. and A. G. Thornton, were established there by the end of the 1870s, and also prominent was Thomas Driver, who marked his rules "Driver's Rule Works".

In Liverpool, the Casartelli family, who started in 1821 and continued until at least the 1960s, and the Dicas family, who started in 1774 and continued until 1837, were the most prolific of the various makers.

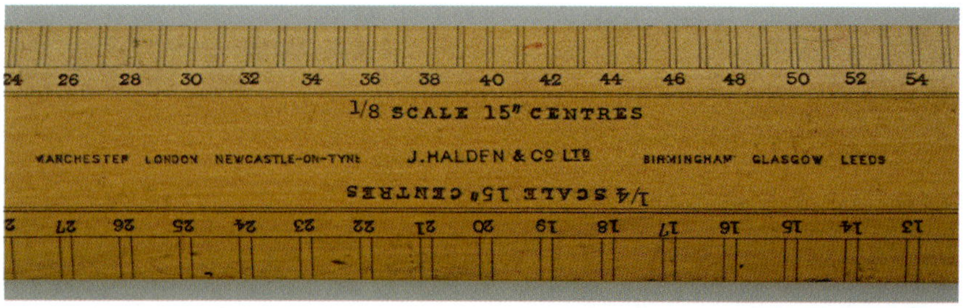

Fig. 4.19. A specialist scale rule for setting out rafters by J. Halden & Co. Ltd. This rule dates from post-1920, when the company had offices in several towns.

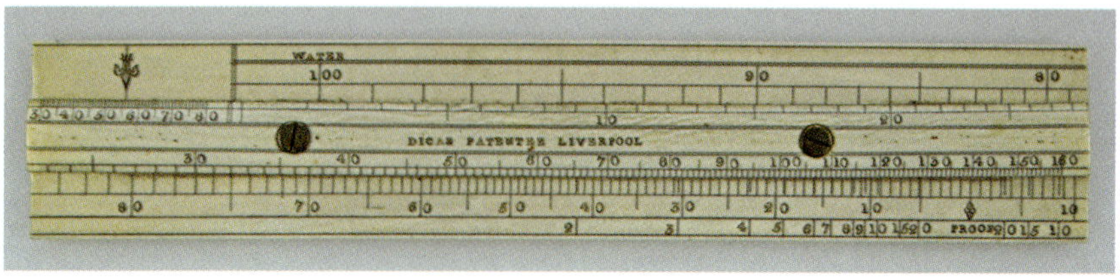

Fig. 4.20. John Dicas took out a patent for this notably different type of slide rule for correcting temperatures in the brewing trade in 1780.

Fig. 4.21. A scale rule by Alfred Shellard (Bristol, 1854-1902), the last Bristol rule maker.

Despite its dominance of the edge-tool business, Sheffield never really competed in the rule making trade until the 20th century. The only two significant makers in the 19th century were James Chesterman and J. P. Cutts who, at various times, were in partnership. Chesterman continued in business until 1963 when they amalgamated with Rabone to form Rabone Chesterman.

Although Bristol and Bath had a small but established trade in the late 18th to the mid-19th century, with 10 makers in Bristol in the mid-century, there were no rule makers in either place after 1902, when the last Bristol maker closed down.

The end of an era

The apocalypse of 1914 ended the era when manufacturers were prepared to list the widest possible range even if hardly any of some items sold, and the days of the trade were numbered. In 1900, there were around 85 rule making businesses, most based in London and Birmingham. By the end of World War II, only 34 remained.

In 1964, one of the last traditional rule makers in London, J. S. Corbishley closed down, The other remaining London rule makers carried on for a short while, with W. F. Stanley ceasing in 1967, Dring & Fage, leading makers of gauging instruments, in 1977 and, in 1986, Threadwell & Sons went out of business, bringing to a close all rule making activity in London.

The only firms then remaining in business in Britain were I. & D. Smallwood and Rabone Chesterman Ltd., both in Birmingham, with the former closing in 1999 when the last owner died and the latter taken over by Stanley Tools in the early 1990s and moved to Sheffield where no wooden rules were made.

Famous makers

Of the numerous rule makers operating in Britain, there are six whose products will be most commonly found by the collector: in Birmingham, John Rabone & Sons, Edward Preston & Sons and I. & D. Smallwood; in Sheffield, James Chesterman & Co. and in London, Dring & Fage and W. F. Stanley. There follows a brief history of each of these firms.

John Rabone & Sons

The first record of Michael Rabone, the founder of this dynasty of rule makers, was in 1784 when he joined with Joseph Hassall in a partnership which operated from 12 Bull Street, Birmingham. (It was also the year that he married his wife, Elizabeth Smith.) In 1791 Hassall also formed a partnership with Zachariah Belcher, which was dissolved when Belcher moved to Sheffield in 1794. Michael Rabone and Joseph Hassall continued to work in partnership at 12 Bull Street until 1800 when Joseph Hassall died. In his will he left the rule making business, and the wood turning business that he also ran, to Michael. Michael moved from Bull Street in 1801 and set up on his own at 8 Snow Hill. However, his time on his own was short-lived and he died

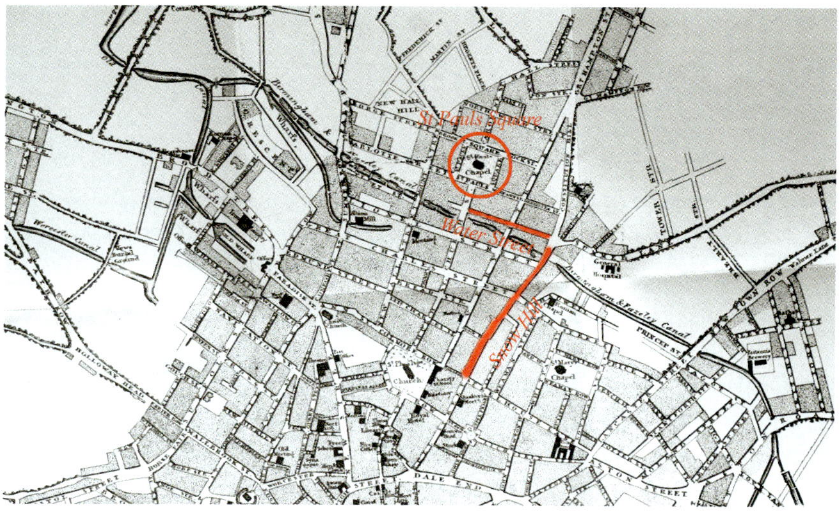

Fig. 4.22. The three Birmingham premises that Rabone occupied from 1801 to 1871: Snow Hill, Water Street and St. Paul's Square, all within walking distance of each other, just north of the modern city centre. Bull Street, where Michael Rabone worked from 1784 to 1800, is the southern continuation of Snow Hill.

Fig. 4.23. After the younger John Rabone joined the firm in 1845, they marked their rules "John Rabone & Son".

in 1803. Michael and his wife Elizabeth had two sons, Ephraim and John. Elizabeth, continued the business and in 1808 Ephraim joined the business and the name became Elizabeth Rabone & Son trading from a premises in Water Street. In 1811, Ephraim started his own rule and thermometer making business at another premises in Water Street. In 1817, John took over the original business which continued to trade under the name of Elizabeth Rabone & Son. (Whether this was due to Ephraim's ill health – he died in 1819 – or because Ephraim wanted to concentrate on his own business has not been established.) In 1823 the business moved to 61 St. Paul's Square and

Fig. 4.24. The Rabone dynasty,. These portraits do not include Michael Rabone, the founder of the firm.

the name was changed to John Rabone. The firm thrived, describing itself as an "English and Foreign Rule Manufactory".

In 1845 John Rabone's son, another John, joined his father and the name of the business became John Rabone & Son.

By good luck many of the books of the firm have survived so it is possible to get an extensive picture of the business from the 1830s onwards.[7] By 1846 it may have been on a firm footing but it was still not large. The accounts for 1846, for example, show that the stock of materials, finished and part finished goods was valued at £128 2*s* 6*d*, and, excluding tools and fixtures, the assets of the business amounted to £292 2*s* 6*d*. "all belonging to J. Rabone" (referring to the elder John). The trading profit was however £226 6*s* 10½*d*. John Rabone had drawn £37 during the year and John, the son, £67.

By 1860 the value of materials, finished and part finished stock had increased to £2026 19*s* 3*d* and by 1863, the total value was £2800 10*s* 3*d*.

An inventory of May 1851 lists the tools and equipment being used in the business. A few items were "kept in the house for safety" confirming that the Rabones were still living on the premises (see Appendix 3).

Expansion was now accelerating and with an increasing change over to working by machine, new premises were needed. In 1871 a purpose built factory, the Hockley Abbey Works was opened. Shortly before this move, John's son, Harry Joseph, joined the firm, being paid the princely sum of £1 per week. In 1872, his younger brother Arthur John also joined, also being paid £1 per week and Harry's wages increased to £3. 3s. In 1873, Harry was made a partner with Arthur becoming a partner in 1877 and once again, the name of the firm changed, to John Rabone & Sons.

The Hockley Abbey and Vulcan trade marks had been registered the previous year, 1876.

Following the move to Hockley Abbey more and more automation was brought in. The firm could afford the cost of these developments – they were making thousands of rules whilst other makers were producing tens. However, despite this increasing mechanisation, there was still a great deal of hand work necessary in assembling, dividing, finishing, inspecting and packing.

Not only were Rabone making rules to be sold under their own name but from the "Specials Pattern Book" in the Rabone archive it is clear that they were making rules for a vast number of tool makers and sellers. These included not only well known tool manufacturers such as Alex. Mathieson in Scotland and Wm. Marples of Sheffield, but also companies such as John Jacques & Sons, sports equipment makers, J. H. Holding, tailors' equipment suppliers, the Army & Navy Stores and even other rule makers, such as Aston & Mander, W. F. Stanley and A. G. Thornton. Rules were also being made for companies in Australia and North America. It is of interest that these "specials" were often noted to be marked with the name of the retailers, or sometimes even the retailer's customers.

The product range was also expanding. By 1892 in addition to the boxwood and ivory rules that the firm had always made, a range of steel and brass rules and a good selection of tapes, linen and steel, in both small pressed metal cases and the traditional rigid leather cases, had been added.

Although the firm prospered overall, there were periods where business was difficult, even to the extent that for a time after World War I the company went into voluntary liquidation.

A few figures will illustrate the output of John Rabone and Sons during the last century. In 1945, a time of high demand, the company was produc-

Fig. 4.25. A small steel tape in a nickel silver case by John Rabone & Sons (Birmingham, 1877–1963). The patent refers to the spring mechanism.

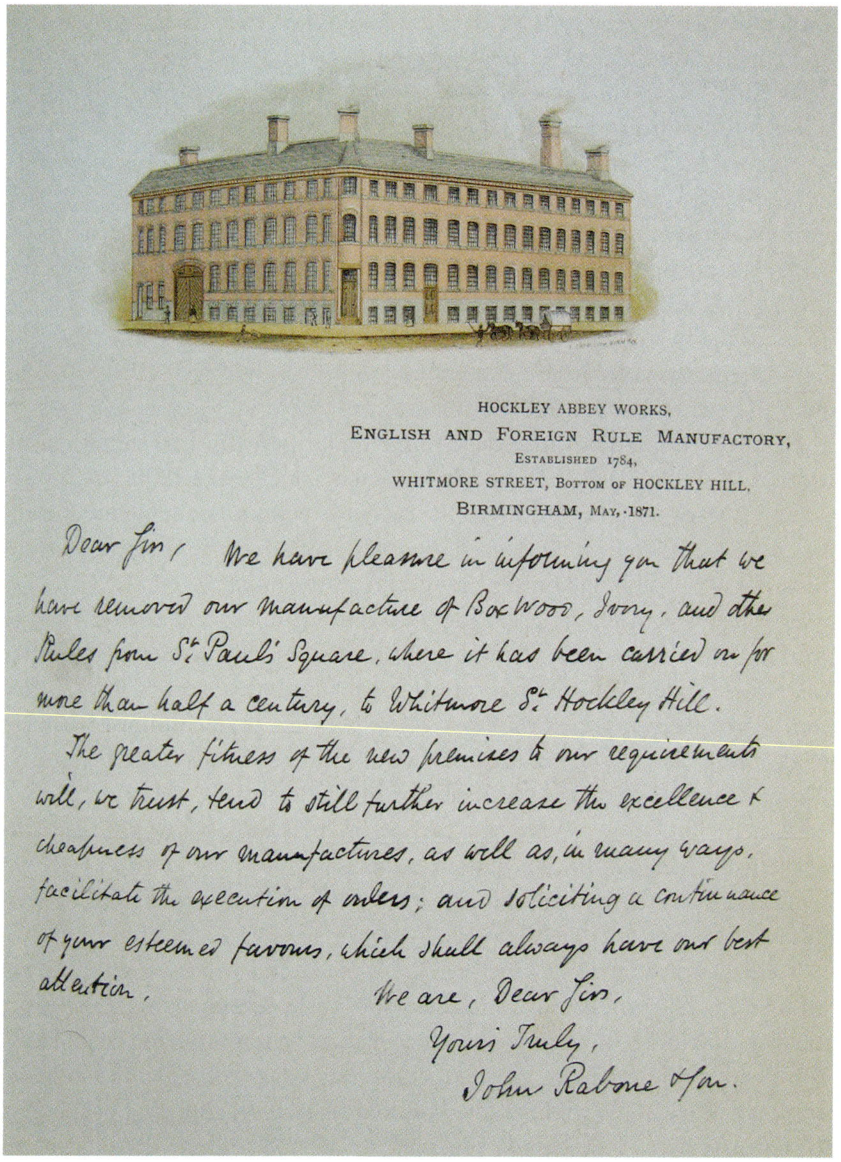

HOCKLEY ABBEY WORKS,
ENGLISH AND FOREIGN RULE MANUFACTORY,
ESTABLISHED 1784,
WHITMORE STREET, BOTTOM OF HOCKLEY HILL,
BIRMINGHAM, MAY, 1871.

Dear Sirs, We have pleasure in informing you that we have removed our manufacture of Boxwood, Ivory, and other Rules from St. Paul's Square, where it has been carried on for more than half a century, to Whitmore St. Hockley Hill.

The greater fitness of the new premises to our requirements will, we trust, tend to still further increase the excellence & cheapness of our manufactures, as well as, in many ways, facilitate the execution of orders; and soliciting a continuance of your esteemed favours, which shall always have our best attention,

We are, Dear Sirs,
Yours Truly,
John Rabone & Son.

Fig. 4.26. Letter from John Rabone & Sons informing customers of the move to the new Hockley Abbey Works.

ing 27,000 rules each week and orders were such that the company needed to increase production. In the early 1950s the company had 1000 tons of boxwood logs in stock.

With the collapse of Edward Preston & Sons in 1932 and its assimilation into Rabone, a major competitor disappeared. After this date I. & D. Smallwood was the only remaining firm of any size competing with Rabone.

The Rabone product line had developed so that there was considerable overlap with that of the Sheffield measuring tool maker James Chesterman, and industrial logic suggested a merger of the two firms. This took place in 1963. The history of both these firms has been documented by Douglas J. Hallam in his history of Rabone Chesterman, *The First 200 Years*.[8]

Manufacture of boxwood rules declined after 1963. The reasons were many: the ascendancy of the boxwood rule had passed; the tradesman preferred the spring steel tape; the supply of boxwood could not be assured; and in any case the quality of the rules had declined, partly due to the quality of the wood, and partly because the faster (stamped) methods of marking produced a product of lesser quality. For the tradesman who must have a folding rule, Rabone offered a white plastic alternative.

Fig. 4.27. A steel tape by Chesterman in a Bakelite case. Bakelite, which had been invented in the early years of the 20th century, was marketed by Chesterman as "a special new type of moulding material".

James Chesterman & Co. Ltd.

The early products of James Chesterman had one thing in common – they all used springs – often in innovative ways. These included such items as clockwork roasting spits, door springs and a tape measure that returned into the case by means of a coil spring. Chesterman, a Londoner, started his business in Sheffield in 1821. In 1829 he took out patent No. 5817 for a spring tape. Landmarks in the development of products were the 1841 production of woven tapes with copper wire warp to prevent stretching and in 1842 the first steel measuring tape with etched markings. The production of long steel tapes is reputed to have been a response to the collapse in the market for crinoline wire (actually a flat strip and another of Chesterman's products) around 1870. Initially the long steel tapes were called "patent steel band measuring chains" as they were intended for land surveyors who had traditionally used wire link chains. The marking was effected with copper rivets.

Like many Sheffield manufacturers, from time to time the firm experimented with selling some other tools and the 1880 catalogue contains some pages of carpentry tools, though none would have been made by Chesterman.

Fig. 4.28. William Chesterman from the Ironmonger *magazine, August 1880. William succeeded his father James as head of the firm in 1867.*

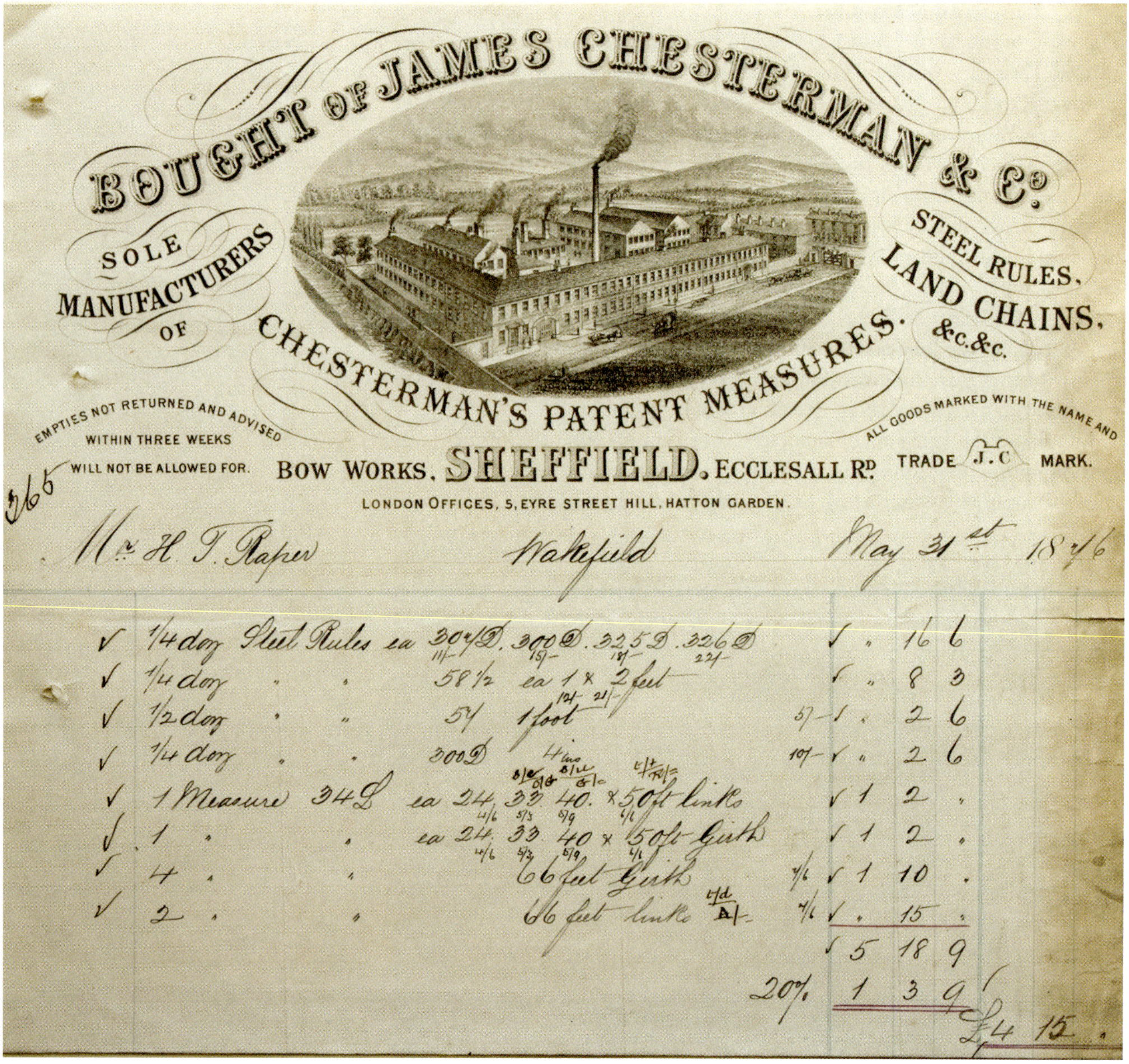

Fig. 4.29. An invoice, dated 1876, from James Chesterman & Co. for steel rules and measuring chains, made out to H. T. Raper, an ironmonger in Wakefield.

However the principal products illustrated are a range of tape measures, made from linen, metallic reinforced tape, and etched steel up to 100 ft. Other products were wire land chains, the Chesterman patent steel band measuring chain, and a good selection of machine divided and etched steel rules. A full page is devoted to the patent spring measuring tape cattle gauge.

After 1900 many engineers' measuring tools were added to the range. Initially mainly such items as feeler gauges, small sliding caliper gauges and some engineers' squares and other engineers' marking tools. In the 1920s the range was further extended with vernier calipers and height gauges. The improved Chesterman height gauge introduced in 1933 had a triangular vertical beam containing a fine adjustment screw. This was a great advance on the previously available height gauges. The range of steel rules and straight edges however continued as a major product line.

In 1963 the firm amalgamated with John Rabone & Sons to form Rabone Chesterman Ltd.

Edward Preston

The business founded in 1825 in Birmingham by Edward Preston initially made wooden planes and little else. In 1864 his son Edward advertised that he was making wood and brass spirit levels and very shortly afterwards he was also offering ivory and boxwood rules, becoming by the end of the nineteenth century, one of the major rule making companies.

In 1888 the well established rule making firm of T. Bradburn & Sons was assimilated into Edward Preston & Sons. Although many were duplicates of rules already made by Preston, the range of Bradburn rules were listed in a separate section in the extensive, well illustrated catalogues published by Preston between 1901 and 1914.

The catalogues issued by Preston contained thousands of different rules which perhaps illustrates one of the factors that was to lead to the demise of the firm in 1932. The product range, with not only rules and levels but an ever expanding range of cast iron planes and shaves and eventually even engineers' small tools, was far more than an undercapitalised family-run business should have considered. On every side they were competing against much larger specialist manufacturers such as the Stanley Rule & Level Co. and John

Fig. 4.30. Edward Preston, son of Edward Preston, the founder of the firm. This photograph is thought to have been taken at the time of his marriage in 1858, when Edward was 22 years of age.

Fig. 4.31. A 2 ft. 4-fold rule by Edward Preston with the Preston mark.

Rabone & Sons. In 1932, after several years of losses, the firm was taken over by Rabone. The factory was swiftly closed with the metal plane making part of the business sold to C. & J. Hampton Ltd. (Record Tools) and the rule and level making being transferred to Hockley Abbey. The Preston business had always made quality goods and indeed the reluctance to compromise standards in a changing world was another factor in the collapse.

The benefit to Rabone came more from the removal of a competitor than from the acquisition of any manufacturing expertise. As mentioned previously, this take-over left only I. & D. Smallwood as a competitor to Rabone.

A history of the firm can be found in the reprint of the 1909 catalogue, published in 1991[9] and further information about the demise of the firm can be found in the Tools & Trades History Society *Newsletter No. 75.*[10]

Fig. 4.32. A printing block for one of the Preston catalogues.

I. & D. Smallwood

This was the last of the traditional rule and level makers of Birmingham, closing in 1999 after the death of the last owner, remaining a family-run firm to the end.

The firm claimed to have been established in 1810. However, Smallwood is a common Birmingham name and the early 19th century trades directories include quite a few following a variety of trades. The earliest record from which a continuity can be proved is in 1823 when William Smallwood, heavy steel toy maker, was at 35 Barford Street. By 1830 he is described as steel toy, tool and iron square maker, and in 1839, iron square maker, working in Moseley Road. In 1858 the name is recorded in the trades directories as John & Daniel Smallwood, the sons of William, and this remained the name until the firm's closure in 1999. However, the style of the name was always shown as "I & D", with the "J" being written in the early form, which by this date was purely an affectation.

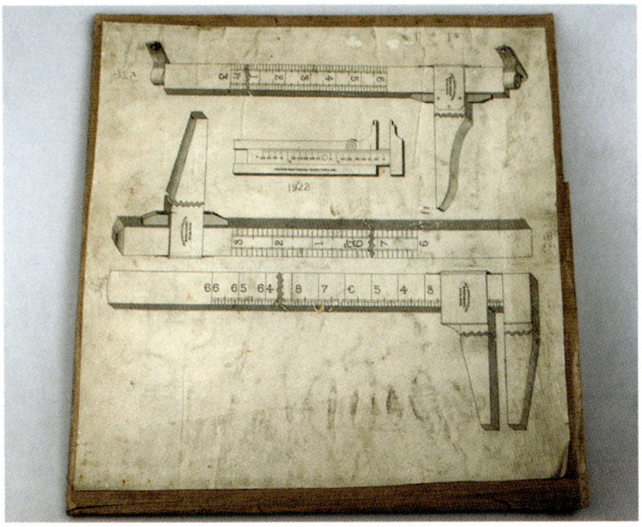

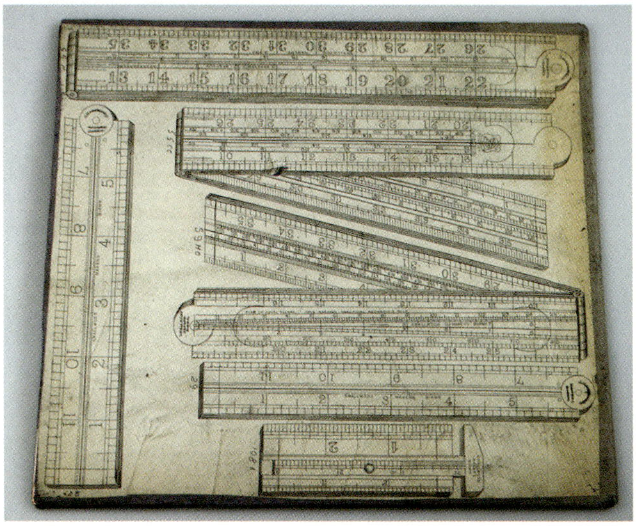

Above left: Fig. 4.33. The sleeve of a printing plate for I. & D. Smallwood's catalogues showing a wall-mounted height measure, a button gauge, free-standing height gauge and a caliper measure.

Above: Fig. 4.34. The sleeve of the printing plate in Figure 4.35 which shows the print as it would appear in a catalogue.

In 1865, the firm moved to premises in Leopold Street where they remained for well over a century; the building is shown in the frontispiece of their 1884 and 1897 catalogues.

The firm made a wide range of products; joiners' squares, metal squares and roofing squares, steel rules, tapes, shoe sticks, plumb bobs and many sorts of levels in both wood and metal were a leading part of the range.

Many of the products shown in the catalogue of 1897 are of the best quality and include Routledge's, Wilkinson's, Hawthorne's and Carrett's engineers' rules.

The 1897 catalogue records the firm taking over the business of T. Aston, described in this catalogue as "The Original Rule Maker", although there are no grounds for this assertion.

Although the earlier products were in no way inferior to those of the other Birmingham makers, in the latter years, their quality seems to have been compromised, with the rules having impressed or printed markings that easily wore off and the finish on many items leaving much to be desired.

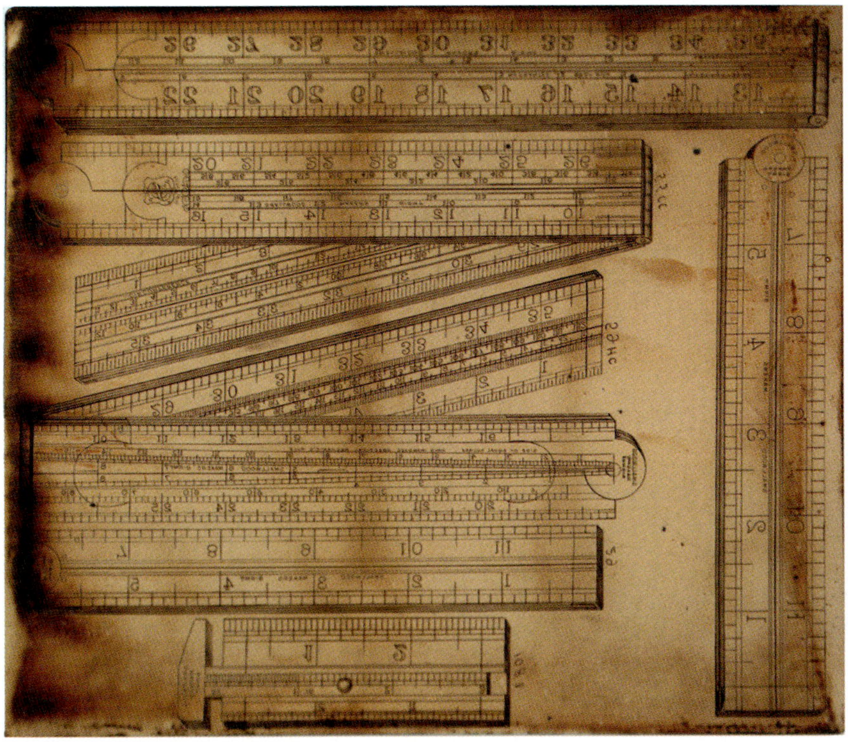

Fig. 4.35. An engraved copper printing plate for I. & D. Smallwood's catalogues showing a button gauge, two 1 ft. 2-fold rules, both sides of a Hay's mechanics' rule and a 3 ft. 4-fold rule.

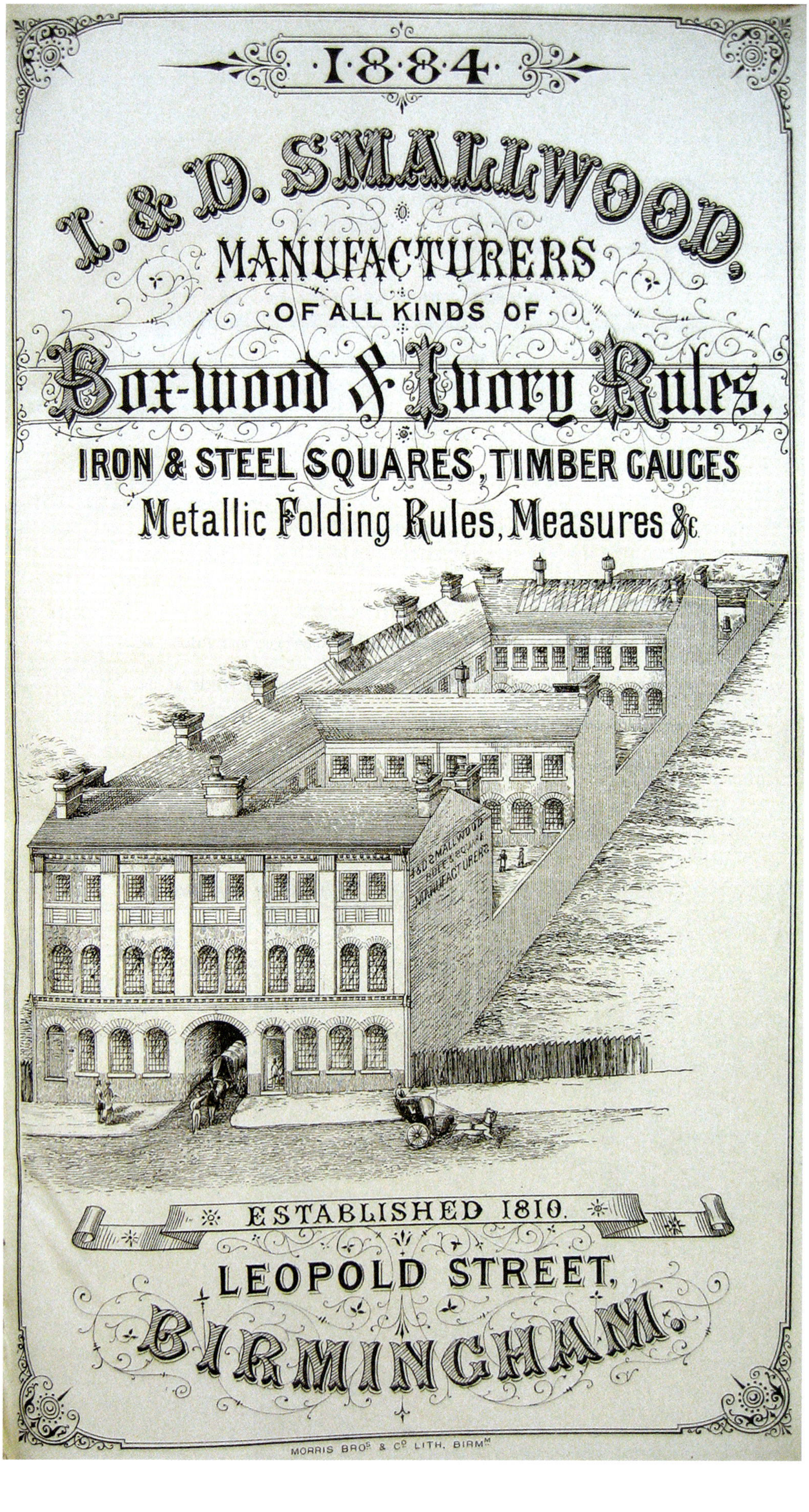
1884
I. & D. SMALLWOOD,
MANUFACTURERS
OF ALL KINDS OF
Box-wood & Ivory Rules,
IRON & STEEL SQUARES, TIMBER GAUGES
Metallic Folding Rules, Measures &c
ESTABLISHED 1810.
LEOPOLD STREET,
BIRMINGHAM.
MORRIS BROS & CO LITH. BIRMM

Opposite: Fig. 4.36. Frontispiece of the I. & D. Smallwood catalogue for 1884 showing the factory in Leopold Street, Birmingham.

Dring & Fage

The origins of this firm can be traced back to 1716 when John Clarke "turner and engine maker" commenced in business. Around 1725 he invented a metal hydrometer which he publicised in his book *The Hydrometer or Brandy Prover,* published in 1746. The business was continued by various members of the Clarke family, eventually passing to John Dring, the brother-in-law of the Clarke in ownership at that time, in 1784. In 1790 Dring entered into partnership with William Fage, creating the firm that was to become a leading maker of hydrometers and gauging instruments. The firm finally closed 1977.

Like many firms trading over such a long period the products varied. An early 19th century trade card gives emphasis to nautical instruments, listing sextants, compasses, telescopes and charts. With the opening of the premises in the Strand in 1883, in addition to the premises in Tooley Street, the style "opticians" was adopted but the core products throughout remained gauging instruments – hydrometers, saccharometers, the necessary accompanying thermometers and the rods, rules, calipers and slide rules used for the measurement of containers and the calculation of their contents. The principal users of these instruments were the officers of the Customs & Excise and from the 1890s the firm claimed "by appointment to the Honourable Commissioners of Customs". Similar claims had been made by several other makers earlier in the century.

The firm seems to have died quietly in 1977. The workshops were just locked up and shortly thereafter were pillaged and set ablaze by vandals. Some papers, including an employee's piecework account, are now in the library of the Science Museum.

A study of this piecework account for the period between 1797 and 1802 gives an indication of the variety of work that the firm was undertaking. There are mentions of a huge variety of rules including "patent slides" (slide rules), ullage rules, gauging rods, Branan's rules, casting rules (perhaps shrinkage rules?), calipers and long calipers, straight rods, glaziers' and gauging rods and rules, yard sticks, glovers' size sticks, cordage rules, plumbers' rules, ivory pocket book rules, and sectors. Extracts form this account are reproduced in Appendix 4.

There were a number of rules where a particular designer of a rule is men-

Fig. 4.37. The hydrometer workshop in the Dring & Fage factory in Tooley Street, London.

tioned by name as with the Branan's rule listed above. These included Gordon's timber rule – no examples of which have been noted to date.

There are numerous other types of rules that appear in the ledgers during the same period that are of interest, such as ordnance rules, starch gauges (probably a duty rule), stationers' rules, diameter rules and diary rules (perhaps a forerunner of the calendar rules that were one of Dring and Fage's specialities in the 20th century).

The firm's trademark, a half moon and dagger, was originally the mark of John Clarke who stated that he put it on his hydrometer. The use of the mark is common on items made after 1880 but does not seem to have been used much in the earlier part of the 19th century.

Fig. 4.38. The Dring & Fage trademark.

W. F. Stanley

In 1853 William Ford Stanley (1829 – 1909) opened a small shop in Great Turnstile, Holborn, London, selling wooden drawing instruments such as T-squares, parallel rules and French curves which were made on the premises. At first progress was slow but the range of items was gradually expanded to provide a wide range of instruments for the draughtsman.

Stanley was clearly an inventive man with drive. Two significant steps in the progress of the firm were firstly the development in 1861 of a linear dividing engine designed and partly made by Stanley himself and capable of dividing scales of any division and secondly, the publication of his book, *A Descriptive Treatise on Mathematical Drawing Instruments* in 1866. This book was regularly republished until 1900 by which time it had been expanded to 364 pages. His style is forthright and even includes some self-criticism, but the message is clear – the products he made are the best – and they were.[11] The dividing engine enabled the production of scales to new standards of a fineness of division and accuracy. The Stanley firm were always leaders in the production of draughtsmen's scales and in 1953 over 1000 types were available.[12]

By 1881 the firm was employing 80 men and in 1898 150. The range of goods expanded to include optical items such as levels and theodolites and almost everything that was needed in the field or office by the design professions. In 1900 the firm converted to a limited company and amalgamation with J. J. Hicks (1911) and Heath & Co. (1926) followed. The firm finally closed in 1967.

Opposite: Fig. 4.39. W. F. Stanley's drawing instrument works, foundry and saw mills in South Norwood, London, and their surveying instrument works in Great Turnstile Road, Holborn, London, from their catalogue of 1919.

References

1. Latham, Robert, *The Shorter Pepys. Extracts from the Diary of Samuel Pepys*, London, 1993.
2. Coggeshall, Henry, *Timber Measure by a Line of more ease, dispatch and exactness, then any other way now in use, by a double scale ...,* London, 1677.
3. Everard, Thomas, *Stereometry, or the Art of Gauging made easy...,* 7th ed., London, 1712.
4. *Ibid.*
5. Flower, W., *A key to the Modern Slide Rule...,* London, 1768.
6. Stock List of John Shaw & Sons, ca. 1805, MSS DB/24/401, records of John Shaw & Sons, Wolverhampton Borough Archives.
7. Records of John Rabone & Sons, MS 1522, Archive Department, Birmingham Library Services.
8. Hallam, Douglas J., *The First 200 Years, A Short History of Rabone Chesterman Ltd.,* Birmingham, 1984.

Stanley's Factories and Workshops

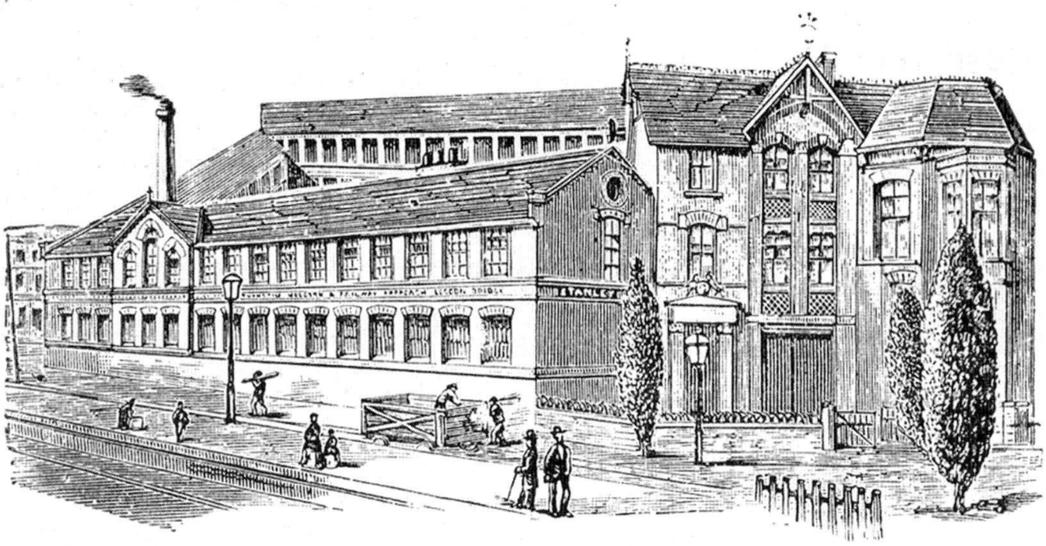

Drawing Instrument Works, Foundry, and Saw Mills, South Norwood.

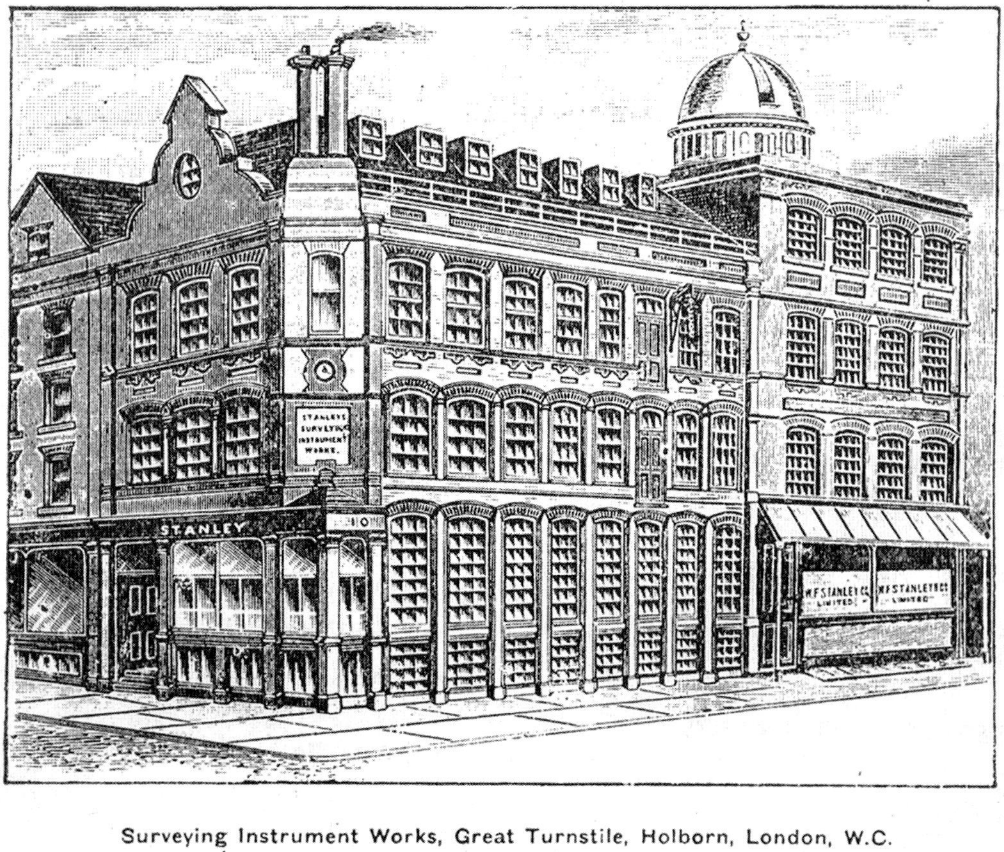

Surveying Instrument Works, Great Turnstile, Holborn, London, W.C.

9. Edward Preston & Sons, *Catalogue No. 18. The "Preston" Catalogue,* 1909 with 1912 supplement, reprinted Mendham, NJ, 1991.
10. Rees, Mark, "Edward Preston & Sons Ltd., The Final Years", *Newsletter No. 75,* Tools & Trades History Society, 2001.
11. Stanley, W. F., *A Descriptive Treatise on Mathematical Drawing Instruments,* 4th ed., London, 1873.
12. Allen, Cecil J., *A Century of Scientific Instrument Making 1853 – 1953,* London, 1953.

Chapter 5
The timber trade

... while dinner was getting ready, he and I practised the measuring of tables and other things, till I did understand measuring of timber and board very well.
Pepys, 1662.[1]

Timber has been brought into Britain for many centuries and by the 13th century there are contemporary references showing that both softwood and hardwood were regularly being imported from the Baltic region and from Norway with the merchants of the Hanseatic League playing a leading role in the trade. The Windsor Castle pipe rolls contain many entries recording the purchase of timber – and quite a lot seems to have been imported via London. In 1252 three thousand "Norway boards" and fifty "great boards" (for making tables) were sent from London – the cost £19 4*s* 6*d*. During the following decade the records show that fir, "waynescoting" and "Estrich" (eastern reich) boards were all being purchased for the king's use at Windsor – the first known evidence of a trade that has continued to the present day.

There is no indigenous softwood in Britain excepting the Scots pine, which has a coarse and twisted grain, suitable for little other than fire wood, so imported pine was a useful material. Whether by the 13th century the population was already overstraining the available resources of hardwood, particularly oak, is not entirely clear but by the 16th century this certainly was the case and from then on there were recurrent concerns about the supply of timber suitable for shipbuilding.

During this early period it would seem that sawn timber was accounted by the board; whether there was an understanding that the board was of a particular size or the equivalent thereof is not known. What is clear from the records is that the timber, whether soft or hard wood, was being imported already sawn, a general practice of the softwood trade that has continued. One interesting record reveals that in 1390 the ship *Haricog* unloaded a cargo of timber in Newcastle made up of "Rygholts" (timber from Riga), "Wanescottes" (oak boards), "Barelholts" (barrel staves), and "Deles" (deals).[2]

The term deal is remarkably difficult to define. In due course it came to mean softwood, just as wainscot came to mean oak, but "**a**" deal is a piece of sawn softwood – within certain size limits though not of any specific section. The general limits were probably between 1½ in. and 3 in. thick by 8 in. to 12 in. deep. Neither is the length specific. If 6 ft. or less it is a "short" but a timber merchant who received a high proportion of pieces only 7 or 8 ft. long would soon complain so typically a deal is in the range of 10 to 16 ft. in length. The timber trade seem to have had a disconcerting habit of referring to deals as if they were a standard item, which they were not. There may have been local understandings between merchant and customer at various times so that to the concerned parties a deal was a fixed item but to someone else it might have a different meaning.

Units of measurement

Timber starts as a standing tree and from there goes through a process of conversion to make it suitable for use. Conversion may entail little change; for example a pit prop needs nothing other than cutting to length – not even de-barking – or it may entail progressive re-working, for example, the sawing,

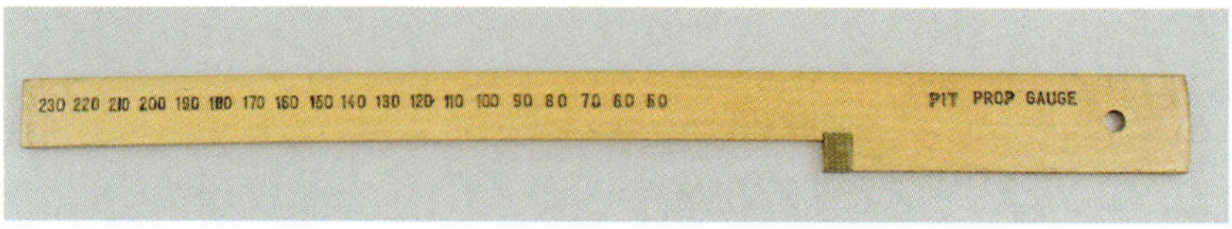

Fig. 5.1. A mine pit prop gauge. Although unmarked, this rule was made by I. & D. Smallwood (Birmingham, 1858–1999).

re-sawing and then selection, needed to prepare a reasonably knot-free batten from which an architrave might be moulded.

The units of measurement used by the timber trade were many, reflecting the changes wrought as the wood progressed through the conversion process and also as it passed from wholesale to retail.

In the past the trade provided a range of products that is perhaps difficult to imagine today. The 1899 list of charges for the Surrey Commercial Docks Company, the proprietors of the principal London dock used for the importation of timber in all its forms, included such diverse items as handspikes, rickers, treenails, "thickstuff", and "balk". The Company's basis for charging was varied. The St. Petersburg Standard Hundred, which was 165 cu. ft., was the basic unit for softwood of normal dimensions but the "load" was used for larger sizes (that is 5 inches and upwards in thickness), although this was not a fixed quantity. Hardwood was charged by the load, whether sawn or round, but if round or "in the tree" the load was stated to be 40 cu. ft. Firewood was charged by the fathom. Smaller items such as staves, box boards and "sashwood" were charged per 1200 pieces and larger items such as putlogs (poles built into walls to support scaffolding), poles and rickers (thin poles or young trees) per 120 pieces. However these are only a few of the units of measurement used in a trade whose origins can be traced back 800 years or more.

The load: The oldest meaningful measure, literally the quantity of timber that can be carried on a cart, this term is probably as old as the use of wheeled vehicles to carry timber. By the 17th century the meaning had become more precise and was thereafter generally taken to be a fixed number of cubic feet. Most frequently this was 50 cu. ft., less commonly 40 cu. ft. and occasionally 30 cu. ft. William Leybourn, writing in 1669, stated that "50 cu. ft. of rough timber is counted a load."[3] Some sources suggest that the 50 cu. ft. load was used for softwood with 40 cu. ft. for hardwood which is heavier so that the weight of a load whether hard or softwood would have been substantially the same.

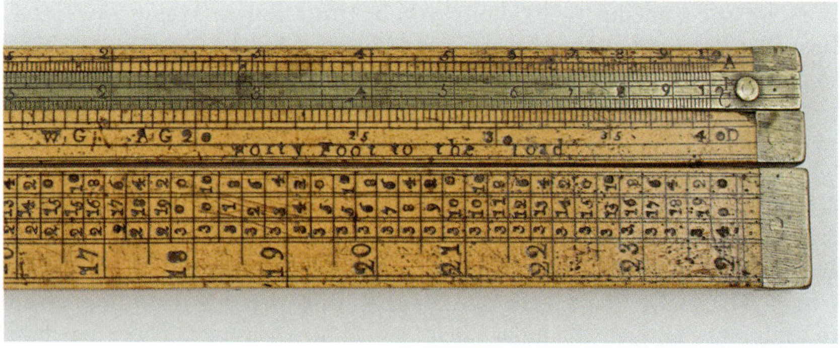

Fig. 5.2. A rule by Thos. Onions (1767–1801) with the size of load, in this instance 40 ft. to the load, marked onto the rule.

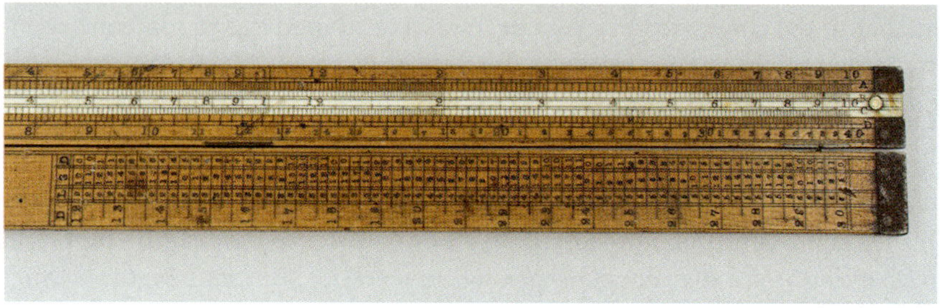

Fig. 5.3. Timber load table from a Coggeshall type rule made by Thomas Tustian (London, 1805–39).

The 1872 list of port charges for Hull Docks Company,[4] a leading port for the Baltic timber trade, accounted "Timber and planks imported – Fir, square, the load of 50 cu. ft., and fir round, the load of 40 cu. ft." This seems to be the generally used definition of a load – sawn timber 50 cu. ft., logs 40 cu. ft. Therefore it is clear that the term had a specific meaning in a particular circumstance although in a different circumstance, it might well have had a different meaning.

As a measure the term certainly had a long usage in the timber trade, and load/cu. ft./money tables are a common enhancement on rules intended for those needing to buy timber.

Cubic measurement: The cubic foot or foot cube is the most usual and perhaps the most understandable unit of measurement for medium and small quantities of timber. Today this unit has been replaced by the cubic metre which is 35 times larger. As a result many orders are for but a fraction of a metre with a price of maybe several hundred pounds per metre.

If finished timber is sold by the cubic foot or cubic metre, it is usual to use the original (nominal) size of the sawn timber when the volume is calculated.

Hoppus foot: This is a cubic foot of wood measured using the Hoppus system (an explanation of this system is given later in this chapter). A Hoppus foot will be somewhat more than a cubic foot: some authorities state that a Hoppus foot is 1.273 cu. ft. but this is not correct in that the cubic content depends on the size and shape of the log.

Board foot: This is one square foot of timber, one inch thick i.e. 144 cu. in. Principally used to measure timber that is in planks, though the planks need not be an inch thick, so for a plank 1¼ in. thick a board foot is only 115.2 sq. in. or for a plank ¾ in. thick, 192 sq. in. This latter statement was not always correct, however, as in some markets, boards under 1 inch were measured as if they were 1 inch thick.

This is perhaps not as illogical as it seems, as the origin of the board foot lies in the size of the area to be cut when sawing timber when, obviously, the thickness has little effect on the work involved. The board foot seems to have been used more extensively in America than in Britain.

The determination of the board foot measure of boards that are of a constant width is just a matter of arithmetic but for tapering or unedged boards it is considerably more difficult. The grade of timber being measured also has to be considered when deciding what is to be included in the measure. To make matters even more difficult, each of the major sources of supply had its own set of rules. A review of them is without the scope of this book

Fathom (piled fathom): When applied to the measurement of timber the meaning is a cubic fathom, 6 ft. x 6 ft. x 6 ft. (= 216 cu. ft.). This is used for goods such as pulpwood, firewood, lathing and some pit props and poles that are not amenable to measurement by other means.

Russian piled fathom: 7 ft. x 7 ft. x 7 ft. (= 343 cu. ft.). This measure was occasionally used for pit props.

Cord: 4 ft. x 4 ft. x 8 ft. long (= 128 cu. ft.). Almost always used for the measurement of firewood, it is more common in America than Britain.

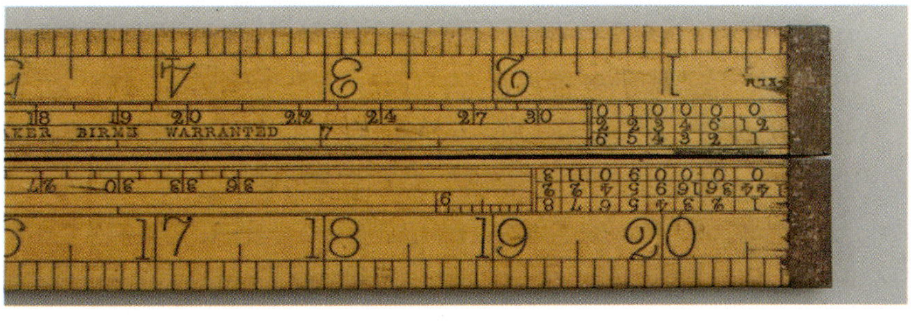

Fig. 5.4. The table for boards narrower than 6 in. at the end of a line on a timber contraction rule by Samuel Albert Smith (Birmingham, 1850–72).

Hundred: As early as the 12th century, references to "a hundred" or "a hundred boards" as a specification of a quantity of timber are commonly found. The presumption must be that even as early as this the hundred was interpreted as a reasonably fixed quantity. With time the description became "a standard" (hundred).

St. Petersburg standard (hundred): 165 cu. ft. Before metrication this was the most widely used unit of quantity in the wholesale timber trade in Europe. The origin is a "hundred" deals 1½ in. x 11 in. x 12 ft. but the hundred is a "long hundred" of 120 pieces. There are actually two St. Petersburg standards: one for sawn goods and one a nominal measure for planed goods. When timber is planed, depending on the size of the section, it is reduced by between ⅛ in. and ¼ in. The actual volume is therefore somewhat variable

Other standards: Many other standards have in the past been used, most only of local significance. The following are but a few examples.

Christiania Standard – 103⅛ cu. ft.
Drammen – 121⅞ cu. ft.
Irish and London – 270 cu. ft.
Quebec – 229⅛ cu. ft.
Gothenburg – 180 cu. ft. (often used for pit props).

Liverpool measure: The practice of allowing the purchaser rebates in the measure that he bought was widespread in the timber trade. Such arrangements, originally intended to compensate for faults, became widespread and institutionalised. For example, in the mahogany trade, Liverpool measure meant that the purchaser only paid for 10½ in. of each 12 inches measured. This system continued until 1940.

Sawyers' measure: In general sawyers were paid by the square foot with the rate expressed per 100 sq. ft. (Hundred Measure). But additional payments were legion. Breaking work – that is the first cut though a log – was frequently measured per foot lineal and sometimes payment for more than one breaking cut was demanded! Small dimension timber, skew cuts and certain species of timber all attracted extras, mostly paid in the form of "over measure".

These quite sophisticated methods of different payment for different work were well established by the last quarter of the 18th century. George Fisher, in the 1799 edition of *The Instructor or Young Man's Best Companion,* sets out the then going rates.[5]

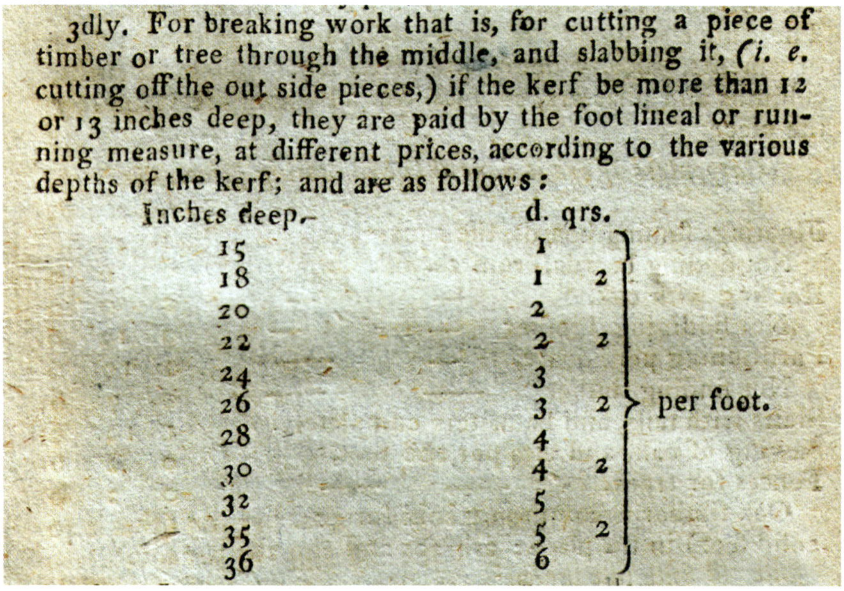

3dly. For breaking work that is, for cutting a piece of timber or tree through the middle, and slabbing it, (*i. e.* cutting off the out side pieces,) if the kerf be more than 12 or 13 inches deep, they are paid by the foot lineal or running measure, at different prices, according to the various depths of the kerf; and are as follows:

Inches deep.	d.	qrs.	
15	1		per foot.
18	1	2	
20	2		
22	2	2	
24	3		
26	3	2	
28	4		
30	4	2	
32	5		
35	5	2	
36	6		

Fig. 5.5. The list of prices for breaking work from the 1799 edition of The Instructor or Young Man's Best Companion.

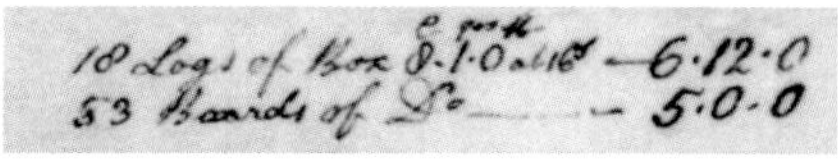

Fig. 5.6. Entry in the Christopher Gabriel 1791 stock list for 18 logs of box

Mille: Small items such as barrel staves, trenails, etc. were sold by the mille. The origin is from the Latin, a thousand, but in many situations the thousand was 1200 pieces.

Bundle: This was the measure by which laths were sold. Leybourn, writing in the 17th century, says that laths are either 4 ft. or 5 ft. long but by the 19th century lengths from 2½ ft. to 5 ft rising by 6 in. increments were normal in the trade.[6] For hand-split laths, the bundle was 360 ft. run but for sawn laths the bundle contained 500 ft. run.

Weight: This was occasionally used for the sale of expensive or exotic woods such as boxwood, ebony, etc. In 1791, Christopher Gabriel listed in his stock, 18 logs of box weighing 8 cwt. 1 quarter 0 pounds at 16 shillings per cwt. and in 1800, he listed 14 cwt and 3 quarters of Turkey box.[7]

Standardisation of sizes: Whilst today most timber is described only by the size "300 m of 50 mm x 50 mm", in the past the description or name given to a piece of timber was very much dependent on the size. The extract from British Standard 565 – 1949, given below, was a late attempt to give standard meaning to terms that had been used for centuries. Even here there is no clarity; a piece 4 in. x 4 in. would fit the definition for a batten, a scantling or a square. The timber trade was clearly no place for the uninitiated. The term batten also had a different meaning in the timber trade from that in the building trade, where a batten is a piece of small section timber not exceeding 1 in. x 3 in.

Plank	2 in. to 4 in. thick x 11 in. and wider.
Deal	2 in. to 4 in. thick x 9 in. to under 11 in. wide.
Batten	2 in. to 4 in. thick x 5 in. to 8 in. wide.
Scantling	2 in. to 4 in. thick x 2 in. to 4½ in. wide.
Board	Under 2 in. thick x 4 in. and wider.
Strip	Under 2 in. thick x under 4 in. wide.
Slating	½ in. to 1 in. thick x 1 in. to 3 in. wide.
Square	1 in. x 1 in. to 6 in. x 6 in.
Baulk	4 in. x 4½ in. or greater cross section.
Flitch	4 in. x 12 in. or greater cross section.
Shook	Set of pieces of sawn and/or machined timber ready for assembly to form a box or crate.

Retail measurements: Today timber is almost always sold by lineal measurement "330 m of 150 mm x 25 mm PAR (planed all round)". With a pocket calculator, any builder should have little difficulty in working out how many metres run he needs to floor a room. Planed timber has, of course, smaller finished dimensions than the nominal size so the calculation of linear feet was not entirely straightforward. It was thus convenient for builders to purchase flooring or sidings by superficial (area) measure. This could be either a laid measure, that is the actual quantity required to complete the area taking into account any overlaps or tongue and groove joints or, less frequently, on a nominal basis. This is the quantity of boards in the sawn state (i.e. of nominal or original size) that would be needed to cover the specified area. The two measures used in the pre-metric era were:

Square: 10 ft. x 10 ft. (100 sq. ft.), a long established area measure much favoured by the building industry.

Square yard: 3 ft. x 3 ft. (9 sq. ft.)

Measurement of round timber

The earliest records of timber sold in the round refer only to trees. One must assume that the price of each tree was adjusted in accordance with the size of the tree.

By the end of the 17th century mahogany and other tropical hardwoods, initially from sources in the Caribbean, began to arrive in Britain, and in the 18th century, South America, Africa and the Far East all became sources for fine or exotic hardwoods. Coming from areas where there were no established saw mills these were generally imported in the log. The assessment and measurement of these imports therefore followed the practices of the home-grown trade which by the 16th century had developed methods of calculation of the cubic content of a log and perhaps more importantly, had grown to understand them.

The measurement of timber in the log would be perfectly easy if trees were round and did not taper. A simple method used in the Scandinavian forest assumes exactly this – the diameter of the *top* of the log is measured and the volume calculated as a cylinder of this diameter. This method is supposed to give a good idea of the *usable* timber that will come from that tree – which it probably does, as up to 30% of a tree will be waste by the time it is converted to square-edge boards or deals.

But trees do taper; indeed some trees, which may taper more in some parts than others, and have large branches, present real difficulties to the measurer which can only be resolved by considering the tree as a number of sections, each separately calculated.

Measuring girth: The best and usual method is to measure without bark but this is not always possible so if the measure is taken over the bark it will be recorded. A customary allowance (usually 7½% to 15% dependent on species) can then be deducted; Hoppus suggests that one tenth or one twelfth of the circumference be deducted before determining according to his method.[8]

Customs (King's) caliper measure: The log is measured with calipers at the mid point and the volume calculated as if it were a perfect cylinder of the measured diameter multiplied by the length. This was a quick and easy method for the Customs & Excise to use. The calculation is therefore the *length x area of cross section.* (All the measurements must be in the same units, so if the length is in feet but the radius is in inches, the radius must be divided by 12.)

The calculation for the area of a circle is πr^2 but as it is simpler to take the diameter than the radius, the formula used is

length (feet) x π x *diameter (inches)*^{2} ÷ 144 x 4, which equation can be reduced to

length (feet) x diameter (inches)^2 ÷ 183.34

which is a quick and easy formula to apply for the calculation of the volume of a log, the result being in cubic feet. In practice the divisor used is 183 without the decimal places. Interestingly Mallinson says that this gives a reasonably true measure.[9]

Customs string measure (113 divisor): The very ancient practice of taking the quarter girth of a log using a piece of string was also used. (This is also at the heart of the Hoppus method.) Once the girth has been taken at the centre point or, if the log is irregular, an average girth, the calculation of

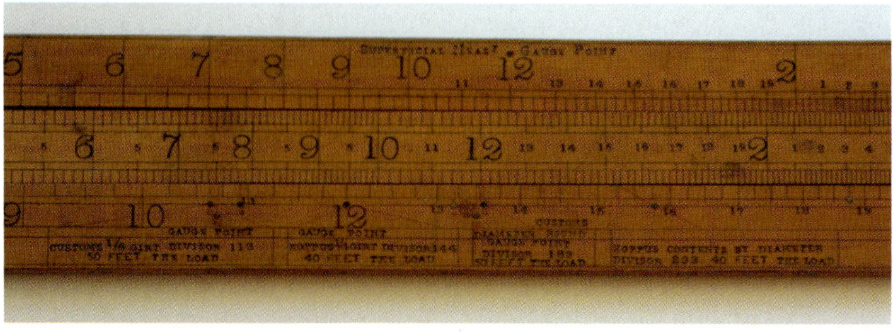

Fig. 5.7. The gauge points marked on a British Standard Cubing rule for the Customs & Excise: The divisors for the following systems are as follows: Customs ¼ girt (string) system = 113; Hoppus ¼ girt system = 144; the diameter round = 183 and the Hoppus content = 233 (this is yet another system for measuring logs). In addition to the gauge points, the cubic feet per load are given. However, the basis for the choice of size of load is not known. Note that the term girt is used, which strictly means to encircle or measure the girth, but in these instances actually means the measurement itself. i.e. the girth.

the volume would be the same as the Customs caliper measure except that the circumference is used instead of the diameter, the formula for an area of a circle using the circumference being $C^2/4\pi$ which produces a divisor of 113 giving the formula

length (feet) x quarter girth (inches)^2 ÷ 113, an easy and quick calculation.

Walter Rose, in his book *The Village Carpenter*, writing about his father in the early years of the 20th century, gives an excellent description of using the string measure for measuring trees.[10]

Hoppus system (144 divisor): In 1736 Edward Hoppus, surveyor to the Corporation of the London Assurance, published *Practical Measuring Now Made Easy to the Meanest Capacity...*[11] A forthright title and he was also, in the fashion of the times, forthright in his condemnation of other books of a similar type. However, the Hoppus method was no more than a restatement of the very long established method of taking the square of the quarter girth as the area of the cross sectional of the log at that point. Hoppus died in 1739 by which time only two editions of the work had been published but after 1751, with a new publisher, there were twelve new editions before the end of the century. With some variation of title, Hoppus's book remained in print, and use, until metrication in the 1970s. It is difficult to think of another technical work that has had an equally long currency.

The purpose of the book, and no doubt the reason why it was used for so long, was to remove the need to do any calculation. The major part is numerous tables from which the cubic content of the piece of wood can be read. There are also some price tables and tables of superficial feet (areas in

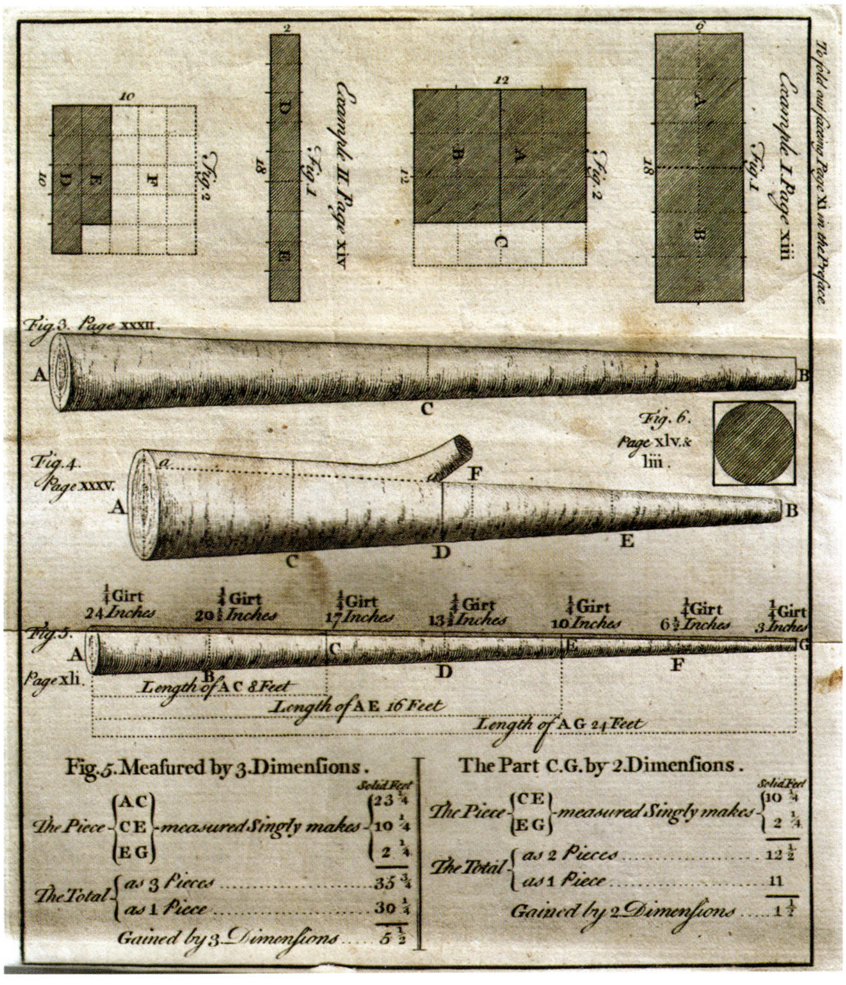

Fig. 5.8. Illustration from the 7th edition of Hoppus (1765) showing how to measure a tree in the round.

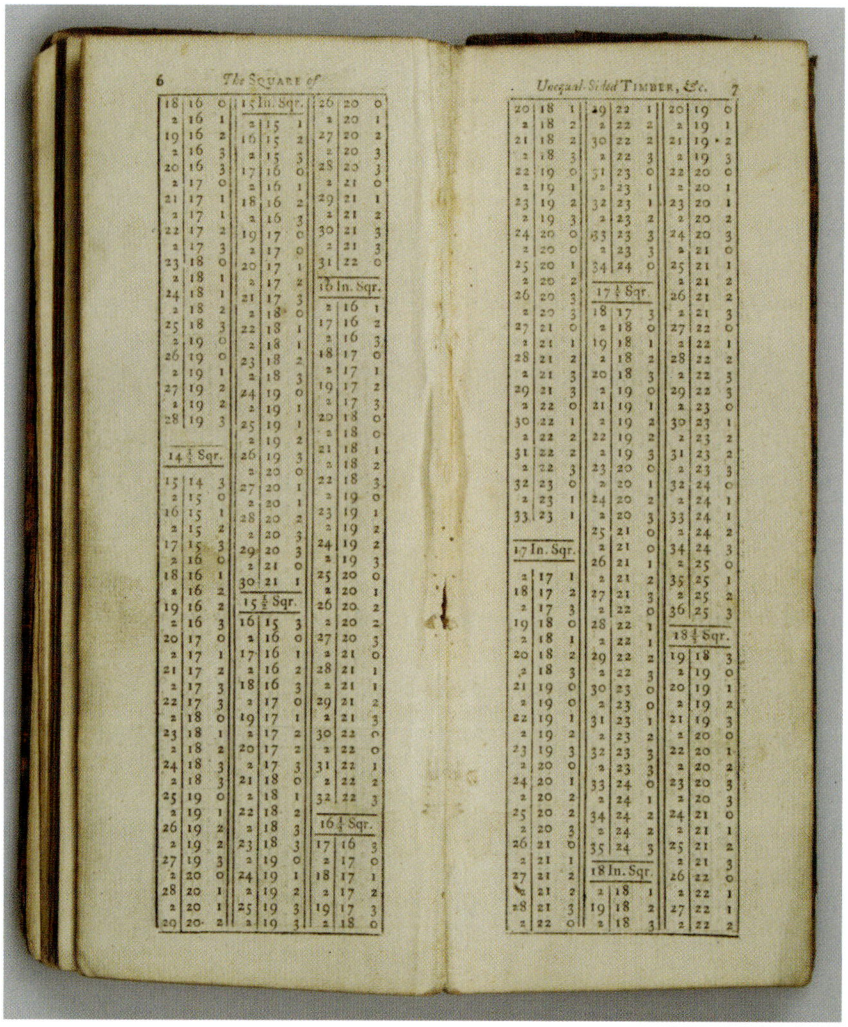

Fig. 5.9. A page of tables giving the square of unequal sided timber from the 7th edition of Hoppus, dated 1765.

sq. ft.). The earlier editions also include some interesting recitals of tools and ironmongery and nails together with prices.

When applied to the measurement of round timber, the Hoppus method gives a considerable degree of under measurement. The timber trade was well aware of this but whether the vendors of trees were is less certain. Figure 5.10 is a diagram of the method applied to a log of 24 in. diameter. It is immediately obvious that the areas outside the square baulk far exceed the small area of the baulk (the square) that lie outside the log (the circle). It would seem inconceivable that Hoppus and the previous writers from whom he had taken the method did not know this! The answer is presumably that the quarter girth method was so well established that he merely followed standard practice.

There is a strong practical reason for the use of the quarter girth in the measurement of trees whether they be felled or standing. Durable tape measures with figured dimensions did not exist before the 19th century, so the use of cord to transfer dimensions to a rule was common practice and by quartering the girth, the majority of trees could be measured against a standard 2 ft. carpenter's rule.

The Hoppus system has many synonyms: string 144 divisor, quarter-girt 144 divisor, Hoppus quarter girt and others. The formula is

length (in feet) x *(C /4)^2 (in inches)* x *1/144* with the answer in cubic feet.

The diameter can be used instead of the quarter girth, in which instance the divisor is 233. This is sometimes known as the Hoppus content divisor. By the late 19th century, the Customs & Excise were well aware of the inaccuracy of the Hoppus system and the Timber Measuring Department of the Custom's Fund and Bill of Entry Office at the Port of London Docks issued a set of calculation sheets covering the measurement of round timber, deals, battens, scantlings, thick stuff, deck planks, Quebec timber and "waney board timber".

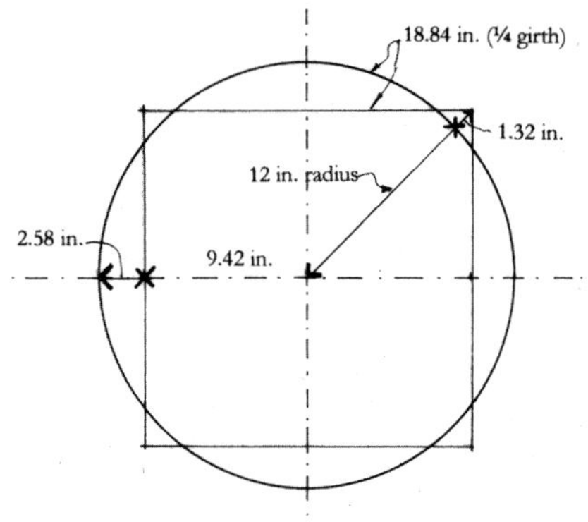

Fig. 5.10. Diagram illustrating the application of Hoppus' method to a log of 2 ft. diameter. The quarter girth of this log is $2\pi r/4$ = 18.84 in. A square with sides the same length as the quarter girth is drawn superimposed. The area of the four slabs outside the square (less the area of the four small triangles outside the circle) gives a graphic illustration of the under measure of round timber using Hoppus' method. The true area of this log is 452.16 sq. in. Using the quarter girt method the area is 354.95 sq. in.!

Measuring devices

Girthing tape: Typically this is a linen tape ¾ in. wide with a ring end all similar to the normal general purpose wind-up tapes. Later examples are likely to be made from metallic thread tape. Preston in 1909 listed these tapes in lengths from 3 to 12 ft. uncased, but examples often appear in leather cases.[12]

The line will give a direct quarter girth reading (in inches) as needed when using the Hoppus and other systems. The tape is divided into 4 in. divisions, marked 1, 2, 3, etc. with each division subdivided and marked ¼, ½, and ¾. If an allowance for bark is included, the divisions will be about 10% larger than 4 in.

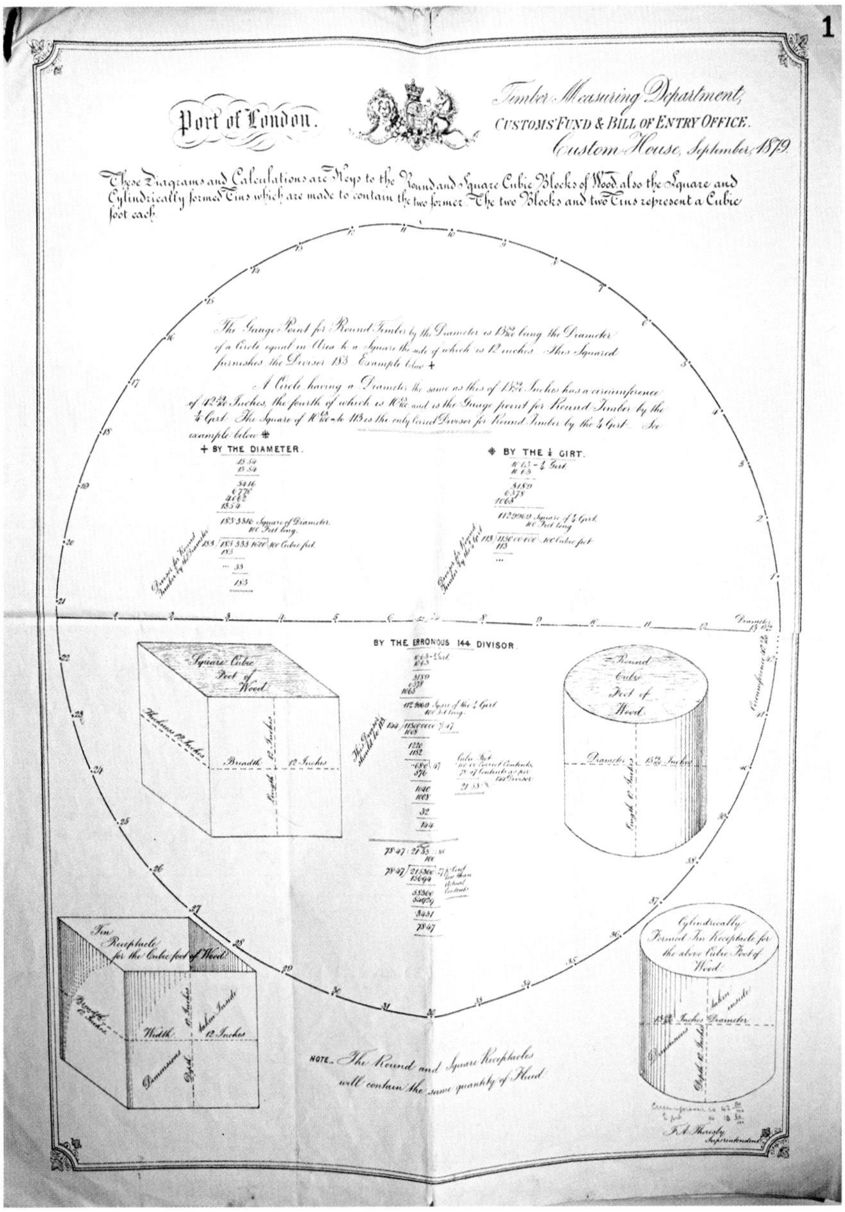

Fig. 5.11. The first page of the Port of London Timber Measuring Department's instructions. This page covers the measurement of round timber.

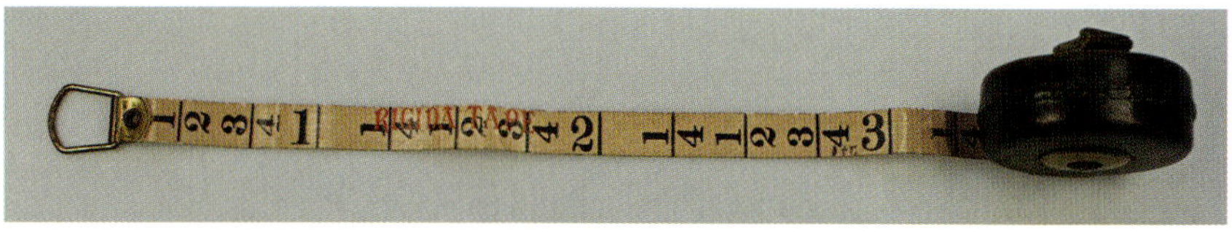

Fig. 5.12. The quarter girth measure on a timber girthing tape by John Rabone & Sons (Birmingham, 1877–1963). A patent was taken out by J. C. Watkins in 1886 for a measuring tape of exactly this form and this tape was sold by Rabone as their "patent girthing tape". It is not known whether this was the same patent, but acquired by Rabone.

Fig. 5.13. The reverse of the above tape marked to give the cubic content of one foot. This enables the user to ascertain the content of any tree by multiplying the cubic content reading by the length.

It is possible to confuse girthing tapes with "circumference" tapes which give the diameter by measuring the circumference of round objects. In such tapes each "inch" mark will actually be at 3.14 in. (π) increments. This is very similar to the girthing tape with allowance for bark. Girthing tapes are easily distinguishable from horse measuring tapes which are similarly marked but with the sub-divisions of each 4 inches marked only 1, 2 and 3, instead of fractions.

Commonly the reverse of the tape is marked with normal feet and inches, though other measures, such as that shown in Fig. 5.13, are also used.

Leather girthing tapes marked with brass studs are sometimes found, presumably made for those who thought the standard tapes not sufficiently durable.

Girthing sword: This adjunct to the girthing tape has a lightly curving metal "blade" with a hook at the end used to catch the ring on the girthing tape and draw it round a log, particularly useful when the log is lying on the ground. The construction of these is quite varied. Most are simple, being made from metal rod, but in some the "blade" is flattened. The version shown in Fig. 5.14 from the W. F. Stanley 1931 catalogue is referred to as a timber needle.[13]

Log (timber) caliper: A large size sliding caliper measuring to 3 ft. or even larger, with jaws of generous projection. The jaws usually have a small upstand at the tips suggesting that they were intended for measuring the diameter of round timber and the dimension the larger sizes of squared timber. It is a direct reading measurement. The frame is usually iron with a wooden handle. Cargo calipers are similar but are of wooden construction.

P1959.

P1959 **Timber Needle** for passing tape under fallen timber 9 0

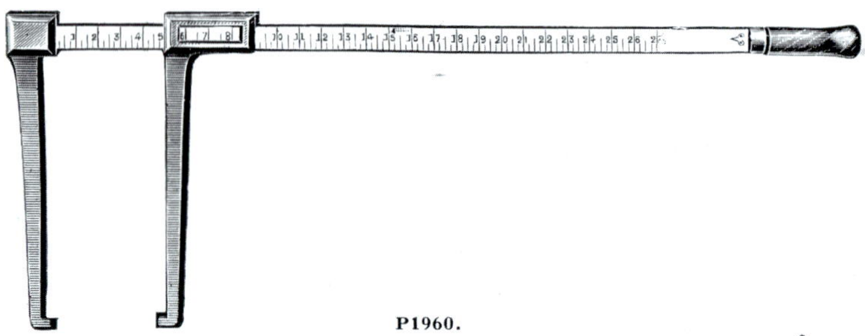

P1960.

£ s. d.

P1960 **Timber Calliper,** iron, with wood handle, measuring inches and quarters up to 27 inches 2 10 0

Fig. 5.14. Girthing sword (Stanley calls this a timber needle) and timber caliper from 1931 W. F. Stanley catalogue.

Timber rod: The 1931 catalogue of W. F. Stanley lists a timber rod for measuring the height of standing timber. It is described as made of bamboo, 25 ft. long, in sections of five 5 ft. which are jointed together. It was sold complete with a canvas case for £3 3*s* 0*d*.[14]

Tree height measures: The 1928 J. H. Steward catalogue lists two instruments for measuring the height of standing trees, the "Auto Simplex" and the "Dendrometer".[15]

Board rules

There appear to be two types of board rule. Both have a hook or lip on the end distant to the user which facilitates rapid use.

Type A: This is used for determining the board feet in a sawn board. The rule is marked with multiple scales, each for a different length of board. The rule is laid across the width of the board and the content read off directly on the scale that corresponds to the length of the board being measured.

Tree Measurers. 54

AUTO SIMPLEX TREE AND HEIGHT MEASURER.

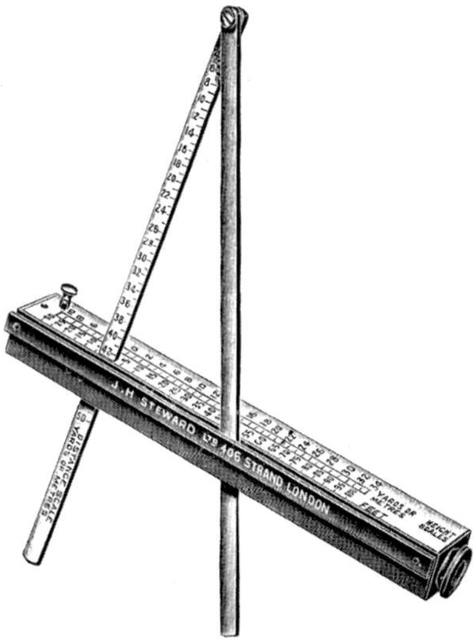

Fig. 69.

An instrument for measuring the height of trees and other objects, consisting of a combination of two scales set at right angles to one another, with an indicating pendulum and sighting tube attached.

The vertical scale is a scale of "distances," which is adjustable, and to which the indicating pendulum is attached.

The horizontal scale is a scale of "heights," and is engraved on the body of the sighting tube.

In use the vertical scale is first set to indicate the distance of the observer from the tree or other object.

With the pendulum swinging free the observer sights the summit of the object, through the tube, and then clamps the pendulum.

The height of the object is read from the "height" scale opposite the indicator line engraved along the pendulum.

Heights are given in feet, yards or metres.

SM 288.—**The Auto Simplex Tree Measurer** in flexible sheath, with instructions .... Fig. 69 **£3 18 0**

SM 289.—**Stiff Leather Case** if preferred instead of sheath. Extra **9 6**

THE DENDROMETER.

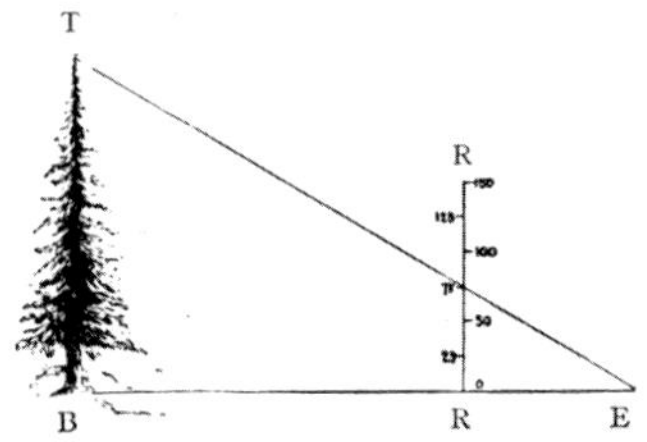

Fig. 70.

SM 290.—**The Woodman's Dendrometer,** for measuring the height of trees, consisting of a rod 24 ins. long, folding to 12 ins. with a cord of definite length attached to its lower extremity. The rod is divided to 100 parts, with bold figures at every fifth division. In use the woodman places himself at E, 30 yards from the tree BT, and with the free end of the cord held between his teeth, holds the rod RR vertically in one hand and stretches the cord taut. The rod is held so that the zero at the bottom R coincides with B the bottom of tree trunk, or with a mark placed on the trunk. The eye is then raised, and the division on the rod RR coinciding with T, the top of the tree indicates the height of the tree in feet, which in the diagram is shown as 75 feet. If it is found necessary to stand at a distance greater or less than 30 yards from the tree, the height indicated will be in proportion to the distance. With canvas case and instructions .... Fig. 70 **12 6**

Timber Girthing Tapes, see page 45.

J. H. Steward, Ltd., 406, Strand, and 457, West Strand, London, W.C.2.

Fig. 5.15. Illustration from the J. H. Steward 1928 catalogue of the two tree height measuring devices

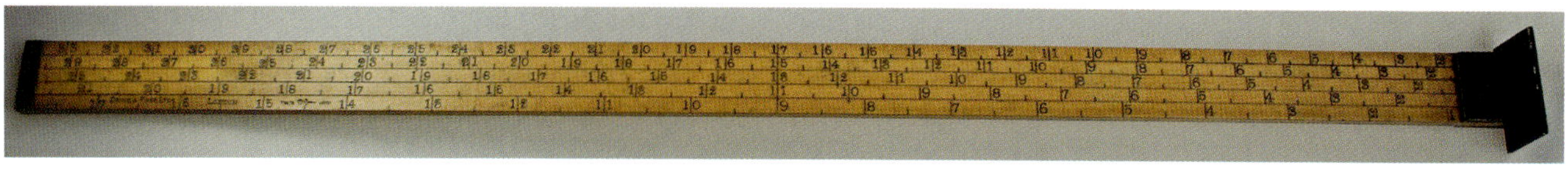

Fig. 5.16. A Type A board rule by Dring & Fage (London, 1790–1977).

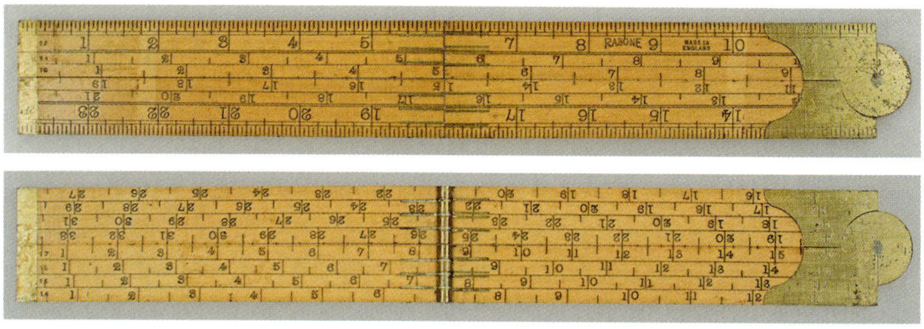

Fig. 5.17a & b. A 2-fold board rule by John Rabone & Sons (Birmingham, 1877–1963). This is a folding pocket version of the Dring & Fage rule shown in Fig. 5.16.

Type B: The rule is applied across the end of a log, and by reference to the appropriate line of figures for the length of the log, the number of board feet that it contains will be shown.

Board rule caliper: The purpose of this device is similar to type B above but the caliper is applied to the log at mid point. Again the reading is in board feet. Devices of this type are commonplace in America but quite rare in Britain. The purpose of these rules is to quickly predict the number of board feet that will be obtained from a log. The yield will be dependent not just on the size of the log but on many factors including the species, the conditions of growth, the form of the tree, and what wood is judged to be acceptable in size and quality.

Nelson[16] has pointed out that the figures on these measures are not all the same, the reason for this being that the marking of these rules was from tables used by different lumber companies to predict the yield – and that there were many companies and many different tables, each intended to take into account the many variables that have been outlined above. However almost all these tables and rules have one thing in common – they under-measure the yield as calculated on a purely mathematical basis or, expressed differently, they all have a built-in allowance for the inevitable waste and the disappointments that occur when sawing takes place. Either way, they produced an answer skewed in favour of the buyer.

Timber shrinkage rules

A rule to measure timber green but show the width of a board when dry would seem to be useful, but such timber shrinkage rules are rare. Perhaps merchants had enough experience not to need help, or exact prediction of

Fig. 5.18. Details from the shrinkage rules by Thomas Nicholls (London, 1860–82).

shrinkage was only of significance for the widest of boards. Whether they were intended for timber or sawn stock is not known.

There are a number of different forms of timber shrinkage rules known. One type is a pair of brass-tipped lancewood laths, with an example shown in Fig. 5.18 which are marked Nicholls Maker Barbican London (working 1860 to 1882). One lath is marked ASH and the other, ELM. The laths are 60 in. long and but only a little over ¼ in. thick at the centre, tapering to approximately ⅛ in. at the ends. Both are identically marked on both sides except that each side commences from the opposite end of the lath, with the line of measure in both cases starting at 6. The shrinkage for ash is the less, with the last marking on the rule being 55, giving a shrinkage of 8.3%, whilst for elm, 53 inches wet measure gives a reduction in the drying of 11.7%.

Another type appears to be an ordinary 2 ft. 2-fold rule. One example was made by Hobcraft, London (1817-52). Close examination shows that on one face, marked ASH, there are only 22 in. marked (a reduction of 8.3%) and the other face, marked ELM, is divided only to 21 in., a reduction in this case of 12.5%.

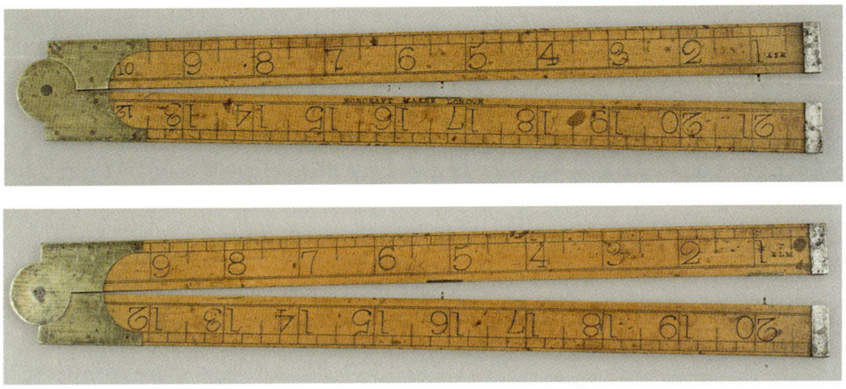

Fig. 5.19a & b. Timber shrinkage rule by William Hobcraft (London, 1817-52). As this rule reads from right to left it dates from before 1824.

Another 2-fold rule is by Wm. Coulsell, 9 Castle Street, Southwark, London. Coulsell was at this address from 1816 to 1855. Again marked ASH on one side and ELM on the other, the legs are 13½ in. long, the ELM face being divided into 24 divisions and the ASH face into 25. It has an ivory slide which is marked to extend the ELM scale to 36 divisions on one side and the ASH scale to 37 divisions on the other.

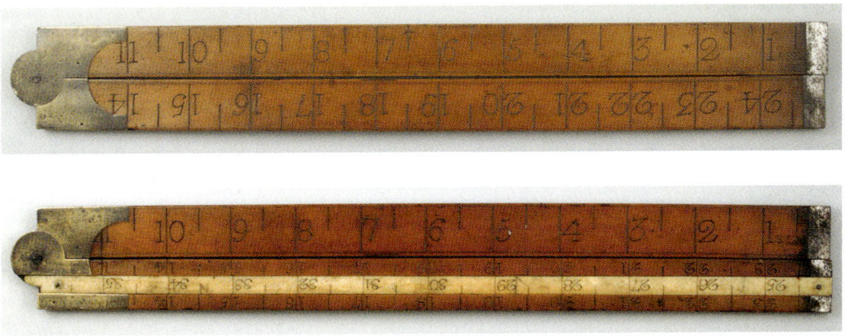

Fig. 5.20a & b. Timber shrinkage rule by William Coulsell, (London, 1809–1934) marked for ASH and ELM

Interestingly, on both the Hobcraft and the Coulsell rules, the ASH divisions are the same size, but the ELM divisions are considerably larger on the Hobcraft rule than those on the Coulsell rule.

A 2 ft. 2-fold timber shrinkage rule by Samuel Albert Smith (Birmingham, 1850-72) is particularly interesting in that it is also marked with a timber line and a board measure. This is probably the latest rule known that is marked

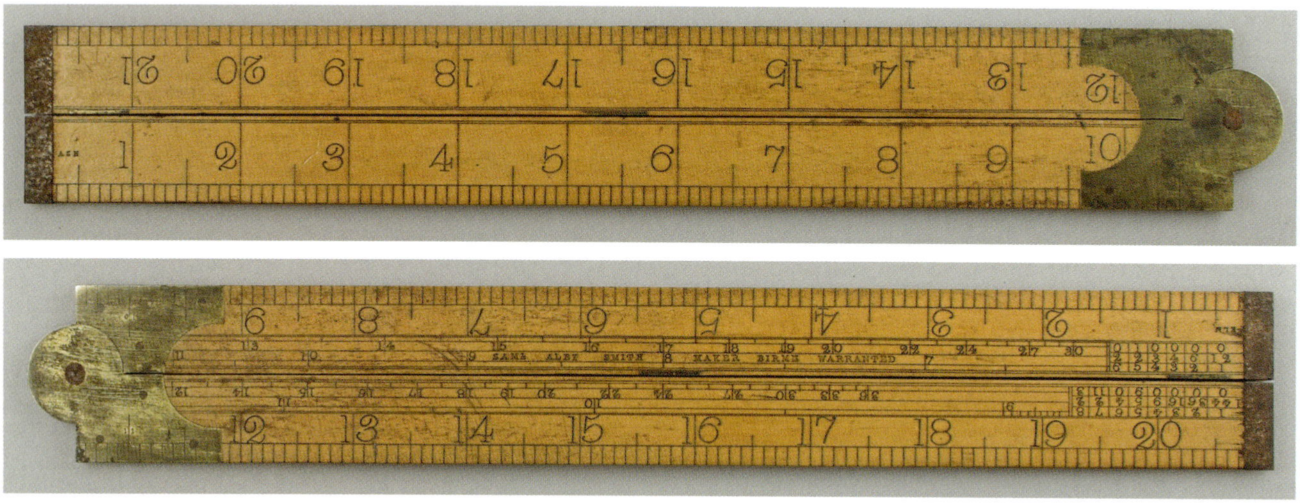

Fig. 5.21a & b. A timber shrinkage rule, also for ash and elm, by Samuel Albert Smith, (Birmingham, 1850–72).

with these two lines. Until this rule was discovered, it was thought that the marking of these lines did not extend beyond the 18th century.

Another 2 ft. 4-fold timber shrinkage rule by J. H. Adams (Liverpool, 1837–1919) has standard inches marked on the inside but on the outside, it is marked with divisions to 21. These divisions are not named but are the same size as the ELM divisions on the Coulsell rule. It is also marked on the face with an owner's name, Ivor Jones, and also Canada Dock. Canada Dock, which opened in 1859, was part of the Port of Liverpool and was a timber dock whose main purpose, as the name implies, was to receive timber from Canada (elm trees grew extensively in Canada until devastated by Dutch elm disease).

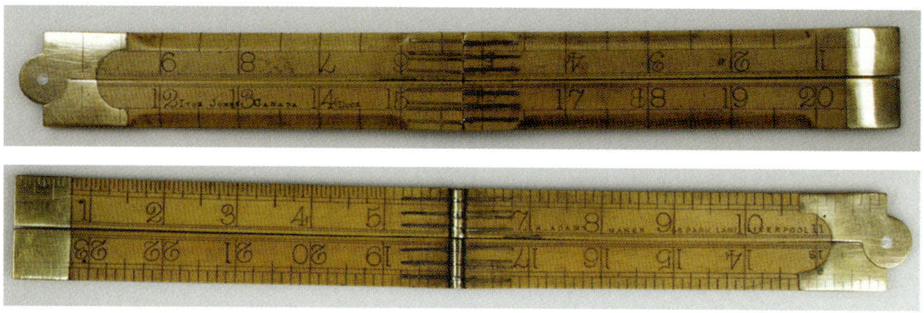

Fig. 5.22a & b. A timber shrinkage rule by J. H. Adams (Liverpool, 1873–1919), not marked for any particular timber but from the amount of shrinkage likely to have been for elm.

Timber rules

A style of rule that was commonly used for timber rules was a 2 ft. 3-fold rule made with two principal legs each 9 in. long, one of which was divided longitudinally to form a folding third leg which thus gives a 2 ft. rule when fully

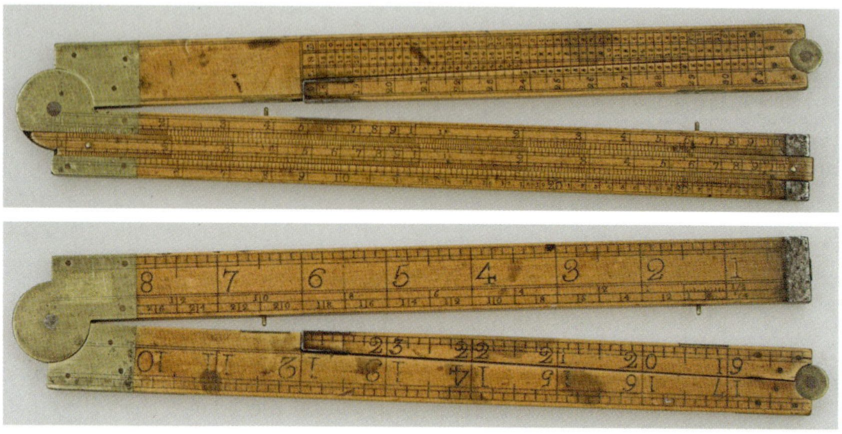

Fig. 5.23a & b. A 2ft. 3-fold boxwood and brass timber rule. Although unmarked it dates from before 1824 as the line of measure reads from right to left. The slide is boxwood. The tips are iron. On one face of one leg are the lines A, B, C, being two cycle log scales, and D, being a one cycle log scale starting at 4 (a girt line). The other leg has a timber load table which spans the extra fold-out leg. From the accuracy of the marking of this table, it would seem likely that the table was marked before the additional leg was created by cutting it out of the main leg. On the reverse of the rule there is a line of inches divided into eighths and also ¼ and ½ in. scales. A line of feet divided into 100 parts is marked on the edge of the rule.

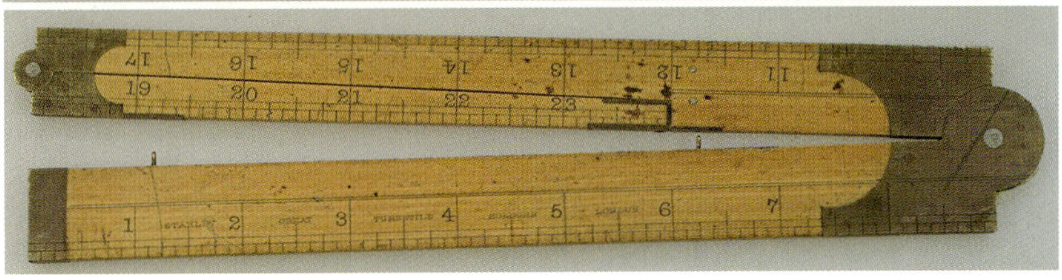

Fig. 5.24a & b. *A simpler style of timber rule by W. F. Stanley (London, 1853–1967) without the timber load table. The addition of an ivory tablet (on which to write notes) is a frequently found feature of this type of rule.*

extended. The rules have the additional features of a log slide in one leg and sometimes an ivory tablet for making notes and calculations. Some examples of these rules were marked with a timber load table. In essence these rules are similar to many so-called carpenters' rules except that they are 3-fold. The 9 in. legs enable a slide of satisfactory length to be fitted whilst at the same time making a rule that will fit into the pocket.

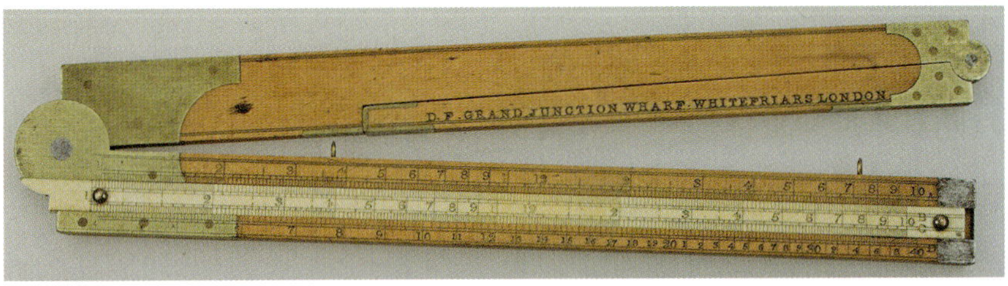

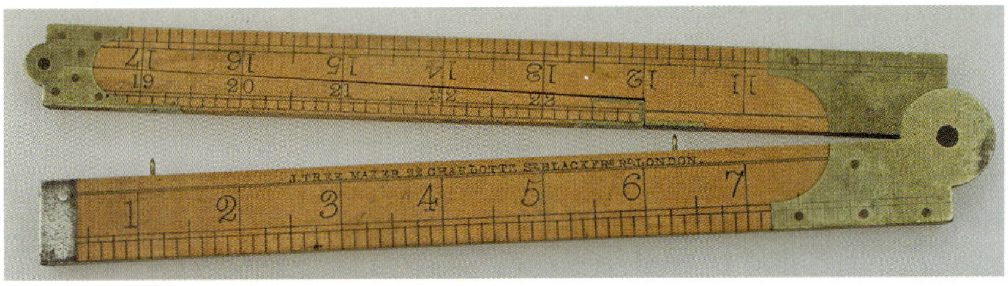

Fig. 5.25a & b. *A still simpler style of rule, by J. Tree (London, 1832–95) which does not have either an ivory tablet or a timber load table. It is also marked with the owner's initials, D.F. and an address, Grand Junction Wharf, Whitefriars, London.*

Most of the known examples are by London makers. This type of rule was listed as late as 1931 by W. F. Stanley.[17]

A most unusual example, shown in Figs. 5.26-28, was made by J. Tree, 22 Charlotte Street, Blackfriars Road and marked with the owner's name, H. Batchelor, Betchworth, Surrey 1842. It has an ivory slide and note tablet and the normal timber load table but in addition is marked with five tables which are headed: THIK STUF 12 IN^{S} BRO^{D}; SQ^{R} SCANTLING TO A LOAD; PLANK IN LOAND

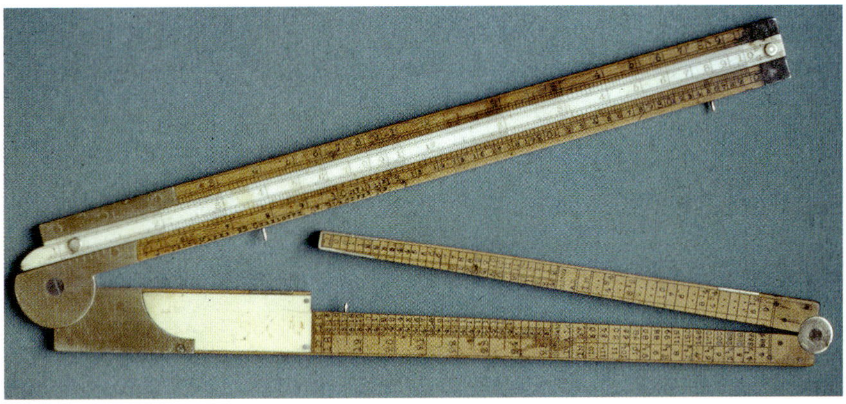

Below left: Fig. 5.26. Timber rule by J. Tree (London, 1832–95) with ivory slide and tablet.

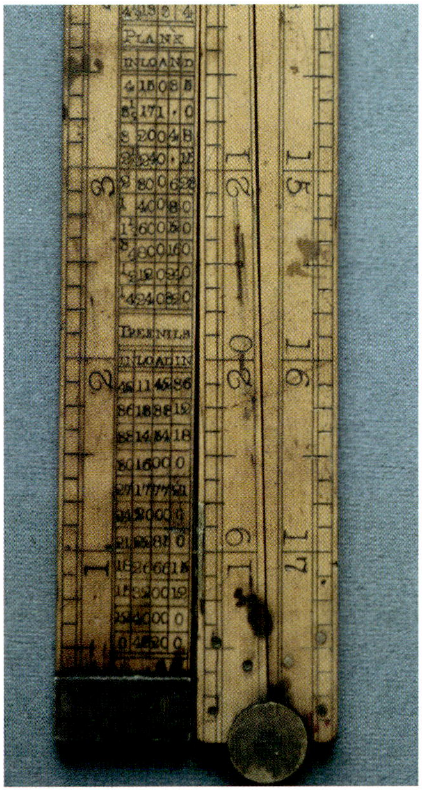

Fig. 5.27. Tables headed "PLANK IN LOAND" and "TREENILS IN LOAD IN".

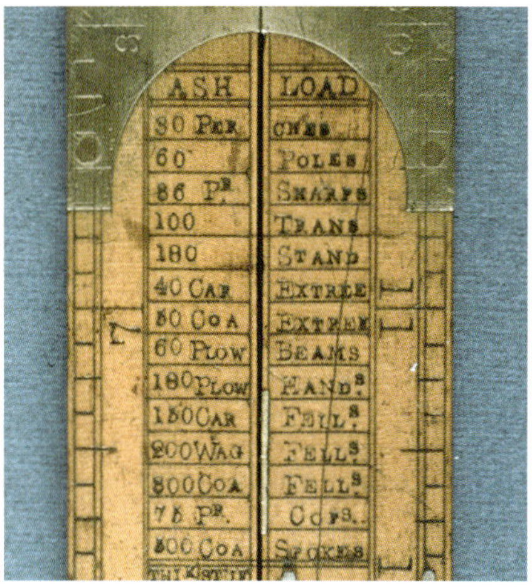

Fig. 5.28. Table headed "ASH LOAD".

(this is an example of the rule maker's error – it should read LOAD); TREENILS IN LOAD IN and ASH LOAD. The latter table lists the number of a variety of coach, wagon, and plough parts to be got from the load.

Slide rules specific to timber measurement

Whilst all types of Customs & Excise slide rule could be, and were, used for calculations related to timber, there exist a few whose only or principal use was for the timber trade. The best known of these is probably the "Vade Mecum" rule patented by George Bousfield on 9 August 1887. The patent description was "Improvements in slide rules specially applicable for reducing from and into the St. Petersburg Standards ..." But this was only one of the various calculations that the slide rule was designed to solve. Others included a scale for taking off quantities from plans and one for calculating the cubic content of round, square and unsquared timber. This was in addition to the ability to undertake all the usual mathematical calculations of any standard slide rule.

Fig. 5.29. George Bousfield demonstrating the use of his rule.

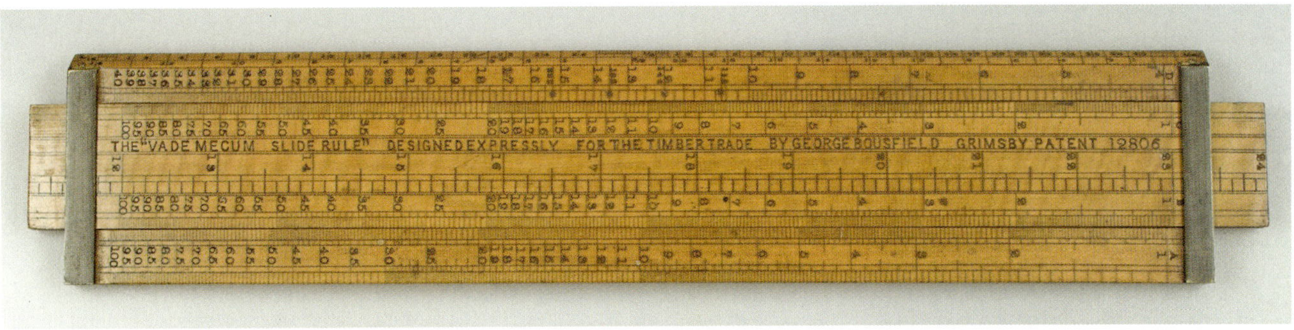

Fig 5.30. The "Vade Mecum" slide rule designed expressly for the timber trade.

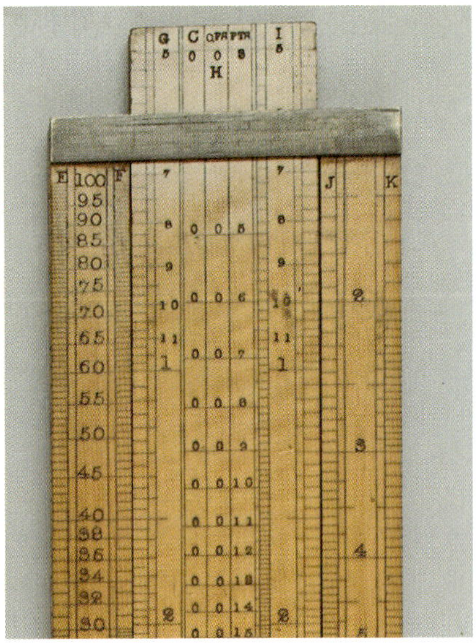

Fig. 5.31. The reverse of the "Vade Mecum" slide rule showing how it was designed to be used vertically.

It is a flat section rule with a single slide; however, the rule is unusual in that the centre slide is very wide, taking up half the width of the whole rule. The other unusual feature is that the rule is marked to be read with the rule held vertically rather than the normal horizontal position. It came with an extensive instruction book stating that for the sum of 1*s* a "T" square could be purchased for reading the small divisions, though no examples of this "T" square seem to have survived.

Although none of the known rules have been found with a maker's name, the 6th edition of the instruction booklet was published jointly by William Rider & Sons and W. F. Stanley. The assumption must be therefore that at least for some period Stanley were making the rule. A full description can be found in the *Journal of the Oughtred Society*.[18]

Aston & Mander also made a very similar rule. This is marked on the edge "DESIGNED REGISTERED AND MADE ONLY BY ASTON & MANDER 25 OLD COMPTON ST., SOHO, LONDON". Aston & Mander worked at this address from 1870 until well into the 20th century so knowing the maker is no help in determining which rule came first.

This rule is rather simpler than the "Vade Mecum" but striking similarities are that it was also designed to be used vertically and the centre slide is again particularly wide, although not as wide as that in Bousfield's rule. One side is marked for timber and has five scales, marked INCHES WIDE, FEET LONG, ST. PETERSBURG STANDARD, FEET CUBE, INCHES THICK. However, the reverse side of the rule is devoted exclusively to brickwork and includes scales for calculating the number of bricks required with relation to the height and thickness of a wall.

Fig. 5.32a & b. Aston & Mander (London, 1870–1930) timber slide rule.

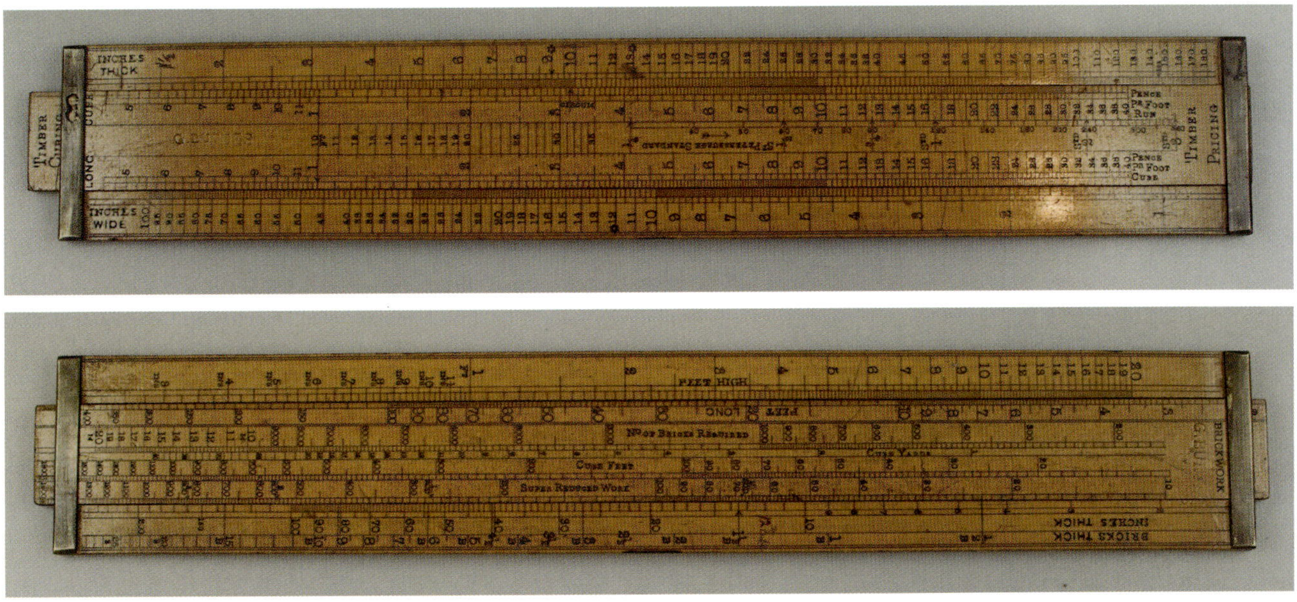

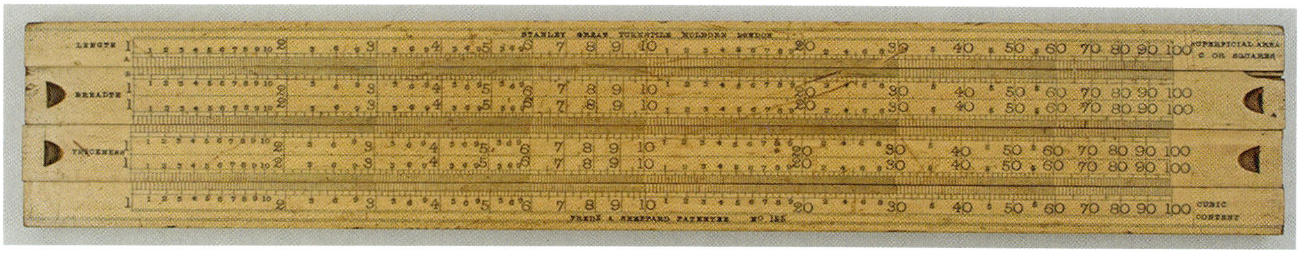

Fig. 5.33. Cubic content rule by W. F. Stanley (London, 1853–1967) marked "Fredk Sheppard Patentee No. 155".

Timber contenting slide rules: A number of slide rules have been recorded intended for calculating the volume of timber. Three of these were written up in the *Journal of the Oughtred Society* in 1997.[19] All were of a similar form, being a double-sided rule, with a single slide. One side is marked with log scales used for calculating the content. The other side is for round, 8-sided and 16-sided timber, the latter two scales being used in making masts.

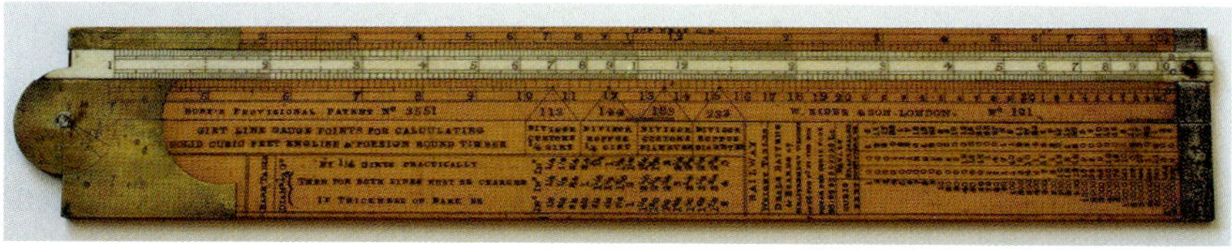

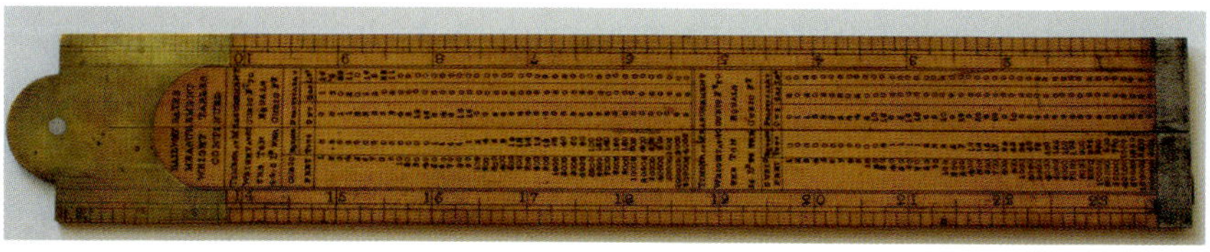

Fig. 5.34a & b. Burt's patent rule for the measurement and weight of timber.

Another slide rule of a different form designed for calculating the cubic content of timber is marked "Fred[k] Sheppard Patentee No. 155". Made by W. F. Stanley, it has four scales, two fixed on the outer edges, and two slides, which work against one another. The four scales are marked "Length", "Breadth", "Thickness" and "Cubic Content". Although it is marked with Sheppard as patentee, there is no record of a patent being taken out, so the intended use of this rule remains a mystery.

Fig. 5.35. The name and patent number on Burt's patent timber contenting slide rule.

In the 1880s, E. A. P. Burt took out a number of patents for rules with regards to both the content and weight of timber. His provisional patent, No. 3551, included Railway Weight tables; this rule was in the form of a 2 ft. 2-fold rule with an ivory slide.

It also included the four divisors for the Customs ¼ girt (string) system = 113; Hoppus ¼ girt system = 144; the diameter round = 183 and the Hoppus content = 233.

Another patent covering content was taken out in 1887. It was intended for "calculating the Superficial Measurement of Planks, Deals, Battens and Boards of every description and size…" This rule was in the form of a slide rule.

References

1. Latham, Robert, *The Shorter Pepys. Extracts from the Diary of Samuel Pepys,* London, 1993.
2. Latham, Bryan, *Timber. Its Development and Distribution, A Historical Survey,* London, 1957.
3. Leybourn, William, *The Art of Measuring: Or The Carpenters new Rule Described and Explained,* London, 1669.
4. Thubron, Robert, *The Port Charges of Great Britain and Ireland,* Newcastle-upon-Tyne, 1872.
5. Fischer, George, *The Instructor or Young Man's Best Companion,* 21st ed., London, 1799.
6. Leybourn, *op. cit.*
7. Rees, J. and M., *Christopher Gabriel and the Tool Trade in 18th Century London,* Needham Market, 1997.
8. Hoppus, Edward, *Practical Measuring Now Made Easy to the Meanest Capacity,* London, 1st ed., 1736.
9. Mallinson, R. F. A., *Timber Trade Practice,* 2nd ed., London, 1953.
10. Rose, William. *The Village Carpenter,* Cambridge, 1937.
11. Hoppus, *op. cit.*
12. Edward Preston & Sons, *Catalogue No. 18. The "Preston" Catalogue 1909 with 1912 supplement,* reprinted Mendham, NJ. 1991.
13. W. F. Stanley, *Catalogue of Surveying and Drawing Instrument,* London, 1931.
14. *Ibid.*
15. J. H. Steward Ltd., *Catalogue of Surveying, Drawing and Nautical Instruments,* London, 1928.
16. Nelson, Robert E., "The case of the missing tool" *Newsletter* No 54, The Tools & Trades History Society, 1996, and "Case solved" *Newsletter* No 56, The Tools & Trades History Society, 1997.
17. Stanley, 1931, *op. cit.*
18. Rees, Jane, "The *Vade Mecum* Slide Rule designed expressly for the timber trade by George Bousfield", *The Journal of the Oughtred Society,* Vol. 8, No. 1, 1999.
19. Bready, James and Babcock, Bruce, "Measuring the Volume of Logs: the Timber Contenting Slide Rule" *The Journal of the Oughtred Society,* Vol. 6, No. 2, 1997.

Chapter 6
Carpenters and builders

Carpenter's joint-Rule, is an instrument usually of box, 24 inches long, and one and a half broad; each inch being subdivided into eight parts – on the same side with these divisions, is usually added Gunter's line of numbers.

On the other side are the lines of timber and board-measure; the first beginning at 82, and continued to 36, near the other end: The latter is numbered from 7 to 36, four inches from the other end. Chambers, 1743.[1]

Any ordinary 2 ft. 2- or 4-fold rule, used by a woodworker, or any other tradesman, can be called a carpenters' rule. Indeed this is the interpretation of the term that most people would subscribe to. However, there is a more specific description of the term, although still imprecise.

The carpenter's joint-rule described in Chambers' *Cyclopaedia*, 1743,[2] is a 2 ft. 2-fold rule, "usually of box" (boxwood), 24 inches long and 1½ inches broad. It is divided into inches, halves, quarters, and eighths and also includes a Gunter's line and on the back, a line of board measure and a line of timber measure, each of which includes the table for timber of small size.

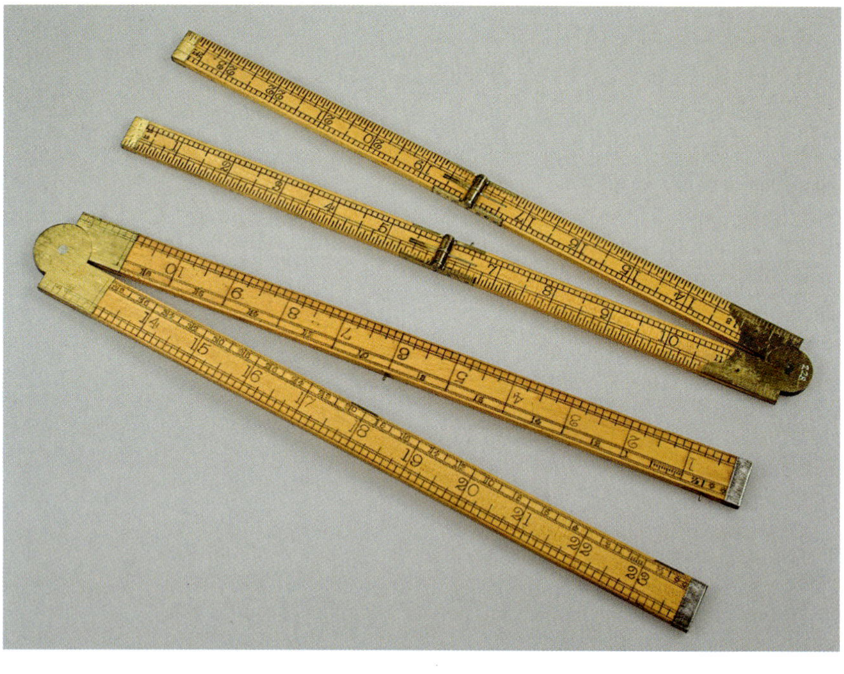

Fig. 6.1. A 2 ft. 4-fold rule and a 2 ft. 2-fold rule, both by John Rabone & Sons (Birmingham, 1877–1963).

Fig. 6.2. A 2 ft. 2-fold rule by William Gibbons (Wolverhampton, 1811–35) with a line of timber measure on the upper leg and a line of board measure on the lower leg.

It is these three latter features which are generally considered to be the defining characteristics of what is known as the carpenters' rule.

Its origins lie nearly two hundred years earlier than the Chambers reference. The 1562 edition of Leonard Digges' *Boke Named Tectonicon*[3] included a description of a carpenters' rule: "wel playned, twelve Inches longe, a quarter of an Inche thicke, and two Inches yn breadth. Truly yt were more commodious if it hath two foote in length". The rule was scaled in inches, half inches, quarter inches and eighths of an inch. It was also marked with a table of timber measure and, on the reverse of the rule, a "Quadrant geometrical" with a 90° scale, and a plummet to hang from the quadrant. A full description of Digges' rule can be found in Paul Kebabian's article on the subject in *Selections from the Chronicle.*[4] In 1602, Richard More published *The Carpenter's Rule.*[5] In this, although More gives credit to Digges for the invention of the rule, he considers many rules inaccurate for board and timber measurement and suggests that a mathematical calculation should be used.

The advancement of the mathematical theory of logarithms by John Napier and other 17th century mathematicians enabled Edmund Gunter, in 1624, to develop a logarithmic line of numbers which, in due course, allowed William Oughtred to invent the slide rule.

The subject of the carpenters' rule continued to exercise the minds of authors. In 1658, George Atwel[l] in his book *The faithfull surveyour ...* included an appendix with instructions for making a carpenters' rule[6] and in 1669, William Leybourn wrote *The Art of Measuring: Or The Carpenters new Rule Described and Explained.*[7]

Lines of board and timber measure

As well as the lines of logarithms, and the line of timber measure mentioned by Digges, early carpenters' rules also often have a line of board measure. In the Tools & Trades History Society *Newsletter No. 20* Richard Knight gives a comprehensive description of two early carpenters' rules that include these lines.[8]

Set alongside the line of numbers, the board measure is a line which is divided in such a manner that the product of any two adjacent numbers on each line gives a total of 144 (the number of square inches in a square foot) so

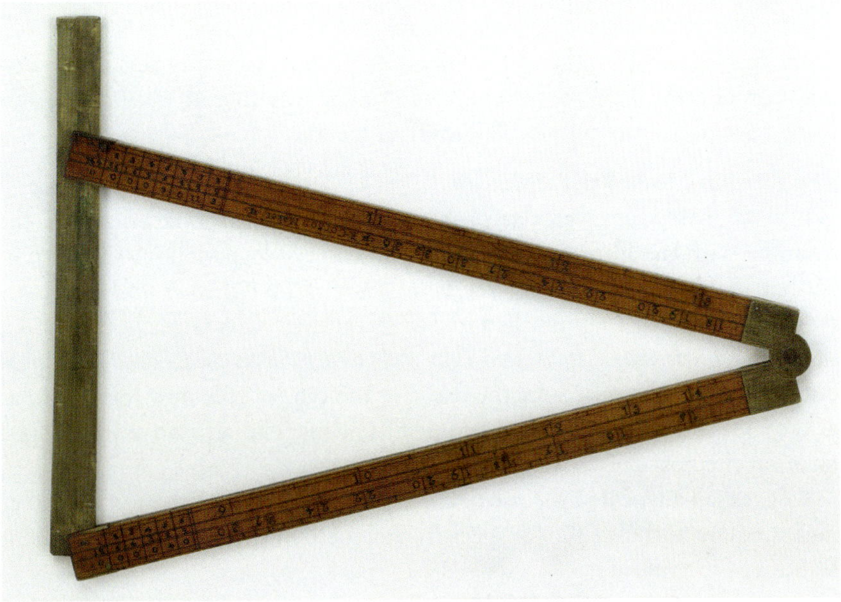

Fig. 6.3. The line of timber measure and the line of board measure including the tables necessary for timber less than 9 in. square and boards less than 6 in. wide on a mast makers' rule by R. Corson (Wolverhampton, 1767–70).

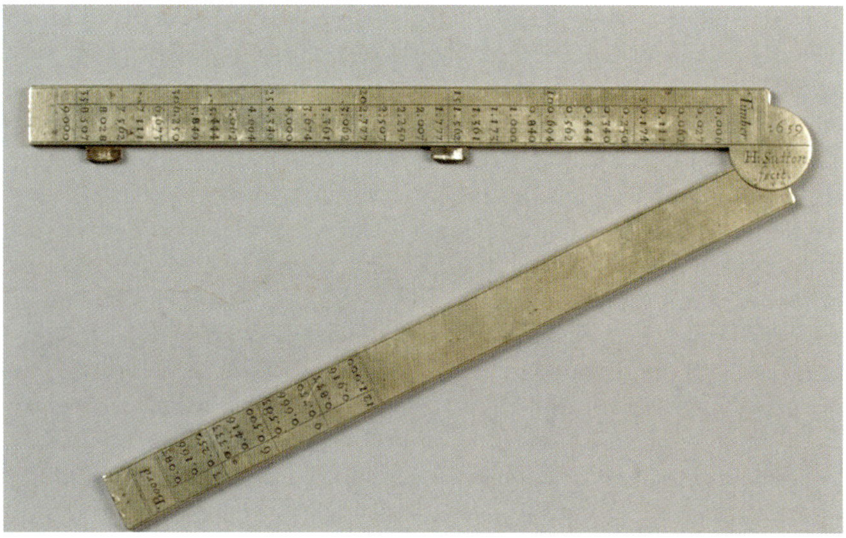

Fig. 6.4. A brass sector, dated 1659 by Henry Sutton (London, 1649–65) with tables for timber and board.

Fig. 6.5a & b. Two timber load tables. Top: on a rule by Thomas Onions (Birmingham, 1767–1801). Bottom: on a rule by William Brown (Birmingham, 1835–52). In the earlier Onions table, the load is 40 cu. ft. (which is also noted on the rule). On the later rule it is 50 cu. ft. It is also interesting to note that on the 18th century rule the table starts at 6 pence per cu. ft. whilst on the later rule it starts at 12 pence.

for any given width, the length of timber needed to measure one square foot can be read off the rule.

The line of timber measure works on the same principle but here the multiplier is 1728, being the number of cubic inches in a cubic foot.

With both these lines, when the board is less than 6 in. wide or the timber is less than approximately 9 in. square, the answers extend beyond the length of a 2 ft. rule so these values are obtained from two small tables to allow the calculations.

An interesting variation to the standard form of timber and board lines can be found on a brass sector made by Henry Sutton (London, 1649-65), and dated 1659. On this one leg is marked "board" and one "timber". These tables show the factor by which the length of a 12 in. board, one inch thick, must be multiplied to give the number of square feet, from the timber table, or cubic content, from the board table.

The majority of 19th century rules do not have either a line of board measure or a line of timber measure, although they are still referred to as "carpenters' rules". They do however, often include a timber load table.

Price per load table

This table is used to establish the price of a load of timber (of a set size) at a given price per cubic foot. The price per foot is given in pence and the size of the load is usually either 40 or 50 cu. ft. Hence, in reality this is a multiplication table of the price per cu. ft. multiplied by either 40 or 50, the result being converted into pounds, shillings and pence.

The introduction of the slide rule

Whilst all these rules were "carpenters' rules", the term took on a different meaning with the improvements made by Henry Coggeshall (1623–90), in his book, *Timber-Measure by a Line of More Ease, Dispatch and Exactness, than any other Way…*, published in 1677[9] and five years later another book, *A Treatise of Measures, By A Two-Foot Rule which Slides to a Foot,*[10] that produced the rule which became the basis for the type of rule now referred to as a "carpenters' rule". Indeed it continued to be made in an almost unchanged form for the next 250 or so, years.

Coggeshall advocated the use of the slide rule for the calculation of timber measurement and his rule was specially designed for this purpose. The salient features of the rule are:

Fig. 6.6. The preface to Coggeshall's Timber-Measure by a Line of More Ease, Dispatch and Exactness, than any other Way...

TO THE

Reader.

IF Thou delighteſt in Timber-Meaſure, or haſt Occaſion for Much; Thou haſt here a Line ſo Fit for thy Purpoſe, that thou canſt not imagine, much leſs, wiſh a better. For what can be more ready and eaſie, then having ſet twelve to the length, to ſee the Content exactly againſt the Girt or Side of the Square. Whereas on Mr. Partridge's *Scale the Content is the Sixth Number, which is far more troubleſome then with Compaſſes. The*

A 2 *Line*

- A 2 ft. 2-fold rule in boxwood.
- The incorporation of a log slide into one leg marked one edge with a single log scale, 1–10, and on the other an offset scale, 4–40 (the girt line).
- The back of the slider is marked with inches thus allowing the effective length of the rule to be extended to 3 ft.
- The edge of the rule is marked with feet divided into 100^{ths}.
- A price/load/cu. ft. table.

Coggeshall clearly understood that the quarter girth squared method of calculating the volume for round logs was flawed, referring to this as the "country measure "or "the common way" but he also describes "true measure of round timber". Clearly the "wrong" method was so established in the timber trade that it was impossible to ignore it.

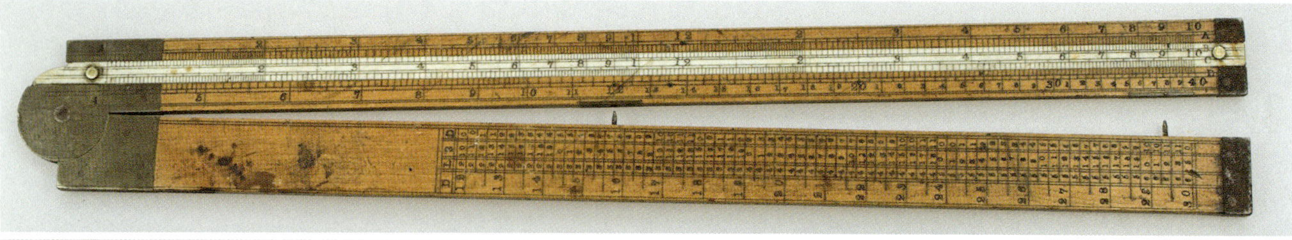

Fig. 6.7. Top, A Coggeshall type rule with an ivory slide by Tustian, London (1803–39). On this face there are a timber load table, three log scales and a girt line.

Fig. 6.8. The reverse face of the above rule. The underside of the ivory slide has a line of numbers marked 1 to 12 to allow the rule to measure to 3 ft. The edge of the rule, not shown, is divided into hundredths.

Coggeshall's improvement

The first three lines on a carpenters' rule, marked A, B and C, are standard logarithmic lines, being the two cycle Gunter's lines as found on any slide rule right up to the 20th century.

Coggeshall's improvement was in the line marked D, which he named the girt line (from its use in calculating the content of timber in the round by means of the quarter girth). This line is marked from 4 to 40, that is, it is "folded" at 4. This means that instead of starting at 1 on the left-hand side of the scale, it starts at 4, progresses to 10 towards the centre of the scale and then on to 40 at the right-hand end. One of the advantages of this was that, as the number 12 was frequently used in calculations concerning timber, this point on the scale was near the centre of the rule and therefore easier and more accurate to use.

Rules including these improvements are strictly speaking slide rules rather than linear measuring rules but Coggeshall's rule became so standard that any rule that included a slide was referred to as a carpenters' rule.

"New Improved Sliding Rule"

Another type of early rule intended for the carpentry and building trades, as well as a number of other trade-related calculations, is described in a generic instruction book for a "New Improved Sliding Rule" published in Birmingham in 1789.

At least five of these rules have been recorded; these are by Robert Corson, 1767–70, and Samuel Davis, 1770–92, both working in Wolverhampton; Wood & Lort (1750–60) and Abraham Hassall, 1773–80, both in Birmingham; and one unmarked. The rules are all 2 ft. 2-fold.

Fig. 6.9a & b. A carpenters' rule by Josiah Culmer (London, 1783–1811) showing the three standard lines of logarithms, marked A, B, and C and Coggeshall's girt line, marked D.

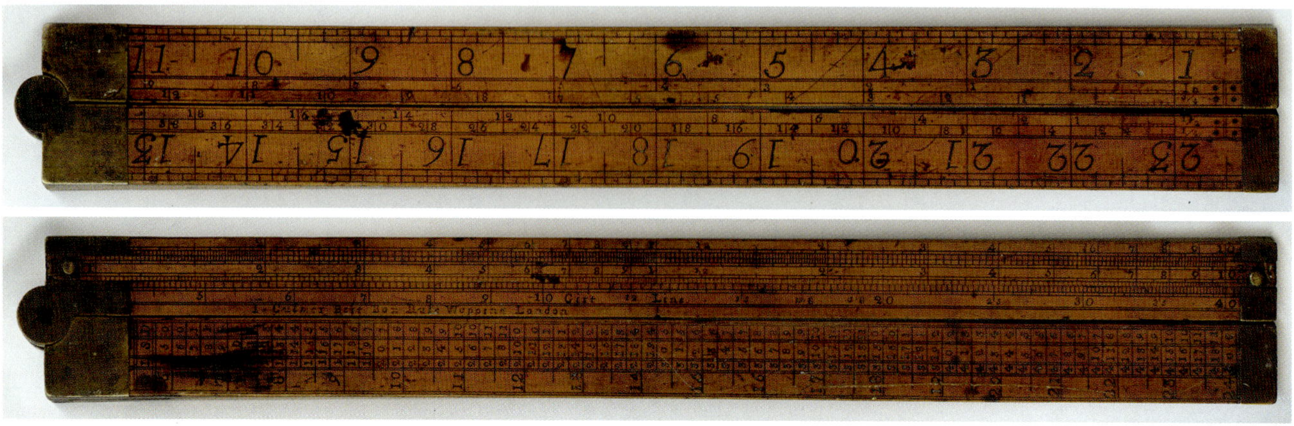

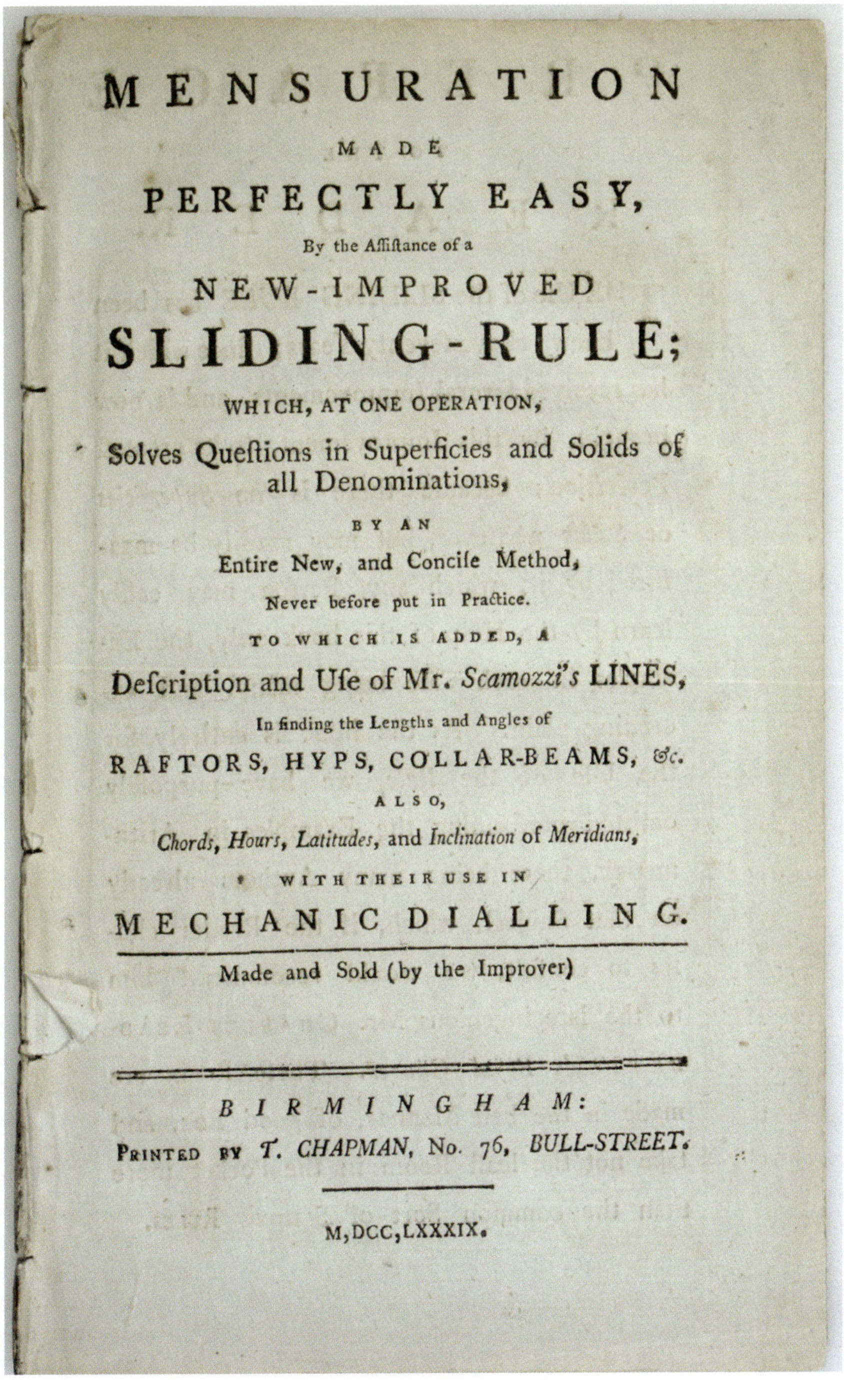

MENSURATION
MADE
PERFECTLY EASY,
By the Aſſiſtance of a
NEW-IMPROVED
SLIDING-RULE;
WHICH, AT ONE OPERATION,
Solves Queſtions in Superficies and Solids of all Denominations,
BY AN
Entire New, and Conciſe Method,
Never before put in Practice.
TO WHICH IS ADDED, A
Deſcription and Uſe of Mr. *Scamozzi's* LINES,
In finding the Lengths and Angles of
RAFTORS, HYPS, COLLAR-BEAMS, *&c.*
ALSO,
Chords, Hours, Latitudes, and *Inclination* of *Meridians,*
WITH THEIR USE IN
MECHANIC DIALLING.

Made and Sold (by the Improver)

BIRMINGHAM:
PRINTED BY *T. CHAPMAN,* No. 76, *BULL-STREET.*

M,DCC,LXXXIX.

Fig. 6.10. The title page of the instruction booklet for the "New Improved Sliding Rule".

Unlike the carpenters' rule, there are two slides set into one face of one leg, sliding one against the other. These are logarithmic scales, marked A, B, C and D, and have two cycles, although D is an inverted line and therefore reads the opposite way to the other three. (They do not have Coggeshall's girt line marked on them.)

Below these lines are a series of numbers. These are gauge points, although unlike later rules, the actual figures are given rather than just an initial to indicate their use. The left-hand point is marked 816. The standard measure for brickwork was a rod, which was 16½ ft. square and 3 half-bricks thick. To calculate the content of a wall you set the length on slide C to the height on slide D against the thickness in half-bricks on slide A and by setting 1 on the lower slide against the divisor 816, the content is given in rods. The other gauge points are for calculations concerning malt, ale and wine.

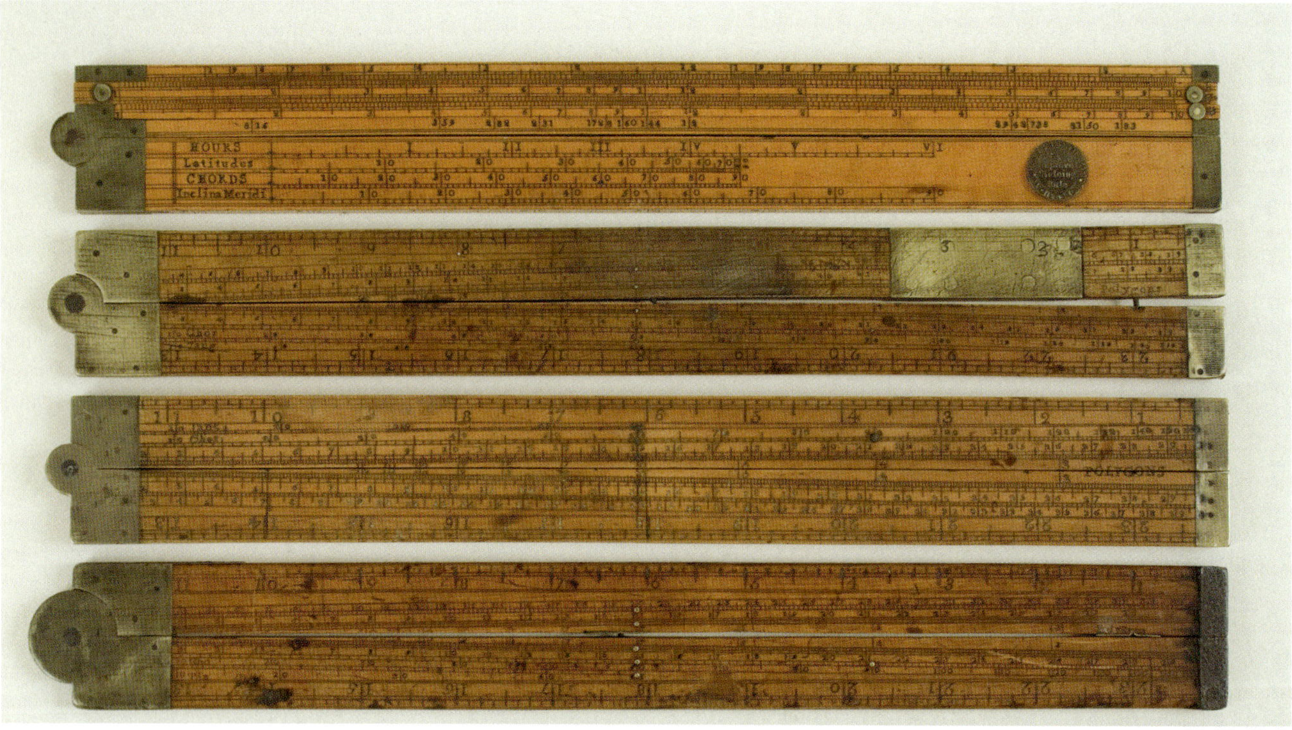

Fig. 6.11. Four different examples of the "New Improved Sliding Rule." From the top: A rule by Robert Corson (Wolverhampton, 1767–70). This rule does not have the architectural scales found on the other rules. A rule by Abraham Hassall (Birmingham, 1773–80). This rule has an early repair made by rivetting on a brass plate on each side of the leg. A rule by Samuel Davis (Wolverhampton, 1770–92) showing the reverse face with Scamozzi's lines and the line of inches. An anonymous rule. This example, together with the rule by Hassall, also has a number of unspecified gauge points marked with brass studs on this face.

On the other leg of this face are lines of chords, hours, latitude and inclination of meridians, the scales necessary for setting out sundials. Three of the known rules also have the standard architectural scales of ¼, ½, ¾, and 1 in. to the foot on this leg.

On the reverse face of the rule are set out Scamozzi's lines, used for "finding the lengths and angles of rafters, hyps, collar beams, etc." These are sectoral lines, that is they are intended for calculations using compasses, rather than direct readings from the rule. The examples given in the instruction booklet all refer to the design of various roof timbers.

Vincenzo Scamozzi was an architect and author, born in Vicenza, Italy, in 1548. His treatise *L'Idea della Archittettura Universale* (*The Universal Idea of Architecture*)[11] was the last of the Renaissance works on the theory of architecture, but he was also interested in the practicalities of building.

There is, in addition, a line marked "Polygons", used for dividing a circle into equal parts. Also on this face is a standard line of inches divided into eighths.

One edge is marked with inches divided into tenths and the other with the lines needed for finding the diameter of a cask. (For further explanation of this line, see Chapter 10.)

Rabone's carpenters' rule

In 1880 Rabone published a booklet describing the history and use of the carpenters' slide rule[12] although in their 1892 catalogue these were not listed

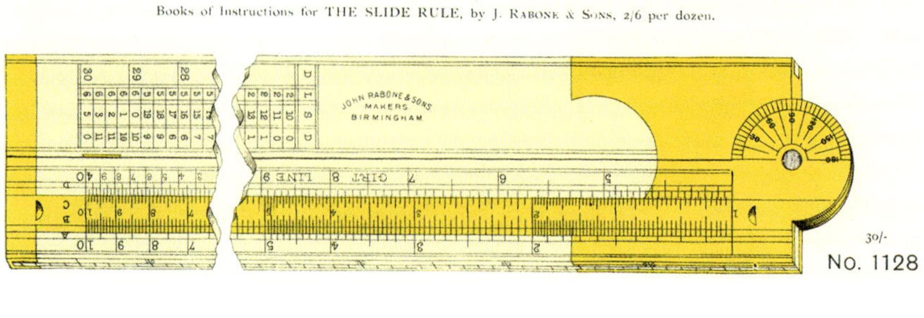

Fig. 6.12. Illustration from the Rabone 1892 catalogue showing one of the "Two Feet Two-fold Boxwood Rules with Brass Slides." The description for this rule No. 1128 is "Arch Joint, Tables for Timber measuring." The timber load table is for a load of 50 cu. ft. The rule was also offered with an ivory slide.

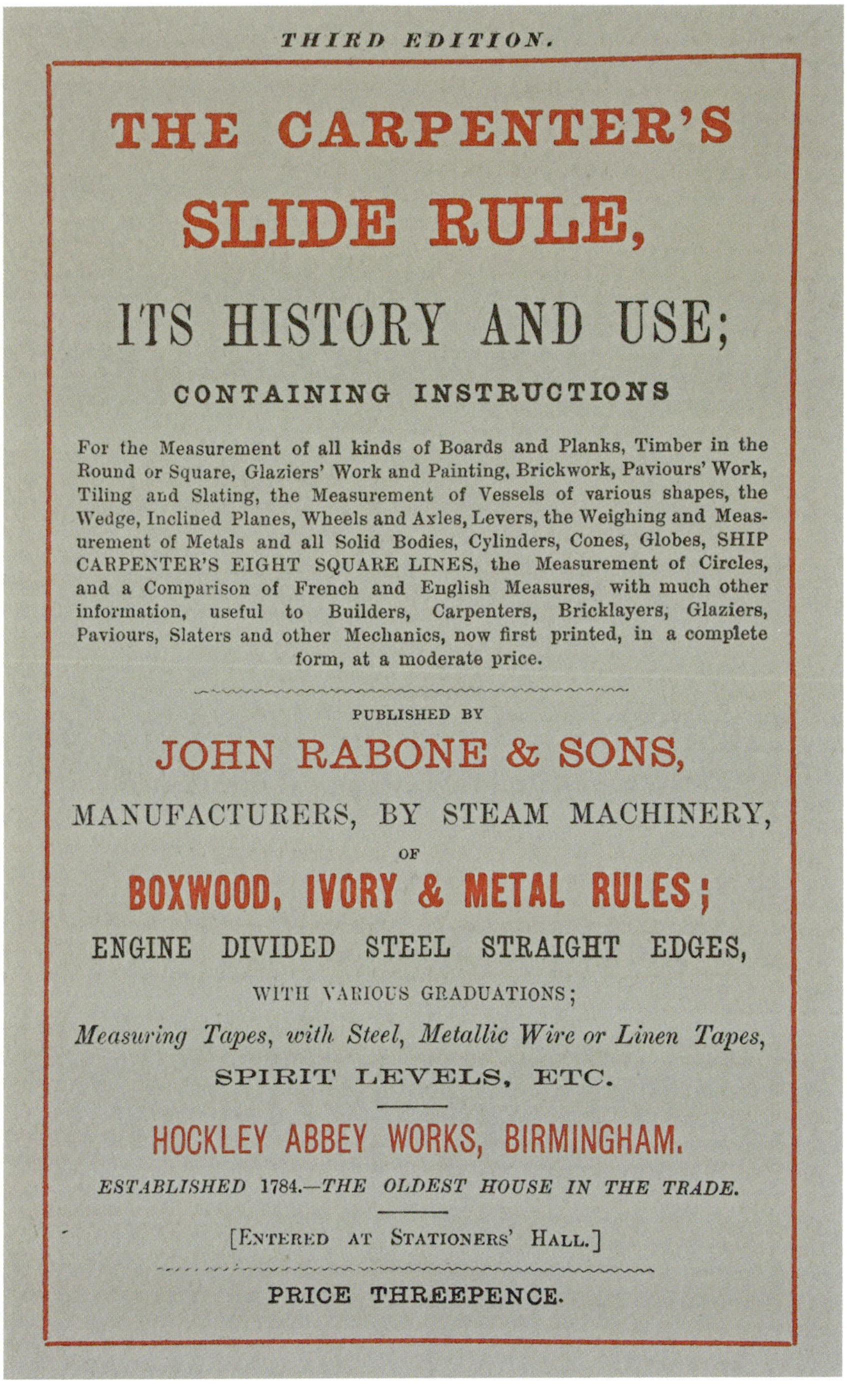
THIRD EDITION.

THE CARPENTER'S

SLIDE RULE,

ITS HISTORY AND USE;

CONTAINING INSTRUCTIONS

For the Measurement of all kinds of Boards and Planks, Timber in the Round or Square, Glaziers' Work and Painting, Brickwork, Paviours' Work, Tiling and Slating, the Measurement of Vessels of various shapes, the Wedge, Inclined Planes, Wheels and Axles, Levers, the Weighing and Measurement of Metals and all Solid Bodies, Cylinders, Cones, Globes, SHIP CARPENTER'S EIGHT SQUARE LINES, the Measurement of Circles, and a Comparison of French and English Measures, with much other information, useful to Builders, Carpenters, Bricklayers, Glaziers, Paviours, Slaters and other Mechanics, now first printed, in a complete form, at a moderate price.

PUBLISHED BY

JOHN RABONE & SONS,

MANUFACTURERS, BY STEAM MACHINERY,

OF

BOXWOOD, IVORY & METAL RULES;

ENGINE DIVIDED STEEL STRAIGHT EDGES,

WITH VARIOUS GRADUATIONS;

Measuring Tapes, with Steel, Metallic Wire or Linen Tapes,

SPIRIT LEVELS, ETC.

HOCKLEY ABBEY WORKS, BIRMINGHAM.

ESTABLISHED 1784.—*THE OLDEST HOUSE IN THE TRADE.*

[ENTERED AT STATIONERS' HALL.]

PRICE THREEPENCE.

Fig. 6.13. The instruction leaflet issued by John Rabone & Sons for their carpenters' rule.

as carpenters' rules, merely as "Two feet Two-fold Boxwood Rules with Brass Slides". The description in the catalogue reads "...marked in Inches in 8ths and 16ths, scales, 8 square lines and Gunter's lines with tables for Timber measuring". There appears from the illustration to be only one table, being a price per load table, giving the cost in pounds, shillings and pence of a load of 50 cu. ft. at up to 30 pence per cubic foot.[13]

The rules included a girt line and were offered with either an ivory or a brass slide. In their ca. 1924 catalogue the tables for timber measure were omitted and only a brass slide was offered.[14] In both the 1892 and the ca. 1924 catalogues, a book of instructions was either offered or included in the price.

By the time of Rabone's catalogue No. 29 (undated but post-1932), one

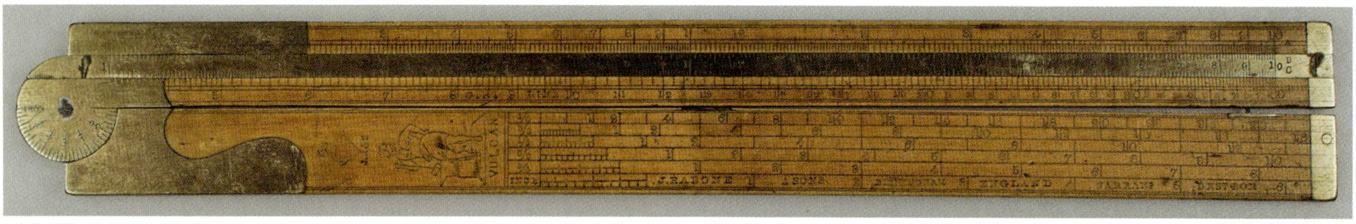

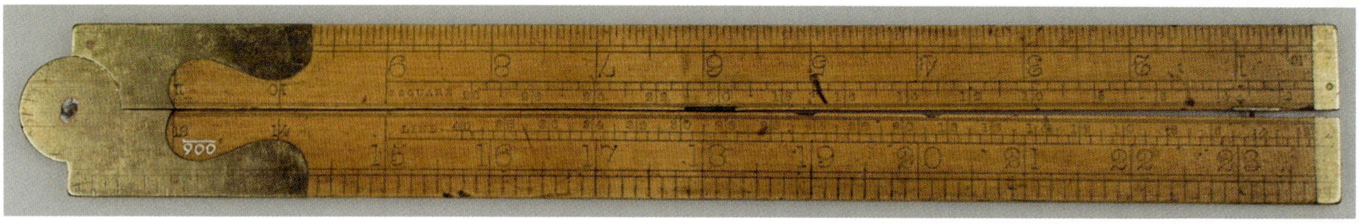

Fig. 6.14a & b. A 2 ft. 2-fold boxwood rule with brass slide and Vulcan joint by John Rabone & Sons (Birmingham, 1877–1963). This rule was not offered with timber load tables but the book of instructions for the carpenters' rule was available.

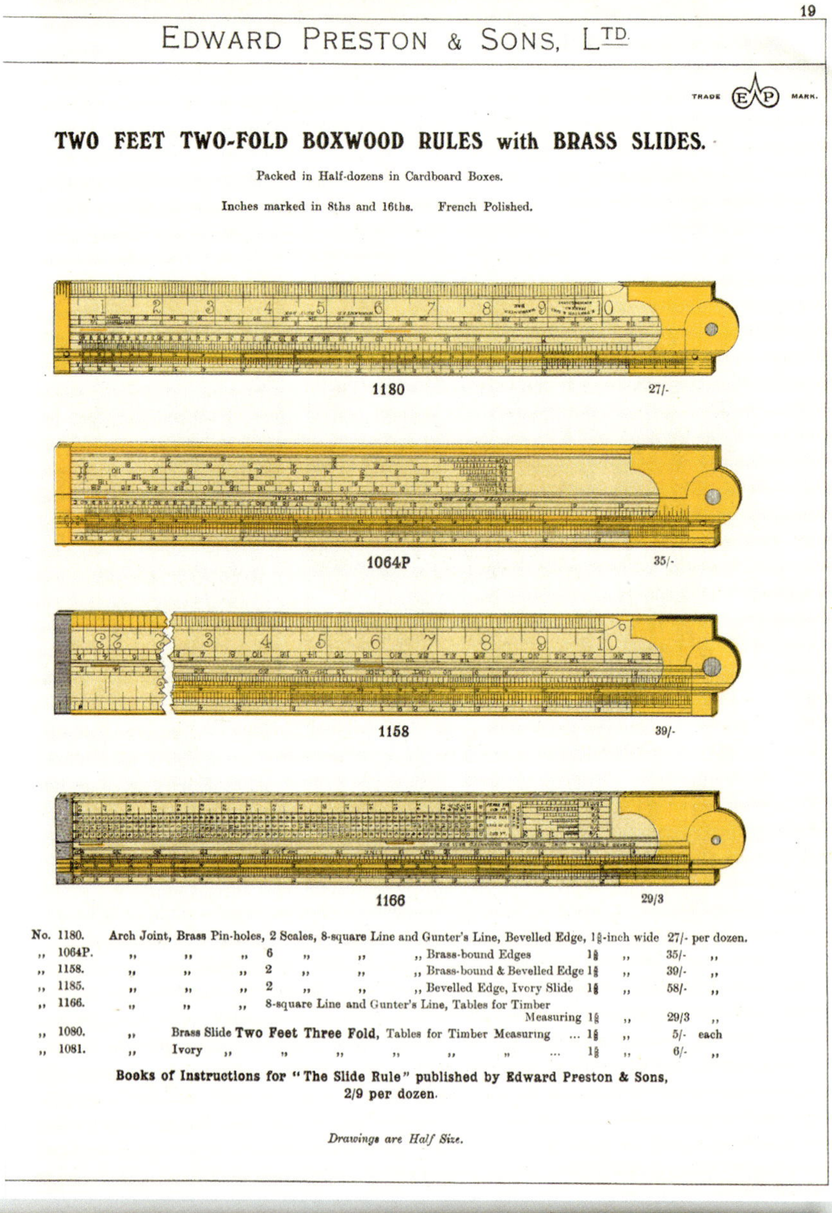

19

EDWARD PRESTON & SONS, LTD.

TRADE MARK.

TWO FEET TWO-FOLD BOXWOOD RULES with BRASS SLIDES.

Packed in Half-dozens in Cardboard Boxes.

Inches marked in 8ths and 16ths. French Polished.

1180 27/-

1064P 35/-

1158 39/-

1166 29/3

No. 1180.	Arch Joint, Brass Pin-holes, 2 Scales, 8-square Line and Gunter's Line, Bevelled Edge, $1\frac{3}{8}$-inch wide	27/-	per dozen.
,, 1064P.	,, ,, ,, 6 ,, ,, ,, Brass-bound Edges $1\frac{3}{8}$,,	35/-	,,
,, 1158.	,, ,, ,, 2 ,, ,, ,, Brass-bound & Bevelled Edge $1\frac{3}{8}$,,	39/-	,,
,, 1185.	,, ,, ,, 2 ,, ,, ,, Bevelled Edge, Ivory Slide $1\frac{3}{8}$,,	58/-	,,
,, 1166.	,, ,, ,, 8-square Line and Gunter's Line, Tables for Timber Measuring $1\frac{3}{8}$,,	29/3	,,
,, 1080.	,, Brass Slide **Two Feet Three Fold,** Tables for Timber Measuring ... $1\frac{3}{8}$,,	5/-	each
,, 1081.	,, Ivory ,, ,, ,, ,, ,, ,, ... $1\frac{3}{8}$,,	6/-	,,

Books of Instructions for "The Slide Rule" published by Edward Preston & Sons, 2/9 per dozen.

Drawings are Half Size.

Fig. 6.15. Page 19 from the Edward Preston & Sons 1909 catalogue.

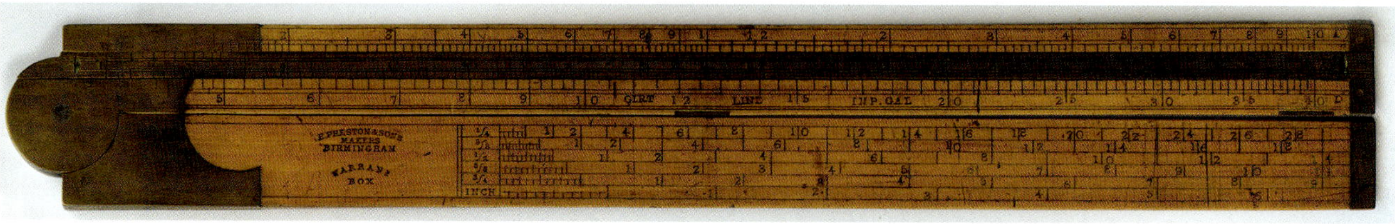

Fig. 6.16. A carpenters' rule by Edward Preston & Sons (Birmingham, 1864–1932).

Fig. 6.17. *Part of the index from the Edward Preston & Sons 1909 catalogue showing the listing for carpenters' slide rules.*

INDEX.

rule with a brass slide was still offered but it had neither timber tables nor girt line and no instruction booklet was available.[15]

Other rule makers, such as Edward Preston, also offered carpenters' rules. It is interesting to note that in the Preston 1909 catalogue, although the rules are listed as carpenters' rules in the index, the page in the catalogue just describes them as "two feet two-fold boxwood rules with brass slide", with an ivory slide offered as an alternative.[16] These rules could be had with or without tables for timber measuring.

They also offered their own book of instructions, *The Slide Rule*, published by Edward Preston & Sons.

The octagon or 8-square line: This line, (actually two lines), is noted in Rabone's booklet and found on many 19th and 20th century carpenters' rules. Sometimes marked E (edge) and M (middle) but more often 8-square line, these enable an octagon to be set out from a square of timber by taking a direct reading off the rule.

Used in conjunction with the line of measure on the rule, the size of the balk of timber is found on the edge or the middle line, depending on whether the octagon is to be set out from the centre or the edge of the balk. The measurement needed is found by looking across the rule to the point on the main line of measure opposite the size of square timber on the E & M scales.

Fig. 6.18. *Below. Diagram showing how an octagon is set out using the 8-square (E & M) line.*

Fig. 6.19. *Below left. Details of the 8-square line from the rule by Edward Preston & Sons shown in Fig. 6.19 showing the direct reading of the offsets for setting out an octagon.*

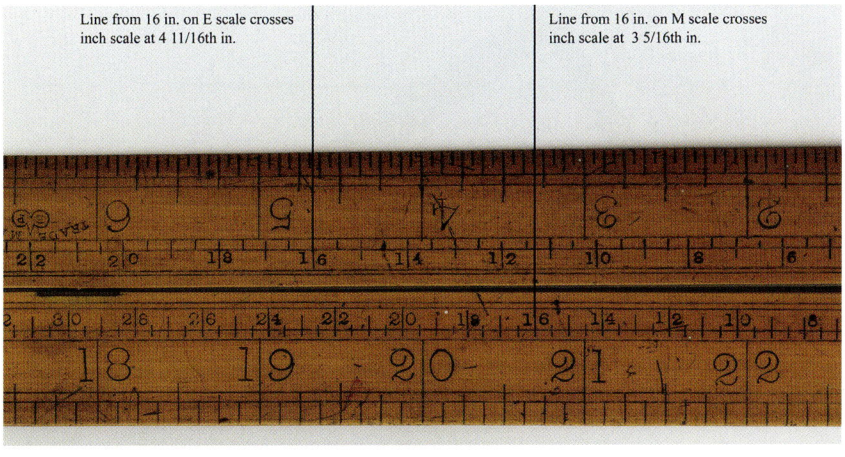

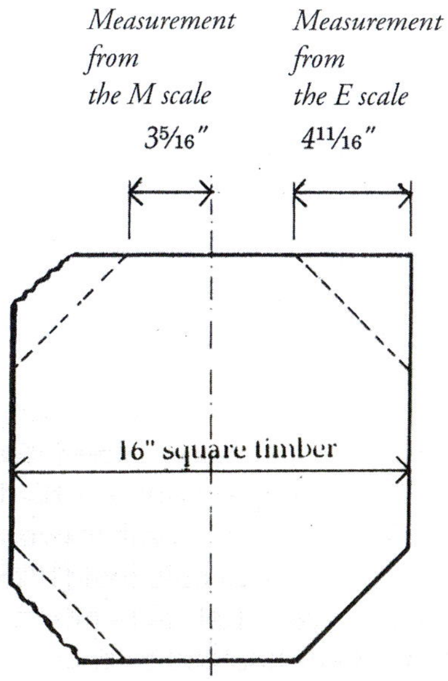

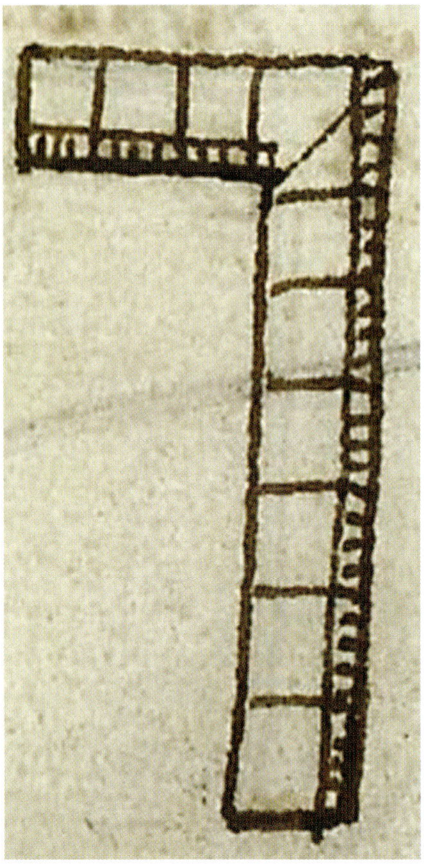

Fig. 6.20. The illustration of the carpenters' square from Randle Holme's Academy of Armory, 1688.

Carpenters' squares

Many mediaeval and Renaissance illustrations show iron squares being used by masons – typically these are made of flat iron, the blades around 2 in. wide and with arms between 12 and 24 in. long. Whether all such squares were marked with inches is open to debate – but many were. The use of iron squares by mediaeval carpenters is not well documented but by 1688 Randle Holme gave a good description of an iron carpenters' square "… is made generally of iron, the shorter part being a foot, and the longer part two foot, both ends being divided into inches, half and quarter inches".[17]

Almost certainly the tool he described had long been used by carpenters. Since Holme's time the iron carpenters' square has remained part of the carpenters' kit, and smiths and some other trades have also used them.

With the lines of inches on the square arranged so that both commence at the corner of the square, it becomes simple to use the square for setting out bevel cuts that match the pitch of the roof by placing the square onto the work and directly marking the cut.

Roofing squares

Also known as a steel, rafter, or framing squares, the modern square is marked with tables that are helpful when framing roofs, though the square also has many other uses. The idea behind such squares is to permit the carpenter to cut all the timbering for a roof on the ground and then erect it – a procedure only for those confident in their use of it!

The development of the modern form of square seems to have taken place in America starting around 1820 when Silas Hawes of Vermont added the table of board measure. By 1880 the fully developed form was being made by the Eagle Square Co. in Vermont. The complexity of the square with its extraordinary range of uses has resulted in the publication of many books on the subject. Perhaps the most garrulous author was an American, Fred T. Hodgson, who managed to write 600 pages on the subject – but he did then produce a pocket version of only 130 pages![18]

In this country too, there was no shortage of instruction booklets for the use of roofing and framing squares produced not only by manufacturers but also many individuals.

The fully developed roofing square, which was eventually available with a variety of finishes, copper plated, nickel plated, blued and others, is marked

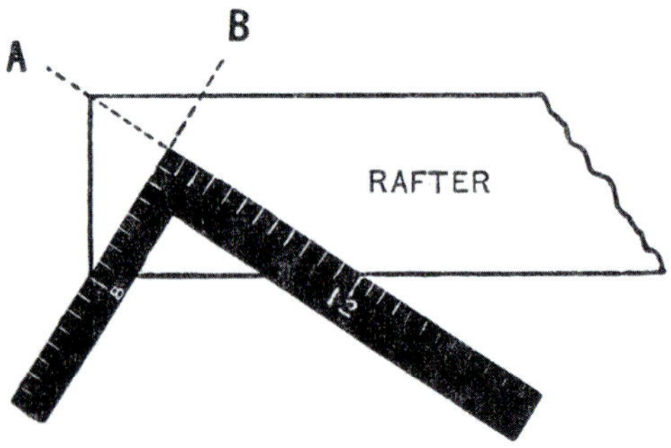

The Cuts or angles of a rafter are obtained by applying the square so that the 12 inch mark on the body and the mark on the tongue that represents the rise shall both be at the edge of the rafter. The illustration shows 8 foot rise, the line A, the cut for the foot end of rafter and B the cut for ridge end.

Fig. 6.21. Explanation of the use of the roofing square from Sargent & Co. Catalogue of Mechanics' Tools, 1911.19

with much information in the form of lines and tables. Not all manufacturers included all the possible lines and information; those likely to be found is briefly described below.

Tables found on a roofing square

Table of (Essex) board measure: A development of the old established line of board measure but now presented as a table of areas for given widths and lengths. The lengths are tabulated under the line of inches on the edge of the square.

Rafter table: The following lines of figures are found on the framing rule as set out by Hodgson.[20]

- The first line of figures gives the length of common rafters for one foot run.
- The second line gives the length of hip or valley rafters for one foot run.
- The third line gives the length of the first jack rafter and the difference in the length of the remainder if spaced at 16 in. centres.
- The fourth line is the same as the third line but for rafters spaced at 2 ft.

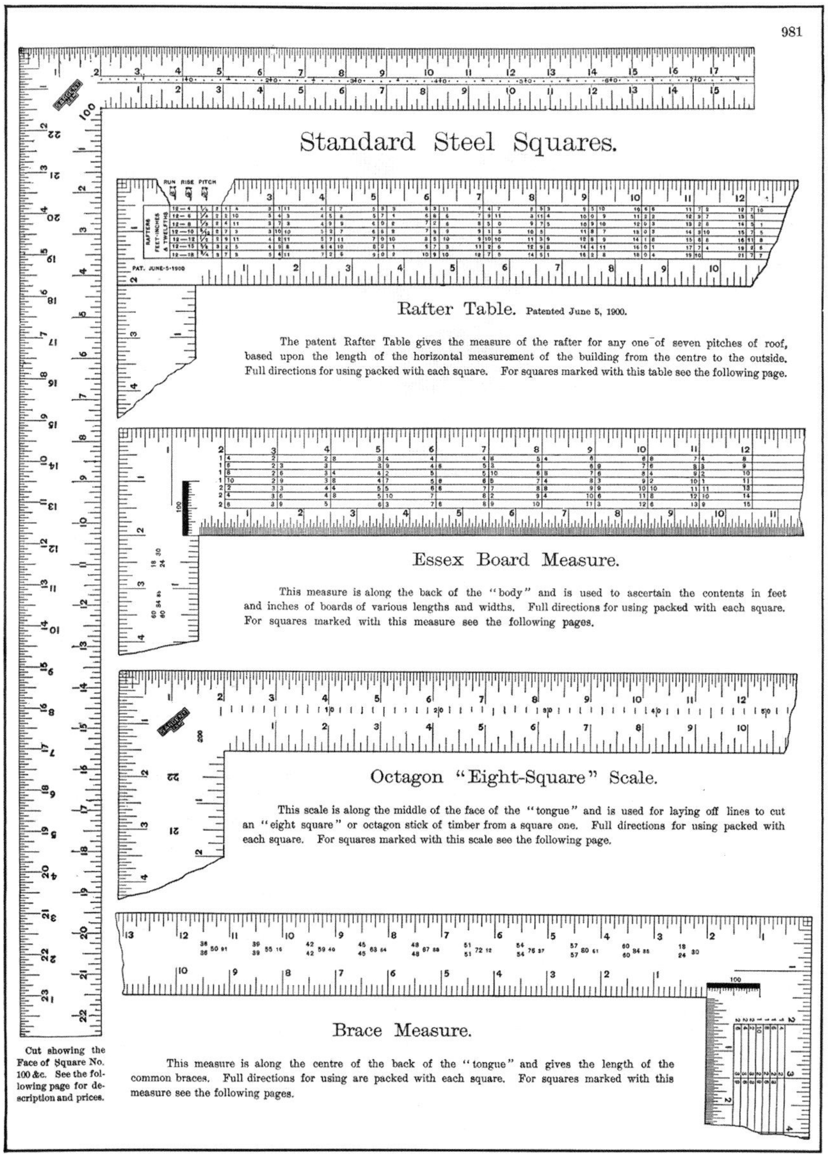

Fig. 6.22. The various tables that were available on the steel squares shown in the Sargent & Co. 1910 catalogue.[21] The squares could be purchased with five different combinations of tables: Brace Measure, 8 Square and Essex's Board Measure; Brace Measure and Essex's Board Measure; Brace Measure, 8 Square and Patent Rafter Table and Brace Measure and Patent Rafter Table and Essex's Board Measure on its own. They could be had in eight different finishes: Polished; Nickel Plated; Electro Copper Plated; Galvanised; Blued White with enameled figures and marks; Blue Yellow with enameled figures and marks; Galvanised Red with enameled figures and marks and Royal Copper

- The fifth line gives the side cut of jack rafters against the hip or valley rafters
- The sixth gives the side cut of hip or valley rafters against the ridge board or deck.
- The seventh line of figures gives the cuts of sheathing and shingle in the valley or hip.

Rafter table, alternative form: Offered by Sargent and other American firms, the table has seven lines with each line being for a stated pitch, 4, 6, 8, 10, 12, 15, and 18 inch rise per foot. The length of common rafters is read (in feet, inches and twelfths of an inch) under the edge line of measure which in the table represents the run, that is, the half span of the roof. These tables, which are easily identified by the three column format, merely reproduce tables of rafter length that had been published in many carpentry books.

THIS PAGE APPLIES TO NOS. 381 AND 383 ONLY

BRACE MEASURE (Fig 23)

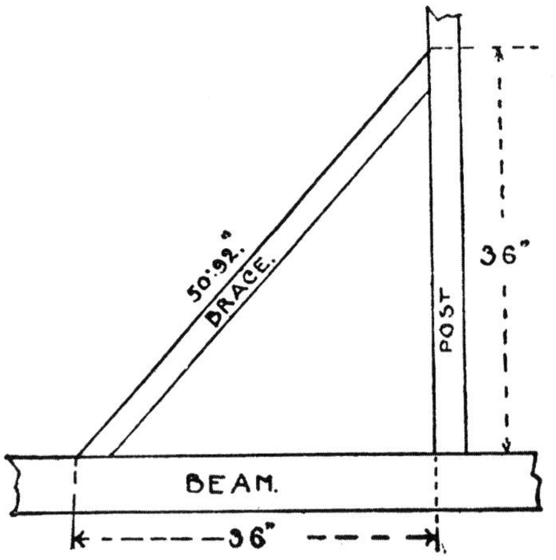

Fig. 23

This measure is given along the centre of the back of the TONGUE and gives the length of common BRACES (Fig. 24).

Example :—Find length of BRACE with run on POST and BEAM equal to 36in.

(i) Look at BRACE TABLE and find expression $\frac{36}{36}$ 50.92 (Fig. 24). Length of BRACE = 50.92 ins.

(ii) If run on post differs from run on beam, then length of BRACE and TOP and BOTTOM cuts may be determined as for COMMON RAFTERS.

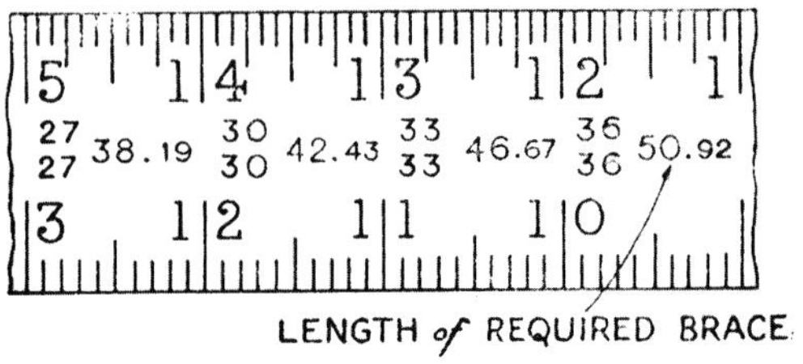

Fig. 24

12

Fig. 6.23. Instructions for the use of the brace measure from the I. & D. Smallwood booklet, Instructions for the Use of Carpenter and Framing or Rafter Steel Squares.[22]

The brace measure: This table does not relate to the line of inches on the square but is complete in itself. This is a table giving the lengths of diagonal braces. It is laid out thus, for example

18 25.45
18

An 18 in. by 18 in. brace (a 45° brace) has a length of 25.45 inches. For small sizes this could of course be measured off the square but the table normally extends up to 60 inches. An inch divided into 100ths is usually included to assist in the use of the dimensions on the table expressed as decimals.

Stair measure: I. & D. Smallwood offered squares that also included a stair or "66" measure. The "66" measure was so called because the depth of the tread multiplied by the height of the rise is sixty-six. They supplied an instruction book to accompany their squares which set out the various ways in which their squares could be used.

STAIR MEASURE

One of the most common is 66 measure because the TREAD multiplied by the RISE is equal to 66—i.e., a stair with tread of 11in. has a rise of 6in., or a stair with tread of 8⅛in. has rise of 8⅛in.

To Set Out TREADS.

(i) Get length of TREAD on outer edge of BODY of square and RISER on the TONGUE (Fig. 25).

(ii) Lay the square on STRINGER as shown and mark along outer edges of BODY and TONGUE.

(iii) Get top and bottom cuts of STRINGER by reversing the square (Figs. 26 and 27).

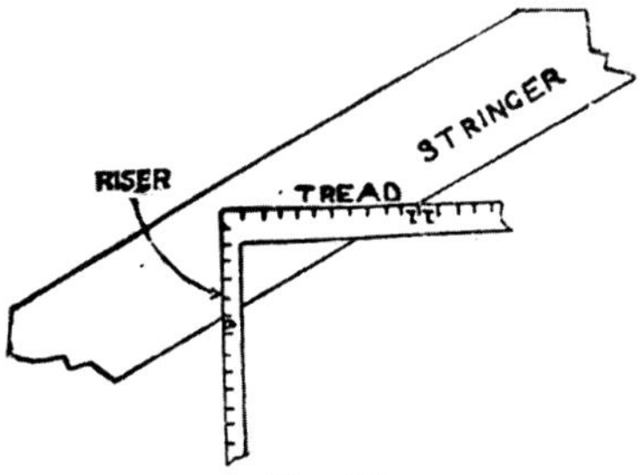

Fig. 25

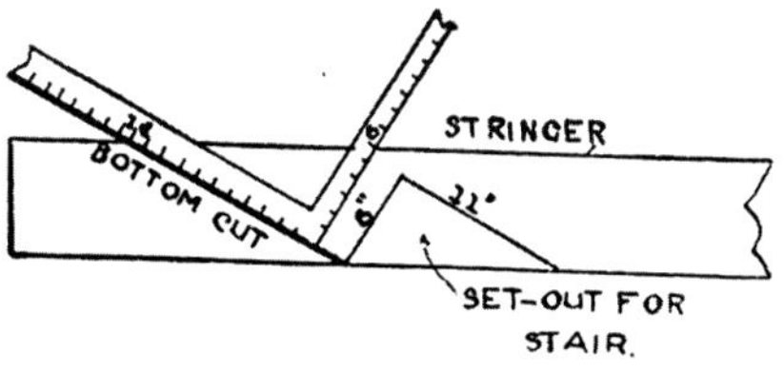

Fig. 26

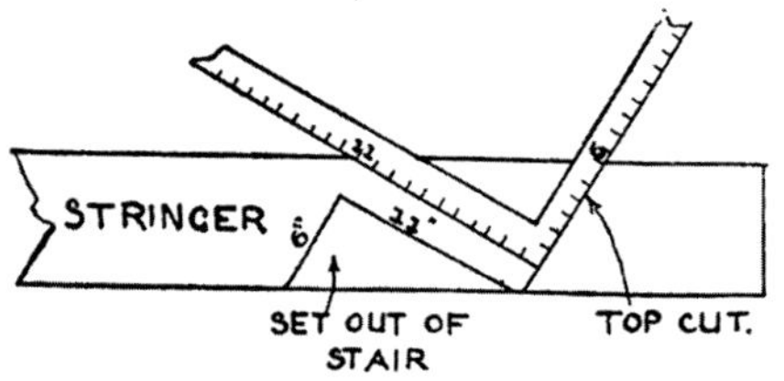

Fig. 27

13

Fig. 6.24. Instructions for Smallwood's stair measure.

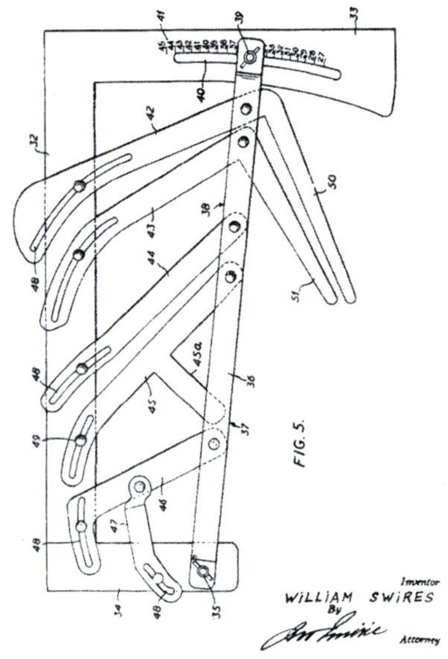

Above: Fig. 6.25. William Swires' patent roofing square.

Above right: Fig. 6.26. William Lanham's "Roofit" bevel. The bevel was both patented and the design was registered. It is crudely made from pasteboard and does not seem very suitable for use on a building site.

Metric roofing squares

Metrication, together with the new practice of specifying of roof pitch in degrees, has caused radical alterations to the layout of the roofing square in the UK. The No. 390 metric square sold by Smallwood has a scale of degrees on the inside edge of the blade. The table for the lengths of common and hip or valley rafters is now no longer keyed to the measurement on the edge of the blade but is an independent table. The ability to calculate the reduction in the length of jack rafters remains but it is now a formula.

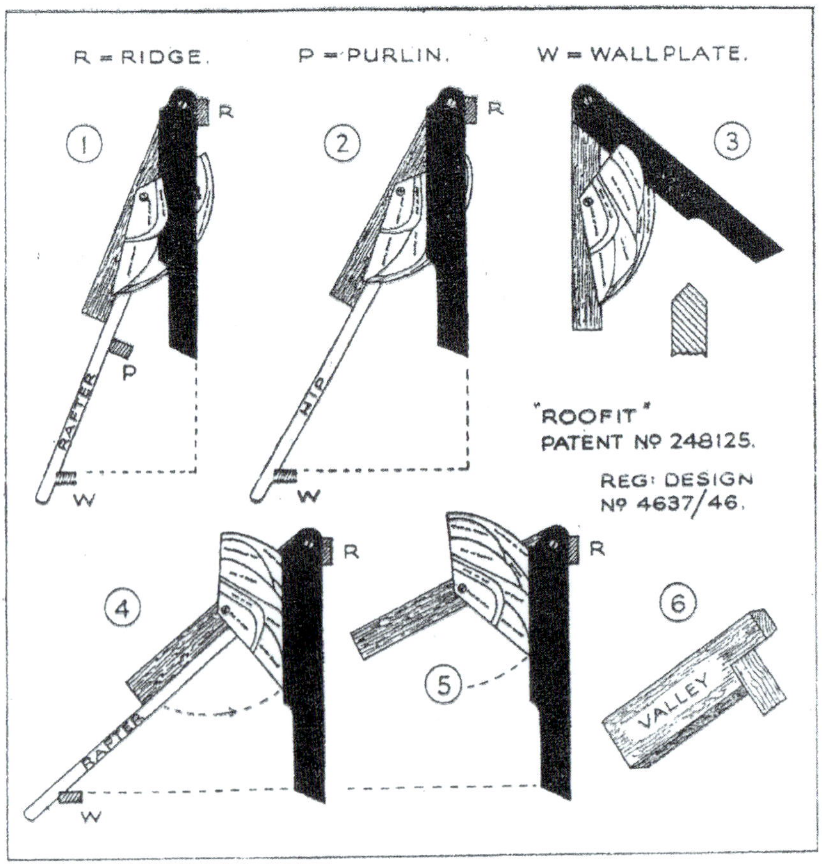

Fig. 6.27. Instructions for using William Lanham's "Roofit" bevel.

Innovations

The roofing square seems to have frequently exercised the minds of inventors on both sides of the Atlantic and there are numerous patent and innovative versions. A particularly complicated one was invented by William Swires in the late 19th century.

Another was the "Roofit" bevel, patented in 1926, by William Lanham of Bath.

Makers

By the end of the 18th century iron squares had become just one of the many products made by the Birmingham "toy" makers.

In later years the principal English maker of carpenters' squares was the Birmingham firm of I. & D. Smallwood.

An early square marked Hildick with crown over has been recorded – this was probably made in Birmingham but as there were Hildicks in the tool making trade in both Birmingham and Sheffield it is not possible to be certain.

Take down squares

The steel square is an awkward tool to accommodate in the tool box. The answer is the "take down" roofing square made by several American manufacturers including Stanley but never made in Britain. This was a square with a separate tongue and blade with a screw that locks the two together at right angles,

Roofing and building slide rules

There were other rules developed to assist with the calculations and setting out both roofing and general building projects.

One inventor who put considerable effort into this was Lala Ganga Ram, who was the Chief Engineer of the Public Works Department in the Punjab

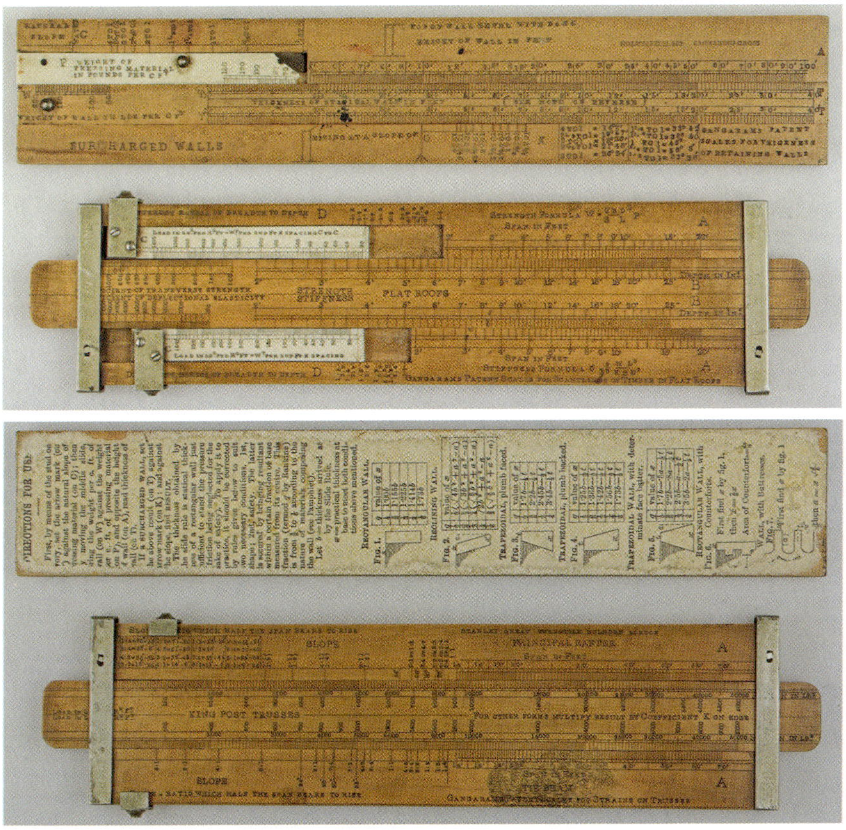

Fig. 6.28a & b. Two slides rules patented by Lala Ganga Ram. Top: Made by Holtzapffel and marked with their "Charing Cross" address (London, 1821–1901), this is for calculations regarding retaining walls. Bottom: This rule, made by W.F. Stanley (London, 1880–1967), is for calculations for trussed rafters. Another example of this rule has been recorded, also made by Holtzapffel.

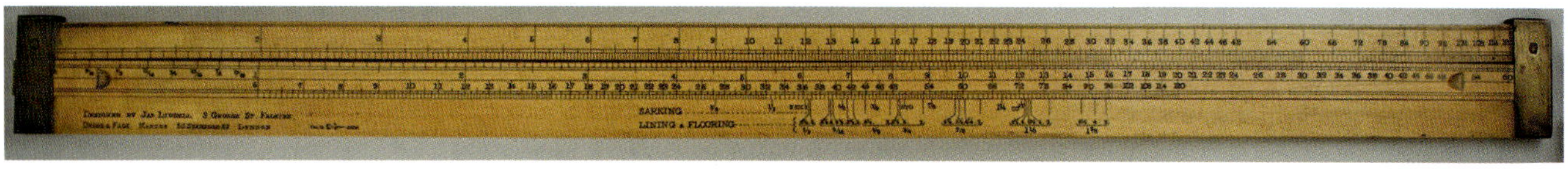

Fig. 6.29. A rule designed by James Liddell concerned with sarking (the waterproof layer between a roof structure and the roof covering), timber linings and flooring, made by Dring & Fage (London, 1790–1977). The exact use of this rule has not been established.

in India. In January 1884 he patented two "Sliding Scales for Engineering Calculations" that he had designed. The patent covered two different rules, both of which are illustrated in the patent, one for calculating timbers in both flat and pitched roofs, the other for calculating retaining walls.

Another calculating rule was designed by James Liddell, which is found in two forms. One rule states that a patent has been applied for but no record of the patent has been found. The other rule gives his address as 3 George Street, Falkirk but no further information has come to light.

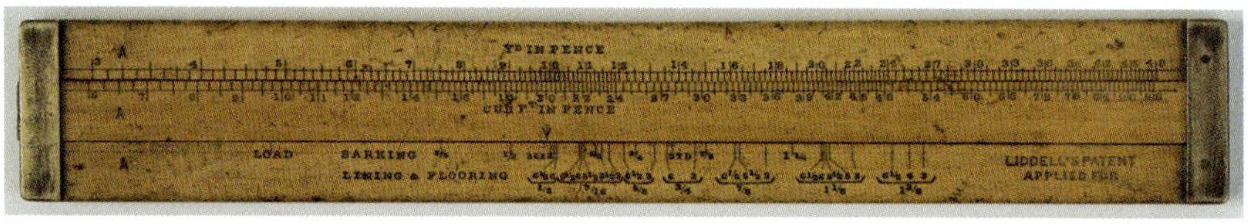

Fig. 6.30a & b. Another of James Liddell's slide rules. This one has sarking, flooring and lining information with the remainder of the rule marked for calculating the cost per cubic foot of timber mouldings if the cost per 100 ft. is known and the cost in cubic ft. per pence if the cost in yards per pence is known, or vice versa. There is no maker marked.

Plumbers' rules and gauges

Many early references to lead roofing are quoted by Salzman in *Building in England down to 1540* but none contain any clear specification of thickness.[23] In the mediaeval period lead roofing was an expensive form of roofing and the records quoted by Salzman suggest that the material was most often supplied to the plumbers by the building owners. The plumbers were paid on the basis of the weight of lead they fixed. The lead was bought by the fother, a weight variable according to time and place, but approximating to a cartload (between 19 and 22½ cwt.). Payment of plumbers' work by weight, whether or not the material was provided by the client, continued until the late 19th century.

No other trade has undergone such a progressive de-skilling of its work as the plumber. In the 18th century the plumber might have been expected to cast the lead sheet he needed for a roof, and then fit it. He would have cast pipes or rolled them up from lead sheet and then soldered them together. Casting the parts of a decorated cistern or hopper head and assembling them

Weight lb./sq. ft.	Thickness in.	Thickness BS 1178 mm	BS 1178 lb./ sq. ft.
3	0.051	1.20	2.91
4	0.068	1.80	4.19
5	0.085	2.24	5.21
6	0.102	2.50	5.82
7	0.119	3.15	7.33
8	0.136	3.55	8.26
9	0.153	–	–
10	0.170	–	–

Fig. 6.31. Thickness / weight of lead sheet and the metric equivalent of the thickness as given in BS 1178 (1969). Traditionally sheet was always specified by weight – the thickness listed in column 2 is only useful as a quick check. Under the British Standard, the old weights became code numbers with the slightly altered weights listed in column 4. Gauges were only a quick method of checking the thickness.

Fig. 6.32a & b. An unmarked 18th century plumbers' rule, brass bound with steel tips, marked with a hundredweight table, a table of weight per square foot of cast lead and a table of weight per yard of lead pipe.

was all part of the trade. First came milled sheet (available by the 1720s), against which the plumbers conducted a long standing campaign of vilification of the material, and then factory-made pipes. The skill needed to make a

CAt		LEAD	
Thick		Wt.	
In	lb	os (z)	dr
1/32	1	15	6
1/16	3	14	13
1/8	7	13	10
1/4	15	11	4
1/32	17	10	10
1/16	19	10	1
1/8	25	8	14
1/2	31	6	8
1/32	33	5	14
1/16	35	5	5
1/8	39	4	2
3/4	47	1	12
1/32	49	1	2
1/16	51	0	9
1/8	54	15	6
In.	62	13	0

Fig. 6.33. Table of weight per sq. ft. of cast lead as found on the plumbers' rule shown in Fig. 6.35. Thickness is expressed in the old-fashioned method of adding smaller fractions to the quarters. Note the fourth column is drams – there are 16 to the ounce. This would seem to introduce accuracy to the point of absurdity!

Fig. 6.34. Table of weight per yard of lead pipe from the rule shown in Fig. 6.34. Until increased water pressures and increased flow rates, the smallest size of pipes used was ¾ in.

LEAD PIPE p^r y^d	
Bore in inches	Lb.
3/4	10
1	12
1 1/2	18
2	24

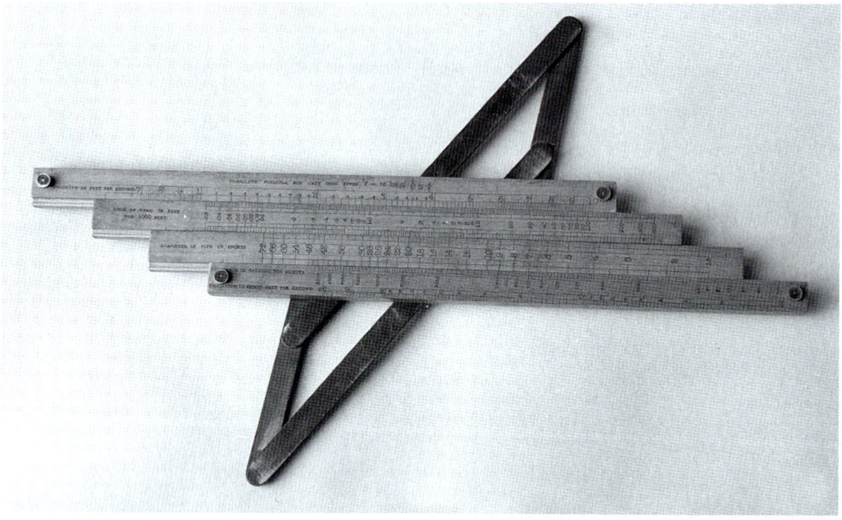

Fig. 6.35. A most unusual unmarked calculating parallel rule probably intended for the design of sewerage pipes.

wiped joint is considerable even when the heat is available from a blow torch – how much more skill was needed when the heat was transferred to the work with a soldering iron.

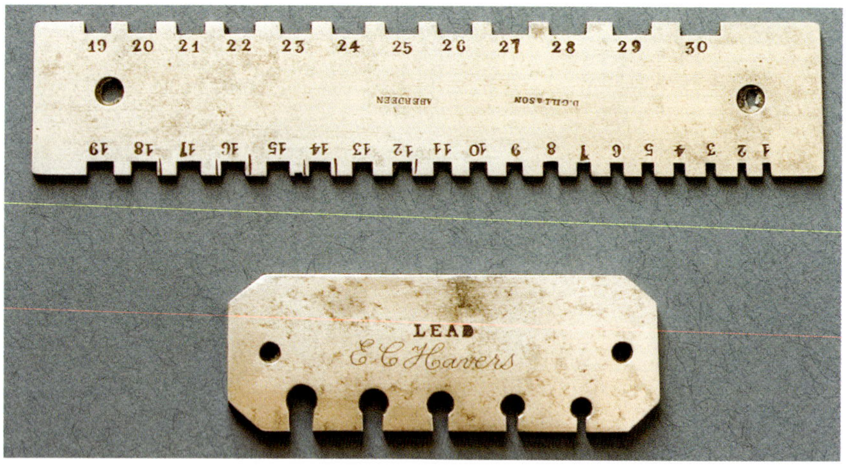

Fig. 6.36. Two lead sheet gauges. Top by D. Gill & Son (Aberdeen, dates not known). Bottom, no maker's name.

Fig. 6.37a & b. A gauge, made in brass, by W. F. Stanley (London, 1853–1967) marked on one side for glass and lead and on the other for standard wire gauge and zinc.

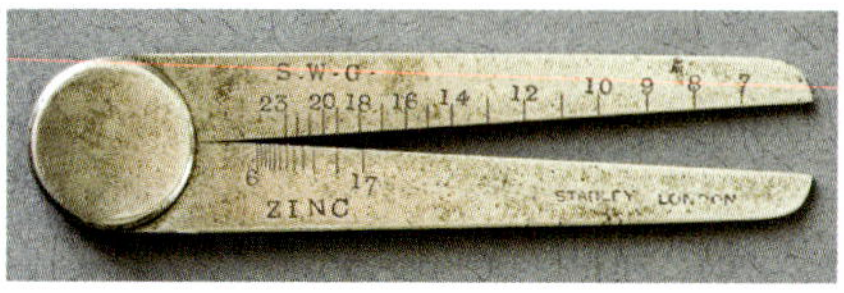

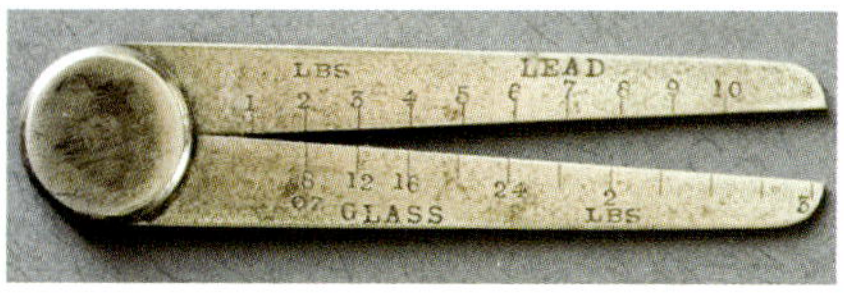

Lead sheet work: Specifying the thickness of sheet lead by the weight of a square foot has been the method used since meaningful specification of building materials started in the building industry – and this was as long ago as the Middle Ages.

The plumbers' rule: A 2 ft. rule with 9 in. legs and a 6 in. brass folding tongue, it is usually marked with three tables – the normal hundredweight table, a table of lead thickness / weight and a table of weight of pipes per yard. The hundredweight table was particularly relevant to plumbers for all leadwork, whether labour only or material and labour, was paid for by the hundredweight. Plumbers' rules were probably obsolete by 1880 – they are not listed by Rabone in 1892.

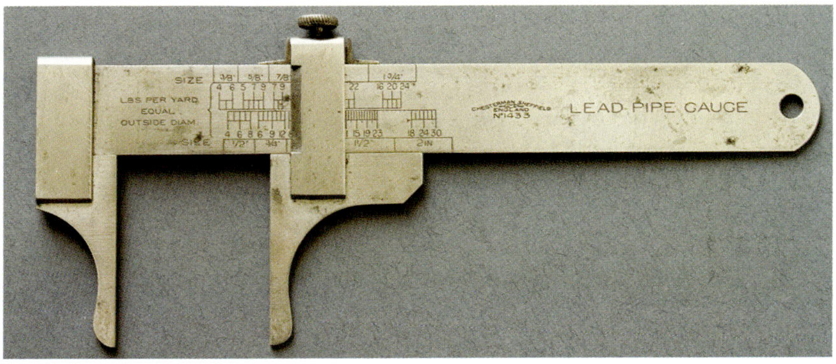

Fig. 6.38. A lead sheet and pipe gauge by James Chesterman & Co. (Sheffield, 1859–1963).

Lead sheet gauges: Most usually in the form of a steel slot gauge but occasionally made as a V-gauge, the normal range of measurement is from 3 to 7 lb. per sq. ft. The gauge is sometimes just a simple plate but often is contained in a brass case, in form similar to a pocket knife. Gauges of this type often contain two or three blades, with leaves for lead, glass, wire gauge or zinc.

Chesterman's lead sheet & pipe gauge: Available in the 1935 Buck & Hickman catalogue,[24] this is an all-steel caliper marked to measure both inside and outside diameters of pipes, weight of lead sheet in pounds per square foot, and the weight per yard of pipe from ⅜ in. to 2 in. bore. Once the bore of the pipe has been established a caliper reading of the outside diameter gives a direct reading of the weight per yard of the pipe. The difference in the wall thickness of lead pipe available is illustrated by the variation from 8 to 17 lbs. in the weight of 1 in. pipe.

Glass and glazing

Before 1845, when the tax on glass was lifted, most windows were glazed with crown glass – glass made by spinning out a thin round "table", between 3 ft. 6 in. and 4 ft. diameter. Crown glass, which got its name from the duty mark with which it was originally marked, could be made thinner than cylinder glass (which was in any case inferior) and thus, as all glass was dutied by weight, carried less tax.

The tables were packed in cases which typically contained 12 to 18 tables but might hold up to 24.[25] To make packing easier and more compact the tables were cut once (usually 4 in. from the central pontil) before being packed in the case.

In the 18th century, as today, it was normal to cut glass in the glaziers' workshop and carry it to site cut to exact size. The London glaziers seem to have done good business cutting both Newcastle- and London-made glass and dispatching panes to sites in the country and even as far as the American colonies. In July 1771, G. Wythe of Williamsburg, Virginia, ordered 400 panes of crown glass and in October of the same year Thos. Everard ordered 100 feet of window glass cut to 11 in. by 9½ in.[26]

Glaziers' rules: Neve in *The City and Country Purchaser and Builder's Dictionary...*[27] writes "...the best and readiest way to compute the Area's, is to take the Dimensions with a Sliding rule, such as the Glaziers generally use; which Rule is divided Centesimally..." Early descriptions of slide rules often describe them as being like a glaziers' rule – having two sticks that are slid alongside one another and held together with brass hoops, and one or two of the earliest slide rules extant are of this form.

Wilson in *The Artist's and Mechanic's Encyclopaedia*[28] (ca. 1830) describes the tools of the glazier in some detail and he describes glaziers' rules as "two

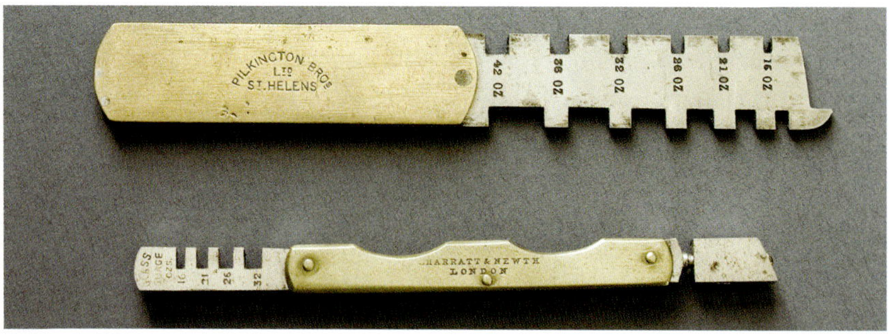

Fig. 6.39. Two folding glass gauges, the lower also incorporating a diamond glass cutter. When the gauge is incorporated with a glass cutter care should be taken to distinguish between breaking slots which are often present on glass cutters, and slots for gauging. The upper gauge is marked Pilkington, St Helens, a well-known glass manufacturer, the lower Sharratt & Newth, makers of glass cutters.

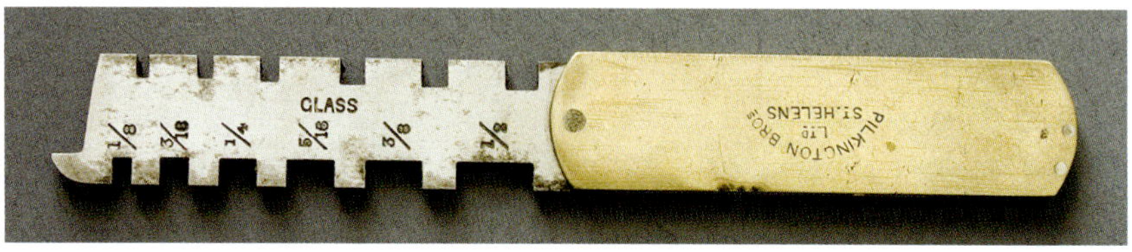

Fig. 6.40. The reverse of the top gauge shown in Fig. 6.39. On one side the glass is measured in ounces per square foot – expressed solely as oz. On this side the measurement given is the thickness of the glass.

feet long in four different pieces" but no rules fitting this description have survived.

Glaziers' work was priced by the square foot. The easiest way of computing the area of a sash pane, $7\frac{3}{8}$ in. by $11\frac{5}{8}$ in., is to measure it in 100th parts of a foot (in this case 65 by 97 = 6014 which is 0.601 sq. ft.) rather than wrestle with fractions and the duodecimal conversion from inches to feet. A decimal result also facilitates the next step which is to multiply by the number of panes in the sash windows. Whether undertaken with a slide rule or by long multiplication there are thus only two quite simple steps in this calculation to decide the area of glass in a whole facade. For this reason it would seem essential that any "glaziers' rule" have both inches and feet in 100 parts lines of measure.

Glass gauge: Tables of crown glass are thick in the centre and, depending on quality, of more or less variable thickness elsewhere. Only when sheet glass became available in the 1840s did the thickness of the pane become a significant matter in the specification of glass. From then until metrication the thickness of sheet glass for building purposes was specified in ounces per square foot. The thinnest glass used in building was 16 oz. and thicker glass could be up to 42 oz. Plate glass was used in a range of thicknesses from $\frac{1}{4}$ in. up to $\frac{1}{2}$ in. and more.

The glass gauge is invariably a slot gauge cut from steel. In most cases it will be found in conjunction with other gauges of use to a builder and often in a case of pocket knife form.

Fig. 6.41. A glaziers' T-square by Edward Preston & Sons (Birmingham, 1864–1932).

Glaziers' T-squares: The traditional type is an all boxwood T-square from 24 to 60 in. long, with square, not bevelled, edges. Used as a straight edge for cutting glass, they are easily distinguished from draughtsmen's T-squares by the inch measure marked on the blade and the ramp in the stock which allows the glass cutter to be brought right up to the stock.

Glaziers' laths: A brass tipped boxwood lath from 24 to 72 in. long the principal use of which is to position the glaziers' T-square. It is usually described as "thin" (to slip under the T-square?) and this is the only characteristic that helps to distinguish it from laths intended for other trades.

The marking of these laths varies but there will always be lines starting from both ends, sometimes on opposite sides, sometimes on the same face. The usual division is inches and eighths.

References

1. Chambers, E., *Cyclopaedia or an Universal Dictionary of Arts & Sciences,* 5th ed., London, 1743.
2. *Ibid.*
3. Digges, Leonard, *A Boke Named Tectonicon*, London, 1562 (first published 1556).
4. Kebabian, Paul B.*, Selections from the Chronicle,* Mendham, NJ, 1991.
5. More, Richard, *The Carpenters Rule, Or, A Booke Shewing Many plaine waies, truly to measure ordinarie Timber ...*, London, 1602.
6. George Atwel[l], *The faithfull surveyour: discovering divers errours in land-measuring; and shewing how to measure all manner of ground ... Teaching likewise the making and use of a new and general instrument, called a pandoron ... With an appendix unfolding errours in board and timber-measure, with directions for making a carpenters-ruler,* Cambridge, 1658.
7. Leybourn, William, *The Art of Measuring: Or The Carpenters new Rule Described and Explained,* London, 1669.
8. Knight, Richard, "The Carpenter's Rule" *Newsletter No. 20,* Tools & Trades History Society, 1988.
9. Coggeshall, Henry, *Timber Measure by a Line of more ease, dispatch and exactness, than any other way now in use ...*, London, 1677.
10. Coggeshall, Henry, *A Treatise of Measures by a Two Foot Rule Which Slides to a Foot ...*, London, 1682.
11. Scamozzi, Vincenzo, *L'Idea della Architettura Universale* (*The Universal Idea of Architecture*), Venice, 1615.
12. Rabone, J., *The Carpenter's Slide Rule*, Birmingham, 1880.
13. John Rabone & Sons, *Catalogue of Rules & Tapes,* Birmingham, 1892.
14. John Rabone & Sons, *Catalogue No. 24,* Birmingham, n.d. but probably 1924.
15. John Rabone & Sons, *Catalogue No. 29,* Birmingham, n.d. but post-1932
16. Edward Preston & Sons, *Catalogue No. 18. The "Preston" Catalogue,* Birmingham, 1909 with 1912 supplement,
17. Alcock, N. W. & Cox, N., ed. *Living & Working in 17th Century England,* CD of Randle Holme's *Academy of Armory, 1688*, London, 2000.
18. Hodgson, Fred T., *The ABC of the Steel Square and its Uses*, Chicago, 1908.
19. Sargent & Co., *Catalogue of Mechanics' Tools 1911,* New Haven, CT.
20. Hodgson, *op. cit.*
21. Sargent & Co. *Catalogue, 1910.* New Haven, CT. Reprinted as part of

The Sargent Tool Catalogue Collection, Mendham, NJ, 1993.
22. I. & D. Smallwood, *Instructions for the Use of Carpenter and Framing or Rafter Steel Squares,* n.d. but probably ca. 1920.
23. Saltzman, L. F., *Building in England down to 1540,* Oxford, 1952.
24. Buck & Hickman, *General Catalogue of Tools & Supplies,* London, 1935.
25. Ayres, James, *Building the Georgian City,* New Haven, 1998.
26. Mason, Frances Norton, ed., *John Norton & Sons, Merchants of London and Virginia,* Newton Abbot, 1968.
27. Richard Neve, *The City and Country Purchaser and Builder's Dictionary...,* 2nd ed., London, 1726.
28. Wilson, Edward James, *The Artist's and Mechanic's Encyclopaedia,* Newcastle on Tyne, ca. 1830.

Chapter 7
Surveyors and architects

Surveying is the art of ascertaining, by measurement, the shape and size of any portion of the earth's surface, and representing the same, on a reduced scale, in a convenient manner, so as to bring the whole under the eye at once. Usill, 1916.[1]

The principal measurements taken when making a survey are angles and distances. Many of the instruments used – levels, theodolites and other optical instruments for measuring angles – are not within the compass of this book but one device for measuring angles and a number of tools for measuring distance do fall within the parameters.

Measuring angles

The simplest tool for establishing horizontal angles is an alidade and plane table. An alidade is a rule-like instrument with a sighting vane at each end and frequently marked with scales on the edges. It is used by pivoting an edge around a centre pin, while looking through one vane to the other, lining up the slot in the farther vane with the subject being surveyed, and then drawing a line along that edge of the alidade. In this method there is no actual measurement of the angles for subsequent plotting, as the sight lines are directly marked onto the paper as the survey is made.

Plane table rules: Plane tables dating from the 18th century and before were often designed to be used with a plane table rule, described by Bion[3] as a box frame. The purpose is to hold the paper onto a plane table and to determine angles between sightings or, if the table is aligned north with a compass,

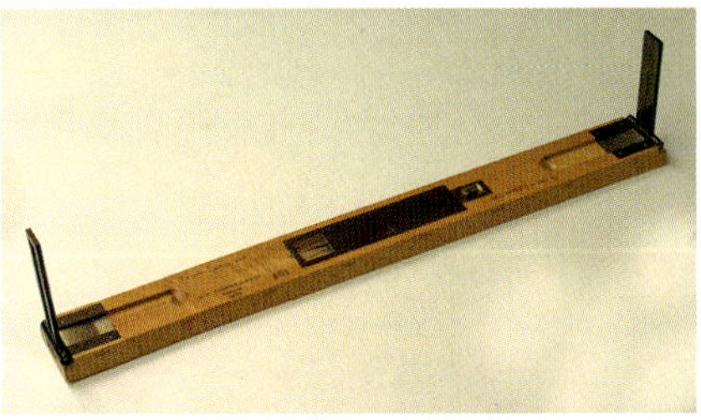

Fig. 7.1. A boxwood alidade by C. T. Cooper & Sons, London (1905–1947>).

Below left: Fig. 7.2. A plane table, with an alidade on the top, in use from Wm. Leybourn's Compleat Surveyor, *1722.*[2]

Below: Fig. 7.3. A plane table and alidade dating from the 1910s. Although these are a military version, civilian versions would have looked no different.

Fig. 7.4. A plane table rule. As with all such rules, it is unmarked but dates from ca. 1700. All four sides are marked in degrees and inches. The inches run from left to right on two (adjacent) sides and from right to left on the other two adjacent sides. The corner joints are formed with brass strips.

to record the bearing of sighted positions. The rule is constructed with square boxwood sections hinged to fold inwards. When erected it forms a rectangular frame which sets into a rebate on the perimeter of the plane table. The rule is marked with degrees projected from the centre point of the table and some scales. None of the examples found have been named – presumably the maker's name was put onto the plane table of which the rule is but a part.

Measuring length

Rods, poles and perches: In the literature on the history of weights and measures several theories have been advanced as to how the English rod, pole or perch – these are different names for the same measure – came to be standardised at 16½ ft. Its importance is that the perch is the unit of measure from which the acre was generated. In early references an acre is specified as an area 4 by 40 perches – 22 yards (a chain) by 220 yards (a furlong).

Perhaps there was a practical aspect. In use a pole of 16½ ft. or thereabouts would have been as long as is easily portable; a shorter pole, although easier to move, would need more repetition when measuring and would thus be less accurate. Whatever the explanation, the pole was used as a measure by practical men measuring out fields and woods from Saxon and mediaeval times.

The king's perch may have been regularised at 16½ ft. but as with so many measures, there were other sizes of perch. Woodland is clearly not as valuable

Linear measure	
16½ feet or 5½ yards	= 1 rod, pole or perch
40 rods	= 1 furlong
8 furlongs	= 1 mile
Gunter's linear measure	
1 link	= 7.92 inches
100 links	= 1 chain i.e. 4 rods or 22 yards
Area measure	
30¼ sq. yards	= 1 sq. rod
40 sq. rods	= 1 rood
4 roods (160 sq. rods)	= 1 acre (4840 sq. yards)

Fig. 7.5. Table of pre-metric measures used for land in Britain.

as farmland so it is quite logical to measure it using a perch (usually known as a rod for woodland) of 18 or 20 feet or even larger – even up to 37½ ft.

Whatever its length the use of a long wooden rod for measuring land could never have been convenient and by the 16th century mention of cords and "wyre" chains were becoming common. In view of the early demise of the rod it is not surprising that no period examples have survived.

Surveyors' cords: These were mentioned by Leonard Digges in 1556 as being 5 perches long with a knot or mark at each perch.[4] He and earlier authors describe how the cord was to be dressed with wax and rosin to stabilise it but there are indications that even by Digges' time the cord was considered obsolete for surveying land.

The wire chains: In terms of stability, durability and accuracy chains composed of wire links were a great improvement over the cord. With the standard perch of 16½ ft. used for the measurement of land, most chains would have been multiples of this length – with 4 perches being the most common. The length of the links however was not standardised and it is clear from other descriptions that there was little or no regularisation.

Gunter's chain: The chain proposed by Edmund Gunter in 1620 was 66 ft (22 yards) long divided into 100 links; thus each link was 7.92 in. The reason for this choice of size for a link was to simplify calculation of the acreage of a piece of land.

The early definition of an acre, 22 yards by 220 yards, was also one Gunter's chain (100 links) by 10 Gunter's chains (1,000 links); the area measured in links is therefore 100 links x 1000 links = 100,000 sq. links. Thus to obtain the area in acres it is only necessary to calculate the area in square links and then move the decimal point five places. For example: 854 links x 1345 links =1,148,630 square links = 11.486 acres.

Edmund Gunter's chain was by no means the only one invented. In 1616, Aaron Rathborne[5] describes not only his own design of chain but mentions that there were several others in common use. In his *Compleat Surveyor*, Leybourn also gives a good description of Rathborne's chain.[6] This differed from Gunter's in that its overall length was only 2 rods (11 yards), with each rod divided into 10 parts called primes, which were then subdivided into a further 10 parts called seconds.

Land chains: The chain known as a land chain today is a hard wearing measure, resistant to damage, but even so surveyors' manuals insist that the

II. Of Mr. *Gunter*'s Chain.

AS every Pole of Mr. *Rathborn*'s Chain was divided into 100 Links, ſo Mr. *Gunter*'s whole Chain (which is always made to contain four Poles) is divided into 100 Links, one of theſe Links being four times the length of the other. Now, if this Chain be made according to the Statute, each Perch to contain 16 ½ Feet; then each Link of this Chain will contain 7 Inches, and $\frac{92}{100}$ of an Inch, and the whole Chain 792 Inches, or 66 Feet.

In meaſuring with this Chain, you are to take notice only of Chains and Links, as ſaying, ſuch a line meaſured by the Chain, contains 72 Chains 48 Links; which you may expreſs more briefly thus, 72, 48. And theſe are all the Denominations which are neceſſary to be taken notice of in ſurveying of Land.

For the ready counting of the Links of this Chain, there ought to be theſe Diſtinctions; namely, in the middle thereof, which is at two Poles end, let there be hanged a large Ring, ſo is the whole Chain by this divided into two equal Parts.

Fig. 7.6. Description of Gunter's chain from Leybourn's Compleat Surveyor.

Fig. 7.7. Description of Rathborne's chain from Leybourn's Compleat Surveyor. *(Leybourne used an alternative spelling without the "e".)*

I. Of Mr. *Rathborn*'s Chain.

THE Chain which Mr. *Rathborn* ordinarily uſed (as himſelf ſaith) contained in length two Statute Poles or Perches, each Pole containing in length $16\frac{1}{2}$ Feet, which is 198 Inches: Then each Pole was divided into 10 equal parts, called *Primes*; every of which contained in length $19\frac{4}{5}$ Inches. Again, every of thoſe *Primes* was ſubdivided into 10 other equal parts, called *Seconds*; ſo that every of theſe *Seconds* contained in length $1\frac{98}{100}$ Inches; ſo that the Pole, Perch, Unite, or Commencement (as he calleth it) was divided into 100 equal parts or Links, called *Seconds*.

This Chain being thus divided and marked, you have every whole Pole equal to ten *Primes*, or 100 *Seconds*; every three-quarters of a pole equal to ſeven *Primes* and an half, or 75 *Seconds*; every half Pole equal to five *Primes*, or 50 *Seconds*; and laſtly every quarter of a Pole equal to two *Primes* and a half, or 95 *Seconds*.

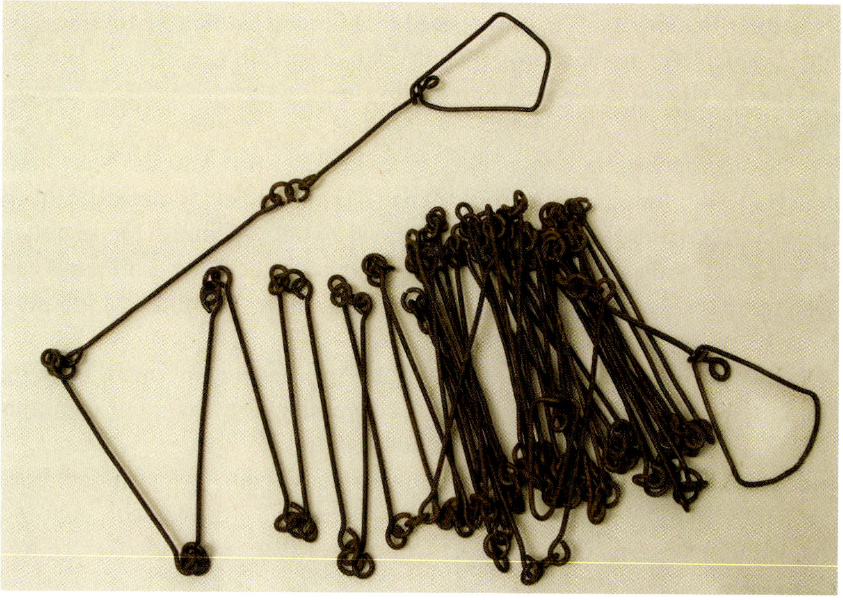

Fig. 7.8. A late 18th / early 19th century land chain. Note that this chain has no tallies – they had not been introduced at this date.

chain should be frequently checked – government surveyors were required to do this every morning before starting work. Although using a steel tape will give greater accuracy than a chain (probably by a factor of about 2) the chain has, until recent years, remained the preferred tool of the land surveyor. The quality of chains did vary. James Chesterman in 1880[7] listed best quality chains made in hardened and tempered steel, with three small links between the main links, whilst cheaper versions were iron with only two small links between the main links. The better quality versions are also of thicker gauge. A further improvement could be made by having the joints brazed shut. All modern chains have specially shaped distance markers called tallies, usually hanging from the links but sometimes inserted into the chain.

Accessories for use with chains include hanging thermometers, handles with adjustment for temperature correction and spring balance handles for ensuring correct tension.

Steel band chains: An alternative to the land chain is the steel band, marked with brass studs at the chosen divisions – links, feet or metres. It is less liable to stretch than a chain but also less robust so is principally used for checking chains. They are most often installed in an open frame (cross) rather

Fig. 7.9. Land chains from the Chesterman 1880 catalogue. They could be bought to measure in links, feet (50 or 100 ft.) or in metric (25 or 30 m.). Note the shaped tallies.

than a full case as is usual with steel tapes and occasionally are to be found with one side marked with etched measures.

Invar bands: T. Cooke and Sons in their 1913 catalogue list "Invar bands, ⅛ or ¼ in. wide, in lengths up to 500 ft."[9] With lines scribed onto a silver strip every 100 ft., invar bands were intended for purposes such as setting out baselines where the greatest accuracy was required. In use they were supported at intervals on tripods and a standard tension was applied by means of weights. Invar is thermally very stable but further steps to increase accuracy included seasoning or ageing of the metal. With such specialised use, few invar bands were sold and are therefore rarely found.

Tape measures: Used by almost every imaginable trade from carpet layer to shipbuilder, both woven linen and steel tapes were, and still are, widely used by surveyors.

STEEL BAND CHAINS.

Fig. 49. On Steel Cross.

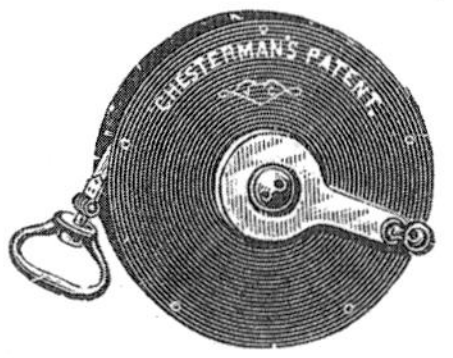

Fig. 50. On Metal Cased Reel.

The Steel Band Chain is as strong as a chain of links. It is lighter, more compact, and easier to clean. It is furnished at the extremities with brass swivelled handles. When not in use the band is coiled on either a steel cross (Fig. 49), which is included in the price, or on a metal reel with handle (Fig. 50) at extra cost, as stated on next page.

The 4-pole (66 feet) bands are divided by brass studs at every link, and numbered at every 10 links, the first and last links into 10ths. The 50 feet and 100 feet bands are divided by brass studs at every foot, and numbered at every 10 feet, the first and last foot into 10ths or 12ths. The 20-metre bands are divided by brass studs at every 5th of metre, and numbered at every second metre.

For prices see next page.

J. H. Steward, Ltd., 406, Strand, and 457, West Strand, London, W.C.2.

Fig. 7.10. Two styles of steel band chain. Although this illustration is taken from the J. H. Steward Ltd. 1928 catalogue, both chains are clearly marked Chesterman's Patent.[8]

Fig. 7.11a & b. A cloth tape in a leather case with a carrying handle attached to the back by Rabone Chesterman Ltd. (Birmingham, 1963–1990s).

The woven tape is easy to handle and maintain but until metallic thread was added into the weave its tendency to stretch limited its use for accurate surveying. The first patent for a woven tape that incorporated strands of copper wire in the warp was taken out by James Chesterman in 1842 and from that date onwards, their "Metallic Woven Tape" was marketed as a resolution to the problem. Woven linen tapes also continued to be sold with Chesterman continually trying to solve the problem of stretching and in 1889 their "Constantia" tape was introduced that was claimed to be the most stable woven tape available, produced by a special patented process. Even this, however, was not infallible. During the 19th and for much of the 20th century the tape was generally contained in a hard leather case with a fold away handle. The longest cloth tapes were 100 ft. but many lengths down to 20 ft. were available.

Surveyors' rods: Although called rods, surveyors' rods are mostly 5 or 6 ft. long lath type rules. Older examples are more likely to be two fold without a locking joint whilst more modern examples are multi-section zig-zag folding rules that do have a locking joint. They are used principally by building surveyors and architects for measuring buildings. The division into feet and inches is boldly marked with the start point on each side at opposite ends. The rigid nature of the rule allows easy measurement of heights to 12 ft. and the rod can be poked into areas otherwise inaccessible. Two part sliding versions were also offered by some suppliers.

Surveying wheels: Named thus by Bion[10] these have more names than is good for any device; the following have all been recorded: hodometer, odometer, waywiser, road measuring wheel, perambulator and there are probably others.

The surveying wheel is an ancient instrument – Hero of Alexandria in the 1st century BC wrote about an example that was fixed to a cart and had a mechanism that released pebbles to count the distance travelled. Today neat aluminium bodied versions with pneumatic tyres are still widely used in the road construction business.

The version shown by Bion has a boldly shaped wooden fork with a spoked wheel and a clock face with two hands moved by gearing. The "minute" hand revolves once each mile and records the number of poles; the "hour" hand

Fig. 7.12. Surveyors' rod by Rabone Chesterman Ltd. (Birmingham, 1963–1990s). This is the only type of British zig-zag rule that is commonly found.

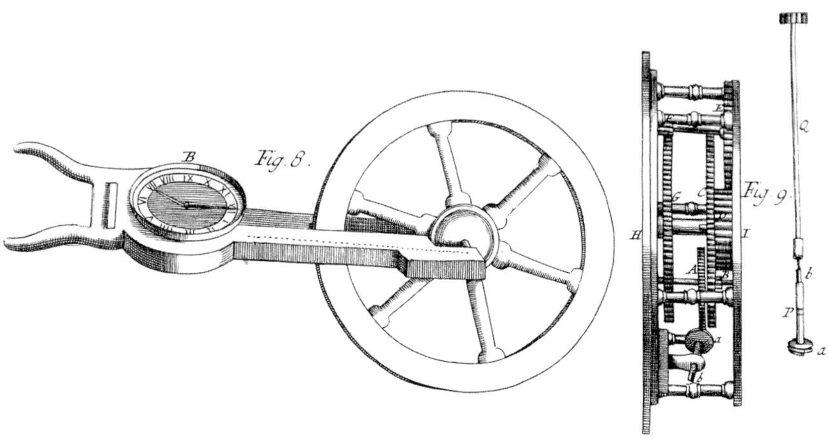

Fig. 7.13. The surveying wheel from Bion. The description is very precise, stating that the wooden wheel, tyred with iron, must be exactly 2 ft. 7½ in. in diameter so that the circumference is 8 ft. 3 in. – half a pole (rod).

286, HIGH HOLBORN, LONDON, W.C.1 157

Road Measuring Wheels (Perambulators).

P1560 to P1562

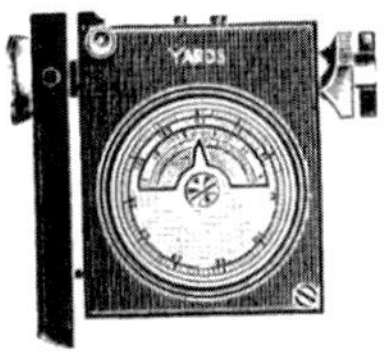

Indicator.

		£	s.	d.
P1560	**Perambulator**, for measuring roads, mahogany wheel, 2 yards circumference, with brass spokes and gun-metal tyre, with detachable handle, registering miles, furlongs, and yards; packed in strong pine case	18	0	0
P1561	Ditto, ditto, in metres	20	0	0
P1562	Ditto, ditto, in links	20	0	0

P1564 to P1566.

P1564	This **Perambulator** or Road Measurer has been designed to provide a specially light, accurate and reliable apparatus. It has a detachable handle, and the whole fits into a light pine case, which is quite portable. The wheel has a heavy gun-metal rim which is turned true, and is exactly two yards in circumference. It runs on ball bearings, and its semi-revolutions are recorded upon a very compact revolution counter, showing the number of single yards traversed by the wheel up to 99,999 yards. It forms the lightest and most simple and accurate apparatus for measuring distances on roads	8	0	0
P1565	Ditto, ditto, in metres	9	0	0
P1566	Ditto, ditto, in links	9	0	0

Fig. 7.14. Two versions of road measuring wheels are shown in W. F. Stanley's 1931 Catalogue.11 A heavy version, top, made in mahogany and recording miles, furlongs and yards and a lightweight version which records the distance solely in yards, with the register reading to 99,999 yards. Both were also available to record in metres or in links.

once every 12 miles and records the number of miles. Handsome wooden examples of this type were made by the London instrument makers throughout the 18th century though they are not common today. Although Bion's example measured half a pole, by the choice of appropriate size of wheel and gearing the measure can be output in whatever unit is required.

Architects' scale drawings

Richard Neve writing in 1726 in *The City and Country Purchaser...* gives an excellent description of scale drawings: "A draught, or draft, is the picture of an intended building described on paper; wherein is laid down by Scale and Compass the devised divisions of every room, in its due proportion to the whole building."[12]

Whether produced by hand or by computer aided design, scale drawings are the mainstay of the architectural profession, though this wasn't always the case. When a Cardiff pub was built in 1905 the four-storied building, which included an elaborate front and bar fittings, was constructed using a single sheet of drawings with the plan and elevations drawn to only a couple of different scales and a specification of six pages. To build it today, the contractor would need probably fifty sheets of drawings, drawn to a whole variety of scales including many full size details, and a 150-page bill of quantities. In the 18th century, documents for the same job would have consisted of a small sketch on the corner of a one-page description.

Drawing to scale: Some undertakings are best worked from full size drawings. The mediaeval mason needed a full size draft before stone could be cut. Evidence of the use of such full size designs for arches and other details is found in the form of inscribed lines on the walls or floors of some mediaeval churches. The practice followed by shipwrights was similar – the full size lines for the ribs of a ship were chalked onto the loft floor before being transferred to the rods used for the shaping of each piece.

On the other hand, by drawing to scale, large objects can be shown in true proportion on manageable sized pieces of paper. The ratio of reduction can be as little as 1:2 (half full size) or may be large – for example 1:10560 (six inches to the mile), a commonly used scale for maps.

Both Roman and mediaeval builders worked to scale drawings for major projects, but rules specifically divided as scales do not seem to have come into use until around 1450 in Italy and not until 100 years later in northern Europe.

Scale rules

A scale rule is marked with divisions which represent true measure but are smaller by the chosen ratio of reduction. All scales can be stated as a simple ratio, but until metrication it was not normal practice to do so. In the building trade "one eighth scale" meant ⅛ in. to 1 ft. (one eighth of an inch on the drawing representing one foot on the building, a ratio of 1:96), **not** $\frac{1}{8}^{th}$ full size (1:8). In civil and mechanical engineering general practice was to express scales in the opposite order, 30 feet (or chains) to the inch; that is one inch on the drawing represents 30 feet, a ratio of 1:360 (30 x 12) or 30 chains, a ratio of 1:23760 (30 x 66 x 12).

Modern practice is to use the scale rule both to mark the dimension on the drawing and to take off dimensions from the finished draft. The accuracy with which dimensions can be measured off from a scale drawing is of course related to the ratio of reduction used. On any scale drawing where exact dimensions are required, these will be figured (written); indeed today almost

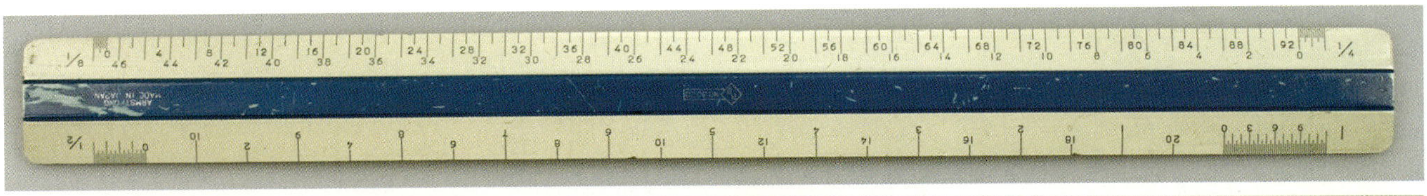

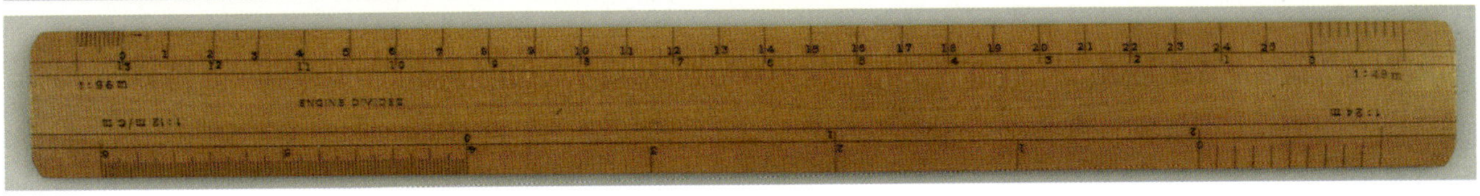

Top: Fig. 7.15. A plastic architects' scale, unmarked but dating from the 1960s. This face is marked with ⅛, ¼, ½ and 1 in. scales, the other side is marked with ¾, ⅜, 1½ in. and 3 in. scales.

Fig. 7.16. A metric conversion scale. Marked "R.I.B.A. conversion scale" this was the design approved by the Royal Institute of British Architects.

all technical drawings are marked with the injunction "do not scale from this drawing", as most modern methods of reproduction do not produce true-to-scale results.

Scales for different users

The division or ratio of scales will in most instances give some indication of the work for which they are intended.

Architects' scales: 19th century and later practice was for architectural scales to be related to the foot and marked ⅛, ¼, ½, etc.

With the change to the metric system in the building industry in the U.K. in the late 1960s, for a short while conversion scales, by which metric dimensions could be taken directly from imperial drawings and vice versa, were produced.

Scale	Ratio	Metric equivalent
1/16 to 1 ft.	1:192	1:200
⅛ in.	1:96	1:100
¼ in.	1:48	1:50
½ in.	1:24	1:20
¾ in.	1:16	–
1 in.	1:12	1:10
1½ in.	1:8	1:5
3 in. to 1 ft.		
(Qtr. full size)	1:4	1:4
Half full size	1:2	1:2
Full size	1:1	1:1

Fig. 7.17. The commonly used imperial architectural scales showing the ratios and the metric equivalents now used. (The metric equivalent of ¾ in. is not used.)

Architects' folding rules: Many rule manufacturers made folding architects' rules. Usually 2 ft. 4-fold, but also in other sizes, they were often in ivory with nickel silver fittings and, as can be seen from many of the inscriptions found, were a favourite retirement or "thank you" present.

They are recognisable by the beveling of their edges on the inside, on both edges or, more rarely, on just the outside. Usually divided with architectural scales, they are also found with surveying scales but are always listed in the catalogues as architects' rules.

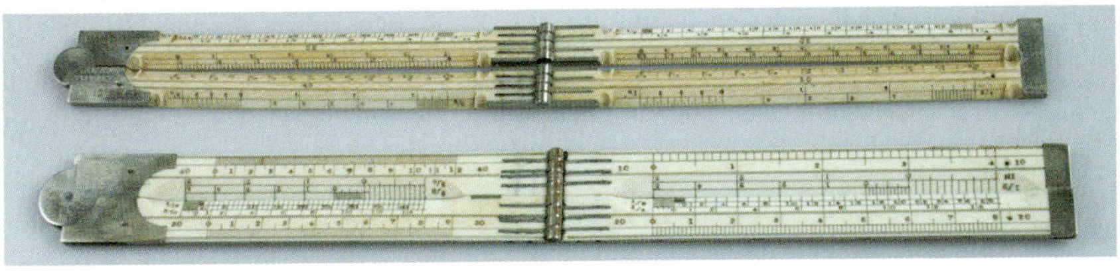

Fig. 7.18. Two ivory architects' 2 ft. 4-fold rules. Top by W. H. Harling (London, 1851–1930s) beveled on both inside and outside edges. Bottom by W. F. Stanley (London, 1853–1967), beveled on just the inside edges.

Fig. 7.19. The pins of the brick gauges are set to equate with 3 in. (one course of brickwork) at the scale marked on the face of the gauge. Also shown in this illustration is a scale rule with a brass inset, marked with smaller divisions than have been marked on the edge of the rule. These can be taken off with a pair of dividers without damaging the rule.

Brick gauges: These little drawing aids are often incorrectly identified. They were usually supplied in sets with gauges to a variety of scales and were used for setting out brickwork on drawings without the need to constantly measure.

Surveyors' scales: Surveying manuals indicate that dimensions were normally plotted to the scales shown in Figure 7.20. However the scale rules were marked as 10, 20, 30, etc. suggesting that they could also have been used at ten times these scales, i.e. 10 chains to 1 in. which equals 1:7920 etc. Less frequently 7, 8 and 9 chains to the inch were also used.

Whilst some surveyors' scale rules are marked only in chains or, very occasionally, links, it is also common to have scales marked with chains on one edge and feet on the other so the same rule can be used whether the survey was made with a 22 yd. or a 100 ft. chain. The ratio is the same, it is only the unit of measurement that differs.

Fig. 7.20. The commonly used surveying scales together with the equivalent ratios.

Scale	Ratio
1 chain to 1 in.	1:792
2 chains to 1 in.	1:1584
3 chains to 1 in.	1:2376
4 chains to 1 in.	1:3168
5 chains to 1 in.	1:3960
6 chains to 1 in.	1:4752

Fig. 7.21. Two surveyors' scales by W. F. Stanley (London, 1853–1967) from a set of six scales and offsets. The top scale is marked 40 on both edges; the bottom scale is marked 1⁄25000 on one edge and LINKS on the other.

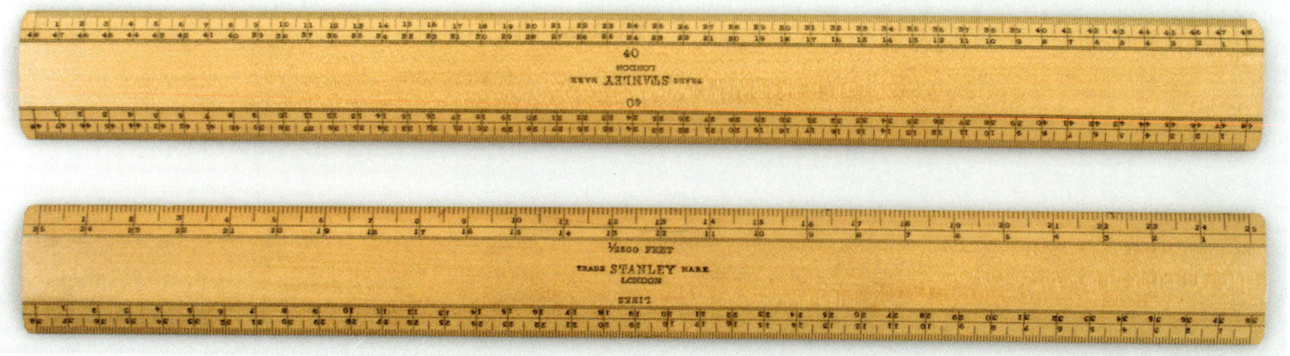

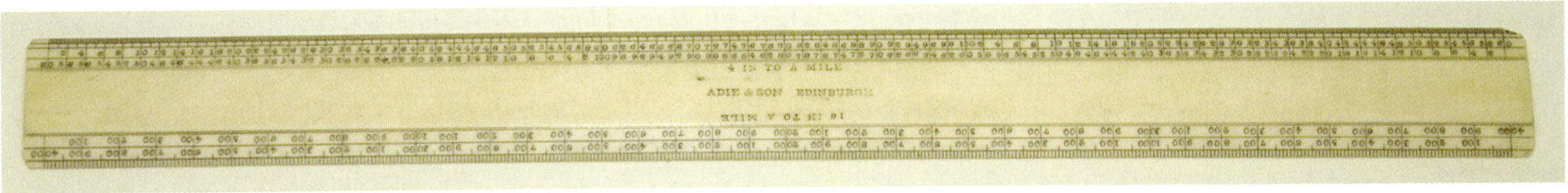

Fig. 7.22. Ivory scale rule by Adie & Son (Edinburgh, <1835–58) marked on the face only with surveyors' scales of 4 in. to the mile and 16 in. to the mile These are unusual scales.

To facilitate plotting either side of a base line some 19th century surveyors' scales are marked with a zero line, either in the centre of the scale, or one inch from an end. The scale is then numbered in both directions from this line.

Offsets: These small scale rules, usually 2 in. but occasionally a little longer, are frequently found. In use they were placed at right angles to the side of a normal 12 in. scale to measure off distances from the edge, thus mimicking the survey method where offsets from a base line are taken. They were normally supplied in cased sets of matching scales and offsets but have often become separated from their rules.

In 1849 Troughton and Simms[13] offered a system with longer offsets sliding in a groove in the long scale. Other improved offsets with a metal "T" end, giving greater accuracy, are also occasionally found.

Engineers' scales: For work related to buildings, engineers normally drew to the same series of scales as architects. However on other works they often used the surveyors' scales but translated to a scale of 10 ft. to the inch (giving a ratio of 1:120), 20 ft. to the inch (1:240), and so on.

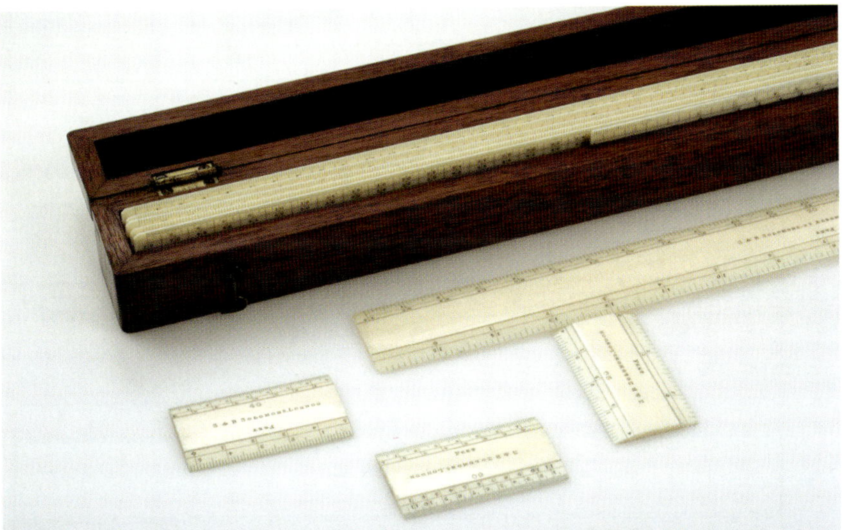

Fig. 7.23. A set of six ivory surveyors' scales with offsets marked S. & B. Solomons (London, 1840–75) contained in a mahogany box. Each scale is divided in chains and feet.

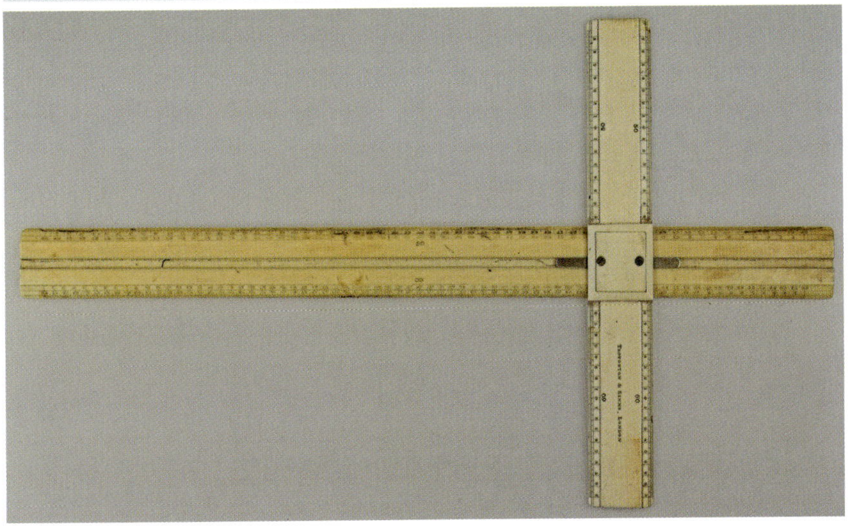

Fig. 7.24. Troughton & Simms (London, 1826–1915) rule with the offsets running in a groove.

Mapping scales: The large scale Ordnance Survey maps, widely used in the civil engineering and construction industries, were produced to the following ratios: 1:1250, 1:2500, 1:10560, which are very close to 50 in., 25 in. and exactly 6 in. to the mile. Scales reading in links, chains, feet, yards and

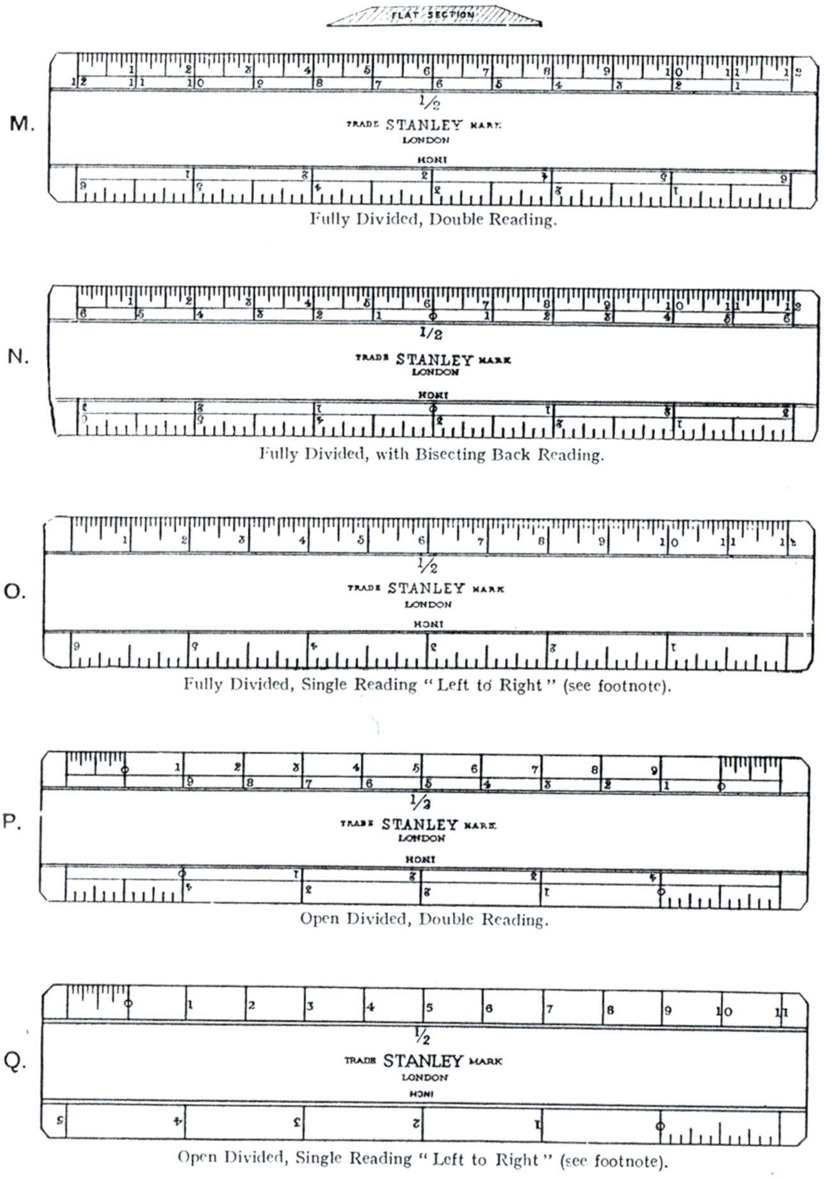

Fig. 7.25. From the 1931 W. F. Stanley catalogue section of engineers' and architects' scales.[14] *This illustration shows the different methods of dividing scales. The scales shown read from left to right but they could also be ordered reading right to left. Taking into account the methods of division and the range of ratios, materials and lengths, the variety of different scales available was almost innumerable.*

Fig. 7.26. A most unusual plotting scale, the exact usage of which has not yet been discovered. Marked "Austin's bi-lateral plotting scale" both sections are divided in the same mapping scales marked 30 and 40.

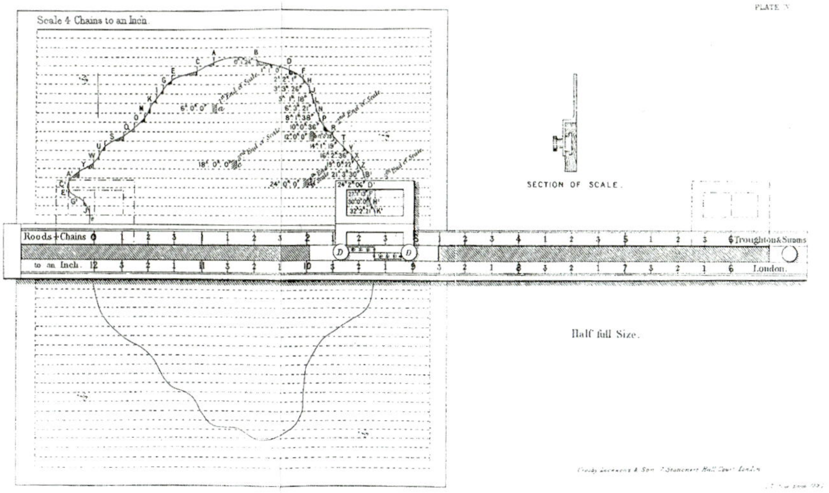

Fig. 7.27. Diagram showing the use of a computation scale from Usill's Practical Surveying, *1916.*[15]

metres for all of these ratios are to be found. These are marked thus: 1:2500 Links, Feet, etc.

Other map makers did not limit themselves to these ratios so scale rules in a wide variety of other ratios are also found.

Computation scales: These consist of a boxwood stock with a sliding cursor offset to the edge. They are used to measure areas on maps by the simple but effective means of overlaying the map with parallel lines and thus dividing the area into strips of a standard width.

The total length of these strips is then measured using the computation scale which is divided to give a direct reading of the area. The scale of the plan, the width of the strips and the scale of the computing rule must of course match or a correction factor will be needed.

Special purpose scale rules: Specially divided scales useful for repetitive setting out of particular objects such as brickwork, joists, and even telephone exchange equipment were occasionally made but such scales are rare.

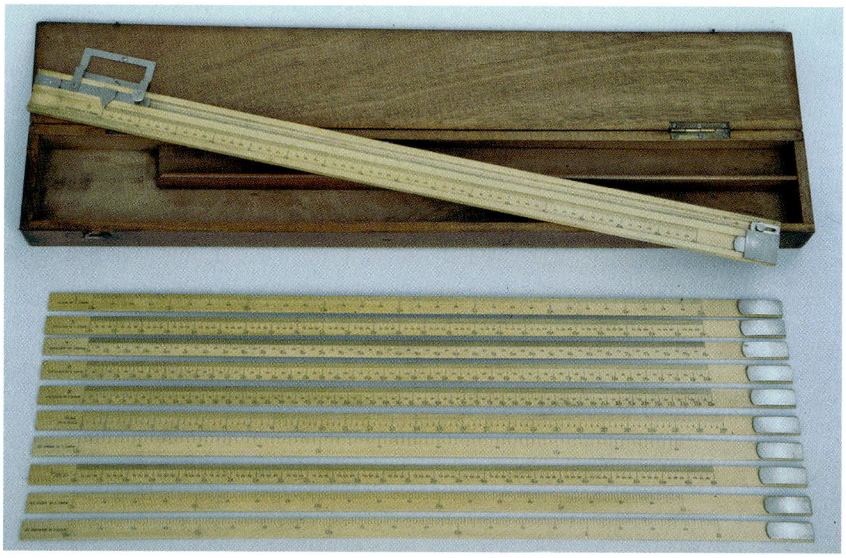

Fig. 7.28. An anonymous computation scale with 11 different mapping scales. Some are defined as chains in one link, some as the standard mapping scales such as 1/25000 and 6 in. to the mile.

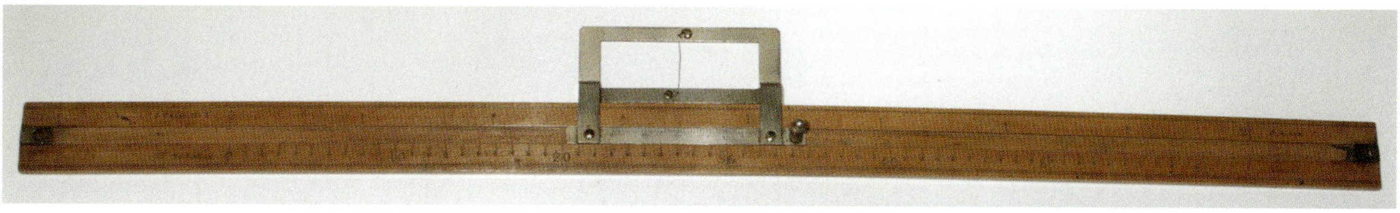

Fig. 7.29. A computation scale by Charles Baker (London, 1851–1947>). The scale is marked for specific scales and it is necessary to match the scale of the map to the scale on the computation scale. This scale is marked for 1 chain to 1 in. and 25 in. to 1 mile.

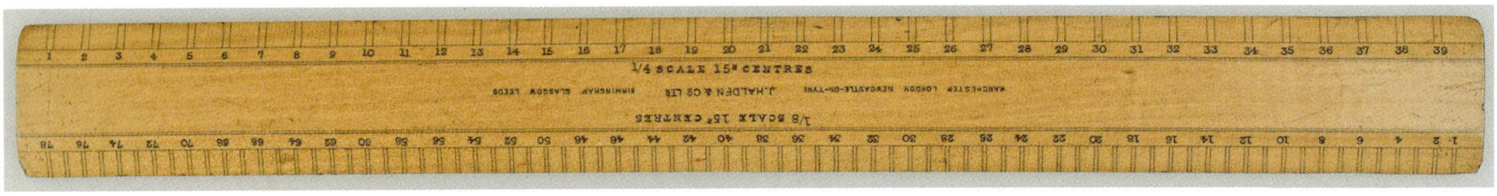

Fig. 7.30. A scale designed to set out rafters on drawings by J. Halden & Co. Ltd. (Manchester, 1877–1969).

Cartographic scales: These are substantial brass or nickel silver scales, up to 4 in. wide and up to 4 ft. long, the face marked with diagonal scales. They should be contained in a wooden case and were used for setting beam compasses when plotting surveys.

Logarithmic scales: These are used for log scaled graphs. The 12 in. scale will usually have a double cycle on one edge and a single on the other.

Steam engine indicator scales: Three or four inch long scales marked 100, 150 etc. to the inch, these were used for interpreting the diagrams made by the indicator instrument. Originally included in the case with the instruments these are now often separated.

Face marked scales

Before the mid-19th century the scales were marked onto the face of the rule so that it was necessary to use dividers to transfer the required dimensions to the drawing. However, whilst the dimensions might be transferred directly onto the drawing from the scale rule it was common practice to mark out a scale on the paper when the drawing was started. This method had the advantage that as the paper expanded and contracted with changes in humidity, the drawn scale would always be correct in relation to the drawing.

The face marked scales of the 18th century and early 19th century are mostly "lines of equal parts", i.e. an inch divided into 30, 35, 40, etc. parts, or diagonal scales with the end division (an inch, half an inch etc.) divided as a diagonal (nonious) scale.

Fig. 7.31. Face marked scales. From the top in both illustrations: A architects' scale by James Gargory (Birmingham, 1835–62); an anonymous architects' scale with a most unusual layout of the scales; a scale by W. Elliott (London, 1817–1853) marked with architectural scales and a protractor on the face and chords, surveying scales and a diagonal scale on the reverse; another unusual scale by J. Tree (London, 1832–95) marked with surveying scales, both edge and face marked on one face but only edge marked on the other. This scale is 10 3/8 in. long.

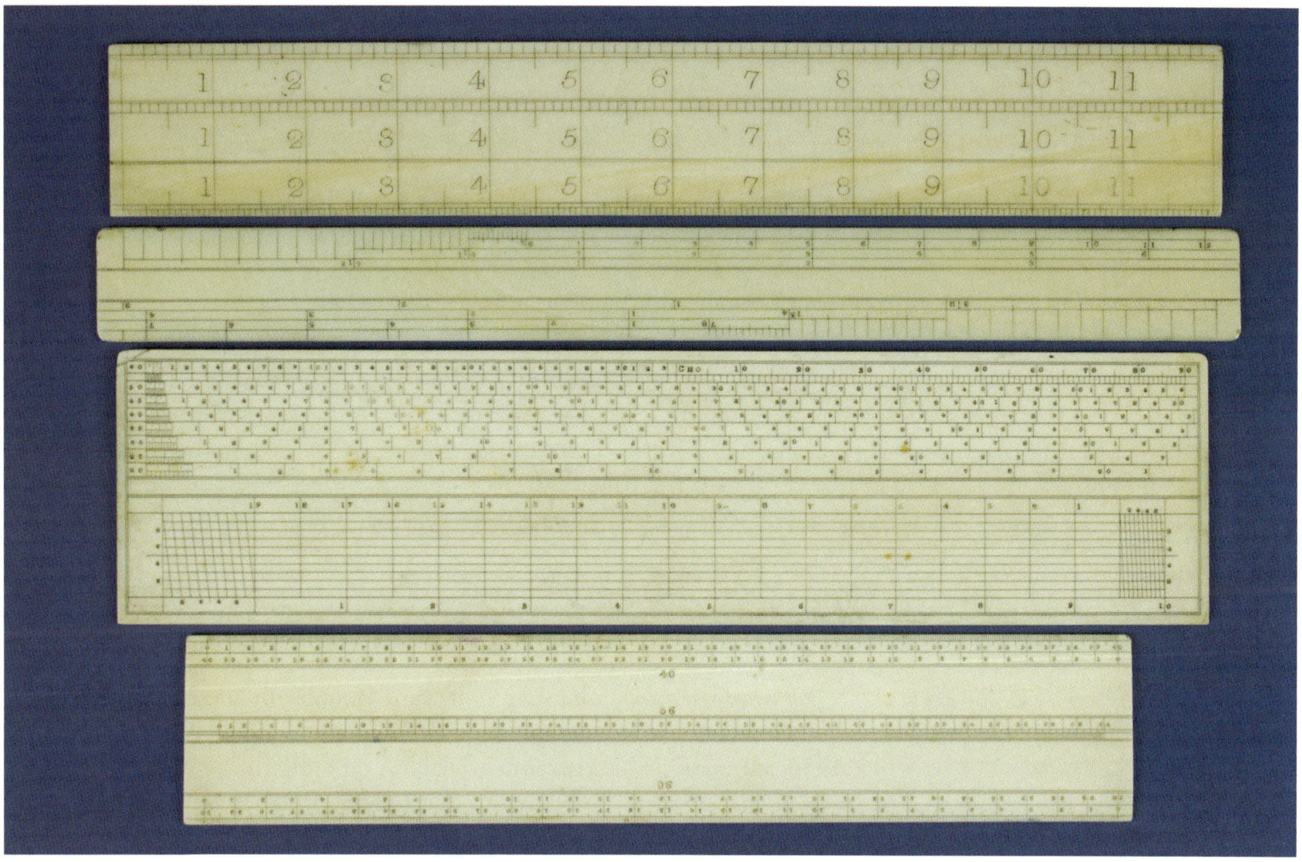

Fig. 7.32. Reverse side of the scales shown in Fig. 7.31.

Edge marked scales

By the 1840s the professional user was working with bevelled scales marked on the edge. These must have given a considerable improvement in speed and accuracy. However, as most professionals would have needed eight or more different scales requiring four or more rules, this was still not an entirely satisfactory solution. For architects the development of the "universal" scale around 1870 and the Armstrong scale which was a design that combined eight different scales on one rule, each edge having two compatible scales, such as ⅛ in. to 1 ft. and ¼ in. to 1 ft. on the same edge, provided some solution to this problem, although surveyors and engineers continued to use separate scales of the inch-divided type.

Materials and construction

Most of the materials used for other types of rule are also used for scale rules with boxwood being the most common. Ivory scales are also frequently found, but as explained in Chapter 2 have a problem with movement and curling.

Fig. 7.33. Two architects' Armstrong scales. Top: Boxwood pre-metric scale, unnamed. Bottom: Plastic metric scale by Blundell Harling (Weymouth, 1961–present).

Fig. 7.34. Scales in four different materials. From the top: Nickel silver by John Crichton (London, 1831–65); steel by James Chesterman & Co. (Sheffield, 1859–1963); ivory by S. & B. Solomons (London, 1840–75); boxwood by W. F. Stanley (London, 1853–1967).

Boxwood with celluloid edges became popular in the early years of the 20th century and continued in manufacture for many years. During the 1950s all-plastic scales became available, some of which were good quality and engine divided, but most were impressed and lacking the fineness and accuracy of scales made of other materials.

Metal scale rules are occasionally found. Unless lacquered, brass scales will mark drawings, a problem that also occurs with steel rules. For the greatest possible accuracy, nickel silver (electrum) was often recommended.

Cheap sets of cardboard rules, intended for educational use, are quite common.

Engine divided: Like all other rules, scales were originally divided by hand but from 1860 W. F. Stanley and other principal makers developed dividing engines which both speeded and improved the accuracy of the process.

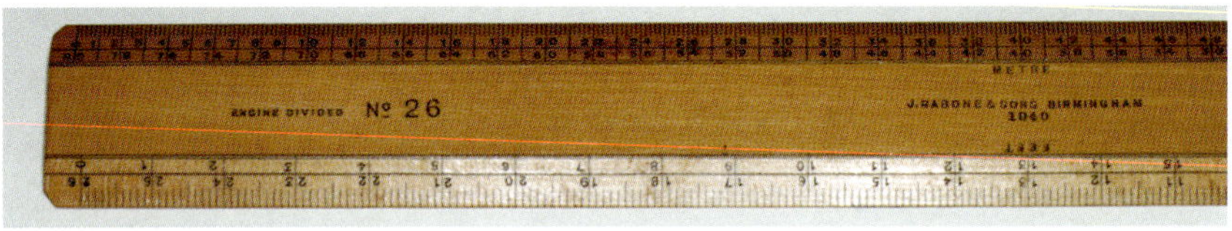

Fig. 7.35 A surveying scale by John Rabone & Sons, dated 1940, marked ENGINE DIVIDED. This scale is a 1⁄2500 scale marked in feet and metres.

Size and form

For architectural and surveying work, the 12 in. scale was the most popular. The 6 in. scale is convenient for the pocket but is too short for general drawing work. This was the size included in the smaller sizes of instrument case but 5 inch, 4 inch and occasionally even shorter scales were made to fit the smallest sets. Longer scales of 18, 24 and 30 inches were also made but are rare.

Oval: Easy to pick up and with four useable edges this shape is preferred by architects and others likely to need several scales on one drawing.

Flat: Having only two bevels and with a plain back, these were mostly used by surveyors. They work well with offsets but have only two useable edges.

Reverse bevel: These are easier to pick up but also have only two useable faces. They are rare.

Fig. 7.36. A boxed set of scales and offsets by W. F. Stanley (London, 1853–1967).

Triangular: Although they sit firmly on the paper, these are awkward for the user who must lean over the scale to see the edge. Made from the 1870s onwards in boxwood, ivory and later, of composite construction. Plastic versions available today are "waisted" to improve visibility.

Boxed sets of scales

From mid-19th century it became usual for makers to supply sets of scales in mahogany boxes – sets of six were the most common although the largest might contain as many as eighteen or even more. Sets intended for surveyors also included matching offsets.

Parallel rules

A common task was drawing parallel lines and the simplest tool for this was a parallel rule. Those intended for architects and surveyors usually had only two sections. They were normally available in ebony and ivory and in the 20th century in various man-made materials. The lengths ranged from 6 in. to 24 in. or even more.

Rolling rules

Where no T-square is available, the rolling rule is something of a substitute – many are plain but rules with edges marked with inches and intended for drafting were available. Such rules, which were included in the larger sets of drawing instruments but were also sold separately, were usually made with an ebony stock and ivory edges marked with inches and/or drawing scales. The most sophisticated examples included indicating wheels showing how much

Fig. 17.37. A pair of matching ebony and brass parallel rules, unmarked.

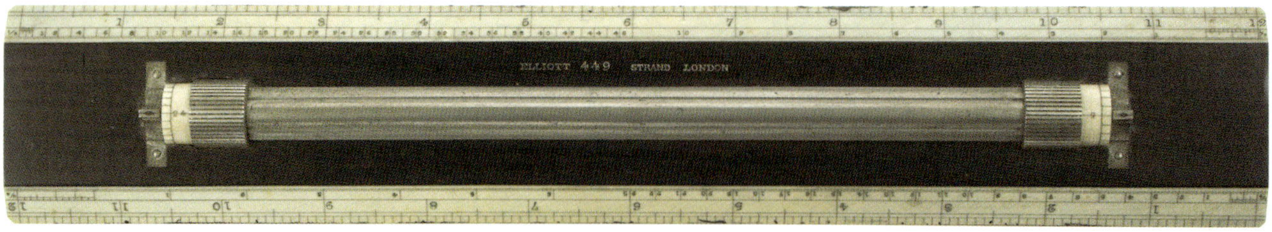

Fig. 7.38. An ebony rolling rule with bevelled ivory edges by Elliott Bros. (London, 1864–1913>) with architectural scales marked on the edge.

the rule had been moved. However, most rolling rules and bar parallel rules were intended for navigational chart work and not for drawing.

Combined protractor and scales

The boxwood or ivory scale rule included in an 18th or 19th century set of mathematical (drawing) instruments is often combined with a protractor. Typically these combined rules are 6 in. by 1¾ in. and are marked with five or six scales (lines of equal parts) together with one or more diagonally divided scales. There may also be a line of chords or tangents. The more sophisticated versions also include either a parallel or a rolling rule.

286, HIGH HOLBORN, LONDON, W.C.1 237

Parallel Rules.

P2615.

		6-in.	9-in.	12-in.	15-in.	18-in.	24-in.
P2615	**Bar,** ebony, well made	3/6	6/6	7/6	9/-	11/6	16/6
P2616	Ditto, transparent celluloid	7/6	—	—	—	—	—
P2617	Ditto, ivory	11/6	—	—	—	—	—

P2620 and P2621.

		6-in.	9-in.	12-in.	15-in.	18-in.	24-in.
P2619	Parallel **Rolling**, ebony, with brass bridge	16/6	18/6	21/-	25/-	30/-	40/-
P2620	Ditto, with divided celluloid edges and brass bridge	—	24/-	30/-	36/-	45/-	55/-
P2621	Ditto, ditto, electrum bridge	—	25/-	31/6	38/-	47/-	57/6

P2620 or **P2621**, with best African ivory, instead of celluloid edges, at an extra cost according to length.

P2622.

		9-in.	12-in.	15-in.	18-in.	24-in.
P2622	Parallel **Rolling**, white opaque celluloid, divided edges, gun-metal bridge, in mahogany case	35/-	42/-	50/-	60/-	72/-

P2623 to P2625.

		6-in.	9-in.	12-in.	15-in.	18-in.	24-in.	30-in.	36-in.
P2623	Parallel **Rolling**, solid brass, in mahogany case ..	30/-	35/-	40/-	45/-	50/-	60/-	95/-	120/-
P2624	Ditto, solid bright gun-metal	32/6	37/6	42/-	47/6	55/-	67/6	100/-	130/-
P2625	Ditto, solid electrum, India pattern	—	37/6	42/-	47/6	55/-	67/6	100/-	130/-

These are strongly recommended. They are made entirely by machinery and supplied largely to the Government. 18, 24, 30 and 36-inch have lifting knobs.

These Rules may be divided to any scale at an extra cost according to length.

For Field's Nautical or Chart Parallels, see page 129; Sectioning Parallel, next page.

Fig. 7.39. The listing of bar parallel and rolling parallel rules available in the 1931 W. F. Stanley catalogue.

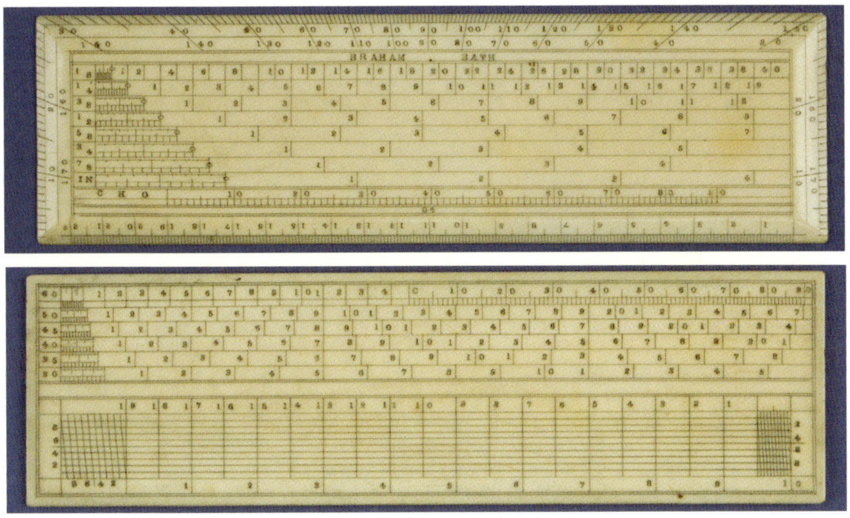

Fig. 7.40a & b. A 6 in. combined protractor and scale by Braham Bros. (Bath, <1852 >).

The combined protractor and scale certainly had a long currency. Although by the early in the 19th century it was obsolete for professional drawing office use, like the sector, it continued to be included in pocket cases into the 20th century and was still listed by W. F. Stanley in 1931.[16] However by then it had been relegated to field use by military officers.

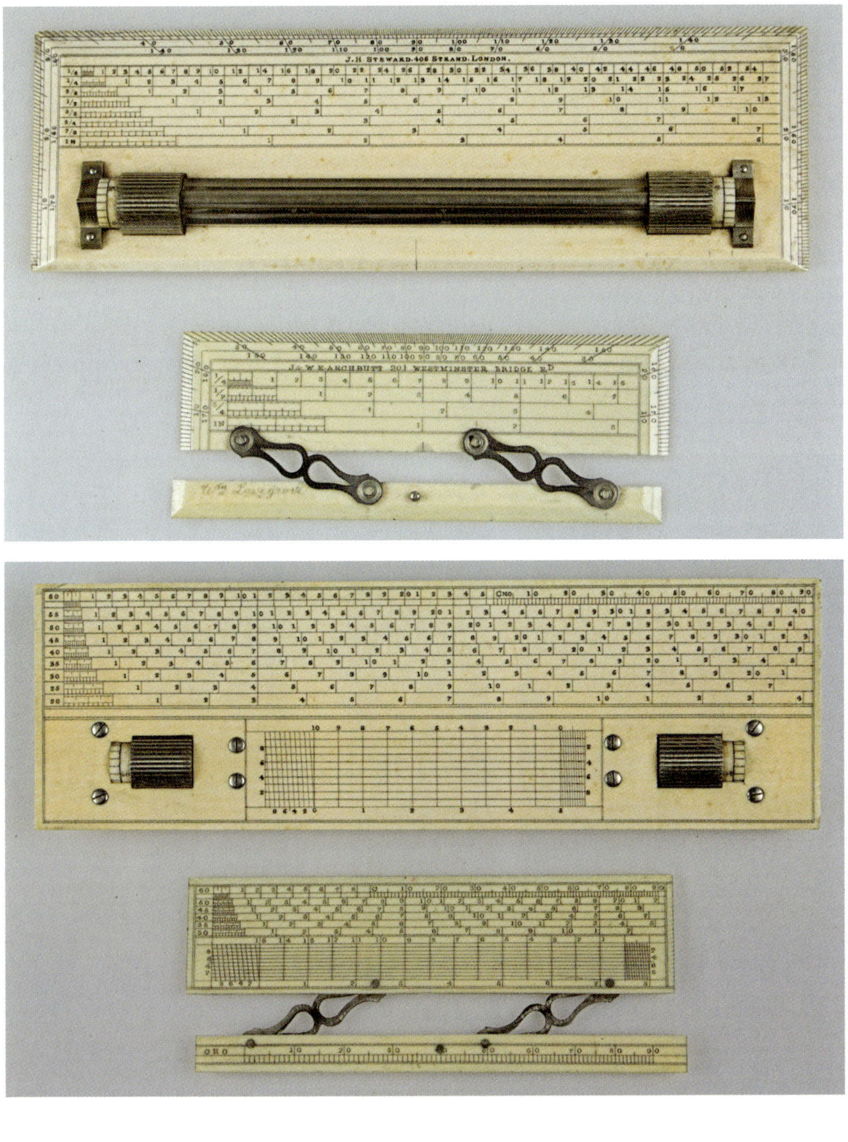

Fig. 7.41a & b. Top: An ivory and nickel silver scale with protractor and architectural scales on the front face and surveying scales and a diagonal scale on the reverse, combined with a rolling rule by J. H. Steward (London, 1867–1982). Bottom: An ivory and nickel silver 5 in. long protractor and scale combined with bar parallel by J. & W. E. Archbutt (London, <1880>). This was probably originally contained in a pocket set. Both these rules are an unusual combination.

Fig. 7.42. An unnamed late 18th or early 19th century étui containing an ivory sector, parallel rule and scale rule.

Pocket sets

Pocket sets of drawing instruments, known as étui, often included scales, parallel rules and sectors, though not all sets included the full range. According to Hambly, the earliest of these date from the 17th century and they continued to be made into the 20th century.[17] Often contained in sharkskin cases, the rules were only 6 in. long.

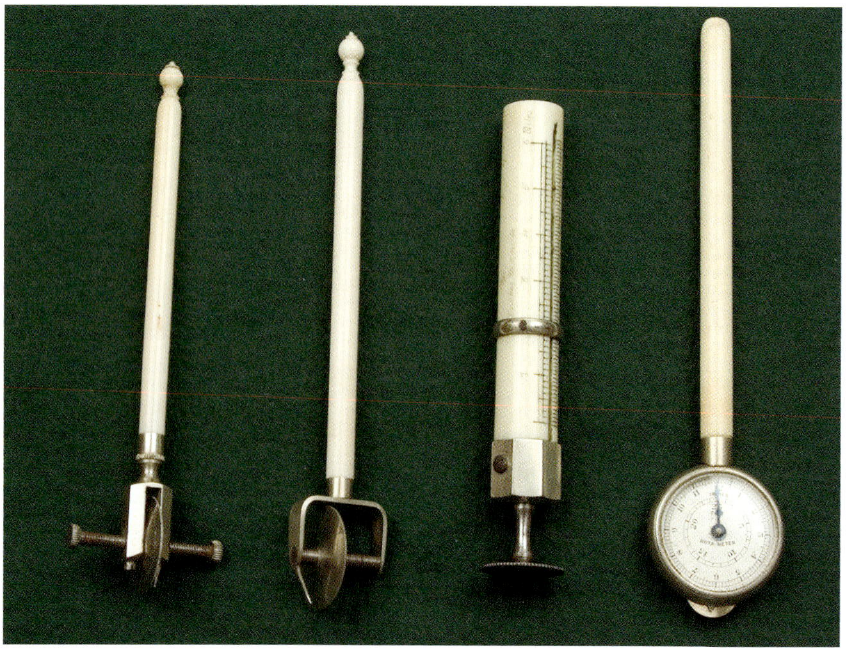

Fig. 7.43. Four different styles of opisometer. The two on the left can be used on any scale of map. The second from the right is marked 6 inches to a mile and can only record off maps of that scale. The one on the right, which was called a rotameter, is marked in inches and feet. It has two hands with one recording inches and the other feet.

Fig. 7.44. Two watch-type map measures. Left: Double-sided measure with miles, kilometres and nautical miles on one side and centimetres to kilometres and inches to miles or verstes (a Russian unit of measurement equal to 3,500 ft.) on the other. There is no maker marked. Right: Sold Keuffel & Esser Co., New York, although the movement is marked "Switzerland", the three dials register 100 ft. in 10 ft. increments, 10 ft. in 1 ft. increments and 12 in. and 1/8 in. increments on the dial and the measuring wheel is marked in 1/32 in. divisions.

Mechanical map measurers

Opisometer: Intended for measuring distances on maps and large drawings – which may, in the case of maps, be along a less than straight road – there are two main forms. The more simple type consists of a wheel running along a threaded axle held in a fork fixed to a handle. The wheel is run along the line to be measured and then the same distance is compared to the scale at the edge of the map by rewinding the wheel along that scale. These are suitable for maps at any scale. More sophisticated examples are divided on the wheel to give a measure in feet and inches. There are also versions that are designed for maps of a specific scale.

Rotameter: The alternative form of the opisometer, known as a rotameter, consists of a small wheel connected with gearing to a hand on a watch type face. Some of these are cheaply made items with little appeal but there are also finely crafted instruments, occasionally with inter-changeable disks allowing use at differing scales.

References

1. Usill, George Wm., *Practical Surveying*, 11th ed. London, 1916.
2. Leybourn, William. *The Compleat Surveyor.* 5th ed. London, 1722. (1st. ed. 1653.)
3. Bion, Nicholas, *The Construction and Principal Uses of Mathematical Instruments.* Paris 1723. Translated by Edmund Stone; ... *To Which are Added, the Construction and Uses of Such Instruments as are Omitted by M. Bion* by Edmund Stone. Second edition; ... *To which is added a Supplement Containing a Further Account of some of the Most Useful Mathematical Instruments as Now Improved.* Supplement by Edmund Stone, London, 1758, reprinted Mendham, NJ, 1995.
4. Digges, Leonard, *A Boke Named Tectonicon*, London, 1556.
5. Rathborne, Aaron, *The Surveyor,* London, 1616.
6. Leybourn, *op. cit.*
7. James Chesterman & Co. Ltd., *Catalogue of Measuring Tapes ...*, Sheffield, 1880.
8. J. H. Steward Ltd., *Catalogue of Surveying and Drawing Instruments*, London, 1928.

9. T. Cooke & Sons, *Catalogue,* London, 1913.
10. Bion, *op. cit.*
11. W. F. Stanley, *Catalogue of Surveying and Drawing Instruments*, London, 1931.
12. Neve, Richard, *The City and Country Purchaser, and Builder's Dictionary, or the Compleat Builder's Guide*, London, 1726, reprinted New York, 1969.
13. Troughton & Simms, *Catalogue,* London, 1849.
14. W. F. Stanley, *op. cit.*
15. Usill, *op. cit.*
16. W. F. Stanley, *op. cit.*
17. Hambly, Maya, *Drawing Instruments 1580-1980*, London, 1988.

Chapter 8
Engineers and mechanics

The Improved Routledge's Rule, a true ready and convenient means of making all measurements and calculations, which no Practical Engineer or other Mechanic should be without. Rabone, ca. 1867.[1]

Although slide rules for the Customs & Excise and for carpenters and for timber measurements had been available since the 17th century, the first slide rule specifically intended for use by "engineers" was that designed by James Watt working with Matthew Boulton at the Soho Manufactory in Birmingham in the 1770s which is known as the Soho rule.[2]

From this developed the so-called engineers' rules intended to assist engineers with their calculations. The basic and most common form is a 2 ft. 2-fold rule with a logarithmic slide and a number of gauge points or constants that are specifically of use to an engineer. The first rule of this type was designed by Joshua Routledge and was not only the first but was clearly the inspiration for the majority of the later designs.

Routledge and his engineers' rule

Joshua Routledge (1773–1829) followed a career which is best described as mill engineer. By 1808 he had designed the rule which bears his name.[3] An early example of the rule, containing a boxwood slide, by Alexander Wellington, Crown Court, London has been recorded – he may well have been the original maker of the rules. In 1824 Routledge took up a job in Warsaw, Poland, supervising the erection of a steam powered mill. He died there in 1829. He does not seem to have protected his design with a patent and probably even before his departure to Poland the rule was being pirated by other rule makers. During his lifetime Routledge published small books of instructions for his rule and after his death, further versions of these were printed and distributed by various rule makers.

Routledge's rule had a long commercial life and was sold for more than a century – it was still available in 1920, with his name and the description "ENGINEER" and usually, but not always, "BOLTON". This marking was even included on several types with considerably modified designs. This is perhaps a commentary on both the rule making trade (who were always prone to slavishly copying from one rule to another) and to the standing of the Routledge rule.

Routledge's rule

The layout of the rule does not seem to vary greatly from maker to maker, though some information is omitted by later makers.

On one face of the rule there are four lines of logarithms, A, B and C consisting of two cycles and D of a single cycle, i.e. starting at 1 on the left, finishing with 10 on the right. (This was the arrangement developed by Boulton & Watt and known as the Soho configuration.) This D line was intended for calculations relating to engineering problems, particularly in steam engine design, and replaces the girt line (Coggeshall's line) on a carpenters' rule.

Below these lines, starting on the upper leg and continuing on the lower

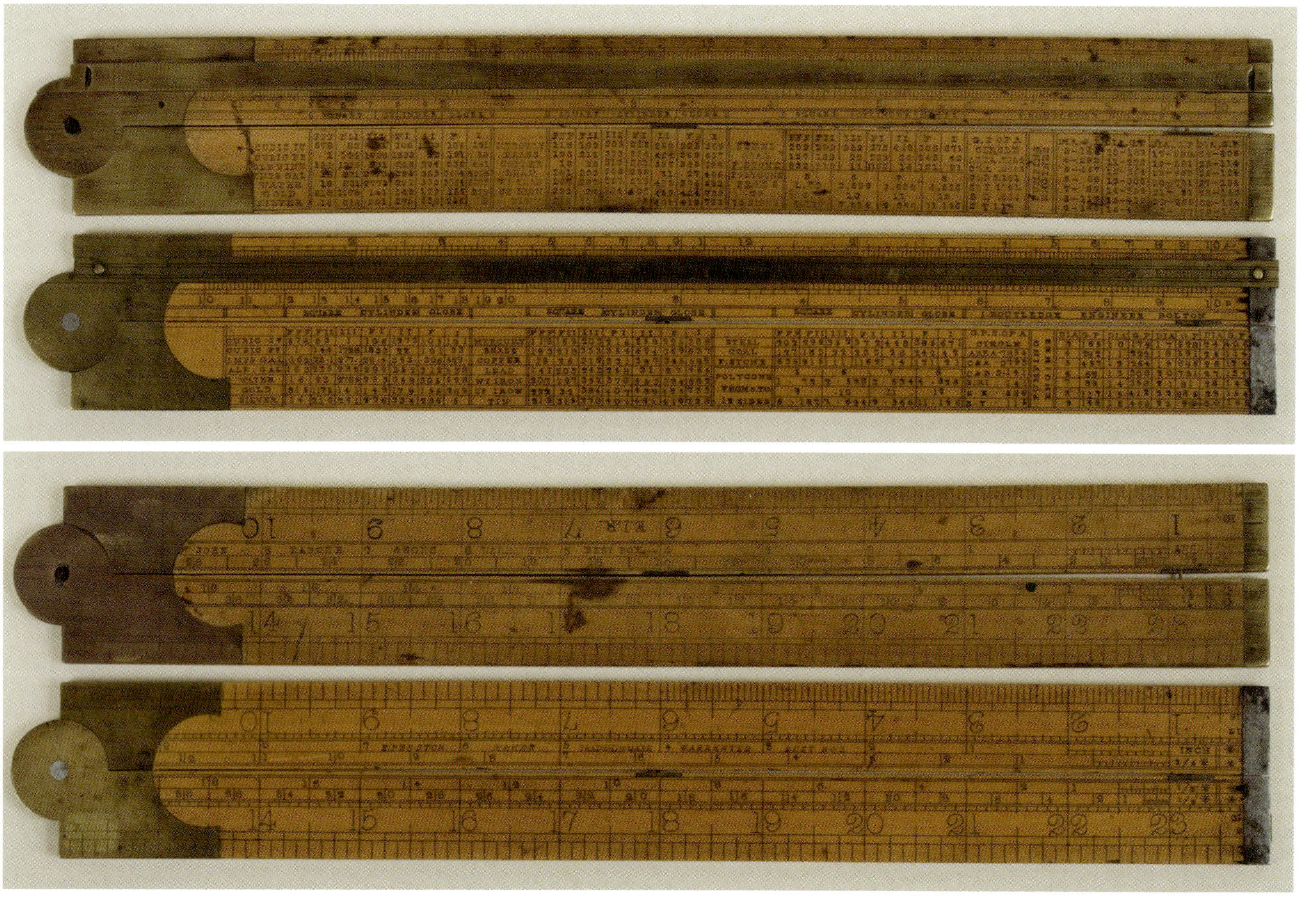

Fig. 8.1a & b. Two Routledge rules. Top: Made by John Rabone & Sons (Birmingham, 1877–1963); Bottom: by Edward Preston & Sons (Birmingham, 1864–1932). There is one difference in the gauge points on these two rules; on the Rabone rule, the gauge point for old wine gallons has been included whereas on the Preston rule this has been replaced by the gauge point for ale gallons. Both of these rules date from after 1824 as they both show the gauge points for imperial gallons, but the old measures have still been included.

leg, are columns of tables. They are in sets of three with each set then subdivided. The first set is headed SQUARE, which is subdivided into FFF (being a cubic solid where all dimensions are in feet), FII (the same but where one dimension is in feet and two are in inches), III (again the same but all three dimensions are in inches). The second set is headed CYLINDER with the subdivisions FI (a cylinder where the length is in feet, the diameter in inches) and II (a cylinder where length and diameter are both in inches). The third set, GLOBE, is divided into F and I (spheres where the diameter is in either feet or inches).

These tables give the conversion gauge points for a variety of materials for the specified geometric solid and also for a number of different types of gallons. The gauge points, often described on these rules as divisors, are intended to simplify calculations by performing some parts of the calculation before the slide rule is used. With fewer movements of the slide greater accuracy could be achieved. It is of note that there are no decimal points shown on the rule. The user of the rule was expected to know the magnitude of the calculation.

The gauge points for wine gallons (231 cu. in.) in a cubic solid measured

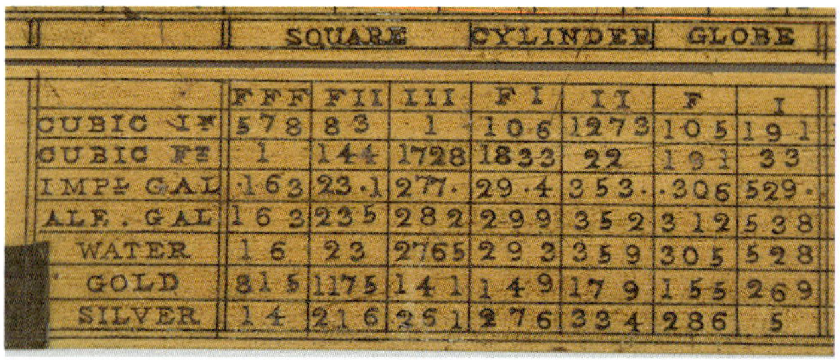

Below left: Fig. 8.2 The first table of gauge points on the Routledge rule by Edward Preston & Sons shown in Fig. 8.1.

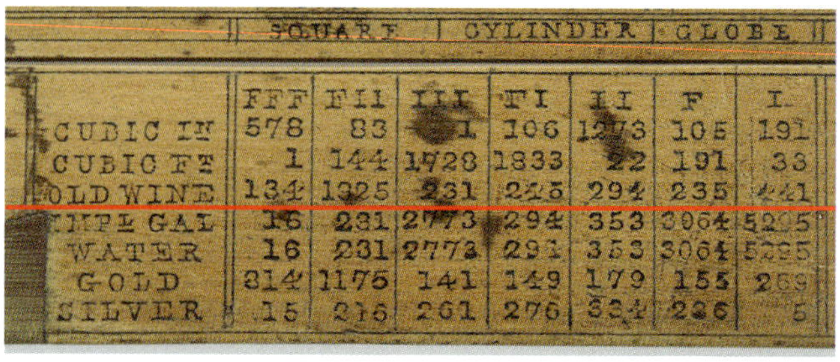

Below right: Fig. 8.3 Gauge points for old wine gallons from the rule by John Rabone & Sons.

SQUARE			*CYLINDER*		*GLOBE*	
FFF	*FII*	*III*	*FI*	*II*	*F*	*I*
134	*1925*	*231*	*245*	*294*	*225*	*441*
0.1337	*19.25*	*231*	*24.5*	*294.1*	*0.255*	*441.1*

Fig. 8.4. An explanatory diagram of the gauge points for old wine gallons (underlined in red in Fig. 8.3), showing the actual values including the decimal points. In the top row is the geometric solid that is to be calculated. The second row gives the choice of combinations of feet or inches for the dimension of the solid, for example FII indicates one dimension is in feet and the other two in inches. In the third row are the gauge points as given on Routledge's rule. The bottom row gives the actual values including the position of the decimal point, which is not shown on the rule. The slide rule user always needed to take care in placing the decimal point – the user of these gauge points would have needed to be doubly careful.

in feet (III) was calculated thus: 12 x 12 x 12 ÷ 231 (1728 ÷ 231) = 7.4805. However the gauge points are all expressed as reciprocals, so the gauge point 7.4805 is expressed as 1 / 7.4805 which is 0.1337 (134 on the rule).

In the booklet sold by John Rabone & Sons to accompany the Routledge rule, published around 1867,[4] there are gauge points listed for a considerably greater number of materials than appear on the rules. They are also set out in the form of a continuous table, so it would appear that the division into three on the rules was purely in order to fit the table onto the rule.

GAUGE POINTS FROM ROUTLEDGE'S RULE.

	SQUARE.			CYLINDER.		GLOBE.	
	FFF	FII	III	FI	II	F	I
Cubic In......	578	83	1	106	1273	105	191
Cubic Ft.	1	144	1728	1833	22	191	33
Old Ale Gal. ..	16	235	282	299	352	312	538
Old Wine Gal.	134	1925	231	245	294	235	441
Imperial Gal...	16	231	2773	294	353	3064	5295
Water	16	231	2773	294	353	3064	5295
Gold	814	1175	141	149	179	155	269
Silver	15	216	261	276	334	286	5
Mercury	118	169	203	216	259	225	389
Brass	193	218	333	354	424	369	637
Copper	18	26	312	331	397	345	596
Lead	141	203	243	258	31	27	465
Wt. Iron......	207	297	357	378	453	394	682
Ct. Iron	222	32	384	407	489	424	733
Tin	219	315	378	401	481	419	723
Steel	202	292	352	372	448	385	671
Coal..........	127	183	22	233	28	242	42
F. Stone	632	915	11	1162	14	121	21
Oak	174	252	303	320	386	382	578
Mahogany	15	217	2605	276	333	286	49
Box	155	243	269	31	342	296	512
Red Deal	242	351	422	458	539	461	805
Marble	591	85	102	116	13	113	195
Brick	795	115	138	147	176	152	263
Oil	174	25	301	319	383	332	574
Bees' Wax	16	231	278	294	355	306	53
Sulphur	8	115	138	146	126	153	264
Alcohol	193	278	333	354	424	369	637
Air	128	1843	22118	2347	28162	244	42243
Malt Bushels..	125	179	2150	2276	2378	267	41

Fig. 8.5. A table of Routledge's gauge points taken from the ca. 1867 Rabone printing of the Instructions for the use of the Practical Engineers' & Mechanics' Improved Slide Rule... *It is interesting to note the choice of materials for which gauge points are given such as Mahogany, Box, Marble, Brick, Bees' Wax and Sulphur. F. Stone refers to free stone.*

The next table to be found on the rule was labelled POLYGONS FROM 5 TO 12 SIDES. This gives the area enclosed in an equal sided polygon when the length of the sides is 1.

Fig. 8.6. The table for polygons from the Edward Preston & Sons rule.

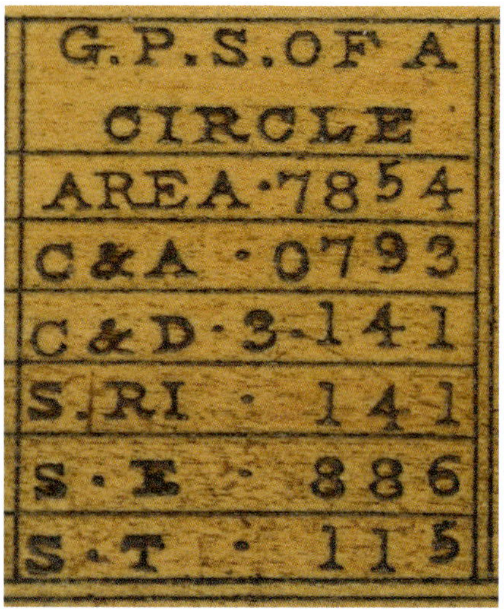

Fig. 8.7. The column for G.P.S. of a circle from the Edward Preston & Sons' rule. Note that once again, some of the decimal points are omitted. The minor difference from those given in the text in some of the values is due to changes in the way decimals are rounded.

There is then a column titled G.P.S. OF A CIRCLE. This gives some gauge points of a circle:

Area of a circle where diameter is 1 (0.7853)

Area of circle where circumference is 1 (0.0795) marked C & A

Circumference of circle where diameter is 1 (3.141) marked C & D

Diameter of a circle to contain a square of side 1 (1.411) marked S.R1

Side of square of area equal to the area of a circle of diameter 1 (0.886) marked S.E

Diameter of a circle within which there is an equilateral triangle of side 1 (1.155) marked S.T

The last table is titled PUMPING ENGINES and sets out the gauge points for calculating the diameter of a steam cylinder needed to power a pump of the specified diameter. When Routledge designed his rule the steam pressures used were low – his gauge points are based on a fixed 7 lb. per sq. in. and intended for beam pumping engines where the steam cylinder and the pump were equidistant from the bearing.

The calculation to find the required diameter of the steam cylinder is the gauge point, G. P, of the diameter (DIA) of the pump multiplied by the lift in yards.

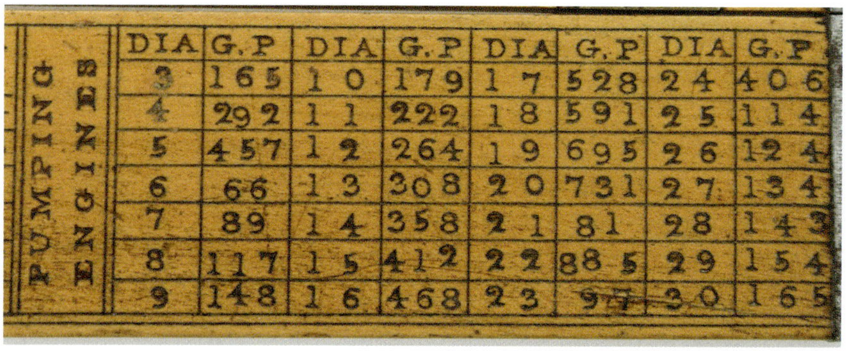

Fig. 8.8. The table for calculating the diameter of steam cylinders from the Edward Preston & Sons rule.

On the reverse side of the rule are a line of measure with the inches divided into 1⁄16ths of an inch and four drawing scales, inch, ¾ in., ½ in. and ¼ in.

There was also a line of a foot divided into 100 parts on the edge of the rule – this made calculation in feet very simple and accurate even if the dimensions were only a part of a foot.

The reverse of the brass slide, as with some other rules, is marked with a line of inches enabling it to be used to extend the rule to 3 ft. length.

Although the majority of engineers' rules are 2 ft. 2-fold, some 4-fold rules have been recorded.

Fig. 8.9a & b. A 2 ft. 4-fold ivory Routledge rule made by Rabone & Sons (Birmingham, 1877-1963). The 6 in. slide would seem to be more of a novelty for gentlemen's pockets than a practical tool for calculations.

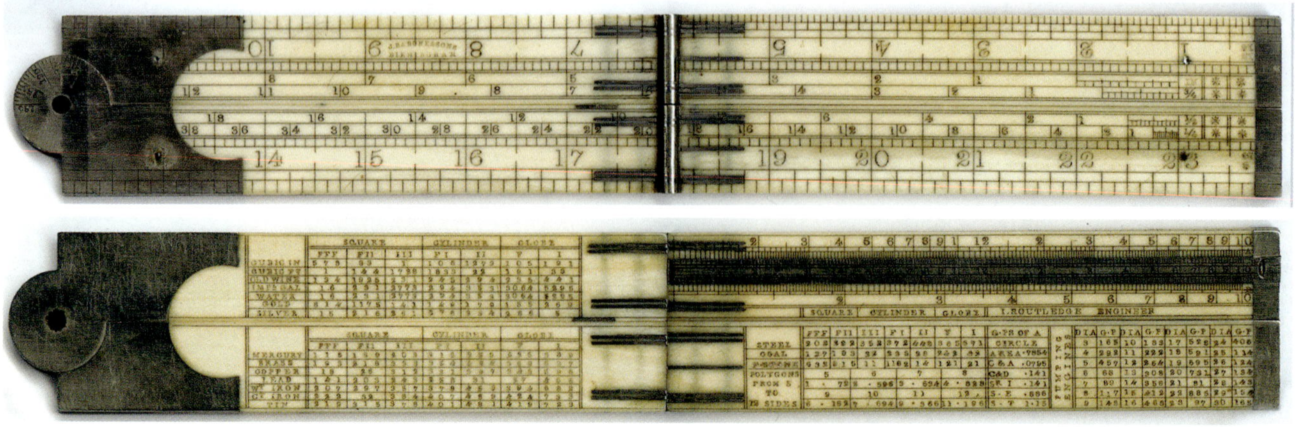

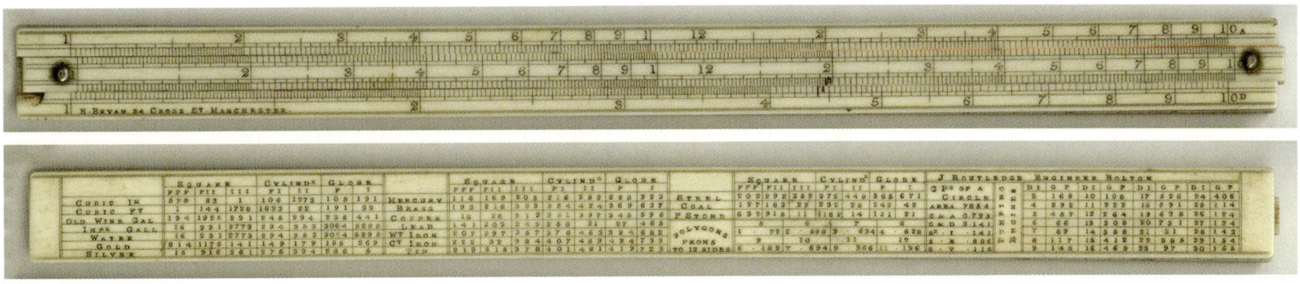

Fig. 8.10a & b. An ivory Soho rule with Routledge's gauge points marked on the reverse, made by E. Bryan (Manchester, 1865–71).

A few engineers' rules have been found that take the form of a rectangular section slide rule rather than a folding rule. These do not include the line of measure on either face, although there is one divided into 10^ths^ on one side by E. Bryan, Manchester. This ivory rule has the Soho lines on one face and Routledge's gauge points on the reverse.

Wilkinson's rule

One of the variations of Routledge's rule that has been found is marked T. WILKINSON IMPROVED RULE. On a rule by John Higgison, the lower rule shown in Fig. 8.11, Wilkinson's name is marked on the face of the rule with the lines of measure and the alteration consists of two additional lines marked NUM and LOG. On the face with the tables, these have been squeezed sufficiently to allow another line of logs to be added.

On a rule by Samuel Albert Smith, the upper rule in Fig. 8.11, Wilkinson's name is marked on the face with the tables, and the alterations are only on this face. The tables have been even further condensed to allow the two additional lines, again marked NUM and LOG, to be added below the gauge point tables. The additional line of logs below the tables has not been included but a line of measure divided into 100ths, marked DECIMAL, has been added above the A line of logs.

Fig. 8.11a & b. Two Routledge rules as improved by T. Wilkinson. The upper rule is by Samuel Albert Smith (Birmingham, 1850–72) and is dated 1862. The lower is by John Higgison (Birmingham, 1833–94).

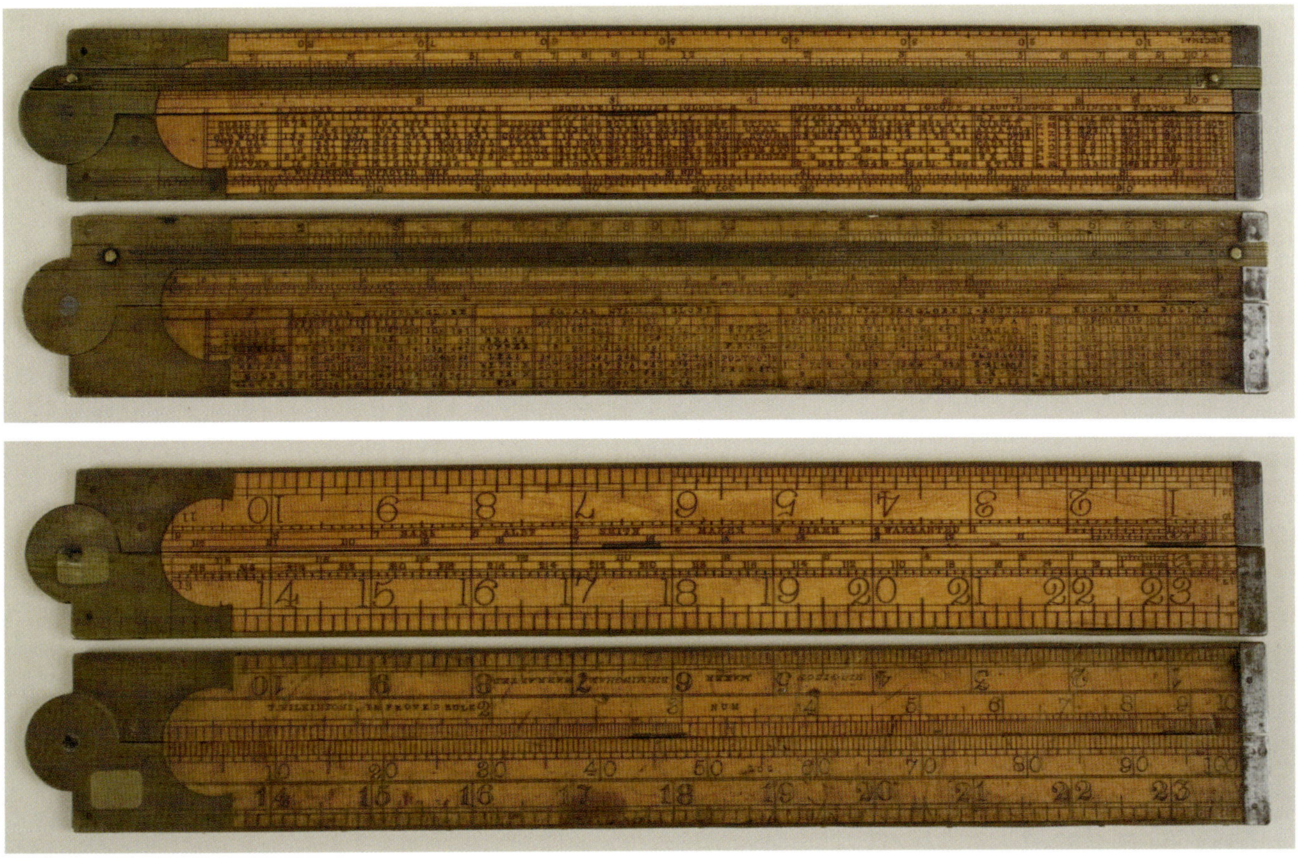

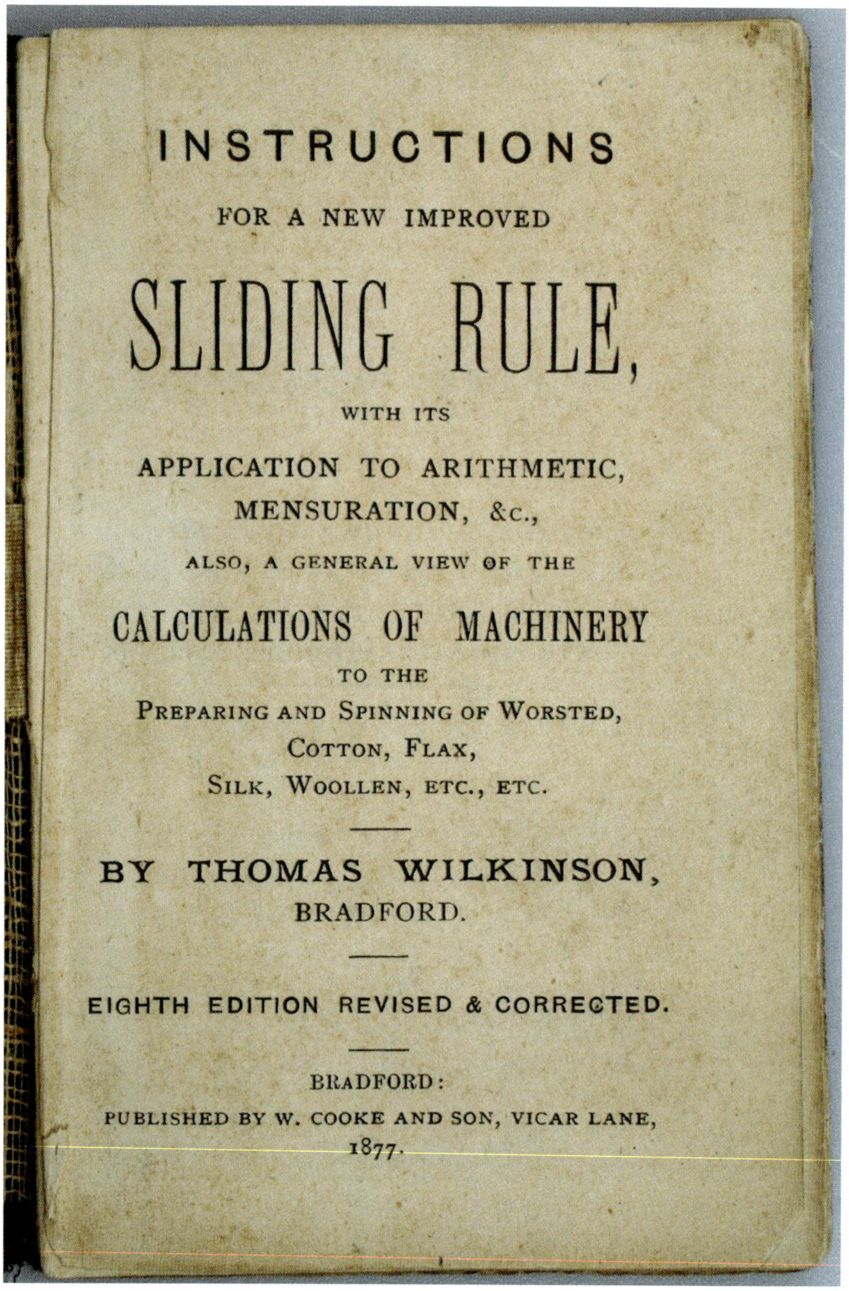
INSTRUCTIONS

FOR A NEW IMPROVED

SLIDING RULE,

WITH ITS

APPLICATION TO ARITHMETIC,
MENSURATION, &c.,

ALSO, A GENERAL VIEW OF THE

CALCULATIONS OF MACHINERY

TO THE

PREPARING AND SPINNING OF WORSTED,
COTTON, FLAX,
SILK, WOOLLEN, ETC., ETC.

BY THOMAS WILKINSON,
BRADFORD.

EIGHTH EDITION REVISED & CORRECTED.

BRADFORD:
PUBLISHED BY W. COOKE AND SON, VICAR LANE,
1877.

Fig. 8.12. The instruction booklet that accompanied Wilkinson's rule. It is a small pocket book, only 2¾ in. by 3¼ in., and contains not only general mathematical information but also instructions for such engineering calculations as finding the power of steam engines and the power of cranes.

Chadwick's rule

Marked J. CHADWICK ENGINEER MANCHESTER, this is a clear derivative of Routledge's rule – the tables for the weights of metals are very similar though gold and silver have been replaced by zinc and aluminium. The inclusion of aluminium indicates a 20th century date for this rule. The pumping engine table of the Routledge rule is replaced by a more modern steam pressure / temperature / volume table.

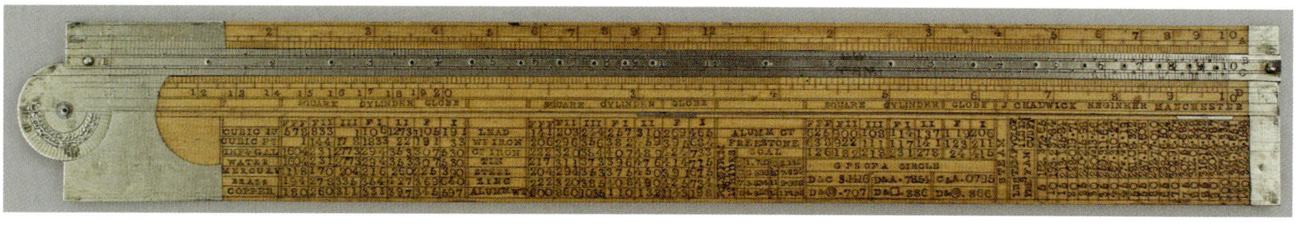

Fig. 8.13. Chadwick's rule. The only known examples of this rule came from the I. & D. Smallwood factory when it was closed in 1999. There is no maker's name marked nor has a catalogue listing yet been found.

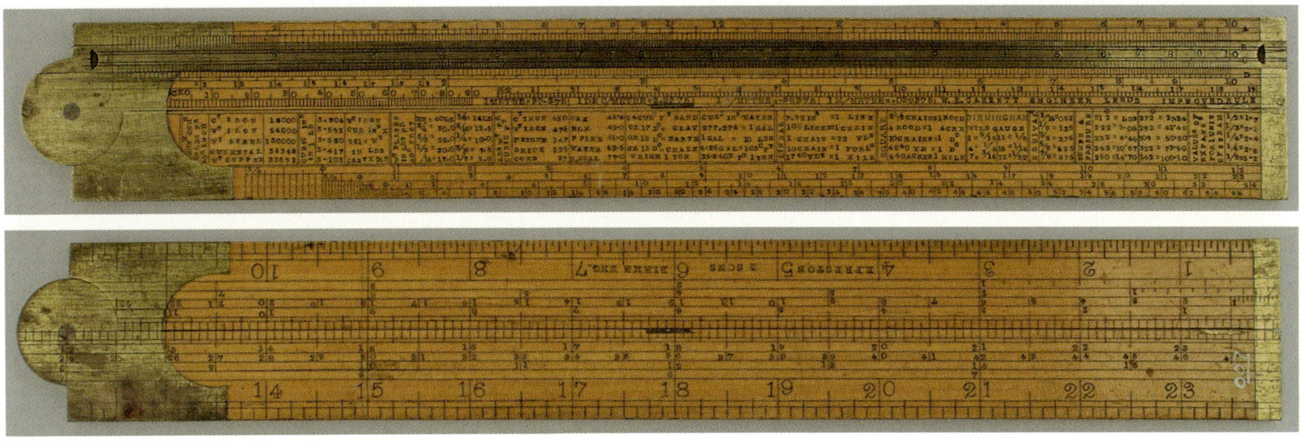

Fig. 8.14a & b. A Carrett's rule made by Edward Preston & Sons (Birmingham, 1864–1932).

Carrett's rule

Although at first glance similar to a Routledge rule, the information is different. The rule is marked W. E. CARRETT, ENGINEER, LEEDS, IMPROVED RULE.

There are still four lines of logarithms, laid out to the Soho pattern but the "magic number" gauge points of the Routledge rule with their partially completed calculations have disappeared. This is clearly a more modern approach. The tables now give values for the breaking strain of various materials (the description is "Ext^{M} Coh^{N} in Lbs P^{R} Sq^{R} Inch"); weights of different thicknesses of iron plate; weights per cubic foot of a range of materials (iron, wrought iron, free stone, bricks, coke, oak, elm, pitch pine, water, coal); the volume of sand, clay, earth and chalk that weighs one ton; decimal equivalents of fractions and tables for steam temperature / pressure and expansion. There are also tables of linear and square measurement and the sizes of Birmingham wire gauges.

The arrangement on the reverse of the rule is also unusual as, in addition to a normal line of measure, there are five adjacent scale measures, marked ½ in., 1½ in, and 2 in., with each division being marked according to the unit.

The rule was listed by Edward Preston & Sons in both their 1901[5] and 1909[6] catalogues.

In the Rabone scrapbooks there are three letters from W. E. Carrett regarding engineers' rules.[7] They are dated 1857 and addressed from the Sun Foundry, Leeds.

Noble's rule

This rule exists in at least six different versions. Three are 2 ft. 2-fold rules with one log slide; two are a 2 ft. 2-fold rule with two slides, one in each leg. The sixth is a brass-bound flat form slide rule and is unusual in having two slides adjacent.

The inscriptions vary somewhat but state that the rule was improved, arranged and manufactured for or by James Noble(s) Engineer, Sheffield. James Noble is listed in the Sheffield trades directory in 1871 as a dealer in

Fig. 8.15. The front face of the rule shown in Fig. 8.18, one of the 2 ft. 2-fold engineers' rules. This face is the same on all five of the Noble style of rule.

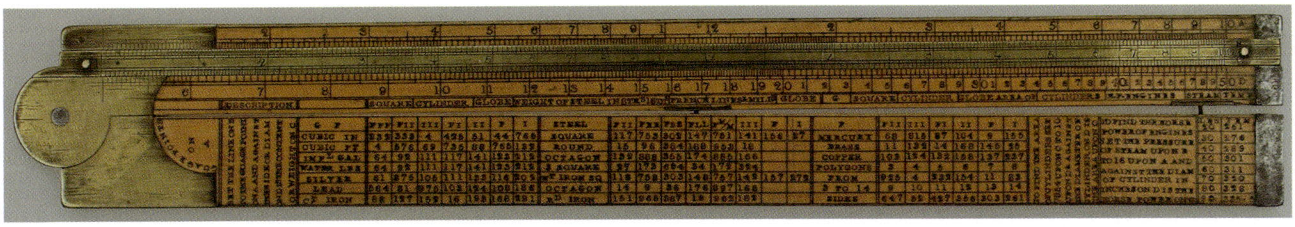

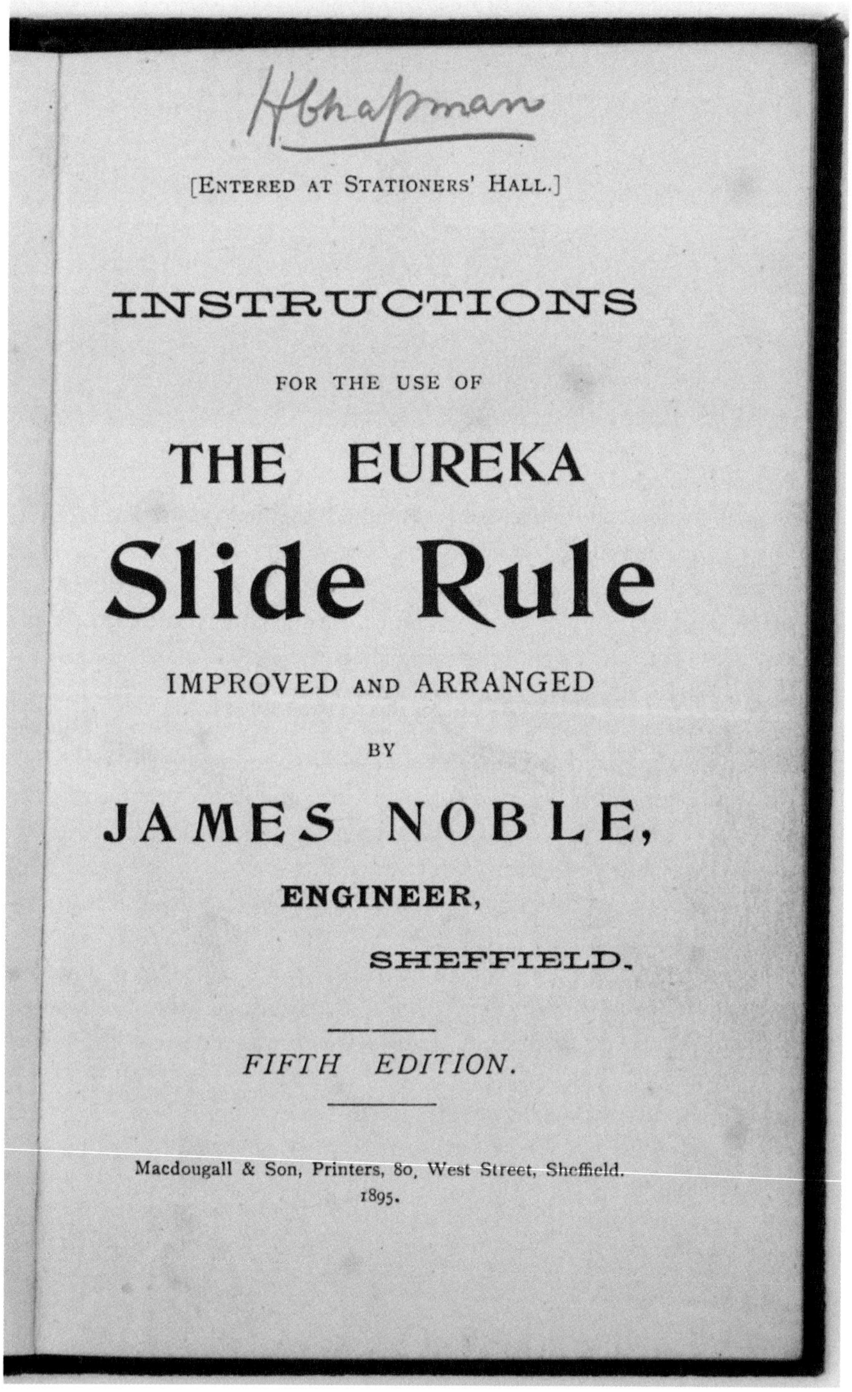
[ENTERED AT STATIONERS' HALL.]

INSTRUCTIONS

FOR THE USE OF

THE EUREKA

Slide Rule

IMPROVED AND ARRANGED

BY

JAMES NOBLE,

ENGINEER,

SHEFFIELD.

FIFTH EDITION.

Macdougall & Son, Printers, 80, West Street, Sheffield.

1895.

Fig. 8.16. Title page from the instruction booklet for James Noble's Eureka slide rule.

engineers' rules and a journeyman engineer. By 1883, he is listed as a foreman engineer after which date he disappears from the directory.

As with many of these calculating rules, an instruction book was available. The 5th edition, published in 1895 names the rule as "The Eureka Slide Rule".

From 1902 to 1909, James Arthur Noble (probably the son of James Noble) is listed in the trades directories as the "Sole maker of the 'Eureka' slide rule". Whether he, or indeed his father, actually made the rules, is doubtful. It is more likely that they were made for them by one of the Sheffield rule makers.

The 2 ft. 2-fold rules take some inspiration from Routledge. On the front face of all five, the first and third tables list the same gauge points as are found

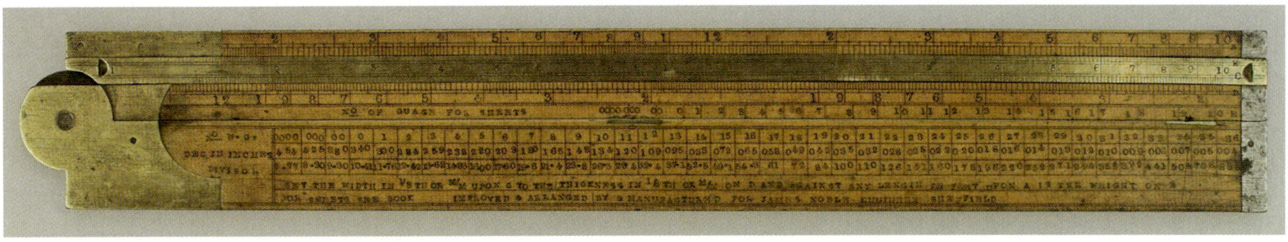

Fig. 8.17. *This is the first arrangement of the rule. The name is marked in a single line on one leg and reads IMPROVED ARRANGED AND MANUFACTURED FOR JAMES NOBLE ENGINEER SHEFFIELD (note that the name does not have an "s" and that the mark reads "manufactured for").*

on a Routledge rule under the same SQUARE, CYLINDER, GLOBE and FFF, FII, etc. headings. However, the second table is headed WEIGHT OF STEEL IN 8TH 16TH; FRENCH LINES & MILLS and another GLOBE. And the headings of the table read FII, FEE, FSS, FLL, FM/M, III, F.I. The steel is listed as square, round, hexagon, octagon, 3 square, Wt. Iron sq. octagon (a second time with different values) and Rd iron. The third table refers to steam and temperature. There are also brief instructions for the use of the rule marked on face.

Unlike the Routledge rule, the D line is folded at 5; consequently, the gauge points are different and the calculation must take account of the automatic division/multiplication that this introduces.

The reverse side of each of the five is different but all have features intended for calculations related to iron rods and sheet.

The first arrangement has a second brass slide, in addition to the one on the front face, which is inset into the opposite leg. The A, B and C lines of the slide on this side are two cycle but the D line is inverted and folded at 1.88. There is an additional line E marked "No. of Gauge for Sheets", although, as this is not against the slide, it cannot be intended for calculations using the slide.

It is marked with an extensive table of wire gauge, marked NO. W.G. decimal thickness in inches, marked DEC IN INCHES and gauge points for weight of sheets, marked DIVISOR.

The edges of the rule are marked with a simple line of numbers divided into ⅛ ths. On the inside edges is a line of numbers divided into 100ths.

There are brief instructions for the use of the rule marked along the lower leg.

The second arrangement, shown in Fig. 8.18, is designed to be read vertically rather than the normal horizontal arrangement. It has four tables, titled S and P; W and G; Square and Round.

The edges of the rule are marked with scales, ¼, ½, ¾ and 1 in., two to each leg, occupying half the length. It also has a simple line of measure with the inches divided into 10ths.

The third arrangement, shown in Fig. 8.19, is also designed to be used vertically. It is marked A TABLE OF MULTIPLIERS TO GET WEIGHT OF STEEL SQUARE below which there is another table marked STEEL ROUND, then STEEL OCTAGON and finally STEEL SQUARE. The instructions for use are also marked.

On this face it has IMPROVED ARRANGED AND MANUFACTURED BY JAMES NOBLES ENGINEER SHEFFIELD with the mark spanning both legs. The edges are marked with the same scales as on the rule shown in Fig. 8.17.

Fig. 8.18. *The second arrangement. This is also marked IMPROVED ARRANGED AND MANUFACTURED BY JAMES NOBLES on one leg of this face.*

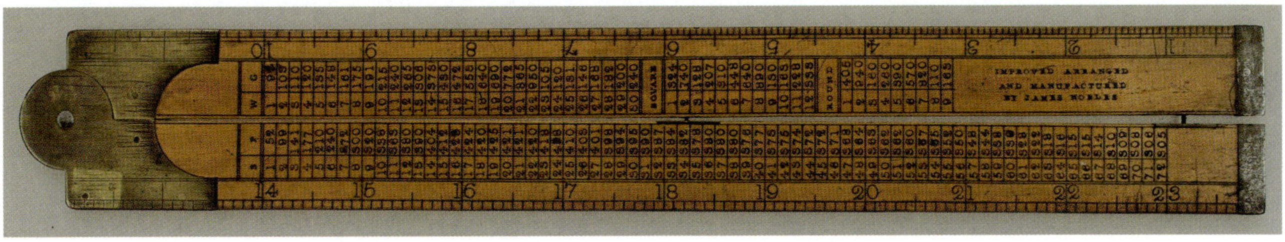

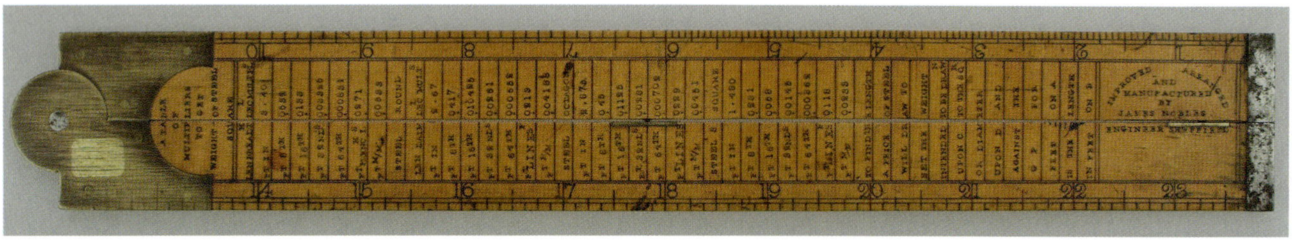

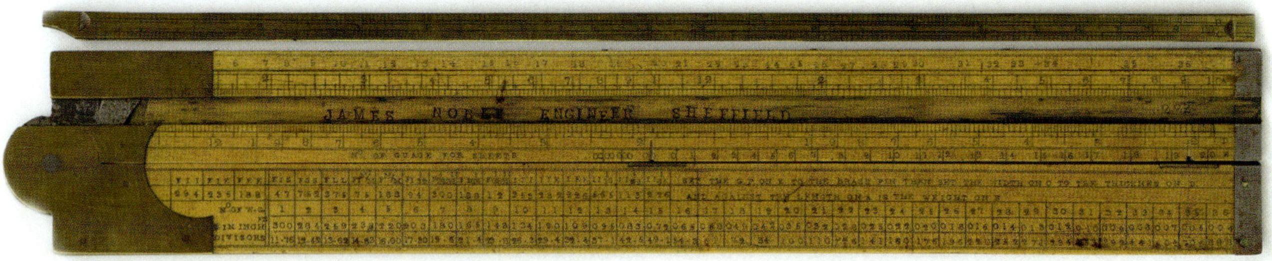

Top: Fig. 8.19. The third arrangement.

Centre: Fig. 8.20. The fourth arrangement.

Bottom: Fig. 8.21. The fifth arrangement is another rule with two slides. On this rule the name, JAMES NOBLE ENGINEER SHEFFIELD, is marked under the second slide.

The fourth rule has a three column table down the centre of the rule, which is the same as that shown in Fig. 8.17, but marked vertically, the first column being the number of wire gauge, the second decimal thickness in inches and the third the gauge points (divisors). A simple line of numbers is marked on each side of this face, as with the rule shown in Fig. 8.18. Again there are brief instructions for the use of the table.

The fifth rule also has a second slide on the reverse face. As with the other rule with two slides, there is an additional line marked E, although on this rule there are instructions for its use. There is also the same three line table for wire gauge, decimal thickness in inches and gauge points and in addition there is a further table for gauge points similar to those on the front face but with different values. There is no indication, however, of the use of these.

The sixth rule is the only one that is not a 2 ft. 2-fold rule, being a 12½ in. open framed, two slide, slide rule. On one face the logarithmic lines are as found on the other Routledge and Noble rules but with the difference that the D line, still folded at 5, is on a second slide. The lines B and C are labeled WEIGHT 1 H P and WIDTH AREA OF PISTON respectively. On the lower part of this face there are also the normal gauge point tables that would be expected, together with some instructions for use.

On the reverse side all the logarithmic lines are two cycles with the D line (not folded) and an additional E line on the second slide. These are labeled

Fig. 8.22. The sixth rule, the only one that is not a 2 ft. 2-fold rule. It is marked on one edge IMPROVED & ARRANGED BY & MANUFACTURED FOR JAMES NOBLE ENGINEER SHEFFIELD.

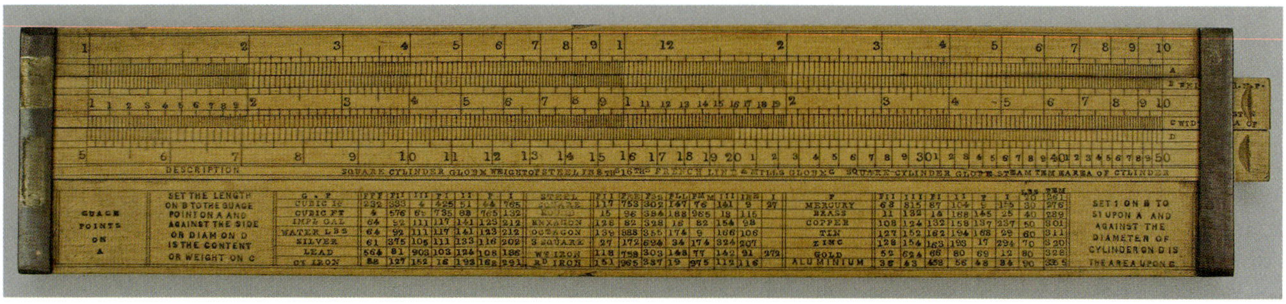

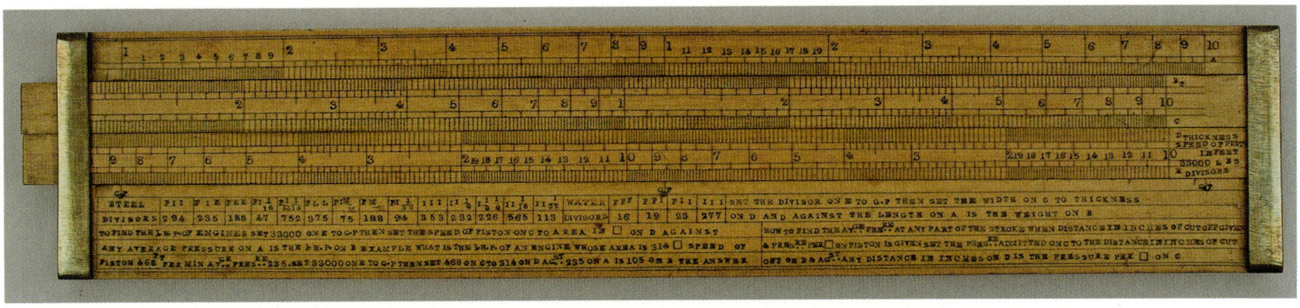

Fig. 8.23. *The reverse face of the slide rule shown in Fig. 8.22.*

THICKNESS SPEED OF PISTON IN FEET and 33000 LBS DIVISORS respectively. On the lower part of this face are gauge points for steel and water. There are instructions for various calculations such as finding horse power and average pressures. The rule does not have any lines of measure.

Fig. 8.24. One edge is marked IMPROVED & ARRANGED BY & MANUFACTURED FOR JAMES NOBLE ENGINEER SHEFFIELD.

Hawthorne's rule

Marked "IMPROVED & ARRANGED BY R^{T} HAWTHORNE, CIVIL ENGINEER, NEWCASTLE TYNE", this rule is of a 2 ft. 2-fold format, the front face being very similar to a Routledge rule.

The rule has a Coggeshall layout with the D line folded at 4. The conversion and materials gauge points are largely Routledge but there are additional gauge points for rope, chain, iron rod, etc. Unlike the Routledge rule it is inscribed with instructions as to the use of the gauge points. The steam pressure / temperature table extends to 50 lb. / sq. in.

The reverse of the rule has tables giving compressive strength of a range of materials, strength of beams, specific gravity and tables of linear square and cubic measures. All of this information is actual values and not partially calculated gauge points.

The Tyneside origin of this rule is displayed in such information in the cubic column as "36 Bushels = 1 Chald[ron] of coals in London & weighs 28 Cwt." and "68 Bushels = 1 Chald[ron] at Newcastle and weighs 58 Cwt." and "a bushel of coal contains 2816 in."

A chaldron was the measure used for coal, the Newcastle and London chaldrons being different sizes. Also, coal which filled a chaldon before loading in Newcastle would have broken into smaller pieces by the time it arrived in London and had been unloaded – so it would not have filled the original chaldron in any case. The Newcastle coal bushel was not the standard bushel either!

Fig. 8.25a & b. Hawthorne's rule.

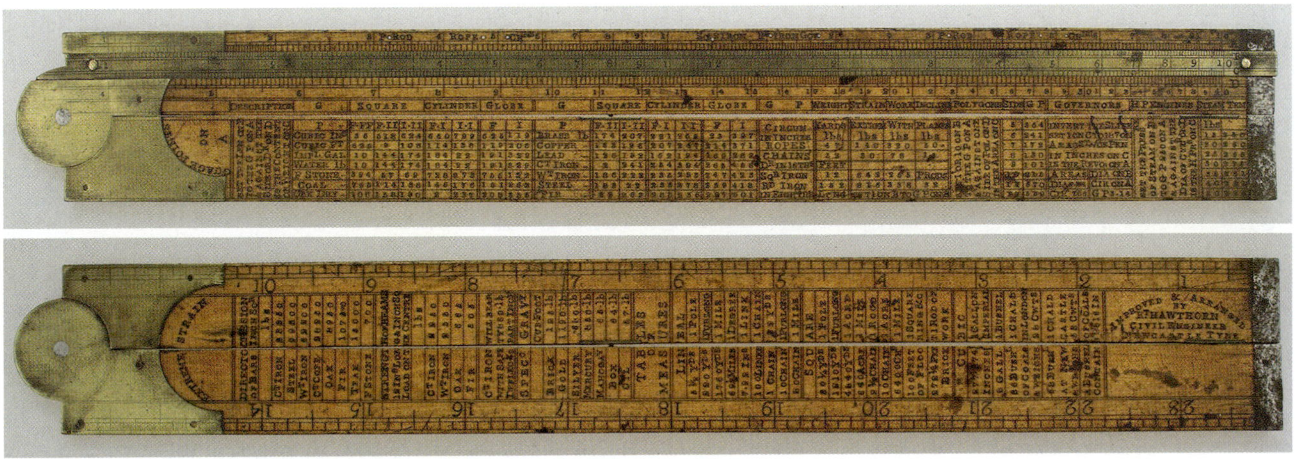

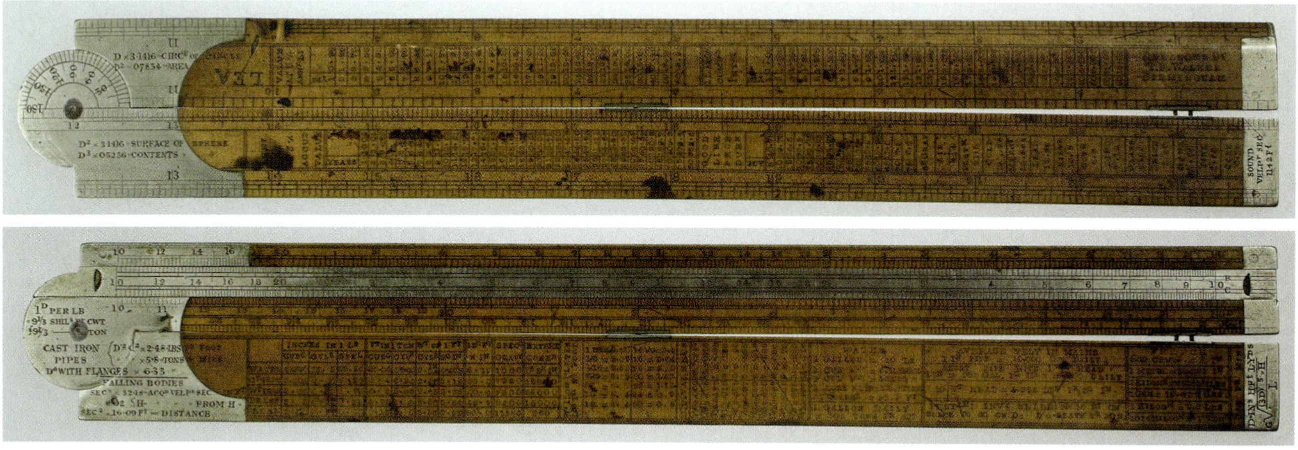

Fig. 8.26a & b. Walker's rule. There is no maker's mark on this rule nor on any others that have been recorded and it does not appear in either the Preston or Rabone catalogues. Walker's rules have been found in both boxwood and ivory.

The rule is listed in Preston catalogues in 1891[8] and 1909[9], Rabone in 1892[10] and Smallwood in 1897[11]. With at least three makers it would suggest that by the 1890s any design protection had run out.

Walker's rule

Marked "ARRANGED BY T. F. WALKER, BIRMINGHAM", this is something of a hybrid – the rule is marked with gauge points and formulae (some particularly directed to water supply pipe work) clearly intended for a civil engineer but also included are tables of factors for financial calculations and a life expectancy table.

It is physically different from other engineers' rules in having a bevelled edge and also some formulae marked on the metal arch and tips. Also included are a line of centimetres and some imperial / metric conversion factors. On the inside edges is a line of inches divided, unusually, into 12ths and on the unbevelled outside edge is a line of inches divided into 100ths.

Simpson's rule

Marked "IMPROVED AND ARRANGED BY ABRAHAM SIMPSON DARNALL SHEFFIELD", this is another rule based on Routledge's rule. The front face is like Noble's rule, with the D line folded at 5. The inclusion of aluminium in the gauge points confirms a 20th century date. Unlike the majority of the other Routledge type rules, this has a maker's name, A. Harper, Birmingham, who was working from sometime before 1900 until 1925.

The rule does not include any of the tables relating to steam engines, horsepower, etc.

Fig. 8.27. An ivory Walker's rule, with the slide removed, showing the data marked under the slide which appears to be financial percentage tables.

On the reverse face, there is another slide with the D line arranged as a Coggeshall girt line, that is, folded at 4.

Below this there is the table found on the two slide Noble's rules, with lines for wire gauge, decimal thickness in inches and gauge points (divisors); in

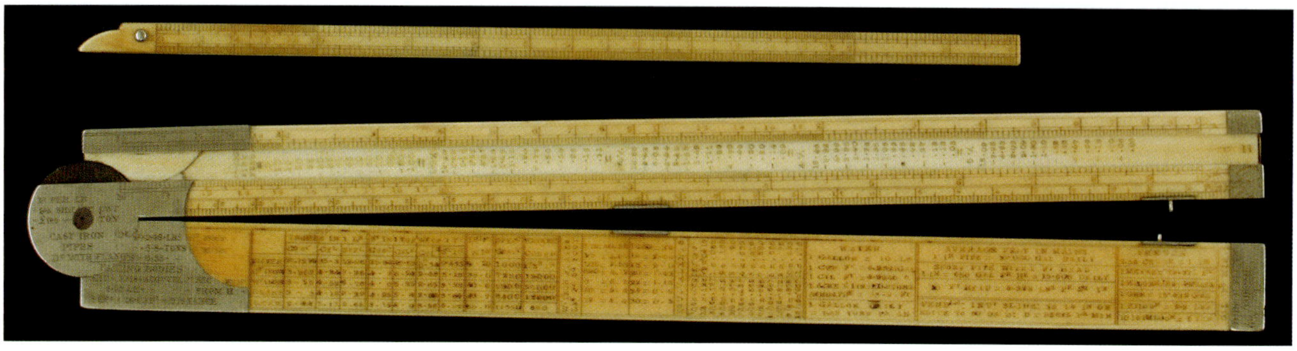

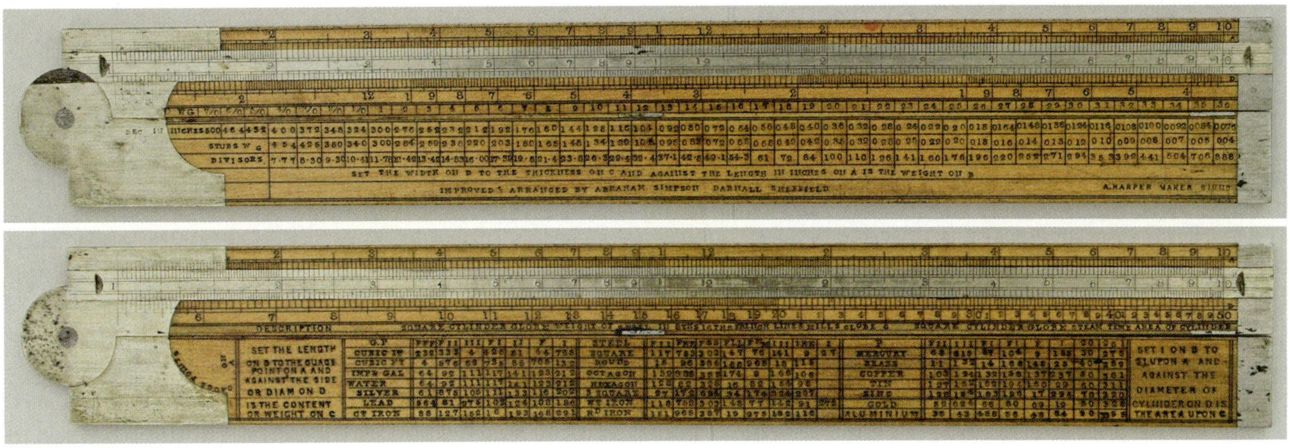

Fig. 8.28a & b. Simpson's rule made by A. Harper (Birmingham, <1900–1925). This is a simpler rule than many of the engineers' rules.

addition there is also a line marked STUBS W G, being Stubs' wire gauge. Once again, there are instructions for the use of the rule marked on this face.

Hay's Improved Mechanics rule

Designed by William Hay of Dumbarton, and patented in America in 1867, this is shown in the 1909 and 1914 Edward Preston catalogues (but not in the earlier ones). Even so, the only examples recorded with a maker's name are marked I. & D. Smallwood.

This rule is different from the other engineers' rules described in this chapter. It does not have a slide or a list of gauge points.

It is a 3 ft. 4-fold rule and in addition to the normal linear measures and two short scales, one ½ in., the other ¼ in., there are four other lines marked, three on the outside face and one on the inside. Those on the outside are a pair of radius lines, marked R^{d}, enabling the user to subdivide a circle into equal parts; a line marked SIDE OF EQUAL SQUARE, parallel to the upper radius line, used in conjunction with it to convert between squares and circle of equal area; and a line of chords marked "C". With the help of a pair of dividers, this line can be used to set the rule open at any angle.

On the inside of the rule, in addition to the normal inch line is a line marked CIR CUM. (On the upper rule this line is only marked CR.) This is a circumference line which, used in conjunction with the inch line, enables the easy conversion from diameter to circumference.

The American patent showed a 2 ft. 2-fold rule which perhaps explains why the markings stop short of the end of the rule, the scales being used to fill the remaining length.

A fuller explanation of this rule can be found in Philip Stanley's *A Source Book for Rule Collectors*.[12]

Textile industry rules

There are a number of engineers' rules that were specifically designed for the textile industry. These include Slater's, Holmes's and Ashenhurst's rules and

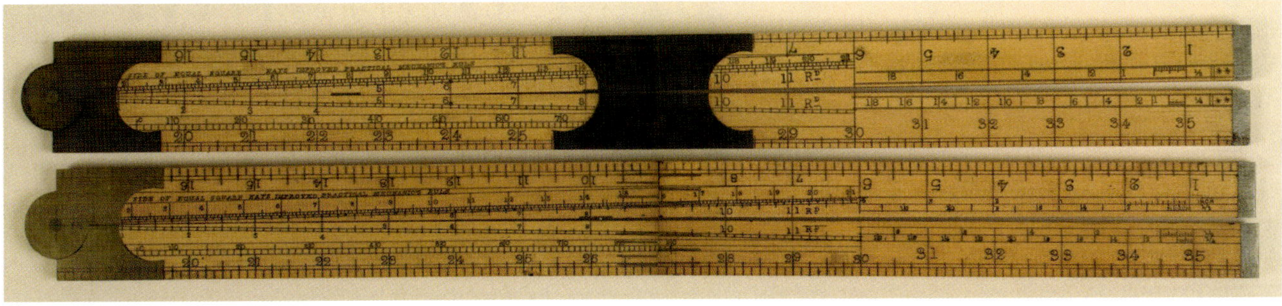

Fig. 8.29. Two Hay's rules. The upper rule is unmarked but a version with three arch joints is listed in the Preston 1909 catalogue. The lower is known to have come from the Smallwood factory, although it is not marked.

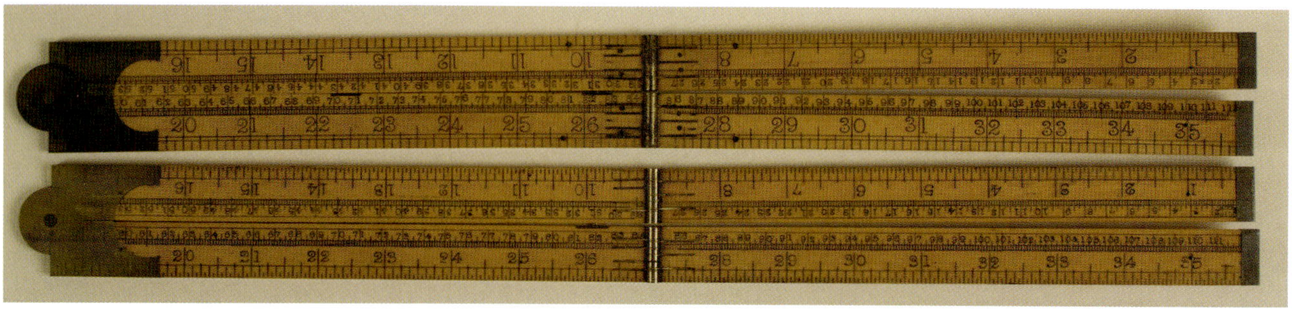

Fig. 8.30. The inside faces of the two Hay's rules.

are described and discussed in Chapter 13.

Engineering gauges

Although engineering measuring lies outside the compass of this book, a brief description of some of the gauges commonly found seems appropriate.

Plate 69 of the Richard Timmins & Sons catalogue of ca. 1845 illustrates 15 designs of slot and hole gauges.[13] Most are made to measure "wire" – which refers to Birmingham wire gauge standard; three are to measure screw gauge and two measure both wire and nail rod. Thus three standards are visible in this one early plate. Gauging systems were plagued with the commonly-found habit of attaching a sizing system to one particular product or material; thus there were different systems applied to steel wire, iron wire, ferrous sheets, zinc sheets and others.

Listed below are brief descriptions of the most frequently found systems. There are doubtless others but the principal purpose of this list is to enable readers to identify those gauges that they may discover. It will also be seen that many of the standards were known by more than one name.

In most series the higher the number the smaller is the size – but there are exceptions. Most systems include 0, 00, 000, etc., often alternatively expressed as 2/0, 3/0, etc.

It is not known if there was any mathematical basis to most of the systems; if there was, it is long forgotten. However, most series increase by a roughly standard percentage, usually 10 or 12 percent.

Before micrometers were available one method of measuring thin wire was to wrap it spirally round a rod and then count the number of turns in an inch.

Stubs' Steel Wire Gauge (Stubs' Wire Gauge, Stubs' Gauge or Lancashire Steel Wire Gauge): This is probably the oldest of the gauges. Whether Peter Stubs, the dominant maker of Lancashire tools in the late 18th and

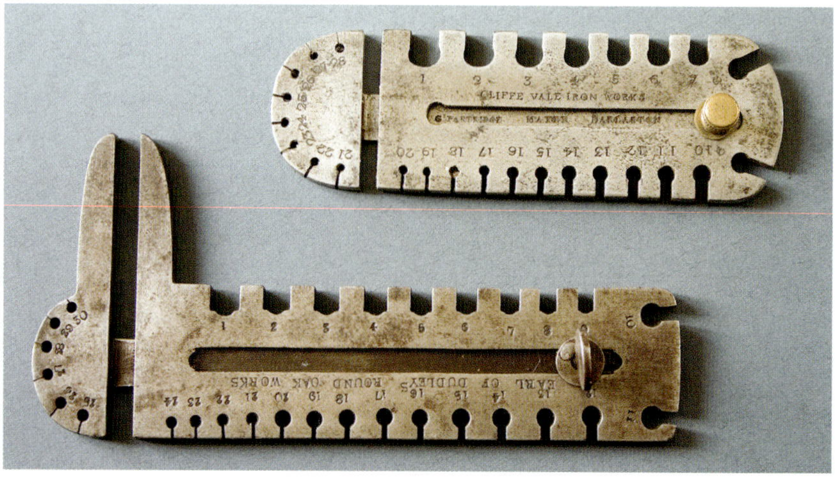

Fig. 8.31. Two wire gauges. The top gauge is similar to one of the designs shown in the Richard Timmins & Sons catalogue of ca. 1845. It is made by Partidge (Darlaston, <1850–96>) and is marked Cliffe Vale Iron Works, which was located in Shelton, near Stoke-on-Trent. The bottom is marked Earl of Dudley Round Oak Works. These were located at Brierley Hill, near Dudley in Staffordshire. They opened in 1855 and continued until at least the 1980s. There is no maker indicated.

early 19th century, devised this system or whether he adopted an already existing Lancashire system is unclear. The gauges were used for steel wire and rod.

Stubs' Iron Wire Gauge: The Brown & Sharpe catalogue of 1926[14] contains a reminder of the difference between Stubs' *Steel Wire* Gauge and Stubs' *Iron Wire* Gauge, which is synonymous with Birmingham Wire Gauge, English Standard Wire Gauge, etc. It states that "In using the gauges known as Stubs' Gauges, there should be constantly borne in mind the differences between the Stubs' Iron Wire Gauge and the Stubs' Steel Wire Gauge. The Stubs' Iron Wire Gauge is the one commonly known as the English Standard Wire or Birmingham Gauge and designates the Stubs' soft wire sizes. The Stubs' steel Wire Gauge is the one that is used in measuring drawn steel wire of drill rods of Stubs' make and is also used by many makers of American drill rods."

Fig. 8.32. A Birmingham wire gauge by W. & C. Wynn (Birmingham, 1795–1869). It is interesting to note, that in the 1893 Wynn Timmins & Co. catalogue, the sizes of Wynn's Birmingham wire gauges differed from the standard BWG by a factor of up to 5%, so it would appear that W. & C. Wynn were still using their own standard although the gauge was marked Birmingham!

Birmingham Wire Gauge (English Standard, Birmingham Standard, or Stubs' Iron Wire): In 1847, Charles Holtzapffel realised that standardisation of the numerous variants of sheet metal and wire gauge was very desirable and set about measuring them. Based on his results, which he published in his catalogue in 1847,[15] he established a standard known as Birmingham Wire Gauge (BWG). There are 40 sizes in the system from 4/0 = 0.454 in. to 36 = 0.004 in., the smaller the number therefore indicating the larger the wire. However both wire and gauges were also made to half sizes. This gauge was theoretically obsolete after 1884 but Starrett was still listing these gauges (Nos. 188 and 189) in 1930.[16]

Birmingham Wire Gauge: Although with the same name as Holtzapffel's BWG, this series was chiefly used around the 1890s by the South Staffordshire rollers of iron plate intended for nail makers in the Black Country.

Birmingham Metal Gauge (Birmingham Plate Gauge): Also described by Holtzapffel, this is intended for sheet metal. The intervals are closer than for wire, going from 1 = 0.013 in. to 30 = 0.167. As can be seen this is one of the few systems where size increases with number. This gauge was used for many sheet metals including gold and silver.

In their 1847 catalogue, Holtzapffel & Co.[17] also list Birmingham brass wire, sheet brass and metal gauges.

Whitworth Standard Wire Gauge: Proposed by Sir Joseph Whitworth in 1857 this was a logical sequence expressed in thousandths of an inch. From 0.001 to 0.050 the advance was by one thousandth of an inch, from 0.055 to 0.100 by 5 thousandths and thereafter by increasing increments to 0.500. The system was never widely used.

Warrington Wire Gauge: Messrs. Rylands Brothers of Warrington, a leading manufacturer, made their wire to their own system which ran from 7/0 (= 0.5 in.) to 22 (0.28 in.). Clarke[18] in 1891 noted that it was then rarely used.

Standard Wire Gauge (SWG) (Imperial Standard Wire Gauge, British Standard Wire Gauge): From March 1st 1884 this became the only legal wire gauge. Others were rendered illegal and were supposed to have gone out of use!

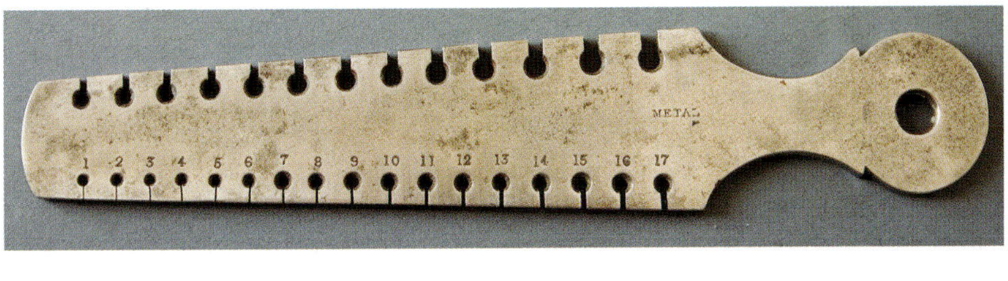

Fig. 8.33. A Birmingham metal gauge, marked only "metal" by Townley (Birmingham, <1910>).

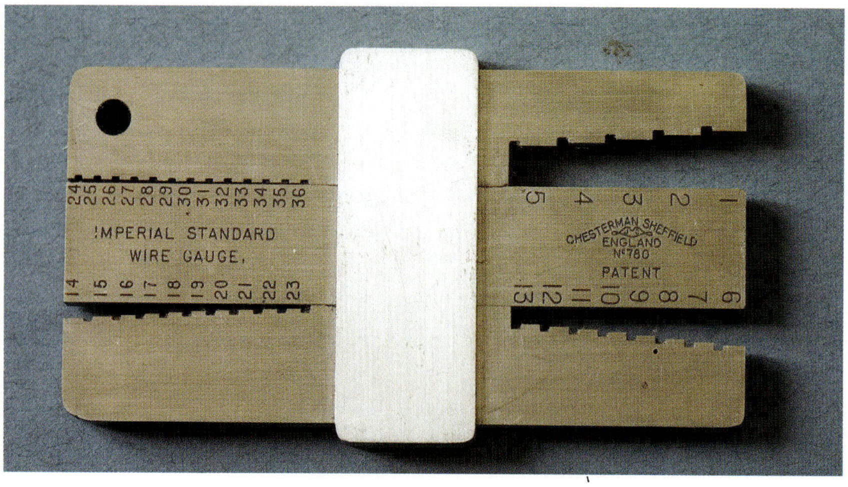

Fig. 8.34. An imperial Standard Wire Gauge, made by Chesterman (Sheffield, 1821–1963).

Nail Rod (Iron Gauge): The Richard Timmins catalogue[19] includes a folding gauge in the manner of a pocket knife that includes a "nail rod". This is a slot gauge with shallow rectangular slots indicating that they must be intended for rectangular rod or sheet metal, though the extent of the gauge cannot be seen. Also shown is a gauge with similar slots, marked to 10, that is described as an iron gauge. Another gauge with rectangular slots has been recorded, which appears similar and is marked with the series 0, 00, and 1 to 8 with additional half sizes of 2½ and 3½.

Fig. 8.35. Wire gauges. Clockwise from top left: Birmingham Wire Gauge; British Standard Wire Gauge, marked legalised 1883; Whitworth Standard Wire Gauge; Birmingham Wire Gauge; Old Birmingham Wire Gauge, made by Wynn Timmins; New Standard (wire gauge).

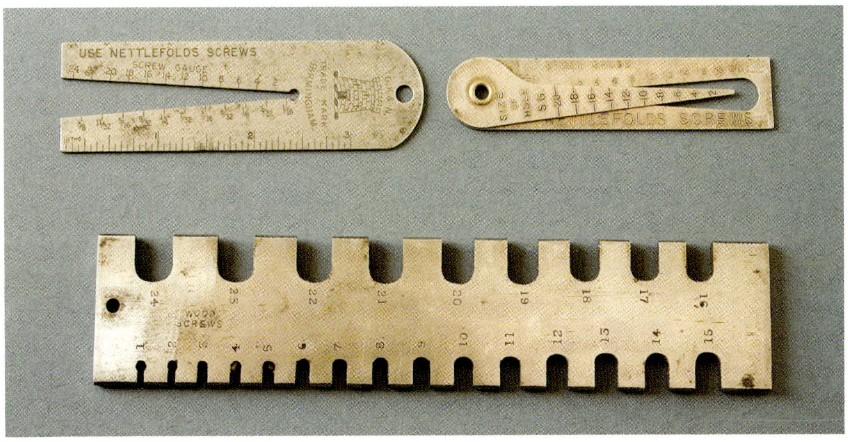

Fig. 8.36. Three screw gauges, the top two marked Guest Keen & Nettlefold, one of the leading screw manufacturers.

Wood Screw Gauge: This differs from most systems in that the higher the number the larger the size. It is still in use today as British-made wood screws are not metricated either in length or gauge.

Zinc Gauge (English Zinc Gauge): These usually extend from Gauge 1 to 21 (0.004 in. to 0.072 in.).

VM (Vieille-Montagne) Zinc Gauge: Extends from Gauge 1 to 26 (0.004 in. to 0.105 in.).

Needle Gauge: As the name suggests, specifically for needle wire, the series extends from 1 to 21 with the sizes being equivalent to Birmingham Wire Gauges sizes 18½ to 36.

Copper Conductors: The [Electrical] Cable Maker's Association specified 13 sizes of copper wire which were used in the production of cables and flexes. Single conductor cables and metrication have rendered these obsolete but small slot gauges marked with the sizes in thousandths were at one time part of the electrician's equipment.

References

1. Rabone, J., *Instructions for the use of the Practical Engineers' and Mechanics' Improved Slide Rule* by J. Routledge, Birmingham, ca. 1867, reprinted Fitzwilliam, NH, 1983.
2. Wess, Jane, "The Soho Rule". *Journal of the Oughtred Society*, Vol. 6, No. 2, 1997.
3. Knott, John V., "Joshua Routledge 1775 – 1829", *Journal of the Oughtred Society*, Vol. 4, No. 2, 1995.
4. Rabone, J., *op. cit.*
5. Edward Preston & Sons. *The Preston Catalogue of rules, measures, levels ...*, Birmingham, 1901.
6. Edward Preston & Sons. *Catalogue No. 18. The "Preston" Catalogue, 1909 with 1912 supplement.*
7. From the collection of scrapbooks kept by John Rabone & Sons, now in the Hawley Collection, University of Sheffield.
8. Edward Preston & Sons Ltd., *The Preston Catalogue of rules, measures, levels...*, Birmingham, 1891.
9. Preston 1909 *op. cit.*
10. John Rabone & Sons, J., *Catalogue of Rules & Tapes.* Birmingham, 1892.
11. I. & D. Smallwood, *Catalogue of Rules, Squares, Spirit Levels, Straight Edges, Gauges, etc.*, Birmingham, 1897.
12. Stanley, Philip E., *A Source Book for Rule Collectors*, Mendham, NJ, 2003.

13. Richard Timmins & Sons, *Pattern Book of Goods*, ca. 1845.
14. Brown & Sharpe, *Small Tools Catalogue No. 30*, Providence, RI, 1926.
15. Holtzapffel & Co., *Catalogue No. 46*, London, 1847.
16. Starrett, *Catalogue,* 1930.
17. Holtzappfel *op. cit.*
18. Clark, Daniel Kinnear, *A Manual of Rules, Tables and Data for Mechanical Engineers*, 6th ed., London, 1891.
19. Richard Timmins & Sons *op. cit.*

Chapter 9
Military matters

The instrument called the Gunners Callipers, consists of two thin rulers or plates, which are moveable quite round a joint, by the plates folding one over the other... On these rulers are a variety of scales, tables, proportions, etc. such as are esteemed to be useful to be known by gunners and other persons employed with artillery. But except for the taking of the calibre of shot and cannon, and the measuring of the magnitude of saliant and entering angles, there are none of the articles with which the callipers are usually filled, essential to this instrument. Robertson, 1775[1]

The practice of officers providing their own clothing, equipment, instruments and even arms lasted until World War II and to a limited extent even continues today. The young officer, depending on the service and regiment he was entering, was instructed to buy certain items which included some measuring equipment.

Military mapping

As the name makes clear, the Ordnance Survey – the first accurate survey of Britain – was a military undertaking produced in response to the Jacobite rebellion of 1745 and the discovery that there were no informative maps of the northern parts of the kingdom. The maps made by the Ordnance Survey, from the 18th century to the present, are produced without pressure of time, using the most accurate methods and instruments available. However the officer in the field may be called upon to map and sketch very quickly and under difficult conditions – even under fire. For this purpose much simpler equipment is required: the sketching protractor, the clinometer, and the sighting rule or alidade.

Military sketching protractors and scales

The simplest examples differ little from the commonly found draftsmen's combined scales and protractors. However, the design of these scales and protractors has exercised the minds of inventors for decades and there are numerous versions with adaptations that the designer, almost invariably a serving officer, considered helpful. The features common to most are markings as a protractor, some mapping scales and information such as a table of horizontal equivalents, useful when assessing gradients. The most complicated versions may include sights and a magnetic compass.

Often identifiable by a scale of inches with diagonal divisions of the end inch, these rules are intended for map work. But from the wild variety of scales observed it would seem that there was not much standardisation. Some scales are expressed in the traditional manner (6 in. to the mile) but the military had a tendency to use ratio based scales (R.F. in army terms) earlier than civilian map makers. The abbreviations found on military protractors are:

- H.E. Horizontal Equivalent: the distance in plan between two contours.
- V.I. Vertical Interval: the difference in level between two adjacent contours.
- R.F. Representative Fraction: the scale of a map expressed as a fraction but always with one as the numerator. For example $\frac{1}{500}$ or $\frac{1}{63000}$.

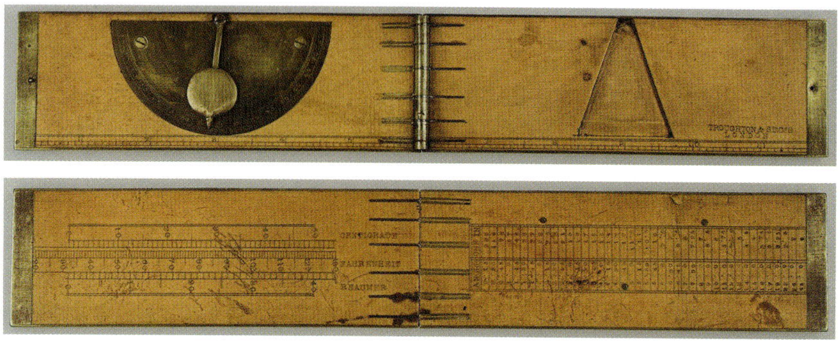

Fig. 9.1a & b. A clinometer in the form of a folding rule by Troughton & Simms (London, 1826–1915). This type of clinometer was supplied to the British Geological Survey and continued in use until the mid-20th century.

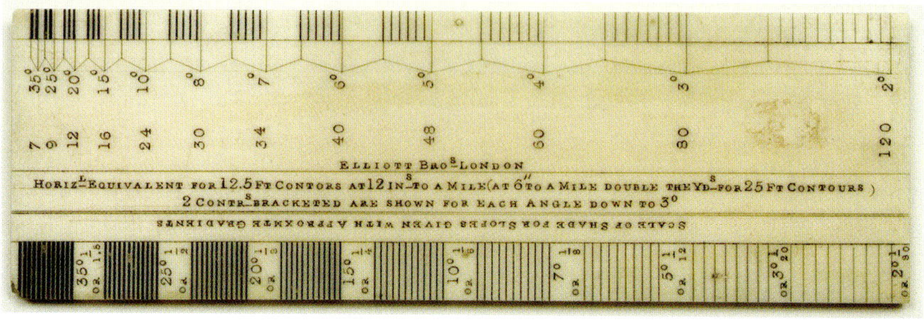

Fig. 9.2. A scale of hachures appears on protractors as blocks of parallel lines of decreasing spacing and increasing thickness. Hachures are a system of delineating the inclination of slopes on maps. This example is on an ivory protractor by Elliott Bros. (London, 1864–1913>).

Below right: Fig. 9.3. Table of horizontal equivalents. (All the examples given in this list are for the same vertical intervals.)

Fig. 9.4. Scale of horizontal equivalents from an ivory rule by Aston & Mander (London, 1870–1930). Also shown is the formula, marked "rule", by which horizontal equivalents are calculated.

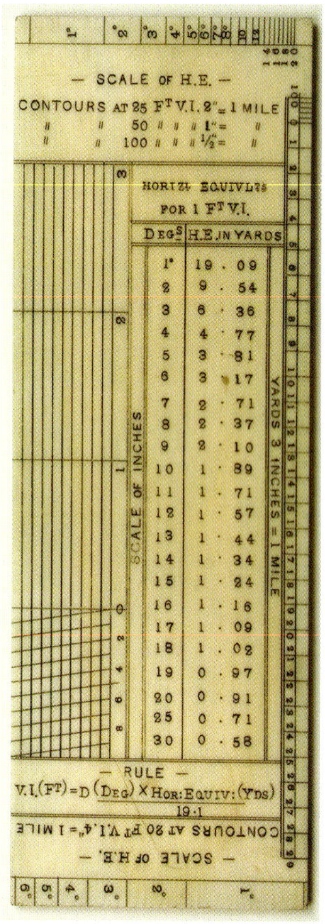

Recording of inclines

In the field a small plummet (weight) attached to the protractor was a simple but not very accurate means of assessing a slope.

Army practice allowed the angle of an incline to be expressed in several ways so tables relating one to the other are a common feature of military protractors.

- In degrees, e.g. 7° slope.
- As a ratio, e.g. 1 in 8 incline.
- Rise in inches per yard, e.g. 4½ in. per yd.
- As a scale of hachures (scale of shade for slopes).
- As a horizontal equivalent.

Contours are unlikely to be plotted at intervals of one foot; twenty-five foot intervals are more usual. Thus, if contours at 25 ft. intervals on the map are measured to be 80 yards apart, the slope will be 80 / 25 = H. E. 3.2 = (by

Horizontal Equivalent for 1 ft V.I.	
Degrees	H.E. in Yards
1	19.09
2	9.54
3	6.36
4	4.77
5	3.81
6	3.17
7	2.77
8	2.37
9	2.10
and so on to....	
20	0.91
25	0.71
30	0.58

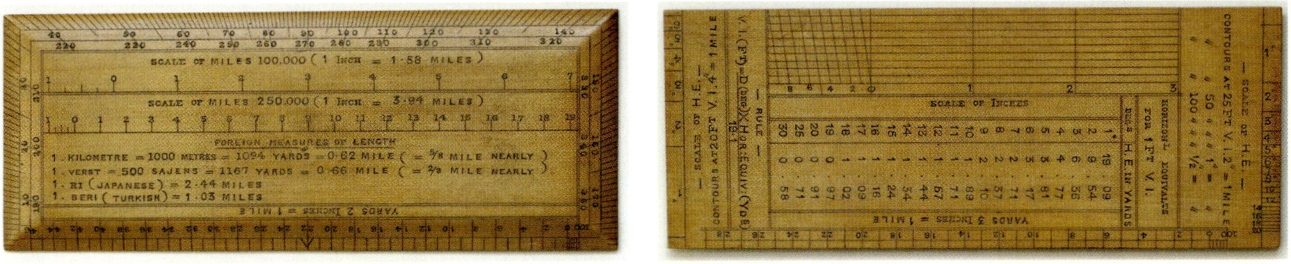

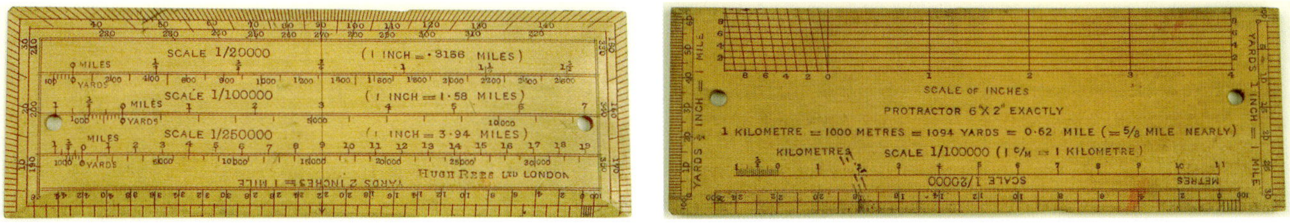

Top: Fig. 9.5a & b. An anonymous military protractor. This protractor is identical to the New Service Protractor illustrated in the J. H. Steward 1895 catalogue.

Below: Fig. 9.6a & b. Another version of a military protractor by Hugh Rees Ltd. (London, <1903–34>).

reference to the table) a slope of a little less than 6°. The formula from which the table is calculated, V.I. (ft.) = D. (deg.) x H.E. (yd.) / 19.1, is also marked on many protractors.

Scale of horizontal equivalent: This is a short cut method of determining the angle of slope in degrees. A scale of this type can only be used when it matches the scale of a map and the vertical interval of the contours are those marked on the scale. For this reason these scales will always be labelled with the scale of the map and the V.I. Some protractors will therefore carry several versions of this scale. In appearance the scale is sharply incremental.

"Tramlines": These are parallel lines, sometimes alternating red and black, inscribed on the face of the protractor that were intended as an aid to sketching onto similarly lined paper or to aid orientation on a gridded map.

Named types

The following are some of the named types of military protractors – there are many others which bear no indication of their origin. Unless noted the size is 6 in. by 2 in.

New Service Protractor: Illustrated in the J. H. Steward 1895 catalogue[2] and listed as the "New" protractor. Presumably there was an earlier version, but the nature of the improvements is not stated.

Service Protractor MK II: Illustrated by Steward in their 1916 catalogue,[3] by which time any foreign measures had been replaced with metric or imperial conversions and an additional scale of 1/80000 (1 inch = 1.26 miles) added. The protractor also had a hole pierced at each end for attaching a cord. The protractor was available in boxwood, xylonite or ivory.

Service Protractor "A" MK III: Illustrated in the War Office *Manual of Map Reading...* 1929[4]; the material is described as ivorine.

The Whitworth Protractor: Designed by Captain D. E. Whitworth, 2nd Lancers, Indian Cavalry and illustrated by Steward in 1916. It is boxwood, 6 in. x 2 ½ in. and includes several time scales. Using these scales the officer can

Fig. 9.7a & b. A Service Protractor "A" MKIII, marked United Chemical Engraving Co. Ltd., Newcastle-on-Tyne and dated 1932.

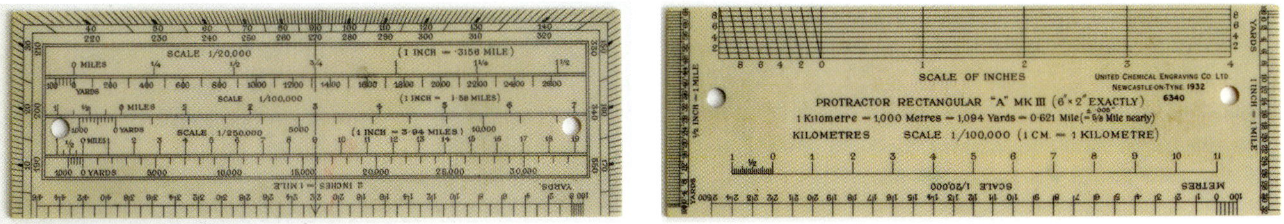

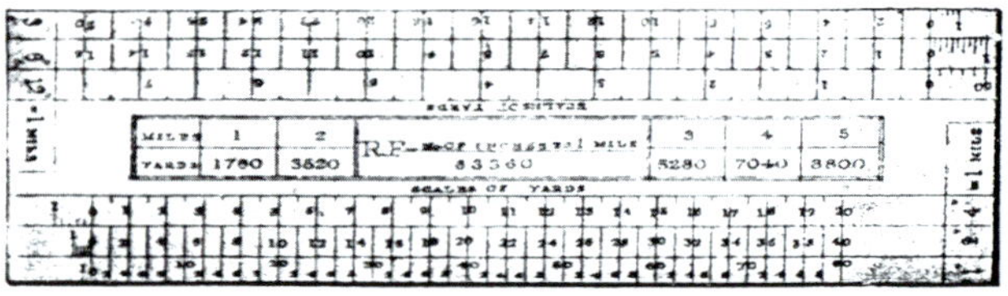

Fig. 77. The "Whitworth" Protractor, 3/6.

Fig. 9.8. The Whitworth protractor from the Steward 1916 catalogue.

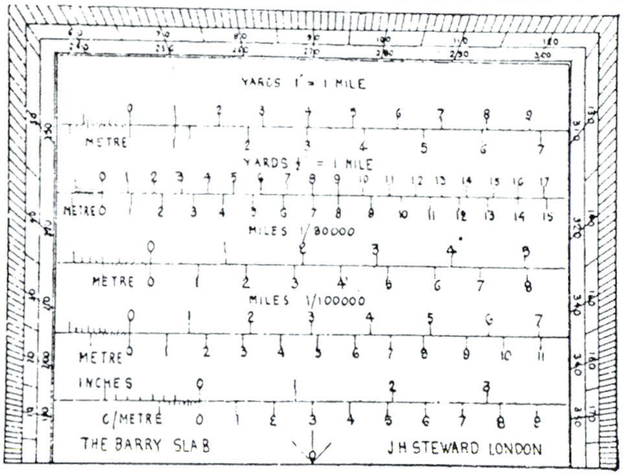

Fig. 75. The "Barry" Slab and Protractor, **5/6.**

Fig. 9.9. The Barry Slab and Protractor from the Steward 1916 catalogue.

measure directly off a map the time that it will take for infantry or horse to march from one point to another. A range of scales allows for different speeds and different scales of map. Available in both imperial and metric versions.

The Barry Slab and Protractor: Invented by Colonel Stanley Barry, an early version of a transparent celluloid protractor intended to fit inside the regulation field book, it is thus 6½ in. by 4½ in. One surface was smooth for writing on.

Colonel Tobin's Time Scale: As the name suggests more a scale than a protractor – the several scales indicate the time taken by infantry and horse to travel the distance measured (Steward, 1895).

Captain Haig's Improved Protractor: "As used at the Staff College". This protractor was available in either boxwood or ivory (Steward, 1895).

The "Verner" Rapid Sketcher: Perhaps the ultimate military protractor, it includes a magnetic compass, extending folding sights and a plummet clinometer, all recessed into the face of the 6 in. x 3 in. boxwood body. The

Fig. 9.10. A "Verner" Rapid Sketcher, marked Major W. Verner Complete Sketching Instrument. On the reverse of the sketcher are various different scales with their representative fractions, degrees of slope, horizontal equivalents, approximate gradients, vertical intervals and a diagonal scale.

Fig. 9.11. An improved version of the rule in Fig. 9.10 with the sights now extending on a sliding track. It was shown in the 1911 Steward catalogue.[5] By this time, Verner had become Lt. Col. W. Verner, which was marked on the rule.

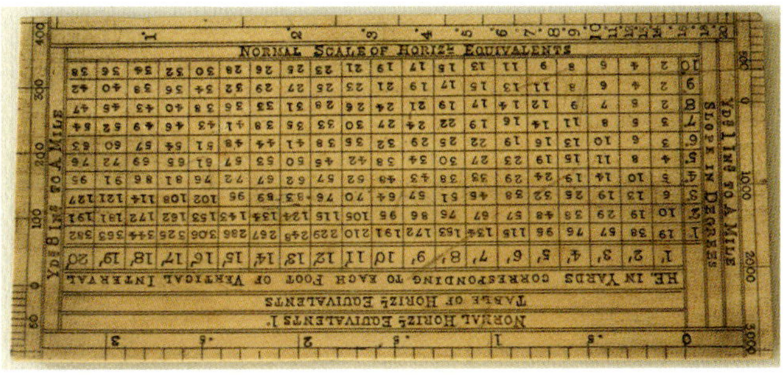

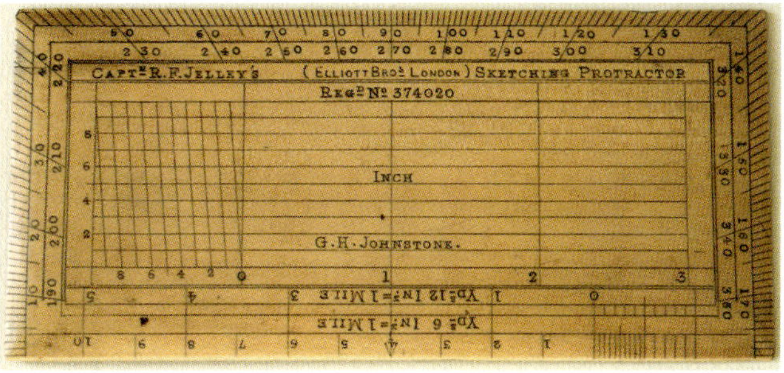

Fig. 9.12a & b. Capt. Jelley's sketching protractor, made by Elliott Bros (London, 1864–1913>) .

inventor was Captain W. W. C. Verner of the Rifle Brigade and he patented his invention in 1891.

The sketcher was listed in the J. H. Steward catalogues from 1895 to at least 1928, and, as Captain Verner moved up the ranks, the name marked on the protractor changes to Major Verner, then Lt. Col. Verner and finally Col. Verner. It was available with either imperial or metric scales.

Captain R. F. Jelley's Sketching Protractor: Reg. Design No. 374020, made by Elliott Bros., London, its size is 5¼ in. x 2½ in. The most unusual feature of this protractor is the extensive table of horizontal equivalents marked on the back. Unlike the standard type of table which records the H. E. only for the V.I. of 1 ft. the table on this rule gives the H. E. for all V. I.s from 1 ft. to 20 ft. Complete as this may be, it would seem most unlikely that the officer in the field should suddenly discover a map with the contours at 14 or 17 ft. intervals! These are rare.

The Sandhurst Protractor: Made by W. F. Stanley and listed in their 1931 catalogue.[6] The most distinctive characteristic is the round flat brass plummet hung on a thread and stored in a housing on the face. The reverse includes a table of horizontal equivalents and a scale of hachures at 25 ft. vertical intervals. The protractor is 5¼ in. x 2¼ in.

Cavalry Sketching Board: Sketching whilst on horseback created its own problems and this was an area that Capt. W. Verner also addressed with his patent cavalry sketching board. The patent was taken out in 1887.

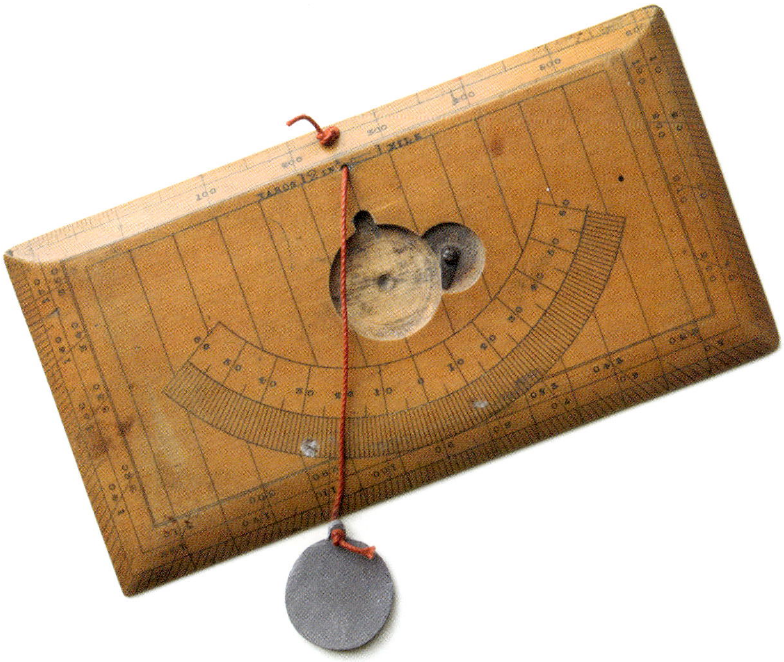

Fig. 9.13. The Sandhurst protractor by W. F. Stanley (London, 1853–1967), shown with its plummet in use.

Fig. 9.14a & b. Front and back of a Verner's patent sketching board. This improved version includes a jointed ruler consisting of a straight edge and an angled arm, which was the subject of an additional patent taken out in 1891. The back view shows the clinometer and the leather arm fastening.

Fig. 9.15. Capt. Verner's patent cavalry sketching board as illustrated in the J. H. Steward catalogues. This example is shown with an alidade held between two elastic bands.

Fig.9.16. The Thomson Artillery Sketcher from the 1911 Steward catalogue.

The Thomson Artillery Sketcher: Designed by Capt. A. F. Thomson, Royal Artillery, this sketcher was listed in the J. H. Steward 1911 catalogue.[7] The rule was held at a standard distance from the eye using a cord held in the teeth. When viewed thus the scale indicates approximate angles in degrees. The rule was supplied with a sketching block marked in squares that corresponded to the graduations on the rule. It also had a protractor projection that also served as a thumbpiece. The description in the catalogue claimed that "…this is the simplest possible method of producing a Panorama sketch, that no artistic talent is required, and that with a little practice an officer or intelligent N. C. O. with no aptitude for drawing can produce a reliable Panorama sketch…"

The De Lisle Cipher Protractor: Although not strictly speaking a measuring device, it is of military protractor form and is marked in the usual manner with degrees and scales on the edges. The purpose of the protractor is as a key to instructions given in cipher. It is made of transparent material (probably celulloid) and the face is divided into ¼ in. squares, marked with letters and numbers. It is 6 in. by 3 in.

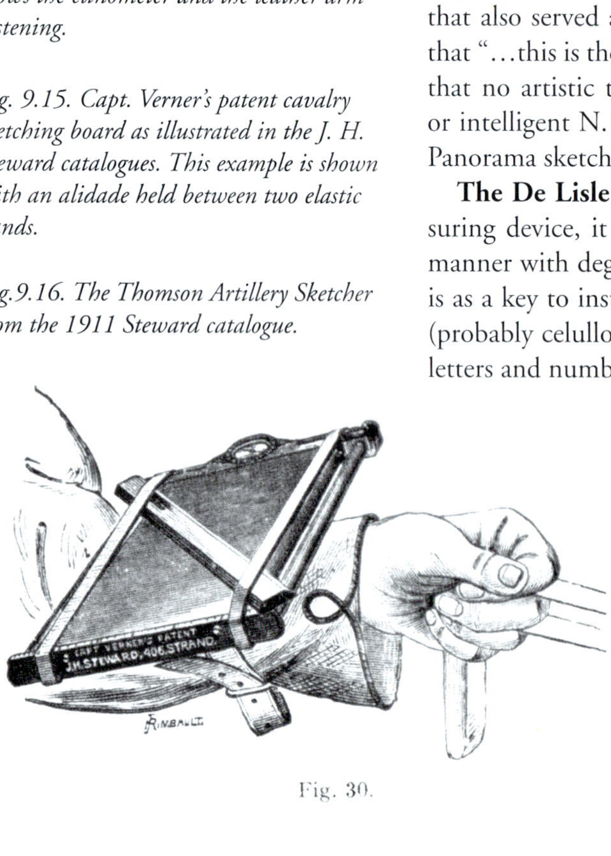

Fig. 30.

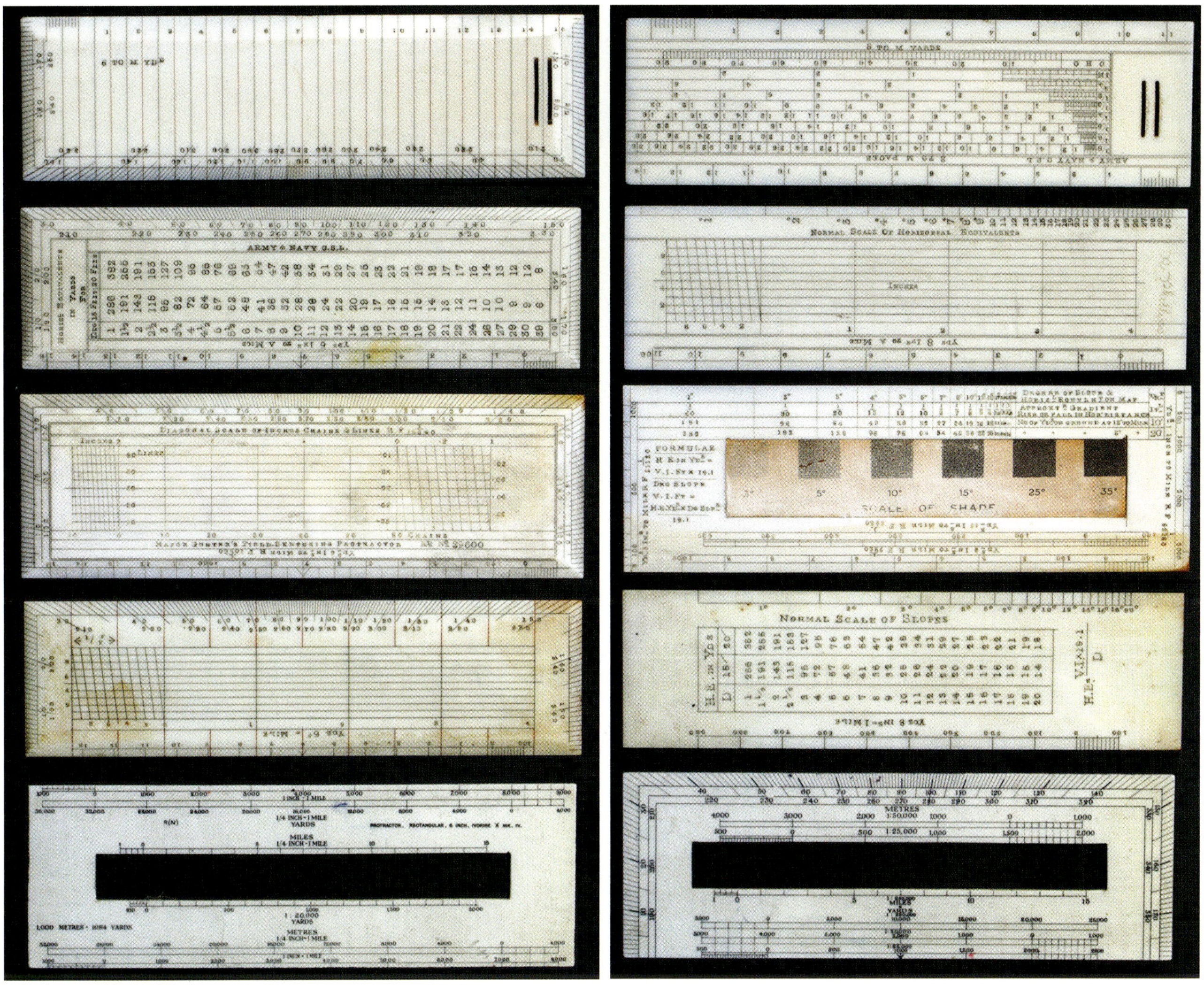

Fig. 9.17a & b. *Five different sketching and military protractors. The top four are ivory, the bottom, ivorine. From the top: 1) A simple type by Army & Navy C.S.L. (London, 1871–ca. 1980). Note the two slots to take a ribbon for suspending round the neck. 2) Another by Army & Navy C.S.L. with a table of horizontal equivalents on the front face and a normal scale of horizontal equivalents on the reverse. 3) Marked Major Gunter's field sketching protractor, but with no maker; this has hachures set into the reverse face. 4) A simple version, with no maker, marked with a table of horizontal equivalents on the reverse. 5) This anonymous ivorine example is marked PROTRACTOR RECTANGULAR 6 INCH IVORINE "X" MK IV.*

Clinometer rules

Only the rule form of clinometer falls within the parameters of this book. The name is also applied to many other forms of instrument – balance circle clinometers, the commonly found Indian pattern military clinometer, and types akin to quadrants or theodolites.

The basic features of the rule type are two thick boxwood legs hinged together with an accurately divided quadrant. This allows the angle between them to be measured with some accuracy. The angle may be expressed in degrees or in inches per yard rise. A level tube is fitted in at least one of the

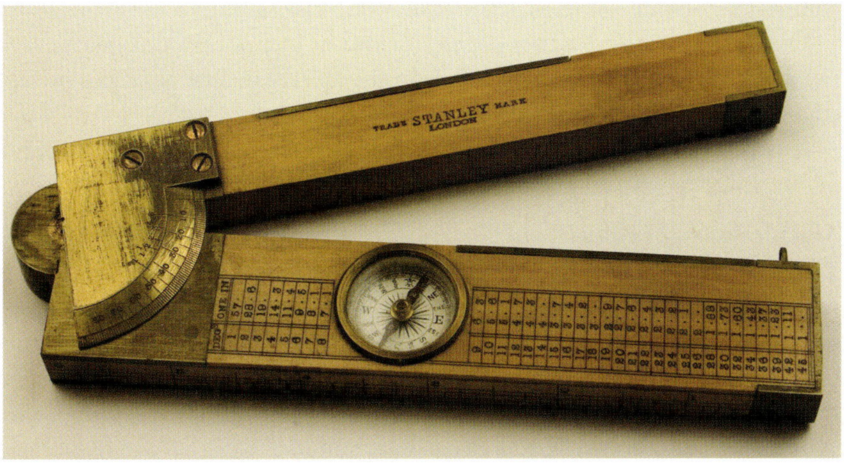

Fig. 9.18. *A clinometer by W. F. Stanley (London, 1853–1967). Of simple form, there is a level in one leg and a compass in the other but no sights.*

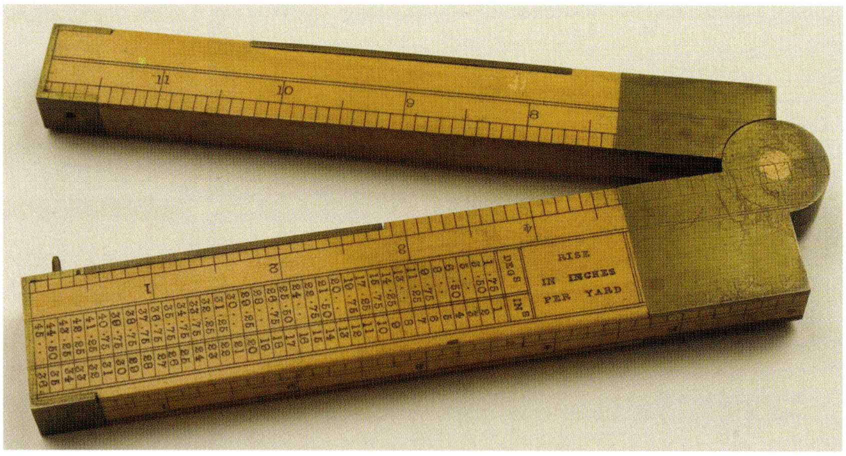

Fig. 9.19. The reverse side of the clinometer shown in Fig. 9.18.

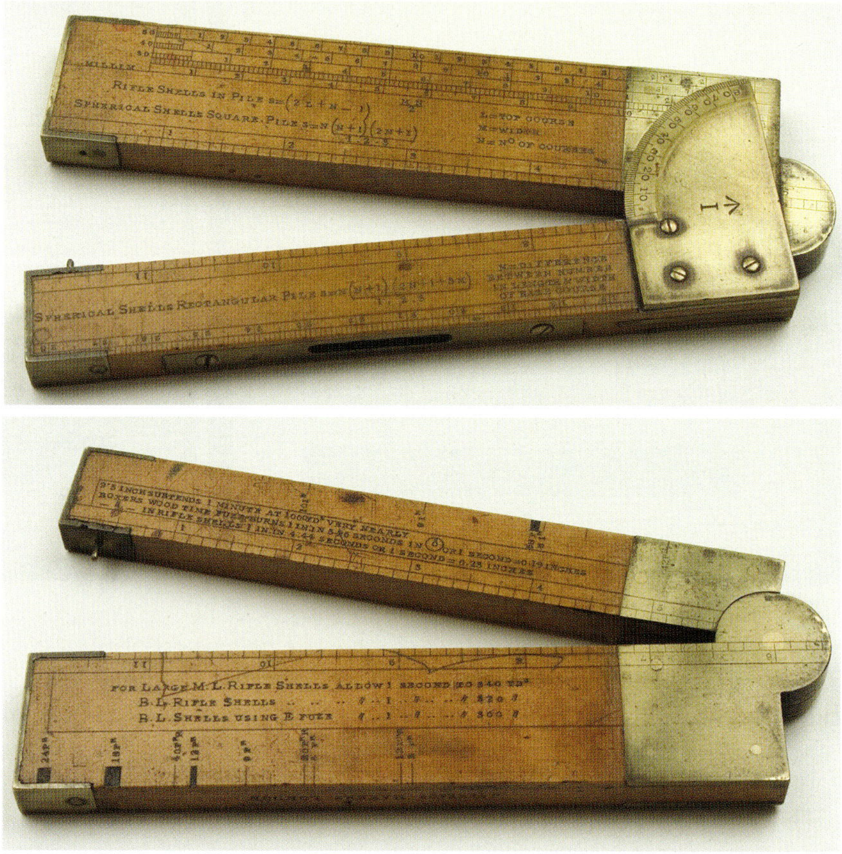

Fig 9.20a & b. A rule type clinometer by J. Tree (London, 1826–95). It is marked with a broad arrow above I indicating that it was the property of the Indian Army – however most clinometers were the property of individual officers. The faces are marked with formulae for calculating the number of "spherical shells in a square pile, rifle shells in pile" and other data relating to Boxers Wood Time Fuze for shells. The sizes of the traditional 3, 6, 12 etc. pounder shot and the sizes of rifled cannon are also indicated.

arms. These are the basics – improvements include levels in both arms, a fixed compass, a rotating compass, sights and even a sighting tube.

One or more tables, either degrees or rise in inches per yard or rise as a ratio (e.g. 1 in 6) are usually marked on the faces. The line of inches on the edge is the only linear measure on these tools.

It is probable that the first clinometers of this type were made around 1850, and they were still being illustrated by W. F. Stanley in 1931.

Alidades or sight rules

The army pattern alidade consists of a stout boxwood rule about 12 in. long with bevelled edges and with folding sights mounted on the top. When making a plane table survey an object is viewed through the sights and the direction of it marked onto the paper using the edge of the rule. One or more mapping scales are marked on the bevelled edges. Examples bearing both World War I and World War II dates are common.

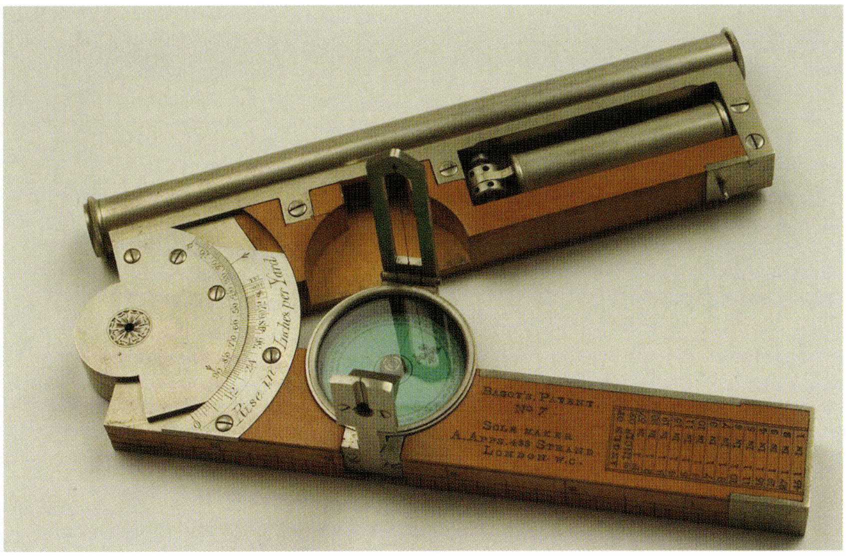

Fig. 9.21. Bagot's patent clinometer by A. Apps, (London, 1866–1910) in boxwood and nickel silver. It is marked No. 7, an indication that this was the seventh example made under the patent. This is perhaps the ultimate in rule type clinometers. The quadrant measures in both degrees and rise in inches per yard. There is a level in each arm; in the upper arm it is set in an adjustable tube, in the lower arm in the inner face of the leg, a sighting telescope and folding and removable sights for taking compass bearings.

Marquois sets

The basic elements of a Marquois set are a 22½ / 67½ / 90 degree triangle, which has an index line marked at the centre of the hypotenuse, and two thick scale rules with square edges. The rules are divided into pairs of "natural" and "artificial" scales along their edges. The natural scales are placed on the inside and divided into actual parts per inch (20, 25, 30, 40, 50 and 60). The artificial scales are on the outside and are divided into intervals three times those of the natural scales. Because of the three-to-one relationship of triangle's dimensions, placing the hypotenuse of the triangle along the edge of the rule and moving the index line according to distances on the artificial scale advances the other two sides of the triangle one-third the distance or

Fig. 9.22. A military style boxwood alidade by C. T. Cooper & Sons, London (1905–1947>).

Fig. 9.23. A Marquois set by J. Coombes (Devonport, <1893>) showing the two rules and set square.

the actual amount indicated on the natural scale. By this means parallel and perpendicular lines, spaced at accurate distances, can be quickly drawn. Thus the set provides all that is needed to prepare and execute accurate drawings in the field in the days before graph paper.

The set was designed by Thomas Marquois, Professor of Artillery and Fortification, around 1778. In 1784 an advertisement was placed in the *New Daily Advertiser* which stated "Marquois Parallel Scales for Drawing all Kinds

Fig. 9.24. A Marquois set in a set of drawing instruments by W. F. Stanley (London, 1853–1967).

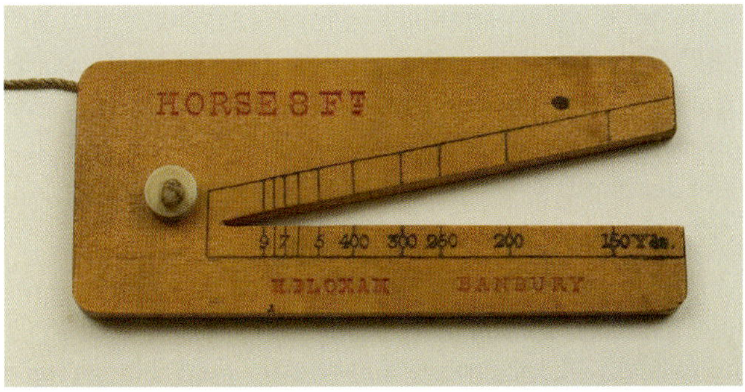

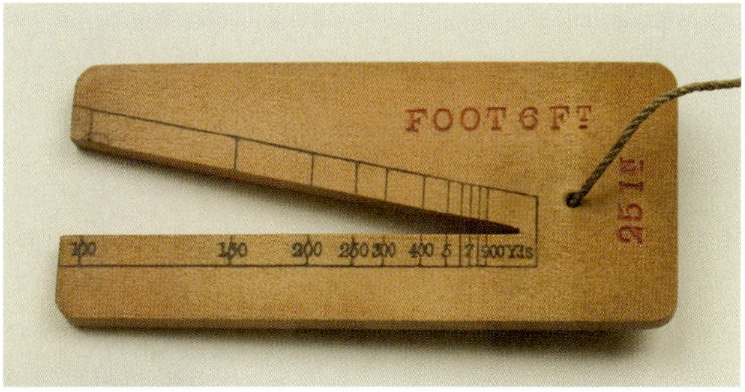

Fig. 9.25a & b. A boxwood range finder made by H. Bloxham, Banbury (unidentified). It is marked on one side with Foot 6 Ft. (for foot soldiers) with a range from 100 to 900 yd. and Horse 8 Ft. (for mounted troops) with a range of 150 to 900 yd. on the other. It is also marked 25 in. being the distance that it should be held from the face, generally considered to be arm's length.

Fig. 9.26. The range finder in use.

of Plans... Sold by the Inventor, at his Military and naval Academy, Charing-Cross, the corner of St. Martin's Lane." Sets were still being sold as late as 1928 by J. H. Steward and are illustrated in their catalogue of that date.[8] The simplest sets contain three pieces – the two scales and a triangle in a mahogany box with a sliding lid. The Marquois pieces were also included in more extensive sets of drawing instruments intended for the military officer.

Range finders

A simple direct measuring form of range finder is shown in Fig. 9.25. It consists of a boxwood plate into which is cut a V-slot. The range finder is held at the correct distance from the eye by means of a cord, with a button on the end, which may be held in the mouth or against the nose. The view of a man or a mounted horseman is then sighted in the slot and the range read from the position in the graduated slot. It reads from 100 yards (for men) and 150 yards (for horses) to 900 yards.

Fig. 9.27a & b. Range tape by J. H. Steward, as illustrated in their 1916 catalogue.

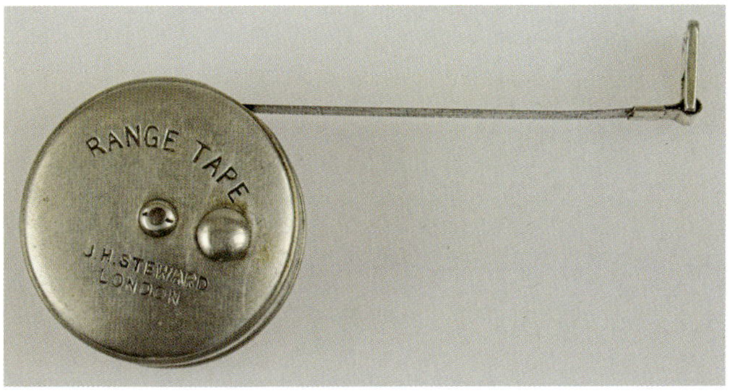

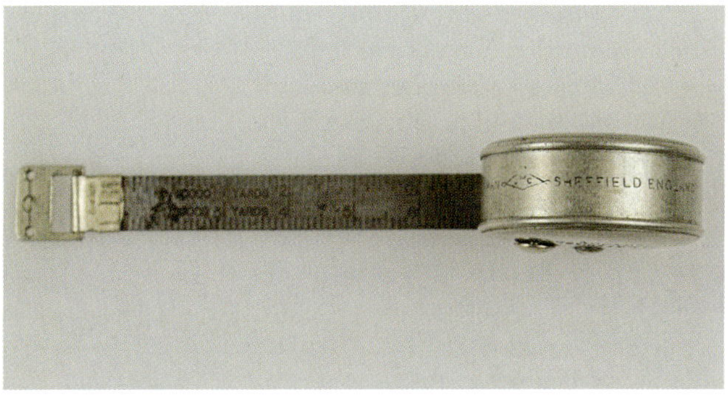

Range tape

Offered by Steward in 1916 this is a metal tape measure in a nickel case. It is divided at the mapping scales of 1/10000, 1/20000, 1/40000 and 1/80000. Rotated round a pin stuck into the map it was a handy method of measuring the distance to a target in any direction.

Gunners' folding rules

A handful of 17th century boxwood rules specifically adapted for use by gunners are known. These are 18 in. 2-fold boxwood rules with an additional brass arm slotted into one leg. In appearance and construction they are very similar to a carpenters' rule of this period, in that they have a wooden mid-joint with minimal metal parts, and thick legs. However they have various adaptations for gunnery.

In use the rule is unfolded into a straight line and the brass tongue is secured at right angles to it. Small sights are fixed into the edge of the rule. Using these an object can be sighted and the angle to it can be measured by means of a small plummet hung from the tongue. They can also be used for setting the elevation of guns.

A gunners' rule of ca. 1680 has been described in the Tools & Trades History Society's *Newsletters No. 56, 57 & 60.*[9] This rule carries tables detailing the quantities of powder for service and proof, length and breadth of powder ladle, and weight of shot for 15 sizes of ordnance; there is also an extensive table giving the weight of shot from 1 in. diameter up to 8 in. diameter in increments of 1/8 in.

The specification of guns by the weight of shot fired was only coming into use at this time – so there were many older guns which were specified by name. The rule bears numerals for some guns and initials for others; for example, WC – whole cannon, BC – bastard cannon, DC – demi cannon, F – falconet, etc.

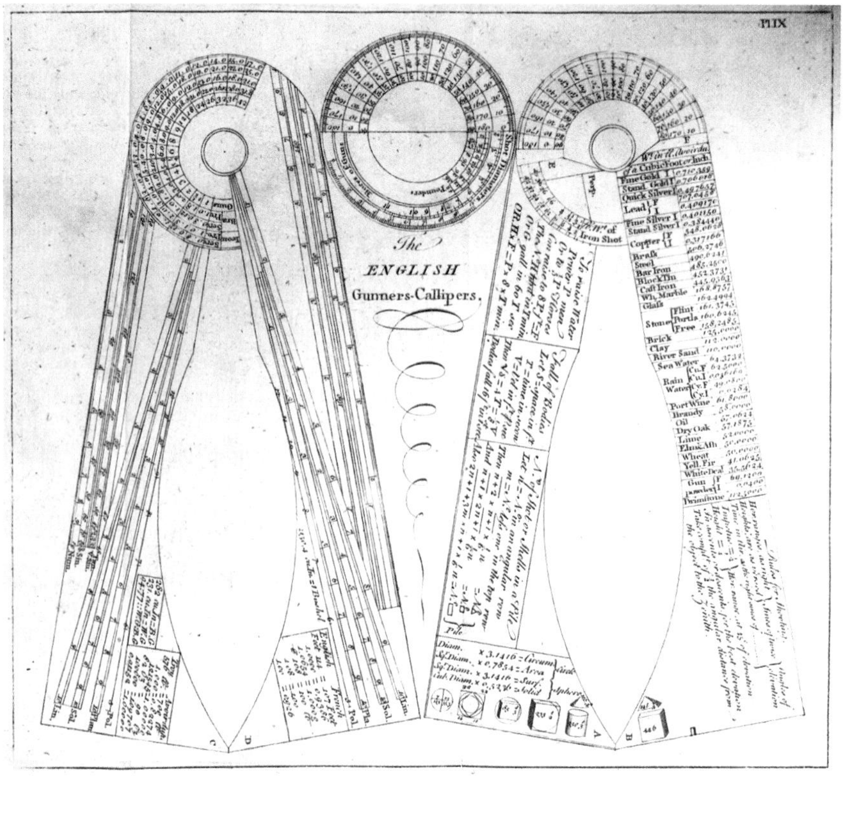

Fig 9.28. Gunners' calipers illustrated in Robertson's Treatise. *Note the sectoral lines and the illustrations of the geometric bodies which are marked with factors and weights. For example, the first two figures are constants relating to a circle and to circles inscribed in squares or squares inscribed in circles. The other figures are weights; for example, the third figure represents a cube of iron set within a sphere of 12 in. diameter – the weight of which is 90¼ lb.*

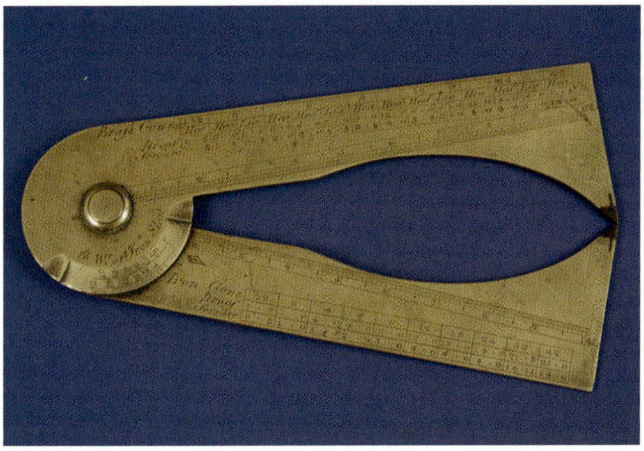

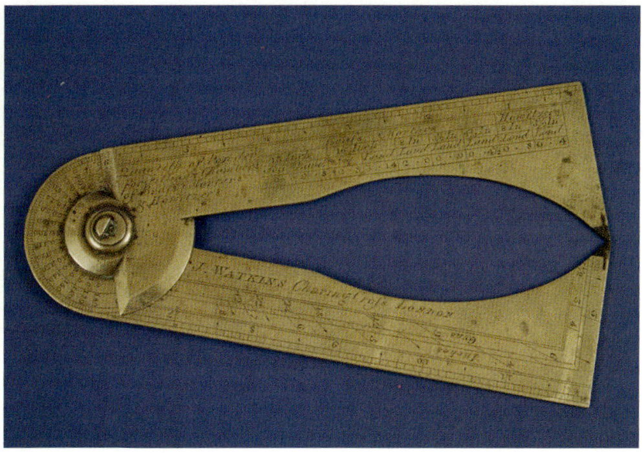

Fig. 9.29a & b. A pair of gunners' calipers by J. Watkins (London 1798–1801). The data and lines on these calipers are fewer than those illustrated by Robertson but there are extensive tables giving quantities of powder for proof and service for iron and brass guns from 1½ to 42 lb., brass mortars and howitzers. The only line for calculation is a line of equal parts.

Gunners' calipers

John Robertson writing in the appendix to *A Treatise of such Mathematical Instruments as are usually put into a Portable Case* in 1775[10] is somewhat disparaging of the gunners' calipers saying that the scales may be put on a sector and the tables put in a pocket book but that if you are desirous of a single instrument to perform many tasks .... well, the gunners' calipers are for you. Amongst the nineteen functions he lists, only the gauging of guns and shot and the taking of angles are essential.

The English pattern gunners' calipers were constructed of two brass plates of distinctive incurved shape joined with a hinge. The points tipped with iron were used to measure the diameter of shot and when passed over one another, the bore of a gun. The design is described by Edmund Stone in his 1758 supplement to the Bion text, but the format had been established 30 years

Fig. 9.30. An exceptionally large pair of gunners' calipers ca. 1775, 23½ in. overall, by Adams (London, 1733-1826).

Fig. 9.31. Detail of one of the tips of the gunners' caliper shown in Fig. 9.30 showing the marking for guns.

earlier.[11] Around 1830 improvements in gunnery rendered this rather simple device of limited accuracy obsolete.

Over the years the basic shape remained constant but the functions and data engraved onto the calipers tended to be reduced – many of the sectoral scales and the tables of specific gravity and weights of bodies of different geometric shape disappeared. The data relating to the weight of iron shot and the weight of powder for service and proofing of guns remained.

The gunners' caliper shown in Fig. 9.30 is very unusual, both for its size and also for the markings. The tips are marked for guns from ½ to 42 lb. On the legs are a diagonal scale of inches and a sectoral line of equal parts and sines. There is also a line of metals (effectively an inverted line of specific gravity taking tin at 1000 as the base). Lines of this type are found on French sectors but rarely on English sectors. The presence of this line however indi-

Fig. 9.32. Designed by Lieut. Cave-Browne of the 23rd Indian Cavalry, the "Visurect" Aim Director was a rule, 1 ft. long, divided into inches for map reading purposes but also with sights at either end. The metal clip was intended to allow the instrument to be attached to either a walking stick or bayonet .

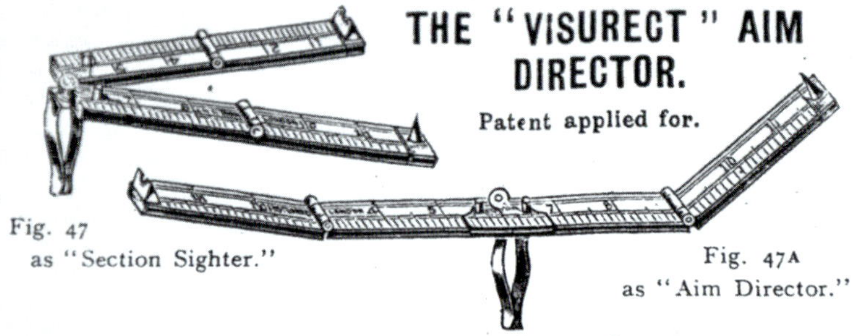

Fig. 9.33a & b. A gunners' aide-mémoire, in the form of an ivory rule by V. & J. Figgins (London, 1835–1910). It is divided on one side with linear distance scales of paces and of yards. For Cases and for Shells, distances are indicated to obtain Good-, Medium-, and Bad-Effect. The reverse has a table of elevations for different types of arms (e.g. Infantry Skirmishers) and various degrees of difficulty.

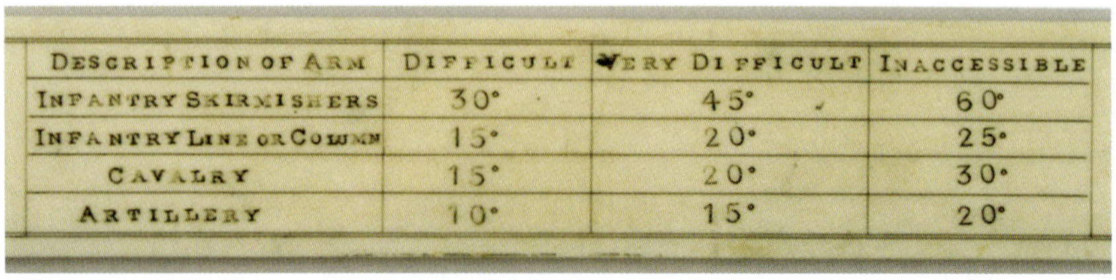

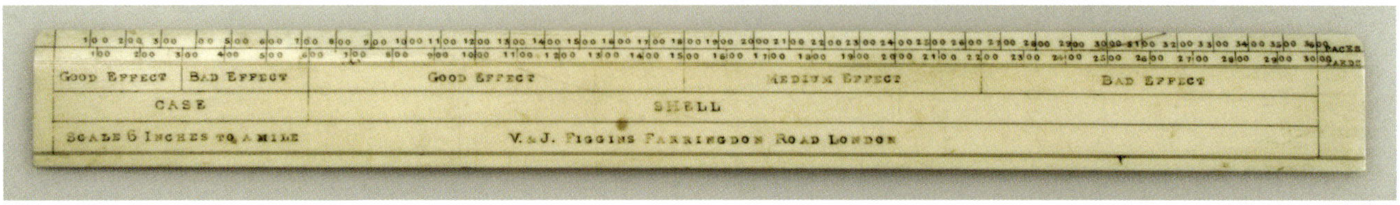

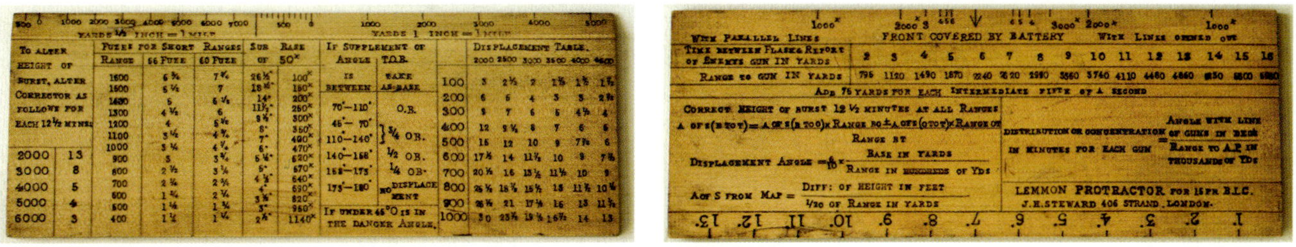

Fig. 9.34a & b. The Lemmon Protractor was designed by Capt. C. H. Lemmon R.F.A. It is 6 in. by 2¼ in. and although listed as a protractor, is not intended for military sketching. It includes scales and data of use by the light gunner in the field. One edge is marked with a sighting line of degrees. When held at arm's length (25 in.) each division subtends one degree. Made of boxwood, it is illustrated in J. H. Steward's 1916 catalogue.

cates something of the origin of the gunners' calipers but it is difficult to see what calculations a gunner might undertake that needed the specific gravity of gold, or even tin, though lead, brass and iron are perhaps understandable.

Artillery rules: During the latter part of the 19th century and early years of the 20th century, numerous rules and slide rules were invented or designed to aid the gunner. Amongst those listed in J. H. Steward's 1895 catalogue are a deflection scale for ranging artillery and the "Visurect" aim director.

By the time of Steward's 1916 catalogue, the artillery rules and slide rules listed included the McLeod Field Artillery Slide Rule, the Machine Gun-

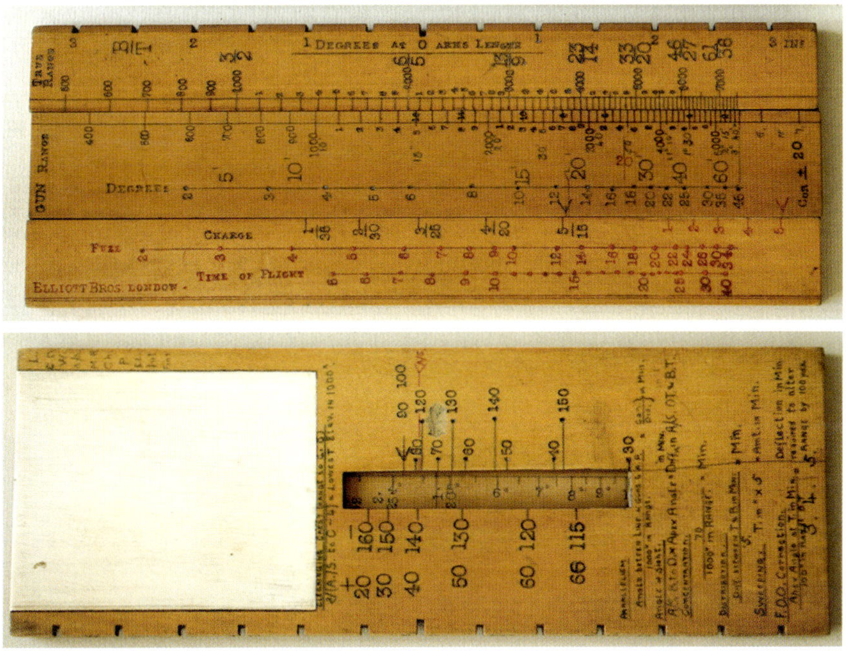

Fig. 9.35a & b. An "improved" version of Burton's 4.5 in. Howitzer Slide Rule, similar to that listed in J. H. Steward's 1916 catalogue, by Elliott Bros. (London, 1864–1913>) which incorporated an ivory tablet for notes.

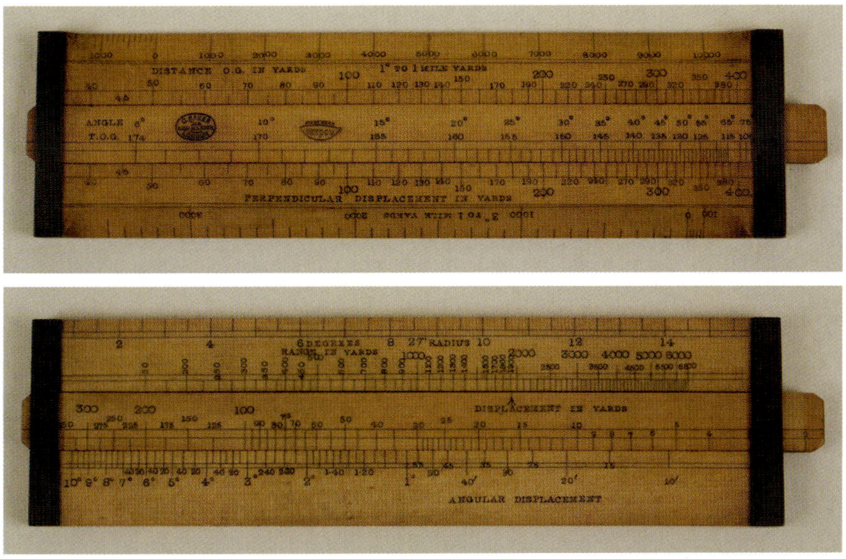

Fig. 9.36a & b. An artillery slide rule by G. Baker (London, 1851–1947).

ners' Slide Rule, the Service "M.G." Slide Rule, the Atmospheric Slide Rule, designed to give instant correction for changes in wind, barometric pressure and temperature, the Lemmon Protractor and Burton's 4.5 in. Howitzer Slide Rule.

References

1. Robertson, John, *A treatise of such Mathematical Instruments as are usually put into a Portable Case,* 3rd ed., London, 1775.
2. J. H. Steward Ltd., *Catalogue of Surveyor's, Mathematical & Nautical Instruments*, London, 1895.
3. J. H. Steward Ltd., *Catalogue of Surveyor's, Mathematical & Nautical Instruments*, London, 1916.
4. The War Office, *The Manual of Map Reading, Photo Reading and Field Sketching,* HMSO, 1929.
5. J. H. Steward Ltd., *Catalogue of Surveyor's, Mathematical & Nautical Instruments*, London, 1911.
6. W. F. Stanley, *Catalogue of Surveying and Drawing Instruments,* London, 1931.
7. Steward, 1911, *op. cit.*
8. J. H. Steward Ltd., *Catalogue of Surveyor's, Mathematical & Nautical Instruments*, London, 1928.
9. Sharman, Peter & Ann, "A gunner's rule & quadrant." *Newsletters Nos. 56, 57 & 60,* The Tools & Trades History Society, 1996, 1997.
10. Robertson, *op. cit.*
11. Bion, Nicholas. *The Construction and Principal Uses of Mathematical Instruments.* Paris 1723. Translated by Edmund Stone;… *To Which are Added, the Construction and Uses of Such Instruments as are Omitted by M. Bion* by Edmund Stone. Second edition; … *To which is added a Supplement Containing a Further Account of some of the Most Useful Mathematical Instruments as Now Improved.* Supplement by Edmund Stone, London, 1758, reprinted Mendham, NJ, 1995.

Chapter 10
Nautical numbers

The sliding rule is generally made of boxwood, and consists of two pieces, or legs, each twelve inches in length, which are connected by a folding joint. Upon its surface and edges are engraved lines graduated into fractional portions, etc. particularly adapted to shipbuilding, carpentry, mastmaking, etc.... On the inside edge are two divisions of lines, which may be used indifferently for the same purpose, namely the eight-squaring of a four-sided piece. These lines are not given on the common sliding rule, but they will always be found on Steel's Improved Rule for the use of Mastmakers and Shipwrights. The Shipwright's Vade Mecum, 1822.[1]

Although today wooden boat building is mostly restricted to small craft, replicas and re-constructions, in the past this was an important woodworking trade and like many other trades, had a number of specialist measuring tools to assist and speed the work.

Fig. 10.1 A boat builders' bevel in boxwood and brass by A. Wellington, (London, 1784–1827).

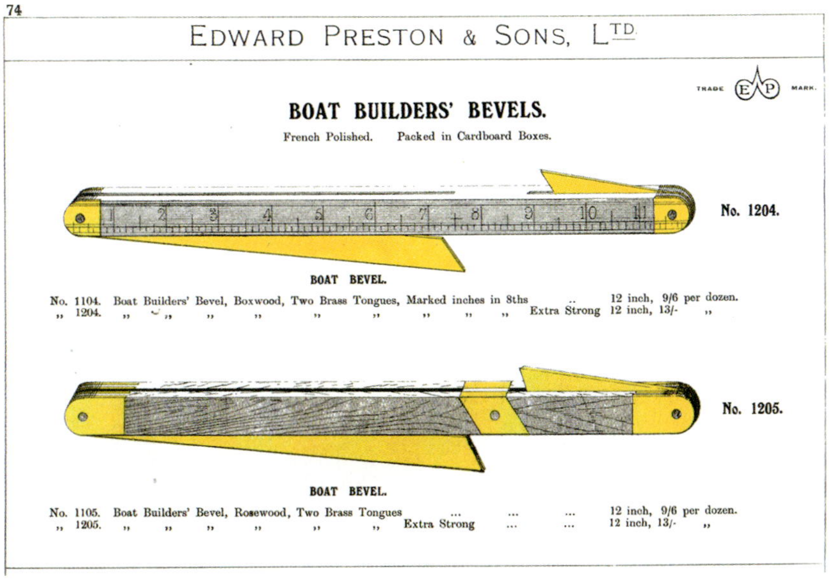

Fig. 10.2. The two styles of boat builders' bevels offered by Edward Preston & Sons in their 1909 catalogue.[2]

Fig. 10.3. Two boat builders' bevels by Edward Preston & Sons (Birmingham, 1864–1932). Top: With a rosewood stock and brass tongue, this style is not shown in any of the known Preston catalogues but is marked with their name. Bottom: A more traditional boxwood and brass boat builders' bevel. Note the similarity to the bevel by Wellington in Fig. 10.1.

Boat builders' bevels

When constructing a wooden boat, the plank edges are bevelled to fit both around the curve of the hull and along the length of the boat. Decks are also laid with a camber to allow water to disperse. The boat builders' bevel had developed to a standard design certainly by the end of the 18th century, as the bevel shown in Fig. 10.1 demonstrates. The style remained unchanged until well into the 20th century, though some minor variations were available. Boat builders' bevels are usually around 12 in. long but considerably narrower than the standard woodworkers' bevel, being only about ⅝ in. wide. Most manufacturers offered them in either boxwood or rosewood and with or without a line of measure. A stock with two brass fold-out blades, of differing lengths, was the more common design but they could also have just a single brass blade.

Mast makers' rules

The traditional method of making a mast was to first shape the timber into a square section, then into an octagon and from this into a round. The edge and middle scales (E & M or 8-square lines) commonly found on carpenters'

Fig. 10.4. A mast makers' rule, unmarked but undoubtedly early 18th century. This rule does not have a Gunter's line marked on it, but a line of board measure and a line of timber measure on this face.

Fig. 10.5. The reverse face of the rule shown in Fig. 10.4. This face has a simple line of inches together with the two sector lines required to set out masts and yards.

rules facilitate the setting out of an octagon from a square so these scales will also be found on rules intended for the mast maker.

Part of, or even all of the mast or spar might be left as an octagon but whatever the cross section it will be tapered. Both the sizes of the spars and the degree of taper were established over the centuries and this knowledge published in many works on shipbuilding. The mast makers' line, which is a sectoral line, gives the information needed to set out the taper, which is not necessarily constant throughout the length.

Kebabian[3] has described a rule of probably 18th century date that is marked with a line for masts and another for spars.

It is a 2-fold rule with each leg measuring 9 in. and an additional brass tongue that folds into one leg which when fully unfolded makes a rule 24 in. long. It has a Gunter's line marked on one face, together with board and timber tables and on the other face a line of measure and adjacent to the inner edges, two sector lines. The inner sector line for masts is marked P (partners – the supporting timbers where the masts pass through the main

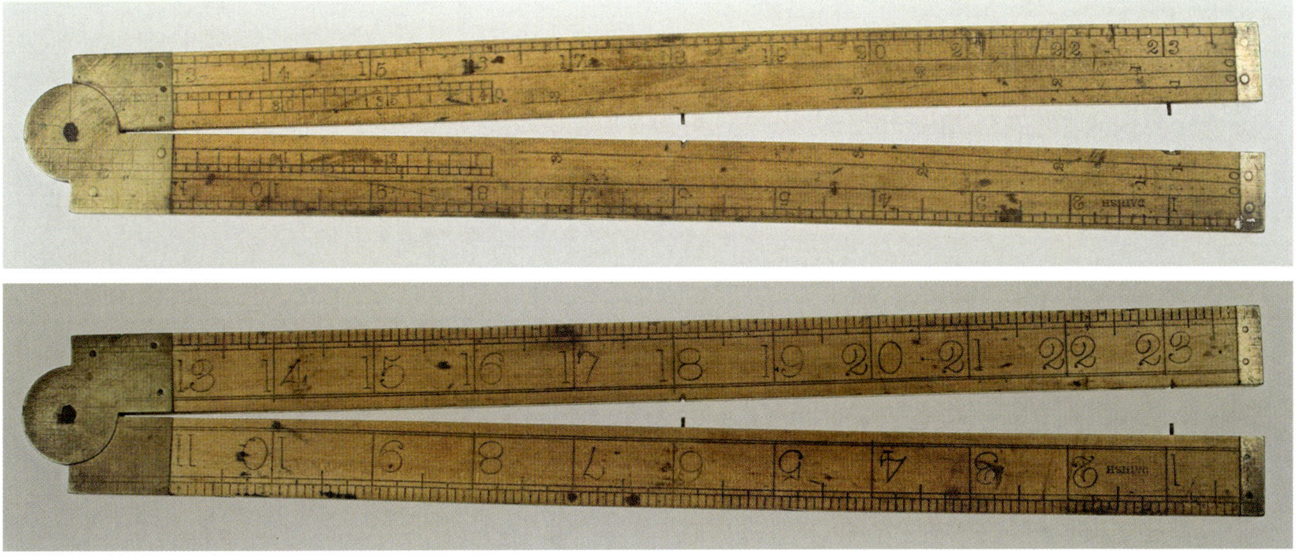

Fig. 10.6a & b. A 2-fold mast makers' rule. There is no maker's name but it is likely that it is English made. The lines of measure on both sides are labeled DANISH, those on the reverse being a simple line of measure. Marked to 24 in., it is actually 24 9⁄16 in. long.

Fig. 10.7a & b. A 2 ft. 2-fold boxwood rule, by Lorkin, (London, 1789–1830), with a brass log slide in one leg. The remaining area of this face is divided into boxes specifying relative sizes of masts and spars. The reverse face has a line of measure along the edges with the remaining area marked to give the sizes of masts, spars, thwarts and other parts of a ship.

deck level); 3Q, 2Q, 1Q, (quarters – the next section of the mast is divided into four lengths with different diameter proportions) and MH, (masthead) and the outer line for yards (the horizontal timbers that hold the sails on a square rigged ship) is marked S (slings), 3Q, 2Q, 1Q and YA. (yard arm, the end portion of the yard). As with masts, the taper along the yards varies throughout their length.

Using the mast rule: The rule is opened until the marks representing deck level on each leg are, say, 3 in. apart. If for example, a mast of 12 in. diameter at this point is required, the compass is walked on the work four times giving the required 12 inches. Without altering the angle of the rule the compass is then used to take the dimension between the next pair of points. This dimension is then also multiplied by four and applied to the work. Thus a slightly smaller dimension will have been marked out. The process is repeated at each of the gauge points on the rule, and then the shaping of the mast can begin. The use of the sectoral lines is, of course, in principle the same as the use of the sector in setting out the taper or other proportions in architecture.

The rule just described is a sophisticated one; other mast makers' rules are simpler, sometimes with the additional brass leg, sometimes just 2 ft. 2-fold rules with the lines only marked with numbers or even inset gauge points with no annotation.

The shipwright faced many problems of size and proportion about which guidance was to be found in the many manuals about ship construction, but the shipyard was no place for a book so recording of the most frequently used data on a rule would seem to be the motive for the rule described in Fig. 10.7.

Fig. 10.8a & b. Another mast makers' / ship builders' rule, by George Cox (Birmingham, 1815–30). Similar to the rule shown in Fig. 10.7 but with slightly different information in the boxes.

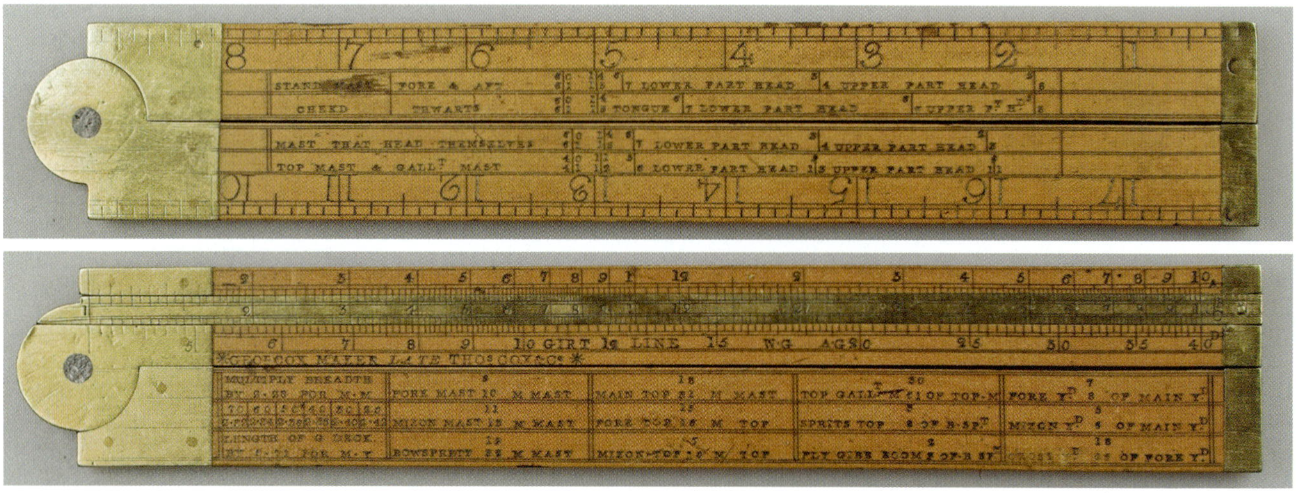

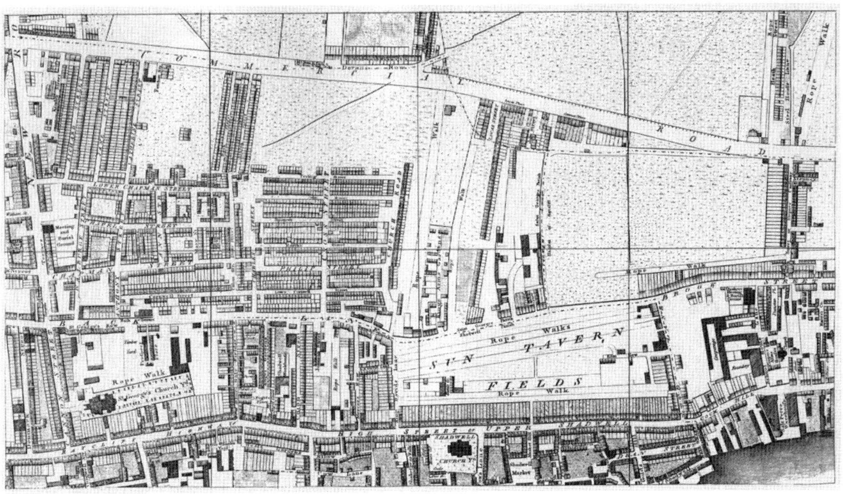

Fig. 10.9. Several rope walks shown on Horwood's 1799 map in the Shadwell area of east London.

Fig. 10.10. The ropewalk at Chatham Historic Dockyard in Kent. This is the only traditional working ropewalk to survive anywhere in the world.

Rope

Sailing ships needed ropes of the best and most durable quality. There is no natural fibre that is durable for more than a few years and even when measures such as tarring were taken, rigging lasted no more than 3 or 4 years. So the demand for cordage was never-ending.

Whilst virtually any fibre can be made into a rope, where the minimum size of rope is important to reduce wind resistance, as in rigging, hemp was the only practical choice. The traditional process of making rope is as follows. The hemp is first hatchelled (cleaned and broken down and combed by being thrown onto and dragged through the hatchel, a ferocious bed of standing spikes). Then the actual making of the rope can commence.

The yarns (threads) are spun. A number of yarns – three and upwards, possibly as many as 20 or 30 if the rope is large – are then twisted together to form a strand; then three strands are twisted together (the term used is laid up, sometimes spelt as layed) to form the rope. For the larger sizes three ropes will be twisted together to form a cable. Other make-ups – four or more strands laid up around a core – also occur but are less common.

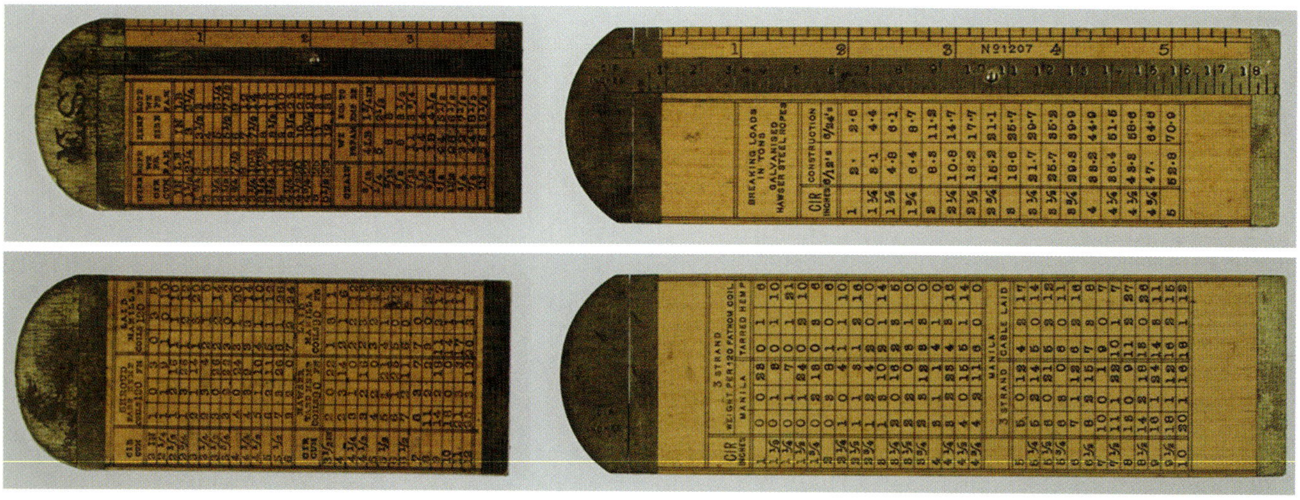

Fig. 10.11a & b. Left: Rope gauge by Edward Preston & Sons (Birmingham, 1864–1932) with tables for wire rope, hemp rope and chain on one side and shroud, laid manilla in 120 fathom coils and 90 fathom coils, and hawser on the reverse. Right: Rope gauge by John Rabone & Sons (Birmingham, 1877–1963) with a table for the breaking load of galvanised hawser steel ropes on one side and tables for 3-strand manilla, tarred hemp and cable laid on the other.

Fig. 10.12a & b. Left: An advertising gauge, unmarked, for a company that manufactured wire ropes. This gauge is notable in both its shape and its lack of any tables. Right: An unusual gauge by John Rabone & Sons (Birmingham, 1877–1963) that measures the weight of rope for round wire rope, hemp and manila. This gauge includes a calculating slide as well as the usual caliper.

Until about 1840 when the first static rope making machines were installed, all rope was made in rope walks. The yarns (threads or twines), the strands and the rope were all made by applying twist at the working end of the walk. The operators worked backwards along the walk to ensure that the twist was put in evenly. There were practical limits to the length of twine, strand or rope that could be made in a rope walk and this is no doubt the reason why the length of a coil of rope became standardised at 120 fathoms (720 ft.), known as the sailors' cable's length. To make a rope this long required a rope walk of over 1000 ft. to allow for shortening as the strands were twisted.

The rope walks adjacent to the river are prominent features on Horwood's map of London of 1799. Rope walks continued to be used long after the arrival of efficient machines; the enclosed facility at Chatham Royal Dock Yard continued to operate until the yard closed in the 1970s.

Rope gauges

These are short wide rules with a caliper at one end and all the major rule makers included them in their catalogues. They are usually in boxwood and brass, normally either 4 in. or 6 in. long, although both Preston in their 1891 catalogue[4] and Rabone in their 1892 catalogue[5] offered a 4 in. ivory version. Ivory rules had been dropped by 1900. Although gauges are found marked with a great variety of tables, in the rule makers' catalogues, they are almost always shown with just the same single table.

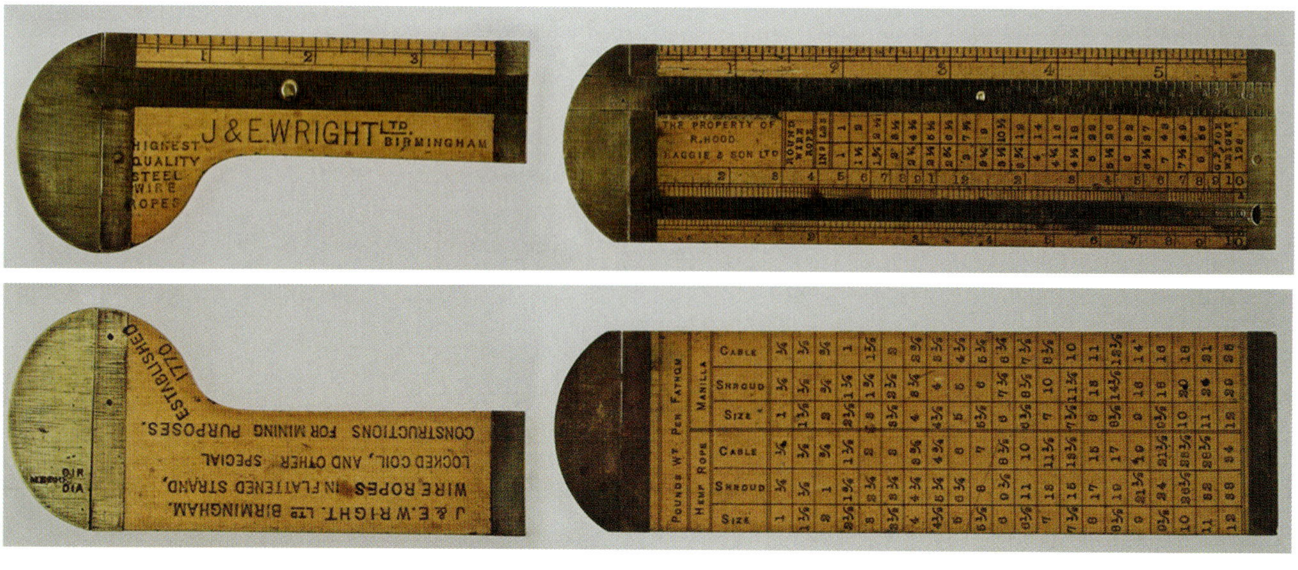

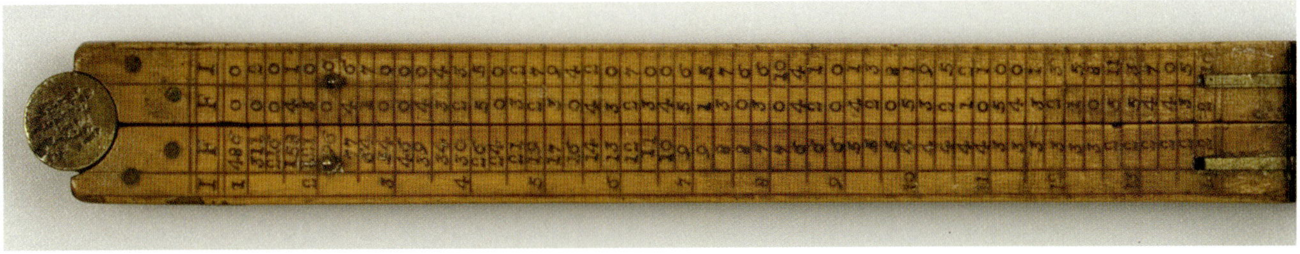

Fig. 10.13. *The IFFI table from a boxwood naval officers' rule dated 1771.*

The caliper is marked to give a direct reading of the circumference of a rope, which is the dimension by which rope is specified. This can give rise to misconception of the size being offered, as an heroic sounding 6 in. hawser is actually a rope of just under 2 in. diameter. Usually, though not always, both faces are marked with tables. Many examples are marked with advertising for rope makers, chandlers or riggers.

It is laborious to measure a long length of rope so it was normal practice to sell rope by weight. The primary purpose of the rope gauge is to facilitate the determination of length from the weight of a variety of types of rope, both natural fibre and wire.

The following are the tables commonly found on rope gauges:

- Size/weight in cwt., qtr. and lb. per coil of 120 fathoms. These are sometimes repeated for different materials such as hemp and manilla. They can also include tarred or white rope and strands and laid.
- Size/weight per coil of 90 fathoms.
- Size/weight per fathom.
- Equivalent strength tables for rope, wire rope and chain.
- Working strength tables.

An early form of rope table is the IFFI table. Tables of this type are not common, but have been observed on a number of 18th and early 19th century rules.

The purpose of the table is to give the length of rope that weighs one hundredweight (112 lb.) of a specified size. The first column gives the circumference in inches and fractions of an inch, the next three give the length in fathoms, feet, and inches.

Rope makers' slide rules

There are two recorded forms of a slide rule made specifically for the rope maker. In addition to the normal general calculations provided by log scales,

Fig. 10.14a & b. Rope makers' slide rule by William Coulsell & Son (London, <1870>).

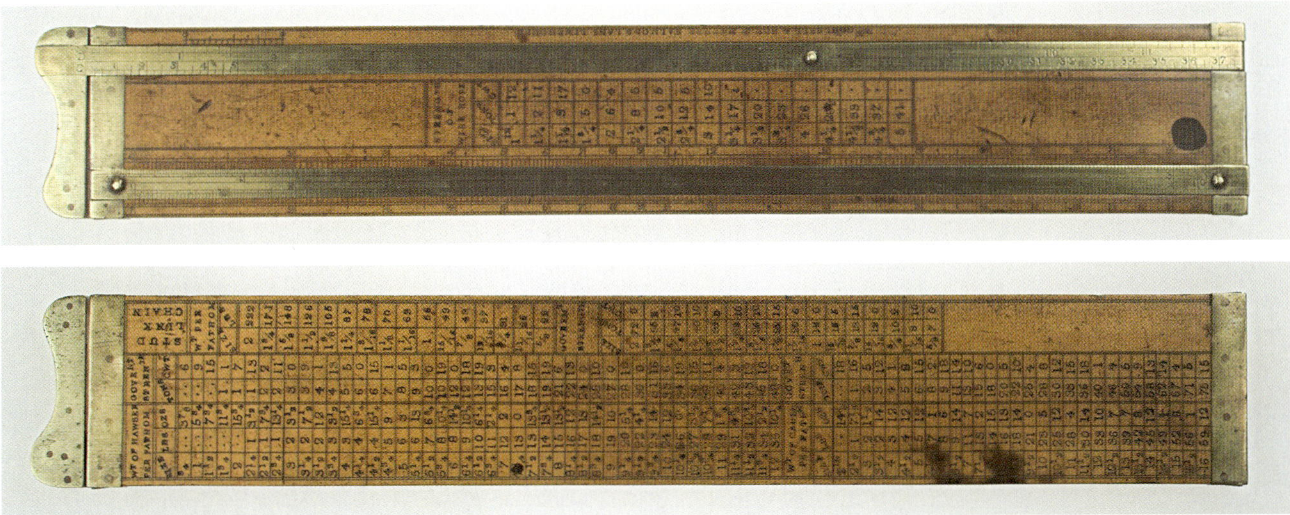

the purpose of this rule was to determine the number of yarns needed to make a rope of the required size.

One is marked Wm. Coulsell & Son, Salmons Lane, Limehouse, and is shown in Fig. 10.14. It is a 12 in. long rule with a caliper. In general, it is similar to a rope gauge but also includes a brass calculating slide marked with two cycles of logs. One edge of the slide is marked with a single line of logs folded at 3.0. The use of this line in rope making is described below.

The other was made by Edward Preston & Sons Ltd. It is 12 in. long and has the same arrangement of slides divided as the Coulsell rule. There is no record of it in any of the known Preston catalogues but a full description of the rule and its use was published in *The Textile Manufacturer*, in 1927.[6]

The rule was used as follows. The principal decision to be made by any rope maker is how many yarns need to be included in each strand in order to make a rope or cable of the required size. As yarns varied somewhat in size the slide rule was used to work this out. Yarn size was specified by the number

Fig. 10.15a, b, c, d, e & f. A rope makers' rule, ca. 1800. The 3 ft. 4-fold rule is marked with a remarkable eight tables, all of which are relevant to rope making. The unusually thick section allows sufficient area for two of the tables to be placed on the inside edges. In addition to the usual lb. / cwt. / money table, IFFI table (the circumference / cwt. / length table), and two versions of the weight per 120 fathom coil table, there are tables whose principal purpose is to list the number of yarns needed to make ropes and cables of different sizes. The precise use of these tables has not been established.

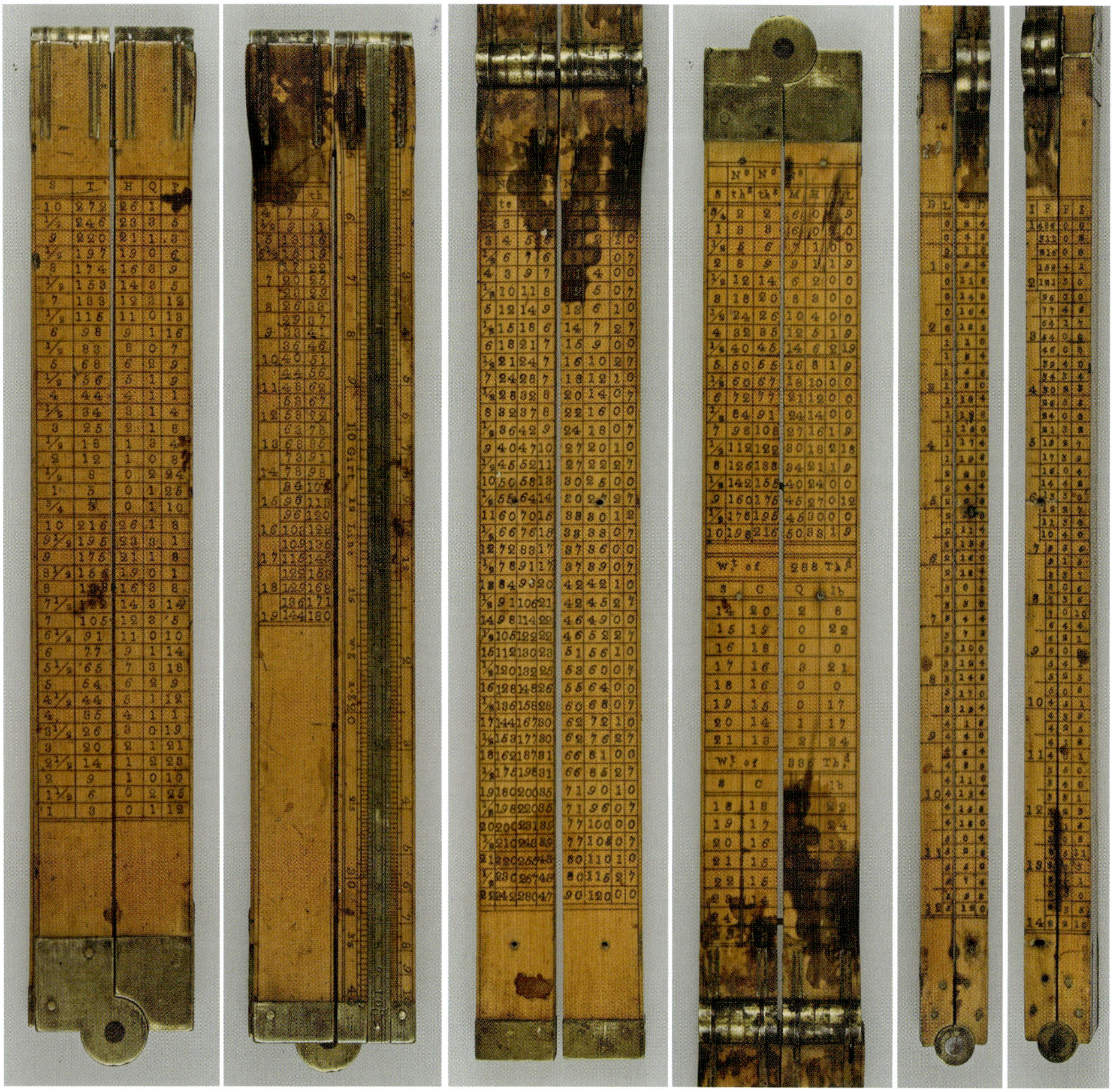

needed to make a rope of 3 in. circumference. This might be from 18 to 60. Thus the yarn sizes were 18, 20, 24, etc.

The special line of divisions on the rope makers' slide rule gives an answer quickly and easily. By setting the 3 on line B against the yarn number on A, the number of yarns to make any size of rope can be directly read. Incidentally a 6 in. rope made with size 24 yarn needed 96 yarns. Similar calculations can be made for four strand ropes and also cables.

Rope makers' rules

Fig. 10.15 shows a rule intended for the rope maker. There were probably others made, but this is the only one known. It contains a log slide (interestingly, with the carpenters' girt line arrangement, not the rope makers' arrangement) and is marked with eight tables, including the hundredweight (money) table, an IFFI rope table and two versions of a weight per coil of 120 fathoms table. The other tables relate to the number of threads needed to lay up ropes of the specified size.

Ships' officers' rules

These rules resemble a four section dip rod but are also marked with lines and tables which would be of use to a ships' officer. The few known examples all appear to date from the last quarter of the 18th century.

In addition to a line of inches all the rules have one or more lines for barrel dipping. The other lines that may be present include a line of diameter of guns calibrated by weight of shot, a similar line for diameter of a shot, a line of foreign inches, and an IFFI rope table. Two such rules are shown in Figs. 10.15, 16, & 17. A fuller description of these rules can be found in *The Chronicle of the Early American Industries Association.*[7]

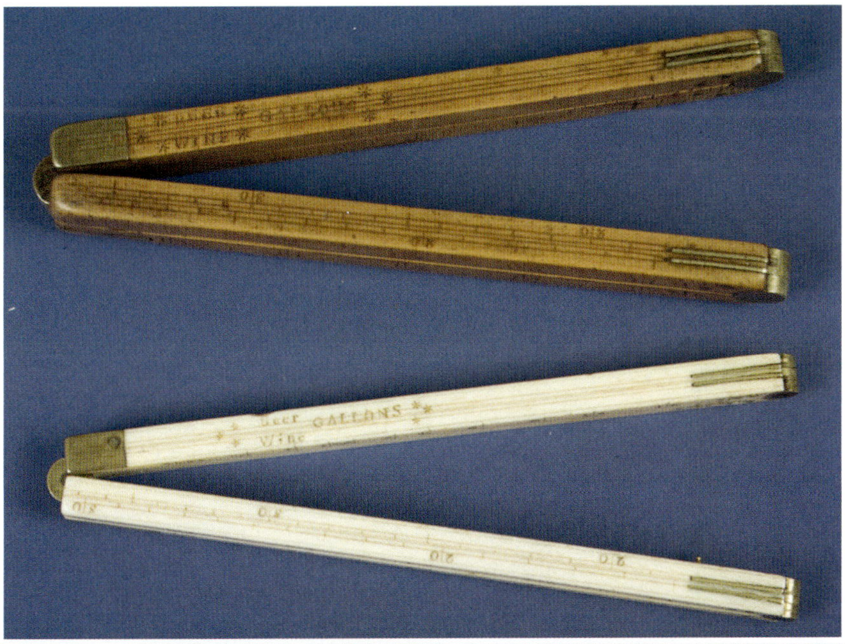

Fig. 10.16. Two 18th century ships' officers' rules showing the dip rod function. The boxwood version is dated 1771.

Fig. 10.17. The inside of the two rules showing the IFFI tables and the hundredweight tables.

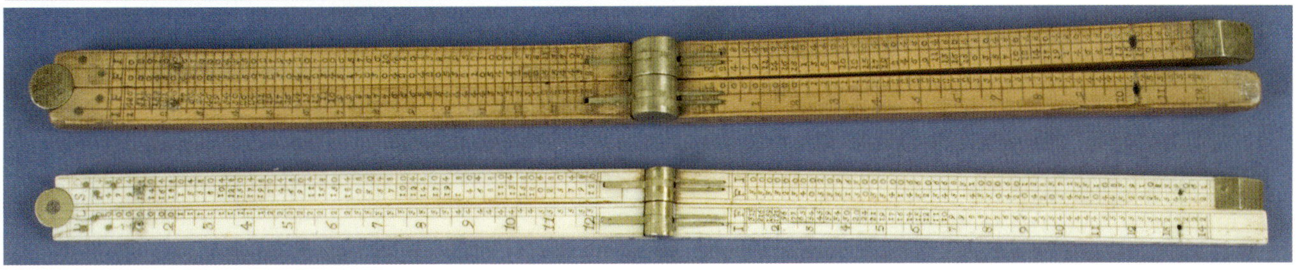

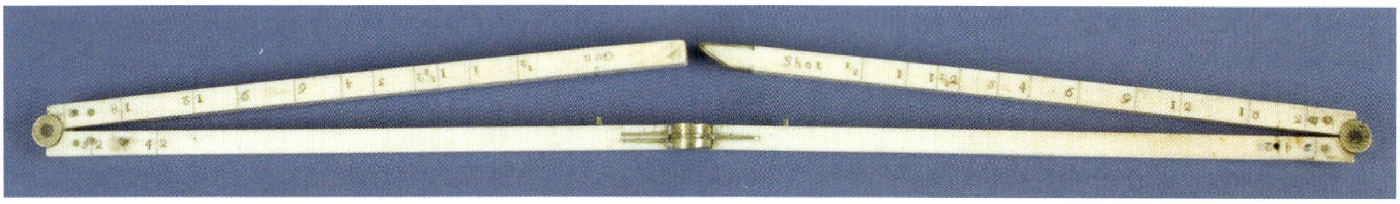

Fig. 10.18. A line of shot calibrated by diameter and a similar line for the diameter of guns calibrated by the weight of shot.

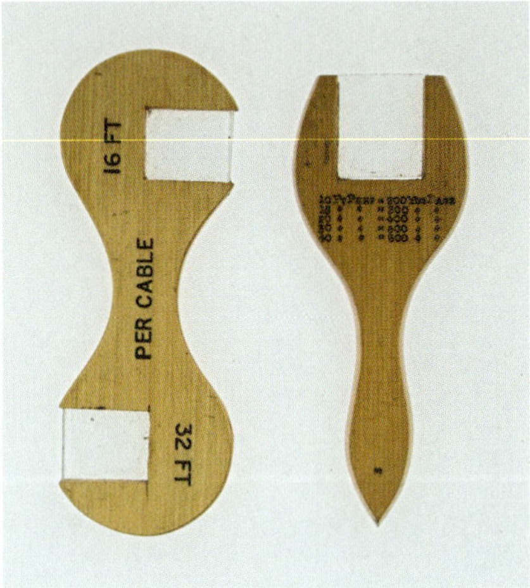

Fig. 10.19. Approximately 3 in. long, so easy to keep in a pocket, the station keeper is held out at arm's length and when the next ship in the formation fills the glass view piece, its distance away can be read off the table. The single ended version is by William Cary (London, 1789–1897). The double ended version is unmarked. This form is rare. The table assumes a knowledge of the size of the ship being viewed.

Above right: Fig. 10.20. An ivory navigation aid that calculates the drift off course, based on the strength of the current by Elliott Bros. (London, 1864–1913>).

Station keepers

Intended for use when ships were travelling in formation or convoy, this is a simple device, in boxwood and glass to ensure that the ship stayed "on station", i.e. in the correct place in relation to the other ships travelling with it.

Navigation

Some of the aids used for navigation are not rules that measure and are not therefore strictly speaking within the compass of this book. However, they are frequently found and sometimes include scales or other lines of measure so seem suitable for inclusion.

Station pointers

Originally used for maritime surveying, for instance when it is necessary to plot where shallows, rocks and other dangers occur, this instrument removes the need for extensive triangulation. A full description of its use can be found in F. W. Simms *Treatise on Mathematical Instruments.*[8]

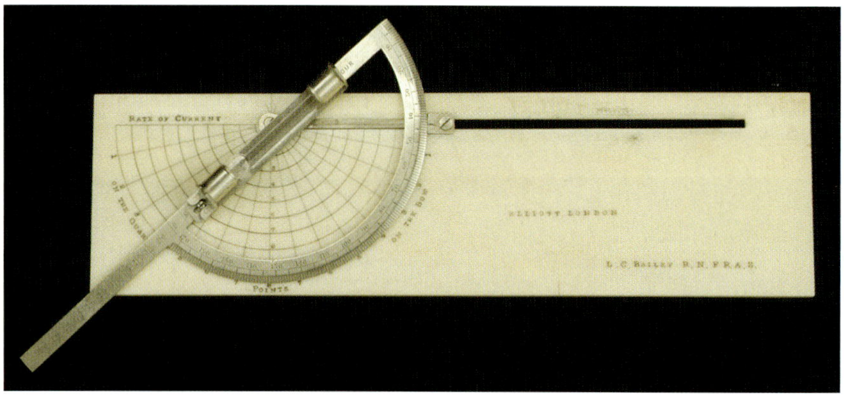

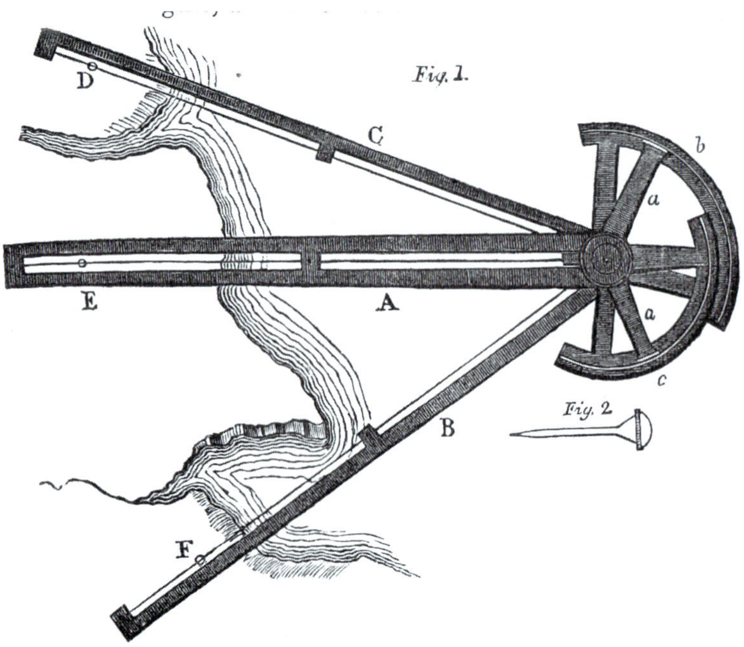

Fig. 10.21. Illustration of a station pointer from F. W. Simms Treatise on Mathematical Instruments, *1849.*

Parallel rules

Parallel rules of both bar and rolling types were widely used for chart work at sea. On land, similar rules were used by all professions that needed to draw – but on land other methods soon rendered them obsolete. At sea the use of parallel rules of both types continues today.

Small 6 in. long parallel rules were a component in many drawing cases, and 9 and 12 in. rules were included in larger boxes. Rules of these sizes would not have been much use for chart work so it is likely that only the larger sizes of bar parallels, from 15 in. upwards, were nautical in intent.

Bar parallel rules

Rules with two, or occasionally three, sections linked with metal bars. On older rules the links are often attractively shaped. Earlier rules are rosewood or ebony but recent examples may be brass or plastic. "Double joint" (three section) rules, which have extended reach, are rare.

Fig. 10.22. Glaholm's patent parallel rule. Patented in 1904 by J. W. Glaholm and G.T. Kent of Nanaimo, B.C., Canada, this is a most unusual form of parallel rule with three arms and additional marking points that run along the top arm to enable measuring or laying off distances without the use of dividers. Although J. W. Glaholm was living in Canada, it is likely that he had family connections with Glaholm, the rule maker in Sunderland.

Captain Field's rule

An improved form of arm type parallel rule, the improvement being the marking of one bar with degrees and the other with points of the compass. These rules have been made in huge quantities over the last 100 years. The

Nautical or Chart Parallel Rules.

P1186 to P1193.

P1186	**Parallel Rule,** Bar, Captain Field's, boxwood, protracted,	15-inch	**10**	**0**
P1187	,, ,, ,, ,, ,,	18-inch	**11**	**0**
P1188	,, ,, ,, ,, ,,	21-inch	**14**	**0**
P1189	,, ,, ,, ,, ,,	24-inch	**16**	**0**
P1190	,, ,, ,, ebony, ,,	15-inch	**11**	**0**
P1191	,, ,, ,, ,, ,,	18-inch	**12**	**0**
P1192	,, ,, ,, ,, ,,	21-inch	**16**	**0**
P1193	,, ,, ,, ,, ,,	24-inch	**18**	**0**

P1196.

P1196	**Parallel Rule,** Rolling, Admiralty pattern, 18-inch, with protracted transparent celluloid base and gun-metal bridge, as illustrated, in mahogany case	**3**	**0**	**0**
P1198	**Chart Dividers,** single handed, electrum, 6-inch		**8**	**0**

Fig. 10.23. Captain Field's parallel rules and Admiralty pattern rolling rules from the W. F. Stanley 1931 catalogue.[9]

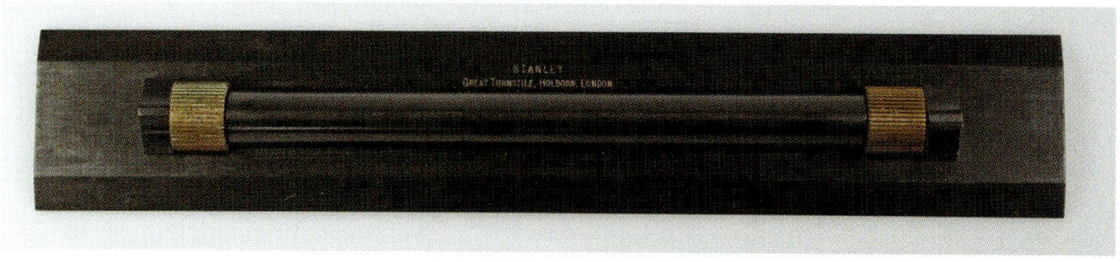

Fig. 10.24. *A blackened brass rolling rule by W. F. Stanley (London, 1853–1967).*

Air Ministry markings found on some examples indicate that they were used for air navigation as well as the more normal use at sea. Many are poorly made in cheap woods with the better rules mostly in boxwood or ebony but opaque and clear plastic models are also common.

Rolling rules

Like parallel bar rules, rolling rules come in the widest range of sizes, materials and markings. Most of the rolling rules intended for navigation are marked with points of the compass or degrees, though plain ones are also found, and many come in boxes – very necessary to prevent them rolling around when not in use – but even so the edges of many are damaged. The larger sizes made in brass or nickel silver are heavy items capable of maiming the unwary in heavy weather.

Gunter's scales

The origins of the Gunter's scale lie in the invention by Edmund Gunter (1581–1626) of a logarithmic line of numbers, but is used with dividers for mathematical calculations. Gunter described his invention in both his *Canon Triangulorum* published in 1620[10] and in his *Description & Use of the Sector, Crosse-staffe & other Instruments* published in 1624[11] and is quoted by Cajori in his history of the slide rule.[12]

By 1652 at the latest, and probably considerably earlier, other lines had been added which enabled the scale to be used for both navigation and dialling (the design of sun dials). The lines on one face of a 12 in. brass Gunter's scale signed "H. Sutton fecit 1652" includes a line of numbers, marked N, lines for sines, S, tangents, T, hour lines, H, rhumbs, R and latitude, L. The other face has two simple lines of measure on each edge and two adjacent diagonal scales.

A very similar 12 in. rule, signed "W. Hayes fecit" (1648–84), shown in Fig. 10.25, has, on one face, lines marked *Numb, Sine, Tan, Mer* and *H*, being respectively a line of numbers, sines, tangents, meridians and hours. There are three further lines which are not labeled. On the other face there are again two lines of measure, marked one on each edge, one being divided into inches, subdivided into 20ths, and one being the length of the rule (one foot) divided into 10ths, again subdivided into 20ths. Between these two lines is a single diagonal scale.

Fig. 10.25a & b. Brass Gunter's scale by Walter Hayes (London, 1648–84).

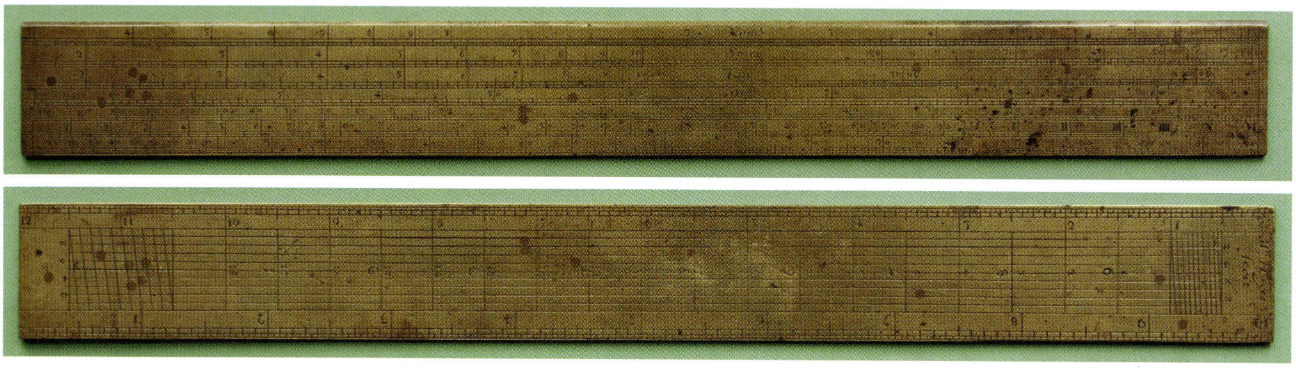

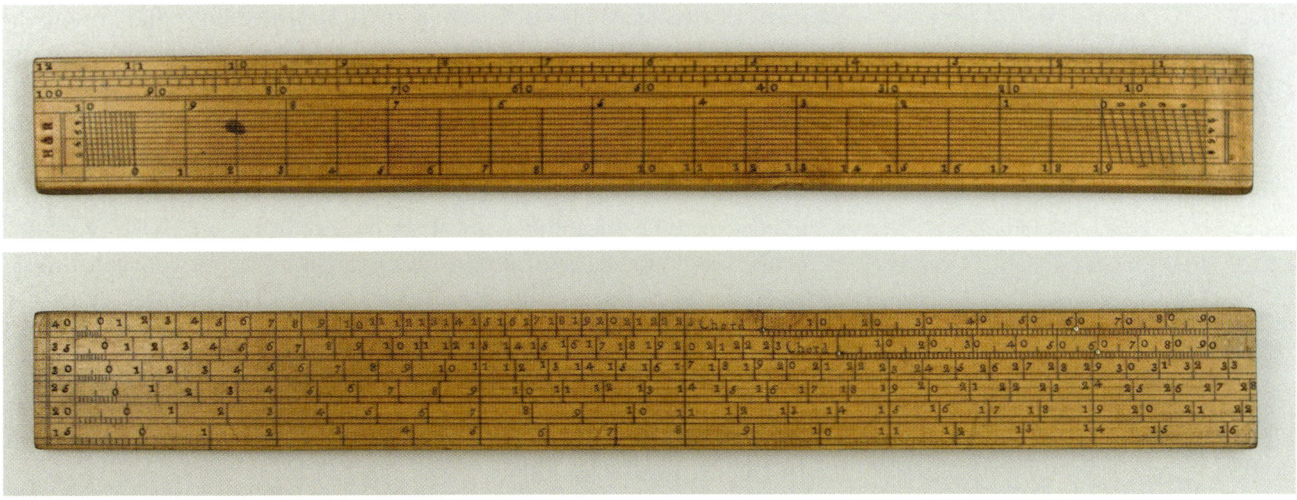

Fig. 10.26a & b. *A boxwood plain scale. This rule is only marked with the initials H&R, which are likely to stand for Hassall & Rabone (Birmingham, 1784–91). It is of rather simpler design than that described by Robertson as it is only marked with scales and chords.*

In Edmund Stone's translation of Bion, published in 1758[13], there is a much more specific interpretation of a Gunter's scale, where further additional lines for navigation, dialling and astronomy have been added. Bion states that a Gunter's scale should be 24 in. long (though he does admit that 12 in. versions also exist). He also mentions a plain-scale. This is 12 in. long (though again he reluctantly admits that there can be 6 in. versions) with a diagonal scale on one face and some of the scales found on a Gunter's scale on the reverse, though he doesn't list which specific lines.

John Robertson in his *Treatise of such Mathematical Instruments ...*" published in 1775[14] however does give a list of the lines.

Bion's[15] description of the lines necessary for a Gunter's scale specifies the following: Line of Numbers marked *Numbers;* Line of artificial Sines marked *Sines;* Line of artificial Tangents marked *Tangents;* Line of artificial versed (reversed, i.e. inverted) Sines marked V.S.; Artificial Sines of the Rhumb marked S.R.; Artificial Tangents of the Rhumbs marked T.R.; The Meridian Line in *Mercator's* Chart, marked *Merid;* and line of equal Parts, marked E.P. These

SECT. VII.

Of the Plain Scale.

THE lines generally drawn on the plane ſcale, are theſe following:

		Marked
I.	Lines of equal parts.	E.P.
II.	——— Chords.	Cho.
III.	——— Rhumbs.	Ru.
IV.	——— Sines.	Sin.
V.	——— Tangents.	Tan.
VI.	——— Secants.	Sec.
VII.	——— Half Tangents.	S. T.
VIII.	——— Longitude.	Lon.
IX.	——— Latitude.	Lat.
X.	——— Hours.	Ho.
XI.	——— Inclinations.	In. Mer.

Fig. 10.27. The lines to be found on a plain scale from Robertson's Treatise.

lines are all two feet long. In addition there are three lines that are only a foot long: Lines of Latitude, Hours and the Inclination of Meridian. On the reverse face are placed all the lines that are put on a plain scale.

Although some of the lines on plain scales can also be used for navigating – and both Bion and Roberston give examples of their use – and are in reality also Gunter's scales, even by Bion's time, the term "a Gunter's scale" had achieved a specialist meaning as a scale intended for navigation.

As previously noted, Bion also insisted that a Gunter's scale needed to be 24 in. long. However, in reality, this is not necessary and both 12 in. and 24 in. scales were made in quantity.

Layout: By the late 18th or very early 19th century the form and layout of a Gunter's scale had become standardised. It is typically a boxwood rule about 1¾ in. wide with one bevelled edge, either 12 in. or 24 in. long, although longer lengths are also found.

The normal layout is 12 or more lines designed for calculation, together with a line of inches, a line of the foot in 100 parts and a ½ in. and 1 in. diagonal scale. On 12 in. rules all the calculating lines are on the back but on 24 in. rules some are on the front alongside the scale. Fig. 10.30 shows a typical 12 in. Gunter's scale.

S R	*Sine of Rhumbs* [a]
T R	*Tangent of Rhumbs*
NUM	*Numbers* [b]
SIN	*Sines*
V S	*Versed sine (inverted sine)*
TAN	*Tangent*
MER	*Meridian line* [c]
E P	*Equal parts*
CHO	*Chords*
SIN	*Sine*
TAN	*Tangent*
S T	*Sines of Tangent*
RU	*Rhumbs*
HOU	*Hours*
LAT	*Latitude*

Fig. 10.28. More modern Gunter's scales have slightly different lines and abbreviations than those quoted by Bion.

Notes:

a) Traditionally the mariners' compass was divided into 32 points each of which was a Rhumb = 11 deg. 15 min;

b) A 2 cycle line of logs;

c) Used in relation to charts of Mercator's projection.

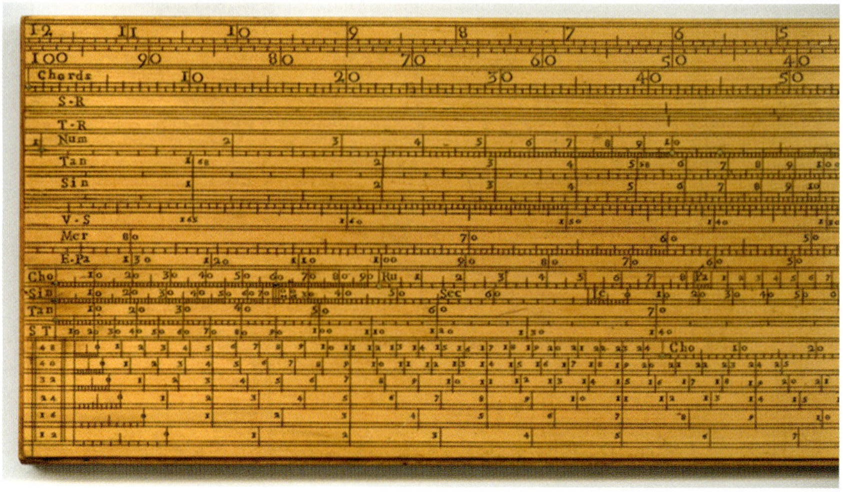

Fig. 10.29. Some of the lines shown in Fig. 10.28 set out on an unnamed but modern 24 in. Gunter's scale.

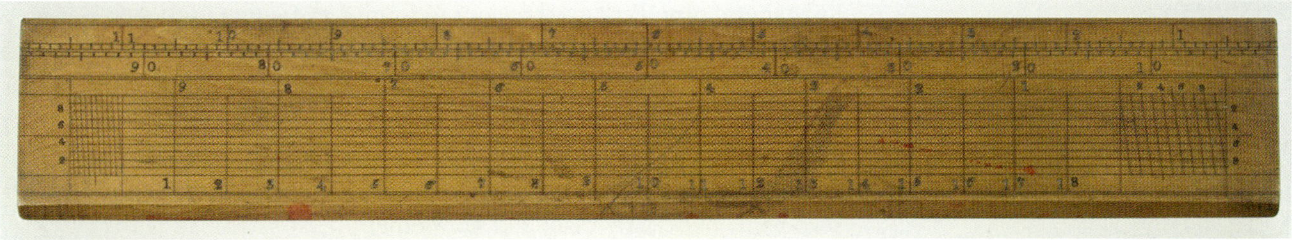

Fig. 10.30a & b. A 12 in. boxwood Gunter's scale by R. & C. Beilby (Bristol, 1809–12).

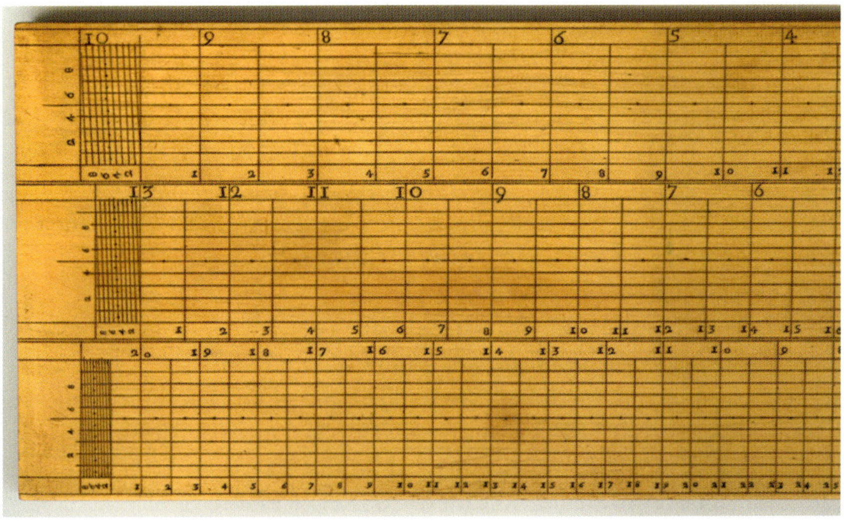

Fig. 10.31. The reverse face of the scale shown in Fig. 10.29.

With a 24 in. rule, the front face is divided with the diagonal scale taking up 12 in. on the left hand side of the face and the "plain" lines taking up the right hand side of the face. The lines on the reverse take up the whole length of the scale.

Use: The rule was used with dividers to perform navigational calculations. As early as 1695 slide rules specifically intended for navigation were available but the Gunter's rule remained the common choice of the navigator for the next 200 years. Writers on the subject have cited a variety of reasons for this – the more convincing include the familiarity of navigators with compasses which were used for chart work and the fact that the 24 in. rule had longer scales than the slide rule, thus giving greater accuracy, whilst the less convincing suggestions include the idea that conditions at sea would have caused a slide rule to swell and jam!

Only around 1880 when the mass produced general purpose slide rule became cheap and widely available was the Gunter displaced – however it was still available until about 1910.

USE V. *Concerning the Line of Rhumbs.*

The Ufe of the Line of Rhumbs is only to lay off, or meafure, the Angles of a Ship's Courfe in Navigation, more expeditioufly than can be done by the Line of Chords: As fuppofe a Ship's Courfe is N N E, it is required to lay it down.

Draw the Line A B, reprefenting the Meridian; take 60 Degrees from the Line of Chords, and about the Point A defcribe the Arc B C. Now becaufe N N E is the third Rhumb from the North, therefore take the third Rhumb in your Compaffes, on the Line of Rhumbs, and lay it off upon the Arc from B to C; draw the Line A C, and the Angle B A C will be the Courfe. Fig. 12.

Fig. 10.32a & b. Instruction from Bion's The Construction and Principal Uses of Mathematical Instruments[16] *for the use of the Line of Rhumbs. Although this instruction was given for a plain scale, exactly the same instruction would be relevant for a Gunter's scale.*

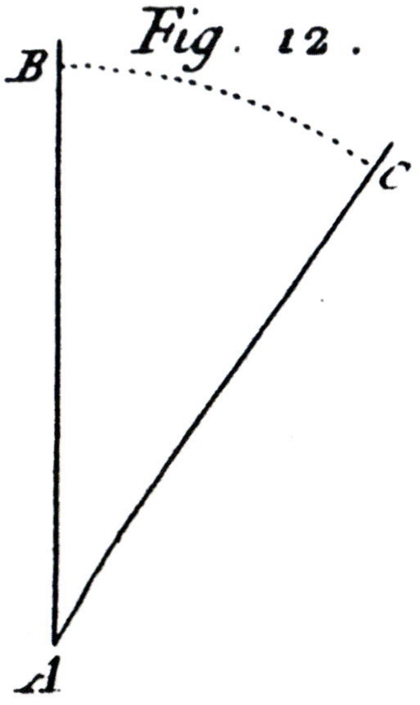

Manufacture: Perhaps three out of four of all Gunter's scales do not have a maker's name. The Rabone records show that by the 1850s the firm was making both 12 and 24 in. rules in quantity but examples marked by this firm are rarely encountered.[17] The rules made by Rabone and the other 19th century quantity manufacturers were often sold by nautical instrument suppliers, naval outfitters and the like, some of whom marked them with their own name but clearly many did not bother.

Gunter's scales continued to be manufactured into the 20th century. The Preston 1901 catalogue[18] lists both 12 in. and 24 in. versions, though the Rabone 1892 catalogue[19] only lists 24 in. versions, described as Gunter's Navigation Scales. In the Preston 1909 catalogue[20], they are still listed in the index, but there are no scales shown on the pages listed.

Donn's scale

As the mark "Navigation scale improv'd by by [*sic*] Donn" indicates, this is an improved version of the Gunter scale. (The double stamping of the word "by" was a mistake – stamping errors are not uncommon on rules.) It was a 24 in. scale and was designed by Benjamin Donn (1729–98), mathematician and public lecturer in science subjects. For a short while after 1769 he taught mathematics and navigation at his academy in Bristol. His rule differs from the normal Gunter's in that some lines were added whilst others were dropped. It was described in his book *The Description and use of the Navigation scale as improved in the year 1772* published in 1772.[21]

The lines on the back of the rule are:

Sine of Rhumbs; Number squared; SC, S Sec.*; Versed Sine; Tangent; Meridian; Meridian Continued*; Numb – Root*;Numb – Cube*

An asterisk marks the lines Donn added. The line of numbers was described as "squared". The lines omitted were the Tangent of Rhumbs, Sines and the line of Equal Parts. Also marked on the scale were some additional formulae concerning calculations for latitude.

The lines on the face of the rule were:

Guns: A line of diameter to 42 lb.; Rhumb; M*Lon; Chord; Sine; Tangent; S. Tangent

There was also a line of equal parts (not labelled) divided in sections 25, 30, 35, 40, 45, 50, and 60, to the inch.

The only additional line on this face was the diameter of guns. Omitted were the second lines of rhumbs and chords.

Donn's rule is not common, but it was made over a considerable period. The Rabone inventory of 1866 records 2 dozen in stock though no example has been recorded with a maker's name.[22]

Fig. 10.33a, b & c. A 24 in. Donn's navigation scale.

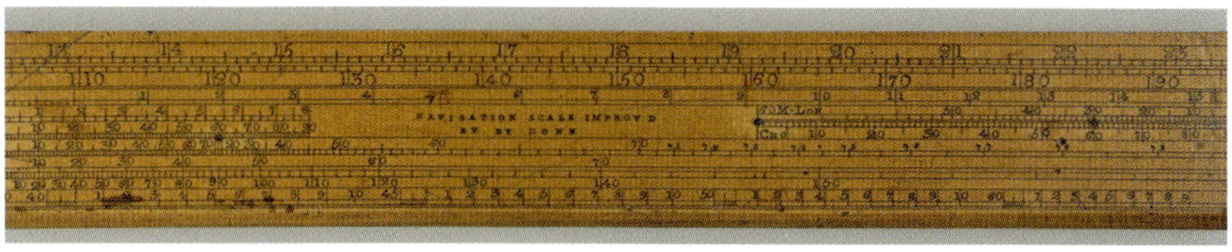

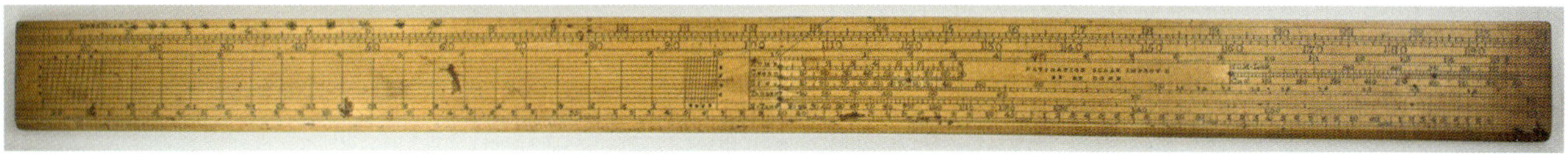

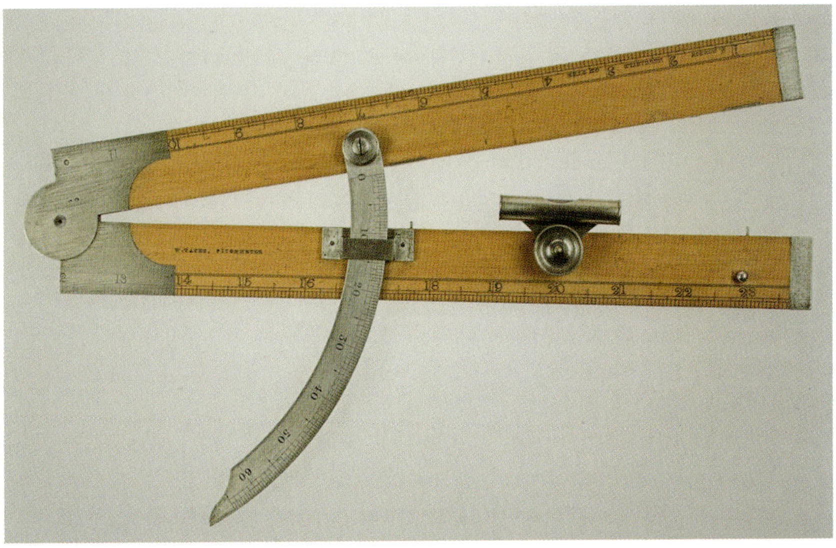

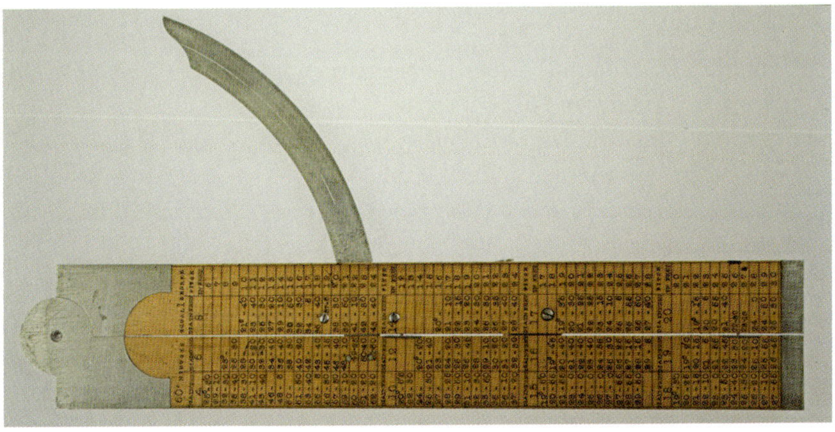

Fig. 10.34a & b. A boxwood and brass rule form pitchometer by F. Robson (Newcastle-upon-Tyne, 1890–present). On the back of the rule is a table giving the pitch angle and pitch length of propellers from four to twenty feet in diameter, thus removing the need for a mathematical calculation.

Pitchometer

The introduction of steam power and a propeller to drive ships through the water gave rise to other specialist rules and slide rules

To get the maximum efficiency from a ship's engine and also to prevent noise and vibration, it is important to match the pitch of the propeller to the engine. The pitch of a propeller is the distance that the blade travels through the water in one revolution and the pitch angle is the angle that any part of the blade makes to the perpendicular of the water flow. This angle relates the pitch length to the diameter of the propeller.

To measure the propeller, a pitchometer was used. These came in a variety of styles, either in brass or in a rule form in boxwood and brass.

Steam power slide rules

A variety of slide rules were designed to assist with the many calculations needed for construction and performance of steam engines, but these are without the compass of this book.

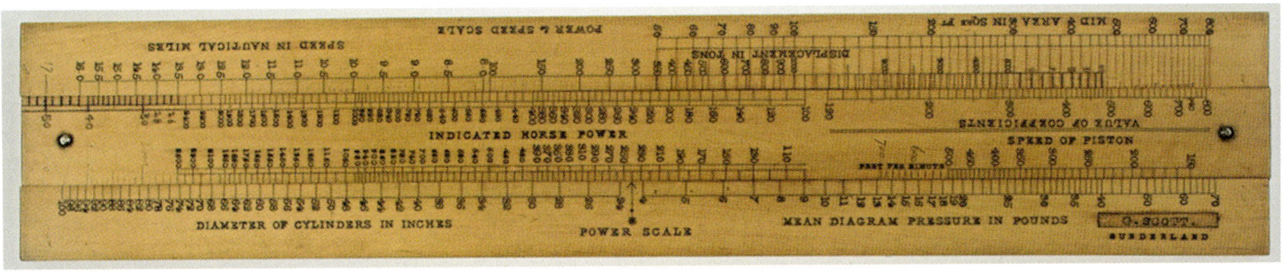

Fig. 10.35. A boxwood steam power calculating slide rule. The maker's name has been overwritten by an owner, but it is likely to have been Glaholm or Glaholm & Robson, Sunderland, who specialised in making rules for the shipbuilding industry.

References

1. Anon., *The Shipwright's Vade Mecum*, 2nd ed., London, 1822.
2. Edward Preston & Sons, *Catalogue No. 18. The "Preston" Catalogue, 1909 with 1912 supplement*, reprinted Mendham, NJ, 1991.
3. Kebabian, Paul B. "Further Notes on the Early English Three Fold Ships Carpenter's Rule." *The Chronicle of the Early American Industries Association*, Vol. 42, No. 1, 1989.
4. Edward Preston & Sons Ltd., *The Preston Catalogue of rules, measures, levels ...*, Birmingham, 1891.
5. John Rabone & Sons, *Catalogue of Rules & Tapes*, Birmingham, 1892.
6. Kilgour, P., "Rope-makers' 12-in. Calculating Slide Rule." *The Textile Manufacturer*, Jan. 15th, 1927.
7. Rees, Mark, "Fathoming it out: The IFFI table". *The Chronicle of the Early American Industries Association*, Vol. 55, No. 3, 2002.
8. Simms, F. W., *Treatise on Mathematical Instruments*, London, 1849.
9. W. F. Stanley, *Catalogue of Surveying and Drawing Instruments*, London, 1931.
10. 10. Gunter, Edmund, *Canon Triangulorum, sive tabulae sinuum et tangentium...*, London, 1620.
11. Gunter, Edmund, *The Description and Use of the Sector, the Crosse-staffe, and other Instruments,* London, 1624.
12. Cajori, Florian, *A History of the Logarithmic Slide Rule and Allied Instruments*, Constable & Co., 1909, reprinted Mendham, NJ, 1994.
13. Bion, Nicholas, *The Construction and Principal Uses of Mathematical Instruments.* Paris 1723. Translated by Edmund Stone; ... *To Which are Added, the Construction and Uses of Such Instruments as are Omitted by M. Bion* by Edmund Stone. Second edition; ... *To which is added a Supplement Containing a Further Account of some of the Most Useful Mathematical Instruments as Now Improved*, supplement by Edmund Stone, London, 1758, reprinted Mendham, NJ, 1995.
14. Robertson, John, *A Treatise of such Mathematical Instruments as are usually found put into a Portable Case ... With an appendix containing the Description and Use of the Gunners Callipers*, London, 1775.
15. Bion, *op. cit.*
16. *Ibid.*
17. Records of John Rabone & Sons, MS 1522, Archive Department, Birmingham Library Services.
18. Edward Preston & Sons Ltd., *The Preston Catalogue of rules, measures, levels ...*, Birmingham, 1901.
19. John Rabone & Sons, *Catalogue of Rules & Tapes,* Birmingham, 1892.
20. Edward Preston & Sons, *1909, op. cit.*
21. Donn, Benjamin, *The Description and use of the Navigation scale as improved in the year 1772,* Bristol, 1772.
22. Rabone archive, *op. cit.*

Chapter 11
Coopers and customs

Gauging made Perfectly Easy, As it is actually practiced by the Officers of his Majesty's Revenue of Excise. Containing … the Established Rules for finding the Contents of all Sorts of Cysterns, Coppers; Backs, Coolers, Tuns, Stills and Casks, when full or Part empty: The Examples being performed here both by the Pen and Sliding Rule: Leadbetter, 1760.[1]

Before the advent of plastic and modern handling equipment, the easiest way to moved goods around was in casks. Colloquially, they are often referred to as barrels but a barrel is just one size of cask. So how big is a barrel? Throughout history a barrel has been used as a measure of quantity for commodities which ranged from beer to honey and even oysters.

Cask sizes

The determination of what quantity was contained in a barrel, or other size of cask, was bedeviled by two variables. Firstly, statutes allowed for different sizes of barrel for different contents and secondly, sizes of gallons also varied.

The tables below give some indication of the variety and complexity of defining the size of a cask.

Anchovies	30 lb.
Nuts or Apples	3 bushels
Pot ash or Barila	200 lb.
Candles	10 doz. lb.
Salmon or Eels	42 gallons
Raisins	1 cwt.
Oil	31 1/2 gallons
Soap	240 lb.
Butter	224 lb.
Herrings	32 gallons

Fig. 11.1. Some barrel contents for different commodities – from The Instructor or Young Man's Best Companion, *1799.*[2]

Cask	Cu. in.	Wine Gal. (231 cu. in.)
Ton	58,212	252
Butt	29,106	126
Puncheon	19,404	84
Hogshead	14,553	63
Tierce	9,702	42
Barrel	7,276	31.5
Rundlet	4,158	18
Ancher	2,310	10

Fig. 11.2. Sizes of casks used for wine in 1760, giving both the size of the cask and the content in wine gallons, taken from Leadbetter.[3]

Fig. 11.3. The sizes of some of the beer and ale casks in use in London in 1760, taken from Leadbetter. The sizes are confusing! The gallons were the same size, 282 cu. in. but there were 36 gallons to the barrel for beer and 32 gallons to the barrel for ale, so the casks sizes were different depending on their content. Ale and beer measures for the country areas were different again as there were 34 gallons to the barrel for both. Other sizes of casks used included tons, butts and pucheons.

Cask	Ale cask cu. in.	No. of gal.	Beer cask cu. in.	No. of gal.
Hogshead	13,536	48	15,228	54
Barrel	9,024	32	10,152	36
Kilderkin	4,512	16	5,076	18
Firkin	2,256	8	2,538	9

At various times in history, efforts were made to standardise measures without much success but by the beginning of the 19th century, the impact of the metric system, introduced in France after the French Revolution, and increasing industrialisation made it imperative. After a series of Parliamentary Commissions the Imperial system was introduced in the UK by the Weights and Measures Act of 1824.

This act replaced both the "old" gallons, wine and ale/beer, with the Imperial gallon. This was intended to be a restatement of the ale gallon, but the definition used resulted in a gallon of 277.27 cu. in., some 1.7% smaller than the ale gallon.

Fig 11.4. Sizes of wine casks and their content in Imperial gallons, taken from Bateman, 1852.[4]

Cask	Cu. In.	Imp. Gal. 277.3 cu. in.
Tun	69,873	252
Pipes or Butts	34,937	126
Puncheons	23,291	84
Hogsheads	17,468	63
Tierces	11,655	42
Barrels	8,734	31.5
Runlets	4,991	18
Ankers	2,773	10

Fig. 11.5. Sizes of some of the beer and ale casks together with their content in imperial gallons, taken from Bateman, 1852. Note that for both wine and beer and ale, although the number of gallons in each size of cask is the same as it was in 1760, the cubic content is not.

Cask	Cu. In.	Imp. Gal. 277.3 cu. in.
Hogsheads	14,973	54
Barrels	9,982	36
Kilderkins	4,991	18
Firkins	2,495	9

Making casks to size

Any experienced cooper knows the three key dimensions of the common cask sizes, being the length, the head diameter and the bung diameter. Working to these dimensions he is expected to make a cask accurate to within 1% of the desired content, which for a barrel meant ± 3 pints. The coopers' diagonals were one aid and the inside could be shaved to add a pint if need be.

To achieve this degree of accuracy in cask content, the bung and the head diameter needed to be controlled to 1⁄10 of an inch above or below the desired size. The coopers' diagonals could be used to determine any adjustments needed to the position of the head but the point to which the trussing hoops could be driven would also indicate whether a cask was over or under size.

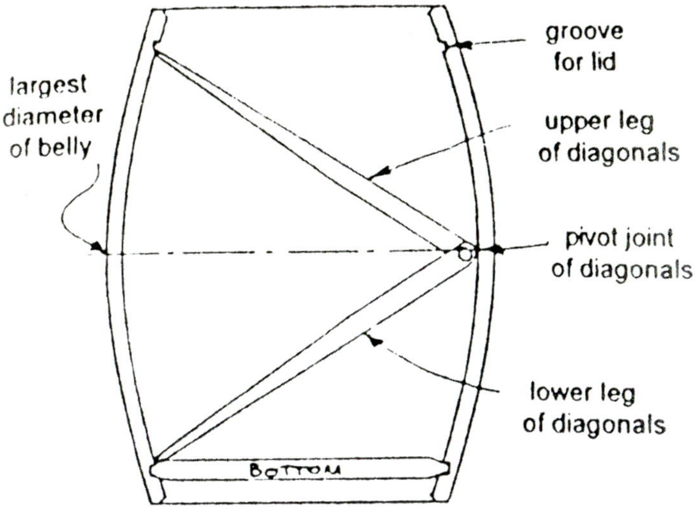

Fig. 11.6. Diagram illustrating the use of the coopers' diagonals, also known as coopers' dividers from "Barrel Gauging" by Jim Packham.[5] The rods, which are hinged together, are placed in the cask and the position of the head marked. The only factor which is not assessed when the diagonals are used is the curvature between the head and the bung, which is the "variety" of the cask.

Coopers' measuring devices

Stave gauge (for width): Seldom mentioned in literature and with only a few examples known, the gauge helped the cooper mark the width of the staves at mid point and also the degree of taper. It may also have been used to check the angle of the side of the stave.

Coopers' diagonals: A pair of rods hinged together, their use is shown in Fig. 11.6.

Coopers' rule: A rule dated 1655, made by Henry Sutton, London, has been described by Richard Knight.[6] It appears to have been made specifically for use by a cooper.

It includes a 4 ft. diagonal line measuring to 300 gallons; however, as it is marked onto a 2 ft. rule it is divided into two sections. This would make it of no use in measuring a complete cask. It would seem therefore that its use was for setting out coopers' diagonals.

Fig. 11.7. A group of coopers' gauges.

The rule also includes a line of board measure extending to a width of 100 inches – far beyond the width of any likely single board. It is suggested that the line was used to measure a number of boards placed alongside each other, for instance the staves of a cask with the circumference of 100 inches.

Passer (passing hoop): A stout steel hoop used both as a trussing hoop and as a ring gauge to check the diameter of a cask. Recorded by Salaman in the *Dictionary of Tools…*[7], it is rarely found, as indeed are the other coopers' measuring items.

Cask varieties

Casks are of no exact geometrical form but for practical purposes are divided into varieties, the term used to describe the shape of a cask. It is generally reckoned that there are four. Each variety approximates to a geometric solid and it is necessary to know this in order to find the correct formula to measure the volume of a particular cask.

Everard, in *Stereometry made easy…* in 1712, describes the four as:[8]

- 1st variety: The middle frustrum of a spheroid.
- 2nd variety: The middle frustrum of a parabolic spindle.
- 3rd variety: The lower frustrum of two equal parabolic conoids, abutting upon one common base.
- 4th variety: The middle frustrum of two cones abutting upon one common base.

CAſks being of *various Curvitures*, Gaugers do uſually ſuppoſe every Caſk to be *in the Form* of ſome one of theſe following *Solids*. [See alſo *Pages* 177, 178, and 179.]

Viz.
1. The *middle Fruſtum* of a *Spheroid.*
2. The *middle Fruſtum* of a *Parabolic Spindle.*
3. The *lower Fruſtums* of two equal *Parabolic Conoids.*
4. The *lower Fruſtums* of two equal *Hyperbolic Conoids.*
5. The *lower Fruſtums* of two equal *Cones.*

First Variety. 1. Now to have a true *Notion* of theſe ſeveral *Caſks*, view *Plate* 4. *Fig.* 3. and *Plate* 5. in which the *outmoſt* Lines, which repreſent the *Staves*, are very much *curved* or *arching*. This repreſents a *Caſk* of the *firſt* Variety, and is what is called the *middle Fruſtum* of a *Spheroid*, and conſequently *will hold more* than a Caſk of *any other Form.*

Second Variety. 2. The *next curved Line* within that I have been ſpeaking of, repreſents a *Caſk* of the *ſecond* Variety; and its *Contents* will be *leſs* than that of a *Spheroid:* This is called the *middle Fruſtum* of a *Parabolic Spindle.*

Third Variety. 3. The *lower Fruſtum* of *two equal Parabolic Conoids* is the *next curved Line*; and holds *leſs* than the *ſecond* Variety, as is eaſy to conceive. This is called a *Cask* of the *third* Variety.

Fourth Variety. 4. The *fourth* Variety, which is the *lower Fruſtum* of *two Hyperbolic Conoids.* The *Staves* of ſuch a *Caſk* are *more ſtreight*, and conſequently holds *leſs* than any of the *other three* before mentioned. This is not repreſented in the Figure, becauſe it would be too much crouded; but will be found in *Plate* 5.

Fifth Variety. 5. A *Caſk* in the Form of *two equal Cones*, *abutting* on *one common Baſe*, is repreſented in *Plate* 5. by the *innermoſt* Line, which is *perfectly ſtreight* from *Head* to *Bung*; that is, from A to B, or from B to I. This *Caſk*, of all others, holds *the leaſt.*

N. B. Some Writers have treated of *another* Variety of Caſks, *viz.* the *Hyperbolic Spindle*, which has its Staves a little *leſs* curved than the *Parabolic Spindle*, and conſequently a Caſk of that Form will hold *leſs* than thoſe of the *firſt* and *ſecond* Variety. How to gauge or find the Contents of a Caſk of *this Form*, I ſhall ſhew by and by.

But as it is difficult, even for expert Gaugers to determine the *Form* of the *Caſk* by the *Eye*, we ſhall lay down the following Method, by which the *true* Form *of any Caſk* may be found.

Fig. 11.8 Description of the different varieties of cask from Charles Leadbetter, The Royal Gauger, *1760.*

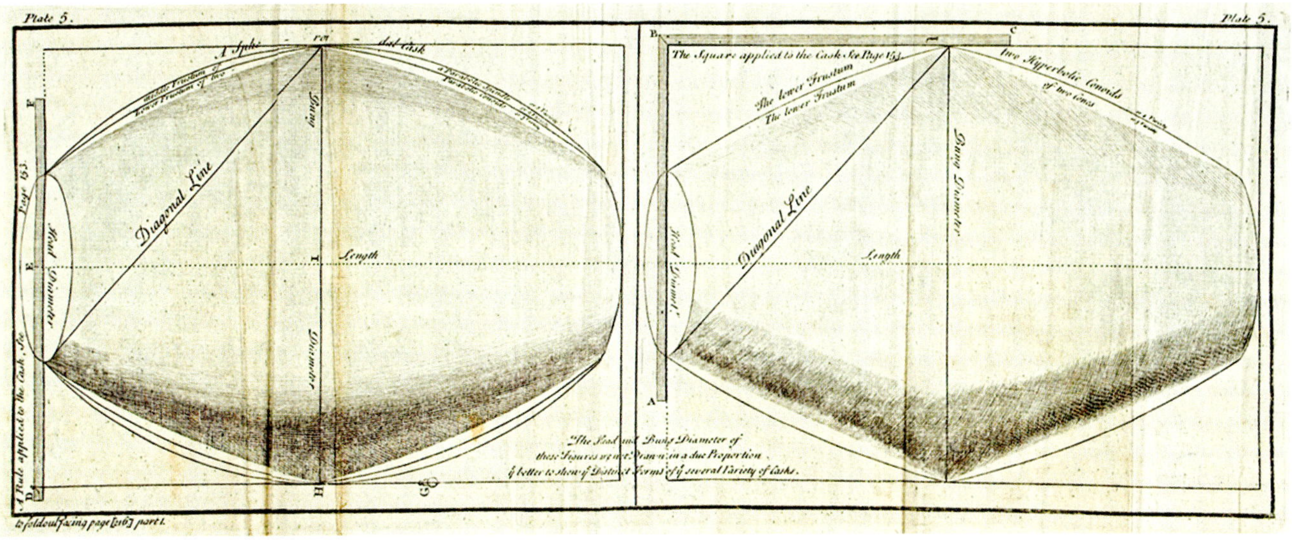

Fig. 11.9. *Diagram of a longitudinal section of a barrel showing the different varieties. From Charles Leadbetter,* The Royal Gauger, *1760.*

Leadbetter in *The Royal Gauger* of 1760[9] gives a considerably more extensive description along with a fifth variety and a diagram, as shown in Figs. 11.8 and 9.

Excise duty

By the 17th century a pressing need for money to fund England's overseas explorations and its active military operations resulted in the introduction of excise duties.

The first attempt to levy these duties in 1626 resulted in riots and administrative chaos but from 1641 duties on beer, cider, perry, butchers' meat, sugar and several other commodities were imposed.

At the time of the Restoration in 1660 the duties on all items except beer, ale, cider and perry were abolished. But the government had learnt that duties were an effective method of collecting money, if the right items were chosen, so in 1685 wines, vinegar, tobacco and sugar were added to the list. Thereafter for the next 120 years more and more items were added but few were removed.

By 1760 the list of dutied goods included all alcoholic drinks, malt, candles, soap, starch, fabrics of many kinds, hops, chocolate, coffee, tea, silver plate, papers and cardboard, leathers and parchment, and glass.

After 1820 the number of commodities subject to duty was gradually reduced – other sources of income had been found. However alcoholic drinks, whether made within Britain or imported, remained prime targets for taxation.

The job of collecting these duties fell to the Customs & Excise who appointed a series of officers to undertake this work.

The measurement and calculation needed for the collection of duties on alcoholic drinks have always been a principal part of the Excise officer's work and, judging from the extensive guidance given in the instruction manuals, the most difficult part of the job. So "gauger" became the common name given to an Excise officer, his job being to "gauge", that is, to measure.

Even for items such as paper, where the gauger needed only count or weigh the finished product to establish quantities, it was not simple. Some types of paper were taxed by the ream, some by the bundle, some by weight or by a percentage of the value (*ad valorem*) so the gauger needed expertise to identify the different types – and some mathematical ability to calculate what was owing.

Fig. 11.10. Title page from the 5th edition of Charles Leadbetter's The Royal Gauger, *published in 1760.*

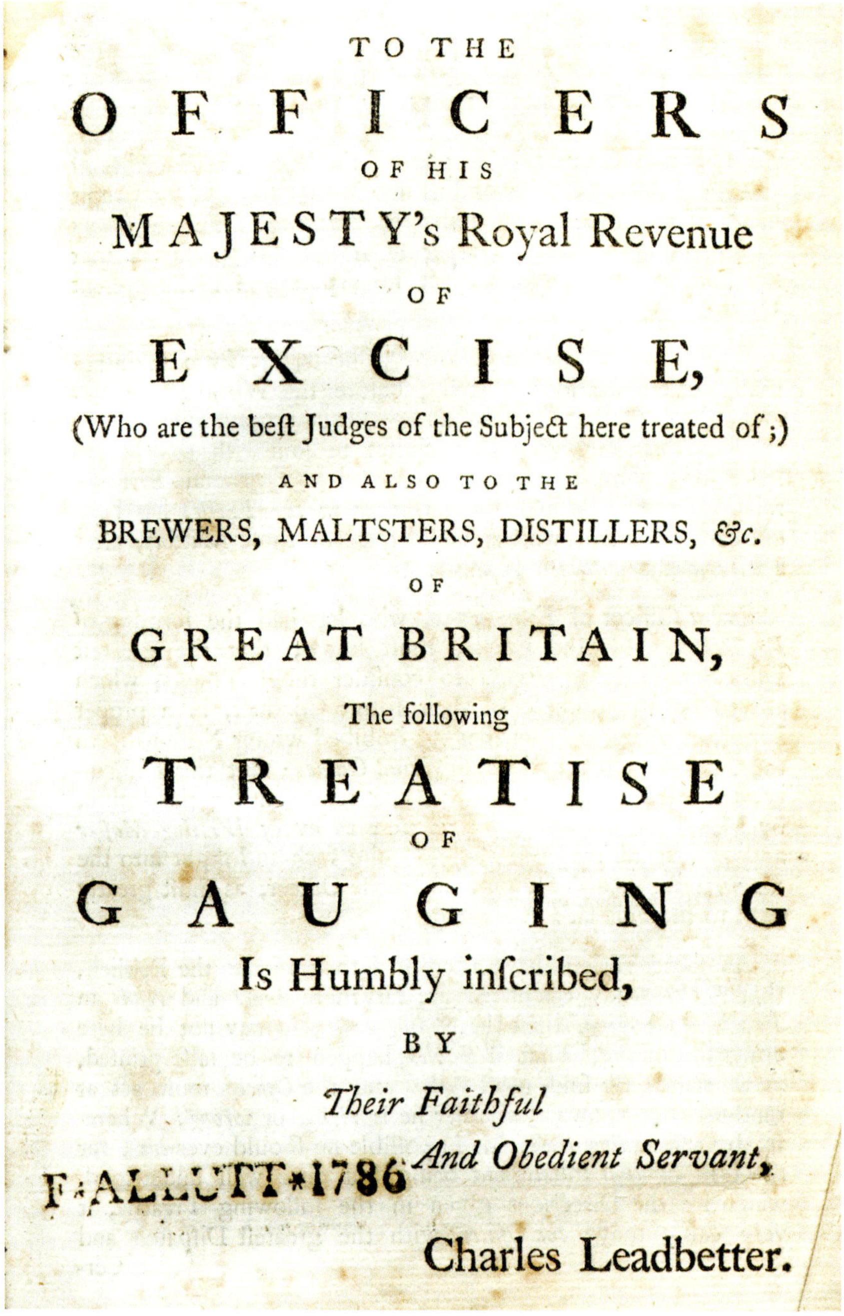

TO THE

OFFICERS

OF HIS

MAJESTY's Royal Revenue

OF

EXCISE,

(Who are the beſt Judges of the Subject here treated of;)

AND ALSO TO THE

BREWERS, MALTSTERS, DISTILLERS, &c.

OF

GREAT BRITAIN,

The following

TREATISE

OF

GAUGING

Is Humbly inſcribed,

BY

Their Faithful

And Obedient Servant,

F·ALLUTT*1786

Charles Leadbetter.

Gaugers' manuals

Probably 100 or more gaugers' manuals were published between 1690 and 1840. These manuals were the textbooks from which employees of the Customs & Excise learnt their trade. Each book describes the equipment current at the time of publication and they are therefore a valuable source for establishing what exactly was being used for, among many other things, barrel gauging at any particular time.

Charles Leadbetter, who described himself as "Many years a Gauger in the Royal Revenue of Excise", graphically describes the establishment and operation of the Excise in *The Royal Gauger*.[10] It is immediately apparent that this was a very well organised operation, alert to the possibilities of fraud by the manufacturers, and of collusion and dishonesty by its officers.

Having shown evidence of suitability, staff entered at the lowest grade and were thereafter promoted by merit. Officers were moved regularly to prevent collusion, and the collection of monies was made by different staff from those issuing the assessments. These were well paid jobs offering opportunity for advancement up a graduated scale of appointments and even a pension when the officer became "old and broken down".

The Commissioners of the Excise sat daily at the head office in London to supervise operations and to determine disputes. They presided over a substantial establishment of clerks and specialist officers in London whilst throughout the country officers, organised into collections, walks and rides, made daily, or even more frequent, visits to the numerous premises where dutiable products were manufactured.

In 1852 the establishment in Bath consisted of 3 officers for the city and 4 covering surrounding areas. They were overseeing 42 maltsters, 184 brewers, 1 soap maker, 1 paper maker, and 76 other premises. In addition there were 187 licensed retailers. This information is taken from Joseph Bateman's *The Excise Officers Manual* which, in common with many manuals, published in the 18th and 19th centuries, lists the more senior staff of the Excise.[11] Such lists can sometimes be a means of identifying the original owners of those rules that are marked with owners' names. Until the end of the 19th century, Customs & Excise officers were required to personally supply their own instruments.

Gauging and ullaging a cask

In order to calculate the duty payable on the contents of a cask, the volume of the cask needed to be established. If the cask was not completely full, it was also necessary to measure the contents (in gaugers' terms "the wet inches") or the empty portion of the cask, known as ullage ("the dry inches").

Gauging rods

This term is the generic description for rods or rules used to internally measure casks to determine the capacity or the contents of casks. Almost all gauging rods are marked with lines that give a direct reading in gallons – thus a cask can be gauged with reasonable accuracy without the need for any calculations.

The names given by writers and makers to these items are varied, and can sometimes be confusing: gauging rod; bung rod; dip rod; dip; dips; dray-

Fig. 11.11. Lines for seven different sizes of cask and for beer and wine gallons marked on a dip rod by William Mole (London, 1750–1811).

Fig. 11.12. Using a diagonal or dip rod.

mans' rule; spile rod; bottle rule; barrel rod; ullage rule, with some rules having more than one commonly-used name. Whilst some of these names indicate a particular purpose, many rods have lines intended for several different types of measurement.

The principle on which all these measures work is that all casks, whatever the size, are "barrel shaped" and that the ratio of length to diameter varies little. Thus the overall volume can be determined from one measurement. However there is some variation in the proportions – some are a little longer and thinner whilst others are a little shorter and fatter – so the most accurate single dimension is the diagonal measurement, from the bung to the base. However, for the gauging rod to measure the actual contents of the cask, it must be marked specifically for that particular variety of cask.

Lines and quantities: The careful observer will see that there are many variations to be found on gauging rods. The markings are usually for some or all of the six London standard sizes of cask but other types also appear. The divisions may indicate wine, ale, beer or imperial gallons but it should be remembered that although the sizes of casks altered when Imperial measure was adopted in 1824, and such changes can be indicators as to the date of a rod, the changeover to the new barrels was a slow process so rods were made with both old and new measures even after this date.

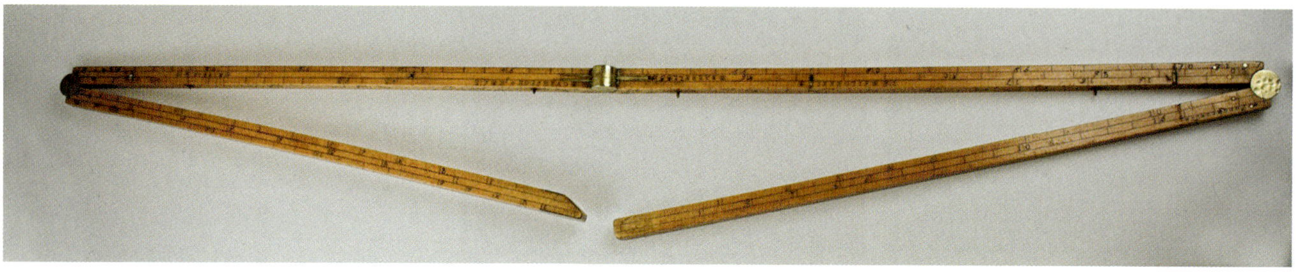

Fig. 11.13. A 4-fold dip rod by William Mole (London 1750–1811).

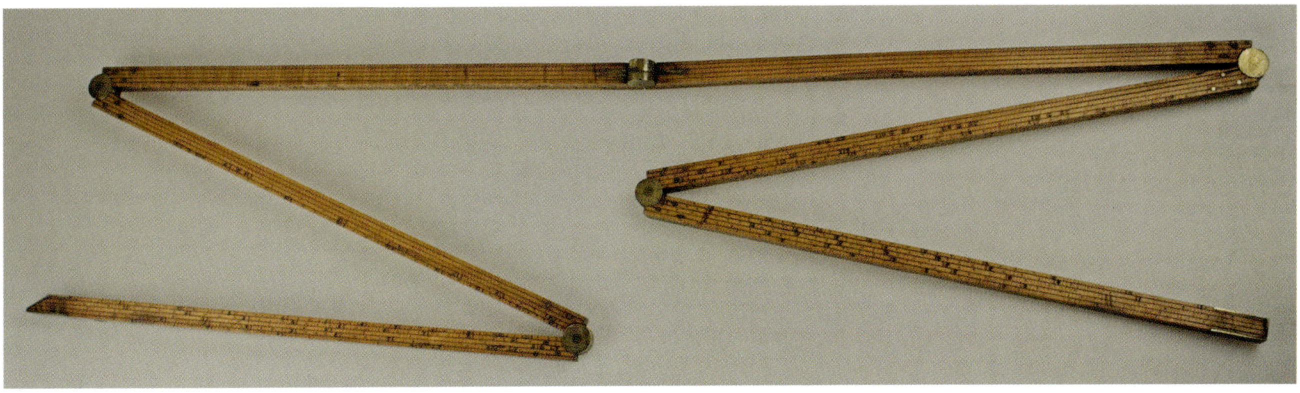

Fig. 11.14. A 6-fold dip rod by Alexander Wellington (London, 1784–1827).

Gauging rod types (dip rods, bung rods): Of the many types of gauging rods, the simplest and earliest form is a solid rod 4 or 5 ft. long, about ½ in. square. Inserted through the bung hole of a cask, it can be used three ways: to determine the diagonal dimension of a cask, for which reason it has a tapering tip; to take the overall length of a cask; and to take the internal diameter at the mid point (in gaugers' parlance, the bung diameter).

The second function of the gauging rod is to ascertain the contents of partly filled casks whether they be standing or lying.

The following scales can be found on gauging rods, though many rods will have only some of these.

- Line of measure in inches, subdivided into tenths but not into feet. Inches and decimal parts are the units used by the gauger in calculations.
- Diagonal line giving a direct reading of the volume of the cask in gallons. Such lines are sharply incremental. A 4 ft. rod will measure to around 250 gallons and a 5 ft. to 400 gallons. There may be two lines graduated for different gallons, for example, ale and wine gallons.
- Lines for casks lying, that is, on their sides, i.e. the axis of the cask is horizontal. Intended for measuring the content (and hence also the ullage), there will be a number of lines, one for each size of cask, firkin, barrel, tun, etc.
- Lines for casks standing. The purpose and marking are similar to that for casks lying but intended for casks standing – that is, on end. As the cross section of a cask, whether standing or lying, is widest at the half point, the spacing of both the standing and lying lines will be symmetrical around the mid point, with graduations closest at that point.

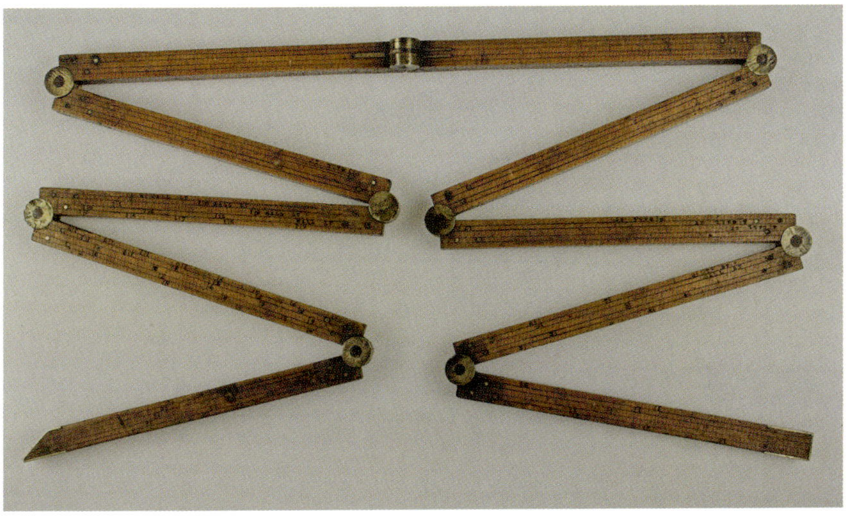

Fig. 11.15. A 10-fold dip rod by Joseph Long (London, 1821–1936).

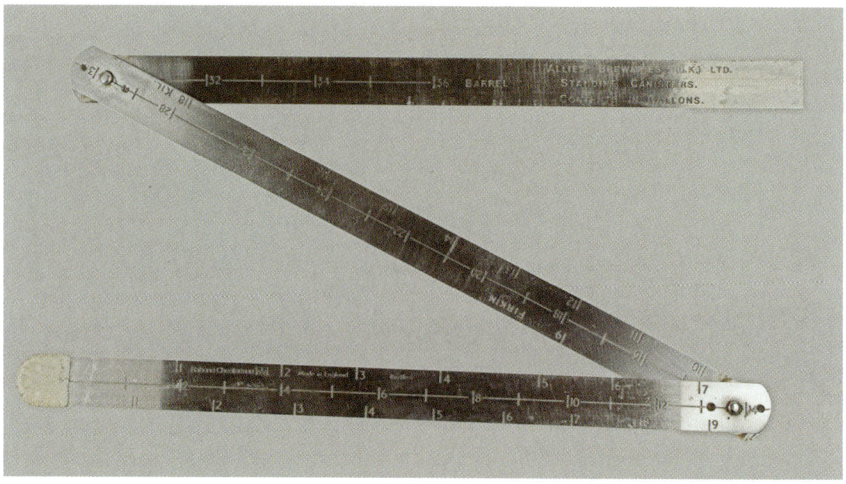

Fig. 11.16. A steel dip rod by Rabone Chesterman (Birmingham, 1963–1990), marked for barrels, kilderkins and firkins, standing.

Occasionally there may also be a line of ale area. If present this will be alongside the line of inches. Placed across the diameter of a cylindrical container, it gives the number of ale gallons in the container per 1 inch depth, the diameter being read from the line of inches alongside. This line is described by Bion in 1723[12] and by Leadbetter in 1739[13] but is only present on a few 18th century rods. After the introduction of Imperial measures, it became the Imperial area line. It continued to be described in the Customs & Excise instructions for officers until their final publication in 1968.

Folding gauging rod: For convenience many rods were made to fold with 4-fold the most commonly found but 6-, 8-, 10-, and even 12-fold versions were also made. Like the rigid rods these will be 4 or 5 ft. long, though occasionally shorter. The sections are thinner than the solid rods, about $3/8$ in. square. Some rods are as short as 2 ft. and therefore only able to measure the smallest cask, the firkin.

As the principal users of gauging equipment were the officers of the Inland Revenue who travelled their circuit daily, compactness was important.

Branan's rule: This is a form of dip rod that takes apart into two sections each of which also folds once. Each section of the rule is marked with logarithmic lines. The purpose of the division is to allow the parts of the rule to be slid alongside each other and thus form a slide rule for the gauger.

At least two pre-1700 dip rods have been observed marked with a line of logs and this would appear to be a regular feature of early dip rods. Branan's rule would thus seem to be a development of this feature.

Fig. 11.17. An anonymous Branan's rule with both halves joined together to form a dip rod.

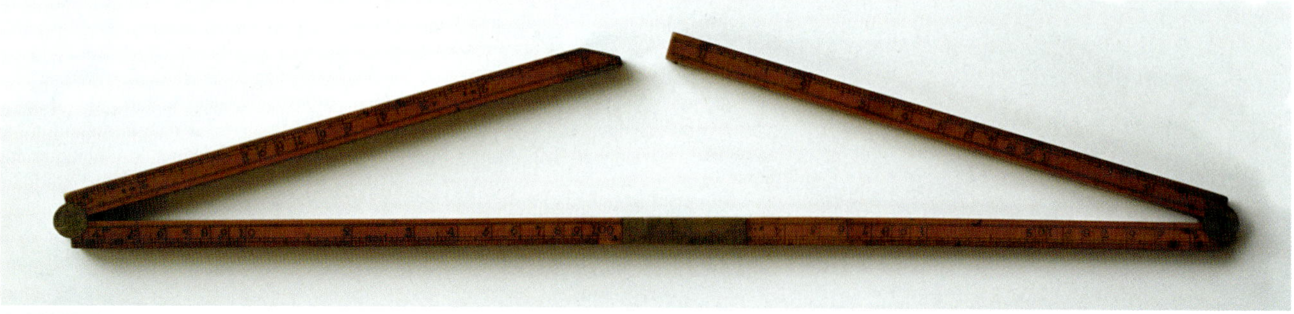

Fig. 11.18. The same rule with the two sections laid side by side to act as a slide rule.

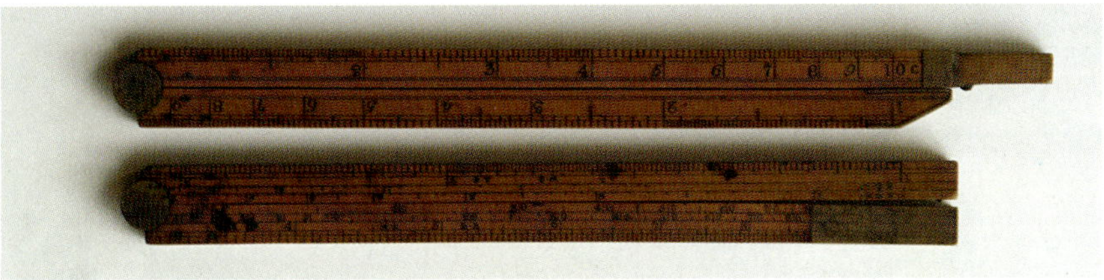

Fig. 11.19. The same rule showing the two separate halves.

The rule and its use have been described in detail by Knight[14]. It is mentioned by Leadbetter in the first edition of *The Royal Gauger* 1739[15] and the Dring and Fage records show that it was being made in some quantity in the 1840s (see Appendix 4). It is however rarer than these references might suggest – the reason may be that parts became lost inside barrels but regardless, it is doubtful if it was ever popular with the users.

Measuring a cask to determine its volume

If a cask is of a known variety or is of standard shape, there are a two methods by which this can be done.

By weighing: This is cumbersome and is only possible if the cask is emptied or filled as part of the process. For this reason it was very rarely used.

By using a diagonal rod: When inserted through the bung, a diagonal rod gave a direct reading of the capacity of the cask in gallons. The cask was measured either in a "standing" or "lying" position, and a correctly divided rod for the size and variety of cask and the type of gallons was required for even reasonable accuracy.

All gauging authorities seem to be agreed that this method was not totally accurate even if the cask were of regular proportions and could be very erroneous if the cask were misshapen or of unusual proportions. This method was thus not considered sufficiently accurate for use by the Excise.

If, however, the cask is not of a known variety or is of irregular shape, the volume can be determined by measuring the three principal dimensions: the bung diameter, the head diameter and the cask length. From these the volume could be calculated. The hurdles to be overcome with this method were numerous – taking accurate measurements; determining the stave and head thicknesses; assessing the variety of the cask and choosing the formula, which included some method of adjustment for the variety.

Having measured the barrel it was necessary to "cast up" the result either by laborious hand calculation or by using a slide rule. Despite its complexities, this procedure was widely used, and still is, when needed, today.

Cask measuring rules

In a treatise on practical gauging, Nesbit and Little[16], writing in 1822, made it clear that there were four instruments used by the port gaugers of the Customs & Excise to measure casks. These were the long caliper, the cross caliper, the bung rod and the head rod. The implication was that all these items, including the head rod, had been in use for many years – indeed reference to the wide variety of gauging manuals endorses this view. The only instrument not mentioned was the endometer but this was always something of an afterthought and was not probably used on a day-by-day basis.

The cross caliper: This is a sliding gauge made in boxwood with arms around 18 in. long at both ends. In use it was placed on the lying cask and the overall diameter was read off directly. Good practice required that, as the cask might have become oval, the caliper was applied several times to obtain an average diameter. Some calipers had built-in factors for different thicknesses of stave but, if not, the thickness of the staves needed to be subtracted from the

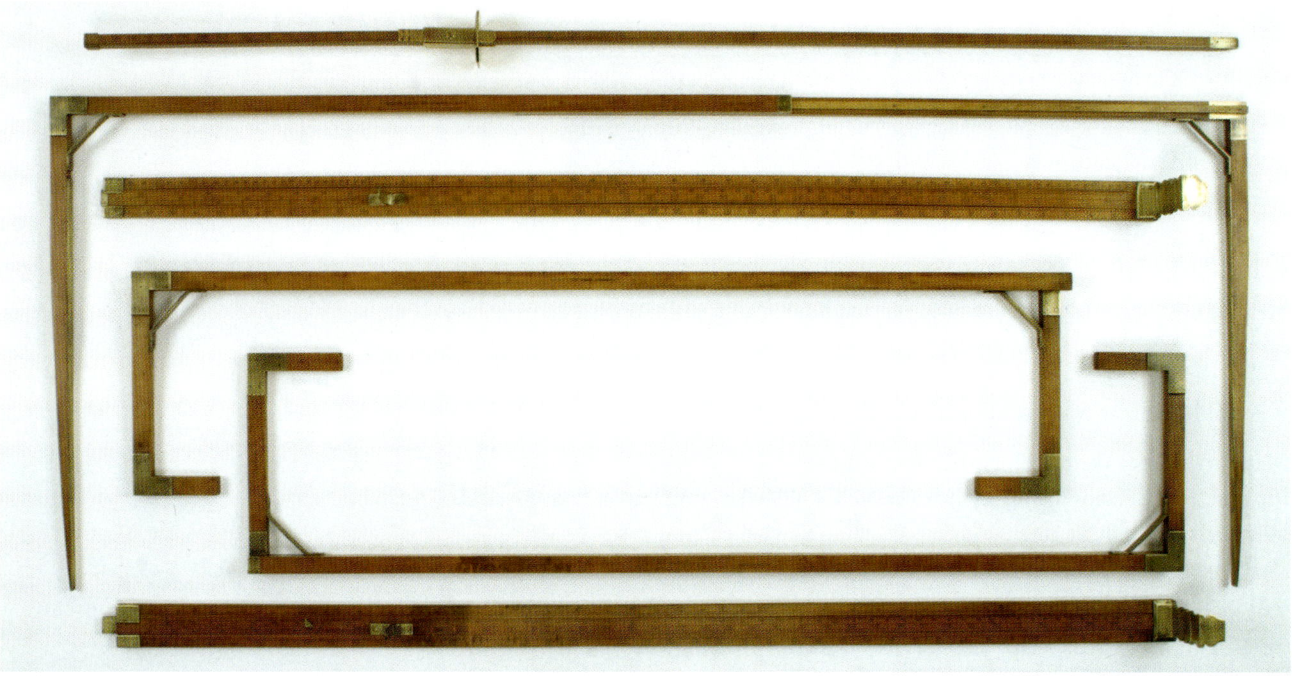

Fig. 11.20. A selection of gauging instruments. From top: Bung rod, cross caliper, head rod, long caliper, another long caliper and another head rod.

measurement. The cross caliper seems to have come into use in the late 18th century – earlier practice seems to have been to use a large square and a rule.

The long caliper: A sliding gauge, usually made in boxwood or lancewood, this rule is for measuring the length of the barrel. It is fitted with short returns at each end so that it measures from one head to the head at the opposite end reaching over the chime (the brim of the cask formed by the ends of the staves projecting beyond the head). Like the cross calipers, some examples have scales to automatically deduct the thickness of the heads.

The head rod (content and ullaging rule): This rule is 44 in. long and with its standing brass feet often presents a puzzle to collectors. In effect it is a combined instrument by which the diameter of the head of a barrel is measured, using the feet to get inside the chimes; and then the content can be calculated using the slide rule facilities; these include lines to calculate both full and partly full casks of different varieties, although frequently only the spheroid and 2nd variety lines are marked.

By the late 19th century the head rod had become the standard instrument used by the Customs & Excise and with the changeover to the issue of instruments rather than personal provision, this became virtually the only rule used for barrel gauging, continuing until metrication.

The bung rod: A straight rod, similar to the diagonal rod but with a sliding collar, this is used to measure the inner diameter of a cask through the bung hole.

The endometer: The rarest of the measuring tools, it is a right-angled gauge with a curving metal arm. The purpose is to measure the thickness of the head of a barrel. In use the curving rod is inserted in the bung hole and the caliper closed until the thickness of the head can be read directly. In reality, it was difficult to use and very capable of giving a wrong answer.

Fig. 11.21. An endometer from Joseph Long's 1922 Catalogue.[17]

Endometer, or Head Calliper

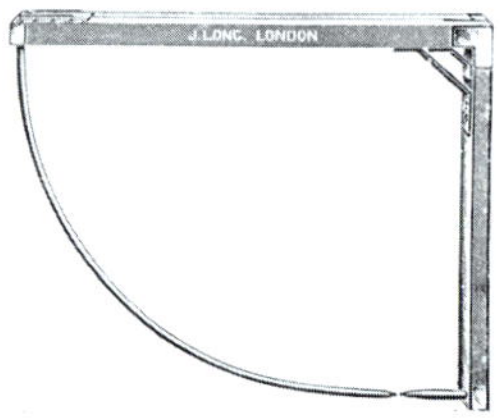

For accurately measuring the thickness of heads of casks.

Measuring the contents of casks

The content of a cask can be measured by using a dip rod provided the size and variety of the cask are known. The rod is divided to give an accurate direct measurement in gallons for both standing or lying positions of a specific type

Fig. 11.22. "The Gaugers at work" in the London Docks from Round London, an album of photographs of the chief places of interest... *published by George Newnes Ltd. in 1896. Key to the instruments illustrated in the photograph: 1) Long caliper; 2) Cross caliper; 3) Bung rod; 4) Head rod; 5) Tin velincher, or samples thief (the Customs & Excise often took samples of the contents of barrels for analysis); 6) Sample jar; 7) Sealed samples; 8) Chalked gaugers' measurements for calculating contents.*[18]

of cask; however most dip rods are only annotated for two or three different types. This method is the quickest and easiest, needing no calculation.

Dip rods: Intended only for dipping, these do not have the diagonal line of the gauging rod. A cask is dipped to establish how much it contains – the *wet* inches or how much vacuity there is – the *dry* inches. Dip rods were made in various forms.

Fig. 11.23. A 4-fold square section dip rod marked only for gallons. There is no maker's name but it is marked with an owner's name, E. Cave, and the date 1788.

Fig. 11.24. An unusual 36 in. round take-apart rod in three sections, unmarked and a 20 in. 2-fold square section rod by Dring & Fage (London, 1790–1977).

- Round take-apart rods, about ½ in. in diameter, which may divide into 4, 5 or 6 parts and may be contained in a leather case.
- Rods of square section. Rarely folding, these are easy to mistake for a gauging rod. However, without the diagonal line they can only be used for dipping.

Spile rods: A slim rod only about ³⁄₁₆ in. diameter and thus thin enough to pass though a spile hole in the bung, the most common type is a round

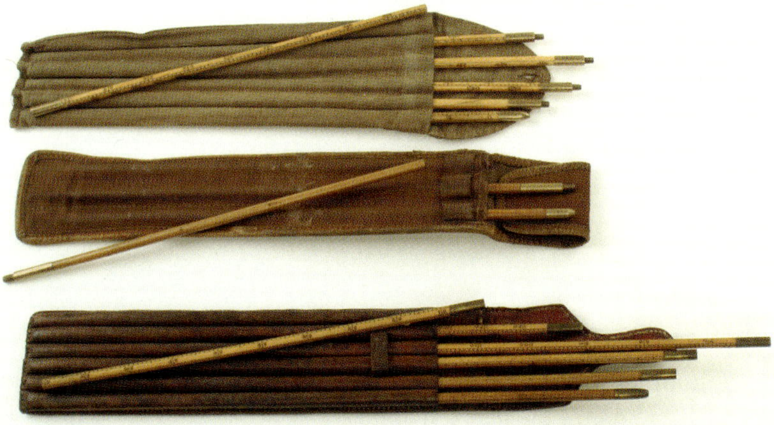

Fig. 11.25. Three sets of spile rods, the top two, unmarked, in canvas cases, the bottom by W. R. Loftus (London, 1855–1947>) in a leather case.

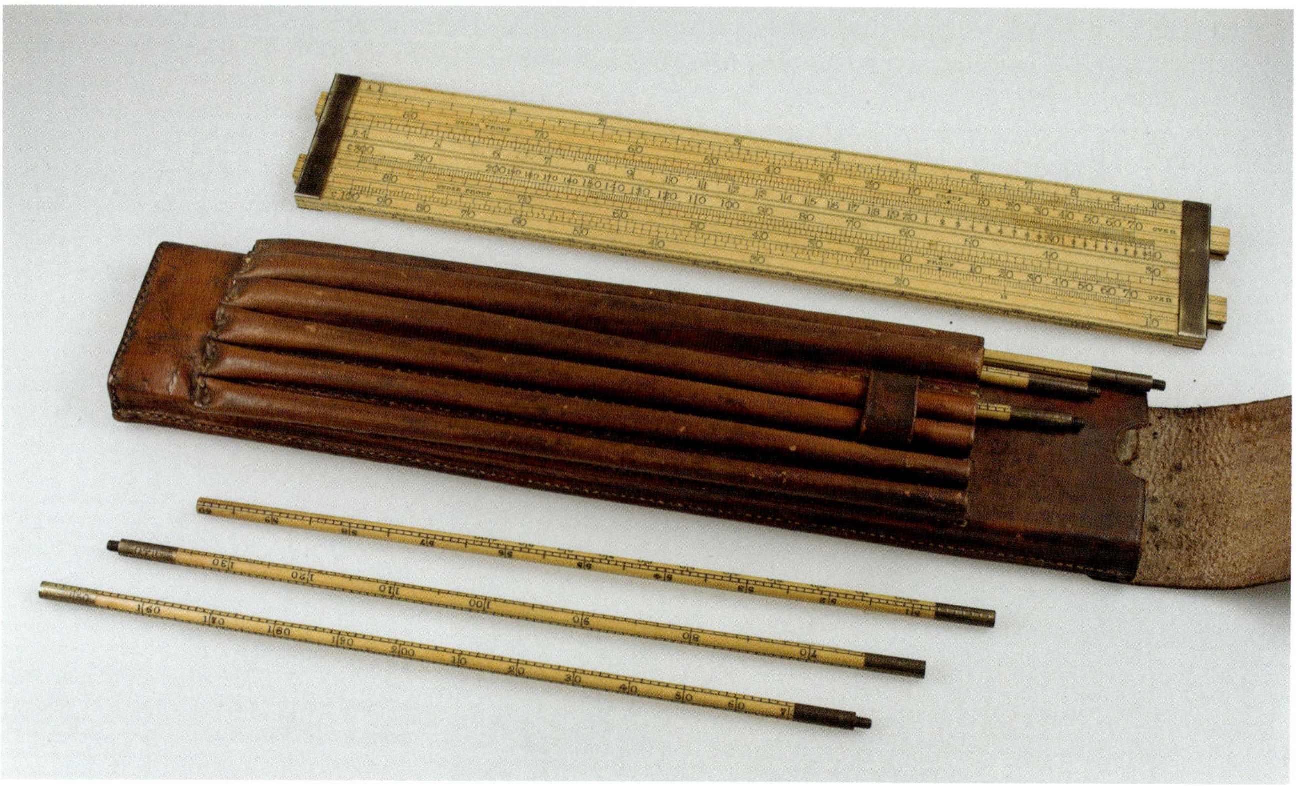

Fig. 11.26. A set of spile rods contained in a leather case with a gaugers' slide rule made by W. R. Loftus (London, 1855–1947>). This set dates to the early years of the 20th century.

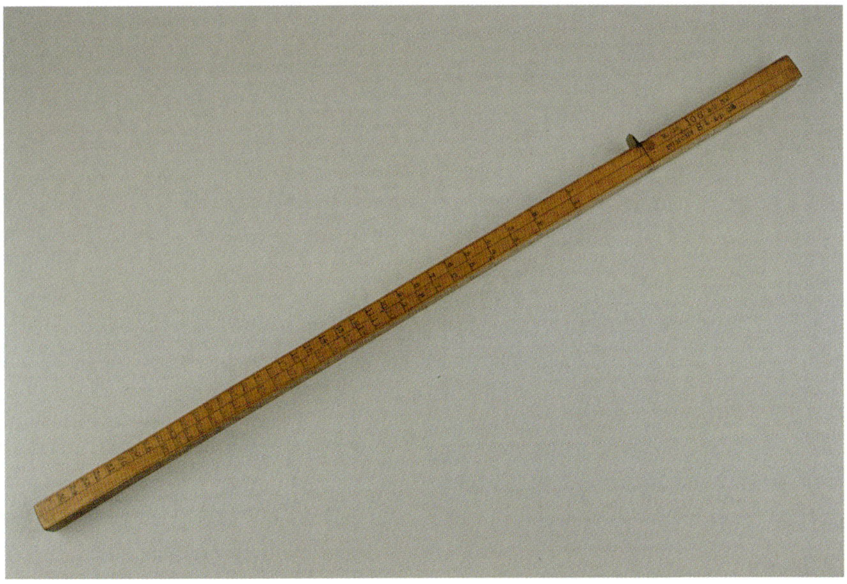

Fig. 11.27. An ullage rule marked for pipes, hogsheads, puncheons, tierces, barrels and half barrels.

boxwood rule, 3⁄16 to 1⁄4 in. in diameter, constructed in 3, 4, 5, or 6 sections, with threaded brass joints. When assembled, the shorter sets dip to 30 in., the longer to 60 in. Many sets are contained in protective leather or canvas cases, sometimes combined with a gaugers' slide rule and occasionally also a gimlet.

As space is limited on a spile rod there is usually only one line of inches and one line of gallons, though some have more. On the shorter rods this is invariably a dip line for a cask lying but on the long rods a diagonal line is usual.

Square brass rods, a more durable type for cellar use, can be found. There are also square section rods of boxwood with threaded brass joints but these are rare.

Ullage sticks: This is a short rule, which, when inserted though the bung hole of a lying cask, measures the ullage i.e. the dry inches. All ullage rules have some form of rest which may be just a small bracket to the side or a wide cross piece.

The line or lines of measure, with a suitable allowance for the thickness of the stave, extend from the rest and give a direct reading of the ullage in gallons. Any ullage line, which is essentially a reversed dip line, will be specific to a particular size of cask so commonly a number of lines are marked on one stick but sticks specific to one size of cask are also found.

When whisky or other spirits are kept in barrels to mature there is a continuing loss due to evaporation through the wood. One form of ullage stick is marked with a line annotated "years". This is a scale of expected loss – if exceeded, the barrel was not tight, or someone was helping themselves!

Draymans' rule: A short form of folding dip rule, available as both 2 ft. 2-fold and 2 ft. 4-fold. This is only long enough for dipping the three smallest sizes of beer casks (4½, 9 and 18 gallons) standing or lying and the 36 gallon barrel lying and is therefore only marked for beer gallons.

Displacers: Hollow brass tubes, each with a volume of 1 pint, strung together on a cord. These were inserted into the cask with additional tubes be-

Fig. 11.28. A 2 ft. 2-fold drayman's rule by Laban Cooke (London, 1826–34).

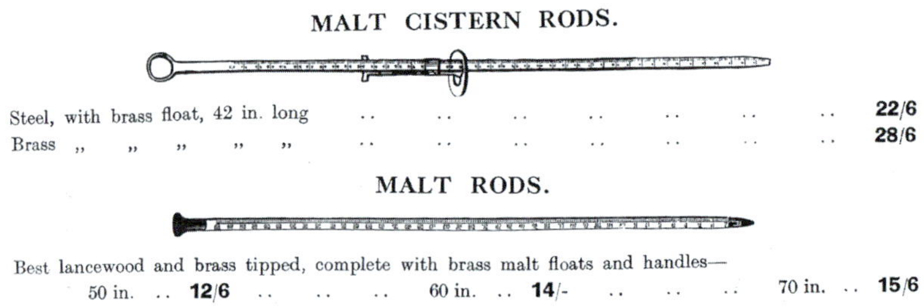

MALT CISTERN RODS.

Steel, with brass float, 42 in. long	22/6
Brass " " " " "	28/6

MALT RODS.

Best lancewood and brass tipped, complete with brass malt floats and handles—
50 in. .. 12/6 .. .. .. 60 in. .. 14/- .. .. .. 70 in. .. 15/6

Fig. 11.29. Maltsters' rules from the Joseph Long 1922 catalogue.

ing added until the liquid in the cask reached the bung. Listed by Loftus[19] in 1875 and described as "originally made for the Inland Revenue".

Brewers' rules and gauges

A number of rules and gauges were used in breweries for measuring the contents of vats during the brewing process.

Float gauge: A rod marked in inches with a cork float fixed to the end that supports the rod above the liquid. Lengths from 20 to 100 inches were listed by Long[20] in 1922. Each fixed receptacle in the brewery would have been previously gauged so that once the measurement had been taken, the contents could be read from a table for that container.

Shoulder rod: Used for measuring down from the top of a vat, it has a shoulder stop with the results being read from tables in the same way as the float gauge. Long[21] lists a "portable float gauge and shoulder rod combined as used by Inland Revenue", a sectional screw-together rod with a brass shoulder which was also screwed on.

Vat rod: This is brass-tipped dip rod of large size, 6 to 12 ft. long, made of lancewood.

Vat tape: A tape with a weighted end, this is used for plumbing the depth of vats.

Maltsters' rules

There are two rods used by maltsters during the process of malting barley.

Cistern rod: Used for measuring the depth of the grain being steeped in a maltsters' cistern, this is a brass or steel rod, usually 42 in. long, with a sliding depth stop (often called a float).

Couch rod (malt rod): This is for measuring the depth of grain in the couch or floor. A brass tipped wooden rod which, like the cistern rod, is fitted with a sliding stop. Some illustrations show a knob handle at the top.

Bottle rules

The contents of stoneware bottles also often need to be measured. There are two shapes – cylindrical, known as cylinder bottles and bulge (or bouge) bottles – and two common forms of the rule:

One form is a square section rule 2 ft. 4–fold with a taper tip. The other

Below: Fig. 11.30. A bulge or bouge bottle, showing the method of measuring.

Below right: Fig. 11.31. A square section bottle rule marked for both bulge bottles and cylinder bottles.

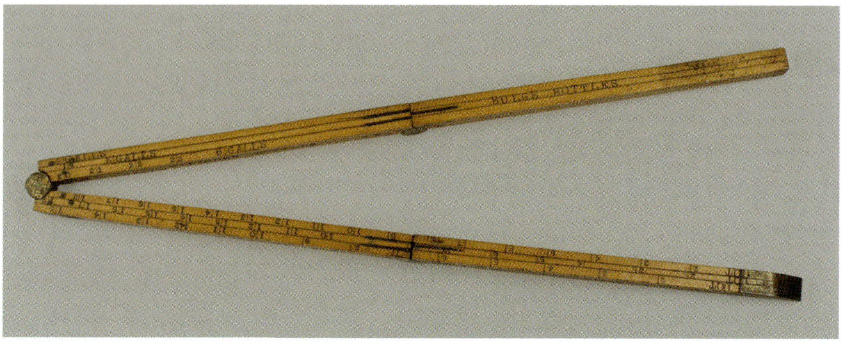

form is a 2 ft. 2–fold rule with one bullet end. The bullet-ended leg is round for part or, in some examples, most of its length.

These rules are likely to be marked with up to six lines for cylindrical bottles (of 1 to 6 gallon capacity) and six for bulge or bouge bottles. The lines give a direct reading in quarts. The angled tips would suggest that the rules were used on the diagonal but it seems unlikely that the angle was sufficient to warrant this.

Fig. 11.32. Jesse's Patent Gauge Rule by C. T. Cooper & Sons (London, 1905–47>). Two slim boxwood rules, each 13 in. by 1³⁄8 in., double sided with brass cursor. Each rule is marked for different liquors.

Bottle gauge rules

Used not only by the bar keeper but also by stocktakers to check pub management, for instance, at the year end or at a change of licensees, these rules are stood upright alongside a bottle to measure the number of shots (measured servings) that have been taken from, or alternatively remain in, the bottle. As spirits and liqueurs are mostly sold in proprietary shaped bottles it is necessary to have a correct scale to measure each type of bottle. The rules generally have 12 or more scales each of which will be annotated with the name of the tipple.

Stave gauges

Thin section brass rules measuring to 3½ in., usually with a ring for the finger at one end and a small hook at the other, but sometimes just with the hook, they are used to determine the thickness of the staves or head of a cask. This may entail boring a small hole and inserting the gauge through it.

Calculating and slide rules

In 1633 William Oughtred published *The New Artificial Gauging Line or Rod* in which he describes his invention of a brass instrument for the gauging of barrels.[22]

The instrument had been made in answer to requests from the Vintner's Company. It consists of two brass rods sliding alongside each other, held together with hoops. It was a multi-purpose instrument: A measuring rod, which, with the addition of caliper arms became a long caliper, and a slide rule. A full description of this instrument and an explanation of the formula used by Oughtred have been published by Otnes.[23]

It is known that sixty of these rules were made, and quickly sold, by the scientific instrument maker Elias Allen (London, 1607–53). Whilst the idea of the sliding calculating scales was Oughtred's, other features in the design came from existing practice. It is significant that these slide rules – almost certainly the earliest slide rules made in any quantity – were special purpose rules for gauging.

Gaugers' slide rules

There is a clear progression in the format of the these rules from the 2-slide rule, described by Thomas Everard in his book *Stereometry or the Art of Gaug-*

Fig. 11.33. Brass stave gauge of the style without the ring.

Fig. 11.34. Brass stave gauge from the Joseph Long 1922 catalogue.

BRASS STAVE GAUGE.

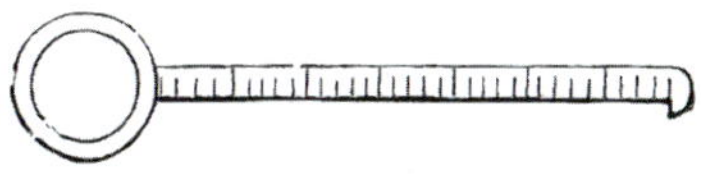

For ascertaining the thickness of the staves, **2/6**

ing, to a rectangular section rule containing 2 slides on opposite faces, then 3 slides and finally 4 slides, one in each face that is recognised today as a gaugers' slide rule. [24]

All these early rules are in the form of a rectangular solid stock into which the slides are inset. The most common length is 12 in. but rules as short as 6 in. and as long as 24 in. were made.

The physical format of the rule is however only one aspect; equally important is the content and layout of the lines and gauge points.

An understanding of the development of the gaugers' rule is difficult as even some of the eminent writers of gauging manuals make assertions that are not correct. For example, William Symons[26] writing in 1803, heads his descriptions of gaugers' sliding rules as follows:

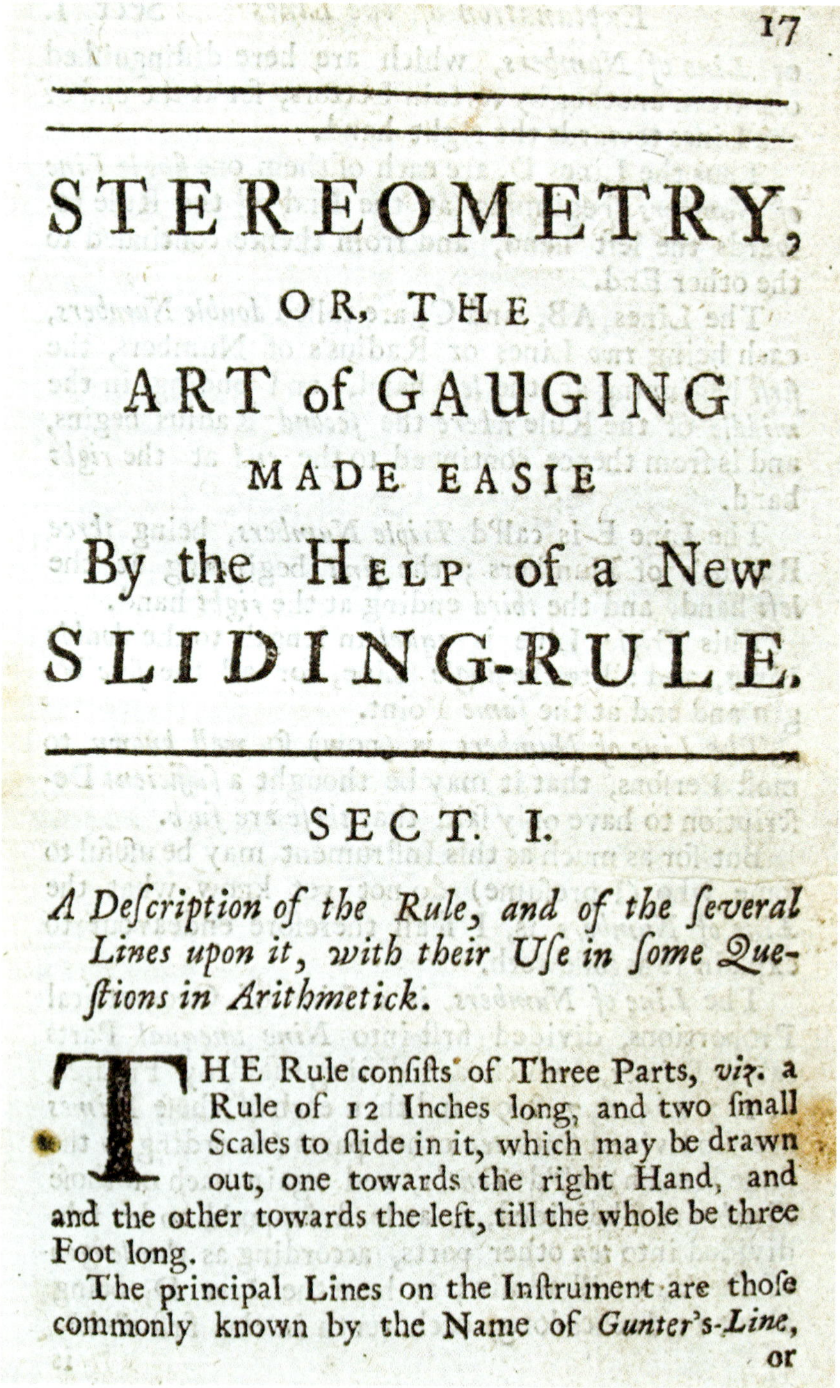

17

STEREOMETRY,

OR, THE

ART of GAUGING

MADE EASIE

By the HELP of a New

SLIDING-RULE.

SECT. I.

A Deſcription of the Rule, and of the ſeveral Lines upon it, with their Uſe in ſome Queſtions in Arithmetick.

THE Rule conſiſts of Three Parts, *viz.* a Rule of 12 Inches long, and two ſmall Scales to ſlide in it, which may be drawn out, one towards the right Hand, and the other towards the left, till the whole be three Foot long.

The principal Lines on the Inſtrument are thoſe commonly known by the Name of *Gunter's-Line,* or

Fig. 11.35. Description of the 2-slide slide rule from Everard's Stereometry.[25]

Fig. 11.36. *Three gaugers' slide rules. Top: 2-slide slide rule by Isaac Carver (London, 1688–1713). Centre: 3-slide slide rule by William Rix (London, <1750>). Bottom: 4-slide slide rule by Edward Roberts (London, 1749–95) dated 1781.*

Everard's Sliding Rule, Having only Two Slides.
Leadbetter's Three Slided Rule, being the preceding rule improved.
Verie's Four Slided Rule, being a further Improvement of the two preceding Rules and that are now principally used by officers of the Revenue.

However his description of Everard's rule includes the inverted malt depth line which does not appear in Everard's description or on the early rules which conform to his description. Similarly Leadbetter[27] illustrates a 4-slide rule which is titled "Everard's Sliding Rule as it is now improved by C. Leadbetter." This also includes the malt depth line (marked MD) and other features which were not Everard's invention. Although claimed by Leadbetter to be his "improvement", it was no more than the then current design.

In essence Everard's rule was a general purpose slide rule with only a limited number of adaptations for the gauger – some gauge points, the line of ale area, and mean diameter lines. Indeed it was very little adapted compared to the post-1750 4-slide rule which had the MD line and a double length line of logs (folded at 3.142). In addition, the slides were marked on the back

Fig. 11.37. Details of the ends of the three rules shown in Fig. 11.36, showing the arrangement of the slides

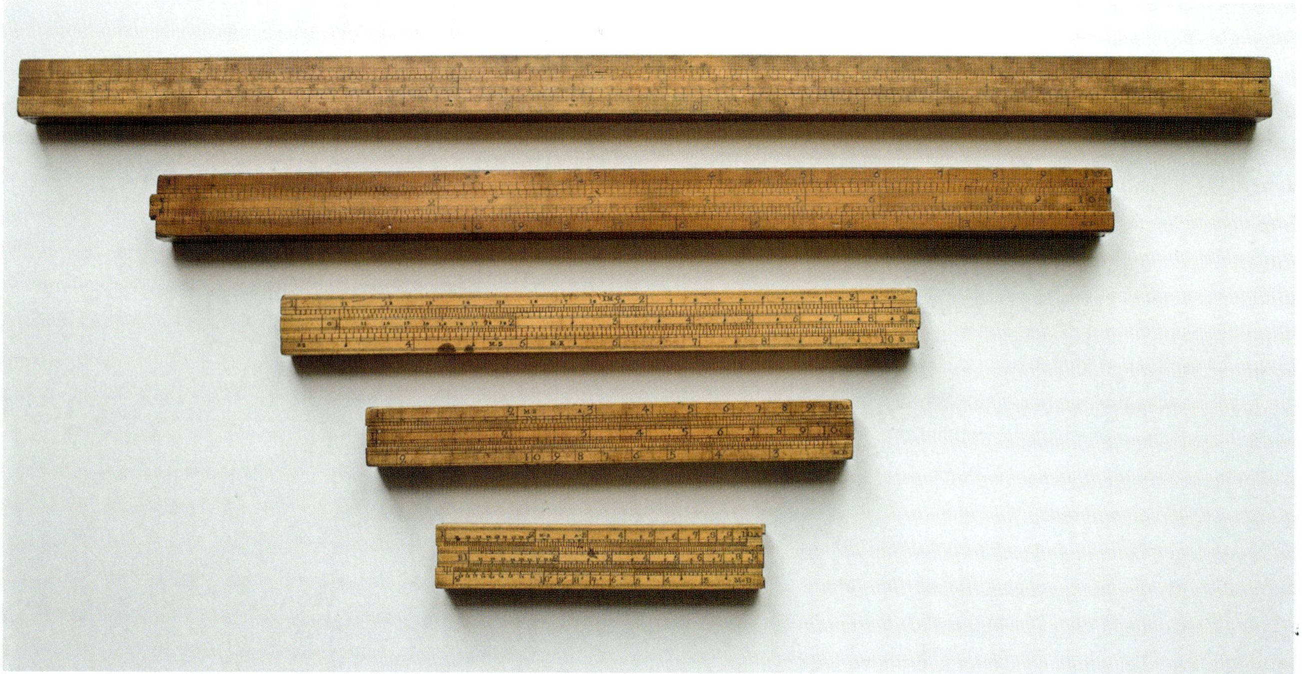

Fig. 11.38. Five gaugers' slide rules, from top: 24 in. rule by Laban Cooke (London, 1826–34); 18 in. rule by Edward Roberts (London, 1749–95); 12 in. rule by William Hobcroft (London, 1817–52); 9 in. rule by Edward Roberts; and 6 in. rule, unmarked.

with numerous gauge points and were capable of being used end to end in the same slot.

Verie's improvement, introduced sometime before 1739, (Verie, whose name is sometimes spelt Vero, described himself as "sometime gauger with the Excise") involved the slides having brass tips riveted to the underside and the positions of the figures on the slide being regulated so that when two slides are placed end to end in one slot, the reading taken will be accurate. The equally important advantage is that it doubles the length of the slide and makes it impossible to "fall off" the end of the slide.

Fig. 11.39. Plate A from Leadbetter's Royal Gauger *showing his improved sliding rule.*

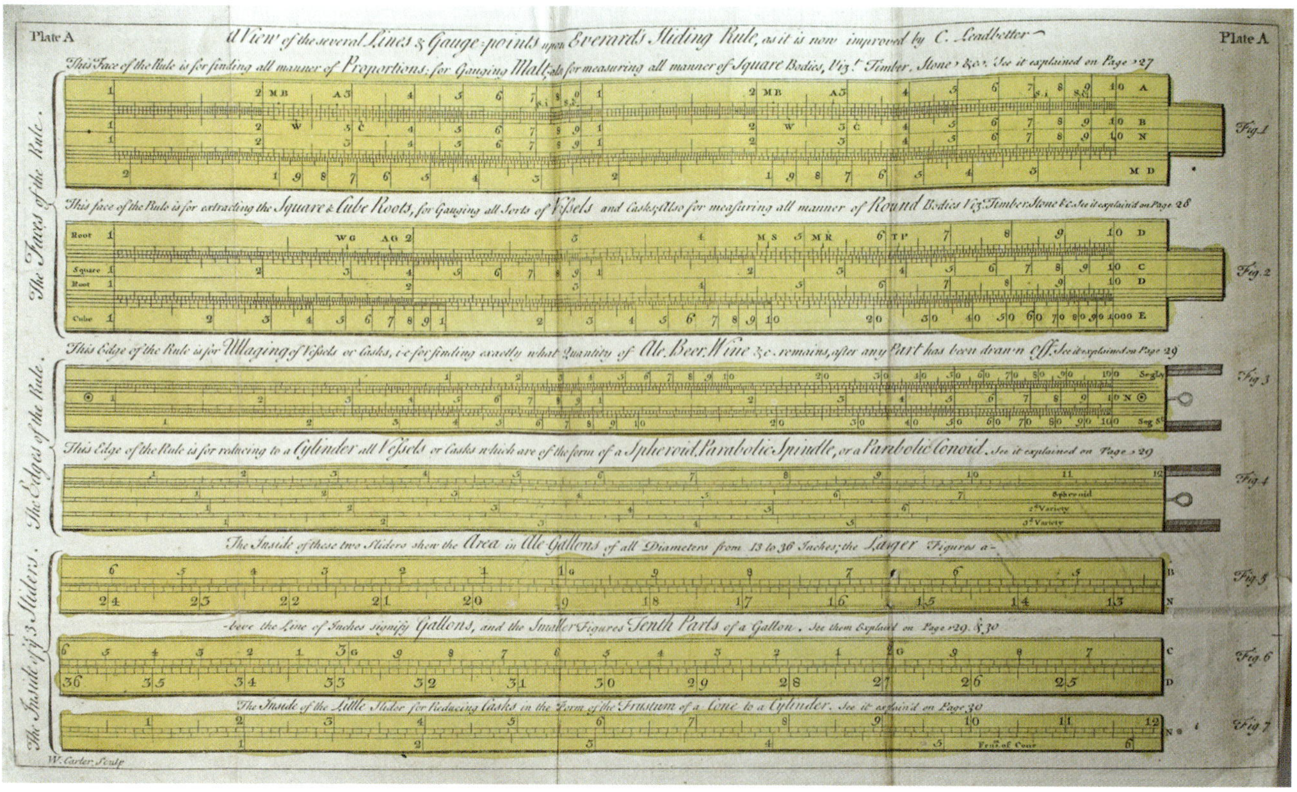

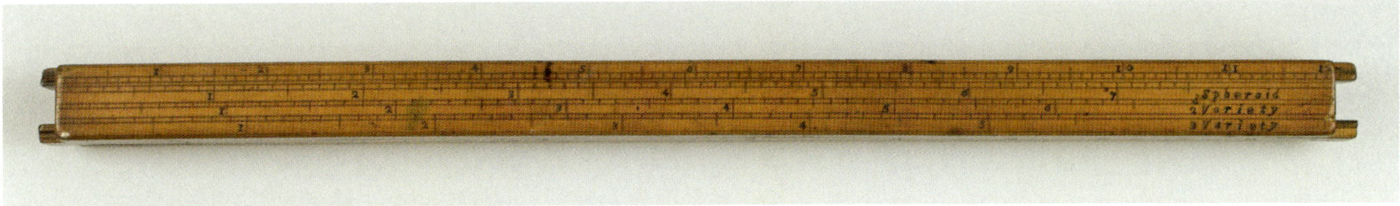

Fig. 11.40. Line of inches and variety lines on a rule by William Rix (London, <1750).

Lines on square section gaugers' rules

The lines are invariably annotated A, B, C, etc., which is helpful when following the instructions in a gaugers' manual but otherwise not significant. On some rules lines may be annotated also with letters indicating the use of the line: N (numbers), MD (malt depth), etc. Listed below are the lines that may be found on rules but it should be noted that not all rules have all the lines.

Lines of measure: Marked in inches, or occasionally feet, but always divided decimally to facilitate calculation.

Variety lines: Two or three lines marked Spheroid (the first variety), 2 Variety and 3 Variety (the fourth variety never seems to have been marked on slide rules). These are used to apply a factor to adjust for the shape of the cask. On early rules made before the third quarter of the 18th century, these are on one edge of the rule, with all three marked; later they are marked on the back of a slide and only two varieties (usually the first and second varieties) are shown.

Calculating lines: One side of a square section rule is marked with a single cycle logarithmic scale, numbered from 1 to 10, marked A, with the slide marked B similarly numbered.

There is a third line, the malt depth line, marked MD. This line is reversed (numbered from right to left) with 1 opposite 2150 on the single line of logs. Using this line within a calculation has the effect of dividing by 2150 – the number of cubic inches in a bushel.

(In 1880 the ingredients used to make beer – malt, hops and sugar – were freed from duty although beer drinkers were no better off as the tax was simply transferred to the wort (the liquid extracted from mashing malted barley when brewing). With malt no longer subject to tax the need for the MD line on the gaugers' rule disappeared.)

Segments standing: On another side there is a line marked Seg. St. This is one of the two ullage lines which, using the wet inches, indicate the percentage of the volume of the cask that is full. This line is used when a cask is standing, i.e. on its end. It is numbered 1 to 8 on the upper scale, with the continuation from 8 to 100 on the lower scale, marked SS (for segments standing). The slide, marked B, is another single cycle logarithmic scale, numbered 1 to 9.

Segments lying: On a third side there is the other ullage line, used when the cask is lying on its side. Marked Seg. Ly., the scale is marked from 1 to 4 on the top scale and 4 to 100 on the bottom scale. Again, the slide is a single cycle logarithmic slide numbered from 1 to 10 and marked C.

On early rules the segments standing and segments lying lines are not di-

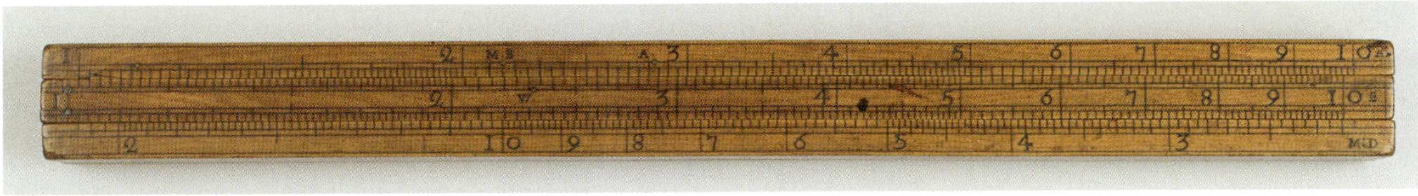

Fig. 11.41. The logarithmic scales A and B and the malt depth line on a rule by Edward Roberts (London, 1749–95) dated 1781.

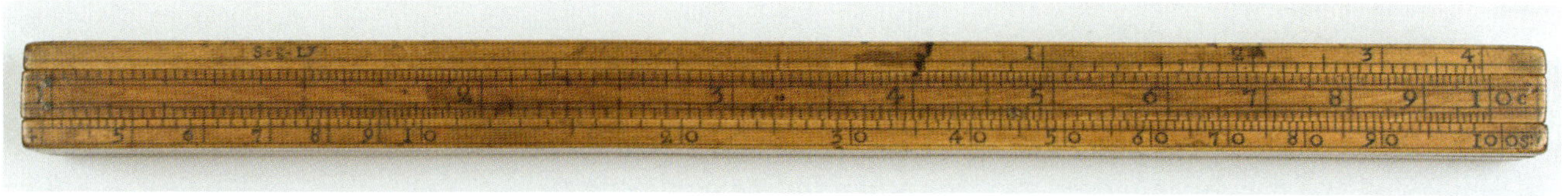

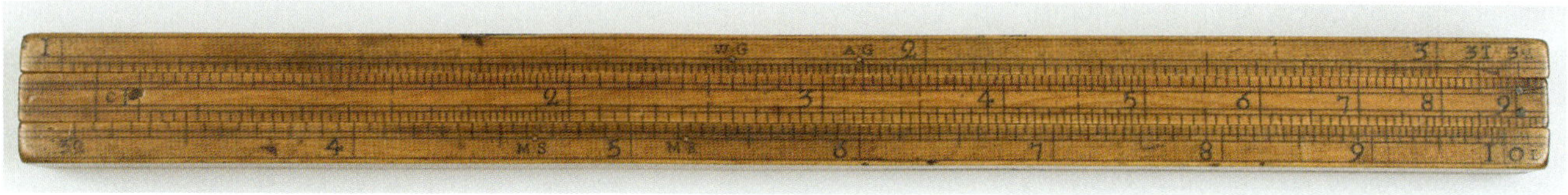

Top: Fig. 11.42. The segments standing line on the rule shown in Fig. 11.41.

Centre: Fig. 11.43. The segments lying line on the rule shown in Fig. 11.41.

Bottom: Fig. 11.44. The square and square root lines on the rule shown in Fig. 11.41.

vided, but on later rules they are divided into two parts, doubling their length and improving accuracy. To use these longer length lines, it may be necessary to put two slides end to end in the slot as some calculations will fall off the end of a single slide.

Square and square root lines: The fourth side of the rule is set out so that the scales can be used to find squares and square roots. The first section of the fixed scale is numbered from 1 to 32, which continues along the lower section from 32 to 100 and is marked D. Again the slide is a simple logarithmic line.

Gauge points

It will be seen in Fig. 11.44 that there are a number of gauge points marked with brass pins. In earlier rules they were often marked only with a stamped dot. These gauge points were used to simplify the complicated mathematical calculations that were needed by part working the equations. On later rules other useful gauge points are annotated on the backs of the slides, as shown in Fig. 11.47 on page 218. Gauge points fall into three categories:

Defining a standard such as the bushel: In Fig. 11.41, a gauge point marked MB can be seen. This is marked at 2150.42, being the number of cubic inches in a malt bushel and W, at 231, the number of cubic inches in a wine gallon.

On the side of the rule shown in Fig. 11.44, the gauge points are WG (Wine Gallon), marked at 17.15, being the diameter in inches of a cylinder that, when filled to a depth of one inch, contains one gallon of wine, i.e. 231 cubic inches; AG marked at 18.95, being the equivalent for ale gallons, i.e. 282 cubic inches. Rules dating from after 1824 will be marked with a gauge point marked IMG for the Imperial gallon, although it seemed that gauge points for wine and ale gallons continued to be marked after that date, so although IMG is an indication of the earliest date a rule thus marked could have been made, WG and AG are not a reliable indication of the latest date.

Also in Fig. 11.44 will be seen gauge points marked MS, at 46.3 and MR at 52.32 These are the sides of a square vessel in inches which contain a malt bushel to the depth of one inch (MS) and the diameter of a circular vessel which contains a malt bushel to the depth of one inch (MR).

Mathematical constants: Most of these relate to a circle or to a circle inscribed within a square. For example Si = .707, the side of a square inscribed within a circle with a diameter of 1.

Fig. 11.45. Flat form of the gaugers' slide rule by Edward Roberts (London, 1749–95) dated 1795.

Conversion factors or part worked calculations: Their purpose is to simplify the number of calculations (movements) that have to be made on the rule. These will often be given as a multiplier or a divisor. Without the explanations contained in a good gaugers' manual these are almost impossible to decipher.

Gaugers' slide rules – flat form

Before the end of the 18th century some makers, including Edward Roberts (London, 1749-95), were making rules with only two slides inset into the face of a solid but flat stock. These rules are invariably longer (18 or 24 in.) than the normal 12 in. length of gauging rules. Having only two slides the lines are fewer – two lines of logs (double) and the two segment lines previously described, indicating that the use was for cask gauging. The absence of any variety lines suggests the rules were intended for use only on British standard spheroid casks. This may have indicated that this type of rule was used more for domestic situations, such as the inspection of breweries, rather than

Fig. 11.46. The illustration of the improved slide rule from Joseph Bateman, The Excise Officers' Manual and Improved Practical Gauger, *1852.*

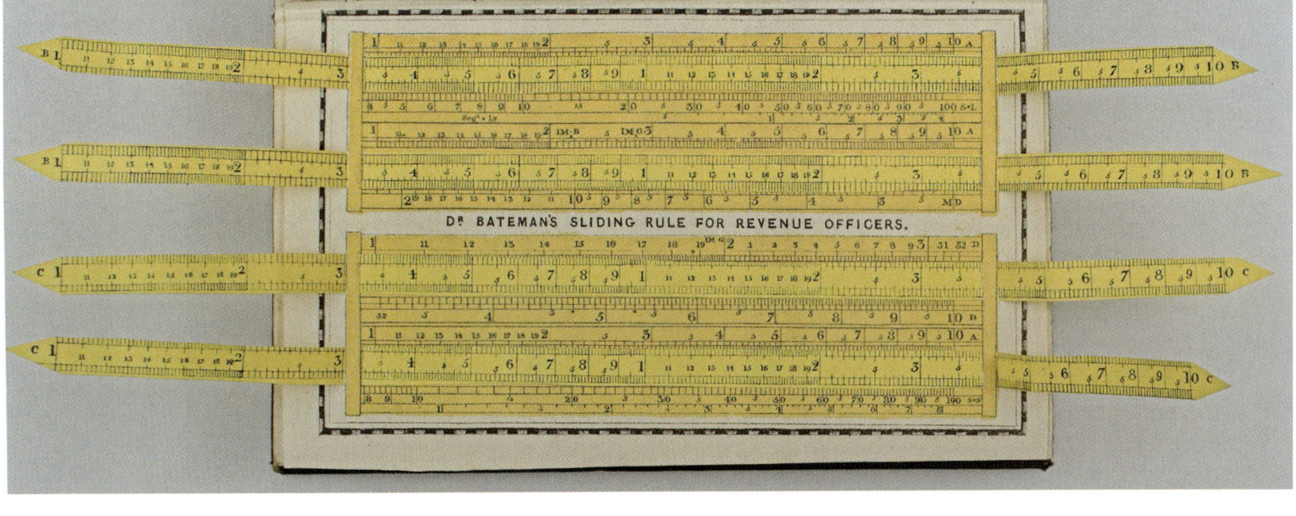

Fig. 11.47. Left: The backs of the slides from a 4-slide gaugers' rule by Edward Roberts (London, 1749–95) dated 1781. Right: The backs of the slides from a 2-slide gaugers' rule by William Rix (London, <1750) with, opposite, an explanation of the abbreviations. Not all gauge points appear on all rules and different gauge points may be found at different dates.

Line Titles:	
A,B,C ...	*Line annotation. Only for identification when following instructions in manuals.*
N, Numb	*Numbers = Line of Logarithms (logs.).*
Root	*Line of logs. (single cycle) positioned adjacent to a double or triple line so square or cube roots can be ascertained.*
Square	*Line of logs. (double). See above.*
Cube	*Line of logs. (triple). See above.*
MD	*Malt depth line.*
Seg. Ly & SL	*Segments lying (casks lying on their sides).*
Seg St & SS	*Segments standing (casks standing on their ends).*
Spheroid	*Mean diameter line for cask of 1st variety.*
2 Variety	*Mean diameter line for casks of the 2nd variety.*
3 Variety	*Mean diameter line for casks of the 3rd variety.*
Standards:	
AG	*Ale gallon, 282 cu. in. Obsolete after 1824.*
WG	*Wine gallon, 231 cu. in. Obsolete after 1824.*
IMG	*Imperial gallon, 277 cu. in. Only marked after 1824.*
MB	*Malt bushel, 2140 cu. in.*
IMB	*Imperial bushel, 2218 cu. in. Only marked after 1824.*
Conversion Factors and Divisors:	
MR	*Malt round cistern. (The vessel used for steeping malt and barley could be either round or square, each requiring different calculations.)*
MS	*Malt square (or rectangular) cistern.*
MT	*Mash tun gallon (227 cu. in.).*
G St. or GS	*Green starch - partly finished product, not dry.*
D St. or DS	*Dry starch.*
HS hot	*Hard soap - hot.*
HS cold	*Hard soap – cold.*
GSS	*Green soft soap.*
WSS	*White soft soap.*
FG or F Glass	*Flint glass.*
PG or Plate	*Plate glass.*
CG	*Crown glass.*
Cr Bro	*Crown and broad glass.*
Ph Bots	*Phial and bottle glass.*
GB	*Glass bottles.*
LW	*Low wine.*
Couch	*Bed or layer of barley in rectangular container. (After steeping malt and barley were transferred to a rectangular container, a couch, where they were gauged. After 36 hours they were then spread as a thick layer on the floor)*
Cyder to couch	*Bed of apples prepared for cider in rectangular container.*
Cyder to floor	*Bed of apples prepared for cider laid onto floor.*
Mum	*Ale brewed using wheat.*
Sweets	*Wines sweetened and flavoured; in 18th century terms, a cordial.*
Ver	*Verjuice: an acidic juice from unripe fruit, used in cooking. The duty on verjuice was repealed in 1818.*
Malls wash to LW	*Mollasses wash to low wine.*
Do LW to spirits	*Low wine to spirits.*
SF	*Not yet identified.*

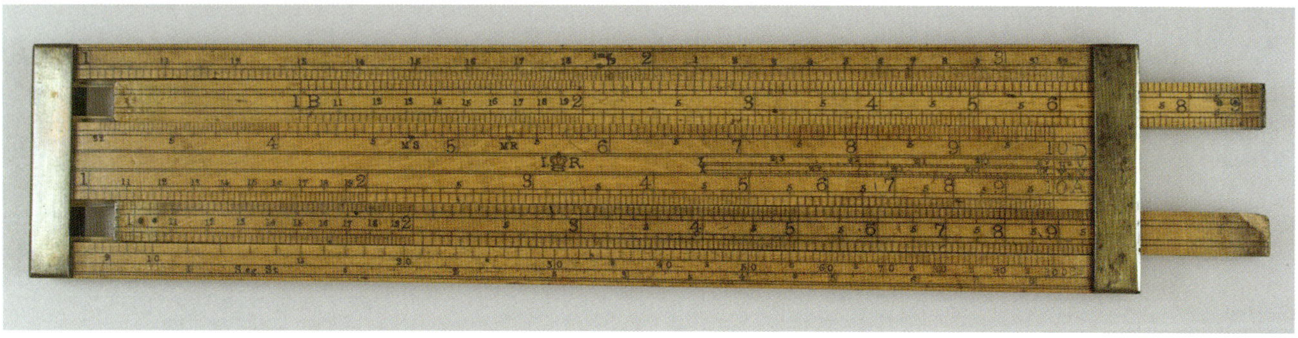

Fig. 11.48a & b. A flat gaugers' rule by Dring & Fage (London, 1790–1977). On one side are scales for 1st and 2nd varieties, the segment lying line and a malt depth line, with gauge points for imperial bushels and imperial gallons. This dates the rule to after 1824, when imperial measures were introduced but before 1880, when the duty was removed from malt. The segment standing line is on the reverse face. Also on the reverse are scales for Hulton's and Young's varieties, marked HsV and YsV, which were alternative formulae for calculating cask volume.

full gauging at ports where foreign casks of variable varieties were arriving.

Around 1840 a radical change took place with the rules becoming a flat double-sided rule the stock being constructed from three pieces, joined at the ends with brass strips.

The illustration of a gaugers' slide rule included in Bateman's 1852 book[28] shows this form of rule but none of its features were new. It is surprising at the solid-stock gaugers' rule remained virtually the only form for so long. The number of surviving examples marked with imperial standards indicate that many were made in the years after 1824.

The lines marked on the new flat rules are, in many examples, entirely similar to those on the solid type. Even the facility to use both slides in one slot was incorporated, as also was the inverted MD line.

Combined gauging, ullaging and reducing rules

The fully developed 4-slide gaugers' rule, with its many scales, some of which were folded for reasons not immediately apparent, had perhaps become excessively complicated. It had features that many users would never need. The simpler "combined gauging, ullaging and reducing rule" actually provided an additional facility as one side was set out with reducing scales for spirits. The format is the same as the flat form of gaugers' rule – a double sided, 2-slide rule usually 12 in. long. The difference is the lines. Ivory versions are relatively common.

Hydrometer rules

Hydrometers measure the specific gravity or density of liquids and are used to measure the alcoholic strength of any spirit. Invented around 1675 by Robert Boyle, these instruments were obviously of interest to the Customs & Excise. In 1787, a hydrometer designed by John Clarke of London was accepted by law for use by the Customs & Excise, even though it was both inaccurate and cumbersome to use.

In 1780, John Dicas of Liverpool designed and patented a different design of hydrometer with a notably different style of slide rule to accompany it. The Dicas system was to became the U.S. standard.

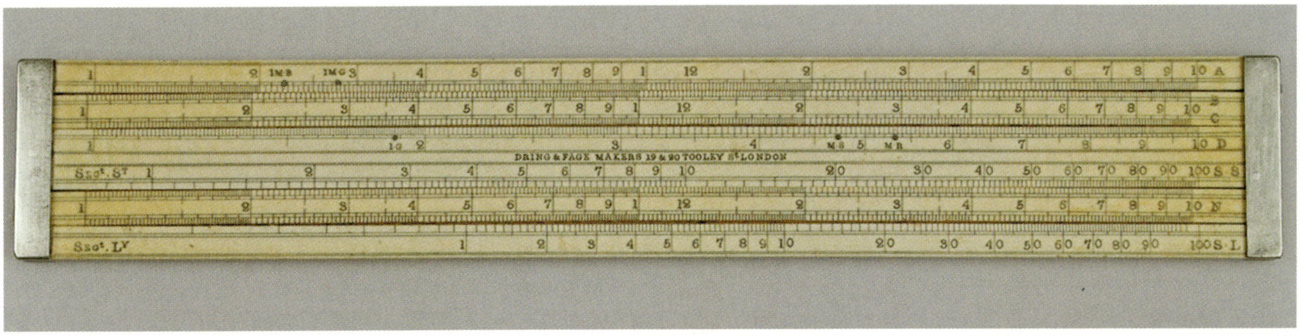

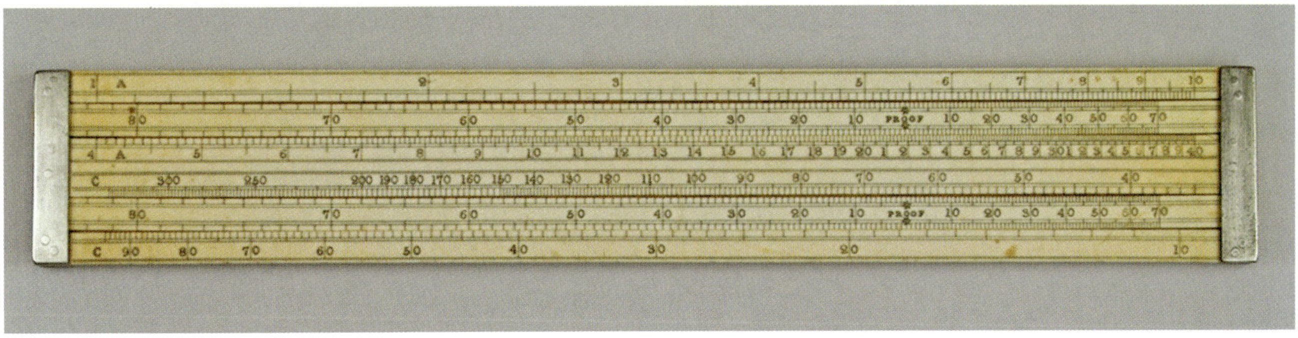

Fig. 11.49a & b. An ivory combined gauging, ullaging and reducing rule by Dring & Fage (London, 1790–1977). On one side, in addition to the normal logarithmic scales, are the lines for segment standing and segment lying and on the other the proof lines.

After a trial of hydrometers in 1803 there followed years of maneuvering until, in 1818, a design by Bartholomew Sikes, an employee of the Customs & Excise, was accepted as the best and the Sikes' pattern hydrometer was adopted by Act of Parliament as the only type acceptable for Excise purposes. By then Bartholomew Sikes was long dead but his nephew Robert Bate, with help of his aunt, Mrs. Sikes, had contrived to get a virtually exclusive contract for the supply of instruments to the Excise, an arrangement that lasted until 1851.[29] With this instrument, it was possible now to make a quick and accurate measurement of the alcoholic strength of any spirit.

The Sikes' hydrometer is very simple to use. The temperature of the sample is taken, the hydrometer is floated in the spirit with as many weights as are needed and the reading is taken. The scale reads from 1 to 100 at 0.2 intervals without named units, which have no actual value until converted

Fig. 11.50a & b. A Clarke's centigrade hydrometer slide rule, another early development, which went into the 1803 hyrdrometer trials, made by Dring & Fage (London, 1790–1977).

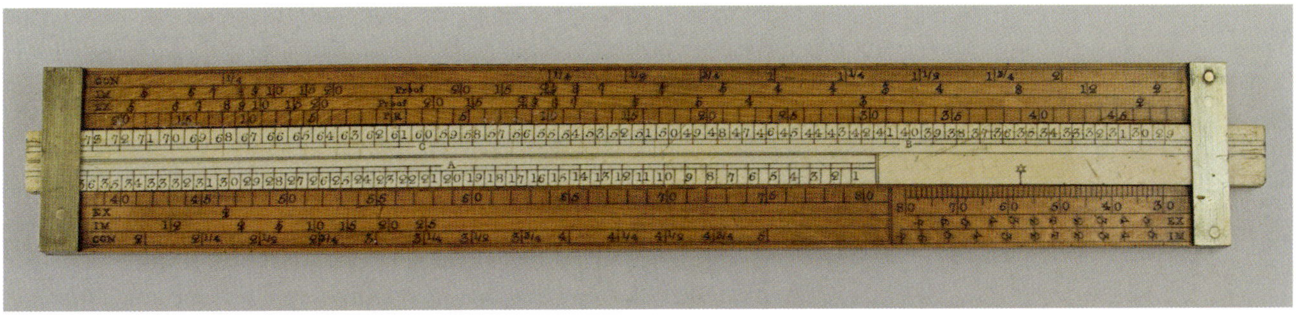

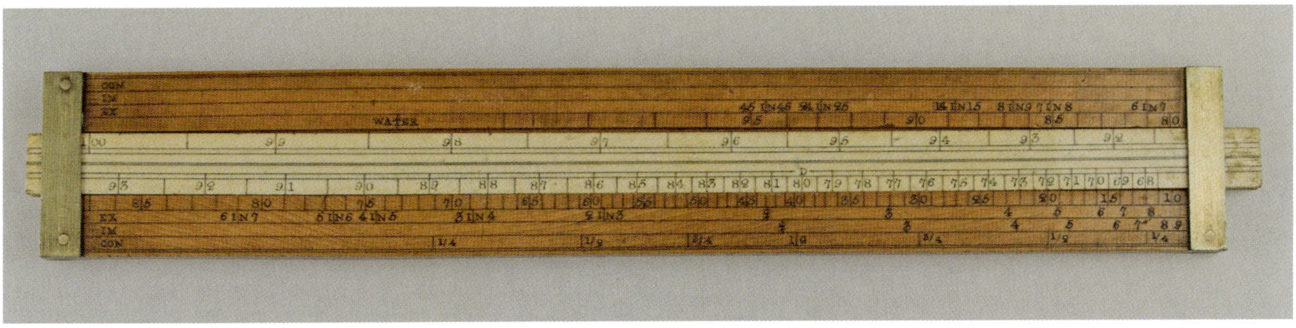

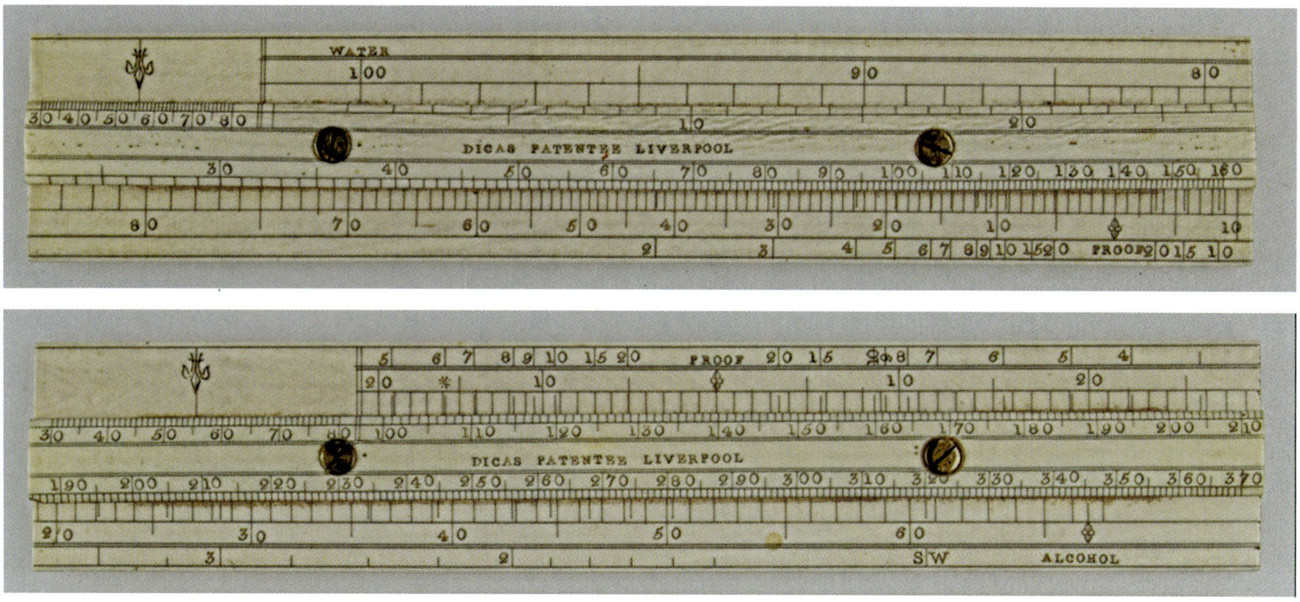

Fig. 11.51a & b. The slide rule designed and patented by John Dicas (Liverpool, 1774–1837).

by the book of tables that accompanied the instrument. Once the reading is taken, the book is opened to the page for the temperature of the sample and the percentage over or under proof can then be read off.

Spirits are deemed proof when the volume weighs $\frac{12}{13}$ths of the equivalent volume of water, measured at 51°F. The alcohol content of proof spirit is actually 49.28% by weight or 57.1% by volume (at 60° F.). The proof scale is direct reading from this point. For example, spirit that is 20% over or under proof will have 20% more or less alcohol than proof spirit. Duty was payable on the basis of proof gallons thus spirits needed to be "compared" or "reduced" to proof strength for calculation of duty.

Proof rules: Many sets of Sikes' hydrometer were sold with either one or two slide rules in the box (both single- and double-sided rules are found), generally 7 in. long, in boxwood or ivory. To many it might seem that these were essential. However, there is no need for the use of a sliding proof rule if the set is accompanied by the book of tables.

To use a slide rule as an alternative to the tables, the temperature is set and the proof strength read off opposite the Sikes' measurement. The back of the rule merely serves to extend the range – thus in most examples, the proof line is broken into no less than four pieces! The slide rule is only accurate over a limited range from 35° under proof to 55° over proof. Outside this range, it was necessary to use the tables.

Comparative valuing and reducing rules: The format is the same as the (Sikes') proof rule, and these are frequently the second rule in the hydrometer box. However, often, the comparative rule is the only rule provided – the user would have used the book of tables for establishing the proof strength.

The purpose of the comparative value side is to compute the "value" of a spirit according to the known value of another spirit. The line is usually marked "aaa"; the value line (which on early rules extends from 1*s* 6*d* to 30*s* and on later rules from 6*s* to 30*s*) is on the stock but is broken in two. The proof line is on the slide.

The purpose of the reducing side is to calculate the amount of water needed to be added to a quantity of spirit at any strength to reduce it to a lower strength. It can also be used in the opposite direction to obtain the yield when liquor of a lower strength is re-distilled to a higher strength. The line is

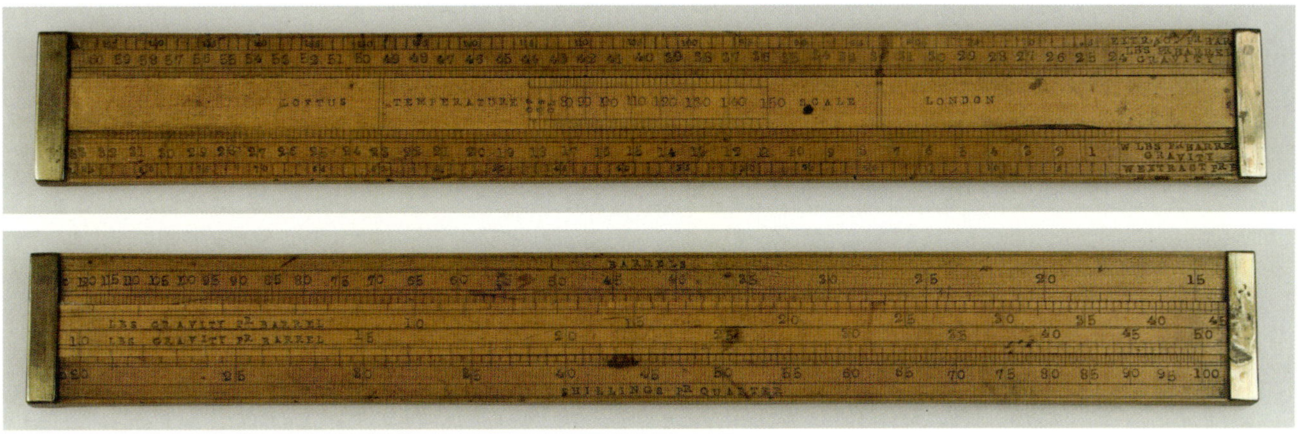

Fig. 11.52a & b. A saccharometer rule by W. R. Loftus (London, 1855–1947>).

usually marked "ccc"; the quantity line, broken into two overlapping parts, is on the stock (marked from 300 or 200 gallons to 20 gallons) The proof line, from 60 under proof to 70 over proof, is on the slide.

Saccharometer rules

After 1880 the taxation of beer was based upon the specific gravity of the wort. The more extractives contained in the wort, the higher will be its specific gravity and the stronger and tastier the resultant brew.[30]

Confusingly the strength of wort was expressed in four different ways: as specific gravity – 1.057; as brewers' gravity – "Gravity 57"; as pounds per barrel (meaning the weight in pounds by which 36 gallons of wort exceeds the weight of 36 gallons of water); and as the weight of extract per barrel, meaning the weight of the residue that would be left if the water was boiled off from 36 gallons of the wort.

Brewers preferred the last two methods of expressing the density so earlier saccharometers (an instrument similar to a Sikes' hydrometer but measuring the sugar content of a liquid, which was needed to obtain the specific gravity of the wort) were scaled to read in pounds per barrel.

There are various forms of slide rule sold to accompany the saccharometer.

Citrometer

Another similar instrument was used for testing the specific gravity of lime juice to assess the duty payable. It was also accompanied by a slide rule.

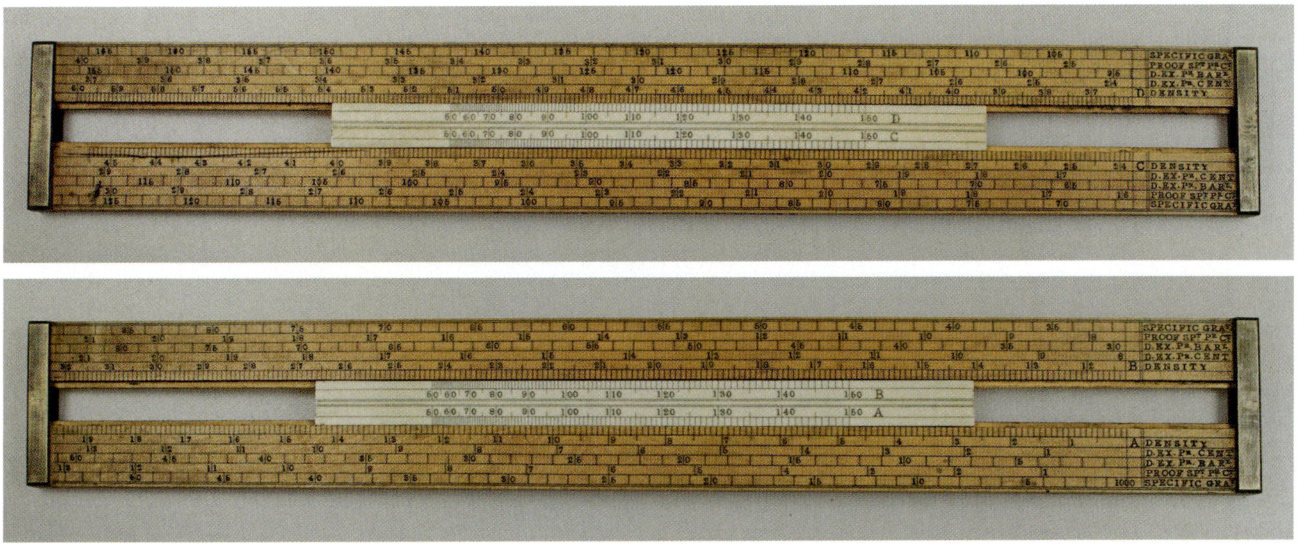

Fig. 11.53a & b. A different form of saccharometer rule by Dring & Fage (London, 1790–1977) to accompany their hydrometer.

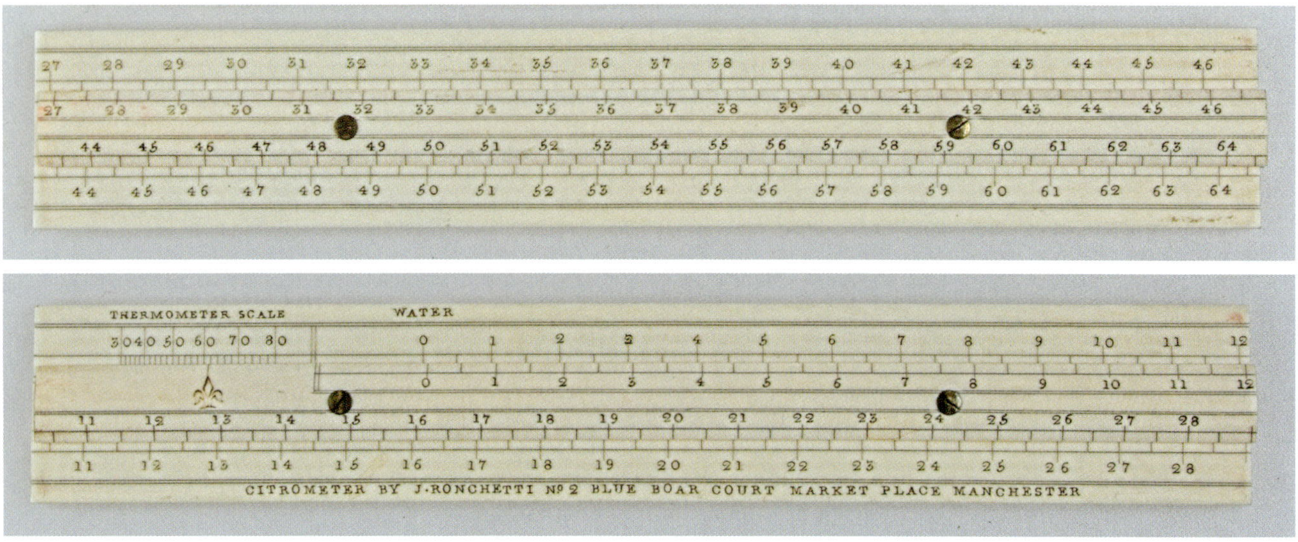

Fig. 11.54a & b. A citrometer rule by Joshua Ronchetti (Manchester, 1817–1841). This is a similar form of rule to the Dicas saccharometer.

Farmar's rules

These were proprietary rules, developed from the gaugers' rule but intended for the "wholesale and retail wine and spirit trade" rather than the Excise, patented by F. C. Farmar in 1908. An earlier less sophisticated version was patented in 1901. The rules are boldly marked with the name, have a cursor, and came in two sizes, 12 in. and 20 in.

Earlier rules are all boxwood but by 1922 they were made with facings of ivorine. Advertisements claimed that in addition to the normal gauging and reducing functions the rules were arranged for blending, pricing, profit-calculating, and stocktaking. Farmar also marketed a bottle gauging rule (an ullage rule) and several other rules for business calculations.

Ure's alcoholometer and slide rule for brewers

Andrew Ure, a Scottish doctor, worked with Joseph Long, London makers of instruments related to the brewing trade, to develop an alcoholometer to ascertain the strength of malt liquors and their respective values. The instrument was supplied with a slide rule. A full description of the rule can be found in Ure's *Dictionary of Arts, Manufactures and Mines.*[32]

Fig. 11.55a & b. A 20 in. Farmar's rule. This rule is not marked with a maker's name, but it was advertised in the Joseph Long 1922 catalogue.[31]

References

1. Leadbetter, Charles, *The Royal Gauger*, 5th ed., London, 1760.
2. Fisher, George, *The Instructor or Young Man's Best Companion,* London, 1799.
3. Leadbetter, *op. cit.*

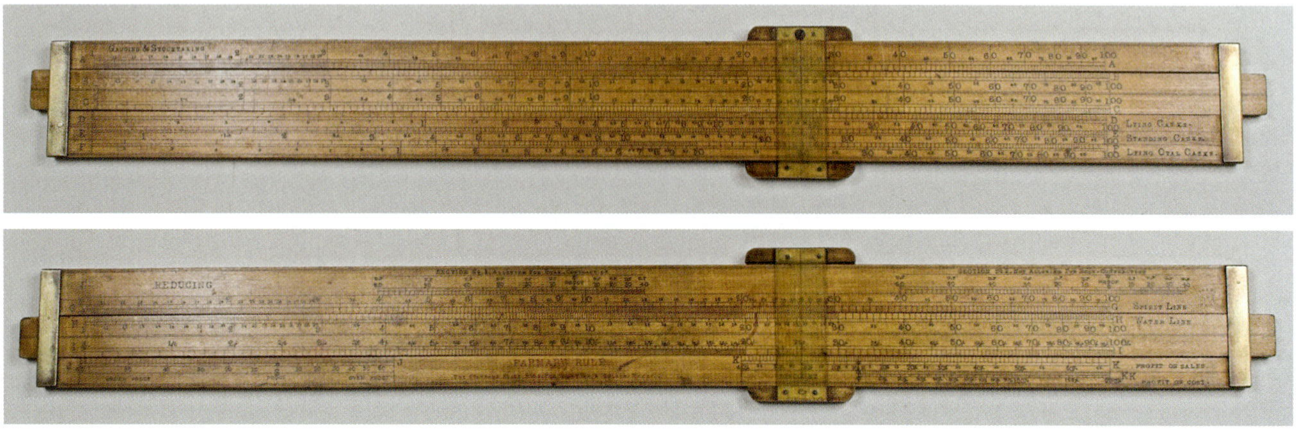

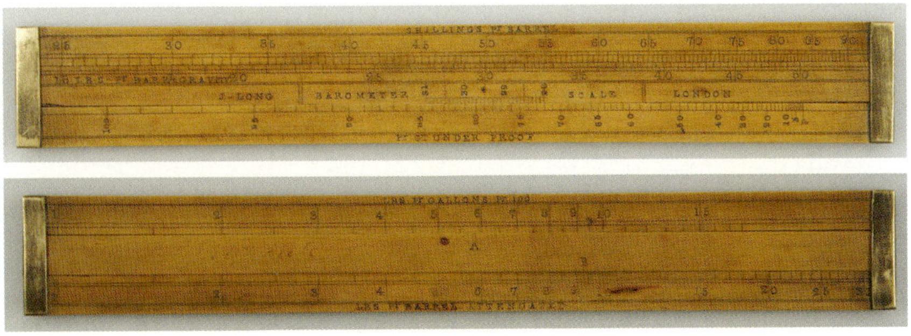

Fig. 11.56a & b. Ure's alcoholometer slide rule by Joseph Long (London 1821–1936).

4. Bateman, Joseph, *The Excise Officers' Manual and Improved Practical Gauger,* 2nd ed. London, 1852.
5. Packham, Jim, "Barrel Gauging", *The Chronicle of the Early American Industries Association,* Vol. 50, No. 4, 1997.
6. Knight, Richard, "An Important 17th Century Rule", *Tools & Trades,* Vol. 10, 1997 and "A Postscript", *Tools & Trades,* Vol. 11, 1999.
7. Salaman, R. A., *A Dictionary of Tools used in the woodworking and allied trades c. 1700-1970,* London, 1975.
8. Everard, Thomas, *Stereometry made easy: or, the description and use of a new gauging rod or sliding rule ... To which is added several new tables with short and plain rules for gauging of Malt.* 7th. Ed. London, 1712.
9. Leadbetter, *op. cit.*
10. *Ibid.*
11. Bateman, *op. cit.*
12. Bion, Nicholas, *The Construction and Principal Uses of Mathematical Instruments,* Paris 1723. Translated by Edmund Stone;... *To Which are Added, the Construction and Uses of Such Instruments as are Omitted by M. Bion* by Edmund Stone. Second edition; ... *To which is added a Supplement Containing a Further Account of some of the Most Useful Mathematical Instruments as Now Improved.* Supplement by Edmund Stone, London, 1758, Reprint, Mendham, NJ, 1995.
13. Leadbetter, Charles, *The Royal Gauger,* 1st ed., London, 1739.
14. Knight, Richard, "Branan's rule: An Undervalued Slide Rule", *The Journal of the Oughtred Society,* Vol. 4 No. 1. 1995.
15. Leadbetter, 1st ed. *op. cit.*
16. Nesbit, A. and Little, W., *A Treatise on Practical Gauging,* York, 1822.
17. Long, Joseph, *Illustrated catalogue,* London, 1922.
18. The instruments in this photograph were identified by P. A. Martin in his paper, "Gauging: the Art behind the Slide Rule" presented at the 8th International Meeting of Slide Rule Collectors, Oct. 2002.
19. Loftus, W. R., *Revised price list, London,* n.d. but ca. 1875.
20. Long, *op. cit.*
21. *Ibid.*
22. Oughtred, William, *The New Artificial Gauging Line or Rod,* London, 1633
23. Otnes, Bob, "Oughtred on Gauging – 1633 AD." *Journal of the Oughtred Society,* Vol. 7, No. 2, 1998.
24. Everard, Thomas, *Stereometry made easy: or, the description and use of a new gauging rod or sliding rule ... To which is added an appendix containing the description and use of another new rule, etc.* 1st ed., London, 1684.
25. Everard, Thomas, *Stereometry, or the Art of Gauging made easy...,* 7th ed., London, 1712.

26. Symons, William, The *Practical Gager*, London, 1803.
27. Leadbetter, 5th ed. *op. cit.*
28. Bateman, *op. cit.*
29. Anita McConnell, *R. B. Bate of the Poultry 1782 – 1847*, London, 1993.
30. H. A. Klein, *The Science of Measurement: an historical survey*, New York, 1974.
31. Long, *op. cit.*
32. Andrew Ure, *Dictionary of Arts, Manufactures and Mines*, 4th. ed., London, 1853.

Chapter 12
Fauna and flora

Chesterman's patent spring measuring tape marked in English feet and inches on one side of the tape and hands on the other, with cattle gauge attached for ascertaining the weight of Cattle by Measurement. To the Farmer and Grazier this is an almost indispensable article, as by its use he may tell the four quarters, either in London stone of 8 lb., the country stone of 14 lb. or the score of 20 lb. Note: ***very fat*** *cattle may weigh a little more than shown by the scale, and* ***very lean*** *ones a little less.* Chesterman, 1880.[1]

There are a number of reasons for wanting to measure animals, plants and other creatures. It may be to ensure that they conform to set standards or it might be a matter of aesthetics.

Horse measures

With regard to horses, the reasons might be to match the size of the rider to the horse; the size of a draught animal determines the amount of work it can do; and in competitions, the size of a horse or pony determines the class in which it can be entered. And appearance is important – a well matched pair drawing a carriage used to be a statement of the owner's position in life.

Horses are measured in hands and inches (1 hand is 4 in.) with the measurement taken at the withers (the highest point on the back of the horse at the top of the shoulder). Thus a modest sized pony might be described as 13.2 (13 hands 2 in.) – the largest horses will measure 17 hands or even more.

Horse measures are often referred to as measuring standards but in the Birmingham rule making trade the term used was horse gauge.[2] There are a number of different types but all consist of a standard marked in hands

Fig. 12.1. Measuring a pony.

with ¼ or sometimes ⅛ inch divisions between, and often on the reverse, feet and inches; the measurement is taken with a sliding arm. Many include a plummet in the standard or a level tube in the arm and superior versions have both.

Common form: With a straight standard and sliding arm, these were made in a wide range of qualities. For the cheapest, inferior wood such as beech was used, for the better quality mostly lancewood or boxwood which may be laminated onto a mahogany core. The best are well shaped with the plumbs and levels integrated into the piece and friction springs to control the movement of the arm.

Walking stick form: Also available in a wide range of qualities, the superior versions have a horn or ivory handle and silver trimmings with the canes made of either rattan or malacca. They are fragile items – the standard, in boxwood, brass or aluminium, sliding out from the cane is less than ½ in. thick with a metal arm folding out from within. The standard is figured in reverse, the measurement being taken at the point where it slides into the cane. Many are unmarked but when they are, the sticks mostly bear the names of riding or gentlemen's outfitters or veterinary instrument makers such as Arnold & Son, rather than the names of rule makers.

Cased tape form: This consists of a brass cased tape measure with a crescent shaped weight intended to drop to the ground. A folding arm with a level tube is attached to the brass case. The whole fits into a leather carrying case. Patented by W. Jones Cooper & Co., Patent No. 13268, in 1895, it is listed as the "Stockwell" Patent Pocket Standard by J. H. Hawkins in their

Fig. 12.2. An 18th century, or earlier, common form horse measure but more sophisticated than many. When fully assembled it measures in hands and inches for horses on one side, and on the reverse, with the lower half unscrewed and removed, it measures in inches for hounds, by W. Newton (Grantham, <1729>).

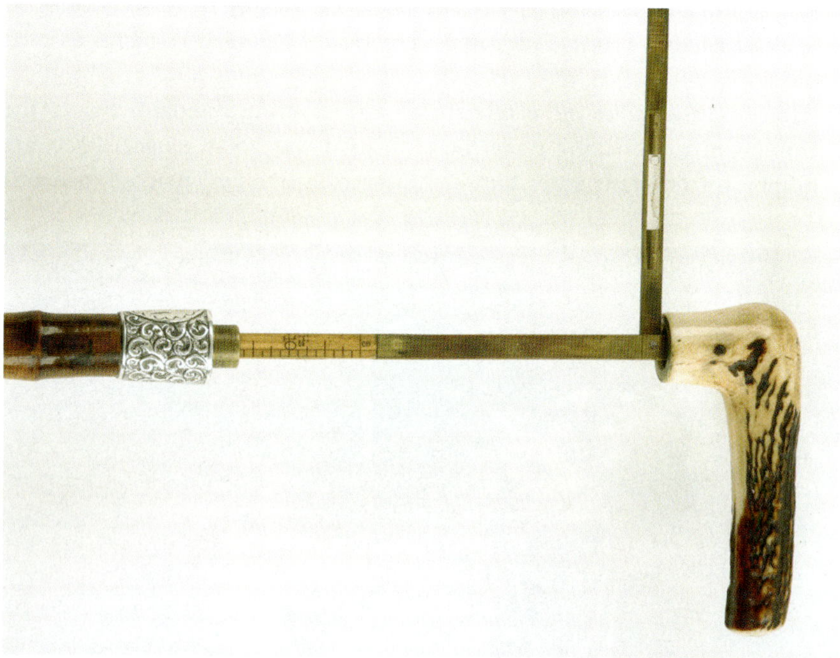

Fig. 12.3. The handle of the cane shown in Fig. 12.4 pulled up to reveal the wood and brass horse measure.

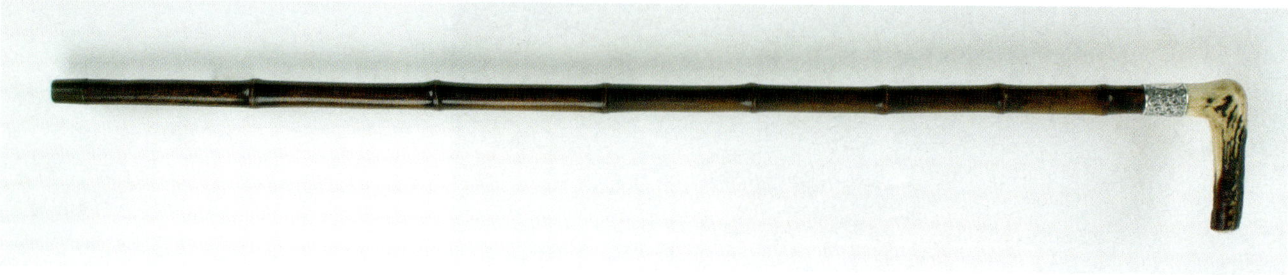

Fig. 12.4. A malacca cane walking stick with silver ferrule and horn handle marked Arnold & Sons (London, 1819 –1938>).

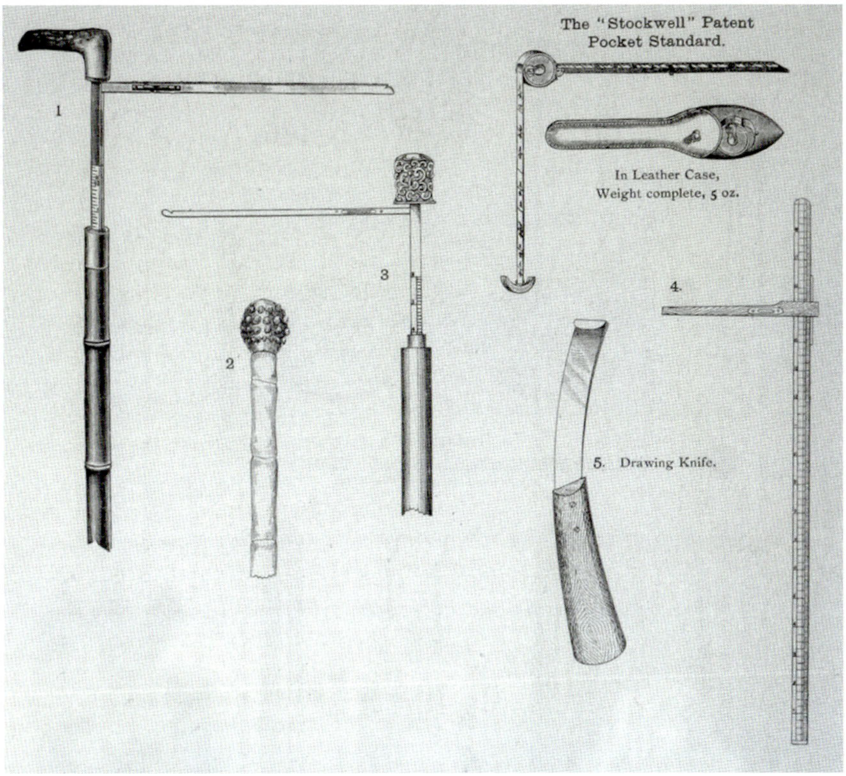

Fig. 12.5. A number of different styles of horse measures including the "Stockwell" Patent Pocket Standard in the J. H. Hawkins Equine Album, *ca. 1922.*

Fig. 12.6 A Stockwell Patent Pocket Standard.

ca. 1922 catalogue, which is called *The Equine Album.*[3] Some examples are marked with Rabone as maker. A fragile construction – many examples have damaged tapes or arms.

Bean's Horse Collar Measure

Used to measure a horse's neck to ensure that the correct size of collar is obtained, a good fit being essential to avoid chafing. The illustration and de-

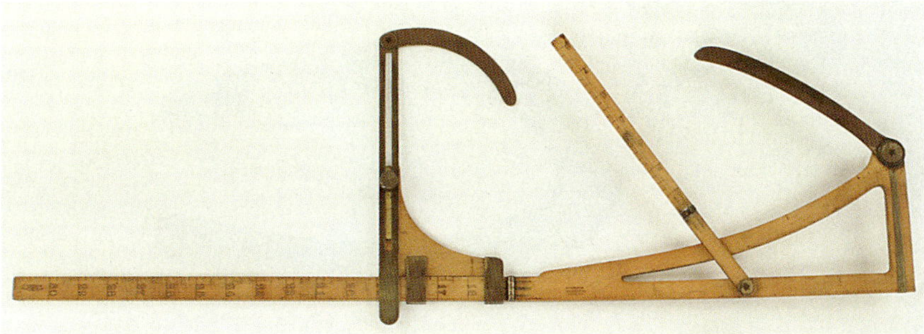

Fig. 12.7. A Bean's Horse Collar Measure, by Edward Preston & Sons (Birmingham, 1864–1932).

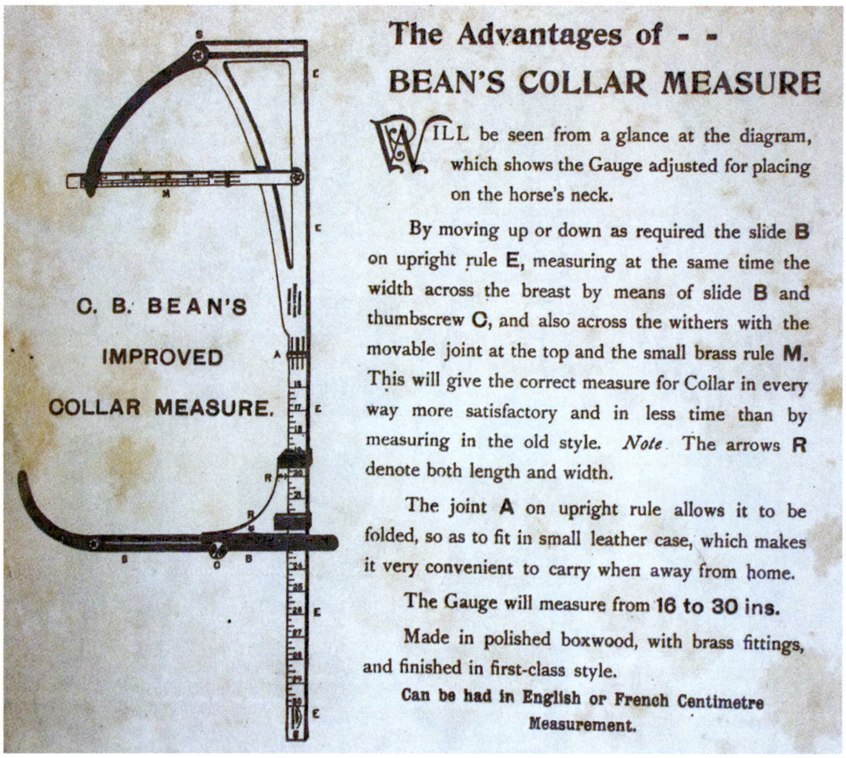

C. B. BEAN'S IMPROVED COLLAR MEASURE.

The Advantages of - -

BEAN'S COLLAR MEASURE

WILL be seen from a glance at the diagram, which shows the Gauge adjusted for placing on the horse's neck.

By moving up or down as required the slide **B** on upright rule **E**, measuring at the same time the width across the breast by means of slide **B** and thumbscrew **C**, and also across the withers with the movable joint at the top and the small brass rule **M**. This will give the correct measure for Collar in every way more satisfactory and in less time than by measuring in the old style. *Note*. The arrows **R** denote both length and width.

The joint **A** on upright rule allows it to be folded, so as to fit in small leather case, which makes it very convenient to carry when away from home.

The Gauge will measure from **16 to 30 ins.**

Made in polished boxwood, with brass fittings, and finished in first-class style.

Can be had in English or French Centimetre Measurement.

Fig 12.8. Illustration and description of Bean's Horse Collar Measure from J. H. Hawkins & Co. Ltd. The Equine Album, *ca. 1922. Much of the content of this catalogue is of earlier date – probably late 19th century.*

scription shown in Fig. 12.8 is from J. H. Hawkins' catalogue.[4] One example, made by Edward Preston & Sons is shown in Fig. 12.7 and it is probable that this firm was the only maker of these. However, they are not listed in any Preston catalogue, suggesting that they may have been made for C. B. Bean who marketed them through saddlers and firms supplying horse accoutrements.

Horse hoof gauge

When shoeing a horse, it is important that the hoof is trimmed to the correct angle so that the weight bearing base of the hoof is as close as possible to the centre of gravity. This ensures the best performance and stamina and helps to avoid damage and injury.

Fig. 12.9. Brass horse hoof gauge. The horseshoe-shaped part is placed under the horse's hoof and the sloping arm aligned with the wall of the hoof.

Most farriers will do this by eye, but if they wish they can use a horse hoof gauge. These tools are more commonly used in America than in Britain.

Hound measures

The display element of hunting – elegant hunters (horses), riders in the correct clothes and a pack of hounds, well matched in size and colour – was, for many, an important part of the sport. Hounds are measured in inches and counted in couples – which in no way indicates the sex of such pairs.

Hound measures come in a variety of forms which are described below. Some types were clearly intended for foxhounds but others were just as likely to have been used for any breed where size was significant. Hound or dog measures of any type are not common.

Basic form: An all-wood height gauge consisting of a vertical standard with a sliding arm, similar to the basic type of horse measure but measuring to about 30 in. in height.

Walking stick form: Arnold 1886[5] lists a dog measure contained in a malacca cane.

Horse and hound – combined type: The only examples of the 18th century type of horse measure shown in Fig. 12.2 that are known are all by either W. Newton (Grantham) or E. Newton (Grantham, <1729>). When the lower part of the standard is removed the upper section with the arm can be used for measuring hounds. It is marked on the back with inches for this purpose.

Arnold's form: A very compact type consisting of a square section brass standard extending from a tubular brass container. A folding arm with a minute level tube is housed in the centre of the standard. It measures to 20 in.

Measuring cattle

The purpose of all cattle measures or gauges is to measure a live beast and from this predict the weight of the carcase from which can be calculated the yield of meat. This is possible because, as a generalisation, all cows are cow shaped, and vary little in proportion. The same is, of course, true of other livestock such as sheep and pigs.

Fig. 12.10. A common form of hound measure by Aston & Mander (London, 1870–1930) measuring in inches on one side, marked to 33 in., and in centimetres on the other, marked to 85 cm.

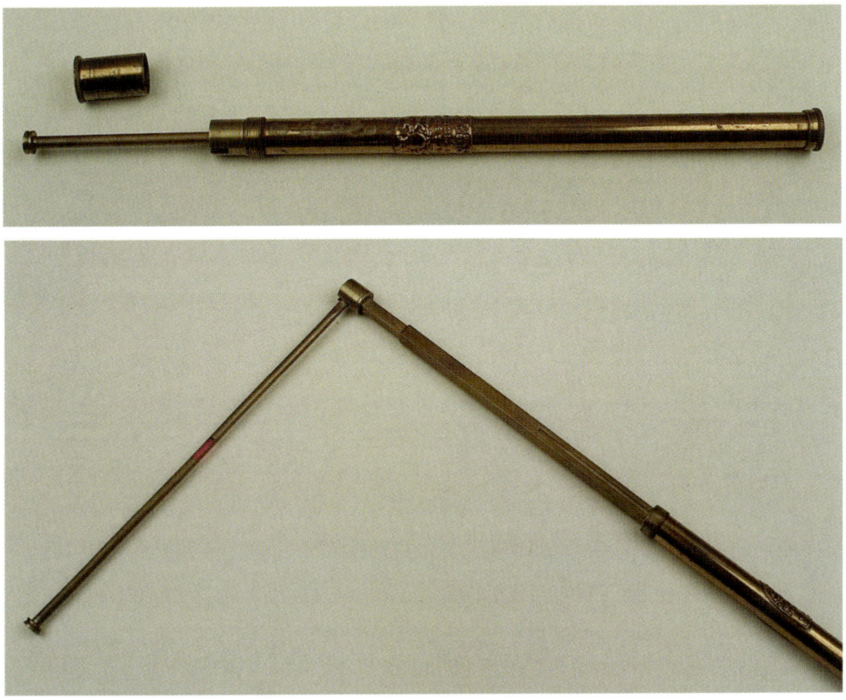

Fig. 12.11a & b. Two views of a brass cased hound measure marked Arnold & Sons (London, 1819 –1938>)

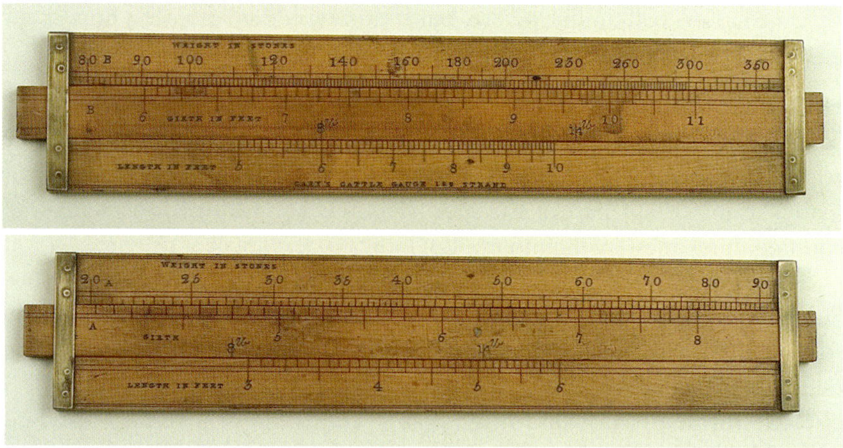

Fig. 12.12a & b. Cary's cattle gauge by William Cary (London, 1789–1897). A double sided boxwood slide rule 8 in. long. The scales extend from a girth of 4 ft. to 11 ft. and a length of 3 ft. to 10 ft. with half of the scale on one side and half on the other. The answer can be in either 8 lb. or 14 lb. stones.

Using this principle it is possible to take just one measurement and from this predict the weight of the carcase. If a tape is used this can be scaled to give a direct reading or tables can be used with a standard tape. However, greater accuracy will be achieved if more than one dimension is taken, but in practice, two – the girth and length – are all that are needed for a reasonably accurate result.

For wholesale purposes, butchers' meat was sold by the stone. The most widely used stone was 14 lb. but the Old London Butchers' stone was 8 lb. This weight was deeply entrenched in the London trade – it was prohibited in 1834 and in many other later regulations but was still being used in 1934 when the Inspectors of Weights and Measures were authorised to continue marking scales with this weight. Edinburgh had its own stone of 17½ lb. A stone or score of 20 lb. was also widely used in the meat trade.

Thus the manufacturers of livestock measures had to present the answer in several different units.

Cattle gauges

By 1820 several books had been published that described the practice of predicting the carcass weight by measurement of the live beast. The method required the use of printed tables which were included in the publications.

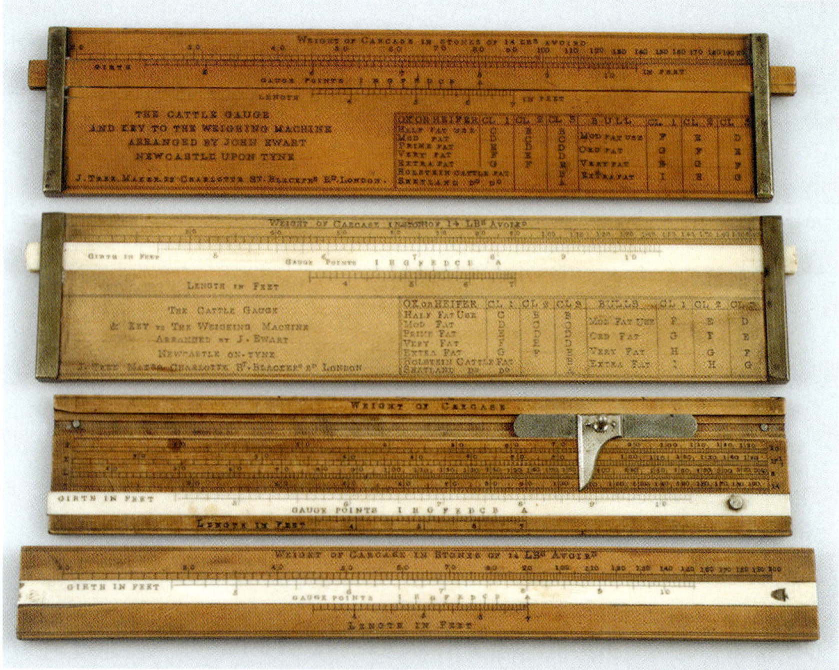

Fig. 12.13. Four examples of Ewart's cattle gauge. The upper three are all 8½ in. long, the bottom rule is 9 in. long. The top two are similar and have the same title "The Cattle Gauge and Key to the Weighing Machine arranged by J. Ewart Newcastle on Tyne" but their layout is different; the third from the top is marked "Ewarts Cattle Gauge arranged for any Market" and the bottom is marked "The Cattle Gauge Arranged by J. Ewart Newcastle on Tyne". All four were made by J. Tree, (London, 1826–95).

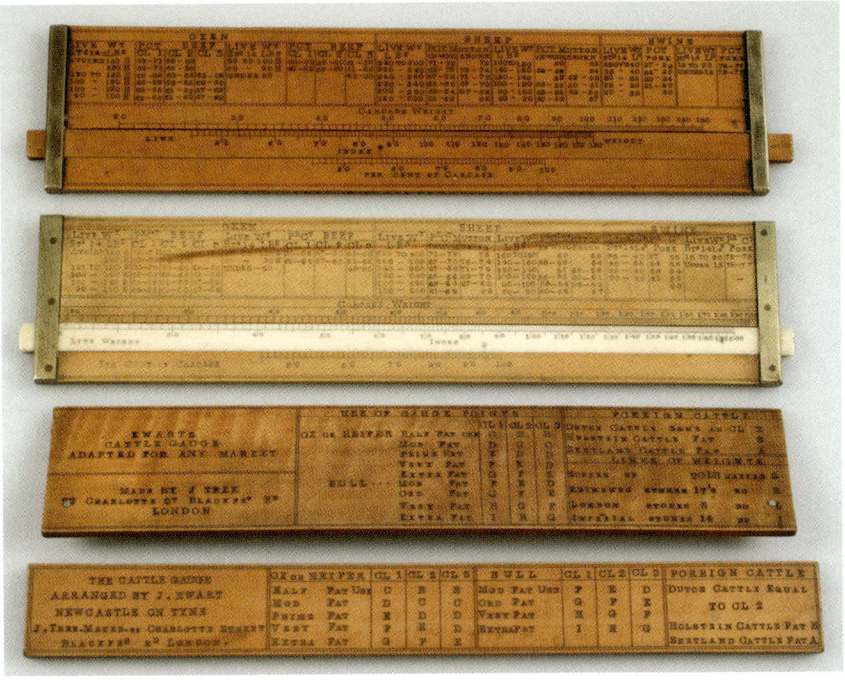

Fig. 12.14. The reverse faces of the rules shown in Fig. 12.13.

The use of a slide rule as an alternative to the tables appears to have been introduced around this time. The Cary's cattle gauge shown in Fig. 12.12 bears the address "182 Strand" indicating that it was made no later than 1822.

Ewart's cattle gauge

In 1844 John Ewart, who is described as a land surveyor, published "A treatise on the produce of beef compared with the live weight of cattle: the method of ascertaining carcase weight of oxen by measurement; and on the use of the sliding rule in computing carcase weight". A publication of 1847 also by Ewart, is titled "Instructions for the use of the slide-rule applied to computations relating to practical husbandry; and of the cattle gauge for ascertaining the carcase weight of oxen".

Ewart's rule in its most developed state has nine gauge points allowing some variation for different cattle. The rules must have sold well for they are fairly common. Arnold in 1886[6] offers seven different versions of the rule, the most expensive being in ivory with an electrum slide, some being sold together with a tape.

The top rule shown in Figs. 12.13 & 12.14 is also marked "sold by J. Buckley, 14, Lower Sackville St. Dublin". Buckley worked from 1832 until 1859.

The third rule from the top in Fig. 12.13 & 12.14 is designed only to provide the basic weight of carcass calculation but the sliding nickel silver index permits the answer to be read from scales for a score of 20 lb. and three different sizes of stones – 17½, 14 and 8 lb. These are marked respectively, "S" for score, "E" for Edinburgh, "L" for London and "I" for imperial. Various different sizes of cattle are covered with gauge points marked for each. Another table gives the gauge points for "Foreign Cattle", being specified as Dutch, Holstein and Shetland. In addition, this rule also includes a table of the different sizes of stone for Edinburgh, London and Imperial and for the size of a score.

Chesterman's cattle gauge

The Chesterman combined measuring tape and cattle gauge is an ingenious

SPRING MEASURES—Continued.

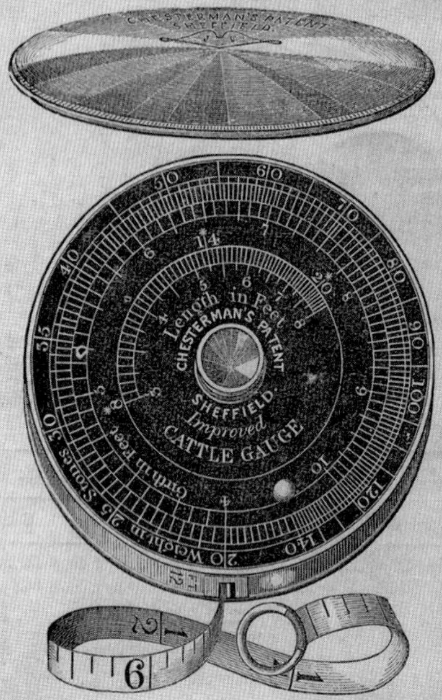

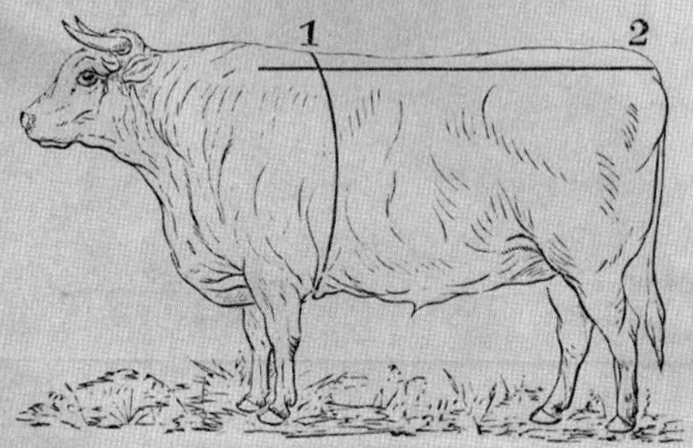

CHESTERMAN'S

PATENT SPRING MEASURING TAPE,

12 FEET LONG,

Marked English feet and inches on one side of the tape and hands on the other, with

CATTLE GAUGE

Attached, for ascertaining the weight of Cattle by Measurement.

To the Farmer and Grazier this is an almost indispensable article, as by its use he may tell the dead weight of the four quarters, either in the London stone of 8lbs., the country stone of 14lbs., or the score of 20lbs.

INSTRUCTIONS FOR USE.

Take the length from the foremost upper corner of the shoulder-blade bone in a straight line to the hindermost point of the rump, by the tail, then take the girth close behind the fore legs.—Suppose the length of the beast to be 5 feet 6 inches, and the girth 6 feet 3 inches.—Turn the slide till the figure $\frac{*}{8}$ is over 5 feet 6 inches on the inner circle, then look at 6 feet 3 inches on the slide, and immediately over that (on the outer circle) will be found the weight, viz., 89½ stones of 8 lbs. to the stone.—If the weight be required in stones of 14 lbs., turn the slide till $\frac{*}{14}$ is over 5 feet 6 inches, and referring as before, the weight will be found 51 stones; or by placing $\frac{*}{20}$ over the length, the weight will be found 35¾ scores.

Note.—*Very fat* cattle will weigh a little more than shown by the scale, and *very lean* ones a little less.

No. 31.	Brass Case, with Linen Tape...........................	7s. 0d. each.
,, 31SS.	,, ,, ,, and Spring Stop......	7s. 6d. ,,
,, 277.	,, ,, Steel Tape	9s. 6d. ,,
,, 277SS.	,, ,, ,, and Spring Stop......	10s. 0d. ,,

With Nickel-Plated Cases, 1s. each extra.

Opposite: Fig. 12.15. Illustration of the patent cattle gauge from the 1880 catalogue of James Chesterman & Co., Sheffield.

Fig. 12.16. Chesterman's cattle gauge opened to show the circular slide rule inside the case.

combination. A tape measure – either linen or steel – is housed in a strong circular case, one face of which unscrews to reveal a circular calculating slide. In principle and practice it is the same as Ewart's rule.

In 1849, Chesterman published a small booklet "A ready reckoner for

Fig. 12.17. Chesterman's cattle gauge with the case closed.

Fig. 12.18. The "Weighband", a direct reading tape measure for calculating the weight of cattle and pigs, unknown manufacturer.

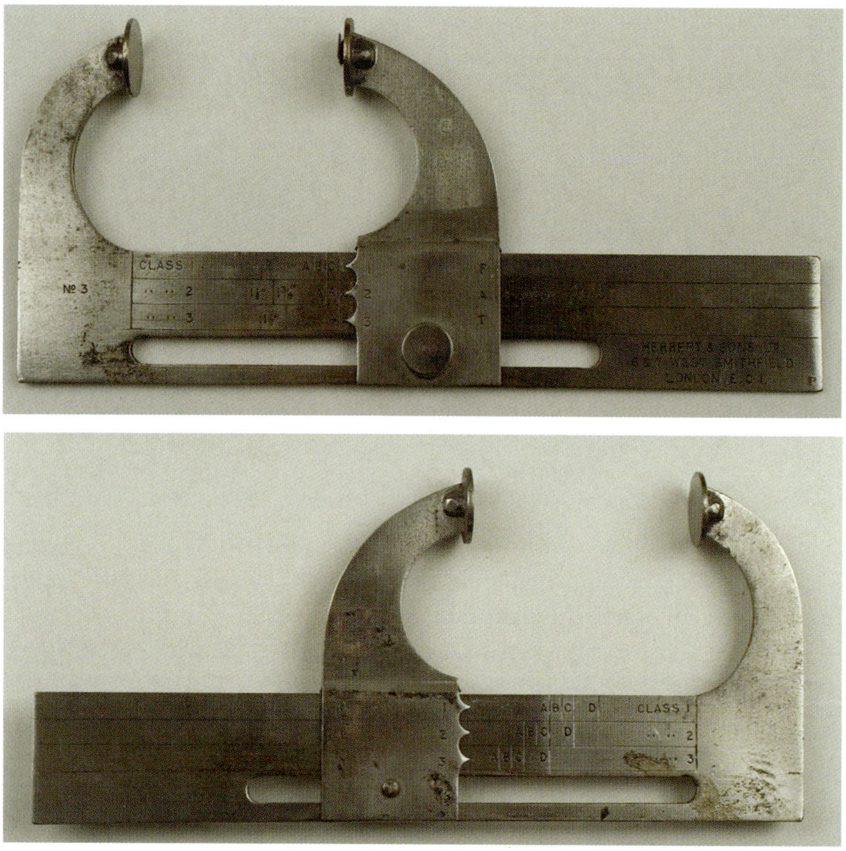

Fig. 12.19a & b. The Smithfield bacon gauge. Although unmarked, it was undoubtedly professionally made.

ascertaining the weight of cattle by measurement" presumably to encourage farmers to buy his tapes. However the cattle gauge combination does not appear in the catalogue until 1880.

Other manufacturers also produced direct reading tape measures for calculating the weight of cattle and pigs, all intended to calculate the carcase weight from the living animal.

Bacon gauge

A sliding calliper gauge, 7¼ in. long, constructed of nickel plated brass, intended for classifying sides of bacon into three classes. One side is marked "FAT", the other "BELLY". It is also marked 1 July 1938, Herbert & Sons Ltd., 6 & 7 West Smithfield, London EC1, although they would not have been the makers.

Bristle rule

A simple solid boxwood rule, always 7 in. long, with brass capped ends and almost invariably marked with the name of a brush maker or bristle merchant. Why always 7 in. long? That is the maximum length to which pigs' bristles grow.

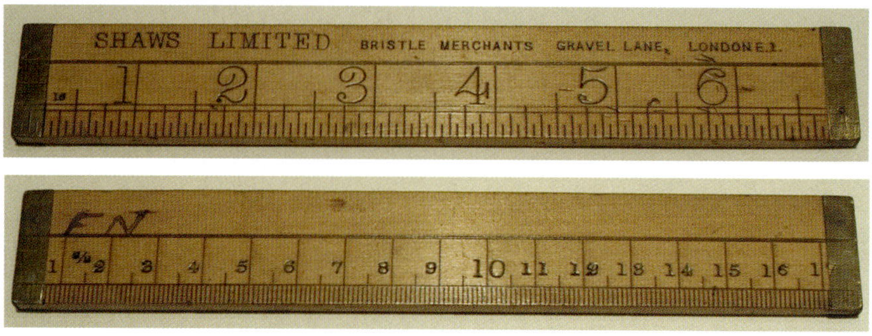

Fig. 12.20a & b. The two sides of a bristle rule advertising Shaw Ltd. bristle merchants, with imperial measures on one face and metric on the other. There is no maker's mark.

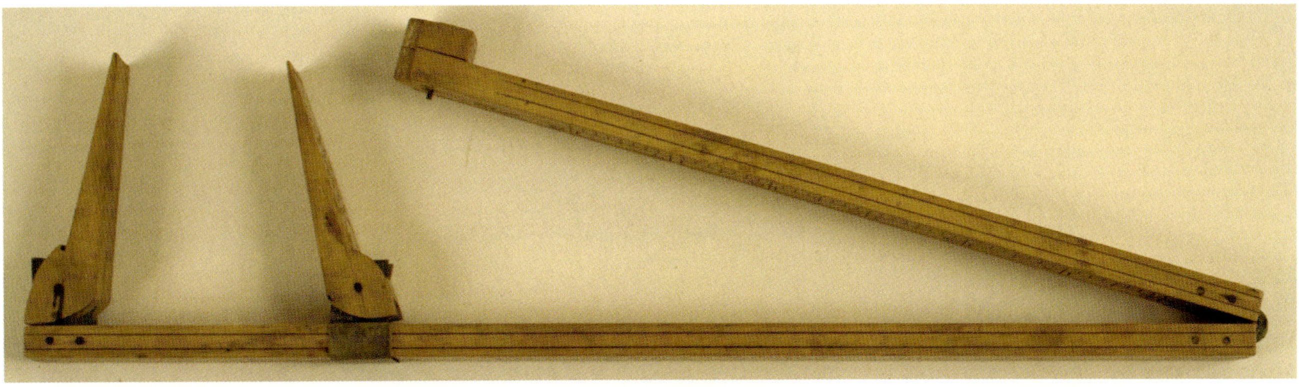

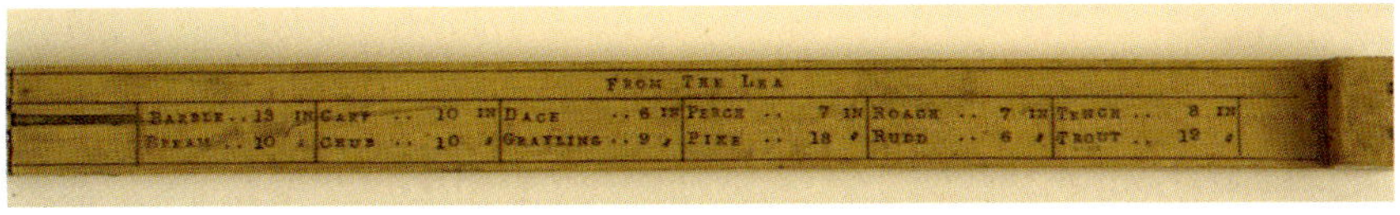

Fig. 12.21a, b, c & d. The Gamages "Thalea" Fish Measure by Edward Preston & Sons (Birmingham, 1864–1932). Sold by Gamages, the London toy and sporting goods store, this rule is adapted from a shoe stick design. It gives the minimum size of various types of fish that can be taken from the river Thames and the river Lea in London.

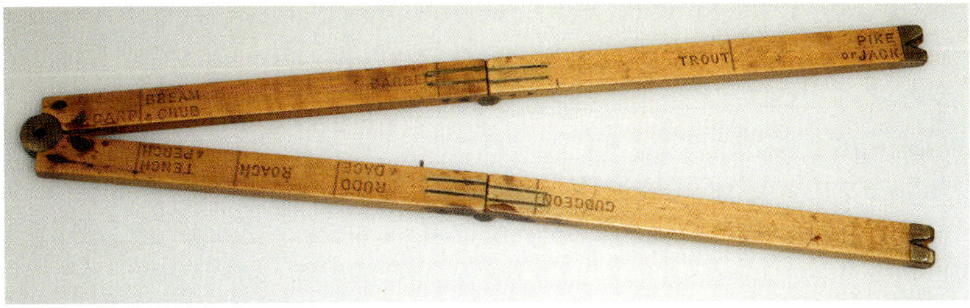

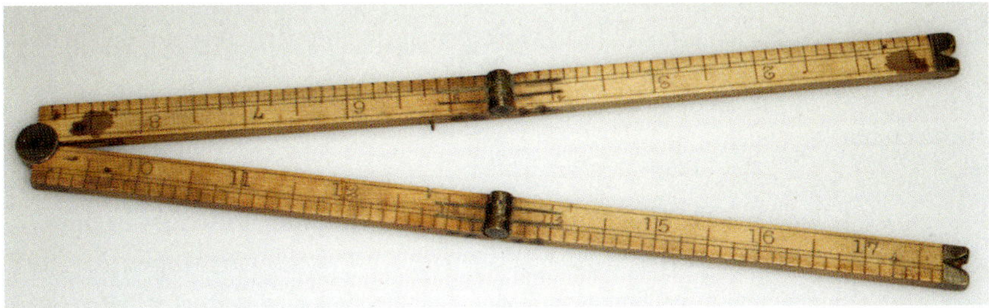

Fig. 12.22a & b. A anonymous rule for checking the size of fish caught. The length of a fish is measured from the dorsal fin to the tail, so to use this rule, the V-slot in one end is placed against the dorsal fin and the fish is measured to the end of the tail. The rule is marked on one face with a line of inches and on the other with the names of various fishes with a mark for the minimum length permitted. As all the species are freshwater fish, the rule was intended for anglers.

Fishermen's rules and gauges

There have always been strict controls on the sizes of fish and shellfish that are permitted to be caught, whether it be commercial fishing or just a pastime. So it is not surprising there were rules and gauges to make it quick and easy to check.

Gardening

In the 19th and 20th centuries, clay flower pots were made in a large variety of sizes which were graded from 1 to 72 depending on how many could be

Below left: Fig. 12.23. A lobster gauge, probably home made. Lobsters are measured by the length of their claws.

Below: Fig. 12.24. An oyster gauge issued by Truro Corporation. Truro, on the River Fal, is the county town of Cornwall. The Fal Estuary, a few miles downstream, was a major oyster fishing area.

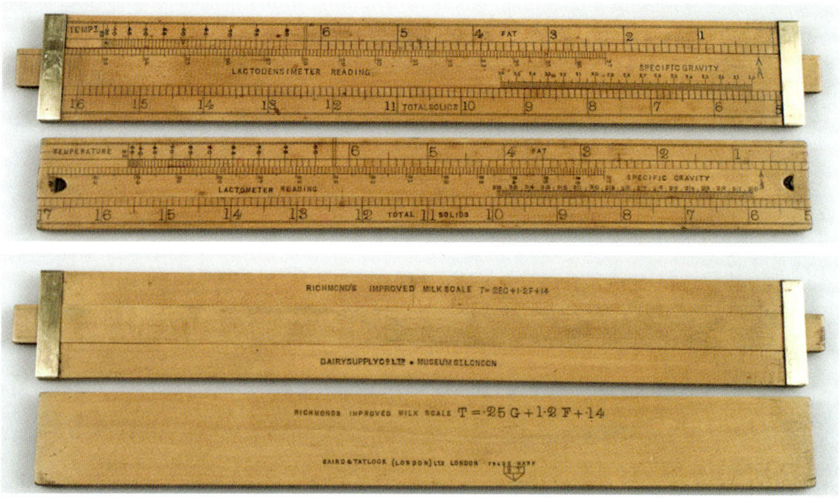

Above: Fig. 12.25. A home-made measure for checking the size of prize leeks. The measurement is taken from the base to the point where the first pair of leaves divides.

Above right: Fig. 12.26a & b. Two lactometer slide rules. The top scale is marked Dairy Supply Co. Ltd. who were the retailers. The lower is marked Baird & Tatlock Ltd. (London, ca. 1905 – 1943>) who were suppliers of laboratory equipment.

Fig. 12.27. Bakers' calculating rule issued by Joseph Rank & Co. flour millers and suppliers, of Cardiff. It would not have been made by them but was undoubtedly professionally made.

made from one cast of clay. Three inch pots were "60s" and five inch pots, "48s".

Leek and onion measures: Garden produce competitions have been a feature of country life for many years, governed by a plethora of rules. A variety of measuring tools, usually home-made, were designed for many different types of produce.

Foodstuffs

A number of rules are found that were used in relation to other food stuffs.

Lactometer slide rules: These were used to calculate the specific gravity of milk, by which means the richness of the milk can be established.

Bakers' calculating rules: Some flour millers and suppliers issued calculating rules to assist bakers. These were not slide rules, merely a rule form of a tables of quantities.

References

1. James Chesterman & Co., *Catalogue of Measuring Tapes, Land Chains, Band Chains…*, Sheffield, n.d. but marked in hand "July 23rd 1880".
2. John Rabone & Sons, *Catalogue of Rules & Tapes,* Birmingham, 1892, and Edward Preston & Sons Ltd., *The Preston Catalogue of rules, measures, levels …,* Birmingham, 1901.
3. J. H. Hawkins & Co. Ltd. *Catalogue, The Equine Album,* Walsall, n.d. but ca. 1922.
4. *Ibid.*
5. Arnold & Sons, *Catalogue of Veterinary Instruments*, London, 1886.
6. *Ibid.*

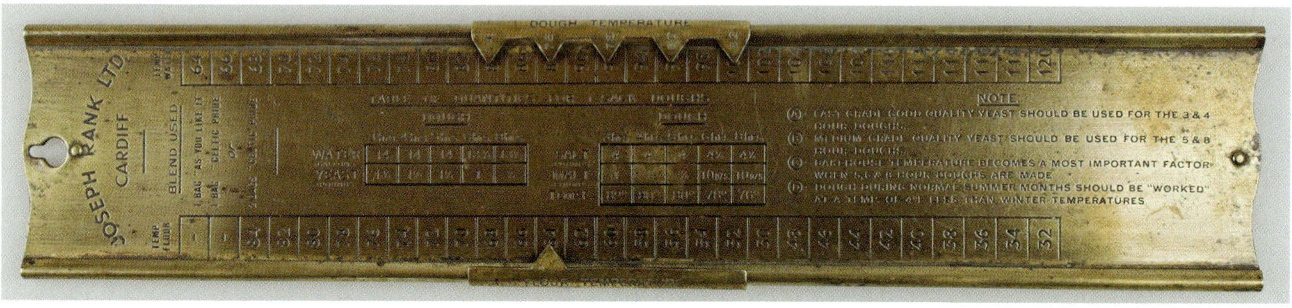

Chapter 13
Wearing apparel

The size of a shooe is the measure off its length, which is in Children divided into 13 parts; and in Men and Women into 15 parts; the first of them being five Inches long before it be taken for a size, what the shoe exceeds that length, every fourth part of an inch is taken for the size 1,2,3. and so forwards to … the size of Men and Women, called the long size, or Mans Size, begins at 1,2,3, etc. to the number 15… Holme, 1688.[1]

The clothing we wear, and for the purpose of this book this includes shoes, hats, wigs and spectacles, and the materials from which it was made were fruitful sources of specialist rules and measuring devices.

The cloth trade

The Magna Carta of 1215 and subsequent charters up to 1251 stated that the width of cloth (which probably referred to woollen cloth) was to be 2 ells. The exact size of the ell referred to is uncertain but it was most likely that the specified width was wider than a single weaver could produce. Until the 18th century invention of the flying shuttle, production of wider material such as woollen broadcloth required two weavers. As might be expected, the trade just ignored a statute that did not take into account the realities of production, and continued to make cloth in narrower widths which were measured, both in length and width, by the quarter (a quarter of a yard, i.e. 9 in.). Traditionally in Britain, until the introduction of metric sizes, the commonly produced widths of cloth varied from 4 to 6½ quarters with the 5 quarters width (45 in.) the most common.

Other units

Until overtaken by metrication, the length of cloth was described by a fraction of a yard rather than by inches until 1/16th of a yard was reached, which length was called a nail.

Unit	Inches	Marking
yard	36	
¾ yard	27	3Q
½ yard	18	2Q
¼ yard	9	Q
⅛ yard	4½	½Q, 2N, HQ
1/16 yard	2¼	N
1/32 yard	1⅛	HN

Fig. 13.1. Cloth measure – subdivisions of a yard and methods of marking these sizes on rules and tapes intended for measuring cloth. Q stands for quarter; N stands for nail; HQ stands for half a quarter and HN stands for half a nail. As can be seen, there are several alternative ways to express ⅛ yard.

Ells

The traditional measure for linen material was the ell. Woollen cloth could also be measured in ells instead of yards. Like many measures used through-

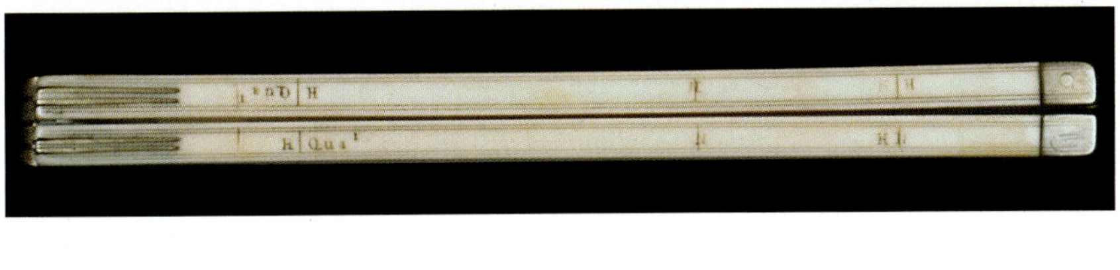

Fig. 13.2. Detail of an 18th century ivory folding ell rule showing the markings for a half quarter nail and a half nail. This is the clue that the rule was intended for cloth.

Fig. 13.3. Table of some examples of European ells. These sizes date from after 1821.[3]

Place	Length in English inches
Nice (France)	46.77
Netherlands	39.37
Munich	32.90
Vienna	31.50
Amsterdam	27.07
Copenhagen	24.71
Zurich	23.62
Hamburg	22.58
Liège	21.71

out Europe, most of the larger continental cities had their own sizes of ell, some of which continued in use until the end of the 19th century.[2] These sizes were highly variable with the majority smaller than the English ell.

The English ell: The English ell consisted of 5 quarters, giving a length of 45 inches. The earliest surviving standard ell is an Elizabethan one of 1588, which when compared with the Imperial standard is commendably similar at 45.04 in. In the following centuries the legitimacy of the ell as a statutory measure was constantly questioned but by custom and usage it hung on and not until the legislation of 1824, introducing Imperial measures, was it finally swept away.

The Scottish ell: Usually quoted as 37 in., occasionally as 37½ in., the Scottish ell had its origins in the old practice of making some allowance in favour of the purchaser – in the case of cloth this was a thumb's width beyond the end of the yardstick. With usage such customary practice became statute and the seemingly odd dimension of 37 in. became recognised as the standard Scottish ell.

The Flemish ell: With Flanders being one of the principal import markets for English cloth and with much linen being exported through the same route, the Flemish (Antwerp) ell also seems to have had some currency in England. In a mercantile guide of 1502 the Flemish ell is described as 27 in. and measurement of a 19th century standard confirms this to within a few hundredths of an inch.

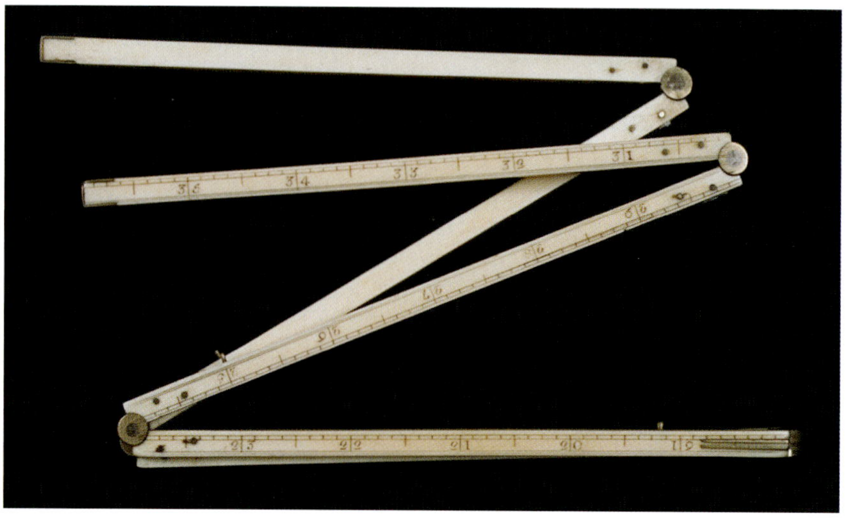

Fig 13.4. A 45 in. (one ell) 4-section ivory rule of mid/late 18th century date (a detail of this rule is shown in Fig. 13.2). On one face the rule is marked with a line of nails, indicated only with points. This and the length of one ell clearly indicate that this rule was intended for the cloth trade – possibly for a travelling buyer. It is contained in a leather slip case but unfortunately is not marked with a maker's name.

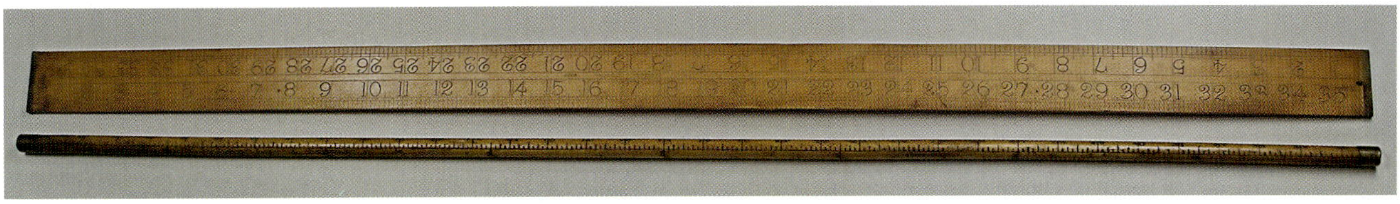

Fig. 13.5. *A round and flat drapers' yard measure, both by Edward Preston & Sons (Birmingham, 1864–1932).*

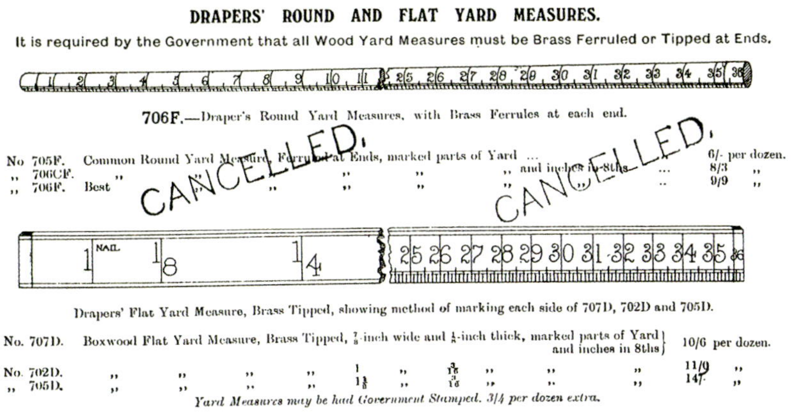

DRAPERS' ROUND AND FLAT YARD MEASURES.

It is required by the Government that all Wood Yard Measures must be Brass Ferruled or Tipped at Ends.

706F.—Draper's Round Yard Measures, with Brass Ferrules at each end.

CANCELLED. CANCELLED.

No 705F.	Common Round Yard Measure, Ferruled at Ends, marked parts of Yard ...	6/- per dozen.
,, 706CF.	,, ,, ,, ,, and inches in 8ths ...	8/3 ,,
,, 706F.	Best ,, ,, ,, ,, ,, ...	9/9 ,,

Drapers' Flat Yard Measure, Brass Tipped, showing method of marking each side of 707D, 702D and 705D.

No. 707D.	Boxwood Flat Yard Measure, Brass Tipped, 7/8-inch wide and 1/8-inch thick, marked parts of Yard and inches in 8ths	10/6 per dozen.
No. 702D.	,, ,, ,, ,, 1 ,, 3/16 ,, ,, ,,	11/9 ,,
,, 705D.	,, ,, ,, ,, 1 1/8 ,, 3/16 ,, ,, ,,	14/- ,,

Yard Measures may be had Government Stamped. 3/4 per dozen extra.

Fig. 13.6. *Drapers' round and flat yard measures from the Edward Preston & Sons 1909 catalogue.*

Ell rules (sticks)

The style of ell rule that usually comes to mind is a measure with some resemblance to a gentleman's (walking) cane, i.e. a tapering stick with a handle at one end. But lath or folding rule forms were also used.

The handled ell rule was commonly made in mahogany or other dark hardwood, the measure boldly indicated by inlaid lines or chequered veneering and not with usual lines and figures. It is probable that most of the rules of this type found today are of European origin – a rule to measure an English ell would need to be 45 in. long plus the handle and we have no record of any handled rules of this length.

Yardsticks and drapers' measures

Although listed in the rule makers' catalogues under "Tailors' Squares and Measures", yardsticks were in reality used not only by tailors but also by those involved in the manufacture of cloth and by drapers (cloth merchants and cloth retailers). They can be distinguished by their division into nails, and were still being offered by Edward Preston & Sons in 1909.[4]

Whilst many styles were offered only as yardsticks, others, described as cloth measures and tailors' measures, were available in lengths up to 60 in.

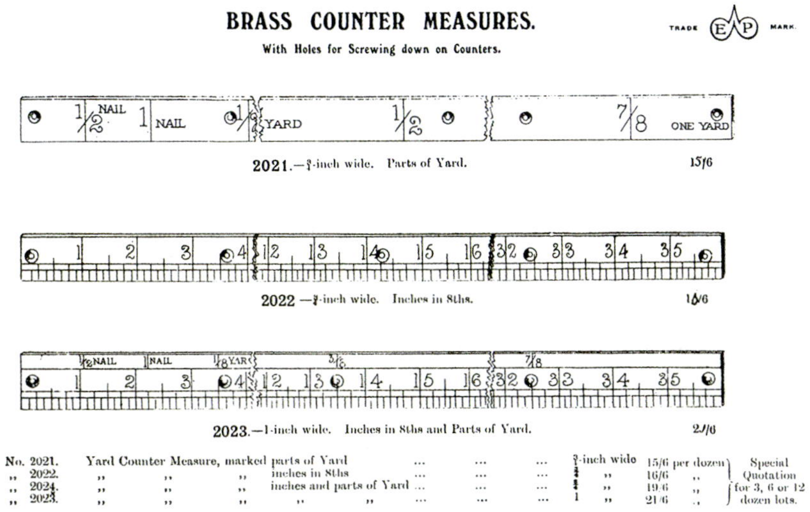

BRASS COUNTER MEASURES.

TRADE MARK

With Holes for Screwing down on Counters.

2021.—5/8-inch wide. Parts of Yard. 15/6

2022.—3/4-inch wide. Inches in 8ths. 16/6

2023.—1-inch wide. Inches in 8ths and Parts of Yard. 21/6

No. 2021.	Yard Counter Measure, marked parts of Yard	5/8-inch wide	15/6 per dozen	Special Quotation for 3, 6 or 12 dozen lots.
,, 2022.	,, ,, ,, inches in 8ths	3/4 ,,	16/6 ,,	
,, 2024.	,, ,, ,, inches and parts of Yard	3/4 ,,	19/6 ,,	
,, 2023.	,, ,, ,, ,, ,, ...	1 ,,	21/6 ,,	

Fig. 13.7 *Brass counter measures from the Edward Preston & Sons 1909 catalogue. Note the countersunk screw holes and the division of the top measure into parts of a yard from half a nail to one yard.*

Fig. 13.8 A unmarked patent "fitometer". This is a example of an innovative tailors' rule.

Counter measures
These were yard-long brass rules, made for screwing to the edge of a shop counter and marked with fractions of a yard down to a nail for the sale of goods by length, and were clearly intended for the draper.

The tailor and seamstress
The tailoring trade required a considerable amount of specialist measuring and there were therefore a wide range of specialist rules and measuring devices. These included rules both for measuring the wearer and for setting out the pattern from which cloth is to be cut. It is not surprising therefore that this was a fruitful area for patents and innovative designs.

Tailors' squares
The basic form of a tailors' square, usually made in boxwood, has unequal arms divided into inches and eighths. These must have had many uses, perhaps both within and outside the tailoring trade, as they are surprisingly common. Whilst they may be of some use when taking a customer's dimensions – for example, assessing the drop of a shoulder – their principal use was for marking out on the cutting table. A variety of these squares were made by all the principal rule makers but many examples are marked only with the names of suppliers of tailoring equipment. There are three main styles.

- **Fixed:** This was the basic form with the arms divided just into inches.
- **Fixed graduated:** These were extra quality squares, usually with one or both edges beveled and an arched reinforcement in the corner. Di-

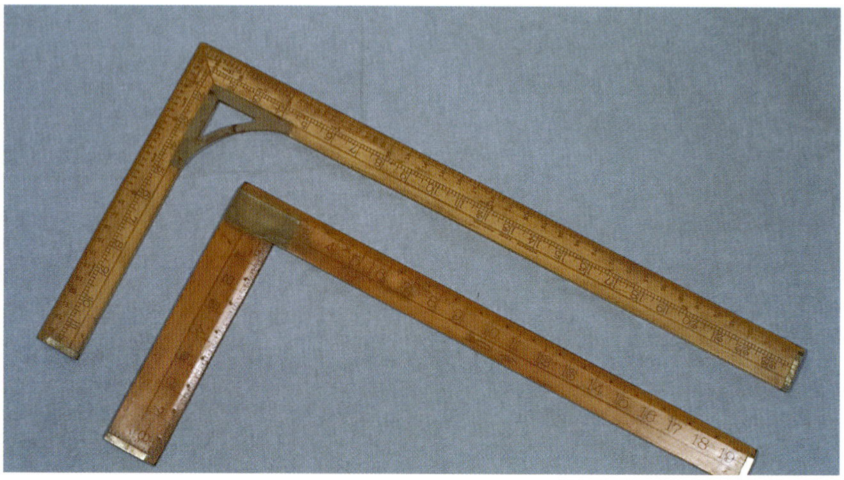

Fig. 13.9. Two tailors' squares by Edward Preston & Sons, (Birmingham, 1864–1932). Preston offered all the styles of squares, fixed, graduated and folding, in a range of five sizes, from 18 in. to 30 in.

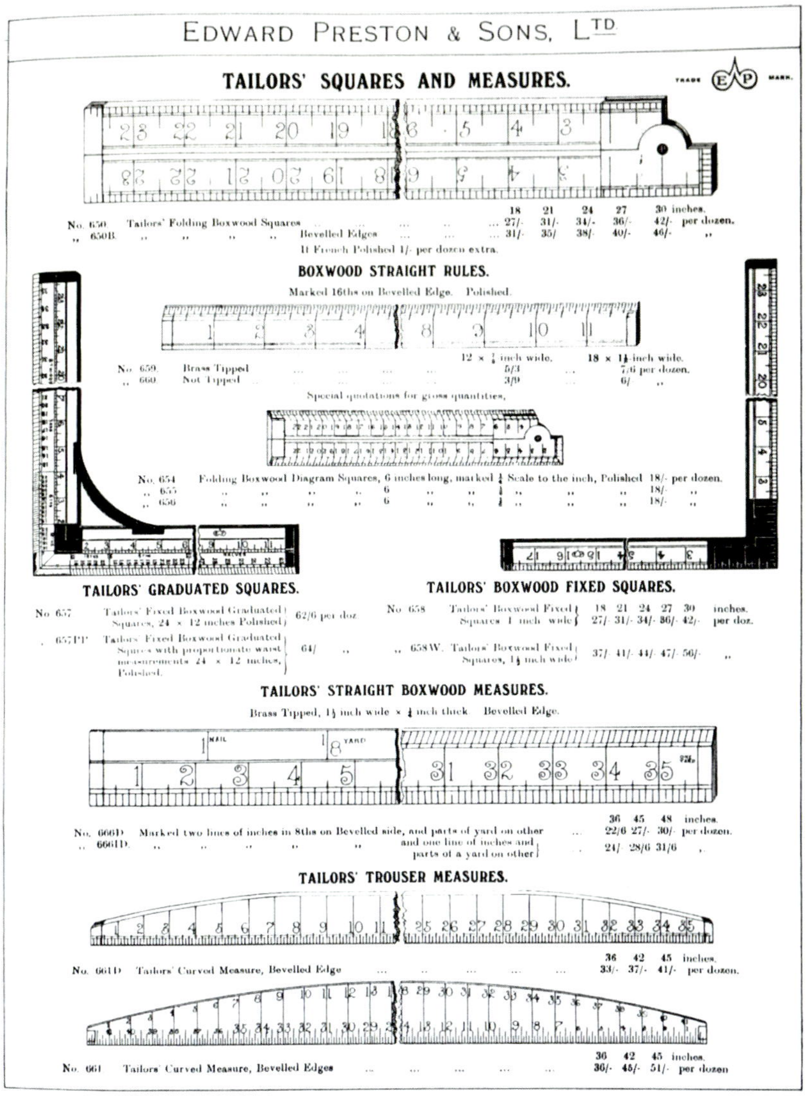

Fig. 13.10. Tailors' squares and measures from the Edward Preston & Sons 1909 catalogue.

vided with the line of inches and also a line of "proportionate waist measure".

- **Folding:** These are commonly found with both bevelled and square edges, and many have arms of equal length.

Trouser and tailors' measures

These were lath-type rules, 36 to 45 in. long, distinguished by having one curved edge. Curved rules of "hockey stick" or other shapes were also used in the tailoring trade, principally intended for marking out on the cutting table.

Fig. 13.11. Four tailors' rules. From top: Trouser measure; tailors' straight boxwood measure; two tailors' measures for use on the cutting table.

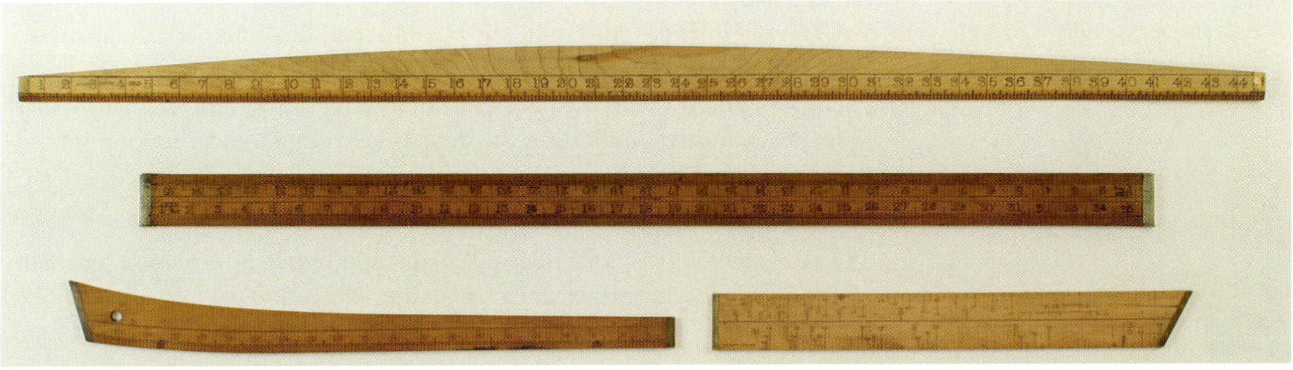

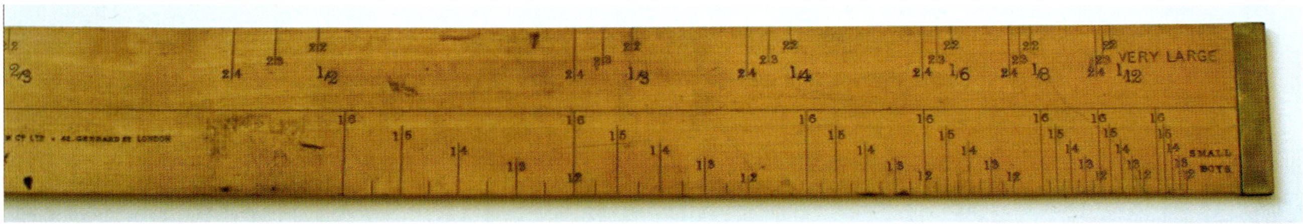

Fig. 13.12. A rule designed to assist in grading garments. It is marked with the name of a tailors' sundries supplier.

Fig. 13.13. A boxwood and brass diagram square, marked ¼ full size.

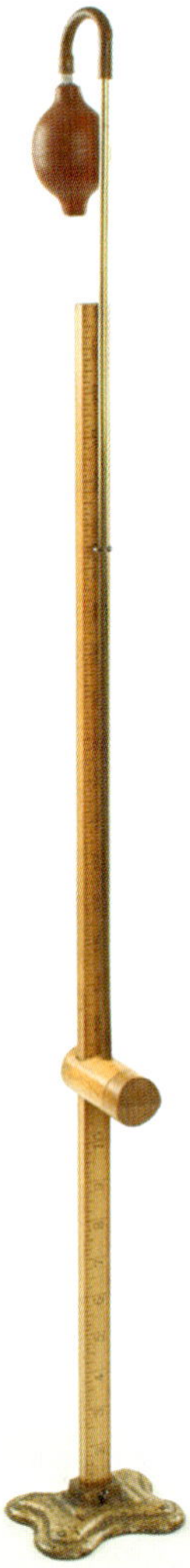

Fig. 13.14. A skirt measure by Edward Preston & Sons (Birmingham, 1864–1932).

Pattern grading rules

Pattern grading is the art of translating the pattern for a garment created at one size into patterns for the same garment in a variety of different sizes. If each dimension is just increased or decreased by a fixed percentage the resulting garment will neither fit nor have the same appearance as the original. Pattern grading rules, which were generally proprietary designs, were intended to aid these translations. Each rule was originally accompanied with instructions, without which their exact use is difficult to understand, and unfortunately these never seem to still be with the rule.

The diagram square

A miniature (6 in. long) version of the tailors' folding square marked with a scale of inches to either ¼, ⅙, or ⅛ full size which, as the name suggests, was used to make sketches in the tailor's work book. They probably came into use when tailoring became more factory orientated and the making of suits started to move away from the tailors' retail shop.

Skirt measures

Generally intended for home use, but also used by professional seamstresses, these measures were for marking the length of a skirt during fitting. French chalk was blown onto the skirt while the wearer slowly turned, giving a mark for the seamstress indicating the desired hem length and checking that the hem was level.

Workbasket rules

These small 1 ft. 4-fold rules are frequently found in boxwood and ivory and sometimes in nickel silver or mother-of-pearl. They are often marked with nail divisions as well as inches and sometimes engraved with a name

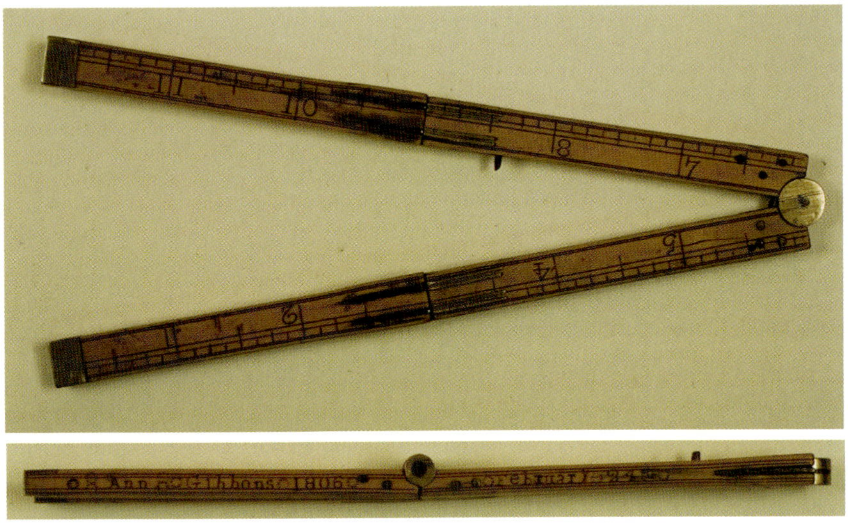

Fig. 13.15a & b. A 1 ft. 4-fold workbasket rule marked in inches and nails, and, on the edge, "Ann Gibbons 1806 February 24th".

and date. Probably intended to fit in a workbasket, they seem to have been popular gifts.

Button gauges

In the late 18th century, Matthew Boulton, the leading Birmingham manufacturer, amongst a variety of products, made buttons. He led the way in standardising the sizes of buttons and produced a new standard that was measured in lines, with 40 lines to the inch.

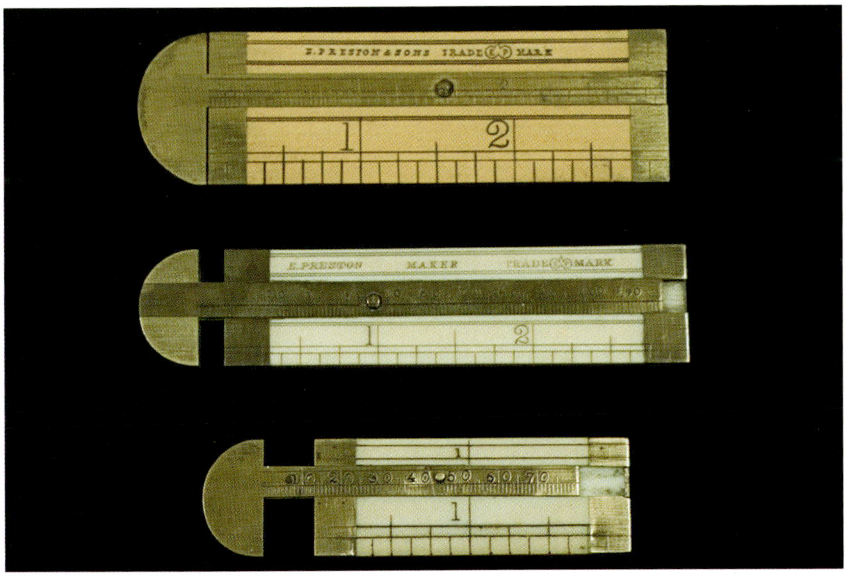

Fig. 13.16. Two 3 in. button gauges by Edward Preston & Sons (Birmingham, 1864–1932) in boxwood and ivory and a 2 in. ivory version. The stocks of the rules are divided into inches with the lines marked on the caliper to give a direct reading of the size of the buttons.

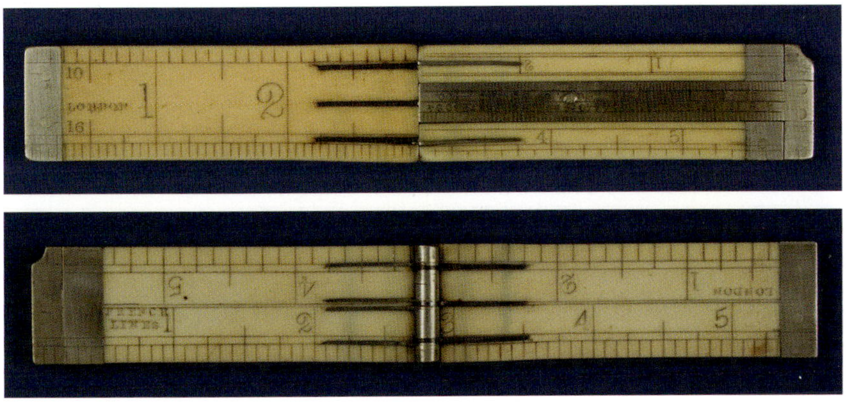

Fig. 13.17a & b. An anonymous folding button gauge in ivory and nickel silver marked LONDON on the outside and LONDON and FRENCH LINES on the inside.

Fig. 13.18. A brass button gauge, probably French, marked in metric and lignes on the body.

Button gauges are usually 3 in. long, though 2 in. versions can also be found. Made in boxwood, brass and sometimes ivory, with a sliding caliper they are recognisable by the markings in lines, either on the caliper or the body of the rule. They are often also marked in metric measurements with the lines labeled ligne.

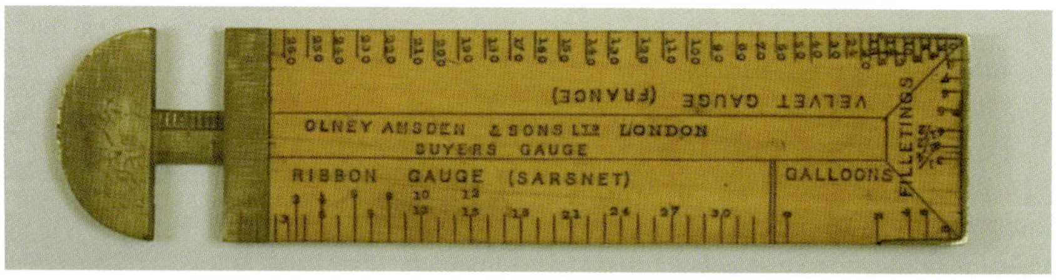

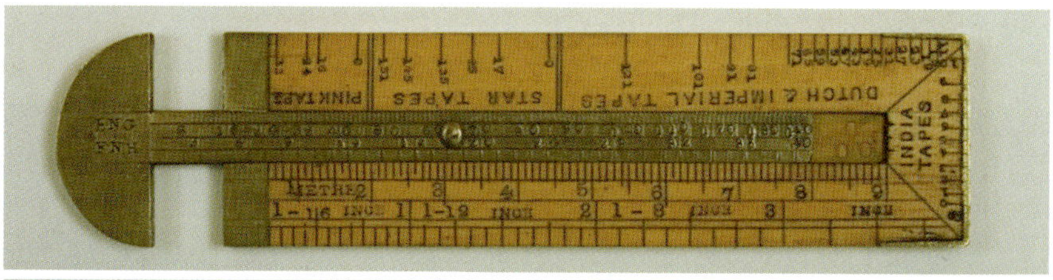

Fig. 13.19a & b. A tape and ribbon gauge. This is marked Olney Amsden & Sons. This company was a wholesale haberdashery supplier and would not have made this rule.

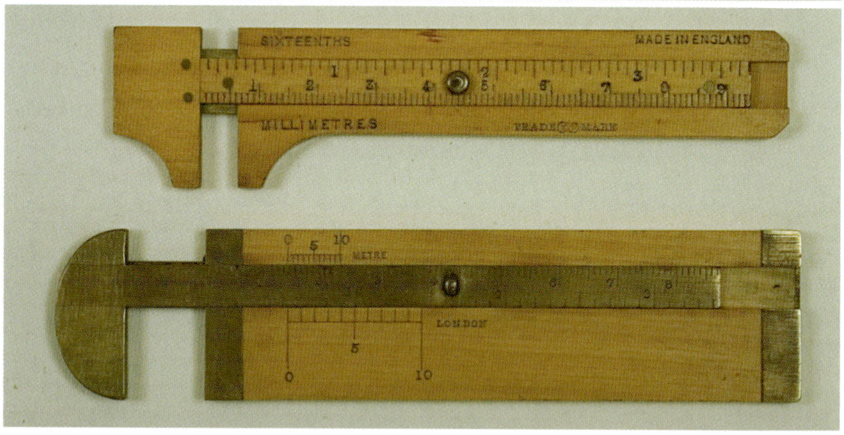

Fig. 13.20. Two different tape and leather strap gauges. Note that both are marked in Imperial and metric measures. The top rule is by Edward Preston & Sons (Birmingham, 1864–1932); the lower is unmarked.

Fig. 13.21a & b. An unusual ivory 4-fold ribbon gauge rule, 14 in. long, by B. Blakeley (London, 1817–45). It is marked on the outside with gauges for SARSNETS (silk) and SATINS ribbons and on the inside with a line of measure.

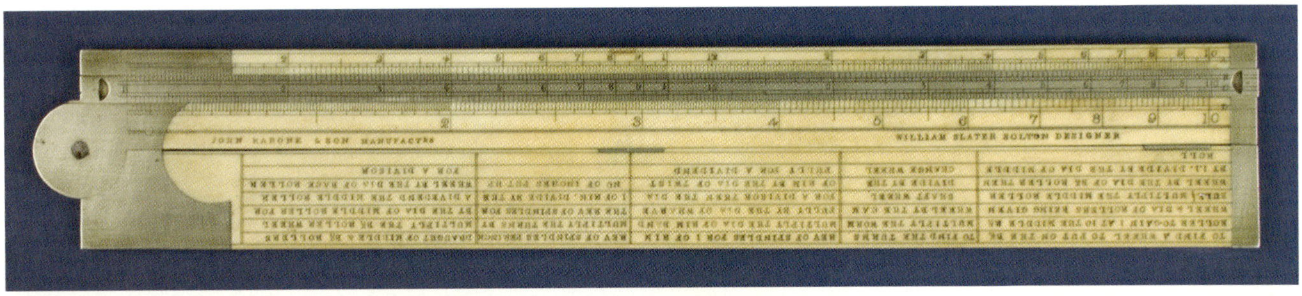

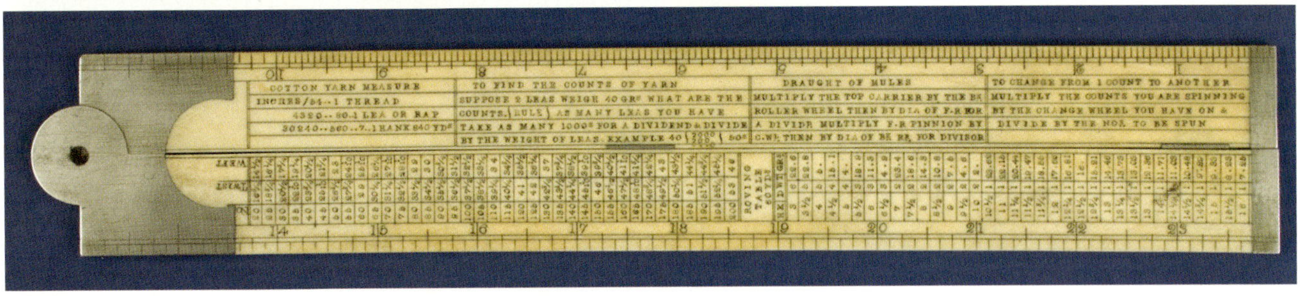

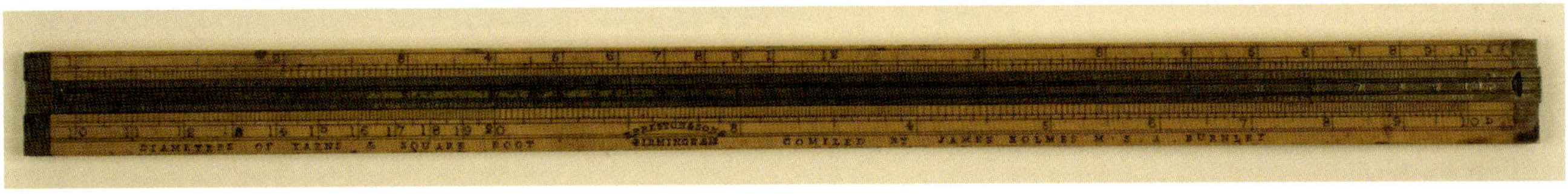

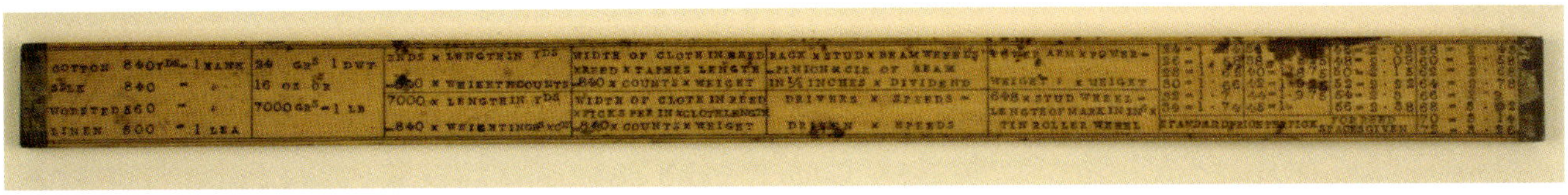

Tape and ribbon gauges

Used for measuring tapes and ribbons, these are also small caliper rules marked for various types of tapes and their sizes. There seemed to have been little standardisation; the rule shown in Fig. 13.19 is marked with eight different systems for eight different types of tape.

The textile industry

Two aspects of the textile industry produced rules that a collector is likely to find. The first concerns the setting up of the machinery; the second the cloth produced.

Slater's rule: The rule, which is a 2 ft. 2-fold rule usually in boxwood (occasionally in ivory) is easy to identify, being marked with the designer's name, "William Slater, Bolton," and with panels giving instructions for setting up spinning machinery. A logarithmic slide is fitted into one leg. It is thus very similar in appearance to other engineers' rules. It would appear that it was never offered other than as a 2 ft. 2-fold rule and Slater's rules were listed by both Edward Preston and John Rabone.

Top: Fig. 13.22a & b. A rare ivory Slater's rule made by John Rabone & Sons (Birmingham, 1877–1963). As the instructions for use were engraved on the rule, no instruction manual was offered.

Fig, 13.23a & b. Another slide rule used in the textile industry for setting up machinery. Made by Edward Preston & Sons (Birmingham, 1864–1932) it was designed by James Holmes M.S.A. from Burnley, a town heavily involved in the textile industry.

Fig. 13.24a & b. Ashenhurst's improved rule, another form of slide rule to assist in the production of cloth. Below is, a detail of the mark. Bradford was also a centre for the textile industry.

Fig. 13.25. Lord's patent calculator for the spinning trade, made by W. Wilson (London, unidentified).

Calculators: In response to the need for managers in the spinning and weaving trades to make many calculations several circular slide rules adapted for this purpose were available. Amongst those commonly found are Fowler's and Lord's.

Weavers' gauges / cloth dissectors: These came in a variety of styles and were used to measure the number of threads in a given width of cloth. This indicated both the weight of the material and the quality. Known as either weavers' measures or cloth dissectors, they usually consisted of two points a set distance apart but could also be in the form of a square hole through which threads could be counted; this latter type was usually made of brass.

The shoemaker

Until recently the English system of shoe sizing consisted of a children's series from 1 to 13 and then, without overlap or break, an adult series from 1 to 14.

Fig. 13.26. Various styles of weavers' gauges. Clockwise from top left: "Smiths Patent Cloth Dissector" with four points at differing intervals and a tape gauge; two fixed pricker style weavers' gauges; a double ended weavers' gauge dated 1850; a multi-leaf pricker style gauge; and two hole style dissectors.

In both series each size advanced by ⅓ of an inch. Today some of the smallest sizes are no longer used and due to the increase in the size of children's feet, boys' shoes now go from children's to adult size 4½ ; girls' to adult size 2½.

The size system described by Randle Holme[5] in the 17th century was similar to this. There were 13 children's sizes and the change point between children's and adults' sizes at 8¼ in. was very close to the 8⅜ in. of the modern series. His system differs in that the size increment is "about" ¼ in. and not the ⅓ in. of today.

Factory production of shoes during the 19th century made standardisation of sizes essential and the ⅓ in. increment became the norm.

French shoe sizes

The metric system consists of a single series for both children's and adults' shoes. The increment (sometimes called a Paris point) is ⅔ cm. The series starts at zero but sizes under 16 do not exist, size 16 being about equivalent to the English children's size 1 and the smallest made. The increment of ⅔ cm. is very similar to the ¼ in. size increment of Randle Holme.

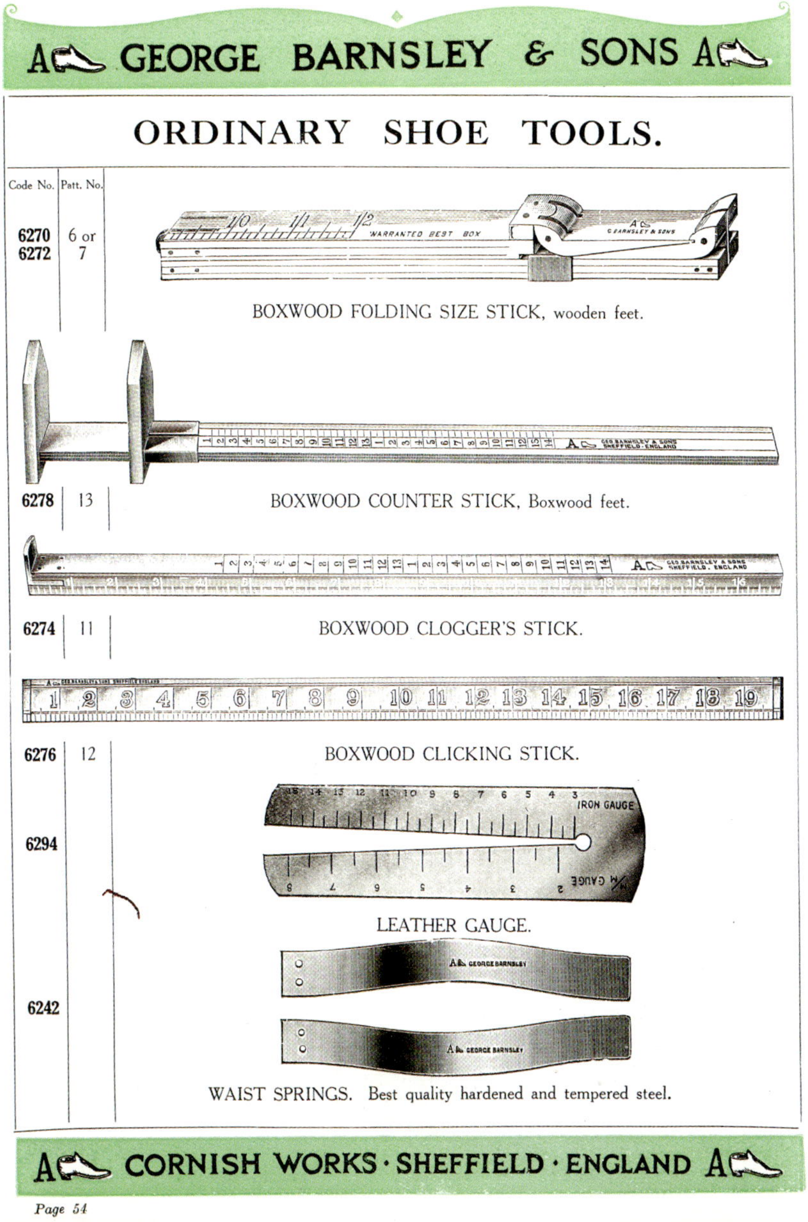

Fig. 13.27. A page from the George Barnsley & Sons 1927 catalogue showing a range of shoe sticks.[7] The name Geo. Barnsley & Sons of Sheffield, the principal English maker of shoemakers' tools, will frequently be found on shoemakers' measures of all types. However evidence suggests that none of the measuring items were made by them but were bought in from rule makers.

Above: Fig. 13.28. The size stick as illustrated by Randle Holme in the Academy of Armory *in 1688.*

Fig. 13.29. The standard type of shoe measure, this example was made by I. & D. Smallwood (Birmingham, 1858-1999). It is still in an unfinished state as it has not yet been divided. This example has brass feet but boxwood feet were also common.

Size sticks

These are used to measure the customer's feet. For convenience the majority of size sticks were made to telescope or fold but the added complexity seems a high price for greater compactness.

The size stick described by Randle Holme[6] (he calls it a shoemakers' measure) is illustrated with a drawing which shows an extending stock which appears to be formed with four triangular pieces.

Early type: This was made as described by Holme – sometimes with the feet carved as a shoe or occasionally a human foot. European shoemakers continued to use this early type of stick long after English shoemakers had switched to the modern folding type. Today continental sticks of late 19th century date are sometimes presented as being early sticks of English origin. The divisions with which the stick is marked will usually indicate whether the item is of English make and sticks made with exotic wood such as ebony are unlikely to be English.

Manufactured types: The Preston catalogue of 1909[8] lists "shoemakers' folding size sticks" in three widths with brass feet and three widths with boxwood feet. This is typical of the range offered by all the principal rule makers of the time. All the sticks fold in the middle with the feet being locked closed when the stock is folded over. Size sticks are mostly simply marked – shoe sizes on the top and inches on the underside.

It is not clear at what date this form of stick was first made but it is probable that they were first produced in quantity as the rule makers' trade developed in the first decades of the 19th century. Pre-1850 examples are rare.

Pocket size sticks: These are a smaller version of the folding size stick, 12 in. long folding to only 6 in. The feet are narrow, run in a channel and are enclosed within the stock when closed. Much less common than the regular size sticks, pocket shoe sticks were illustrated by George Barnsley in 1927.[9]

Fig. 13.30. A folding foot measure also by I. & D. Smallwood (Birmingham, 1858-1999). This measure is also unfinished.

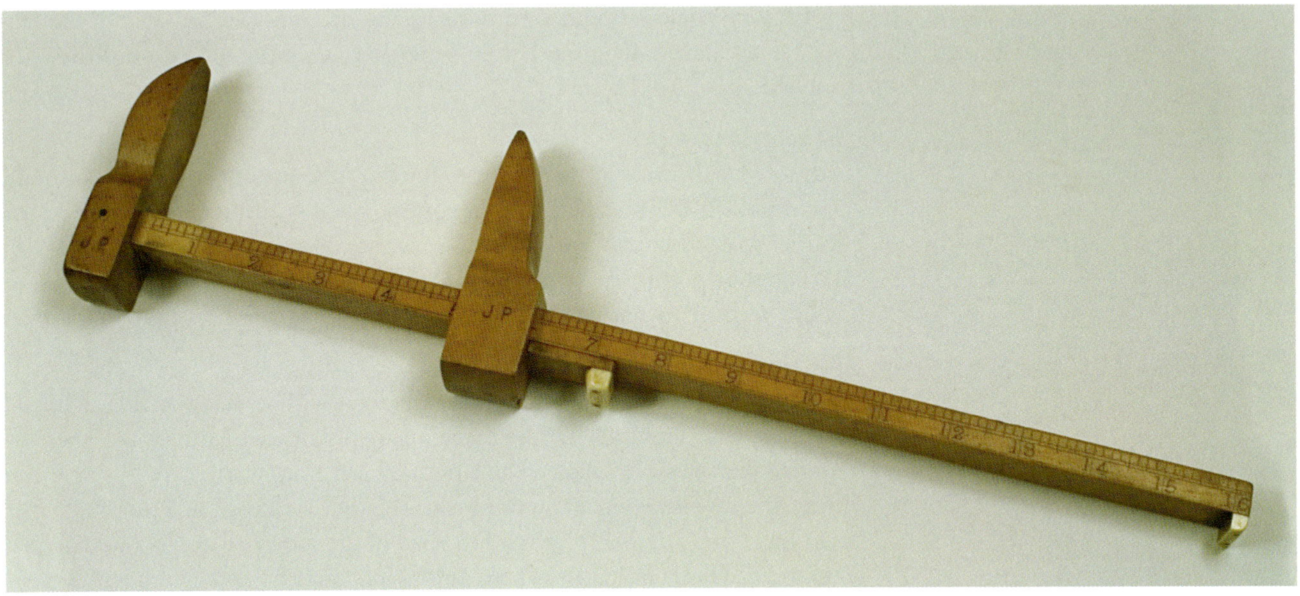

Fig. 13.31. *A counter size stick by Edward Preston & Sons (Birmingham, 1864–1932).*

Counter sticks: Intended for use in the shoe shop, they do not fold and are of heavier construction than the folding size stick. The most sophisticated examples have locking devices on the sliding foot.

The principal use was to measure the size of the customer's foot but most counter sticks were also marked with other lines including a scale to indicate the size of a shoe, French sizes, and inch or metric measurement. There were many variations, including measuring the width of the foot either with a second sliding gauge or with a tape and "improved" versions with a concave heel rest. A version made by I. & D. Smallwood marked "sizes of army boots" has been recorded.

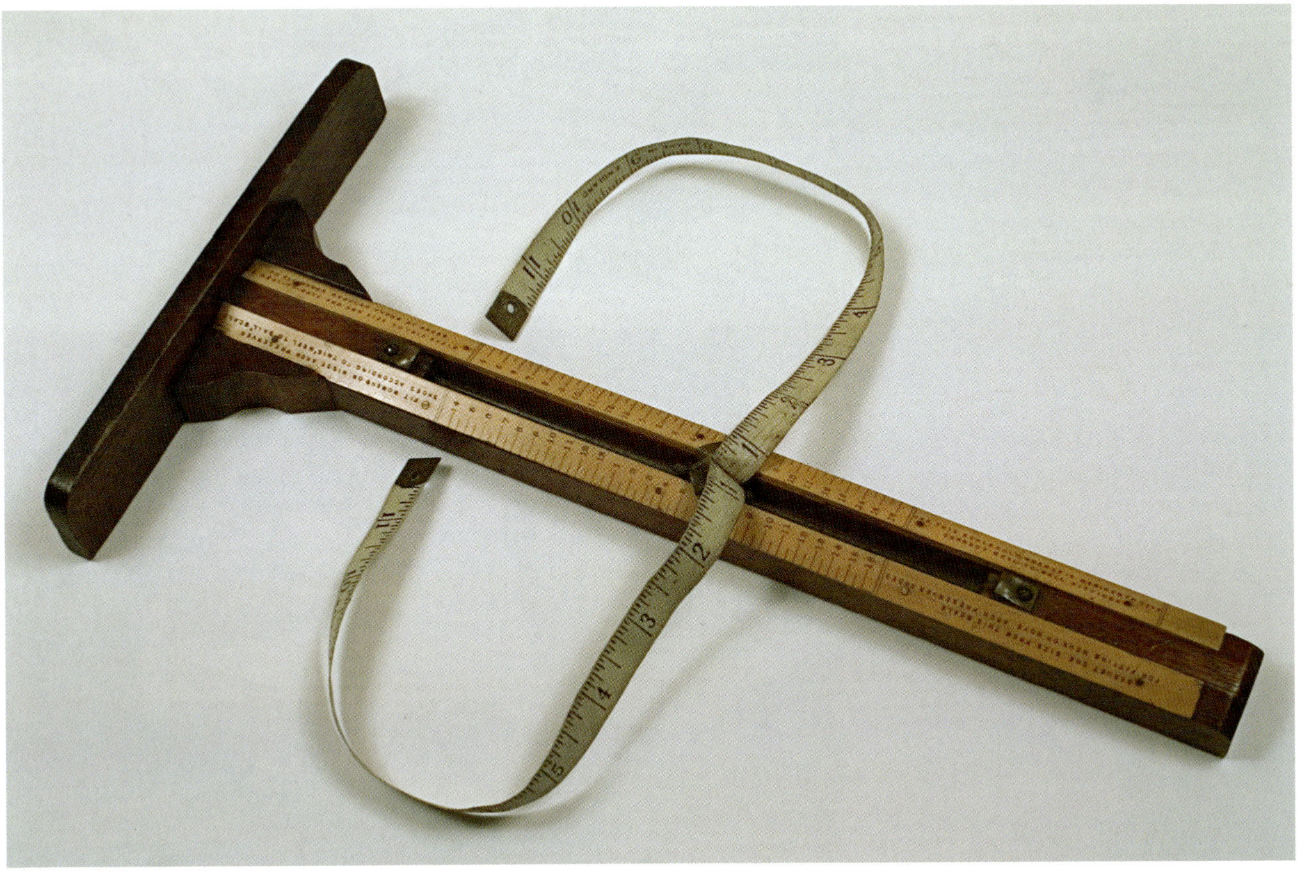

Fig. 13.32. *A more sophisticated counter size stick that measures the width of the customer's foot. Instructions for fitting different types of shoe and customer (man or woman, boy or girl) are included on the stick.*

Factory sticks: Intended to measure the size of shoes or boots in the factory. They have only a fixed stop – the shoe is simply placed onto the stick and the size judged by eye.

Shoemakers' tapes

Edward Dean in 1919[10] lists the following tape measures for the boot and shoe trade: a 24 in. tape made in cotton, linen with metallic thread or red leather with shoe scale and inches on reverse: a French measure marked with French shoe points and cm., a spring pocket tape and a last measure, ½ in. in width with a ring and 1⁄16 in. divisions.

Measuring for last making

Whether a shoe is bespoke or factory made, it will be made on a last. Before a pair of bespoke shoes can be made a pair of lasts, which replicate the customer's feet, must be made. The length of each foot is measured using a size stick; then, with a pencil, an outline of the foot is drawn on a piece of paper and three measurements, around the ball of the foot, the instep and diagonal around the heel, are taken. These dimensions may be taken with a shoemakers' tape but more often are just made with a strip of strong paper marked with small tears. From this information the last is made.

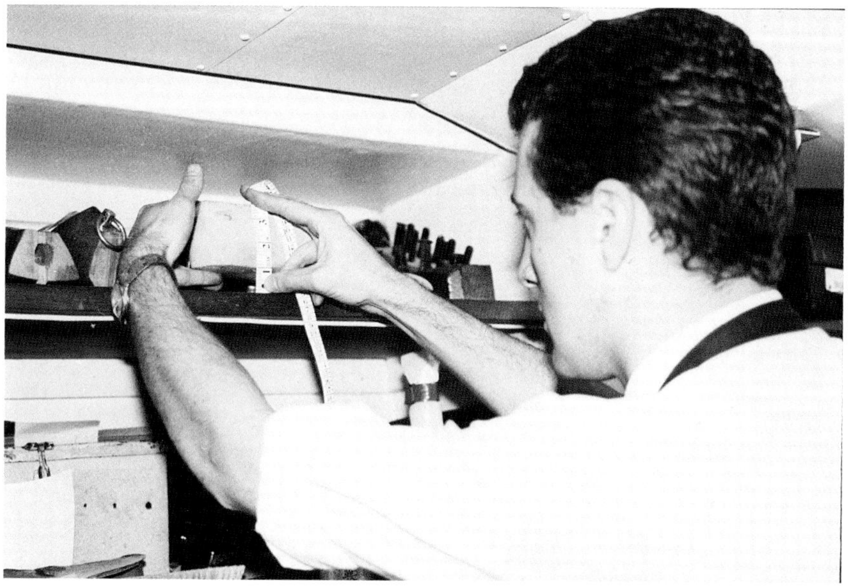

Fig. 13.33. Checking the dimension of a last using a tape measure. Both this photograph, and that shown in Fig. 13.34 were taken in the last making workshop of John Lobb Ltd., bespoke shoemakers, of St James, London.

Fig. 13.34. The last is also checked using a shoe size stick.

Fig. 13.35. A metal last makers' measure used by Edwin Hale, master boot maker.

Clog makers' sticks

The English clog is a shoe or boot with a wooden sole and a leather upper. Clogs are much simpler to make than leather-soled footwear – the leather is simply nailed round the edges of the sole and the only sewing involved is likely to be a seam at the heel. Much cheaper than shoes they were the footwear of the poor and for work and were still being worn into the 1930s.

The cloggers' stick shown in George Barnsley & Sons 1927 catalogue, illustrated in Fig. 13.27, is a boxwood stick about 1 in. square with a simple brass end stop. There is no sliding foot. It is marked with shoe sizes, a line of inches and a line for measuring clogs. Salaman quotes one source which stated that "old" clog sizes were two sizes bigger than shoes (clog size 10 = shoe size 8).[11]

Substance gauges

The substance, which is the thickness of leather for the soles of boots or shoes, is measured in irons with 48 irons to the inch. The common form of substance gauge is a V-gauge around 4¼ in. long, cut from thick steel (these gauges can be up to ⅛ in. thick) and marked Iron Gauge (sometimes abbreviated to I/G) and MM for millimetres. They are thus often mistaken for some type of engineers' gauge. Substance gauges were also made in the form of round and rectangular slot gauges.

These gauges were often given away for advertising, so many are marked with the names of suppliers of leather. Some are marked "Crispin's Gauge" or "St. Crispin", the patron saint of cobblers.

Salaman[12] suggests that the iron as a unit of measurement developed from the width of the cobblers' irons that were used to burnish the edges of the soles and that, surprisingly, this unit of measurement only became usual around 1920. Before that date leather had been specified by thickness using fractions of an inch down to 1/16 in. but this was too coarse a measure so every thickness also existed in "stout" and "light".

Below left: Fig. 13.36. An anonymous substance gauge for shoe leather marked St. Crispin.

Below right: Fig. 13.37. Another anonymous substance gauge for shoe leather, this time marked IRON gauge and MM gauge.

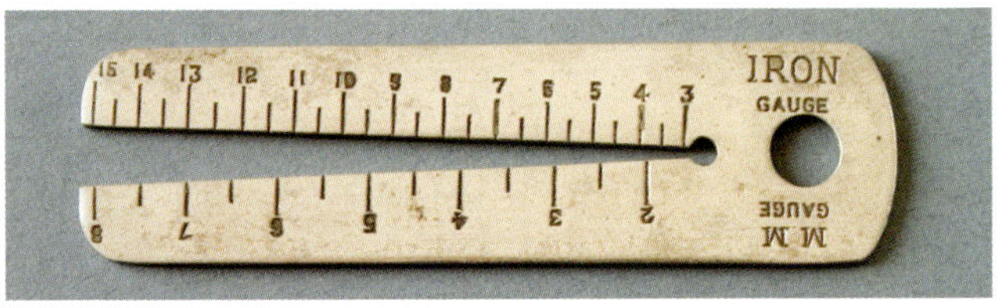

Clicking sticks and bits of rules

The clicking stick (clicking is the process of cutting out the leather for a shoe) illustrated by George Barnsley & Sons (see Fig. 13.27) is just a simple brass tipped boxwood lath, the same as any other lath – but only 20 in. long.[13]

Both steel rules and boxwood laths up to 24 in. were in general use for shoemaking. Salaman[14] suggests that these were marked with shoe sizes as well as inches and fractions. References to a "Bit of rule" are a description of the shoemaker's use of pieces of broken rule – generally the shoemaker only needed to measure small lengths.

The glover

The size of gloves is determined by measuring the circumference of the hand at the knuckles. A typical man's hand might measure 8½ in. round. The glove size is a direct representation of this; it would be size 8½.

Glovers' hand gauges: The patent hand gauge shown at the top of Fig. 13.38 is clearly inspired by the shoemakers' size stick, the design including even folding feet which engage together. The hand is placed palm down between these feet with the knuckles at the points of contact; the result is read directly in glove size.

Glovers' rules (sticks): The centre rule shown in Fig. 13.38, these are square in section and easily identified in that they taper from about ⅓ in. to ¼ in. with a length of either 14 or 16 in. They are marked with inches divided down to ⅛ in. generally with all four faces marked the same. Some examples also marked with French inches have been found – an interesting example of the continuing use of old measures long after France's metrication in the early 19th century. No doubt French glovers in the couture trade saw no reason to change their system of sizes.

The reason for the thin end on the rule is to allow it to be inserted into the finger of a glove either to measure or maybe just to push out an in-turned finger. The glovers' stick would seem to be an unsuitable rule for general use in the workshop – it was more likely used to check the size of finished gloves.

Fig 13.38. Top: Boxwood glovers' hand gauge bearing the patent no. 29036. The gauge is marked with three scales – men's from size 7 to 10 but also marked SM (small), M (medium), OSM (outsize medium) and XOS (extra outsize); ladies from 5¼ to 7½; and boys & girls from 0000 to 7. Centre: A glovers' rule (stick) by John Rabone & Sons (Birmingham, 1877–1963). This example is marked only in English inches. Bottom: An unmarked French glovers' size stick, marked in Paris inches on one side and English inches on the other.

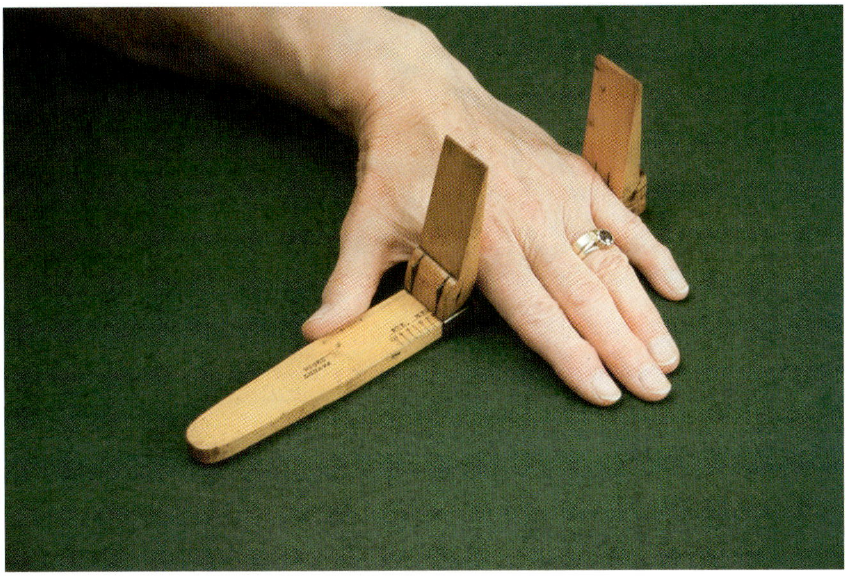

Fig. 13.39. The patent glove measure shown in Figure 13.38 in use.

Glovers' sticks made by Rabone, Preston and Smallwood have been noted though there are few listings in the catalogues. All the named sticks are likely to be of 20th century date.

French size sticks

The lower rule in Fig. 13.38, French size sticks are lath-type rules marked with French inches. They would have been used in retail shops rather than the glovers' workshop and are further evidence of the French hold on the top end of the glove trade. Salaman[15] quotes an example which is 13 in. long; the one illustrated measures 15 Paris inches, which equates to 16 English inches. Another example by Rabone, also 16 in. long and marked with both French and English inches, has also been recorded.

The hatter

Hat makers had a number of measuring devices to fit hats to heads. They frequently used conformatures, of various complexity, for taking measurement of hats. One of the simpler forms worked on the same principle as scissors. These were for measuring the inner circumference.

Hatters' rules: These are short inside-measuring rules with a central slide, used for measuring hard hats, such as bowlers and top hats. Usually 5 in. long, they are marked on the slide side with inches divided into ⅛ in. in both

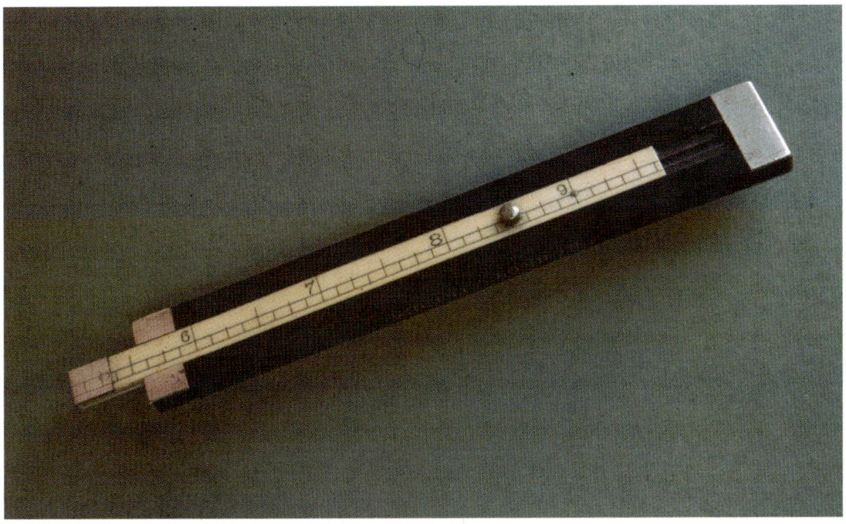

Fig. 13.40. A hatters' rule in ebony and ivory, marked Christy, London. Christy are a well-known London hat makers. They commenced in business in 1773 and are still in business today. They would not have made this rule.

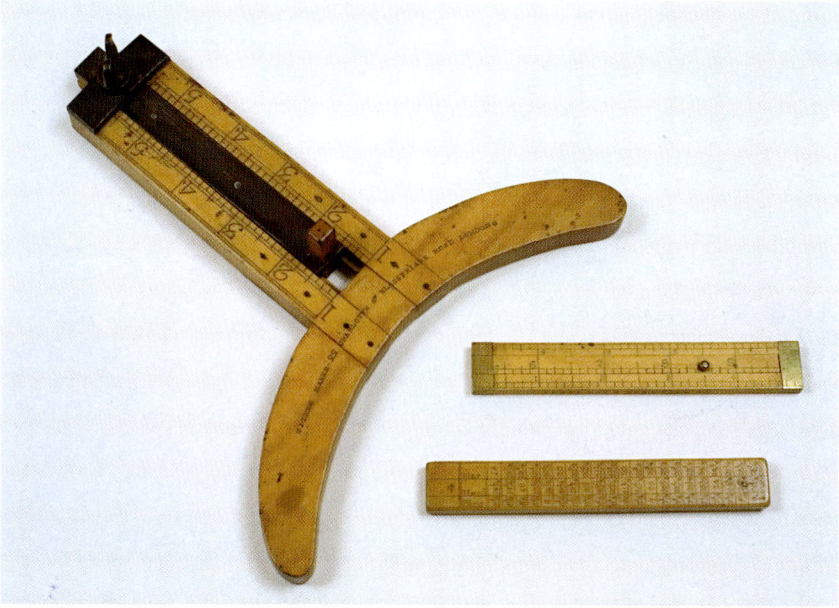

Fig. 13.41. Three hatters' rules. Left: A hat brim gauge by J. Tree (London, 1826–95). Top right: An all-boxwood hatters' rule with brass ends by Edward Preston & Sons (Birmingham, 1864–1932) Bottom right: A plain hatters' rule, unmarked. This has a boxwood slide on the reverse side.

directions on the stick and from 5 to 9 inches on the slide. When fitting a hat, three measurements were taken, heads not being round, from front to back and side to side and the circumference, using a tape. On the reverse of the rule is a table giving the relationship between the hat size, marked SIZE, the circumference of the head, marked HEAD, and the circumference of the hat, marked HAT. These are usually all boxwood, but can be in other woods and were sometimes offered with a brass or an ivory slide.

Two other hatters' tools sometimes found are a hat crown and brim measure and a hat brim gauge. This latter is not strictly a measuring tool but a cutting gauge, used to trim the brim to a constant width.

Wigmakers' rules

Wigmakers' rules are described and illustrated in both Diderot's *Encyclopédie*[16] and in *The Art of the Wigmaker* written by de Garsault in 1767[17], although no example has yet been identified.

Diderot's description is of a rule of lath type around 2 feet long divided into inches and twelfths of an inch. The description of the divisions is not entirely clear but is as follows: The first division is 2½ in. from the end of the rule. The next six divisions increase by ½ in. The seventh to the twelfth divisions are 8⁄12 in. long; the twelfth to the sixteenth increase from 8⁄12 in. to 11⁄12 in.; the sixteenth to the eighteenth are increased by 1 in.; from the

Below left: Fig. 13.42. A hat crown and brim measure in use for measuring the crown of a top hat at Locke & Co., London.

Below right: Fig. 13.43. The same measure shown in Fig. 13.42 measuring the brim of a top hat.

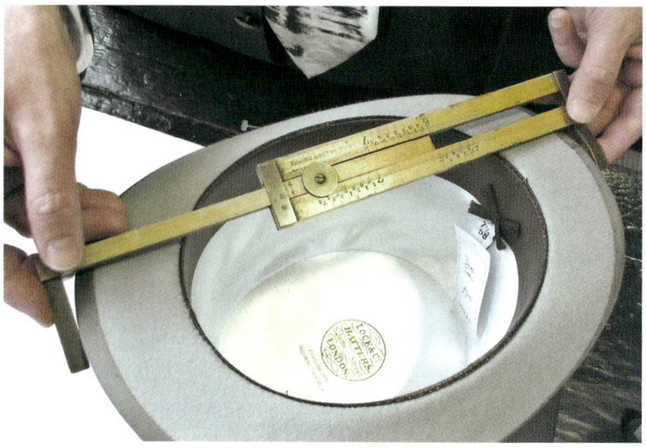

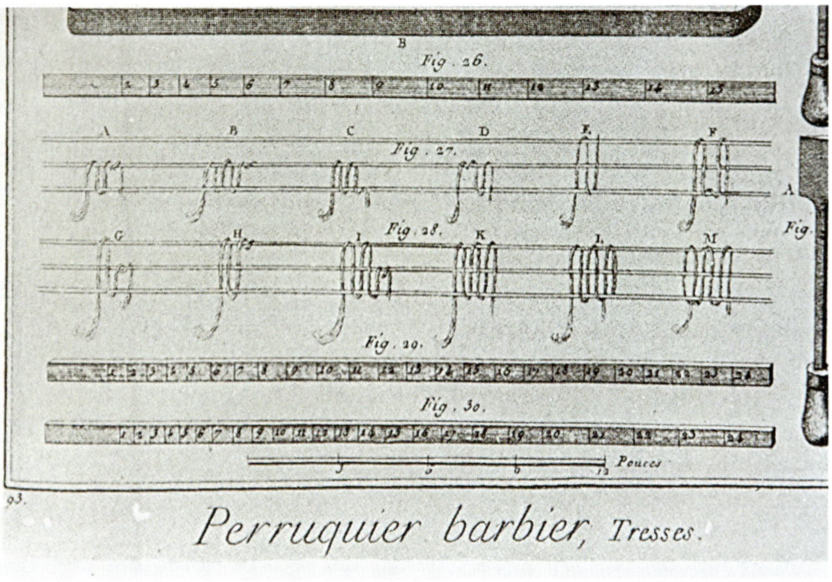

Fig. 13.44. Wigmakers' rule from Diderot's Encyclopédie, ou Dictionnaire raisonné des sciences, des métiers et des arts.

eighteenth to the twentieth by 1⅙ in.; the twentieth to the twenty-fourth are 1½ in. and the final length should be ¾ of an ell, this being the maximum length of hair that can be used.

Garsault's description is slightly different. The rule is still a lath type rule but is only 19 in. long. His first division is 2 in. from the end marked with a line and numbered 2. The spaces between the second and the fifth divisions have 8 lines each – though the size of the lines is not specified – the sixth and seventh divisions have 9 lines, the eight division has 10 lines and all the remainder are one inch.

The optician

In a correctly made pair of spectacles the axis of the lens should be opposite the centre of the eye. Without today's technology this had to be achieved by the proper size of frame whereas today it is possible to adjust the shape of the lens so that the axis in the centre of the eye – even if this is not near the centre of the frame.

A number of rules were developed to assist the optician in the proper fitting of spectacles.

The dimensions that might be taken by the optician were the distance between the pupils, the height of the nose (the distance from the level of the pupils to the point of rest on the nose), the width of the nose and distance from ear to the front line of the spectacles.

Multi-purpose rules: These were intended to take all the dimensions listed above and also measure frames. The most commonly found type in

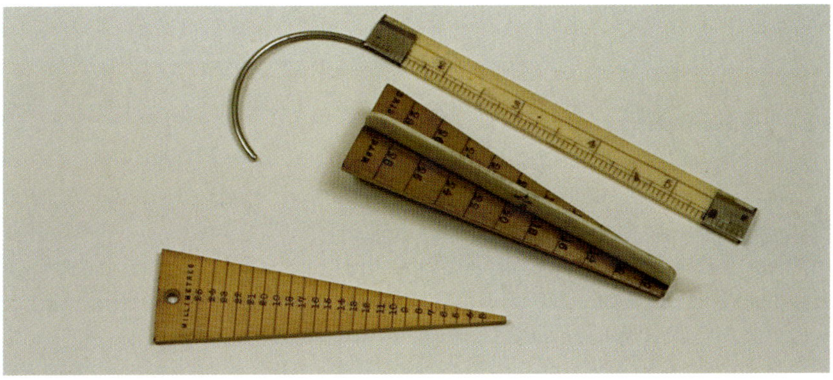

Fig 13.45. Some opticians' rules. From the top: An ivory and nickel silver side rule. A registered design bridge gauge by the Willesden Optical Works Ltd. marked in millimetres radius and width. A simpler boxwood bridge gauge marked J & H Taylor, Birmingham and London.

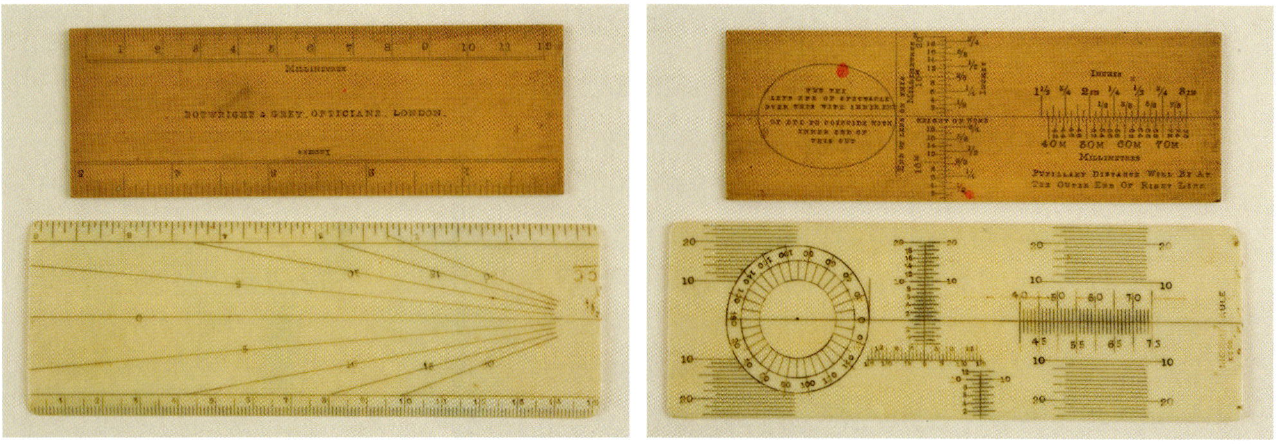

Fig. 13.46a & b. Two opticians' rules for measuring spectacles. Neither are marked with a maker's name. Top: A boxwood rule marked Botwright & Grey, Opticians, London. Bottom: A celluloid rule marked "Higway" rule, Reg^{d}.

Britain is the Orthops rule made by Raphaels, a London wholesaler of opticians' supplies.

Most are boxwood but white celluloid and even ivory versions were also made. Nor was the design of the rule standard. Over the years many different versions were sold.

Freeman's refractionists' rule is in many respects similar to the Orthops rule but without the notches for width of nose measurement. It was listed by Hirst Bros. in their ca. 1914 *Wide Awake* catalogue in both boxwood and ivory.[18]

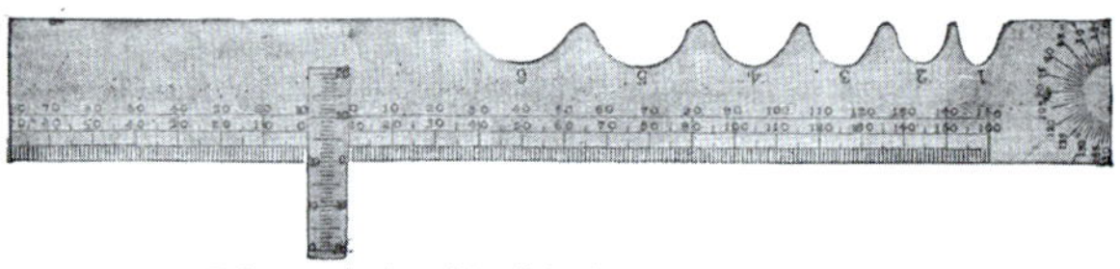

No. O2348. Each 4/- The Orthops Rule serves for determining all facial measurements for spectacle and eyeglass frames, and the dimensions of frames themselves. Full instructions sent with each. Made in Boxwood.

No. O2353. Each 7/- The Orthops Rule. The same as No. O2348, but made in White Celluloid.

Fig. 13.47. Orthops rule from Hirst Bros. "Wide Awake" catalogue, ca. 1914. This rule was available in either boxwood or celluloid.

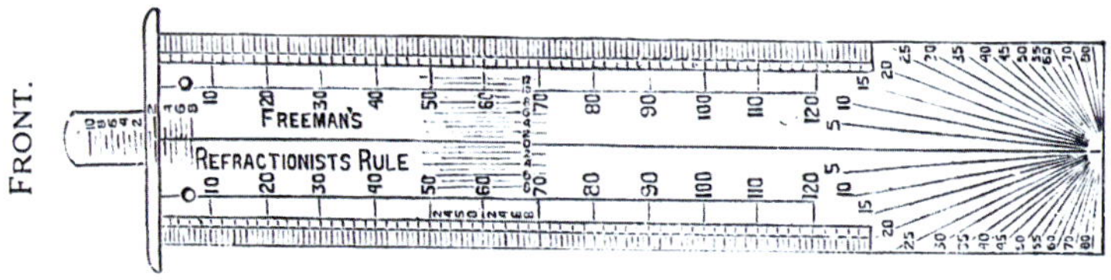

Freeman's Rule with metrical and inch scales, for measuring centres, heights of bridges, insets and outsets of bridges, and angles of sides. On reverse side is a protractor for giving the axis that cylindrical lenses are set in frames.

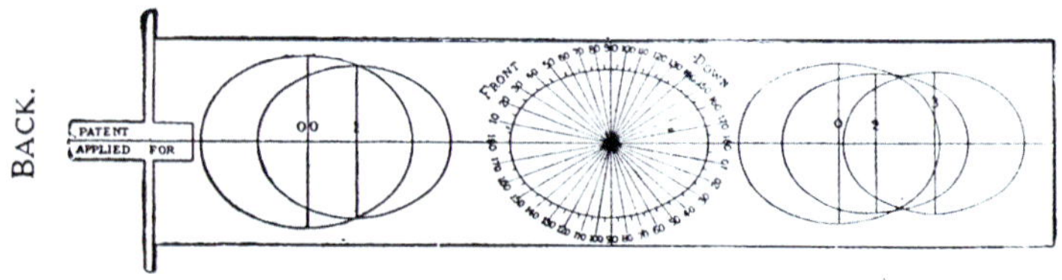

No. O2343. Each 4/6. Boxwood.
No. O2344. „ 10/6. Ivory.

Fig. 13.48. Freeman's improved refractionists' rule also from the Hirst Bros. "Wide Awake" catalogue, ca. 1914.

Face rules: Suitable only for measuring the face of the customer and not suitable for measuring frames.

Near point measures: Easily distinguished by the forked end piece which was placed against the face. A suitably marked card held in a sliding carrier was slid up and down the rule and the near distance at which the eye could

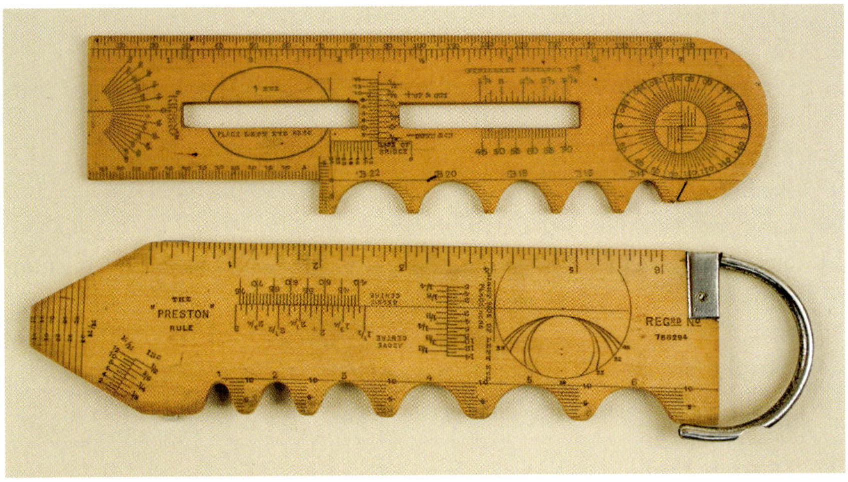

Fig. 13.49. Two boxwood face rules. Top: Marked only Reg^d. No. 689606. Bottom: Marked "The "Preston" rule" and Reg^rd No. 756294. From this is might be inferred that it was made by Edward Preston & Sons of Birmingham; however, it does not appear in any of the known Preston catalogues nor was the design registered by Preston.

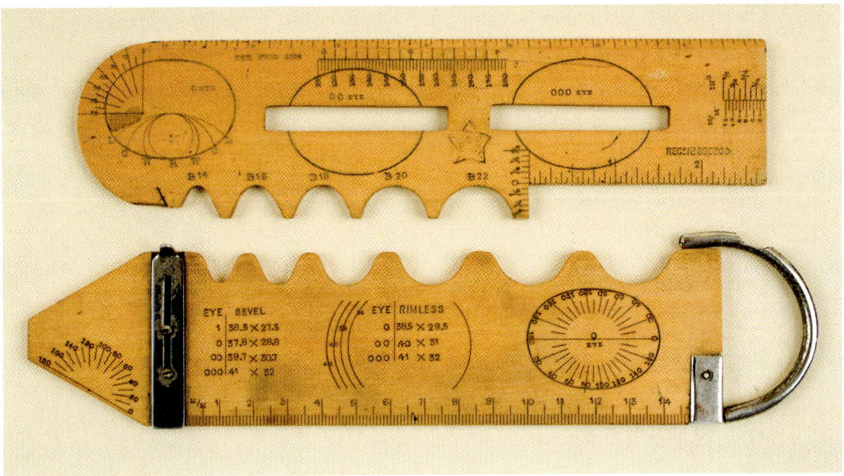

focus was measured. The slider is often missing in those found today. Another version of this tool was the optometer – this has a ring through which the card in the sliding carrier was viewed. With this device a card marked with parallel lines also allowed astigmatism to be diagnosed.

No. O2347. Each 4/6. A Boxwood Instrument for proving the near point, and for determining and measuring various conditions of the eye, with improved slide.

Fig. 13.50. A boxwood near point measure from the Hirst Bros. "Wide Awake" catalogue, ca. 1914.

Bridge measures: These are tapering gauges used to measure the size of the bridge on a pair of spectacles. Two examples are shown in Fig. 13.45.

References

1. Alcock, N. W. & Cox, N. ed., *Living & Working in 17th Century England*, CD of Randle Holme's, *Academy of Armory 1688,* London, 2000.
2. Gilliland, William N., ""Some Measures of History"", *Mensuration*, Vol. 2, No. 1, 1990.
3. *Ibid.*
4. Edward Preston & Sons, *Catalogue No. 18. The "Preston" Catalogue*, 1909 with 1912 supplement, reprinted Mendham, NJ, 1991.

5. Alcock & Cox, *op. cit.*
6. *Ibid.*
7. George Barnsley & Sons, *Illustrated catalogue of Shoemakers' Tools*, Sheffield, 1927.
8. Preston 1909, *op. cit.*
9. Barnsley, *op. cit.*
10. Edward Dean, Lerro & Co., *Catalogue,* London, 1919.
11. Salaman, R. A., *Dictionary of Leather Working Tools, c. 1700–1950,* London, 1986.
12. *Ibid.*
13. Barnsley, *op. cit.*
14. Salaman, *op. cit.*
15. *Ibid.*
16. Diderot, Denis and d'Alembert Jean Le Rond, *Encyclopédie, ou Dictionnaire raisonné des sciences, des métiers et des arts,* 3rd ed., 1776.
17. de Garsault, François Alexandre, *The Art of the Wigmaker*, 1767.
18. Hirst Bros & Co. Ltd., *The Fourth "Wide Awake" Catalogue of Tools and Materials for Watchmakers, Clockmakers, Jewellers and Opticians,* Oldham, n.d. but ca. 1914.

Chapter 14
Miscellaneous measures

Measuring, defined popularly, is the using of a certain known measure, and determining thereby, the precise extent, quantity or capacity of any thing. Chambers, 1743.[1]

At some time, in every walk of life there is a need to measure. This last chapter is a review of a number of measures and measuring tools that have not been covered elsewhere in this book and that the collector might find and wish to identify.

Paper

Historically, there have been considerable variations in the measurements applied to both the paper and printing trades, which makes for much confusion.

Paper quantities: Two dozen sheets of paper are a quire and twenty quires, that is 480 sheets, are a ream. Our office supplies catalogue describes the packets of copy paper that we use as reams – the packets actually contain 500 sheets (maybe this is the photocopier ream) – but there is also the printers' ream of 516 sheets which is, however, sometimes 510 or even 500 sheets, calculated from 20 printers' quires each of 25 sheets. Unreliable paper making machines and printing presses resulted in substandard products at every stage, so one must assume that the larger reams were originally intended as compensation for sheets that had to be discarded. Larger but seldom used quantities of paper are the bundle (a double ream) and the bale (five reams).

Paper sizes: The sizes of sheets of paper of a particular name varied, depending on whether it was intended for drawing, writing or printing.

The most commonly found sizes were foolscap, crown, demy and double elephant, although there were actually, in Britain, around 40 different sizes. The sizes of paper also varied over periods of time, so the size of a particular name in, say the 1850s, might be different from that in the 1880s.

In Britain the adoption of metric paper sizes quickly followed the publishing of BS 3176 in 1959. This is a rational system founded on the A0 sheet, 1189 x 841 mm, which has an area of $1m^2$ and which, when cut in half, produces sheets of the same proportion. These are designated A1, A2, A3, A4, etc.

	Sheet size (in inches)
Large foolscap	17 x 13½
Crown	20 x 15
Large post	21 x 16½
Demy	22½ x 17
Medium	23 x 18
Royal	25 x 20
Super royal	27 x 20
Double crown	30 x 20
Imperial	30 x 22
Double demy	35 x 22½

Table 14.1. Some pre-metrication paper sizes used for printing taken from Molesworth, 1851[2]. Even from the few examples in this and the next figure it can be seen that sheets with the same name were not always of the same size.

Fig. 14.2 Table showing the pre-metrication sizes of writing and drawing paper as shown in Tomlinson's Cyclopaedia, *1852.*[3]

	Sheet size (in inches)
Antiquarian	52½ × 30½
Double elephant	39½ × 26½
Atlas	33 × 26
Columbier	34½ × 23
Elephant	28 × 23
Imperial	29½ × 21½
Super royal	27¼ × 19¼
Royal	23½ × 19
Medium	22¼ × 17¼
Demy	19½ × 15¼
Extra large post	22¼ × 17¼
Large post	21 × 16½
Post	19 × 15¼
Copy	20 × 16
Sheet-and-half foolscap	25½ × 13¼
Sheet-and-third foolscap	22 × 13¼
Foolscap	16½ × 13¼
Pott	15½ × 12¼

Substance: Traditional British practice in the paper trade was to express the substance of paper in pounds per ream. Thus the weight of a ream of paper depended not only on the substance of the paper but also on the size of the sheet. For example a ream of crown sheets weighing 18 lb. has the same substance as a ream of quad demy weighing 95 lb., both being 85 grams/m^2, expressed in the modern system of stating paper substance in grams/m^2. So it can be seen that sheet size matters not.

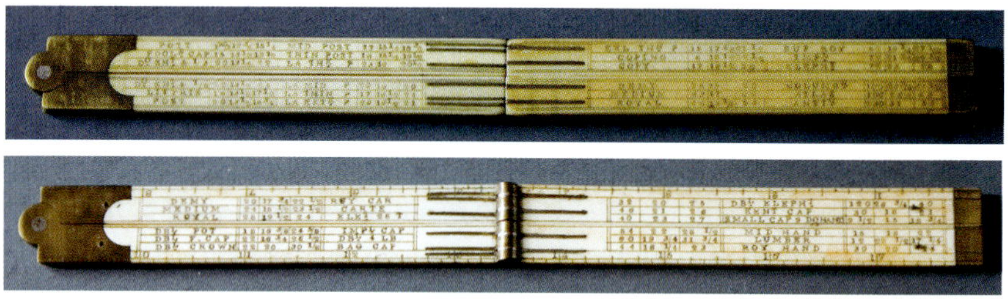

Fig. 14.3a & b. An ivory paper rule with tables for various paper sizes. Some, though not all, can be identified from the preceding figures and include writing and drawing papers, and printing paper. The first of the three columns is the weight in pounds of 1 ream. The second and third columns give the dimensions of the paper.

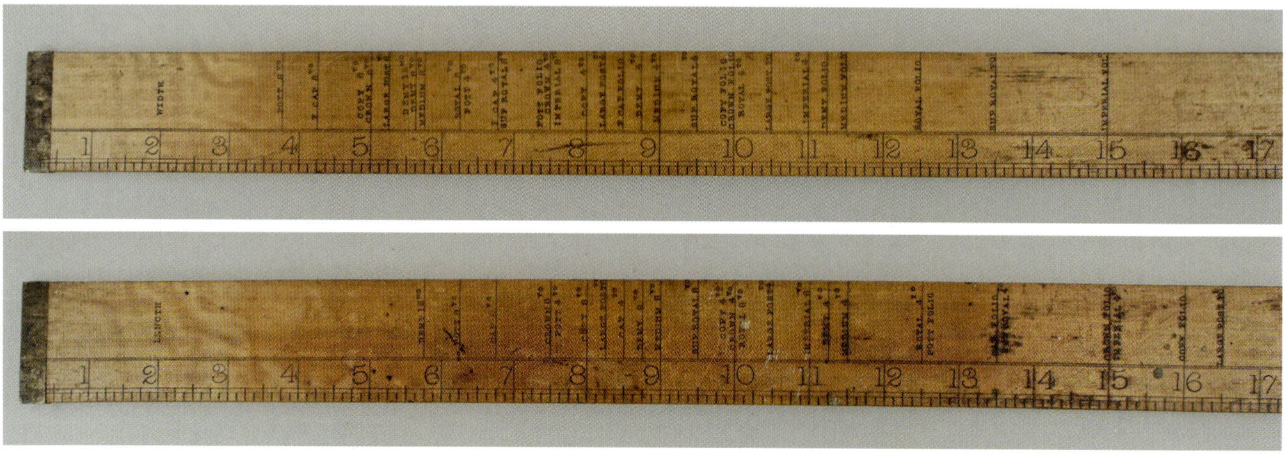

Below: Fig. 14.4a & b. A different style of paper rule by Aston & Mander (London, 1870–1930) marked with a range of sizes, the width being read off one side of the rule as a direct reading dimension and then the length off the other side.

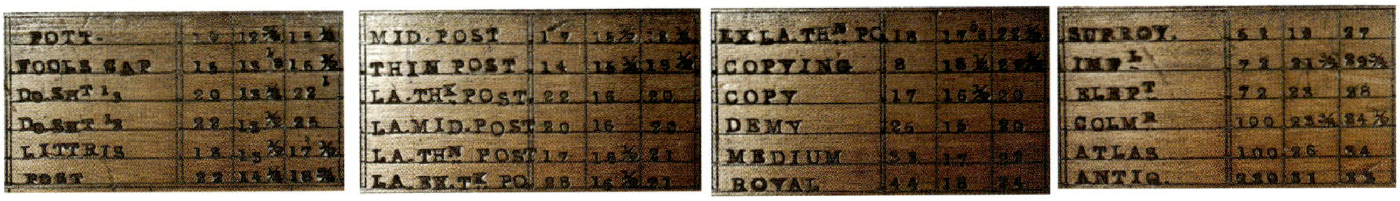

Top: Fig. 14.5. An anonymous paper trade rule.

Fig. 14.6. The first four tables of paper sizes as marked on another anonymous paper trade rule. The first column gives the weight in pounds per ream. The second and third columns give the dimensions. The abbreviations in these tables stand for pott, foolscap, foolscap sheet and a third, foolscap sheet and a half, littris, post, middle post, thin post, large thick post, large middle post, large thin post, large extra thick post, extra large thin post, copying, copy, demy, medium, royal, super royal; imperial, elephant, colombier, atlas, antiquarian.

Paper micrometers: A quick method of checking the substance is to use a paper micrometer. These are small micrometers, opening to ½ in. or 15 mm, with pad anvils. This type of micrometer was made by Moore & Wright, Starrett and most of the other precision tool makers. Smaller nickel silver, German-made examples often bearing advertising for paper merchants are also to be found.

To determine substance the micrometer reading needs to be converted using tables to give the weight of a ream according to sheet size. However this method can never be very accurate as various factors including density and surface texture will affect the result.

Paper trade rules: These are lath type rules, between 18 and 48 in. long, distinguished from other lath type rules by being thick (⅜ in. or more), having one edge bevelled and one or usually both edges protected with a brass strip. Apart from general measuring these rules were intended to act as a straight edge when cutting up to make smaller sheets – indeed many bear scars where the knife slipped. A few examples marked with tables of standard paper sizes and pounds per ream or substance have been found. A rule by William Hobcraft (London, 1817–52) is one of the few named examples recorded and they have not been found listed in any catalogue.

The printer or paper merchant needed a lot of information about the "standard" paper sizes, the weights of reams and much else but perhaps this information was so vast that it did not lend itself to inscribing on rules; rules with this sort of information are a rarity. An example of a rule clearly intended for the user of paper is known marked "arranged & sold by Underwood, 34 Gopsal St., Birmingham". This 2 ft. 4-fold boxwood rule has tables for size/weight for writing paper, printing paper and sheet sizes for writing, printing, brown papers and cards and sizes of foolscap and small post when folded into sheets.

Its most unusual table is "waste allowed". This is reproduced in Fig. 14.8.

Paper merchants' slide rules: There can be few industries where the conversion from Imperial measurement to metric has taken place over a longer

Fig. 14.7. The fifth to the eighth tables on the same rule as that in Fig. 14.6. The sizes given in these tables include some of the same sizes as in Fig. 14.1 & 2 but of different weight per ream. The abbreviations stand for double elephant, demy, medium, royal, double pott, double foolscap, double crown, royal cap, cartridge, elephant, imperial cap, double four pound, draught cap, kent cap, small cap, small cap hand, middle hand, lumber and royal hand.

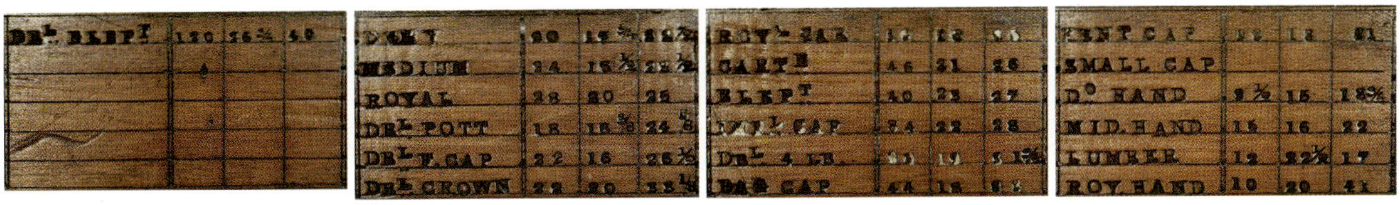

Fig. 14.8. Table of "Waste Allowed" from a paper trade rule. The waste is calculated on the basis of one in 2, 4, 6, 8, 12 or 16 being waste. The quantities in the first column are given as hundreds and thousands of sheets but the waste is expressed in quires and sheets.

	2		4		6	
Quan	Q	S	Q	S	Q	S
100	2	6	1	3		18
250	5	9	2	16	1	19
500	10	15	5	8	3	14
1000	21	2	10	14	7	1
2000	41	22	21	0	14	0
3000	62	18	31	11	20	23
4000	83	16	41	21	27	21
5000	104	17	52	7	34	21
10000	210	12	105	6	70	0

	8		12		16	
Quan	Q	S	Q	S	Q	S
100		13		9		7
250	1	8		22		17
500	2	16	1	19	1	8
1000	5	5	3	18	2	16
2000	10	12	7	0	5	6
3000	15	17	10	11	7	20
4000	20	22	14	0	10	11
5000	26	4	17	12	13	22
10000	53		35	3	27	0

period, before finally being resolved in Britain in 1959. For over a century the trade had to wrestle with inch and metric manufacturers and users. Several slide rules designed to make the conversions between systems were widely used. The most commonly found is Tomlinson's Equivalent Paper Slide Scale which exists in a number of variants.

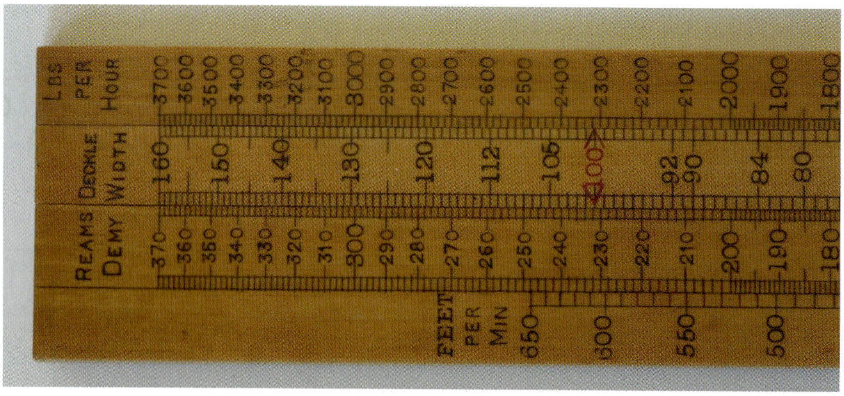

Fig. 14.9. A slide rule for the paper manufacturing trade. As with many specialist slide rules, without the instructions, it is difficult to interpret the use.

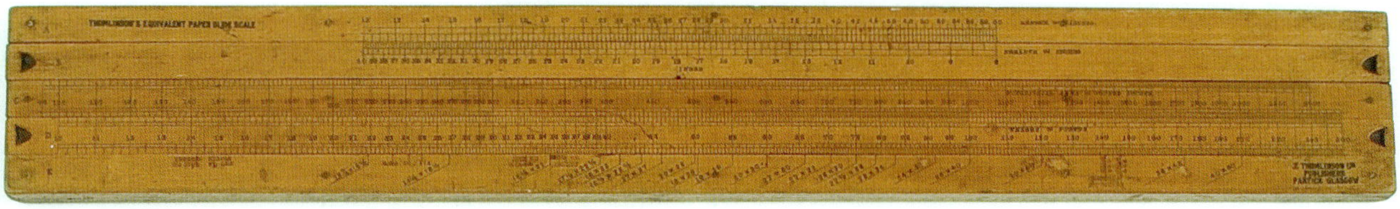

Fig. 14.10. A Tomlinson's equivalent paper slide scale. It is marked "J. Tomlinson, publisher, Partick, Glasgow." Tomlinson has not been identified but is unlikely to have been the maker of the rule.

Printing

Joseph Moxon[4] writing in 1683 lists 10 bodies of type and specifies the sizes. Each body of type originated from a set of steel punches made by a punch cutter. The type resulting from these was therefore of a particular size and form (face) and this was the original meaning of body. However by Moxon's time the principal meaning of the word was size of type.

Incidentally Moxon uses the term "pica sized" when describing small metal parts – meaningful to his intended readers though less easy to understand today.

Pearl	*184*
Nonpareil	*150*
Brevier	*112*
Long Primer	*92*
Pica	*75*
English	*66*
Great Primer	*50*
Double Pica	*38*
Two Lin'd English	*33*
Great Cannon	*17½*

Table 14.11. Joseph Moxon's list of type bodies. The figures show the amount of type contained in one linear foot.

Type

Early printers applied names to the different sizes of type that they used. These names, pica, brevier, cannon etc., are thought to have originated from the work in which they were originally used.

Fournier, a leading French type founder, produced the first numerical system in 1737. This used his "typographical point" of 0.0137 in. and in 1784 Didot, another Parisian, started to make type to his system of Didot Points (0.0148 in.) which gradually became widely used on the continent and remains the standard today.

In Britain names continued to be used until at least 1898 when the American point system (0.0138 in.), which had only developed after 1872, was adopted. To printers, named type is "old body".

Definitions: A piece of type has three principal dimensions. The body (this is the point or named size); the front, which varies with the width of the letter, "M" being the widest; and the type height, that is the height from the printing surface to the foot – 0.918 in. a dimension standardised since the earliest years of printing.

The point size defines the body on which the typeface is cast, not the size of the letter as printed – different faces will occupy more or less of the available space.

The widest letter in any font will be the "M" (em) – with very few exceptions its width will be equal to the size of the type. Pica was the most commonly used type size in old body and its ems (0.166 – approximately 12 to

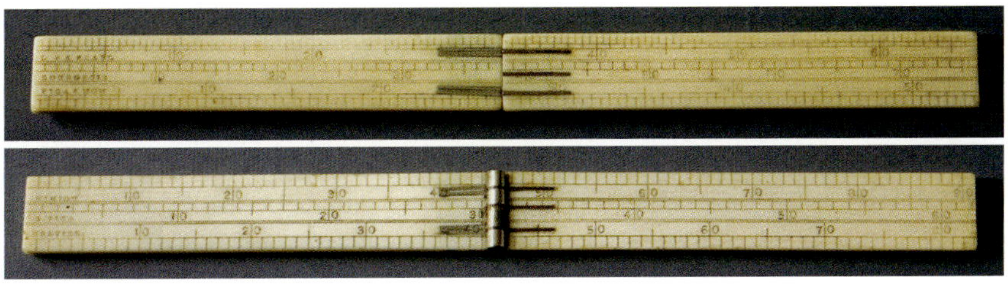

Fig. 14.12a & b. An ivory printers' rule marked for Pearl, Bourgeois and Pica and Nonpareil together on one side and Minion, Small Pica and Brevier on the other, giving the number of ems contained in the length of the rule for each body.

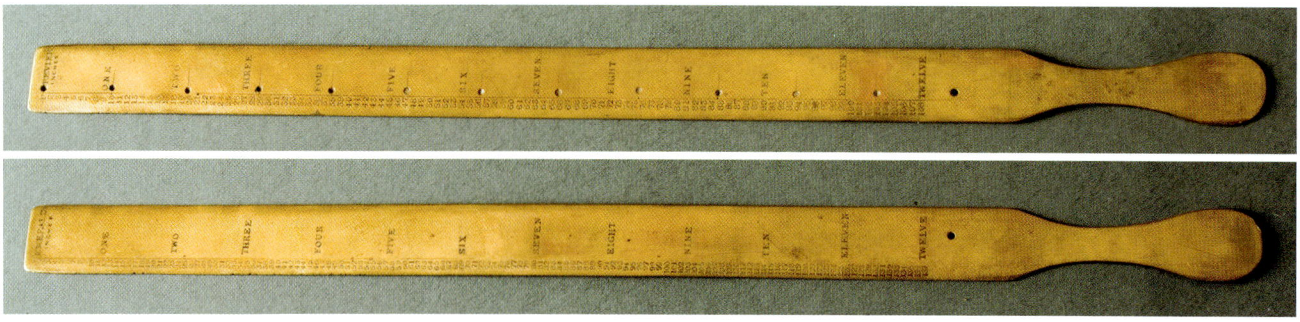

Fig. 14.13a & b. A brass rule marked for Emerald on one side and Brevier on the other, giving the ems per inch.

the foot) became a widely used standard of measurement in the printing trade and the size of the printed page might be given in ems even if it were printed in a type of different size.

There are thus two types of ems – the standard pica em, often referred to as a pica, and the em of whatever size type is being set.

The measurement of ems was important to the printer because it enabled him to work out the number of words that could be set on a page.

Printers' rules: The traditional folding form is a 9 in. 2-fold rule, folding face to face and made without tips thus allowing the divisions to be clearly marked right to the ends and usually made in boxwood though ivory examples are also found. The lines of measure are ems (or body size; they are the same) for old body type; inches are rarely marked. Two lines will often be superimposed, for example Pica and Nonpareil (which is half the size of Pica). The naming of the lines is often contracted, sometimes to a single letter.

Brass rules, old body: These usually appear to be craftsman made and are marked for only one or two sizes of old body type.

Brass rules, new body: Brass rules, about 12 in. long, marked in ems in modern point sizes often with 6 and 12 point, and 5 and 10 point superimposed. Many have a shoulder at the index end and often carry advertising for type founders.

Composing rules: These are brass strips with small ears at each end, often 40 or more of increasing length, contained in a box. They are sometimes marked with numbers (the lengths in picas). They are often supposed to have a measuring purpose but this is not the case. They are placed on top of a row of type in the composing stick to make it easier to slide in the type and are then, using the ears, removed. The marked numbers are merely a help when selecting the appropriate size.

Modern rules: Since the introduction of point size type many different sorts of printers' rules have been available and are still being made today in a variety of materials.

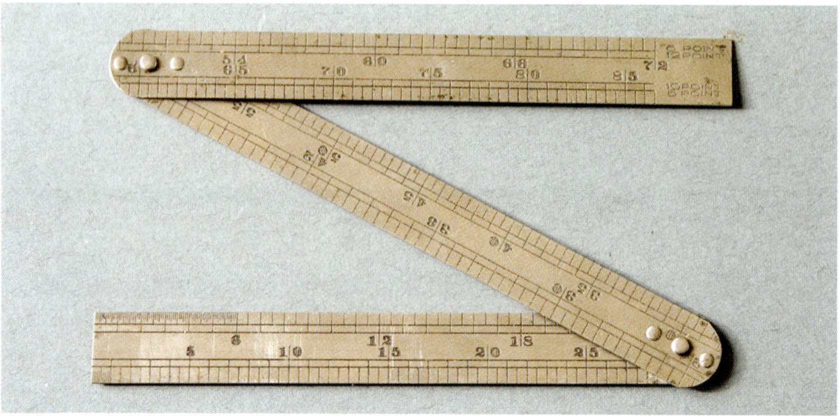

Fig. 14.14. A modern steel 3-fold rule marked with scales for 6 and 12 point (combined) and 10 and 5 point. The reverse face of the rule is blank.

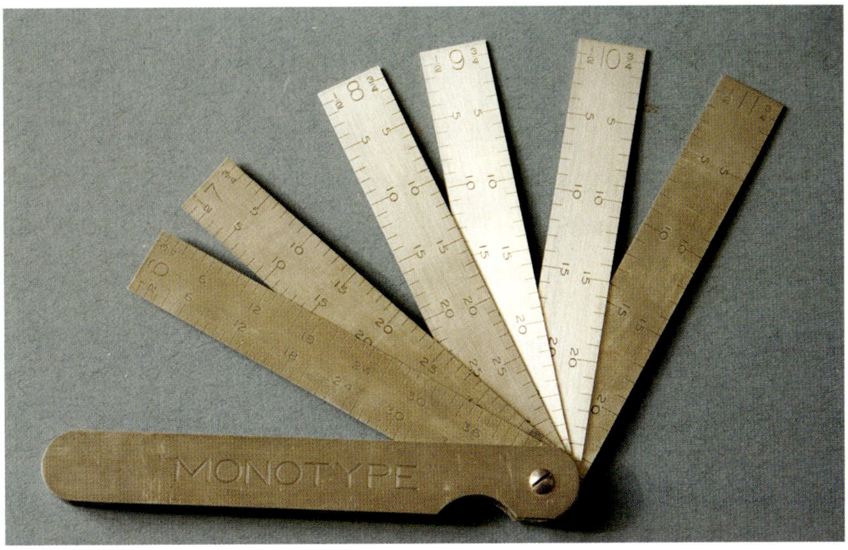

Fig. 14.15. Another style of modern type rule, by Chesterman (Sheffield, 1821–1963). Each tongue is marked for a different point size, from 6 to 11, each size with a separate scale for the whole number, plus a quarter, plus a half and plus three-quarters.

Advertising rules: Type rules were frequently used for advertising, such as the Linotype rule which is a 12 in. boxwood rule with a metal cursor, marked with point sizes on the face and an advertisement on the reverse.

Clock and watch making

The *Catalogue of Tools for Watch and Clock Makers* issued by John Wyke of Liverpool, ca. 1770[5] contains illustrations of hundreds of tools. Included are many tools such as calipers, screw operated gauges and the like which could be used to take or compare sizes but there is not one item in the catalogue that is capable of measuring and giving the result in a figured size for the watch or clock maker. The fit and free movement of the mechanism were achieved by feel.

As late as 1851 Peter Stubs in his *List of Prices of Lancashire Tools etc.* reproduced by Surrey E. Dane[6] still contains little equipment that will give a figured measurement apart from one page showing a few slot gauges for other metal working trades.

Only later in the century, when the mass produced watch with interchangeable parts had arrived, does the trade start to use much measuring equipment to give actual sizes. But thereafter a whole range of specialist equipment comes into use. The difference in the range and accuracy of the tools available to the watch maker or repairer of 1890 compared with 1850 is truly surprising.

Perhaps the most perplexing aspect of all watchmakers' items is the identification of the units in which the measure is made. The small units used may

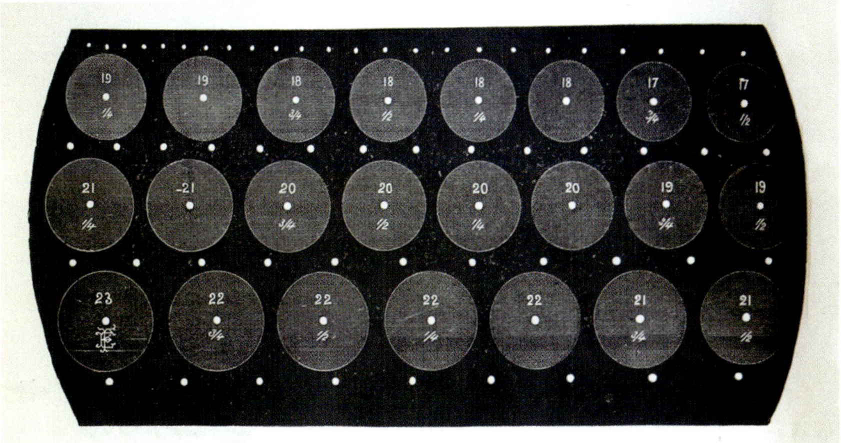

Fig 14.16. A wheel gauge. The whole number sizes are in millimetres with either quarter or eighth millimetre intermediate sizes. From the Henry Isaacs catalogue ca. 1900.[8]

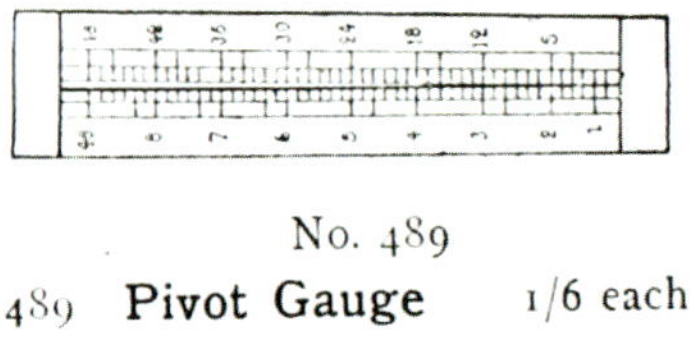

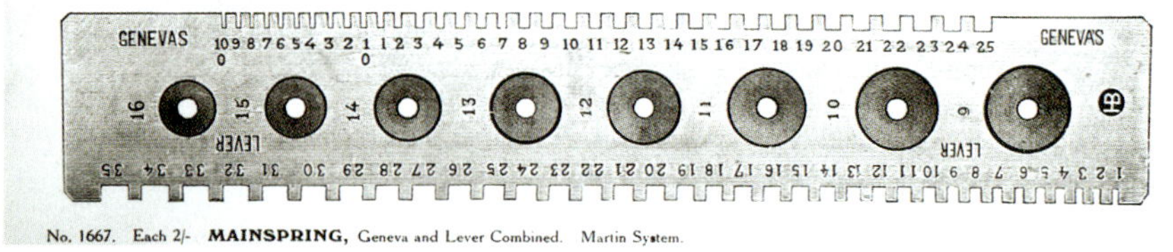

Above right: Fig. 14.17. A mainspring gauge. Hirst Bros catalogue ca. 1914.[9]

Above: Fig. 14.18. Pivot gauge: This is marked with two measures though the basis for either of these is uncertain. Henry Isaacs catalogue.

be fractional parts of a millimetre but fractional divisions of the French, Swiss or English inches may also be the basis of the measure.

Numerous gauges for the use of clock and watchmakers are listed in catalogues, many of which are rarely found; only the more common have been described below. A full description of watchmaking tools and gauges can be found in the books written by T. R. Crom.[7]

Wheel and barrel gauges: A plate machined with circular recesses, each recess being drilled through at the centre. This allows wheels or barrels already mounted on pinions to be measured. Every watchmaker must have had several of these as they are very common.

Mainspring gauges: Two measurements are necessary – the width of the spring and the thickness. The gauge for the latter is usually called a mainspring force gauge. The two items are sometimes combined.

Pivot gauges: A small closed end V-gauge set to a fine angle and in most respects similar to a mainspring strength gauge but distinguished by not having an aperture to enter the spring.

Watch glass gauges: When a watch glass or "crystal" needs replacing, it is necessary to determine its size. There are two systems used for the measurement of these – Geneva or Lunette. Geneva measurements are expressed in units and either 8ths or 16ths of the unit; Lunette in units and 4ths.

The normal form of watch glass measure consists of a boxwood slip (though occasionally ivory or nickel silver) marked with a scale. The glass rests on a raised strip, usually brass, fixed to the face of the gauge and the diameter measured by eye against the scale. If the system is the Geneva system, the scales consist of a series of vertical divisions, either 15, 9, 7, or 3, which are crossed

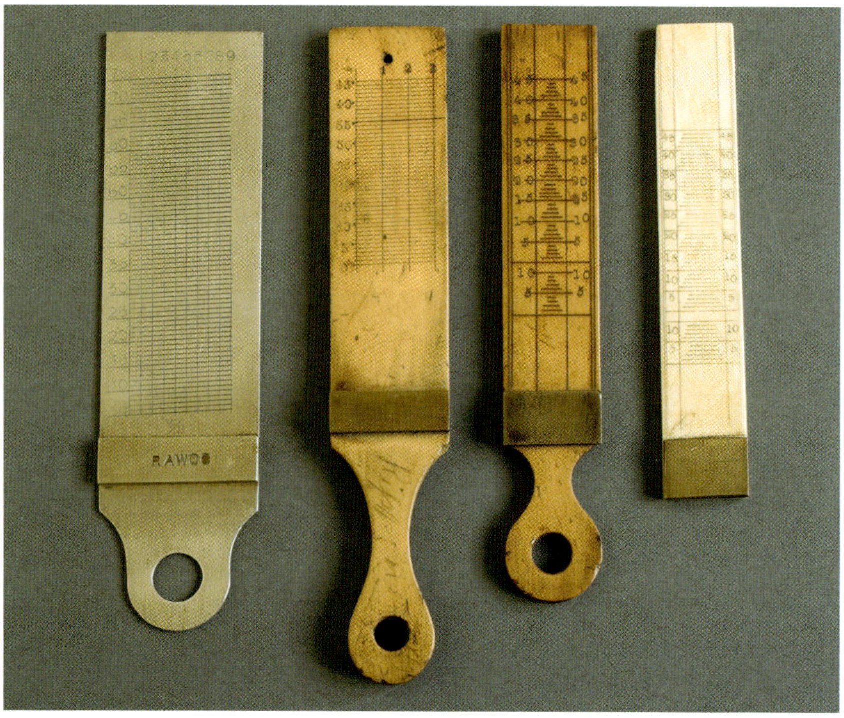

Fig. 14.19. Watch glass gauges. The two on the left for the Geneva system, the two on the right for the Lunette system.

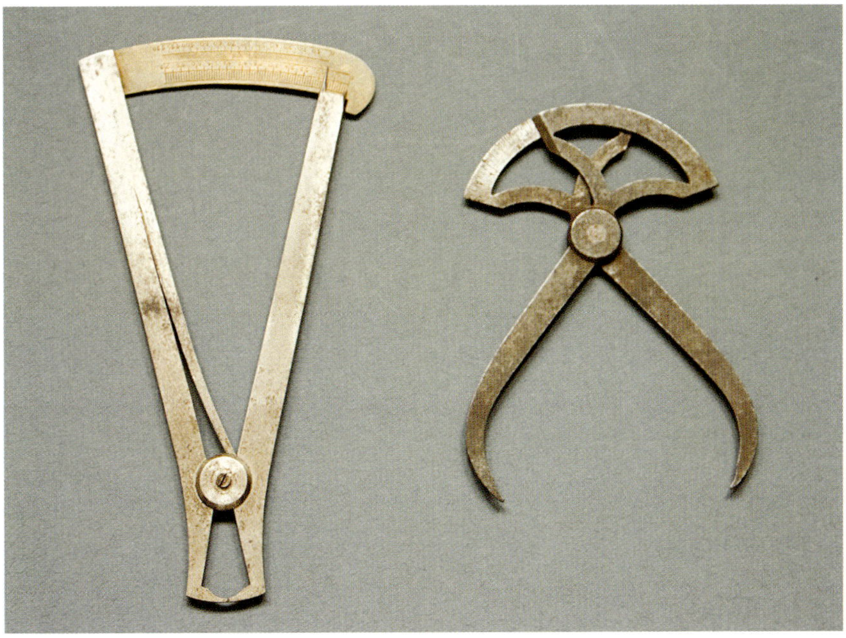

Fig. 14.20. Two different styles of douzieme gauge. The one shown on the left is the more common style.

by a series of diagonal lines. The glass is placed with its lower edge against the strip and the size determined by the graduation intersection reached.

With the Lunette system, the diameter is read off directly.

Watch key gauges: Until keyless watches, which are the modern type of watch with a permanently fixed winding knob, became the norm a removable key was needed to wind the watch. Needless to say the winding squares were of differing sizes. The watch key gauge consists of a range of steel squares fitted to a steel or brass back plate. Circular versions with the squares projecting radially are also known.

Douzieme gauges: The name suggests that this gauge multiplies by 12 and this may have originally been the case when the measure was French inches but the more recent types measure in millimetres and multiply by 10. In America these are known as degree gauges. These are a quick tool to use and must have found many uses for they are common.

Musical instrument string measuring

The traditional material for the strings on the violin family of instruments is catgut, though steel is sometimes used for the higher pitch strings. For pianos, harps and a range of other instruments such as banjos and guitars, steel wire is used. Catgut strings are specified by instrument and note - for example, for the violin, E, A, G, and D strings; metal strings are usually specified by gauge.

British music wire gauge: This extends from 00 (sometimes expressed as 2/0) to 30, the largest. Although this is already a series with small increments, surprisingly many gauges also include half sizes.

String gauges for violins, violas and cellos: Catgut strings are variable in thickness; indeed it is desirable that they are, so that the best thickness can be selected to achieve the desired tone. The most appropriate type of gauge is therefore a brass or nickel silver V-gauge with some latitude allowed for each string. Many players carried gauges of this type. The markings are (from thin to thick) for a violin, E, A, G, D; for a viola A, D, G, C; for a cello A, D, G, C.

These gauges are occasionally combined with some other item of use to the player such as a pitch pipe, a mute or a tuning fork.

Fig. 14.21. Music wire gauge with the half sizes shown between 12 and 22 gauge only. It is marked J. & J. Goddard, Tottenham Court Road, London. Goddard was one of the leading suppliers of tools for musical instrument making but they would not have manufactured the gauge.

Left: Fig. 14.22. A selection of music string gauges. Top: A combined catgut string gauge for violin, viola and cello. Second row: Two music wire gauges by G. Buck, a firm of London tool makers and sellers; two gauges marked with a large range of string sizes, probably for harp strings; an unidentified gauge for C and F strings. Bottom row: A combined mute, pitch pipe and violin string gauge; two violin string gauges and a combined music wire gauge and tuning fork.

Jewellery measuring

Diamonds and other precious jewels are specified by carat weight. The carat is an ancient weight derived from the weight of the seed of the carob plant (*Ceratonia siliqua*) which grows in the Mediterranean area. This was chosen because it produced seeds of very regular size and absorbed or lost moisture slowly. In British metrology the carat was defined as 4 grains (1 grain was approximately 65 mg.) but in 1887 it was agreed by leading jewellers to be

Fig. 14.23. A combined stone and pearl gauge. Unmarked, it dates from the early 20th century and has 44 holes marked from ½ to 40.

205 mg. The metric carat, in use since 1913, is only 200 mg. Small though the carat is, for small diamonds an even smaller system of measurement is convenient; the point system of 1 point = 2 mg. = 1⁄100 of a carat being an alternative.

However once stones are mounted it is no longer possible to weigh them so gauges that will indicate the weight by measurement are useful. Clearly all these devices will only give an approximate indication of the weight of a stone; the cut, i.e. the shape of the stone, is also of significance.

Gauges for precious stones: Mostly multi-leaved, these hole gauges are intended to be placed over a cut stone. Gauges in the form of calipers to measure the depth of the stone are also available. As might be expected the result is expressed as carat weight. The gauge holes are usually round but can be square with heavily rounded corners. Gauges for stones and for pearls are frequently combined in the same case.

Pearl gauges: These are not a new invention. The Edmund Stone translation of Bion,[10] in 1723 illustrates a five leaf pearl gauge, a form which is almost identical to those sold today. As with many items in this work, in detail the description is somewhat jumbled. The weights of the smaller sizes of pearls are given in grains, but the larger ones are given in carats. Bion's gauge will measure pearls from ½ grain (⅛ of a carat) to 8½ carats (34 grains). The sizes of holes are given in lines – but how many lines there were to the inch (certainly more than eight) and whether these are French or English inches is not clear.

In measuring pearls by carats Bion would seem to be out of step with 18th century practice – Chambers' *Encyclopaedia* of 1743[11] cites grains as the unit of weight – and modern practice still uses grains though the pearl grain is now standardised at 50 mg.

Ring gauges: In Britain two systems, a letter series and a number series, were commonly used for the sizing of rings. As can be seen from ring sticks, which are cone shaped, the size increment is a regular amount in both series.

The letter system extends from A (the smallest) to Z. The smaller sizes, below G, only fit children and are usually omitted on ring sticks. The increments in this series are greater than in the number series so half sizes are sometimes marked.

The number system extends from 1 (the smallest) to 33.

Ring sticks: These were available in a variety of materials. The most appealing are the boxwood examples but for heavier use brass or wood sheathed

Fig. 14.24. Two anonymous boxwood ring sticks, the top marked with the letter system, the lower with the number system.

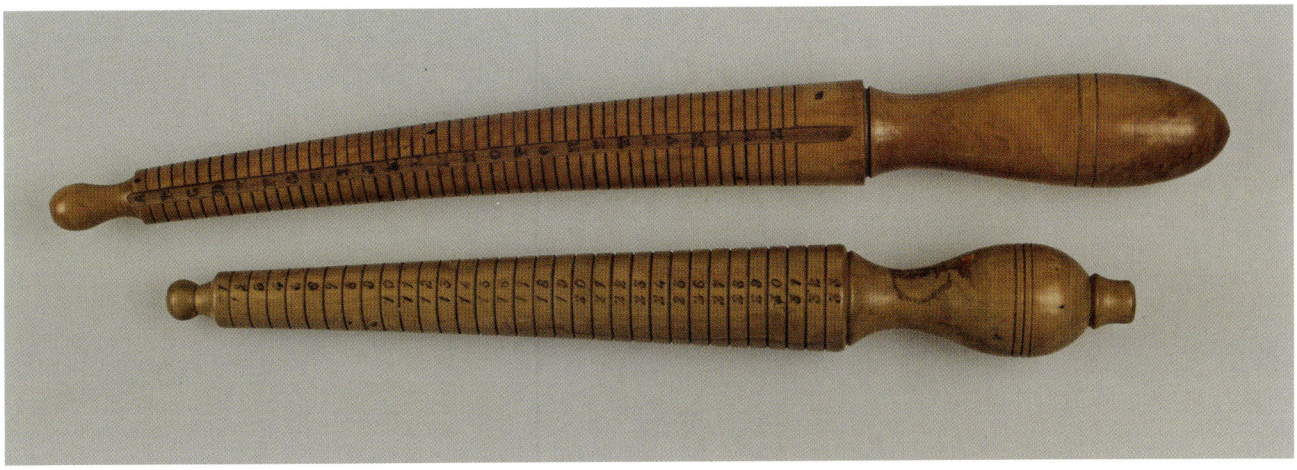

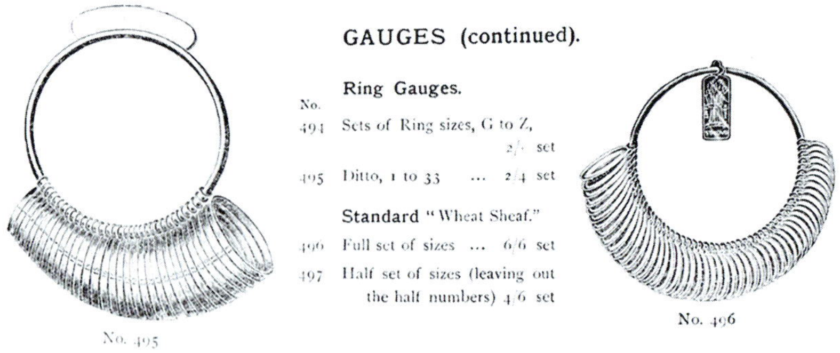

Fig. 14.25. Ring gauges from the Henry Isaacs catalogue, ca. 1900.

with metal were more durable. Also available were hardened steel mandrels which were marked in sizes though these were more commonly used for reshaping a ring or fractionally expanding a ring.

Finger ring sizes: The size of a finger is taken using a set of ring gauges. The quality of these varied enormously. For professional use, there were heavily-made sets of metal rings as well as the more usual sets with lighter metal rings – all captive on a chain or large wire ring. Cheaper sets were in the form of leaf gauges folding into a pressed metal cover. Cheaper still were similar sets made from plastic sheet – these often carried advertising for ring manufacturers. Cheapest of all were those made of light card.

Sports and games

Many sports and games required measuring devices to apply rules and regulations. Although the majority of these do not strictly come within the orbit of linear measurement, this book would not be complete without a brief overview of some of the variety to be found.

Bowls: The winner in a game of lawns bowls is the player whose bowl finishes closest to the jack and accurate and fine measurements were needed to determine the winner.

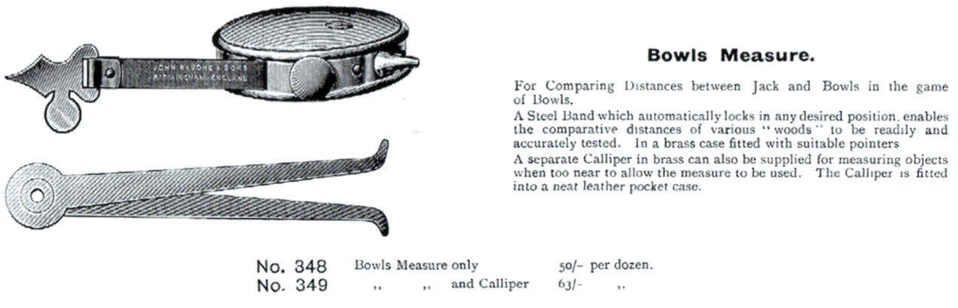

Fig. 14.26. A bowls measuring set consisted of a pull-out tape with a small horn on the side of the case and a pair of inside calipers from the John Rabone & Sons Ltd. catalogue. Undated but probably 1924.[12]

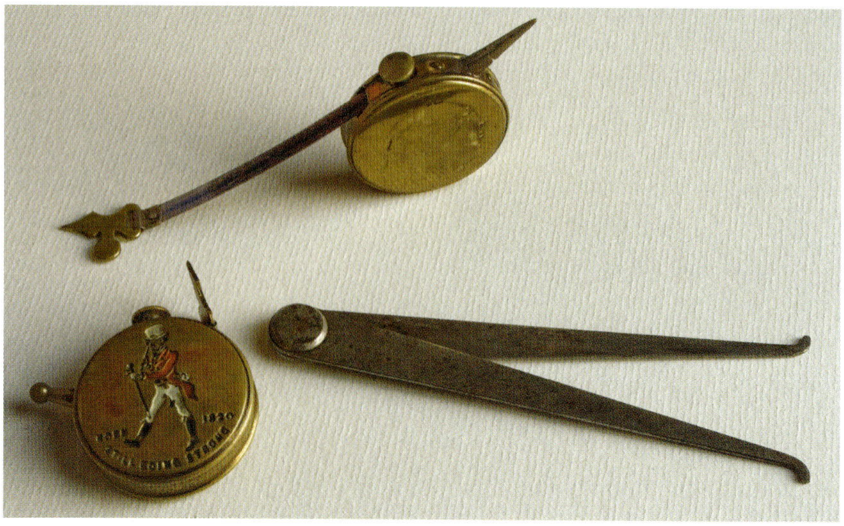

Fig. 14.27. Two bowls measures, one advertising Johnny Walker whisky, and a pair of dividers of the style shown in the Rabone catalogue. Unmarked but likely to have been made by either John Rabone & Sons (Birmingham, 1877–1963) or James Chesterman & Co. (Sheffield, 1859–1963).

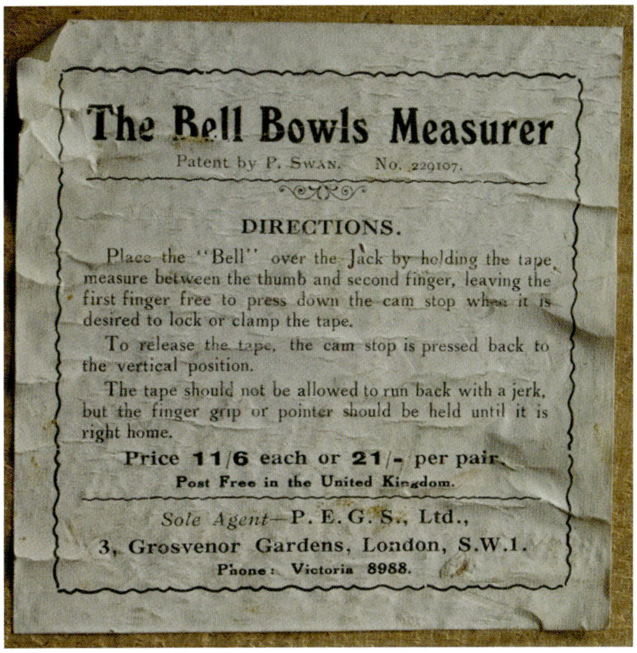

The Bell Bowls Measurer

Patent by P. Swan. No. 229107.

DIRECTIONS.

Place the "Bell" over the Jack by holding the tape measure between the thumb and second finger, leaving the first finger free to press down the cam stop when it is desired to lock or clamp the tape.

To release the tape, the cam stop is pressed back to the vertical position.

The tape should not be allowed to run back with a jerk, but the finger grip or pointer should be held until it is right home.

Price **11/6** each or **21/-** per pair.

Post Free in the United Kingdom.

Sole Agent—P. E. G. S., Ltd.,
3, Grosvenor Gardens, London, S.W.1.
Phone: Victoria 8988.

Above left: Fig. 14.28. A Bell Bowls Measurer, patented by P. Swan, incorporating a metal tape in a spring case. Although unmarked, it is likely that the tape was made by Rabone Chesterman.

Above right: Fig. 14.29. The instructions for the Bell Bowls Measurer pasted onto the inside lid of its wooden box.

Ball games: In most ball games the size of the ball is specified by the ruling body of that game and measuring tools were produced to check these. Many, but not all, were of the "go – no go" type.

The equipment used by the players was also subject to regulations as well as the pitches on which the games were played and the measuring devices for

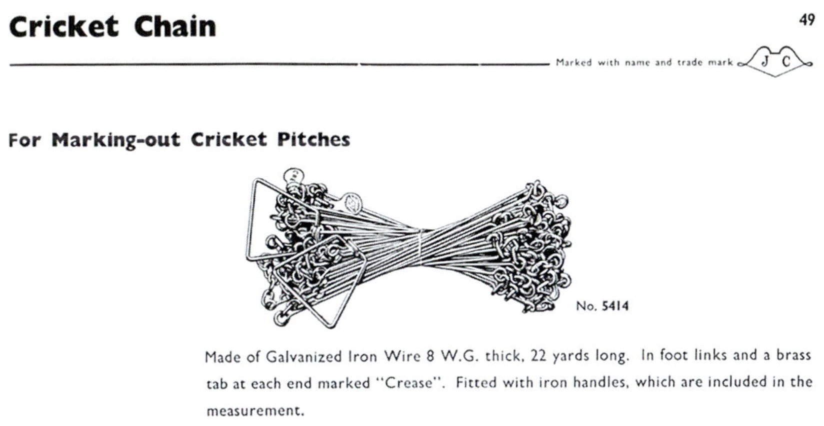

Cricket Chain 49

Marked with name and trade mark J C

For Marking-out Cricket Pitches

No. 5414

Made of Galvanized Iron Wire 8 W.G. thick, 22 yards long. In foot links and a brass tab at each end marked "Crease". Fitted with iron handles, which are included in the measurement.

Fig. 14.30. A chain made specifically for marking out cricket pitches by James Chesterman & Co. Ltd. (Sheffield, 1859–1963) from an undated but certainly 20th century catalogue.

Fig. 14.31. Below left: An unmarked cricket ball gauge.

Fig. 14.32. A patent cricket ball gauge, marked Duke & Sons, a Kent cricket ball manufacturer.

Above right: Fig. 14.33. A croquet hoop measure. This is used to check that the hoops are set in the ground squarely to ensure that the two arms are exactly 4 in. apart. John Jaques & Sons Ltd. were sporting goods suppliers.

Above: Fig. 14.34. Hockey sticks are measured by the diameter of the handle.

Fig. 14.35. A leather-cased tape measure for setting out lawn tennis courts with the dimensions of the court on the face of the case, unnamed.

checking the sizes of these are usually marked with the names of the suppliers of the equipment rather than the makers of the measuring devices; many rule makers kept name stamps for their principal customers.

Calendar and money rules

There are a number of rules used to calculate the number of days between two given dates for rent and other financial matters. As with many other rules, without an instruction book, it is often difficult to understand their exact intended use.

Fig. 14.36. A billiard ball caliper by Edward Preston & Sons (Birmingham, 1864–1932).

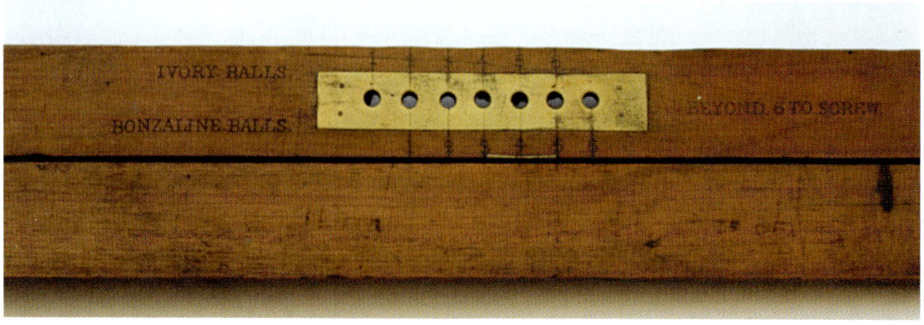

Fig. 14.37. The exact use of this rule has not been established but it is marked for billiard balls. There is no maker marked.

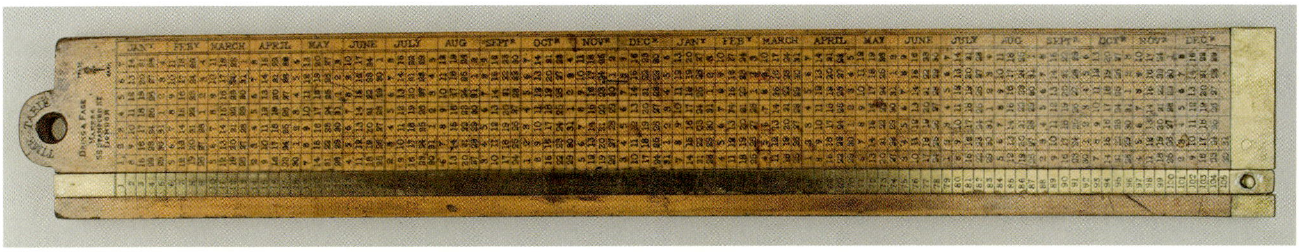

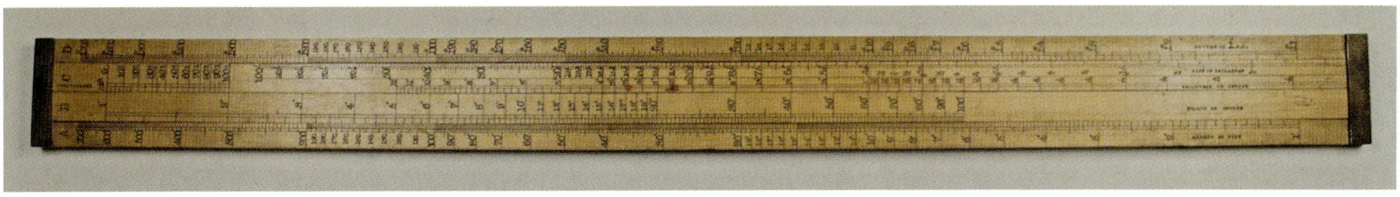

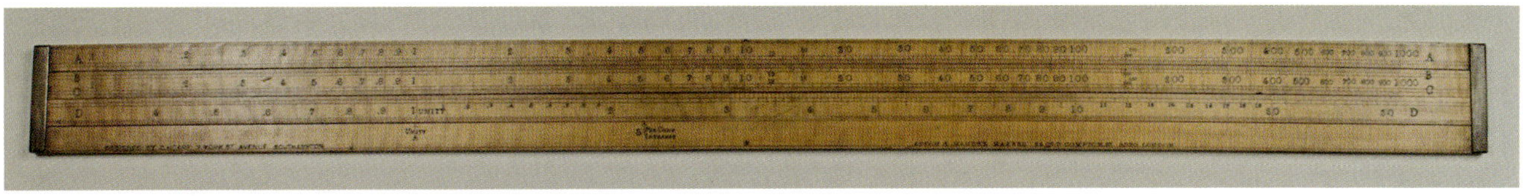

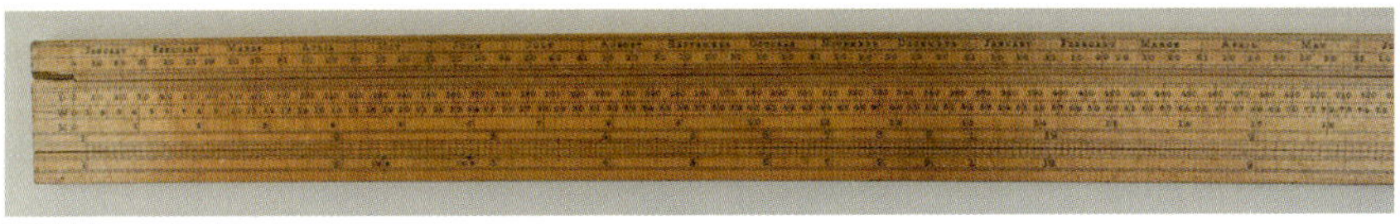

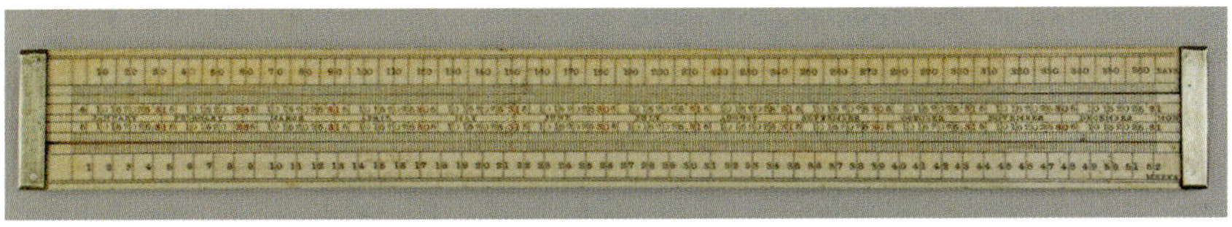

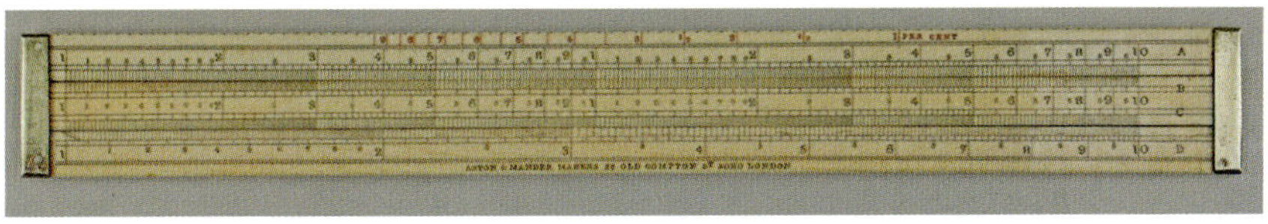

From the top: Fig. 14.38. A dock rent rule by Dring & Fage (London, 1790–1977). This calculates the number of weeks between any two chosen dates.

Fig. 14.39. A rule by Aston & Mander (London, 1870–1930) marked with scales to calculate prices for a given pound price per foot or a rate in shillings.

Fig. 14.40. A rule designed by C. Hoare, 7 York Street Ave., Southampton and made by Aston & Mander (London, 1870-1930) with scales for calculating percentages.

Fig. 14.41 An anonymous calendar rule with months, days and weeks marked.

Fig. 14.42a & b. A calendar rule by Aston & Mander (London, 1870–1930) with days, weeks and months on one face and a slide rule on the other with an additional scale for percentages.

In their 1960s catalogue, Dring & Fage[13] advertised their Time Table & Rent Rule. The advertisement included the words "...once again available." indicating that this must have been made previously, though the date has not been established. This rule enabled the user to calculate the number of days and the number of whole weeks between any two dates over a period of two years.

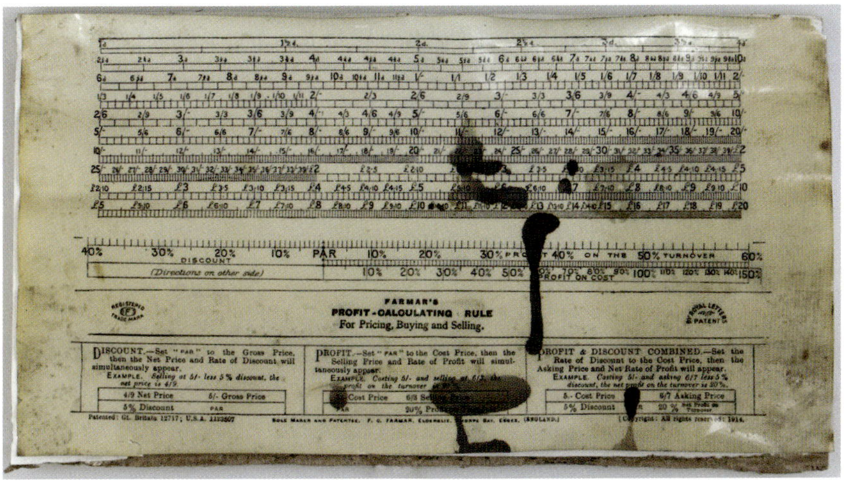

Fig. 14.43. An advertisement for Farmar's Profit Calculating rule, designed by F. C. Farmar who also designed several rules for the wine and spirit trade, described in Chapter 11.

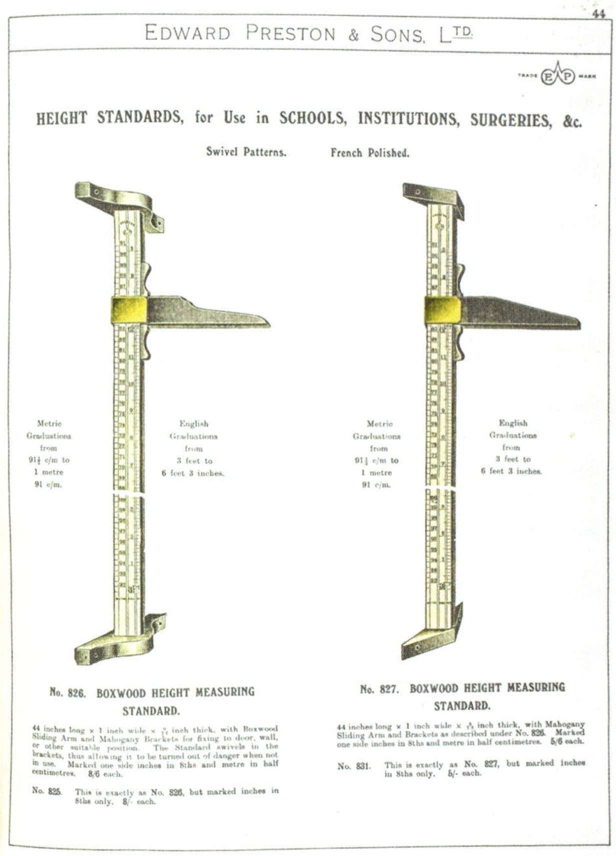

44

EDWARD PRESTON & SONS, LTD.

TRADE EP MARK

HEIGHT STANDARDS, for Use in SCHOOLS, INSTITUTIONS, SURGERIES, &c.

Swivel Patterns. French Polished.

Metric Graduations from 91¼ c/m to 1 metre 91 c/m.

English Graduations from 3 feet to 6 feet 3 inches.

Metric Graduations from 91¼ c/m to 1 metre 91 c/m.

English Graduations from 3 feet to 6 feet 3 inches.

No. 826. BOXWOOD HEIGHT MEASURING STANDARD.

44 inches long × 1 inch wide × 9/16 inch thick, with Boxwood Sliding Arm and Mahogany Brackets for fixing to door, wall, or other suitable position. The Standard swivels in the brackets, thus allowing it to be turned out of danger when not in use. Marked one side inches in 8ths and metre in half centimetres. 8/6 each.

No. 825. This is exactly as No. 826, but marked inches in 8ths only. 8/- each.

No. 827. BOXWOOD HEIGHT MEASURING STANDARD.

44 inches long × 1 inch wide × 5/16 inch thick, with Mahogany Sliding Arm and Brackets as described under No. 826. Marked one side inches in 8ths and metre in half centimetres. 5/6 each.

No. 831. This is exactly as No. 827, but marked inches in 8ths only. 5/- each.

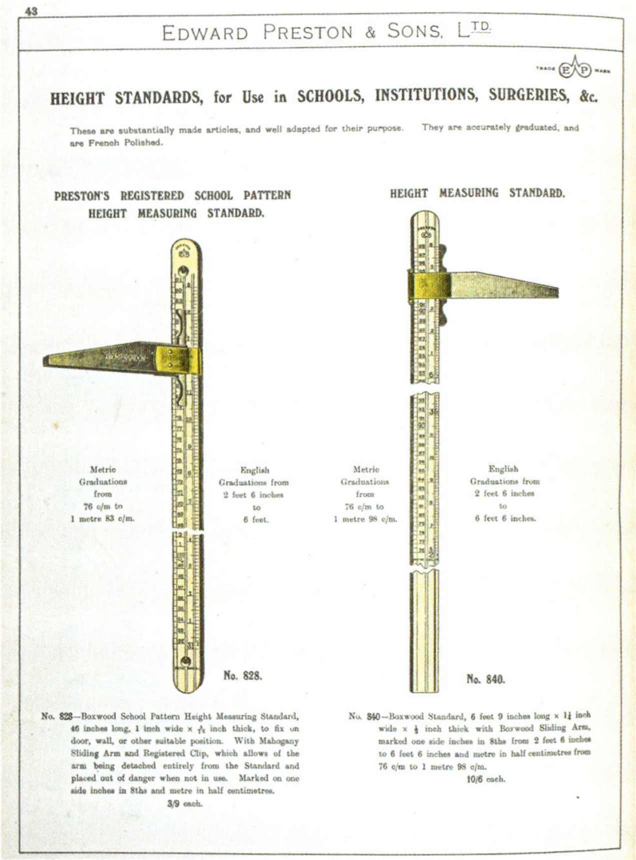

43

EDWARD PRESTON & SONS, LTD.

TRADE EP MARK

HEIGHT STANDARDS, for Use in SCHOOLS, INSTITUTIONS, SURGERIES, &c.

These are substantially made articles, and well adapted for their purpose. They are accurately graduated, and are French Polished.

PRESTON'S REGISTERED SCHOOL PATTERN HEIGHT MEASURING STANDARD.

HEIGHT MEASURING STANDARD.

Metric Graduations from 76 c/m to 1 metre 83 c/m.

English Graduations from 2 feet 6 inches to 6 feet.

Metric Graduations from 76 c/m to 1 metre 98 c/m.

English Graduations from 2 feet 6 inches to 6 feet 6 inches.

No. 828.

No. 840.

No. 828—Boxwood School Pattern Height Measuring Standard, 46 inches long, 1 inch wide × 5/16 inch thick, to fix on door, wall, or other suitable position. With Mahogany Sliding Arm and Registered Clip, which allows of the arm being detached entirely from the Standard and placed out of danger when not in use. Marked on one side inches in 8ths and metre in half centimetres.
3/9 each.

No. 840—Boxwood Standard, 6 feet 9 inches long × 1¼ inch wide × ½ inch thick with Boxwood Sliding Arm, marked one side inches in 8ths from 2 feet 6 inches to 6 feet 6 inches and metre in half centimetres from 76 c/m to 1 metre 98 c/m.
10/6 each.

Above left: Fig. 14.44. Edward Preston & Sons (Birmingham, 1864-1932) wall mounted height standards from their 1914 catalogue.

Above right: Fig. 14.45. Preston's Registered School Pattern Height Measuring Standard, also from their 1914 catalogue.

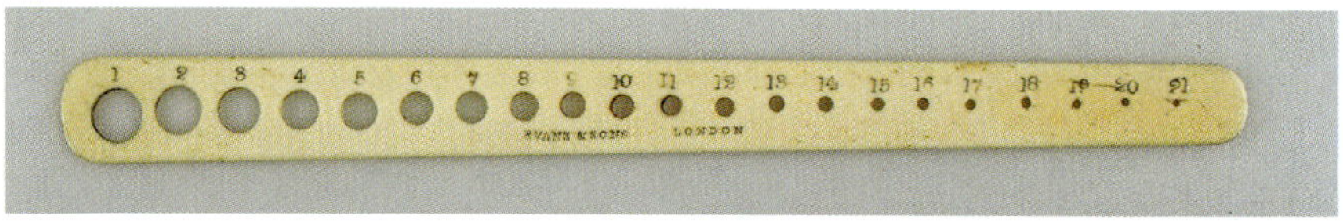

Fig. 14.46. A catheter gauge in ivory. It is marked Evans & Sons, London, who were suppliers of medical equipment.

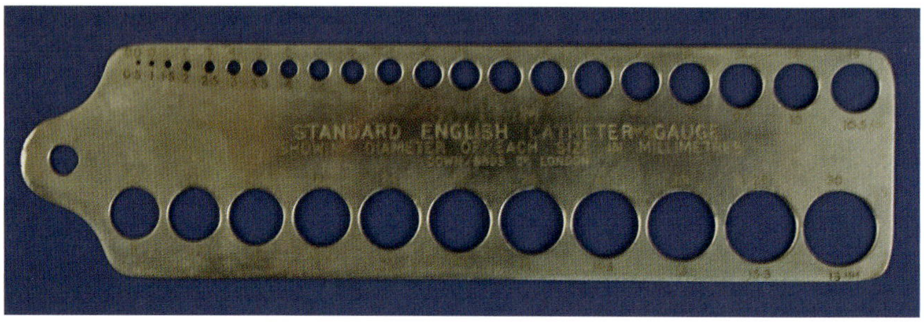

Fig. 14.47. A stainless steel catheter gauge. It is marked Standard English Catheter gauge and bears the name of Downs Bros. Ltd., London. Down Bros. were suppliers of medical equipment.

Bottom: Fig. 14.48. An anonymous 2 ft. 2-fold boxwood rule marked CONTRACTION FOR CLAY 1" TO FOOT. This may have been used fir brick making, tile making or flower pot making.

And everything else …

Medical measures

"Height standards for use in schools, institutions, surgeries, etc." first appeared in the 1912 supplement of the Preston catalogue.[14]. They also figured in Rabone's Catalogue No. 24.[15]

Catheter gauges can also be found. These are sometimes named but not always and can be confused with other hole gauges.

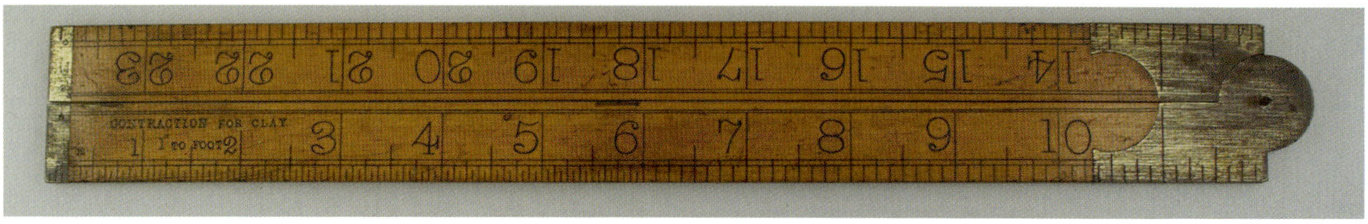

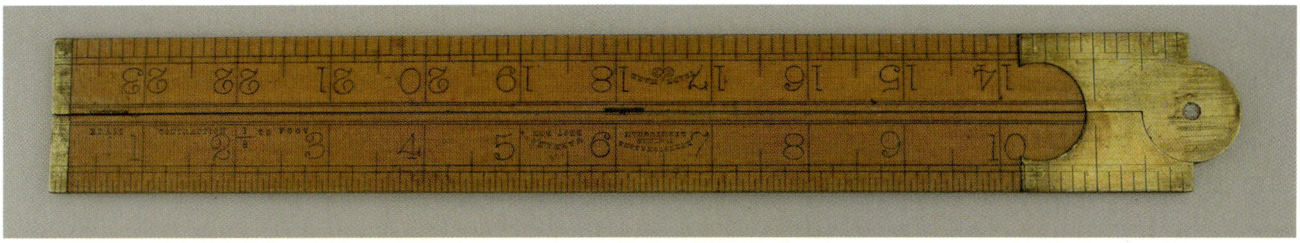

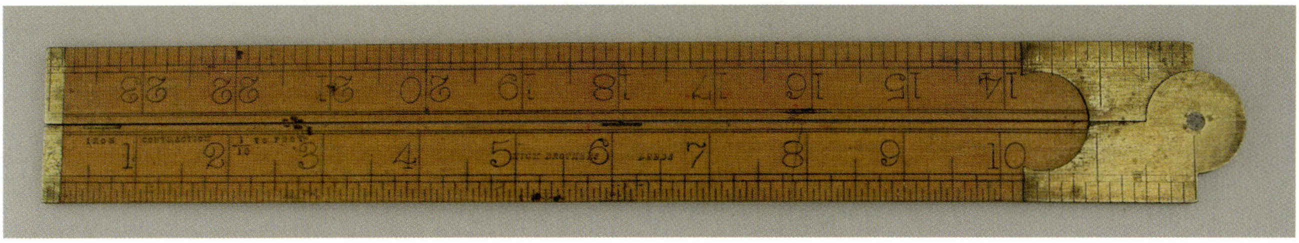

Top: Fig. 14.49a & b. A 2 ft. 2-fold boxwood contraction rule by Edward Preston & Sons (Birmingham, 1864–1932), marked BRASS CONTRACTION 1/8 TO FOOT on one side and IRON CONTRACTION 1/10 TO FOOT on the other.

Fig. 14.50. Gunstockers' bend board by Edward Preston & Sons (Birmingham, 1864–1932).

Fig. 14.51. Detail of the bend board shown above.

Contraction rules

Timber contraction rules have been discussed in Chapter 5. There are others that will be found, the most common being 3 ft. lath type steel rules intended for pattern makers to compensate for the percentage shrinkage when casting. Some boxwood folding contraction rules for a variety of substances also occasionally appear.

Gunstockers' bend boards

Although clearly marked with the name Edward Preston & Sons, this tool does not appear in any known catalogue.

It is used to measure a customer when the stock of a custom-built shotgun is ordered. It measures both the offset of the stock and the drop of the stock to enable the owner to sight down the barrel accurately.

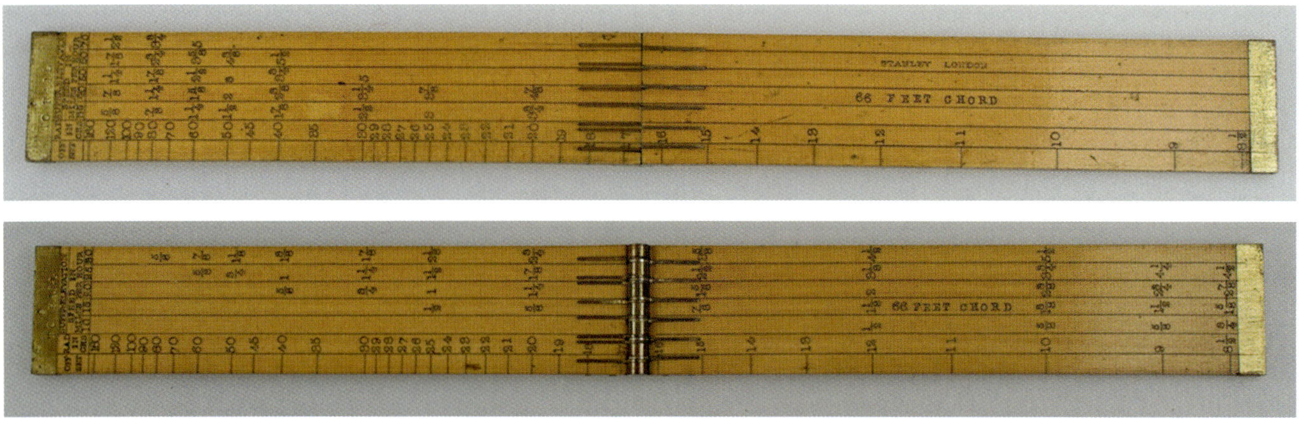

Fig. 14.52a & b. A rule by W. F. Stanley (London, 1880–1967) used for calculating the elevation when setting out railway tracks.

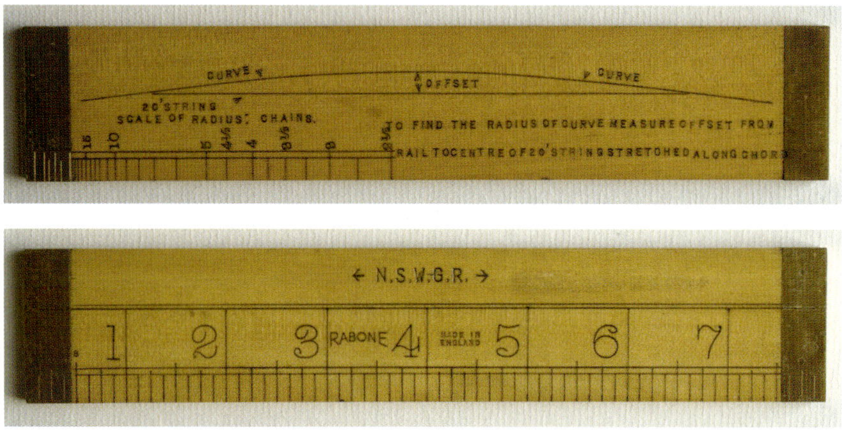

Fig. 14.53a & b. A rule by John Rabone & Sons (Birmingham, 1877-1963) to determine the radius of the curve in a railway line. This may have been to check that the curve was not excessive or for setting speed limits. It is also marked N.S.W.G.R. which was the New South Wales Government Railway in Australia which operated from 1855 to 1972.

Other measuring tools

A small selection of unusual measuring tools that are occasionally found are shown in Figs. 14.52 to 14.56.

Fig. 14.54. A hole gauge for measuring glass tubing or rods.

Fig. 14.55. A caliper rule intended for use in walking stick manufacture.

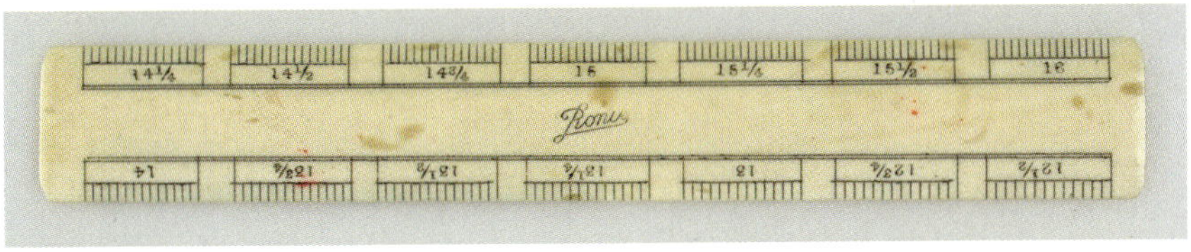

Fig. 14.56. A rule for measuring the perforations on postage stamps, intended for the stamp collector.

References

1. Chambers, E., *Cyclopaedia or an Universal Dictionary of Arts & Sciences,* 5th edition, London, 1743.
2. Thurston, A. P. ed., *Molesworth's Handbook of Engineering Formula and Data*, London, 1851.
3. Tomlinson, Charles, *Cyclopaedia of Useful Arts and Manufactures*, London, 1852.
4. Moxon, Joseph, *Mechanick Exercises on the whole Art of Printing* (1683), reprinted New York, 1962.
5. John Wyke, *Catalogue of Tools for Watch & Clock Makers,* Liverpool, ca. 1770, reprinted Winterthur Museum, 1978.
6. Dane, E. Surrey, *Peter Stubs and the Lancashire Hand Tool Industry*, Altrincham, 1973.
7. Crom, T. R., *Horological Shop Tools 1700 to 1900* and *Horological and Other Shop Tools 1700 to 1900,* Melrose, FL, 1980 and 1987.
8. Henry Isaacs, *Catalogue & Price List*, 5th ed., Leeds, n.d. ca. 1900.
9. Hirst Bros & Co. Ltd., *The Fourth "Wide Awake" Catalogue of Tools and Materials for Watchmakers, Clockmakers, Jewellers and Opticians*, Oldham, n.d. but probably ca. 1914.
10. Bion, Nicholas, *The Construction and Principal Uses of Mathematical Instruments.* Paris 1723. Translated by Edmund Stone;... *To Which are Added, the Construction and Uses of Such Instruments as are Omitted by M. Bion* by Edmund Stone. Second edition; ... *To which is added a Supplement Containing a Further Account of some of the Most Useful Mathematical Instruments as Now Improved.* Supplement by Edmund Stone, London, 1758, reprinted Mendham, NJ, 1995.
11. Chambers, E., *op. cit.*
12. John Rabone & Sons, *Catalogue No. 24*, Birmingham, n.d. but probably 1924.
13. Dring & Fage Ltd., *Catalogue of Hydrometers, Saccharometers, Thermometers, Gauging instrument...*, London, n.d. but probably 1960s.
14. Edward Preston & Sons, *Catalogue No. 18. The "Preston" Catalogue, 1909 with 1912 supplement,* Birmingham, 1912.
15. Rabone, 1924, *op. cit.*

Appendix 1

Directions for making a carpenter's rule. George Atwel, 1662

The following is a transcript of part of the appendix to the 2nd edition of George Atwel's book The faithfull surveyour: discovering divers errours in land-measuring; and shewing how to measure all manner of ground ... Teaching likewise the making and use of a new and general instrument, called a pandoron ... With an appendix unfolding errours in board and timber-measure, with directions for making a carpenters-ruler, *published in Cambridge in 1662. The appendix contains the instructions for making a carpenters' rule. The mathematical examples and exercises that accompanied these instructions have been omitted from this transcript.*

Chap. I: Of making the Rule

First, I would have the Rule,(whether it be of box, or of brass, whether joined in the middle, or streight out) to be just two-foot-long by some standard of brass, kept by the *Clerk of the Market* and to, as I have seen some, that have been half an inch too long. Let it be an inch and an half broad at the least, and a third part of an inch thick with a square stroke struck round about it just in the middle of the length thereof. Let one edge be befild off: which serves that if you have occasion to draw lines with a pen if you turn that side downward, you need no fear of blotting; if your rule chance to be black with inke, if you rub it well with sorrel, that will fetch it out. Through the midst of the befill strike a Gage-stroke; another along the midst of the other edge; divide the rest of this side, beside the befill, into eight equall parts with seven Gage-strokes. In the next 4 columnes save one to the befill, you may place all the under-measure of this Table of board-measure following, which will not fall in a scale upon the rule, *viz.* all inches, halves, and quarters from one inch to six, or, if you will to ten inches, in small spaces the inches of the breadth of the board, in the column next save one to the befill: the feet required to a foot foreward at the breadth in the next; the odd inches in the third; and the *Centesmes* in the fourth. And adjoyning to this Table toward the middle of the Rule, in the first of those four columns set one inch divided into ten equall parts, and each of those into halves, and each of those halves into five, or suppose them so divided; so is divided into 100 parts or *Centesmes* from which inch you shall take all you *Centesmes* with your compasses, that are to be set in any of your scales.

For making the Scale of board-measure

Before you can make this scale, you must have one column, on the other side of the Rule next to the befill, parted into three small parts with Gage-strokes, and divided in the middle of the length of the rule into two equall parts or feet; whereof divide one of them into ten equall parts, and each of them into ten more, and each of them suppose at least to be divided into ten other; so shall that foot be divided into 1000, and this *Gunther* calleth foot measure: which must be reckoned both wayes, first from the beginning of the rule to the middle, thus 1, 2, 3, &c. and backwards again, and thus 11, 12, 13, &c., and because the other foot makes ten of these inches, and these ten make twelve of them, therefore divide the other foot into twelve equall parts or inches, and each inch into eight parts, and number it from end towards the middle with 1, 2, 3, 4, &c. but from the middle to the end with 13, 14, 15, &c. and this he calleth inch-measure. By help of this inch-line and the inch aforesaid, and by help of your Tables for board and timber-measure, are made your scales for board and timber-measure. And this Table of board-measure is this made: First, for all whole inches divide 144 by the inches of the breadth, and you have the inches forward to a foot. If any remain after division, it is the Numerator of a common Fraction, whose Denominator is the Divisor; to which remain annex two ciphers on the right hand, and divide again by the same Divisor, and you have the *Centesme* desired. *Example*

Let a board be seven inches broad. I desire to know how many inches forward makes a foot. Divide 144 by seven, it gives twenty inches, or one foot eight inches 4/7. Now bring 4/7 into *centesme*, annex two ciphers to the remain four, it makes 400 which divide again by seven, it gives 57/100. But for half-inches reduce the breadth into an improper Fraction, as 6½ is 11/2; then multiply 144 by the Denominator 2, it gives 288: so you must always divide 288 by the Numerator, or number of half-inches of the breadth of the board, which is 13; so have you 22, or one foot, 10 inches, 15 *centesme*. But if your breadth be an odd quarter or three quarters: First, reduce it into quarters, and divide 576 by it: so 6¾ is 27 quarters, therefore divide 576 by 27, it gives 21 inches; or one foot nine inches, 9/27, or 33 *centesme*.

Now to place this Table upon the rule, divide the second, third, fourth, and fifth columns next to the befill, at one end into small squares that may hold two figures a piece, in which set over-most the inches of the breadth, in the second the feet required in length, at each inch, half inch, and quartern. In the next the odd inches, and in the next the odd centesme: and this you must do to fix inches, you may do it to ten inches if you will. Then at the end of ten inches, set one inch divided into ten equal parts, and each of them into halves, and suppose each halve into five, so will it be supposed to be divided into an hundred parts, as before. Then from six inches to 36 you shall set all in the column next to the befill, with small strokes in this manner: First I begin with six inches and a quarter, to which I finde in the Table there belongeth one foot, eleven inches, four centesme, that is eleven inches, four centesme from the middle cross stroke of the rule. But because my compasses will not reach so far, I onely take 56 centesme from the former inch, which makes it just two foot from the same end, which I set under measure at.

Another example let be 9¼, for which I find in the Table one foot, three inches, 56 centesme. First I take with my compasses 56 centesme from my inch of centesme, and prick it down upon a line upon a paper. Also with my compasses I take three inches in the foot-line of inch-measure on the other side of the Rule: set that distance also on the paper at the end of the 56 *centesme* in the same line; then take with your compasses the whole length of both, set one foot in the middle-cross-line of the Rule, and in the same scale, and the other toward the beginning of the Rule, and it gives the length correspondent to nine inches and ¼, from the stroke to the end of the Rule. Thus do with all the rest; marking each whole inch with its proper number to 24, also 30 and 36.

And now, before we proceed to shew you the making of the Table of timber-measure, we will first shew the measure of boards.

Chap. II: Of measuring of boards with the Rule

[There are two paragraphs describing to two arithmetical calculations for measuring board feet.]

3. A third way of measuring board is by this rule. Measure the breadth of the board, if it be less than six inches, your Table of under-measure will shew you how much forward you must take to a foot forward. If it be broader, and under 36 inches, then the strokes on your scale give it.

4. Some measure all the breadths of the boards with a line, then stretch the length on a block, and so measure the breadths of all the stock at once, then measure the length of a board, then multiply the length in feet and parts, by the breadth in feet and parts: so suppose the breadth of all the boards is ten foot, nine inches, and the length 154 inches, instead of nine inches I take ½ ¼ of a foot and instead of four inches I take 1/3 or ¼ one inch, and the work will be thus, and it makes 164 feet ¾ , 1 inch and an half.

And this is a very good way in case of a block be hewn eight-square, before it be sawn: which if it be fir for boards, it is pitty it should be hewn any other way; so will it be no loss of timber , the boards will be all straight-edged. If it be sold in timber and measured as eight-square (as shall be shewn) there will be no loss either to buyer or seller.

Chap. III: Of making of a Table of timber-measure for square timber, to make the scale of square timber-measure by; as also the under-measure.

First know that a foot of timber is twelve inches every way breadth, length and thickness, and therefore containeth 1728 square inches, for 12 times 12 is 144, that is a foot of board or a *supersicies*, and twelve foot of board make 1728 inches; therefore proceed to the table. First, For the whole inches: square the square of the piece, that is multiply the square by it self and by that product divide 1728. *[There follows an example.]*

To place this Table on the Rule

Begin at the other end of the Rule, taking those 4 columns next to the thick edge save one, and divide them into little spaces, as you did for the board-measure, setting on them all under measure to 8 inches and an half square, yet you may do it to 12 inches, if you will; setting the square inches of the block in the column next save one to the edge: then the feet required to make a foot forward in the next: then the odd inches in the next to that, and the *Centesme* in the last of the 4. Then from 8 and ½ to 36 you may take off your inches from your line of inch measure, and your *Centesme* from your inch of *Centesme*, as you did in board-measure, and place it backward or forward, according as it shall be more or less than a foot.

Chap. IIII: Of measuring solids, as stone, timber, &c. and first of square timber

For measuring all kinds of solids the fundamental or general way is to multiply the inches breadth by the inches depth, and that product by the inches of the length and divide the last product by 1728. This is so plain it needs no example; this is the best way for stone of all others.

2. A second way of measuring square timber is by this Ruler. Having the square of the piece given look on the Rule, and see how often you finde the length required at that square between that and the end of the Rule in the length of the block, so many foot of timber is in that block.

[There follow three paragraphs on doing this calculation mathematically by extracting the square root.]

Chap. V: Of round timber

Because to every circle there belogeth 3 squares, first the square without the circle, or the square of the diameter; secondly, the square equal to the circle, not in Peripherie, but in the *area*; for if the *area* of the circle of a mile round, and a mile about in a squre be compared, we shall finde the square to contain just 40 acres, whereas the circle of the same Peripherie containeth 50 acres; 3 roods, 25 poles 5/11; and thirdly the side of the square within the circle; therefore we will first shew the manner of making these 4 scales, then the measure of round timber; yet before we shew the making of them our best way is to take *Virgil's* advice, and do as he doth with his Bees. *Principio sedes apidus statioque petenda.*

So before we shew the making of them we will first finde a seat for each of them, then the making of them one after another. First, in the beginning of the first chapter, we shewed that we would have one of the edges on one side befild off: the rest of that side divided length wise into eight equal columns with 7 Gage-strokes upon the befill, ½ the length of the Rule, you may set a scale of 20 in the inch dividing each into halves and quarters. Numbering each half-inch with 10, 20, 30, etc. save that half-inch next the beginning, which must not be accounted for in any of the tens, but must be divided into 10 equall parts by it self, to take the odd inches above even ones, that any round block or circle is about.

Besides this, you have three other scales that are for round measure, that shew the three squares belonging to the circle: and any of these four being known, all the rest are known onely by taking the number thereof upon its proper scale with your compasses, and apply that distance to the scale proper to the thing desired: and these three scales for these squares one for the Diameter, or side of a square without the circle, and that each side thereof toucheth the circle. Another is the side of a square within the circle, or of the chords of 90 degr. And the other is the side of a square, whose content is equal to the content of a circle. For *Example.* Let a block be girded about with a nealed wyer, and then that wyer laid along the block, being found to be 88 inches. I set one foot of the compasses in 80 of the said circle scale, and the other foot in 8 of thosel 10 odd parts next the beginning of the Rule, reckoned from ten upward, being the contrary way to the other 80. If you then desire to know the Diameter of the circle, or side of the square including the circle, you shall finde it just 28 inches, by setting one foot of the compasses in 25 of the Diameter scale, and the other will fall in three odd part, which added make 28: for all these three last scales must be divided into fives, and number 5, 10, 15, etc. and five odd ones, at the beginning. Likewise if you apply the same wideness of the compasses to the scale of the square within the circle, that is, to the square, that a block being round, will be, being hewed just to the four edges: then set one foot of the compasses in one of those great divisions by fives, so that the other may fall amongst the odd small divisions, and it gives you 19 ¾ *feré.*

And lastly, if you apply the same wideness of the compasses to the scale for the square equal, setting one foot in the great divisions, so that the other may fall in the five odd small ones, it gives 24 and about 2/3.

And in like manner if any of the other three scales be given, as if the Diameter 14 be given; if you take 14 upon the Diameter, and carry that to the circle, it gives 44; if to the square equal, it gives about 12 1/3, and so to the rest.

Chap. VI: Of the proof if these scales by Arithmetical calculation
[This chapter, as it says, goes through the arithmetic.]

Chap. VII: Showing the manner of placing these upon the rule
First, To set out the Diameter, you may take the nether part of the third column of the befil'd side, to set it on the middle square stroke of the Rule. Then *Gunther* (in his *Use of the line of numbers in broad-measure, Prop. II*) hath this proportion. Having the Circumference of a circle, to finde the Diameter: As 3143 is to 1000, so is the Circumference, suppose it 47 /13 to the Diameter 15: so that you take 47 /13 in your circle-scale, and set in that column from the middle square downwards, so shall you set out 15 in that distance, run that distance as oft as you can to the bottom of the Rule, which will be 4 times more, divide each of them into 3 equal parts, and the uppermost third into 5 equal, and number all the other great parts, save that with 5, 10, 15 etc., or if you will you may double 47/13, that is 94, 26, and take it from the circle-scale, set it there they will be 30; then half it and they will be 15, then third it into fives.

2. To finde how to proportion the square within the circle by the Diameter. Let the Diameter be the *Radius* 1000, then will the chord of 90 degrees, which is the side of the square included, be the natural sine of half 90: *viz.* 45 degrees, the sine whereof is 707, therefore then because I would divided my scale into even sines, if therefore I take 7 times 5, that is 35, the proportion will be 707, 1000: : 35. 49/50 or 49 ½ : therefore if you take 49 ½ on the Diameter, and set it on the scale of chords, and divide it into 7 equal parts, and that next the end into 5 small parts, numbring all but that with 5, 10, 15 etc. you have your scale of chords, or square within a circle. Or (if you think it troublesome to divide it into 7 equal parts) you may take 6 times 5, that is 30. and say 707. 1000: : 38, 42/43, so you may take 42/43 of the Diameter, and set it on your scale of chords and then divide each of them into halves and each half in to 3 parts.

Otherwise thus, The content of this circle according to *Archimedes* is just ½ the content of the square of the Diameter. Suppose the Diameter 24, the square thereof is 576, the half thereof is 208, the root thereof is 17 *fere*? *Answer* 56½: therefore take 56 ½ of the Diameter and set it on the scale of chords, which gives 8 times 5, first divide it into halves, then into quarters, then into eight.

3. It also may be made by this Rule of his. The *area* of the square within the circle is to the content of the circle as 11 is to 7, so that the circle begin [sic] known, the content is thus found: ½ the Diameter multiplied in ½ the Circumference gives the content of the circle, which if you multiply by 7, and divide the product by 33, it gives the content of the square within: whereof take the square root and you have the side desired; therefore 19/8.88 : : 20.88 / 9 or as M^{r}. *Wingate* hath it (in *Problem 33,* of his *Appendix* to his *Rule of Proportion*) 225 . 1000 : : 20.88 / 9. So that take 88 / 9 from the Circumference and set it on this scale, and divide it into four fives and this scale may be set on the lower half of the befil'd edge.

4. Having the content of the Circumference, to find the side of square equal. Take the square root thereof: so we found before that the Circumference being 88, the content is 616, whos square root is 24 / 40/49, that is more the 24 ¾ . or more easily, because, *Gunther* hath it, the Circumference is to the of a square equal as 1000 the *Radius* to 282, therefore say, 282. 1000: : 20.70 / 9. Therefore take 70 / 9 of the Circumference, and set it in the scale

of the square equal, it gives 20 of that scale; with which distance set out all the twenties the side will bear, dividing each 20 into four fives, and the last into five little ones, and numbring them by five as afore: and this scale may be set in the over part of the third column next the square edge.

[Chap. 7. then continues with five pages on the errors found in measuring round timber by the quarter circumference (i.e. quarter girth) method. There is also a reference to Mr. Stirrup's Plain-scale or Carpenters new Rule. *This is Thomas Stirrup's* The Artificers plain Scale, or the Carpenter's new Rule, etc., *London, 1651, 2nd edition ,1657.]*

Chap. VIII: Of taper timber, whether Conical or Pyramidal
[This is a chapter of mathematical calculations only.]

Chap. IX: Of the making of four other lines on the flat-sides, whereof three are Mr. Gunthers lines, of numbers, sines and tangents; and instead of the Meridian- line, which is onely useful for Navigation, whereas Carpenters make little or no use, we have added a sextant of chords.
Although *Mr. Wingate* (in his book *The Rule of Proportion,*) hath set down the making of them, yet for that he hath done them after another manner then here is shown, neither will an ordinary Rule bear all those lines, we will therefore content our selves with *Mr Gunther's* & this line of chords only. You shall divide the rest of the Rule beside the columns of feet and inch-measure before spoken of, into four other great columns, and divide each of them into equal, and one of them into two also, so the great shall be for figures, the other two for strokes. These two of *Mr Gunther's* you may set in the three middle columns, and the line of chords on the other outside.

First, for making the line of numbers.
I told you before that I would have you strike a stroke round about cross the Rule, I would also have another at each end of the Rule so close as possibly you can, onely to set one point of the compasses on. Then first set out your great divisions in each foot; *viz.* the thousands, if your numbers consist of four figures, or howsoever they are to be the left hand figures of any number, as in 3 in 3. 32. 346. 37046, etc. and must be marked with the 9 digits in either foot, and the first last and middle-most with one so that you may understand as many ciphers with it as shall be requisite, so that it may figure 1. 10. 100. 1000. and then if one signifie 10, the next two will naturally signifie 20, but not always. Now take and set the number 2 in his right place, take a Table of Logarithmes of absolute numbers, and look either the Logarithme of 2. 20. or 200. and take the three next figures to the Characteristick, which are 301 then with your compasses take 301. *viz.* three inches, no tenth part of an inch and 1/10 of a tenth part or *Centesme* of an inch, and set one foot in the nether-most cross stroke, where you set the first one, and turn the other upward in the same column, and there set your 2 likewise with the same numbers, set one foot in the middle cross stroke where you set the middle one , and turn the other upward toward the uppermost one, and there set your 2 also: likewise, do with 3 whose Logarithme is 477 (*id eft*) 4 inches, 7 tenths, 7*Centesme*: also with 4. And these figures for the making of this line we will call hundreds, the next subdivision tens, and the least *Centesme*. But now because we will suppose you compasses will not well reach beyond the figure 4, whose Logarithme is 602, that is above 6 of those inches: therefore first set let us set on the tens so far on both feet, and then the rest of each foot afterwards. Next set out each fifth tenth so far: because you must mark them

with longer strokes, then each single ten: so you must not account the next of those fifths to 1 as 5. (for you will account one for nothing) but must account it for 15. or 150. and so take the Logarithme thereof, which is 176. Likewise 25, or 250, is 398, which you must take with your compasses, and set in their places in both feet, and in like sort shall you do with all your single tens, accounting the next 1 not for 1, nor 2, but for 11. Or instead of taking them off with your compasses, strike out all the first foot with a fine small striking squire of brass, laying it upon the Log. in the line of foot-measure, then set out the other foot with your compasses by this.

Now for the rest of each foot , look out the Logar. of your numbers, and take the distance between it and the middle cross-stroke, and with that wideness set one foot in the upper 1, and where the other falls there is the place of that number. *Example.* I would set out 70, the Log. is 845; I take the distance between it and the middle-stroke of the Rule, or the Arithmetical complement of it, 154, and set it both from the upper stroke and middle-stroke downward, and you set out seventy, But your over-foot may bear unites to 29, and from thence to 40, divide each tenth into five, and from thence to the end into two.

To make the line of sines

First you must know that neither the line of sines, nor tangents, enter the Rule until 35 minutes: where you see the two next figures to the characteristick 8, are both ciphers; there also the characteristick changeth from 7 to 8; for your characteristick shews what foot you are in; therefore since we reckon the minutes onely by tens, our first number or division upon the Rule will be at 40 minutes of the first foot, shewn by the characteristick 8: for 9 is the last, and therefore belongs to the last foot; so that whereas you see that the Log. of one minute hath 6 the characteristick , & 463 the three next figures: therefore one minute would be above a foot and a half before the entrance on the Rule, and likewise would the first minute of the tangent be. Now the Logar. of 40 minutes has beside the characteristick 8, the three first figures 066 *fere* : therefore take off 0 inch, 6 tenths, and 6 *Centesme*, or 5 *Centesme,* and 7 *millesmes*, if you can guess so near, and set them from the nethermost cross-stroke at the beginning of the line of sines forward. And thus do for all under two degrees, be it sine or tangent: but from thence to sine 5 degr. 45 min. or tangent 5 degr. 43 min. (As suppose the sine of 4 degr. whose Logar. beside characteristick is 843) you shall take the distance between 8 inches, 4 tenths, 3 *cent.* and ten inches, and apply that distance from the middle-stroke downwards: and so of the rest of the quarter. But for all both sines and tangents in this first foot: you may by their Logarithmes strike with a square, as you did the line of numbers.

Now for the upper part shewed by the characteristick for all sines and tangents to 20 degr. as suppose the tangent of 20 degr. the Logarithmes of 20 degr. tangent is 56: set it from the middle-stroke forward, but from thence to the sine if 90, and the tangent of 45 degr. as the sine of 40, whose Logar. is 808; take the distance between it and the middle-cross line, and apply it in the line of sines from the upper cross-stroke downward: then number all the whole degrees to ten, with 1, 2, 3 and after that in the sines with 20, 30, 40 etc. to 90, and the tangents with 10, 20, to 45, and back with 50, 60, to 80 degrees.

Lastly, for making the sextant of chords

Set a pair of beam compasses, with a beam of willow, deal, or sallow, near half an inch thick and ¼ broad; make a little nut of good tough wood with a mortes in it, that the beam may slide in it to and fro, indifferently stiff, and in all places alike, with a short prick, or a little piece of an aule-blade in one end, so long from the beam as the other point is. If it goeth not stiff enough to stand and tran with at any place, make the mortes a little deeper one way to put in a wedge, or else help yourself with a screw-pin, then go to some smooth loft boards, opening your compasses to 23 ½ inches, and with that wideness tran an arch, that may be two foot long at the least, and with each foot of the compasses make a prick in the said arch, and set it likewise upon the Rule, then divide space in the arch into two equal parts, which will be 30 degr. a piece, and each of them into three pieces, which will be 10 degr. apiece, and each of them into two, which will be five apiece, and each of them into five simple ones. Then take them off from the floor, and set them on the Rule, one after another, and number them 10, 20, 30, 40 50, 60, and this will be wonderful beneficial in Dialling, and also in many other things, as to divide a circle into any number of equal parts, or to make an angle of any number of degrees, or to finde the quantity of any angle, and so by the line of foot-measure you may also divide a streight line into as many parts as you will.

Now I have shewed the use of all the line on the other side of the Rule, and also of both the out-side lines on this side, so for the other three, I must content myself to shew you the use in general: for if I should descend to particulars, all the paper in *Cambridge* would be too little to hold them. First therefore, you see already, that as by the line of foot-measure, and Table of Logarithms these lines are made, so may you by these lines find the Logarithme of any absolute number, tangent or sine, as if it were by the Tables of Logarithms.

Secondly, By these two lines of numbers and foot-measure may be resolved all questions whatsoever, that common Arithmetick can resolve. And more; for hereby may be resolved all questions of interest, Purchases, Annuities, etc.

Thirdly, by these three lines of numbers, sines, and tangents is resolved the whole doctrine of Triangles and whatsoever may be performed by them, either in Meauring, Dialling, Geography, Geometry, Arithmetick, Navigation, Cosmography, Astronomy, etc.

But, because (gentle Reader) I would have thee learn now to go alone; I will commit thee to thine own consideration, knowing that the chicken that will peck up never a corn, but what the hen puts in its mouth, will never be a fat chicken. Now if the Rule of three is accounted of all men worthy of its excellency of the name of the *Golden-Rule* (which is the least part if the use if one of the lines of this Ruler) then justly may this Ruler be called the *Golden Ruler.*

Appendix 2
Measuring Rules by John Rabone, Junior, 1866

From Timmins, Samuel, ed., "Measuring Rules" by John Rabone, *Birmingham and the Midland Hardware District: The Resources, Products and Industrial History*, London, 1866.

Of the early history of this manufacture it may be sufficient to state that until the early part of the seventeenth century, at which time Edward Gunter invented the line of logarithms graduated upon a sliding scale, which solves problems instrumentally in the same manner as logarithms do arithmetically, the trade never assumed sufficient importance to cause it to be followed by persons who had no other occupation, and to make it worthy of being designated a craft. Up to that time the best measures had been made by the mathematical instrument makers; but this ingenious invention of Gunter, by reason of its universal applicability to measuring purposes, called into existence another class of workmen, superior to those who had hitherto chiefly made the notched sticks similar to those used in many rural districts at the present day, but still somewhat distinct from the opticians and makers of such instruments as quadrants, sextants, and the finer kind of optical and mathematical instruments. The first men who were worthy of the name of rule-makers were to be found only in London; but after a time the trade gradually extended itself to Wolverhampton and Birmingham. Under the date of the 10th of August 1664, Mr. Pepys records in his diary his experience of the scarcity of competent rule-finishers (as the men who make the rules arecalled) in London. He writes:

> "Abroad to find out one to engrave upon my new sliding rule with silver plates, it being so small that Browne, that made it, cannot get one to do it. So I got Cocker, (the celebrated arithmetician and famous writing master), to do it, and I sat an hour by him to see him design it all; and strange it is to see him, with his natural eyes, to cut so small, at his first designing it and read it all over, without any missing, when for my life I could not, with my best skill, read one word or letter."

And then on the next day Pepys writes:

> "Come Cocker, with my rule, which he hath engraved to admiration, for goodness and smallness of work; it cost me 14s. the doing."

From the fact that Browne, the mathematical instrument maker, was unable to find anyone to engrave Mr. Pepys's sliding rule, it is to be supposed that the art had not then attained to much perfection, or rather the proficiency in it had been made but by a few. In 1775, nearly a century later, it is recorded that even in London "it was a most difficult matter to get good rules, there being only one man who could make them perfectly well, and he had lately taken to other work." In July, 1755, James Watt found his first employment in London as a rule-maker, under Mr. John Morgan, of Finch Lane, Cornhill, of whom Watt writes to his father, that "though he works chiefly in the brass way, yet he can teach me most branches of the business, such as rules, scales, quadrants. &c." In less than two months after his arrival in London Watt says he had made a brass parallel rule eighteen inches long, "and a brass scale of the same length." By December he "could work tolerably well," and "he expected that by April he would understand so much of his business as to be able to work for himself." When June had again come round, and within twelve months of his first attempt at rule-making, Watt wrote to his father, that, "he could now make a brass sector with a French joint, which is reckoned as nice a piece of framing-work as is in the trade." Watt, however, having doubtless acquired proficiency in the art of rule-making did not "work for himself," as it appears he intended to do, but turned his attention to the improvement of the steam engine, and thus conferred greater benefits upon the world than he would, had he continued to manufacture "parallel rules," or even "sectors with French joints." It may be interesting to know that the sliding rule, long used by James Watt, is still preserved in Birmingham in the possession of an eminent founder; and is still used as a check in testing calculations. Is it not probable that this rule may be the handiwork of its former great possessor? Five and twenty years later the finer kind of measures were still made in London by the instrument makers; but the more ordinary kinds of carpenter's rules were manufactured almost exclusively in Wolverhampton. At a latter part of the past century only three or four rule masters, each employing a few apprentices and men, were to be found in Birmingham, and one at Harborne adjacent; but now the trade has almost deserted Wolverhampton, which numbers only four or five persons employed in it, while Birmingham affords employment to as many hundreds. With the exception of three or four makers scattered throughout the country the trade is now entirely confined to Birmingham and London. Many of the rules sold as London-made are produced in Birmingham, and many are framed in Birmingham and sold to London makers, who mark and finish them themselves. The joints of the earliest made rules were cut from the wooden pieces which formed the legs, and the brass plates were attached to the outside to strengthen the otherwise weak and wooden joints. Afterwards the joints were composed of metal, and attached to the wooden sides, instead of being made out of them. The joint, so well known as the arch-joint, and which is seen upon so great a number of joiners' and other rules, was first made about sixty years ago; yet in a celebrated modern painting representing the early printers at their labours in Westminster Abbey, we find an accurate representation of a two-foot rule of this kind with an arched joint, the artist being unaware that it was not invented till more than 300 years after those early printers had distributed their last form.

The general shape of rules has been but little varied, the convenient arrangement of two or four legs having been early introduced. Seventy years ago, the ordinary thickness of the boxwood carpenter's rule was not less than a quarter of an inch, but that size has gradually reduced from time to time in order to give greater flexibility. Many rules have been made of only the sixteenth of an inch in thickness, this small size of wood having been cut away further to admit the insertion of the joint, which is inlaid so as to lie level with the surface of the wood. To remove this weakness of the joint, which was the common fault of all thin rules, a joint was produced by Messrs. John Rabone & Son, which being rivetted upon the surface of the boxwood legs instead of any portion of them being cut away to make place for it, adds strength to that part of the rule which had formerly been the weakest. These joints are also enriched by many artistic and ornamental designs stamped upon them, which renders the rules, whether regarded for the greatest strength combined with extreme thinness, or for their elegant appearance, appropriately designated by the appellation of "nonpareil." The processes of sawing, rivetting, filing, and planing, are too well known to need reference here; but the numbering and graduation of rules is an operation which, to those unacquainted with it, is generally regarded as the most interesting of the processes. Since the invention of the dividing engine by Ramsden, and its improvement by Troughton, Aide, Reichenbach, Gambey, and others, all the most accurately and finely graduated astronomical and other instruments have been divided and marked by means of it; but the process is much too slow and costly to be employed upon the rules in general use by mechanics. Straight scales and rules are usually divided by placing the article to be divided and the original pattern side by side, then passing a straight-edge with a shoulder fixed at right angles to serve as a guide, along the original, and pausing at each division; then a corresponding line is made on the copy by the dividing knife. Segments of circles are also graduated in the same way, by making a straight-edge revolve on the centre of the circle, and marking off the divisions as on the straight scale. This method in skilful hands, admits of much accuracy, and was applied to the graduation of theodolites and other instruments prior to the invention of the dividing engine.

About eighty years ago much labour was spent by a Wolverhampton rule-maker in the construction of a pair of steel rolls, having upon their surfaces in relief, the various lines, divisions, and figures necessary to impress, at one operation, the two sides of a scale or rule. Its costliness may be guessed at, when it is stated that the scale it was intended to impress contains 2205 lines, divisions and figures. But the attempt was, as might have been supposed, a failure. Owing to the inequality in the density of the wood, and the slight unevenness of its surface, the work performed by it was defective and its use was abandoned. Marking by pressure produces a thick, shallow, and ill-defined line; whereas percussion, when applied to the steel stamps which bear the figures or letters, gives a sharp, clear, deep, and lasting impression. The filling up of the marks on the rules is effected by simply smearing over them a mixture of oil and charcoal, and wiping the surface clean.

In the year 1858, the late Dr. William Church, of this town, invented and patented a number of machines for the marking and graduating of rules. These inventions displayed a vast amount of thought and ingenuity; but he was unaware of what had been done by a few persons in the trade before, and unfortunately for him all that was good and practicable in his inventions had been in use for many years, and his marking apparatus was too complicated ever to remain long in working order. And when it is known that 540 divisions in the Gunter's line, which may he seen on the brass slide of any carpenter's slide rule, are cut singly and accurately by hand in ten minutes by the process above described, it will not be surprising in this instance manual labour should be cheaper and more to be relied on than machinery, designed as Dr. Church's was, to draw all those divisions at a single operation. In the working of several hundred teeth or cutters all connected together, the blunting or breaking of any one of them would render the work imperfect, and would consume more time in the filing up than the entire process when performed by the usual method.

The box-wood, of which rules are made, is chiefly grown in Turkey, the English grown box-wood not being sufficient size and quality. Formerly, large quantities of rules were exported to the United States of America; but during the past thirty years, they have been manufactured in the States, a prohibitive tariff having prevented the introduction of English goods. The way in which some nations – or rather – some Governments – regard the importation of rules into their territories is rather strange. Some allow, at fair and equitable rates of duty, the importation of English made rules, the measures being suitable to use in their respective countries. But while one country will not admit them at all unless the measures of the country are surrounded by other measures, so as to make the rule a scale of varied measures, instead of being the only legal standard of the country, another Government looks upon all rules, coming to its ports as illegal measures, because they have not the Government stamp upon them, and all so found are seized. The makers do not put a resemblance of the official stamp upon them; and, as there are no native rule makers, and all rules are imported, measures, a little different from the legal standard are made, and being marked with the name of another country than that for which their use is intended, are allowed to be admitted; thus, the greater part of the people are necessarily using an incorrect measure, at the same time that the true standard might be supplied to them as cheaply and more readily. The various countries and states of the world have, at the present time, about 150 different measures; many small places, each containing only a few thousand inhabitants, have their own peculiar measures. At various times many of these measures have been changed from duodecimal to the decimal division, still retaining the original standard of length; but within the past few years, the use of the French metrical measure has become more common in all parts of the world. Many nations now have a large proportion of their rules marked with the French metre, in addition to their own standard; and are thus, doubtless, gradually tending to the general adoption of so convenient and perfect a measure of length as the French metre is admitted to be.

Appendix 3
Inventory of John Rabone & Son, May 1851

This appendix reproduces the inventory of tools and the stock list, taken in May 1851, taken from the Rabone Stock Book for that year. There are stock books which cover several decades in the middle of the 19th century. However, they are generally lists of materials and work in progress. Whilst these general stock lists are of considerable interest in that they contain contemporary terminology relating to the rule making process and if analysed would no doubt be helpful in assessing the production of the firm, the numbers of items listed are very large and an in-depth study of these would be an extensive undertaking. Therefore only one year, 1851, has been selected for reproduction here.

One observation that might be made is that the number of prepared woods or sticks, that is the wooden portions of rules, are very numerous compared with the number of framed, that is assembled, but not finished, rules. The number of completed rules, divided, polished and packed ready to go, is even smaller.

The inventory of tools is of particular interest in that it allows some assessment of the degree to which the rule making process at John Rabone & Son had been mechanised by this date which is considerably less than the highly mechanised process that followed the move to the Hockley Abbey Works in 1871. It is worth noting too that those mechanical aids that were employed were of a comparatively simple, but no doubt effective, nature. Other rule makers of the same period were likely to have been even less mechanised than Rabone.

The first section is the inventory of tools. The second section is the inventory of stock. There were a considerable number of abbreviations used in the stock list; the more frequently used were: Dbl or D = double; Jt = joint; Dz = dozen; comm = common; bwd = boxwood; GS = German steel. Many entries are also indicted with ditto marks. To clarify these, the duplicated information is stated in brackets.

The entries in the right hand column are some notes explaining the less familiar terms. They were not in the original inventory. The value is given in pounds, shillings and pence.

This inventory is taken from the records of John Rabone & Sons, MS 1522, Archive Department, Birmingham Library Services.

Inventory of tools May 1851

Closet in Shop over Warehouse

	£	s	d	
3 Screw Carriers 2/- ea		6	0	
Box of Drills, Rhymers & Brace	1	5	0	
Tool Cupboard		5	0	
Sliding Rest Box & Stops		7	0	
Slide Rest in D° & Gouging out Holder on Top	7	10	0	
Sawing off Table & Slide [a]	1	15	0	*a. For cutting woods to length.*
Sawing in Table [b]	2	5	0	*b. Slotting jig for quarter joint.*
25 Turning off Tools used in Lathe 3/ea	3	15	0	
3 Screw Brasses for Turning off Joints		10	0	
3 Drilling Pegs for Joints		3	0	
4 Drilling Pegs 4d		1	4	
1 Each Large 5/6 & Small 2/- Screw Wrench		7	6	
1 Ea Rose Countersink 1/- Fork 1/- to Lathe		2	0	
46 Brass Chucks for Drills 4/- per Doz [c] of these are in use at Lathes		15	4	*c. Small drill bit fitted permanently into chucks (holders). (No number entered.)*
8 screw Spindles without Saws 1/4		10	8	
1 Long [spindles without saws]		2	0	
3 Loose Saws 10d		2	6	
1 Long Spindle with 1 Saw		3	0	
10 Spindles 1 Saw Ea 2/1 [d]	1	0	10	*d. The small sawing work was being carried out by multi-bladed saws fixed on spindles that were mounted into the lathes.*
12 Spindles 2 Saws Ea 3/- ea	1	16	0	
11 Spindles 3 Saws Ea +3/10 ea	2	2	2	
2 Spindles 4 Saws Ea 4/8 ea		9	4	
2 Spindles & Saws for Thick Squares 2/3ea		4	6	
1 Spindle Six Saws for Tips [e]		5	10	*e. 1 spindle with 6 saws for sawing tips.*
1 Spindle 19 Saws for Setting down	1	5	0	
1 Long Spindle 8 Saws for Arching		8	6	
1 Long Spindle 8 Saws for Bevils		8	6	
Bowl containing Lathe Centers etc.				
2 Drilling Plates	1	1	0	
1 Ea Large & Small Stocks & Dies	2	0	0	
	£ 31	16	0	

Closet in Shop over Warehouse Continued

	£	s	d
Brought forward	31	16	0
8 Screw Gouging out Tools 5/ ea	2	0	0
Setting down Table & Slide & Wood Table for Squares	5	5	0
Arching Table & Slide & Guide for Bevils	8	0	0
Handvices in Handle		2	0
Set of Gimbletts		2	6

Those Marked X are kept in the House for Safety

Shop over Warehouse

	£	s	d	
Turning off Lathe [f]	7	0	0	*f. There were six lathes in total.*
Lathe Board		2	6	
2 Shelves		4	0	
2 Vices 6/ea & 1 Small Table Vice 3/-		15	0	
Bow Saw & Back Saw		2	6	
27 Turning Tools for Iron & Brass in Rack in Closet 6d		13	6	
Stove		5	0	
Wilkinson's Tools [g]		8	0	*g. Tools issued to one of Rabone's employees.*

Top Shop adjoining the above

	£	s	d
Sawing off & Letting down Lathe	8	6	0
Lathe Board		3	0
3 Shelves		5	0
3 Vices 6/- ea		18	0

Brace		2	0	
Oil Stone		2	0	
R. Linforth's Tools [6]		8	0	
S. Weaver's [tools] [6]		8	0	
R. Stensons [tools] [6]		8	0	
Stove		5	0	
Large Top Shop				
Sawing in & Gouging Lathe	8	6	0	
2 lathes adjoining the above £8 6s 0d [e]	16	12	0	
Lathe Board for 3		9	6	
	£ 93	8	6	
Large Top Shop Continued				
Brought forward	93	8	6	
Wood Bed Lathe	5	10	0	
Lathe Board		4	0	
Set of Heads here & heads & Old plan Turning off Tools in Shop under		14	0	
Holes partitioned for Work & Shelving under Middle Bench		16	0	
7 Vices 6/ ea	2	2	0	
Large Vice		10	0	
Stove		5	0	
3 Countersink Chucks, Brooches &		6	0	
16 Wrenches 1/- distributed in the different shops		16	0	
Stenson Tools [8]		8	0	
Reads [tools] [8]		8	0	
Womens [tools] [h]		8	0	*h. Some of the employees were women.*
Hand Saw, Planes & Odd Carpenters Tools		15	0	
Lower Framing Shop				
3 Pairs Shears (1 of them in Top Shop)		12	0	
Edging Tables & Planes [i]		17	6	*i. A jig and planes*
Slide & Size Stick Planes [j]	1	1	0	*j. Special narrow planes for cutting the grooves to receive a slide*
Brace		2	0	
3 Pairs Pliers & 2 Pairs Tongs		5	0	
Stove		5	0	
Oil Stones		3	0	
7 Vices 6/- ea	2	2	0	
Steady & Block, Vice Steady's & Stool's[?] (Part in Top Shop)	1	0	0	
Riddle Iron Pan & Brushes		5	0	
2 Presses 1 at 20/- 1 at £3 5s 0d [k]	4	5	0	*k. Two fly presses*
Hidson's Tools [6]		8	0	
Hope's " [6]		8	0	
Finishing Shop				
Vice		6	0	
Stove [l]		5	0	*l. French polishing must be carried out in a warm atmosphere.*
	£ 118	14	0	
Finishing Shop Continued				
Brought Forward	118	14	0	
Guaging Blocks (*sic*)		2	6	
Finishing Stones	1	0	0	
Dividing Patterns [m]		10	0	*m. Patterns for hand dividing*
New Dividing Boards not made up		17	0	
Foreign Standards & 4 feet Guaging (*sic*) Rod Patterns &	1	0	0	

43 English Punches 1/6	3	4	6	
3 Alphabets	1	1	0	
Large New Figures 1 Set		13	6	
17 Sets Figures. 8 Sets at 3/6 9 Sets at 4/6	3	8	6	
Dividing knives, Hammers & Squares		10	0	
Guages & Dividers		7	0	
Shelf		1	6	
Warehouse				
High Stack of Drawers for Joints	1	10	0	
Middleplate Drawers		16	0	
Large length of Small Shelves wide underneath	1	5	0	
Cupboard with 1 Door		5	0	
Counter with Drawers & Shelf under		12	0	
Desk, Inkstand & Ruler		10	0	
Painted Counter	1	5	0	
Scales & Weights		8	0	
Scizzars, Large Knife, String Can & Shears		5	0	
Ladder		4	0	
Pans & Boxes for Joints		14	4	
Nest of Wide Shelves by shop Door	1	15	0	
[Nest of Wide Shelves] adjoining towards Closet	1	10	0	
Shelf over Fire Place		2	6	
Work Box for Work people		2	6	
Gunter Punches, Box & 57 Punches	8	0	0	*n. Set of punches for Gunter's scales*
Sector Punches & Boards	2	15	0	
	£ 153	9	4	
Warehouse Continued				
Brought forward	153	9	4	
Foreign punches 87 1/6 ea 35 Names 2/- ea	10	0	6	
Ribbon Punches	0	10	0	
43 Pr Press Tools 22 at 3/- 21 Pr at 5/-	8	11	0	
New Nest of Drawers for Joints	1	12	0	
4 Iron Squares		10	0	
12 Guaging Rod Punches [o]	1	4	0	*o. Each type had a set of punches.*
Casting Patterns for Joints [r]	1	10	0	*r. Joints in gauging rods were cast.*
Bolsters & pullers off for Press		12	0	
Steelyards & Weighing hoops		7	0	
Benching in all the Shops	4	0	0	
Gas Fittings 35 Lights	10	10	0	
Bells	1	0	0	
Hammering Machine	8	0	0	
Wheelbarrow		10	0	
Signs	3	10	0	
Grinding Stone		16	0	
Baskets		5	0	
	£ 206	**16**	**10**	

Stock 12th May 1851

98	Thick 2 feet Gunters cut out		1	12	.
4	2 ft Sadlers rulers	1d	.	2	.
4	3 ft Laths	6d	.	2	.
12	[3 ft] Rulers	8d	.	9	.
21 Dz.& 2	2 ft Gunters	3/-	3	3	.
100 Dz	Broad 2 ft Commn	7d	2	18	4
36 Dz	2 ft	1/3	2	5	.
103 Dz	Slides Best	1/3	6	8	9
100 Dz	3 ft Dbl Jts?	6d	2	10	
6 Dz	Iron Guages	9d	.	4	6
29 Dz	Inch Wide 3 ft	8d	.	19	4
48 Dz	Best Slides	1/3	3	.	.
120 Dz	Middle quality [Slides]	9d	4	10	.
53 Dz	5/8 wide Size Sticks	5½d	1	4	3
144 Dz	2 ft Flat folding	4d	2	8	.
100 Dz	Broad 2 ft Comm.	7d	2	18	.
15 Dz	Sliding Size sticks	7d	.	9	.
4 Gro 4 Dz	Red Bevils (cost price) abt. 6d Dz		1	9	.
13 Cwt 2 qrs	Boxwood in stick	20/-	13	10	.
2 Planks	Satin wood		2	15	.
Piece of Ebony			.	3	.
12 lbs 8 oz	Ivory uncut	6/9	4	4	4
5½Dz	4 feet Rods	4/-	1	2	.
Odd ruler stuff satin wood etc			.	2	6
Rolled metal in cellar & w'house 120 lb 8 oz 7¾d for bevels			3	17	9
89 lbs	Slide wire 11d & 9 ½ lb ditto	11d	4	10	6
9 Dz	Best Slides	1/3	.	11	3
16 Dz	Best Slides	1/3	1	.	.
10 Dz	Large 4 ft Dbe & Edged	1/-	.	10	.
4 Dz	Thin 2 ft archd	1/8	.	6	8
4 Gross	Thin 2 foots from Hills	18/-	3	12	.
			£ 72	17	6

		Forward	72	17	6
Pattern Cards in cupboard & rules on			.	15	.
Patterns in Bag			2	.	.
Patterns out 19 cards out unpaid for		5/-	4	15	.
Invoice heads &			1	.	.
Routledge & other Books			.	15	.
Scrap metal 45 lbs		6¼d	1	3	5
Rule Dust 5 cwt 20 lbs		9/-	2	6	6

Ivory Rules cut

8	Ivory Slips for rulers		.	2	.
5	Thin 2 ft & Slides	2/8	.	13	4
4	Broad Folding 3 ft	4/-	.	16	.
4	Inch & ⅛ [3 ft]	3/-	.	12	.
3	[Inch & ⅛ 3 ft]	thick for metres 3/-	.	9	.
6	2 ft Broad Folding	2/8	.	16	.
14	Inch & ⅛ 2 ft	2/3	1	11	6
2	Inch wide 1 Foot	11d	.	1	10
4	[Inch wide] 2 ft	1/10	.	7	4
6	small 1 Foot folding	4/dz (small)	.	2	.
½ Dz	6 inch Scales	3/-and odd scales 2/6	.	5	6
5	Small D. Jt 3 ft	2/-	.	10	.
4 Dz & 8	Button Guages	3/-	.	14	.
4 Dz	small Double Jt 2 ft	16/-	3	4	.

8	2 ft Flat Folding	10^{d}		.	6	8
13	1 ft Small Dble Jt	7/-		.	7	7
5½ doz	1 ft Flat folding	5/-		1	7	6
14 lbs	String mixed	8^{d}		.	9	4
	Brown paper 1⅔ ream			3	7	.
½ ream	Sand paper			.	5	.
87 lbs	Brass edging	10^{d}		3	12	6
53 lbs	Brass wire	$9½^{d}$		2	2	.
16 lbs	German Silver in ??	3/2		2	10	8
				£ 110	5	2
			Forward	110	5	2
3 Doz	Com^{n} Broad Folding	10^{d}		.	2	6
66 Dz	Broad Folding 3 ft not Edged	1/9		5	15	6
56 Dz	2 ft Dble Jt	4^{d}		.	18	8
36 Dz	[2 ft] Flat folding	4^{d}		.	12	.
11 Dz	Thick Square	$5¼^{d}$		.	3	.
17 Doz	[Thick] 2 ft Dble Jt	4^{d}		.	5	8
1 Doz	Large Size Sticks sawn in			.	.	6
5 Doz	[Large] 2 ft Dble Jt not Edged 5^{d}			.	2	1
20 Doz	Small 3 ft [Dble Jt not Edged] $6½^{d}$				10	10
17 Doz	Large Wood Feet Size Sticks	$9½^{d}$		.	13	6
3 Doz	Inch wide 3 ft not Edged	7^{d}		.	1	9
52 Doz	Mid Size Sticks	$4½^{d}$		.	18	9
Things on the Top Shelf in W'House				.	15	9
20 Doz	Inch ⅛ 3 Ft	$9½^{d}$		.	15	10
8 Doz	[Inch ⅛] 2 ft not edged	$5½^{d}$		.	3	8
22 Doz	Small Dble Jt 3 ft not Edged	$6½^{d}$		.	11	11
17 Doz	Best Slides Edged	1/4		1	2	8
29 Doz	[Best Slides Edged]	1/4		1	18	8
33 Doz	Small Dble Jt 3 ft Edged	$7½^{d}$		1	.	7
13 Doz	[Small Dble Jt] 2 ft not Edged	4^{d}		.	4	4
8 Doz	Inch wide 2 ft Edged	$5½^{d}$		.	3	8
11 Doz	Inch ⅛ [2 ft Edged]	$6½^{d}$		.	6	.
3 Doz	Flat Folding [2 ft Edged]	4^{d}		.	1	.
36 Doz	Middle Size Sticks long legs not Edged counted single reckond 18 Doz Double $4½^{d}$			6	9	.
48 Doz	small Dble Jt 2 Ft Edged	5^{d}		1	.	.
8 Doz	Com^{n} Bwd 2 ft [Edged]	8^{d}		.	5	4
9 Doz	[Com^{n} Bwd 2 ft] Let Down Squ^{r} 1^{d}			.	7	6
15 Doz	[Com^{n}] Slides Edged	8^{d}		.	10	.
14 Doz	Best [Slides Edged]	1/4		.	18	8
1 Doz	Bwd Folds 3 ft & Inch ⅛ flush jointed (odd)			.	2	6
24 Doz	Single mid Size Sticks long legs (12 Doz Dble) $4½^{d}$			.	4	6
				£ 131	9	3
			Forward	131	9	3
25 Doz	Large Dble Jt 2 ft not Edged	5^{d}		.	10	5
15 Doz	[Large Dble Jt 2 ft] Edged	6^{d}		.	7	6
29 Doz	Com^{n} Size Sticks (long & Short legs) sawn in 4^{d}			.	9	8
3 Doz	2 ft Bwd folding Drilled	1/-		.	3	.
2 Doz	Com^{n} Squares Edged	$3¼^{d}$ (half of $6½^{d}$ Doz)		.	7	.
1 Doz	Flat folds 3 ft [Edged]	6^{d}		.	.	6
10 Doz	Com^{n} Size Sticks long legs Edged 5 Doz Dble $3½^{d}$			.	1	6
2 Doz	Thin 2 ft Edged	1/4		.	2	8
13 Doz	Best Slides [Edged]	1/4		.	17	4
28 Doz	[Best Slides Edged]	1/4		1	17	4
6 Doz	2 ft Dble Jts [Edged]	4^{d}		.	2	.
20 Doz	Large [Dble Jts] 3 ft	9^{d}		.	15	.
110 Doz	2 ft flat folding Edged	4^{d}		1	16	8

5 Doz	Sqr folding 2 ft Edged	4^{d}		.	1	8
14 Doz	Large Dble Jt 2 ft	6^{d}		.	7	.
22 Doz	Narrow 2 ft let down & sawn in 8^{d}			.	14	8
In Closet	34 2 ft Gunters	3/- Doz		.	8	6
12	2 ft Laths	3^{d}		.	3	.
13	3 ft [Laths]	3^{d}		.	3	3
10	Sliding Gunters 2 ft not Edged	6^{d}		.	5	.
2½ Doz	Tailors Square woods	2/-		.	5	.
15	3 ft Rulers	6^{d}		.	5	.
15	2 ft [Rulers]	4^{d}		.	5	.
1 Doz	Odd Gunters	3/-		.	3	.
8½ Doz	Slides Let down	1/1		.	9	.
5½ Doz	Arch [Slides Let down] (Com^{n})	1/3		.	7	.
4½ Doz	Com^{n} Thin 2 ft Let down	10^{d}		.	3	9
5 Doz	Best [Thin 2 ft Let down]	1/6		.	7	6
2 Doz	Rnd Jt [Thin 2 ft Let down] (Com^{n})	1/-		.	2	.
	Narrow 2 ft pinned on			.	10	.
6 Doz	Inside narrow 2 ft pinned together			.	6	.
				£ 143	19	9
			Forward	143	19	9
3½ Doz	Box Bevils Slit down & c	1/6		.	5	3
2 Doz	Best wood [Bevils] Let down	1/6		.	3	.
4 Doz	Bwd 2 ft Gouged out	9^{d}		.	3	.
10	Guaging Rods Finished	1/-		.	10	.
18	4 ft Double Jts 1st jtd	2/6		.	3	9
11 Doz	Dble Jts & folds various pinned together	2/9		1	10	3
2 Doz	[Dble Jts & folds various pinned together	2/6		.	5	.
2½ Doz	[Dble Jts & folds] Framed	3/6		.	8	9
3 Doz	Large D Jt 40 in [Framed]	3/6		.	10	6
7 Doz	2 ft Dble Jt pinned on middles	2/9		.	19	3
1 Doz	Large Dble Jt 40 inch [pinned on middles]	3/6		.	3	6
7½ Doz	2 ft flat folding Framed	3/-		1	2	6
1½ Doz	Com^{n} Squ Dble Jt 40 in [Framed]	3/6		.	5	3
2 Doz	2 ft D Jt [Framed]	3/6		.	7	.
21	various pinned together	2/9		.	4	9
1 Doz	in Dble Jt & folding 2 ft Finished	3/9		.	7	6
2½ Doz	2 ft Dble Jt Framed	3/6		.	8	9
2 Doz	[2ft] Large [Jt Framed]	3/6		.	7	.
1 Doz	Box Bevils [Framed]	4/-		.	4	.
19 Doz	1st jointed various Dble Jts & folding	2/3		2	2	9
10 Doz	Archines finished unsaleable at present	2/6		1	5	.
21	2 ft flat folding Framed	3/-		.	5	3
2 Doz & 5	3 ft Dble "	3/6		.	8	6
1½ Doz	large Dble Jt 2 ft "	3/6		.	5	3
3 Doz & 9	various 1st jointed	2/-		.	7	6
4½ Doz	folding pinned on middle	2/3		.	10	2
5½ Doz	Large D Jt 3 ft Framed	3/6		.	19	3
28 Doz	2 ft D Jt [Framed]	3/6		4	18	.
3 Doz	3 ft [D Jt Framed]	3/6		.	10	6
3½ Doz	Dble Jt 1st jointed	2/3		.	8	.
2 Doz	Com^{n} Squares jointed	8^{d}		.	1	4
				£ 164	10	3
			Forward	164	10	3
2 Doz	folding & Dble Jt jointed average price	2/6		.	5	.
4½ Doz	2 ft folding 1st jtd	1/9		.	8	.
4 Doz	[2 ft folding 1st jtd]	1/9		.	7	.
½ Doz	Mid Size Sticks finished	3/6		.	1	9
10	Dble Jt 2 ft [finished]	3/9		.	3	6
3½ Doz	3 Flat folding Framed	3/-		.	10	6

9	rules jointed various	2/9	.	2	3
2 Doz	2 ft Dble Jt pinned on middle	2/9	.	5	6
1 Doz	[2 ft] Folding Framed	3/-	.	3	.
8 Doz	3 ft D Jt [Framed]	3/6	1	8	.
14 Doz	2 ft [D Jt Framed]	3/6	2	9	.
1½ Doz	[2 ft D Jt Framed]	3/6	.	5	3
9	Folding D Jt jointed	2/6	.	2	3
13 Doz	Box bevils Framed	4/-	2	12	.
2 Doz	3 ft D Jt pinned on middles	2/9	.	5	6
1½ Doz	4 ft [D Jt] Framed	4/-	.	6	.
4 Doz	2 ft [D Jt] pinned on middles	2/9	.	11	.
3 Doz	[2 ft D Jt] Framed	3/6	.	10	6
9	Small 3 ft Squ. mid framed	6/-	.	4	6
2½ Doz	Flat foots [Framed]	1/-	.	2	6
3	Inch wide and mid [Framed]	7/6	.	2	.
9	3 ft D Jt finished on middles	2/9	.	2	.
1½ Doz	Flat foots jointed	9^{d}	.	1	1
2½ Doz	Com^{n} Squares jointed	8^{d}	.	1	8
2 Doz	Flat folding Framed	3/-	.	6	.
1½ Doz	Squ. [folding] 1 ft [Framed]	2/6	.	3	9
2 Doz	[Squ. folding 1 ft] pinned in middles	1/9	.	3	6
5 Doz & 9	Mid S. Sticks Framed	3/-	.	17	.
13½ Doz	Com^{n} [Mid S. Sticks Framed]	2/6	1	13	9
2 Doz	Squ Folding 1st jointed	1/6	.	3	.
3 Doz	Box sectors framed	2/-	.	6	.
			£ 179	13	0

		Forward	179	13	.
½ Doz	3 ft Dble Jt Framed	3/6	.	1	9
3½ Doz	outside narrow 2 ft [Framed]	1/10	.	6	3
3 Doz	3 ft D Jt [Framed]	3/6	.	10	6
2 Doz & 3	Flat folds 3 ft 1st jtd	1/9	.	4	.
1½ Doz	Large D Jt 2 ft Framed	3/6	.	5	3
3 Doz	1 ft D Jt [Framed]	3/6	.	10	6
1 Doz	[1 ft D Jt] pinned on middle	2/9	.	2	9
3 Doz	odd D Jts Framed	3/6	.	10	6
8 Doz	Large 3 ft D Jt outside & arch Framed	12/6	5	.	.
2 Doz	[Large] 2 ft [D Jt] Squ middle [Framed]	4/6	.	9	.
1 Doz	Com DJt 2 ft [Squ middle]	4/6	.	4	6
8 Doz	Flat folds 3 ft Framed	3/-	1	4	.
1 Doz	Com^{n} arch Jt [Framed]	6/-	.	6	.
11 Doz & 3	3 ft D Jts [Framed]	3/6	1	18	11
4½ Doz	Flat folds 2 ft [Framed]	3/-	.	13	6
3 Doz	2 ft D Jt [Framed]	3/6	.	10	6
5 Doz	flat foots let down for tips	10^{d}	.	4	2
4½ Doz	Inch⅛ 2 ft arch Framed	4/3	.	19	.
5½ Doz	Arch Jt Slides [Framed] medium	8/-	2	4	.
8	Brass edge arch Jt Slides Finished	18/6	.	12	4
½ Doz	Ivory 1 ft folds Framed	10/-	.	5	.
½ Doz	Bwd Folds 2 ft Best [Framed]	12/-	.	6	.
4 Doz & 4	Brass Edge Arch Jt 2 ft Framed	10/-	2	3	4
½ Doz	Inch⅛ arch middle 2 ft [Framed]	8/0	.	4	.
2 Doz	Arch Jt 2 ft Framed Best	6/9	.	13	6
2½ Doz	Flat folding 2 ft 1st jointed	1/9	.	4	4
5½ Doz	Broad Squares let down	10^{d}	.	4	7
8½ Doz	Flat foots Framed	1/1	.	9	3
4 Doz	Large DJt 2 ft pinned on middles	2/9	.	11	.
½ Doz	Best wood Bevils Framed	4/-	.	2	.
½ Doz	odd rules		.	2	6
			£ 201	15	11

				£	s	d
			Forward	£ 201	15	11
1 Doz	Brass Edge arch Jt Slides Framed	14/6		.	14	6
1 Doz	Inch arch 3 ft [Framed]	8/-		.	8	.
1 Doz	Ex thin arch Jt 2 ft [Framed] Best 7/3			.	7	3
1	Ivory 2 ft arch GS Inch ⅛ with edge [Framed]			.	5	3
1 Doz	Brass edge arch Jt Slides [Framed] 14/-			.	14	.
2 Doz	head Jt 2 ft [Framed]	6/9		.	13	6
22 Doz	Large Jt [2 ft Framed]	2/9		3	.	6
1 Doz	[Large Jt 2 ft]Finished	3/-		.	3	.
6 Doz	Small head Bwd 2 ft Framed	2/6		.	15	.
3½ Doz	[Small head Bwd 2 ft] Finished	3/-		.	10	6
11 Doz	Thin 2 ft Com Framed	3/6		1	18	6
9 Doz	[Thin 2 ft] Best [Framed]	3/6		1	11	6
FILES						
4 Doz & 8	10 inch Bastard	7/-		1	12	8
1 Doz & 4	[10 inch] Rough	7/-		.	9	4
5	10 Inch Bastard Smooth	9^{d}		.	3	9
7	[10 inch] Smooth	10^{d}		.	5	10
5 Doz & 8	Safe edges	4/6		1	5	6
3 Doz	3 Squ. Smooths	6/-		.	18	.
8	Rough 3 squ. Recut	3/-		.	2	.
2 Doz & 7	Second cut 3 Squares	5/-		.	13	.
251 lbs	Rolled metal 7¾^{d}	£ 8.2.0¾		8	2	.
279 lbs	Rough casting 11^{d}			12	15	9
271 lbs	casting got ready for use	1/3		16	18	9
65½ lbs	of Hollows cut out	1/3		4	1	10
41 Doz	2 ft for various 4 Fold 6^{d} Best Edged			1	.	6
20 Doz	Waste Bwd 2 ft	6^{d}		1	.	.
5 Doz	4 ft Dble Jt Best Edged	8^{d}		.	3	4
7 Doz	[4 ft Dble Jt Best]	8^{d} Edged		.	4	8
6 Doz	3 ft flat folding best edged	5^{d}		.	2	6
6 Doz	2 ft Dble Jt [best edged]	3^{d}		.	1	6
				£ 263	12	4
			Forward	263	12	4
8 Doz	4 ft Dble Jt not edged	8^{d}		.	5	4
7 Doz	[4 ft Dble Jt not edged]	8^{d}		.	4	8
36 Doz	3 ft flat folds Edged	6^{d}		.	18	.
31 Doz	2 ft Dble Jt [Edged]	4^{d}		.	10	4
50 Doz	[2 ft Dble Jt Edged]	4^{d}		.	16	8
28 Doz	[2 ft] Squ. Folding	4^{d}		.	9	4
Tips iron 12 lbs cut & not turned		4^{d}		.	4	.
17½ lbs	Iron tips turned	6^{d}		.	8	9
2 lbs	Size Stick bitts	3^{d}		.	.	6
Iron in shop not cut				.	15	.
101¼ lbs	Middle plates various	5^{d}		2	3	6
133 lbs	Brass tips turned	1/3		8	6	3
11 Doz	Mid head Bwd 2 ft Framed	2/9		1	10	3
2 Doz & 9	[Mid head Bwd 2 ft] Finished	3/-		.	8	3
½ Doz	Sliding Size Sticks [Finished]	7/6		.	3	9
9 only	[Sliding Size Sticks] Framed	6/6		.	5	.
6 Doz	Bwd wood feet [Size Sticks Framed] 6/8			2	.	.
1 Doz	Mixed Size Sticks Finished	4/-		.	4	.
5 Doz	Bwd folds 3 ft Com^{n} Framed	8/6		2	2	6
1½ Doz	[Bwd folds 3 ft Com^{n} Framed]	8/6		.	12	9
½ Doz	Inch ⅛ Slides Framed	10/6		.	5	3
¼ Doz	[Inch ⅛ Slides Framed]	10/6		.	2	7
4	Ivory Slip Slides Finished	1/6		.	6	.
3	Dble Jt [Ivory Slip Slides Finished]10^{d}			.	2	6

2 Doz & 3	Squ. Jt Brass edge 2 ft Framed	8/-		.	18	.
3 Dble	Edge arch Jt [2 ft Framed]	15/-		.	3	9
½ Doz	Brass Edge new Jt [2 ft Framed]	10/-		.	5	.
2½ Doz	[Brass] Com^{n} Arch Jt Finished	6/9		.	16	9
½ Doz	[Brass Com^{n} Arch Jt] Framed	6/-		.	3	.
½ Doz	Odd Rules			.	3	.
½	Small head Bwd folds Finished			.	3	.
				£ 289	10	.
			Forward	289	10	.
½ Doz	Best 2 ft Bwd folding Framed	12/-		.	6	.
2½ Doz	Com^{n} New Jt 2 ft [Framed]	6/-		.	15	.
2½ Doz	Best [New Jt 2 ft Framed]	6/9		.	16	9
½ Doz	[Best New Jt 2 ft] Finished	7/6		.	3	9
1 Doz	Small Dble Jt arch [Finished]	7/6		.	7	6
1½ Doz	Extra thin arch Jt 2 ft Framed	7/3		.	11	.
1 Doz	[Extra thin arch Jt 2 ft] Finished	8/-		.	8	.
½ Doz	Small Dble Jt Arch Framed	7/6		.	3	9
16	odd rules			.	6	6
1½ Doz	Inch ⅛ Arch mid Framed	8/-		.	12	.
1½ Doz	[Inch ⅛] Squ [Framed]	6/6		.	9	9
2 Doz	[Inch ⅛] Insides [Framed]	4/3		.	8	6
4 Doz	Inch Arch 3 ft [Framed]	8/-		1	12	.
2 Doz	Inch Squ. 3 ft [Framed]	6/6		.	13	.
2 Doz	Small D Jt [3 ft] arch [Framed]	7/6		.	15	.
1 Doz	[Small D Jt 3 ft] Squ [Framed]	6/-		.	6	.
1½ Doz	[Small D Jt 3 ft] Inside [Framed]	4/3		.	6	4
¼ Doz	[Small D Jt 3 ft] Arch Finished	8/-		.	2	.
1 Doz	Inch wide 2 ft arch Framed	8/-		.	8	.
1 Doz	[Inch wide 2 ft] Squ [Framed]	6/6		.	6	6
1 Doz	Small D Jt [2 ft] arch [Framed]	7/-		.	7	6
½ Doz	[Small D Jt 2 ft] Squ [Framed]	6/-		.	3	.
IVORY						
2	Ivory Edge Parallel Sticks			.	2	.
1	Box Drawing Scales			.	.	6
3	12 In Ivory [Drawing Scales]	2/6		.	7	6
1 ea	2 $^{9/-}$ 4-fold$^{11/-}$ Ivory Engineers Finished			1	.	.
1	Bwd folding 2 ft [Finished]			.	7	.
½ Doz	Inch ⅛ 2 ft GS Framed	4/6 ea		1	7	.
½ Doz	[Inch ⅛ 2 ft] Brass [Framed]	4/-		1	4	.
1 ea	[Inch ⅛ 2 ft Brass] with Guage in	5/6 & 5/-		.	10	6
				£304	16	4
			Forward	304	16	4
IVORY						
1 ea	Bwd fold$^{7/-}$ & Inch$^{6/6}$ ⅛ Small yd D$^{3/6}$ Jt 3 ft Framed			.	17	.
1	bwd fold 2 ft Framed	4/8		.	4	8
5	Inch wide [2 ft Framed]	3/3		.	16	3
11	Small D Jt arch 1 ft GS 22/- Framed			1	2	.
1½ Doz	[Small D Jt arch 1 ft] Brass1 9/- [Framed]			1	8	6
3	[Small D Jt arch 1 ft Brass] with Guage [Framed]2/6			.	7	6
3	[Small D Jt arch 1 ft Brass with Guage Framed] 2/6			.	7	6
2 Doz & 8	Small D Jt 2 ft inside	23/- [Framed]		3	1	4
2	[Small D Jt 2 ft] arch	28/- [Framed]		.	4	8
5	[Small D Jt 2 ft] 1 ft inside	14/- [Framed]		.	5	10
3 Doz & 7	Flat Folds 1 ft	10/- [Framed]		1	15	10
2	Inch⅛ 2 ft arch D	4/3 [Framed]		.	8	6
2 Doz & 7	1 ft Dble Jt	14/- [Framed]		1	16	2

1 Doz & 3	squ. Folds 1 ft	10/- [Framed]		.	12	6
2	2 ft Squ Folds GS Best	3/3 [Framed]		.	6	6
3	Squ. Folds 2 ft	15/- doz [Framed]		.	3	9
1	Inch⅛ metre outside & arch	[Framed]		.	8	.
2	Small 1 ft GS with Guage	3/- [Framed]		.	6	.
5	1 ft Small D Jt	14/- [Framed]		.	5	10
5	Button Guages	10d [Framed]		.	4	2
8	1 ft various			.	11	.
5	[1 ft various]			.	6	6
1 Doz	[1 ft various]			.	6	.
1	2 ft Small D Jt arch	2/4 [Framed]		.	2	4
½ Doz	1 ft [Small D Jt] GS Finished	25/-		.	12	6
7	[1 ft Small D Jt] Brass [Finished]	22/-		.	12	10
2	Small D Jt 1 ft [Finished]	16/-		.	2	8
3	Small D Jt 2 ft [Finished]	26/-		.	6	6
4	18 Inch Framed	19/-		.	6	4
7	18 Inch folding [Framed]	12/6		.	7	3
½ Doz	1 ft [folding] Finished	11/-		.	5	6
				£ 323	16	5
			Forward	323	16	5
1 Doz & 7	Button Guages Framed	10d		.	15	10
1 Doz	[Button Guages] Finished	1/1			.	13
5	Flat foot callipers Finished			.	5	.
1 ea	6 in Scale 18d & Sector [Finished]	1/9		.	2	7
7	Box Iron Guages [Finished]	18/-		.	10	6
5	odd rules				.	2
1 Doz	Plotting Scales [Finished]				.	1
1 Doz	Sectors [Finished]				.	4
Patterns rules, Ullages, Hawthornes etc.				.	10	.
Parallel Rule					.	1
1 Doz	Engineers Slides [Finished]	19/-		.	19	.
2	Hawthornes Slides Finished]	2/1			.	4
4 s	Routledge Dble Brass Edges	2/3		.	9	.
5	odd rules				.	2
½ Doz	Com arch Jt 2 ft Framed	6/-		.	3	.
2	Engineers Slides Finished	1/7		.	3	2
4	Arch Jt Slides [Finished]	12/3			.	4
3	[Arch Jt] 2 ft [Finished]	7/6			.	1
5	Inch ⅛ arch [Finished]	1/7			.	6
1	Bwd folding Slide 3 ft [Finished]	1/6		.	1	6
7 Doz & 4	Large Jt. Slides Framed	5/-		1	17	.
2	Dble Bra edge arch Jt Slides [Framed]	20/-		.	3	4
2	New Jt Slides "	9/-		.	1	6
7	Brass edge Squ. Jt Best Finished	16/-		.	9	.
½ Doz	Arch Jt Slides [Framed]	12/3		.	6	2
4	Slides Finished [Framed]	10d		.	3	4
4	21 Inch Tailors Squares Framed	12/-		.	4	.
Pattern makers Rule				.	.	.
1	Dmn Scales	1/9			.	1
2	2 ft Squares Finished	1/2			.	2
				£ 333	8	9
			Forward	333	8	9
9	2 ft Sadlers Rules Framed	9d		.	6	9
1	Straight Size Stick [Framed]	1/-		.	1	.
1 Doz & 3	Large Jt Slides Finished	8/6		.	10	9
7 Doz	Comn Arch Jt Slides Framed	7/6		2	12	6

2 Doz	[Com^{n}] Large Jt [Slides Framed]	5/-		.	10
1 Doz & 9	Large Jt Slides Bitted [Framed]	6/-	.	10	6
9	Large Jt Slides Finished com	8/-	.	6	.
6½ Doz	Large Jt 2 ft Iron tips Framed	2/9		.	17
2 Doz	Large D Jt 2 ft Squ. Mid Finished	5/-	.	10	.
14	40 Inch Portuguese D Jt [Finished]	4/6	.	5	3
7	Brass Edge arch Jt Slides [Finished]	18/6	.	10	10
2 Doz	2 ft Dble Jt Framed	3/6		.	7
5 Doz	3 ft Bwd folds arch^{d} only	4/-	1	.	.
5 Doz & 9	2 ft [folds arch^{d} only]	4/-		1	3
3 Doz & 11	3 ft Dble Jt Middle joints)				
8 Doz & 7	[3 ft Dble Jt Middle joints])	16 Doz 2/9		2	4
4 Doz	[3 ft Dble Jt Middle joints])				
3 Gro 3 Doz	2 ft [Dble Jt] past first jointed in hand		1	2	.
Rules in Cupboard (which will not match)			1	10	.
100 Doz	2 ft Dble Jt	4^{d}		1	13
9½ lb Rolled metal		7$^{3/4 d}$	.	6	.
½ Doz	new Jt Slides Framed	9/-		.	4
4½ Doz	Dbl Palms sawn in		.	2	6
			350	2	5

Debts in Day Book not posted to Ledger to May 12 inclusive	163	6	.
Ledger (outstanding debts $^{£}$199 ^{s}8 0^{d})	36	2	.
Cash in hand	149	15	9
	699	6	2
Tools & Fixtures as valuation May 12 /51	206	16	10
Gross capital, stock, cash tools & fixtures	906	3	.

Appendix 4

Dring & Fage: Journeyman's ledger, 1797 – 1818

The ledger of a journeyman rule maker for work done between July 1797 and July 1818 has been preserved in the archives of Dring & Fage, deposited in the Science Museum Library in London when the firm ceased trading in 1977. The value of this ledger lies in the descriptions of the work being carried out and also the terminology used to describe the rules.

The ledger has STRONG stencilled onto the front cover – perhaps the name of the journeyman – and some payments have been intialled JH.

The entire ledger is too long to be included in this appendix so only a few extracts for individual days have been reproduced here. The first table include the full entries for the day in question. The second table contains part entries for the day quoted to give an indication of the great variety of rules being produced and the prices paid for the various jobs undertaken. Where no prices are listed, these were not noted in the ledger.

In addition to the rules listed in this appendix, other types found in the ledger included cordage rules; sectors, both box and ivory; casting rules; carpenters' squares; diary rules; plumbers' rules; verniers for thermometers; stationers' rules; yard sticks; hatter's rules; glovers' sticks; pocket book rules; butt gauges and scale rules.

As in the Rabone inventory, a number of abbreviations were used, the most common being D^{oz} of D^{n} for dozen, Do for ditto and Box for boxwood.

Full daily entries

		£	*s.*	*d.*
1797				
July 29th	1 large double joint yards		1	0
	2 D^{oz} thin large head wood slides iron tips and plated	1	7	0
	Buttoning Etc. 2 doz slide rules [a]		1	0
	18 Four sect. gauging rods		15	9
		2	4	9
	Received	1	1	0
		1	3	9
Aug 12th	2 D^{n} 3 head Double jnt Yards	1	0	0
	Received		15	0
	JH		5	0
	Lent		10	6
	August 12th 97 in debt 10/6	0	15	6
Aug 26th	38 Dipping rules		17	5
	Buttoning 2 D^{oz} slides		1	0
	1 D^{n} Ivory sectors after out		5	0
		1	3	5
	Received		10	6
			12	11
Sept. 16th	2 D^{oz} Ivory sectors after planed up		10	0
	2 large 5 foot gauging rules		4	0

a. Buttoning = French polishing

		£	s	d	
	1 large 3 foot gauging rule		0	8	
	1 middle plate		0	2	
	4 steel Jt Ivory folding rules		3	10	
	4 Doz 3 head double jts @ 0/10	2	0	0	
	1 doz 2 feet gauging Rules large size		8	0	
	1 doz Sundry lengths etc		8	0	
	13 large head 2 fts iron tips etc		9	9	
		4	4	5	
	Received	1	1	0	
Sept 30th	JH	3	3	5	
Oct. 7	2 five feet Gauging rules		3	4	
	6 large head Patent Brass Slides		7	9	
	7 D° wood slide		7	10 ½	
	Settled JH		18	11 ½	
Nov. 11th	1 Doz Brass jts Ivory folding rules		10	0	
	½ Doz Box Double Jt			6	
	1 side & a middle plate			5	
	An Ivory slip			6	
	7 5 ft Gauge Rules		11	8	
		1	3	11	
	Received	1	3		
				11	
1803					
April 30th	1 Doz Large Hd 2 feet		7	0	
	6 Thumb pieces etc. Hd Rods		2	0	
	Jobs 2 Guaging Rods & Brass slides [b]			9	*b. Jobs = repairs*
	2 Doz Thin large hd wood slides	1	7	0	
May 14th	1 Brass rod		1	0	
	2 Doz thin Large hd Brass slides	1	11	0	
	1 Thin large hd Ivory slip		1	6	
	Jobs to Callipers & Head Rod		2	9	
	2 Rods			6	
May 21st	7 Taper leg wood slides Thin		7	0	
	6 Cross Callipers	1	13	0	
	Jobs to Callipers & Hd Rods			6	
	3 Sides and slip side to Broad slide		1	3	
May 28th	1 Thin wood slide iron tip etc.		1	0	
	6 Box guaging Rods		1	0	
	6 Head Rods	1	1	0	
June 11th	Butt Jts 6 2 feet Ullage Rules		1	0	
	2 Long Callipers mounted with Iron	1	1	0	
	1 Doz Large hd 2 feet Iron tips etc.		9	0	
	1 Doz D° Brass Tips		7	0	
	2 Slips and two Jobs			6	
		9	0	0	
1805					
July 26th	Tipping 1 Doz Station Staves		2	0	
	1 40 Inch Bobber [c]			10	*c. Not identified*
	6 Diameter		7	0	
	1 Doz 40 Inch Brannans [d]		13	0	*d. Brannan's rules*
	6 40 feet D°		6	6	
	Thumb piece for Head Rod			4	

		£	s	d	
	5 Sides and 4 Mid Plates		1	11	
	Job Sliding size stick & Guage Rod			10	
	Easing etc. 5 Calendar Rules			5	
July 27th	6 40 feet Brannans		6	6	
	1 Doz Dippers		5	6	
	Getting out wood for Long Callipers			9	
	2 Ullage Rules		2	0	
	2 Five Foots		3	4	
	6 Diameters		7	0	
	Job to Head Rod			9	
	Easing 2 Ullage Rules			4	
	6 Diameters		7	0	
		3	6	0	
	Reced	2	2	0	
	Settled July 27 1805 JH	1	4	0	
1809					
Aug. 19th	14 13 inch brewing rules ivory slides		17	6	
	2 Doz Dippers		11	0	
	1 Large Cross Calliper		10	0	
	Half a Long Calliper		2	0	
	Side slide etc. Cross Calliper		1	8	
	Reparing 6 Old Callipers		4	0	
	Easing etc. 17 Vero's [e]		2	10	*e. Vero's improvement to slide rules invented before 1739, was to let in small slips of brass to the end of the slides.*
	Butt Jts etc. 14 Ullage rules		2	4	
	Slide to Vero & Tab Long head			4	
Aug. 26th	15 12 Inch Vero's		17	6	
	1 9 Inch Do		1	0	
	2 2 feet Ullage Rules		2	0	
	1 18 inch Do			9	
	Repairing Cross Calliper		1	3	
	6 Dippers [f]		2	9	*f. Probably dip rods but could also have been sampling containers.*
	Butt Jts etc 8 Ullage Rules and 12 Vero's		3	4	
	3 2 feet sliding Gunters		5	0	
Sept. 9th	5 2 feet sliding Gunters		5	0	
	2 18 Inch Ullage Rules		1	6	
	3½ Doz Dippers		19	3	
	Easing etc 10 Vero's		1	8	
	5 12 Inch Vero's		5	10	
		5	16	6	
Sept. 16th	10 12 Inch Vero's		11	8	
	Easing Etc Do		1	0	
	6 Long Callipers	1	5	6	
Sept. 23rd	21 2 feet Ullage Rules	1	1	0	
	1 18 Inch Do			9	
	Butts Jts Etc 21 Ullage Rules		3	6	
	Easing Etc 6 Vero's		1	0	
Sept. 30th	5 2 feet Ullage Rules		5	0	
	8 18 Inch Do		6	0	
	7 12 Inch Do		3	6	
	2 Slides to Vero's			6	
	Easing Etc. 8 Vero's		1	4	
	8 Jewkers Slide Rules [g]		10	8	*g. Not identified.*

Oct. 7th	Butt Jts Etc. 7 Ullage Rules		1	2
	1 Long Calliper		4	3
	5 Cross Do	1	5	0
	1 Do Do		5	0
Oct. 14th	1 Cross Calliper		5	0
	7 12 Inch Vero's		8	2
	1 9 Inch Do		1	0
	1 Dzn Dippers		5	6
	Butt Jts etc 9 Ullage Rules		1	6
		7	8	0

Partial daily entries

1798					
Jan 6th	Job to patent slide		1	0	
Mar 24th	Making 4 rule jts		0	9	
May 19th	Buttoning etc. Ullage rules & patent slide			4	
Jun 9th	Buttoning and easing to Ullage rules				
Jun 23rd	Tabs to 2 Gauging rods				
Sept 8th	½ a Branan & side & springs to Do			10½	
	Side etc. to thin 2 fts			4	
	1 Dn Casting rules [h]		10	0	*h. Probably contraction rules.*
Sept 29th	1 Branan with thumb springs 4 ft		2	6	
Oct. 6th	Job to Calliper			6	
1799					
Jan 12th	Planing up 10 Dn long calipers with tips	1	10	0	
Feb 9th	Tab to Glaziers & Gauging rods				
Mar 9th	Tab to straight rods				
Apr 6th	1 small falt rule [i]			4	*i. Not identified.*
	1 ft heads & Butt Jnt Rules			6	
Apr 20th	1 pr heads Tongue to Bevil & job to rule			7	
	Job to Box square & slips to Rule		2	2	
Aug 29th	1 long Caliper after planing up		1	9	
Oct 12th	Log to Gauging rod with springs			10	
Nov 9th	2 Jobs to Floats			10	
	Making a box square [j]			9	*j. Boxwood drawing square.*
	Finishing 6 Yard sticks		2	0	
Dec 21st	Glovers size sticks & job to Dimension rules			3	
1800					
Mar 15th	1 Doz Ordy Steel jted Squares		4	0	
Apr 5th	1 Steel jted Cordage Rule			9½	
May 2nd	Tongue to plumbers rule [k]			3	*k. 6 in. brass extension.*
Jun 14th	2 Doz Ivory Pocket Book rules		5	0	
Jun 21st	6 2 feet sliding timber rules		6	0	
Jul 5th	1 Dzn Gordon's 2 ft timber rules [l]		14	0	*l. Not identified.*
Aug 30th	1 Ordnance rule [m]		1	0	*m. Artillery rule.*
1801					
Feb 7th	Box Square and 18 inch Scale			10	
Apr 4th	Making a 2 feet bevel		1	6	
Apr 15th	Job to machine		1	0	
1802					
Mar 13th	Filing up Steel Rods		5	0	
May 29th	Filing up starch gauges		3	0	
Jul 3rd	17 Jewker's Gauging Rules	1	2	8	

Aug ?	3 thin 2 Jnt wood slides taper leg		3	0	
1803					
Mar 19th	Fitting cross band Parallel rules		2	6	
Apr 2nd	14 Gordon's timber rules		16	4	
	Thumb pieces etc. to 10 Jtd Rules		3	4	
May 24th	7 Taper leg wood slides thin		7	0	
Aug 6th	Furniture to 6 Patent slides		1	6	
1804					
Feb 25th	Job to T Squares		1	3	
May 12th	4 Comb makers size sticks		2	0	
May 26th	2 Ivory Ordnance Rules		4	0	
Aug 3rd	Making Patterns			8	
Sep 1st	Repairing long Calliper		2	0	
Dec 15th	10 5 inch Hatters rule wth slides		4	0	
1805					
Nov 2nd	1 Dzn Jewker's Gauging rules		16	0	
Nov 30th	1 2 feet Leadbetters Slide rule		2	0	
1806					
Jun 21st	3 Half Branans		1	9	
Jun 28th	A stout double slide broad slide rule		2	0	
Jul 12th	Repairing Mr Wood's Rule		2	0	
Jul 26th	Scale & Slider for Hydrometer work			9	
Oct 4th	1 30 inch Bobber		1	0	
Oct 25	Finishing off 4 five feet Gaugg Rules		3	6	
Nov 1st	2 Branans wh additional halves & Thumb springs		4	4	
1807					
May 16th	Brazing 2 large Squares 3 Days		15	0	
Aug 8th	Long Hydrometer slide & Lath		1	6	
Sep 26th	1 Dzn stout L.H. wood slides		10	6	
Dec 5th	6 steel Jnt Ivory Folding Foots		5	9	
1808					
May 28th	Box pattern for Hydrometer work			8	
Jun 11th	Repairing thumb spring socket rod		2	6	
1809					
Feb 7th	2 cross Callipers 30 inch shanks		12	0	
1810					
Feb 24th	Repairing 11 Callipers 3½ days		17	6	
Mar 17th	4 Saccharometer sliding rules		5	0	
Mar 24th	2 shoulder pieces to Long Rod		1	3	
Jun 9th	Repairing 10 Callipers 2.5 hours		12	6	
Oct 27th	14 13 inch Brewing Rules		17	6	
1811					
Jan 7th	Long Calliper mounted with iron 3½ days	1	1	0	
Sep 28th	6 Stout Large Jtd Brass slides		6	9	
	6 Do Wood slides		5	9	
	2 Jointed wood slides thin iron tips		3	5	
Oct 19th	2 small sliding rules wh extra slides		3	0	
Nov 2nd	Easing etc 31 Veries & Ullage rules n		5	2	*n. An alternative spelling for Vero.*
Nov 30th	5 Box Hydrometer rules		3	4	
	Tipping measuring rods & slide Veries		1	8	

1812				
Jan 9th	4 12 Inch chain scales		8	
Mar 14th	Altering 2 small casting rules	1	0	
1814				
May 14th	Plain table frame [o]	3	0	*o. A plane table rule.*
	Job triangles		3	
1816				
Apr 13th	Satin Wood Ruler		6	
1818				
Apr 25th	Edging 5 Dzn & 8 Gunters	1	5	

Checklist of rule makers and dealers

Introduction

The purpose of this checklist is to enable the reader to identify the names found on a broad spectrum of mostly linear measuring equipment made or marked in Great Britain.

A survey of several hundred Post Office and trades directories produced all those who entered themselves as rule makers and this formed the basic checklist whether or not examples of their work have been found. However, there were many other manufacturers who made measuring equipment but described their trade differently, for example as mathematical instrument makers, opticians or even naval or military outfitters, or, often, a combination of these.

Information was also extracted from a variety of other books, catalogues and databases. Wherever possible, this information has been checked, sometimes expanded and, occasionally, corrected. These sources are listed at the end of this introduction.

Another source was the names found on the rules themselves. These are included where evidence indicates that the name is that of a maker or a retailer, rather than an owner. In many cases inclusion was on the basis of probability resulting from the examination of many rules.

Criteria for inclusion in the checklist

a) Any firm or person described as a rule maker either in a trades directory or other literature.
b) Any firm that is known from other sources to have made rules and linear measuring devices even if they did not describe themselves as rule makers.
c) Vendors and specialist suppliers to particular trades whose names are regularly found on rules. Some may have made the products they sold but it is likely that most "bought them in" for resale. Sometimes there are clear indicators that the firms were simply vendors but frequently it is not so.

Criteria for exclusion from the checklist

Hundreds of names have been recorded on rules and other measuring equipment that have not been included.

a) The name is not a recognised maker or retailer and no evidence has been found to suggest that the name is that of a maker or retailer. It is most probably therefore that of an owner.
b) The name is that of an American and other overseas maker. There are brief entries identifying the major American firms such as Stanley Rule & Level Co., Starrett, Brown & Sharpe, Lufkin etc. who sold their products extensively in Britain.

Style of entries

Those firms for whom there is clear evidence that they made rules have the details of town or city, address, date and descriptions as comprehensively as it has been possible to establish.

Those firms who were general tool dealers are only listed in outline, even if their own names were stamped on the rules they obtained from the manufacturers.

Names in bold italics indicate cross-references to other entries in the checklist.

Nomenclature

Surnames and other names, where known, are shown in full and are usually given as found in the trades directories. However, errors in directories are quite common and if there is verifiable information, the correct spelling is given. This may differ from the name marked on their rules.

Businesses have been entered using their normal trading name. The marks used by manufacturers are often simplifications and may not accord with the firm's title. It is also worth noting that marks were not necessarily changed every time there was an alteration of title or ownership. Care should be taken in using a particular form of a name as an indication of date.

Where a business became a company (Co.) or limited (Ltd.) some time during the period of its operation this information has been included in parenthesis.

Order of listing

Strict alphabetical order has not been followed in all cases in order to allow the listing of firms or family members in chronological order or in a manner that best explains their trading relationships.

Addresses

Addresses are given as in trades directories and other sources. In some instances addresses have been compacted, indicated by the use of brackets in which are given alternative premises numbers and addresses.

Street numbers often differ in the directories from one year to the next. This may be due to errors in the directories or may be due to the renumbering of streets, which occurred on a regular basis as development took place. Street names also changed frequently.

In the 18th and 19th centuries the practice of street naming and numbering was casual. Prior to 1750 few streets were numbered and the location of a shop was indicated by a trade sign hung outside the premises. It was common practice to name buildings, terraces, rows, courts, etc. distinct from the street in which they were situated. This might indicate a row of buildings named after the developer who built them and might be given as the address even though the building stood in a well known street. Alternatively, it might be a court off the street or even a side street.

Confusion may also arise when the same addresses are expressed in different ways over the passing years. "Crown Court, Soho" may become "Crown Court, Berwick St." or "Crown Court, St. Martins" or later on "No. 100, Berwick St."

Dates

The dates given in the checklist are the period during which the maker or firm was in business.

Directories, particularly in the earlier years of the period covered by this checklist, often were published at infrequent intervals until annual directories became normal in the later 19th century. Dates cited are only as accurate as that provided by the directory.

Trade catalogues were also used to help in dating but these too were often produced at infrequent intervals.

Dates have therefore been expressed in the following way.

1816 – 1843 Evidence exists to support these dates. They should either be correct or not be more than a year or two out.

<1816 – 1843> Indicates that the firm may well have been in business before and after the dates shown.

Description of makers' trades

Many of the makers of rules and other measuring devices did not describe themselves as rule makers because rules formed only a part of their production. Many of them were makers of scientific instruments in a variety of different categories. The description of a maker's trade given in the listing is as found in the trades directories. The most commonly found trades for rule makers were:

Mathematical instrument makers: This covered all those instruments to do with calculation and measurement and the practical application of mathematics. These included rules of all description, surveying equipment, compasses, levels and theodolites and the drawing instruments and other equipment needed for recording such information.

Philosophical instrument makers: These were instruments concerned with the natural sciences and the study thereof. Included were orreries and globes, barometers, thermometers and other apparatus for investigating and demonstrating the natural world.

Optical instrument makers: Optical instruments were concerned with the nature and quality of light whether direct, refracted or reflected. These included telescopes, microscopes, mirrors and magic lanterns as well as spectacles.

Nautical instrument makers: These included all those instruments related to ships and the sea, including instruments for navigation.

Scientific instrument makers: This term, which covered all of the above trades, did not come into use until the last few decades of the 19th century and even after this date, many makers continued to call themselves mathematical, optical or nautical instrument makers, though the term philosophical instrument maker appears to have been dropped.

Inventors and designers

Also included in the checklist are the inventors and designers of a variety of specialist rules and slide rules for a number of trades. The inventors would not have made the rules that bear their names.

Reproduction of marks

Where available, illustrations of the makers' and retailers' marks found on rules have been included. It should be noted, however, that these are not to scale and, although it is often possible to identify the size of the mark from the layout of the rule, the illustrations should not be taken as an exact indication of the size.

Where several marks for one maker are shown, they have been arranged in what is likely to be the chronological order; however, this is frequently a matter of judgement and should not be taken as totally proven.

Sources

Shown below are the various books and databases consulted in the preparation of this checklist. When researching a checklist of this magnitude, it is impossible to delve too deeply into any individual maker and there is undoubtedly more information that could be found. It would be good to think that this checklist will be the inspiration for others to continue with this research.

Anderson, R. G. W., Burnett, J., Gee, B., *Handlist of Scientific Instrument-Makers' Trade Catalogues* (National Museums of Scotland Information), Edinburgh, 1990.

Clifton, Gloria and Turner, G. Le E., *Directory of British Scientific Instrument Makers 1550–1851,* London, 1995.

Goodman, W., *British Planemakers from 1700.* 3rd ed. Edited by Jane & Mark Rees, Needham Market, 1993.

Hopp, Peter M., *Slide Rules, Their History, Models and Makers*, Mendham, NJ, 1999.

Newcastle University, *Biography Database 1630-1830,* CD, Issue 1, 1995; Issue 2, 1998; Issue 3, 2000.

Roberts, K. D., Reprint of *John Rabone & Sons 1892 Catalogue*, Fitzwilliam, NH, 1982.

Webster Signature Database: An internet database of scientific instrument makers set up by the Adler Planetarium and Astronomy Museum, Chicago, USA. Web site: www.adlerplanetarium.org

Published histories of firms have been noted in the entries for those firms.

A

ABC Rules & Measuring Ltd. — **LONDON**
63 Old Compton Street — 1931 – 1947>

Rule makers.

ABRAHAM, Abraham (& Co.) — **LIVERPOOL**
20 Lord Street — 1818 – 1878
ABRAHAM & DANCER — **MANCHESTER**
(13) (43) Cross Street — 1841 – 1844

Opticians and mathematical instrument makers. Abraham & Dancer was a partnership between Abraham and ***John Benjamin Dancer.*** Abraham Abraham also had retail premises in Glasgow. The business was taken over by ***G.S. Wood & Co.***

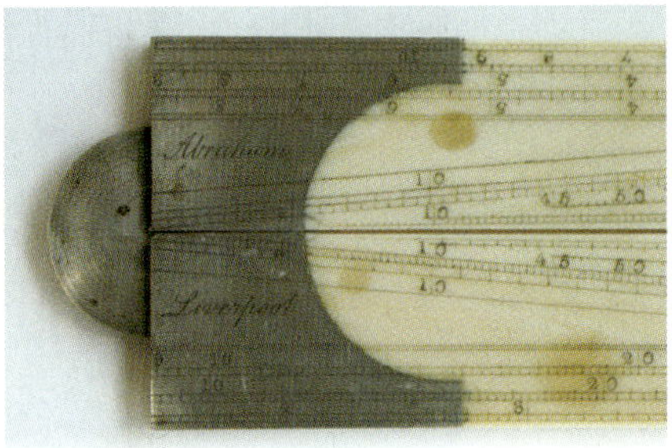

ABRAHAM, Jacob — **BATH**
St Andrew's Terrace — 1805 – 1811
7 Bartlett Street — 1830 – 1851

Optician. He also had premises in Cheltenham. ***John Davis*** was apprenticed to Jacob Abraham and in the 1820s was working at the Cheltenham premises.

ADAMS, George (I) — **LONDON**
60 Fleet Street — 1733 – 1772
ADAMS, Ann & George (II)
60 Fleet Street — 1772 – 1774
ADAMS, George (II)
60 Fleet Street — 1774 – 1795
ADAMS, Hannah
60 Fleet Street — 1795 – 1796
ADAMS, Dudley
53 Charing Cross — 1788 – 1796
60 Fleet Street — 1796 – 1826

Mathematical instrument makers. After George (I)'s death his wife, Ann, continued the business with their son George (II) for two years. George (II) continued until his death, after which, his wife, Hannah, carried on for a year before the premises were taken over by George (II)'s brother Dudley. Although Dudley was made bankrupt in 1817, he appears to have stayed at the premises in Fleet Street until 1826. However, in 1820, the stock in trade was sold to ***W. & S. Jones.*** The premises at 60 Fleet Street were known as Tycho Brahe's Head (Tycho Brahe was a celebrated 16th century Danish astronomer), which has been recorded on rules. A full history of the firm can be found in Millburn, John R., *Adams of Fleet Street,* Oxford, 2000.

ADAMS, James H. — **LIVERPOOL**
66 Park Lane — 1873 – 1919
and 31 Basnett Street — 1873 – 1875

ADAMS, Herbert D.
66 Park Lane — 1920 – 1932

Rule makers.

ADDENBROOKE, Edward — **BIRMINGHAM**
39 Hurst Street — 1850

Rule maker.

ADDINGTON, Thomas — **WOLVERHAMPTON**
Cock Street — 1818

Rule maker. His name was also spelt Agginton.

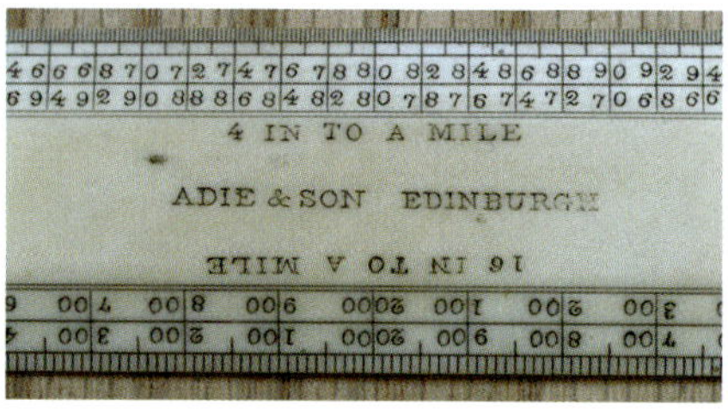

ADIE, Alexander James — **EDINBURGH**
(94) (96) (8) Nicholson Street — 1807 – 1818
(35) (58) Princes Street — 1818 – 1835
ADIE & SON
(58) (70) Princes Street — 1835 – 1880
ADIE & WEDDERBURN

Opticians. Alexander traded with his uncle under the name Miller & Adie from 1804, although this name has not been found on any rules. His uncle died in 1815 but the name of the business remained unchanged until 1822. Adie & Son was a partnership between Alexander and his son John. In 1857 the business was managed by Richard Adie (John's son?). Thomas Wedderburn was employed by Adie & Son from 1835-1880 and was foreman under Richard. To date only one rule has been recorded marked Adie & Wedderburn.

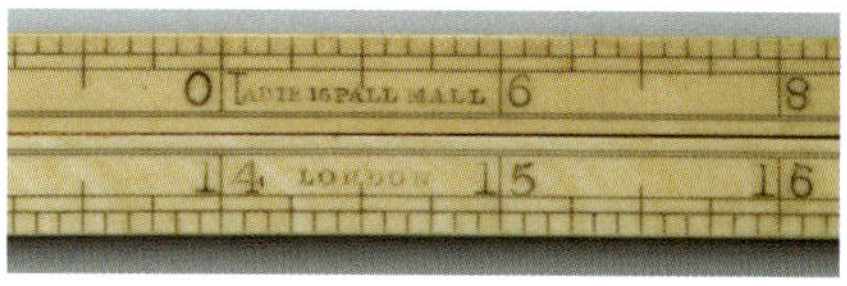

ADIE, Patrick — **LONDON**
395 Strand — 1848 – 1868
15 Pall Mall — 1869 – 1885
29 Regent Street — 1869 – 1870
Tothill Street — 1875
Broadway Works, Westminster — 1890 – 1913
45 Horseferry Road — 1917 – 1931 >

Mathematical instrument maker.

ADLAM, George — **BIRMINGHAM**
Emily Street — 1869
(60) (18) Dymoke Street — 1870 – 1880
6 Court, Prospect Row — 1884 – 1888
13½ Woodcock Street — 1892 – 1907
ADLAM, George, & Son
27 Ravenhurst Street — 1921 – 1948 >

Rule makers. 1902-05: Adlam also had an address in Oxford Street, London.

AIRD & ANDERSON — **LIVERPOOL**
12 Whitechapel — 1876 – 1935>

Tool manufacturers. They did not make rules and any found marked with their name would have been bought in from a rule maker. See *British Planemakers from 1700,* 3rd ed.

AJS & Co.

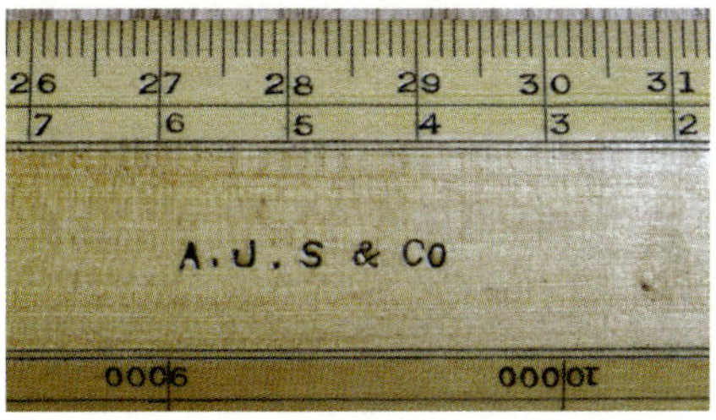

Unidentified but probably a retailer. The name has been recorded on a boxed set of scales.

ALDOUS, J. — **LONDON**
1 London Street — 1834 – 1856
67 Berwick Street — <1837 – 1860

Tool sellers and pawnbrokers. They did not make rules and those found marked with their name would have been bought in. See *British Planemakers from 1700,* 3rd ed.

ALLBUT(T), Isaac — **WOLVERHAMPTON**
(105) (39) Salop Street — 1811 – 1902 >

Rule maker.

ALLEN, Elias — **LONDON**
Blacke Horse-alley — 1607 – 1611
Without Temple Bar, Strand — 1618 – 1653

Mathematical instrument maker in brass and book plate engraver. A variety of addresses are known, but as they are all described as Against St. Clement's Church, they are likely to be in the Strand, just without Temple Bar, and are probably all the same premises.

ALLEN, Jacob — **BIRMINGHAM**
No. 13 Court, Howe Street — < 1852 >

Rule maker.

ALLEN & ROWE — **BIRMINGHAM**
12 St. Mary's Row, St. Mary's Square — 1849 – 1850
ALLEN, Thomas
12 St. Mary's Row and 6 Legge Street — 1850 – 1854
45 Woodcock Street — 1855
ALLEN, William
55 Lombard Street — 1876

Rule makers.

ALLOTT, JONES & Co. Ltd. — **LIVERPOOL**
1–5 Bixteth Street — 1894 – 1910
22 Pall Mall — 1911 – 1946 >

Stationers. One rule recorded with the front view of a steam engine marked on it to signify that it was engine divided.

APPELBY, Edward — **BIRMINGHAM**
146 Summer Lane — 1856

Rule maker.

APPS, Alfred — **LONDON**
433 Strand — 1866 – 1910

Opticians.

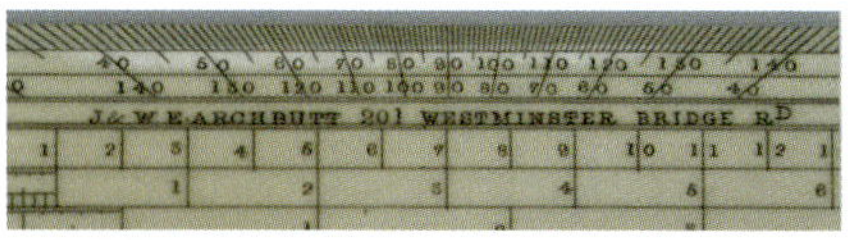

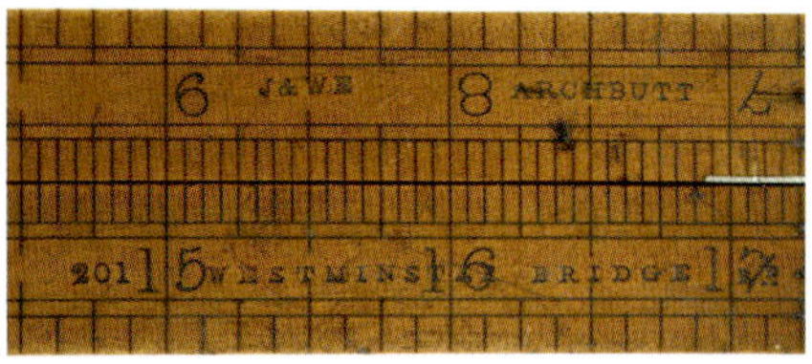

ARCHBUTT, John (& Sons) **LONDON**
20 Bridge Road, Lambeth 1838 – 1864
ARCHBUTT, John & William E.
201 Westminster Bridge Road 1864 – 1892
8 Bridge Street, Westminster

Opticians, mathematical instrument makers and pawnbrokers.

ARCHER, William H. **LONDON**
(11) (45) Googe Street 1842 – 1863

Tool maker and dealer. See *British Planemakers from 1700,* 3rd ed.

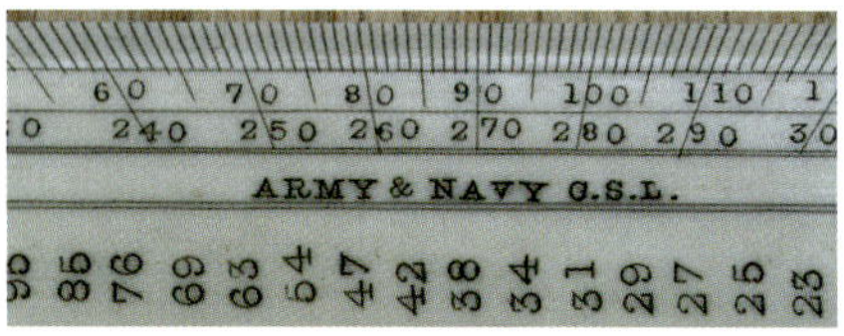

ARMY & NAVY CSL **LONDON**
105 Victoria Street SW1 1871 – ca. 1980

The Army & Navy Co-operative Society Ltd. was founded in the late 19th century to provide equipment and clothing for service personal before overseas postings, expeditions, etc. The Society also had depots in Plymouth, Bombay and Calcutta. They are listed as customers of ***John Rabone & Sons.***

ARNOLD, Thomas **LONDON**
Parish of St Bartholomew < 1730

Mathematical instrument maker. A Gunter's scale has been recorded signed "Thomas Arnold 1700".

ARNOLD & Sons **LONDON**
31 West Smithfield < 1906
52 Wigmore Street 1924 – 1938 >

Established in 1819, and continuing until the mid 1900s the firm were makers and suppliers of medical and veterinary equipment, whose name is found on specialist measuring equipment related to these subjects.

ARONSBERG, Maurice (& Co.) **LIVERPOOL**
4 Manchester Street 1862 – 1878
(17) 39 Castle Street 1867 – 1890
40 Manchester Street 1894 – 1946 >
ARONSBERG, Charles (& Co.)
24 Manchester Street 1890 - 1946 >

Opticians and mathematical instrument makers.

ARSTALL, George **LIVERPOOL**
1 Temple Court 1809 – 1815
50 Castle Street 1821
ARSTALL, Widow, & Son
37 Moorfield 1835 – 1851
and 26 & 28 Cable Street < 1851

Scale and hydrometer makers. George Arstall was married to ***Mary Dicas*** and traded under the Dicas name.

ASHENHURST **BRADFORD**

A rule has been recorded marked "Copyright No. 1 Improved Cotton Calculating Rule". It was unlikely to have been made by Ashenurst.

ASSHETON, Richard

1659

A carpenters' joint rule, in the Science Museum Collection, London, dated 1659, bears this name but nothing further is known.

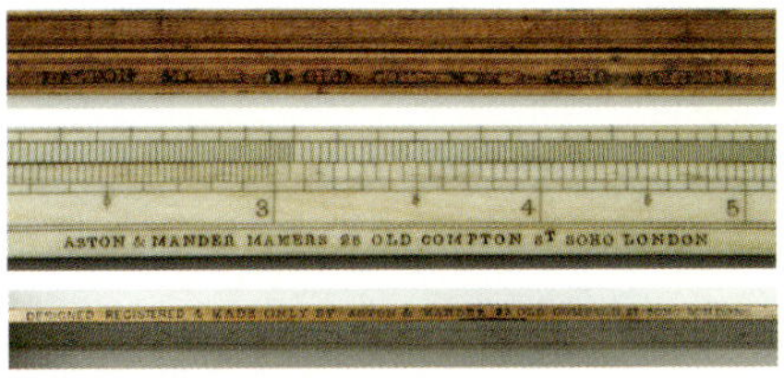

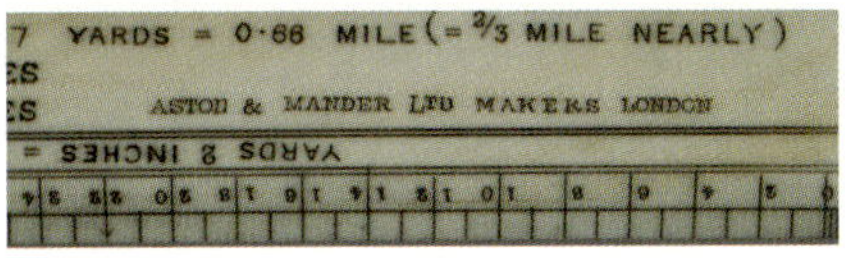

ASTON, Isaac (I)	**LONDON**	
23 Prince's Street, Soho	1811	
25 Compton Street, Soho	1814 – 1840	
ASTON, Mrs		
25 Old Compton Street	1841 – 1842	
ASTON, Isaac (II)		
25 Old Compton Street	1851 – 1870	
ASTON & MANDER		
(25) (61) Old Compton Street	1870 – 1916	
ASTON & MANDER (1917) Ltd.		
(25) (61) Old Compton Street	1917 – 1930	

Rule makers. Aston & Mander was a partnership formed from the businesses of Isaac Aston and ***George Mander.*** The business was renamed in 1917 and some time after 1930, they relocated to Margate but the details have not been established.

ASTON, William	**BIRMINGHAM**
	1770
ASTON, Samuel	
Cheapside	1785 – 1788
5 Water Street	1800 – 1801
Birchole Street	1809 – 1818
Cheapside	1821
ASTON, Thomas (& Sons)	
Coleshill Street	1818 – 1821
20 Bartholomew Row	1823 – 1830
17 Jennens Row	1835 – 1870
25½ Willis Street	1845 – 1890
216 Heneage Street	1858 – 1882
16½ Lupin Street	1895 – 1897
ASTON, Sampson (& Co.)	
1 Jennen's Row	1833 – 1841
23 Masshouse Lane	1850 – 1872

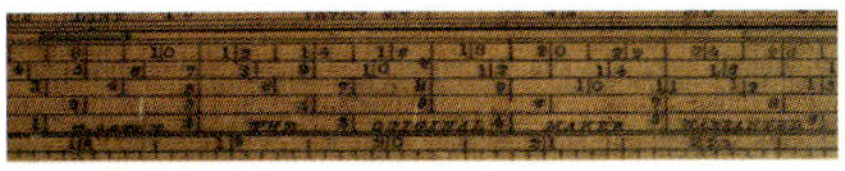

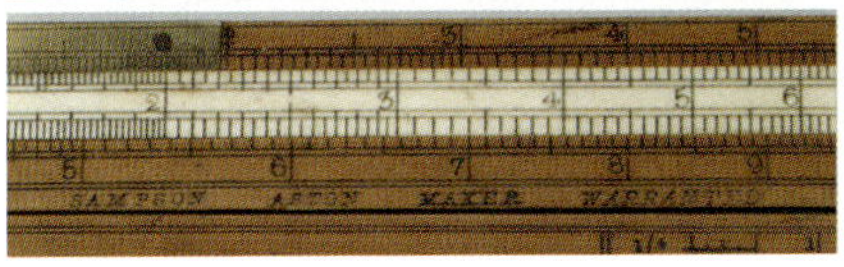

1, Jennens Row,
corner of
Masshouse Lane, Birmingham, 186
M
Bo.t of Sampson Aston & Co
Box & Ivory Rule Manufacturers.
Gross Nett

Whitmore Street, Hockley 1875

ASTON, John

8 Princes Row 1854 – 1860
and (94) (63) Gt. Brook Street
(37) (27) Scholefield Street 1862 – 1882

Rule makers. 1861: Sampson Aston stated that "woods could be cut to any size and prepared for engravers." 1850s – 1870s: There are two separate entries in the directories for T. Aston and John Aston at different addresses but it is likely that all the Aston family were part of the same firm. Jennens (also spelt Jenner's) Row and Masshouse Lane are different versions of the same address. 1876: John Aston at 27 & 28 Scholefield Street, states that "this is also the business name and trade mark of Thomas Aston, the original rule maker of 17 Jennens Row." 1897: The firm of Thomas Aston & Sons was taken over by ***I. & D. Smallwood.*** Advertising posters record the sale of the Sampson Aston business, held on 11 February 1873. Annotations record that amongst the purchasers were ***William Dowell, Samuel Parkes, Edward Preston & Sons, John Rabone & Sons*** and ***I. & D. Smallwood***, all of whome were Birmingham rule makers. The valuable goodwill of the firm together with about 300 pattern books, rules, samples, cases, etc. was bought by John Rabone & Sons for £100.

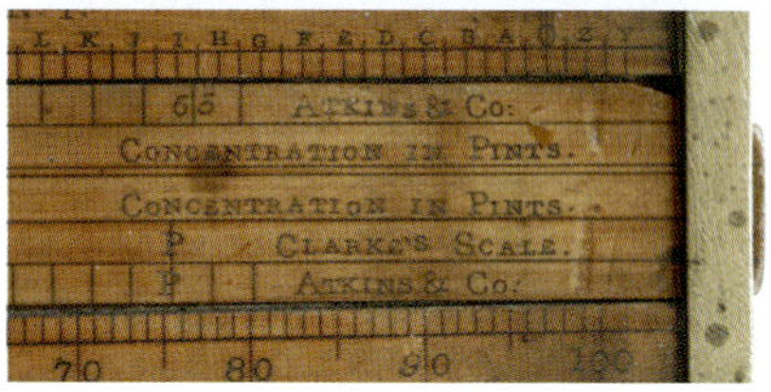

ATKINS, Robert **LONDON**

136 Fenchurch Street 1801 – 1806

ATKINS, George

136 Fenchurch Street 1807
8 Little Charlotte Street, Blackfriars Road 1811 – 1812

ATKINS, George, & Co.

Whitehorse Court, Borough 1813
4 Churchyard Alley, Borough 1815 – 1822
Tooley Street, Southwark 1817

Mathematical instrument makers. George Atkins & Co. also described themselves as opticians.

ATKINSON, James (I) **LONDON**

St. Savior's Dock < 1665
Redriff Wall 1665-1711

ATKINSON, James (II)

21 Cherry Garden Stairs 1700 – 1728

Mathematical instrument makers.

AUSTIN WOOD & Co

A firm of London typefounders, in business towards the end of the 19th century. The name has been recorded on an ivory printers' rule but it is unlikely that they made the rule.

AVERY, W. & T., & Co. **BIRMINGHAM**

1817 – 1960>

AVERY DE GRAVE

1900 >

Well known makers of weighing equipment including equipment for weights

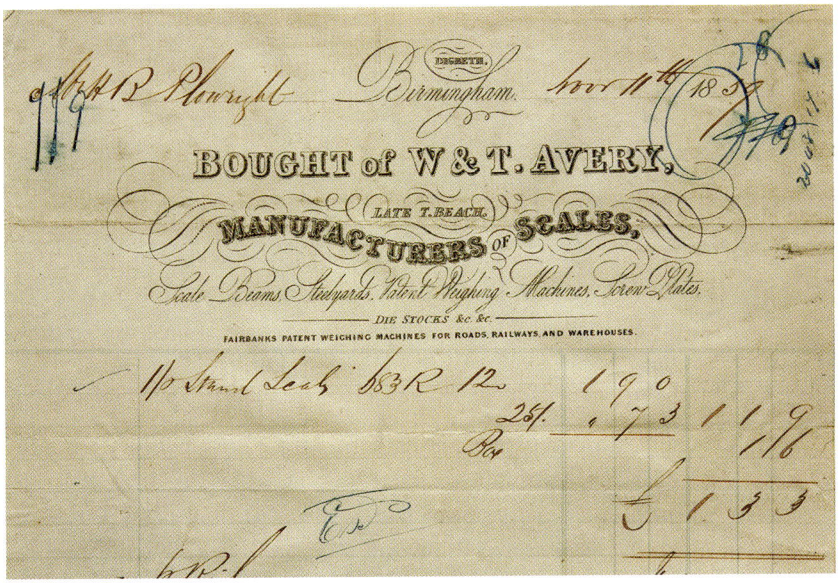
DIGBETH.
Birmingham
BOUGHT of W & T. AVERY,
LATE T. BEACH
MANUFACTURERS OF SCALES,
Scale Beams, Steelyards, Patent Weighing Machines, Screw Plates,
DIE STOCKS &c. &c.
FAIRBANKS PATENT WEIGHING MACHINES FOR ROADS, RAILWAYS AND WAREHOUSES.

and measures inspection. Some standard lineal measures have been recorded marked with their name. ***De Grave, Short Fanner & Co.*** joined the Avery group in the 1900s, with Avery De Grave being used as one of the trading names.

B

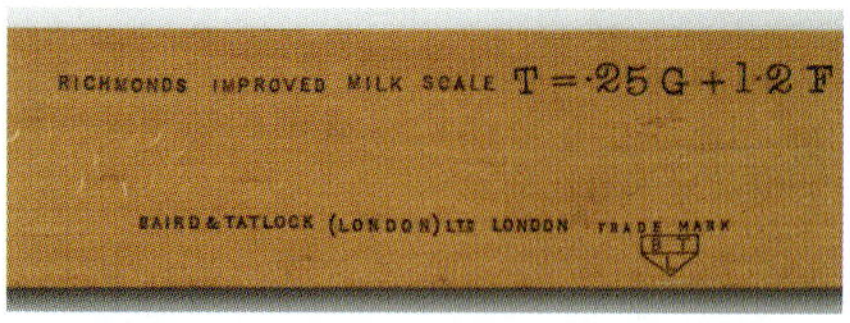

BAIRD & TATLOCK Ltd. — **LONDON**
14 Cross Street — < 1905 – 1943 >

Scientific instrument makers, laboratory furnishers, etc.

BAKER, Charles — **LONDON**
244 High Holborn — 1851 – 1947 >

Mathematical and optical instrument makers. By the 20th century the business was described as scientific instrument makers. They also traded at various times as Curtis, Charles, Lee and Charles, Lees, Curtis and T. J. Offer. Although they described themselves as "makers", according to the reminiscences of one of the instrument makers employed by ***J. J. Threadwell***, many of the drawing instruments sold by Baker in the later years came from Threadwell.

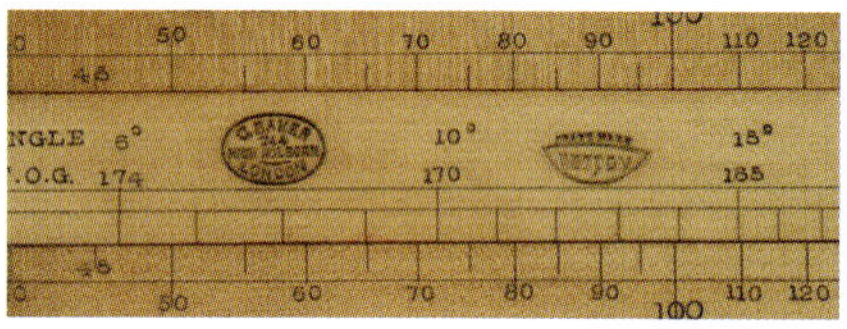

BAKER, Henry — **BIRMINGHAM**
Court, 61 Cecil Street — 1839 – 1842
41 New Town Row — 1842 – 1862
BAKER, Arthur
Upper Sutton Street — 1870
BAKER, Thomas
(10) 36 Theodore Street — 1876 – 1886

Rule makers.

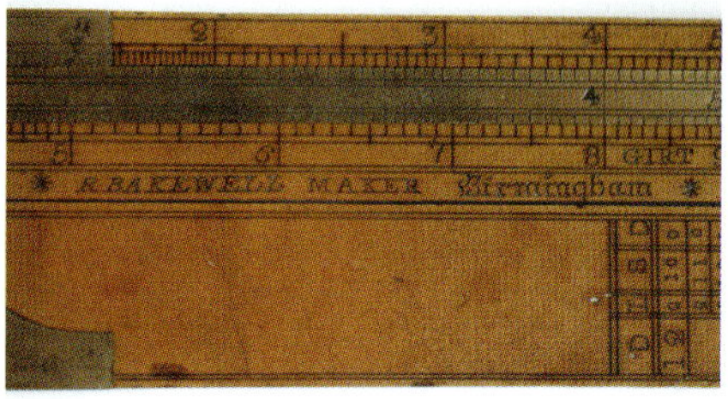

BAKEWELL, Richard — **BIRMINGHAM**
49 Loveday Street — 1792 – 1826
BAKEWELL, Isabella
42 Loveday Street — 1828 – 1829
BAKEWELL, W.
1854

Rule and mathematical instrument makers. 1815: manufacturers of mathematical instruments, brass compasses, plated and brass dog collars, ivory and box rules, land surveyors' measuring tapes, etc. Richard Bakewell's business was taken over by ***Isaac Trow***.

BALL, Thomas — **GARSTANG**

Unidentified, but unlikely to be a maker. The name has been recorded on a tennis court tape.

BAMES & MORRIS — **LONDON**
Aubrey House, Ely Place, EC1

Unidentified but probably a retailer. The name has been recorded on a protractor.

BANKS, Anthony & Robert — **LONDON**
25 Piccadilly — 1792 – 1795
440 Strand — 1765 – 1804

441 Strand 1805 – 1830
BANKS, Robert, & Son
119 New Bond Street 1831 – 1834

Optical and mathematical instrument makers. The name is often found on paper rules and straight edges. It was also spelt Bancks.

BARKER, Francis & Sons **LONDON**
12 Clerkenwell Road 1890 – c. 1965

Wholesale makers of ships', pocket and prismatic compasses, sextants, telescopes, marine and pocket aneroids, drawing and surveying instruments, Barker's patent altitude instruments, sun dials etc. A continuation of ***Groves & Barker***.

BARLING, W.H. **LONDON**
Navigation Warehouse, Limehouse 1822 – 1824
BARLING, Elizabeth
Navigation Warehouse, Limehouse 1826
BARLING, Frederick
Navigation Warehouse, Limehouse 1836 – 1840

Mathematical instrument makers.

BARNETT, Thomas **LONDON**
(6) (61) (East) Tower Street 1789 – 1796
East Street 1799 – 1816
No. 4 Mores Yard

Mathematical instrument makers.

BARNSLEY, George, & Son **SHEFFIELD**
Cornish Works 1836 – 1927 >

The leading makers of tools for shoemakers, their name is found on many measuring items for this trade. They would not have made any of the rules bearing their name.

BARRATT, Jonathon **BIRMINGHAM**
Court 1, Price Street 1855 – 1858

Rule maker.

BARRINGTON, Philip **BIRMINGHAM**
19 Court, Lower Hospital Road 1839
176 (178) Unett Street 1845 – 1865
57/58 Barr Street 1867 – 1901
BARRINGTON, Philip, & Son
57/58 Barr Street 1904 – 1915
BARRINGTON, Samuel
8½ Guest Street 1904

Rule makers.

BARRISHES-ROBERTS Ltd. **LONDON**

Unidentified but probably a retailer. The name has been recorded on a dip rod.

BARRON, T.G. — **WEST HARTLEPOOL**
30 York Road
BARRON & HARDING
Unidentified. The name has been recorded on a calculating rule also marked with "pounds to the inch" and "ordinate" and a printers' rule.

BARROW, Henry — **LONDON**
1790
18 Crown Court, Soho — 1824 – 1826
26 Oxenden Street — 1849 – 1864
BARROW, Henry, & Co.
26 Oxenden Street — 1850 – 1869

Mathematical instrument makers. Henry worked as an "outworker" for ***Dollond*** and ***Troughton***. He was appointed mathematical instrument maker under the Surveyor General of India and worked in Calcutta between 1830 and 1839. In 1842 he succeeded to the business of ***Thomas Charles Robinson*** and traded under the name of ***Robinson & Barrow***.

BARROW, Joseph — **BIRMINGHAM**
11½ Branston Street — < 1852 >
BARROW, F.
< 1863 >

Rule makers.

BARRODALE, T. H. — **BIRMINGHAM**
<1888>

Unidentified, but could be a maker. Rules marked "Patent R^d 97038" have been recorded. The name is also spelt "Barrowdale".

BATE, Robert Brettell — **LONDON**
(17) (20) (21) The Poultry — 1807 – 1842
33 Royal Exchange — 1846 – 1850

Mathematical instrument makers. Bate was the nephew of Bartholomew Sikes, the inventor of the hydrometer accepted by the Excise Board in 1802 as the standard instrument. By 1808 Bate was the official supplier to the Excise Commissioners. After their introduction in 1824, Bate was commissioned to make the new Imperial Standard Weights & Measures. His trade cards state: "Chart Agent to the R^t. Honble the Lords Commissioners of the Admiralty" and "…to the Honourable Board of Excise Hydrometer maker for the Revenue of the United Kingdom." After Robert Bate died in 1847, the business was continued by his widow until finally closing in 1850. The hydrometer and gauging instrument part of the business were then taken over by Edward Ladd and John Streatfield, two of Bates' workmen, who traded as ***Ladd & Streatfield***. The Chart Agency was taken over by ***J. D. Potter***, who was also one of his workmen. For a history of the firm, see McConnell A., *R. B. Bate of the Poultry, 1782 – 1847,* Scientific Instrument Society, London, 1993.

BATES, Edwin — **BIRMINGHAM**
3 Marshall Street, Holloway Head — 1850

Rule maker.

BATES, Richard **LONDON**
1730
Quadrant without Newgate 1740
St. Sepulchres 1744
1750

Mathematical instrument maker. He took ***Richard Bates Gearing*** and ***Thomas Rolph*** as apprentices (amongst others). He advertised in *The Young Mathematician* in 1740 and the *Daily Advertiser* in 1744.

BAYLIS, William **LIVERPOOL**
(56) (67) (75) (85) (81) Dale Street 1834 – 1851
27 South John Street 1851 – 1864
81 Dale Street 1864 – 1867
67 Park Lane 1870 – 1873

Rule and mathematical instrument makers, opticians, brass founders. There are a variety of other Liverpool addresses in the directories but some of these are likely to have been residences.

BECKER, Charles **LONDON**
30 Strand 1861

Mathematical instrument maker.

BEDFORD, Hilkiah **LONDON**
Hosier Lane 1654 – 1666
Fleet Street 1666 – 1689

Mathematical instrument maker and engraver.

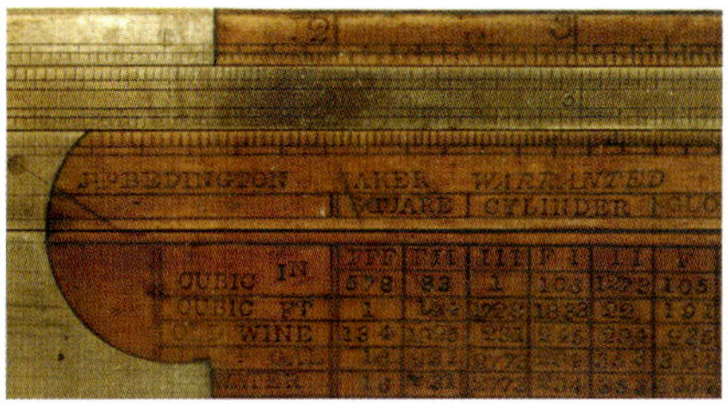

BEDDINGTON, John **BIRMINGHAM**
40 Digbeth, St Martins Lane 1845 – 1852
BEDDINGTON, Mrs Emma
40 Digbeth 1854 – 1858
25 Ravenhurst Street 1863

Mathematical instrument makers, opticians, manufacturers of rules and measuring tapes.

R. Bedington & Sons trade mark

BEDINGTON, Robert & Sons **BIRMINGHAM**
107 Lancaster Street 1841
BEDINGTON, George
10 Court, Snow Hill 1845
BEDINGTON, James
10 Russell Street, St. Mary's Square 1845 – 1870
BEDINGTON, James, & Son
10 Russell Street 1876 – 1890

Measuring tape makers. 1858: Mathematical instrument makers. 1860: "late Cutts, Chesterman & Beddington". 1884: Mathematical instrument makers. James Bedington was also a partner in the firm of ***Cutts, Chesterman, & Bedington***, Sheffield. The name is sometimes spelt Beddington.

BEDWELL, Geo. **LONDON**
4 Little Alie Street 1850 – 1870

54 Great Awe(?) Street

Hydrometer and saccharometer maker.

BEDWORTH, John **BIRMINGHAM**
49 Great King Street < 1852 >
BEDWORTH, Thomas
68½ Smith Street, Gt. Hampton Row 1863

Rule makers.

BEES, Charles H. **BIRMINGHAM**
BEES, HIGGISON & Co.
10 Bishop St 1895
135 Park Road 1900 – 1909

Rule makers. In 1894, ***John Higgison*** and Charles Bees combined to form Bees, Higgison & Co. In 1910 the company was renamed ***Rulers Ltd***.

BEILBY, Richard & Charles **BRISTOL**
2 Clare Street 1809 – 1812
BEILBY, Charles
2 Clare Street 1814 – 1819

Optical, mathematical and philosophical instrument makers. R. & C. Beilby took over the business of ***Joshua Springer***. The Beilby business was taken over by ***John King***.

BELCHER & HASSALL **BIRMINGHAM**
12 Bull Street 1791 – 1798

Rule makers. A partnership between ***Zachariah Belcher*** and ***Joseph Hassall***. Although listed in the directories until 1798, by 1794 Zachariah Belcher had moved to Sheffield.

BELCHER, Zachariah **SHEFFIELD**
Sheffield Moor 1794
4 Bright Street 1811 -1825
BELCHER, Zachariah & Sons
4 (1) Bright Street 1828 – 1837
270 (272) Bright Street 1841 – 1871
(270) 272 Fitzwilliam Street 1876 – 1877
74 Mary Street 1879 – 1900
Wallace Works, Furnival Street 1902 – 1906
101 Napier Street 1909
BELCHER, John
(1) 2 Bright Street 1833 – 1837
Ecclesall New Road 1841
270 Bright Street 1852 –1864
BELCHER, Henry
270 Bright Street 1841

Ivory and box rule makers. In 1794 Zachariah Belcher moved from Birmingham and set up as a rule maker in Sheffield. (In the 1797 directory the name is given as Zachary, but this is probably a diminutive.) Zachariah purchased his Freedom of the Cutlers' Company in 1797 as a knifemaker. 1852: John

Belcher's business listed as Zachariah Belcher & Sons. 1868: Manufacturers of English and foreign rules. In 1820, one of Zachariah's sons, Thomas, emigrated to America, followed in 1822 by his brother William and sister Martha and then in 1824, another brother, Charles. In 1822, Thomas Belcher set up as a rule maker in New York. When he was joined in 1824 by his brother William, the firm was re-named T. & W. Belcher and in 1827, Belcher Bros. In 1851, when the sons of William and Thomas took over the New York business, the name became Belcher Bros. and Co.

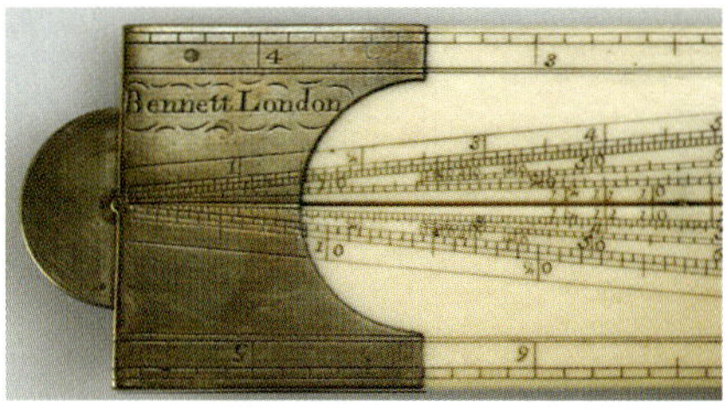

BENNETT, John — **LONDON**
The Globe, Crown Court, Soho — 1735 – 1770

Mathematical instrument maker. He took as an apprentice James Fitch, the father of James Fitch of ***Fitch & Jones***, and ***James Search***. He made Fowler's Sliding Rules and Halfpenny's Tangent Rules. On the back of a trade card in the Heal Collection in the British Library, there is a receipt, dated May 24, 1760, for a Hadley's quadrant costing £2 10*s* 0*d* and a Gunter's scale and long compasses costing 4*s* 0*d* which were bought by Ch. Younge. The business was taken over by ***James Search***.

BENNETT, Wm. — **BIRMINGHAM**
Top of Hunter's Lane — 1854

Rule maker.

BENTLEY & TAYLOR — **LONDON**
105/107 Endwell Road, SE4 — 1950s – present

Rule makers.

BERGE, Matthew — **LONDON**
199 Piccadilly — 1800 – 1819

Optician and mathematical instrument maker. He took over the business of ***Jesse Ramsden***, by whom he had been employed. Berge's business was taken over by ***Worthington & Allan***.

BERRICK, Ralph — **WOLVERHAMPTON**
Queen Street — 1770

Rule maker.

BESWICK, W. — **BIRMINGHAM**
Albert Buildings, Camp Hill — 1862

Rule maker.

BETTERIDGE, John — **WOLVERHAMPTON**
2 Charles Street — 1792

Rule maker.

BETTS, Thomas — **BIRMINGHAM**
Top of Bromsgrove Street — 1816
Inge Street — 1818 – 1821
(42) (125) (128) Great Brooke Street — 1835 – 1862
BETTS, William

102 Bromsgrove Street 1828 – 1830
1 Court, Exeter Row 1835
BETTS, James
31 Horsefair 1823 – 1826
228 Bristol Street 1828 – 1830
32 Duke Street 1829 – 1830
7 Court, Ashtead Row 1839 – 1842
BETTS, John
9 Blucher Street, Exeter Road 1823 – 1826
46 St. James Place 1850
Vauxhall Road 1852
BETTS, Edwin
2 (3) Marshall Street 1849 – 1882
BETTS, Peter
5 Court, St James Place, Vauxhall Road 1860 – 1865
BETTS, Richard
2 Marshall Street 1875

Rule makers.

BIRCHMORE, William **LONDON**
Without Newgate 1672

Ironmonger.

BIRD, John **LONDON**
New Exchange Buildings 1747-1763
Berwick Street 1771
29 Little Marylebone Street 1772-76

Mathematical instrument maker. He made a dividing engine for the Royal Observatory, Greenwich. He also made the 36 in. brass rule used by James "Athenian" Stuart to measure the Acropolis for his book, *Antiquities in Athens*, published in 1751.

BIRD, Joseph **BIRMINGHAM**
239 Heneage Street 1882

Rule maker.

BIRMINGHAM RULE Co. **BIRMINGHAM**

Unidentified but possibly a maker. The name has been recorded on a 2 ft. 4-fold boxwood rule with brass slide and marked in inches, gamla and nya (Swedish old and new inches).

BLAINEY, William **BRISTOL**
5 Lower Maudlin Street 1847 – 1853

Rule and gauging instrument maker. The business was taken over by ***Alfred Shellard***.

BLAKE, T. O. (Ltd.) **LONDON**
(57) (58) (40) Hatton Garden 1905 – 1947 >

Hydrometer makers. The initials T. O. stand for Thomas Odempsey, which were also the forenames of ***T. O. Buss***. The assumption must be therefore that there was a connection between the two firms – perhaps they were alternative trading names for the same firm.

BLAKELEY, Benjamin **LONDON**
(15) (125)Lambeth Lower Marsh 1817 – 1840
21 Lambeth Upper Marsh 1841 – 1845

Rule and mathematical and philosophical instrument maker. Other addresses in Lambeth are also recorded.

BLEULER, John **LONDON**
27 Ludgate Hill 1790 – 1822

Optician and mathematical and philosophical instrument maker. The address on a trade card in the Heal Collection in the British Library is given as 11, 12 & 13 Ludgate Street.

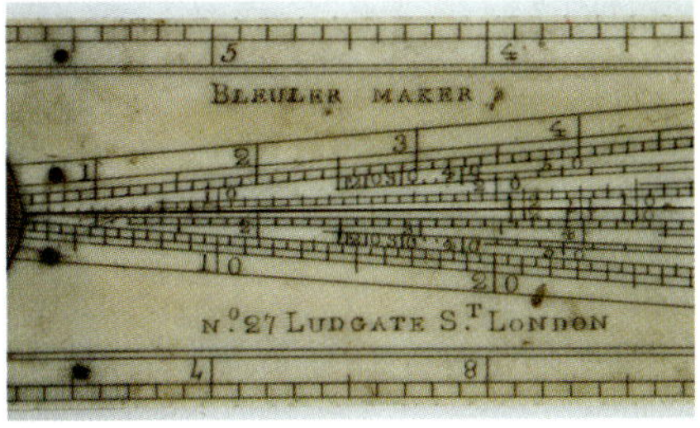

BLEYGHTON, John **LONDON**
1626
Tower Street 1630 – 1631
1654

Mathematical instrument maker.

BLOOMER **BIRMINGHAM**
18 Court, Staniforth Street 1852

Rule maker.

BLOXAM, H. **BANBURY**

Unidentified but probably a retailer. The name has been recorded on a range finder.

BLUNDELL RULES Ltd. **LUTON**
Chaul End Lane 1948 – 1956
WEYMOUTH
Regulus Works, Lynch Lane 1956 – 1964
BLUNDELL HARLING
Regulus Works, Lynch Lane 1964 – present

Makers of drawing scales, slide rules and other draughting equipment. The firm was form to create ongoing post-war work for a machine shop created during WWII. Early rules are marked B.R.L for Blundell Rules Limited. Later rules are marked BLUNDELL. In 1964 they took over ***W. H. Harling*** to form Blundell Harling. For a description of rule making at Blundell Harling, see David Goodison, "Crafting Scale Rules by Hand – An Historical Note", *Journal of the Oughtred Society*, Vol. 7, No 2, 1998.

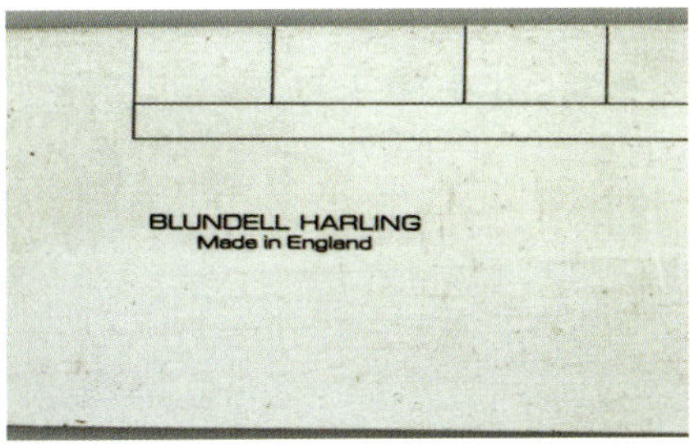

BLUNT, Thomas **LONDON**
22 Cornhill 1794 – 1822

Apprenticed to ***Edward Nairne*** in 1760, he was admitted to the Company of Spectacle Makers in 1771. He formed a partnership with Nairne trading as ***Nairne & Blunt*** from 1774 until 1793 and continued the business after Nairne's death.

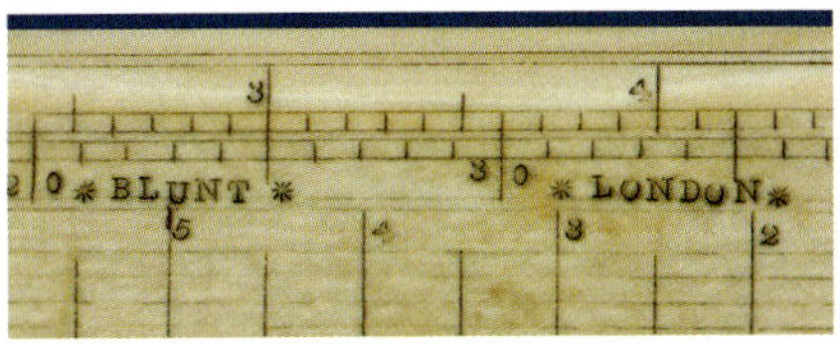

BLUNT, Thomas — **BIRMINGHAM**
Griffiths Bldgs. Carver Street — 1852
49 Fordrough Street — 1860

1852: Rule cutter. 1860: Rule maker.

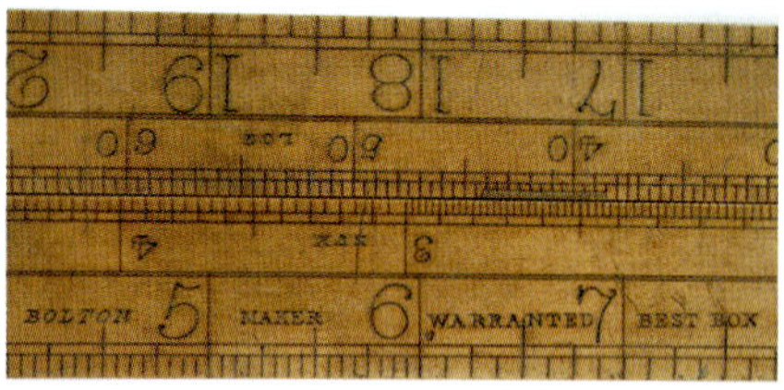

BOLTON, Thomas — **BIRMINGHAM**
6 Colmore Row — 1809 – 1818
61 & 62 Loveday Street — 1828 – 1835
BOLTON, Frederick
7 Vauxhall Grove — 1839
BOLTON, Frederick & Henry
61 & 62 Loveday Street, Steelhouse La. — 1841 – 1858

Rule makers.

BOTWRIGHT & GREY — **LONDON**

Opticians. They would not have made the rules that bear their name.

BOURNE, John — **WOLVERHAMPTON**
7 Wadhams Hill — 1792

Rule maker.

BOUSFIELD, George — **GRIMSBY**
3 Sylvan Crescent, Hainton Street — < 1886
250 Hainton Street — 1890
Eccleston Villa — 1891 – 1902

Commercial traveller in the timber trade. In 1886, George Bousfield patented the "Vade Mecum Slide Rule", designed for the timber trade.

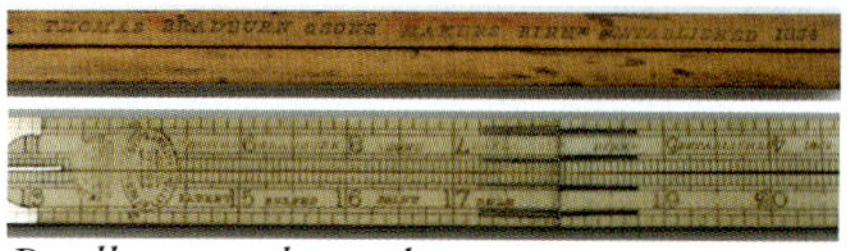

Bradburn trade marks

BRADBURN, Thomas — **BIRMINGHAM**
7 Court, Digbeth — 1835
BRADBURN, George & Thomas
18 (19) Alcester Street — 1839 – 1852
BRADBURN, G.
18 (19) Alcester Street — 1858
BRADBURN, Thomas
111 Gt. Charles Street — 1852 – 1861
BRADBURN, Robert
Back of 22 Bishop Street — 1862 – 1863
BRADBURN, Thomas & Sons
111 Gt. Charles Street — 1862 – 1882
BRADBURN, George
32½ Bell Barn Road — 1876

Rule makers and manufacturer of rules, measuring tapes, joiners' and edge

Memorandum.
From Thomas Bradburn & Sons
GREAT CHARLES STREET.
BIRMINGHAM.
Aug 29th 1872.
To Messrs J. Rabone & Sons
Hockley

tools, general hardware factors. The firm claimed to have been established in 1834. In 1860 an address at 201 City Road, London was also included in the trades directory. In 1876 the firm was assimilated into Ed. Preston who continued to make rules bearing this name until 1914.

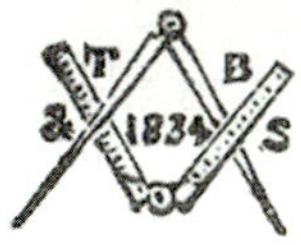

T. BRADBURN & SONS.

BRADFORD, Isaac & John — **LONDON**
(69) (87) Bell Dock, Wapping — 1795 – 1815
136 Minories — < 1822 >
BRADFORD & Son
136 Minories — < 1822 >
BADFORD, George
99 Minories — 1817 – 1843

Mathematical, nautical and optical instrument makers.

BRADLEY, John — **BIRMINGHAM**
(8) (10) 5 Hurst Street — 1829 – 1842
3 Horse Fair — 1852
159 Bromsgrove Street — 1858 – 1861

Rule maker.

BRADLEY, Richard — **LONDON**
20 (23) Compton Street — 1838 – 1839

Rule and tool maker.

BRAILSFORD, John — **LONDON**
St Martin's Court, Leicester Fields — ca. 1750

Cutler. His trade card advertised drawing instruments, rules, etc. "Pawson & Brailsford" has ben recorded on a set of drawing instruments.

BRADSHAW, Alexander — **BIRMINGHAM**
Court, 4, Communication Row — 1854 – 1861
14 Bell Barn Road — 1875
295½ Bell Barn Road — 1876

1858: Box & ivory rule maker. 1875: Rule finisher. 1876: Ebony rule maker.

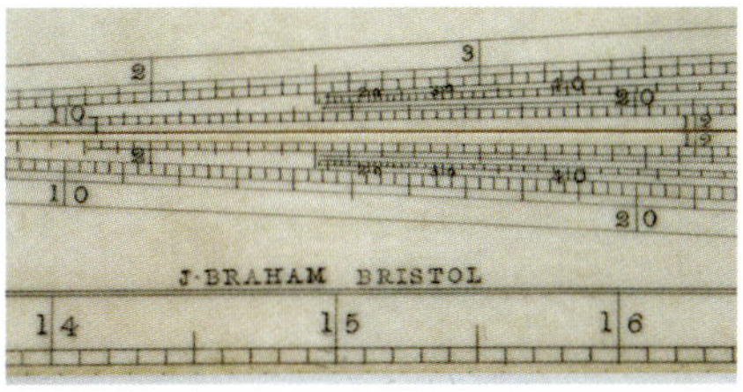

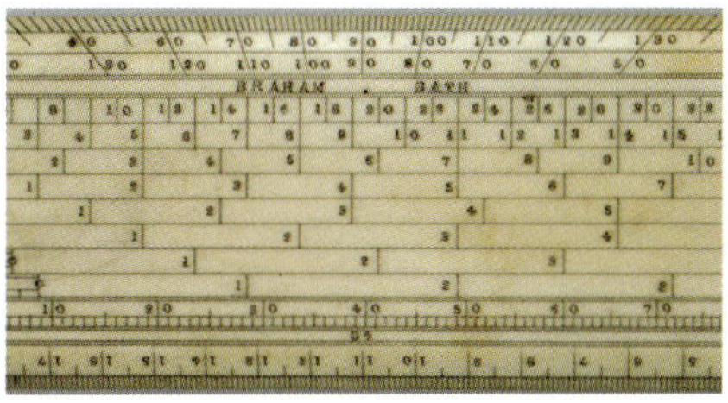

BRAHAM, J. — **BRISTOL & BATH**
12 Clare Street, Bristol — 1828
42 College Green, Bristol — 1829
10 St. Augustine's Parade, Bristol — 1830 – 1838
8 Pulteney Bridge, Bath — 1833
5 York Buildings, Bath — 1837
BRAHAM & Co.
7 St. Augustine's Parade, Bristol — 1840 – 1851
BRAHAM Bros.
7 St. Augustine's Parade, Bristol — 1852
BRAHAM, J.
7 St. Augustine's Parade, Bristol — 1853 – 1863
BRAHAM, J. & J.
7 St. Augustine's Parade, Bristol — 1864 – 1865

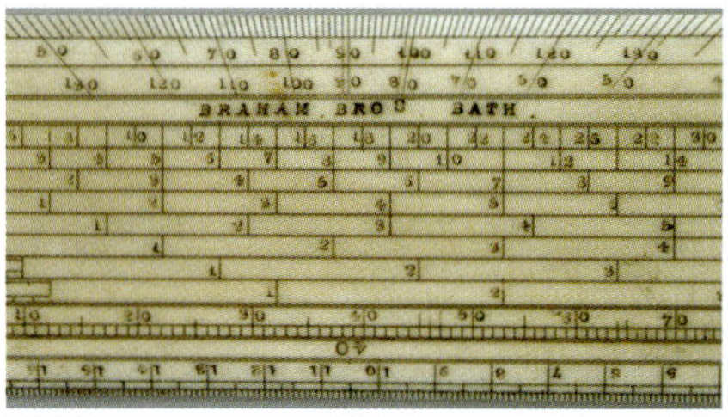

BRAHAM, Joseph
24 College Road, Bristol — 1866
BRAHAM, James D.
30 Triangle Road, Bristol — 1866 – 1867
BRAHAM, L., & Co.
20 St Augustine's Parade, Bristol — 1868 – 1874

Optician and mathematical instrument makers. Although not listed in the directories, Braham Bros. also had premises in Bath. The business was continued by ***M. W. Dunscome***.

BRAILEY, John — **BIRMINGHAM**
142 Ashtead Row — 1880 – 1882

Rule maker.

BRATT, Edwin — **MANCHESTER**
3 Harris Street, City Road — 1861 – 1871
18A Piccadilly — 1875
146 London Road — 1886
46 Buxton Street, London Road — 1890

Rule maker.

BRIGG, T. & Sons — **LONDON**
St. James Street — 1860 – 1938 >

Manufacturers of umbrellas, walking sticks, riding whips, etc. They also had a branch in Paris.

BRITTLE, Joseph — **BIRMINGHAM**
Palmer Street — 1835 – 1839
46 Key Hill — 1854
Ledsam Street — 1855
BRITTLE, John
43 Lower Windsor Street — 1852
81a Brearly Street — 1882

Rule makers.

BROCKHURST, William — **BIRMINGHAM**
(6) (9) Chesher Street — 1880 – 1886

Rule maker.

BROOKE(S), George — **BIRMINGHAM**
12 St James St — 1852 – 1863

Rule maker.

BROUNCKER, Joseph — **LONDON**
1719

He was known to have sold a rule to Christ's Hospital School.

BROWN G.
St Mary's Street — < 1841 >

Unidentified. The name recorded on Customs & Excise slide rule dated 1841 but it could be that of an owner.

BROWN & SHARPE Mfg. Co.

Providence, Rhode Island, U.S.A. Leading manufacturers in the U.S.A. of machinists' tools, the company was founded in 1833 by David Brown and his son Joseph R. Brown. In 1853 Lucian Sharpe joined Brown to form J.R. Brown & Sharpe. In 1866 S. Darling joined the partnership and the company traded under the name of Darling Brown & Sharpe until 1892. The Browne & Sharpe Mfg. Co. was incorporated in 1868. Their name is frequently recorded on engineering measuring devices. The company is still in business.

BROWN(E), Thomas (I) — **LONDON**
The Globe, Fenchurch Street — 1626 – 1657
BROWN(E), John
Sign of the Sun Dyal, Minories — 1654 - 1697
BROWN(E), Thomas (II)
1676 – 1679

Mathematical instrument makers. Thomas (I)'s apprentices included ***Henry Sutton*** and ***Walter Henshaw***. He is known to have made a circular wooden slide rule in 1631. The "new sliding rule" that Samuel Pepys that sought to have engraved on 10 August 1664 is thought to have been made by John Browne.

BROWN(E), William — **BIRMINGHAM**
1 Vittoria Street — 1835
13 Court, Great Hampton Row — 1852
BROWN(E), William Henry
11 Caroline Street — 1835 – 1870
and 11 Castle Street — 1839
BROWN(E), William Henry, & Co.
10 Fisher Street — 1875 – 1886

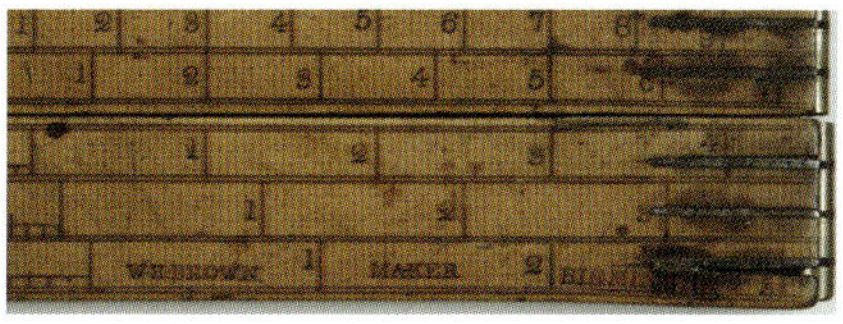

Rule makers. 1876: Manufacturers of box and ivory rules; chain scales; surveyors' dipping rods; plain, ivory edged and metal parallels; cattle gauges; clinometers; size sticks and most of the articles connected with the trade. The name is also spelt Browne. From 1828 to 1830, 1 Vittoria Street was the address of ***George Cox***.

BROWN, William — **MANCHESTER**
Chorley Lane, Chorley

Unidentified. He designed a cotton cloth calculator slide rule.

BRUETON, John — **BIRMINGHAM**
252 Heneage Street — 1882

Rule maker.

BRYAN, E. — **MANCHESTER**
(24)(25) Cross Street — 1865 – 1871

Mathematical instrument maker.

THE BUCK FAMILY — **LONDON**

There were several branches of the Buck family in the tool making and selling business working from 1826 to the present day. The firms' names included

Joseph Buck, George Buck, Buck & Ryan and Buck & Hickman. See also ***George Ryan***. See *British Planemakers from 1700,* 3rd ed.

BUCHANAN BROS. **GLASGOW**

Unidentified. The name has been recorded on a boxwood scale.

BULL, John **LONDON**
Exchange Gate 1582

Mathematical instrument maker (in metal).

BURCHETT, R. & W. **RUDGWICK**

Unidentified, but probably a retailer. The name has been recorded on rules marked "Best Box Rules".

BURROUGHS & WATTS **LONDON**

A firm of billiard table makers and suppliers of equipment for billiards, operating from the middle of the 19th century. Their name is found on billiard cue calipers. It is unlikely that they made the products they sold.

BURROWS, W. & J. **LONDON**
15 Seething Lane

Unidentified but probably a retailer.

BURT, E. A. P. **LONDON**
Surrey Commercial Docks <1887 – 1906 >

Chief measurer and designer and patentee of a variety of rules and slide rules for the timber trade. He also published a books and tables for both timber and corn.

BUSBY, George **BIRMINGHAM**
14 Aston Street < 1849 – 1879 >
BUSBY, Edwin & Sons
Dartmouth Street < 1896 >

Ironmonger, wire worker, weaver and improved land surveyors' measuring chain makers.

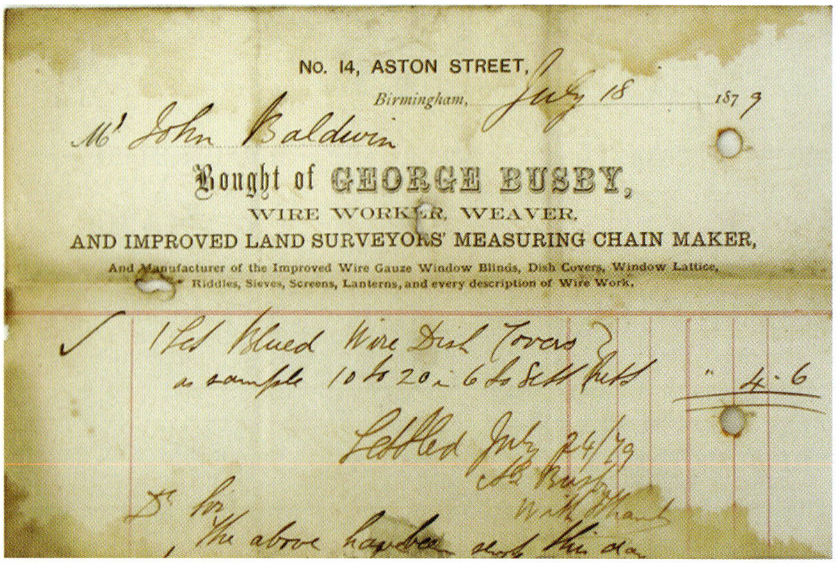

NO. 14, ASTON STREET,
Birmingham, July 18 1879
Mr John Baldwin
Bought of GEORGE BUSBY,
WIRE WORKER, WEAVER,
AND IMPROVED LAND SURVEYORS' MEASURING CHAIN MAKER,
And Manufacturer of the Improved Wire Gauze Window Blinds, Dish Covers, Window Lattice, Riddles, Sieves, Screens, Lanterns, and every description of Wire Work.

1 Set Blued Wire Dial Covers
a sample 10 to 20 ... 4·6
Settled July 24/79
With thanks

BUSS, Thomas Odempsey	**LONDON**
3 Upper Street	1864
33 (48) Hatton Garden	1865 – 1911
77 Clerkenwell	1921
33 Layton Street	1930 – 1947 >

Sykes hydrometer, saccharometer, and steam salinator manufacturers to the Admiralty, Customs and H.M. Dockyards. Agent by direct appointment from the Lords of the Admiralty for the sale of charts, plans and views as issued by them. In 1870 he claimed to be the Successors to the Inventor of the Sykes hydrometer. By 1947 the business was described as scientific instrument makers. See also ***T. O. Blake***.

BUTLER & POWELL	**BIRMINGHAM**
	1812
Lionel Street	1816 – 1823
BUTLER, Benjamin (I)	
4 Court, Lionel Street	1828 – 1829
65 Brierly Street	1829 – 1830
Snow Hill Grinding Mill	1835 – 1842
& (88) (86) Brearly Street	1842 – 1855
(2 ½) (1) (1½) Brearly Street	1858 – 1865
and 1 Hospital Street	1862
BUTLER, Benjamin (II)	
New Street, Aston New Town	1865 – 1876
BUTLER, William	
2 Phillips Street	1876

Rule makers. Butler and Powell was a partnership formed with ***Robert Powell***, but it has not been established whether this was with Benjamin (I) or a predecessor.

BYWATER & Co.	**LIVERPOOL**
BYWATER, John, DAWSON & Co.	
49 Gloucester Street	1816
(18) (20) Pool Lane	1819 – 1832
44 Seymour Street	1832 – 1835

Mathematical instrument makers.

C

CAIL, John	**NEWCASTLE UPON TYNE**
2 New Bridge Street	1823 – 1837
44 Northumberland Street	1838 – 1839
61 Pilgrim Street and 45 Quay Street	1841 – 1853
21 Grey Street	1855 – 1858
CAIL, John & Septimus Anthony	
45 Quayside	1851 –1853
CAIL, Septimus Anthony	
21 Grey Street	1857 – 1884
CAIL & Sons	
28 Quayside	< 1890 >

Opticians and mathematical instrument makers. John Cail worked for ***Edward Troughton***.

CALVERT	**LONDON**
Fleet Street	

Unidentified but probably a retailer. The name has been recorded on a 9 in. ivory rule.

CALVERT, John & Sons	**MANCHESTER**
97 & 99 Great Jackson Street	1886 – 1895
CALVERT, A. E. B.	
69 Sackville Street	1890

Steel rule makers.

CARLILE, John	**GLASGOW**
Bell's Wynd	1755

General merchant, mathematical, optical and surgical instrument maker. No rules have been found bearing his name to date but his advertisements included rules amongst his products.

CARPANI	**LONDON**
Goodge Street	

Unidentified but probably a retailer. The name has been recorded on an ivory rule.

CARPENTER & WESTLEY	**BIRMINGHAM & LONDON**
111 New Street, Birmingham	1835
(21) (24) Regent Street, London	1838 – 1914 >

Mathematical instrument and drawing instrument makers.

CARR, Jonathan — **BIRMINGHAM**
146½ Latimer Street — 1855 – 1858
CARR & WISEMAN
147 Latimer Street — 1861

Rule makers.

CARRETT, W. E. — **LEEDS**
Sun Foundry — < 1857 >

Engineer and designer of an engineer's slide rule.

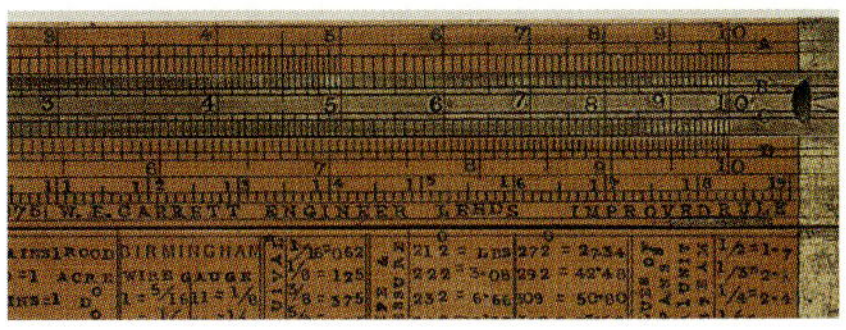

CARTWRIGHT, C., & Son — **PRESTON**
ca. 1840

Optical instrument makers.

CARVER, Isaac — **LONDON**
The Globe Dial, Horsely Down — 1668 – 1713
CARVER, Jacob
Globe Dial, Horsely Down — 1717
Atlas & Quadrant, Rotherhithe Wall — 1721 – 1742

Mathematical instrument makers. Isaac Carver placed an advertisement in the 7th edition of Thomas Everard's *Stereometry, or the Art of Gauging made easy*...published in London in 1712.

CARY, William (& Co.) — **LONDON**
(272) 181 & 182 Strand — 1789 – 1894
7 Pall Mall — 1895 – 1897
CARY PORTER Ltd.
7 Pall Mall — 1897 – 1920
22 Charing Cross — 1923 – 1925
29 & 30 Charing Cross — 1930
41 & 43 Whitehall — 1931 – 1933

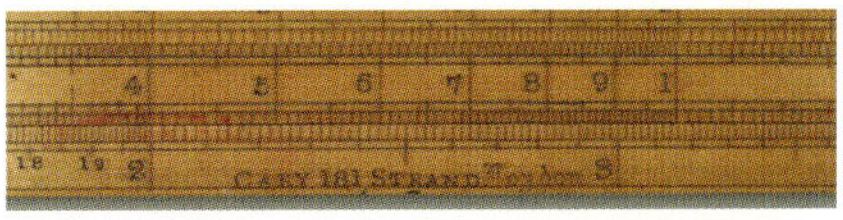

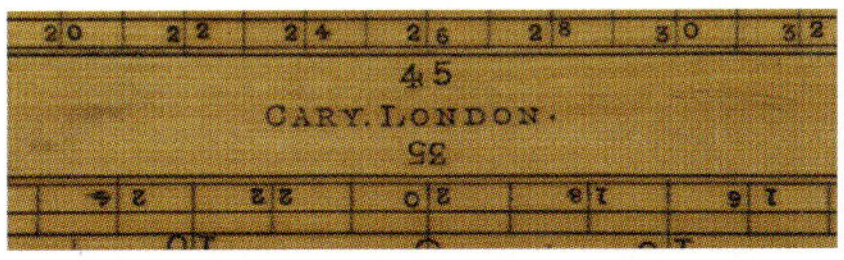

Optician, nautical instrument and telescope makers. William and his brother John, who was a globe seller, map seller, chart seller and engraver, sometimes worked in partnership. William was apprenticed to ***Jesse Ramsden***. William died in 1825 and the business was continued by his sons, George and John, still under the name William Cary. John died in 1852 and George in 1859 and they were succeeded by Henry Gould. Gould also continued to use the name William Cary until 1899. The premises at 181 Strand were also occupied by another firm of opticians, Gould & Porter. Both firms relocated to 7 Pall Mall in 1895. In 1896, Henry Porter was operating without Gould and in 1897, Cary and Porter combined to form Cary Porter & Co.

CASARTELLI, Louis (Lewis) — **LIVERPOOL**
King Street — 1821 – 1832
20 Duke Street — 1830 – 1848
CASARTELLI, Anthony & Joseph
20 Duke Street — 1845 – 1849
CASARTELLI, Anthony (& Co.)
20 Duke Street — 1849 – 1864
11 Castle Street — 1865 – 1872

54 (80) Old Hall Street — 1873 – 1925
and 4 Street Johns Lane — 1879 – 1925

CASARTELLI, Joseph
20 Duke Street — 1849
6 Street John Street — 1851

CASARTELLI, Louis
20 Duke Street — 1849 – 1900

CASARTELLI Bros.
20 Duke Street — 1901 – 1919

CASARTELLI, J., & Co.
89 Old Hall Street — 1930 – 1932

CASARTELLI, J., & Son — **MANCHESTER**
43 Market Place — 1851 – 1989

1821: Opticians. 1849: Opticians, mathematical instrument makers, scientific instrument makers. 1851: Casartelli took over the business of ***J. B. & J. Ronchetti***, Manchester.

CASE, W. — **LONDON**
22 Crown Street, Finsbury Square — 1794 – 1836

Rule maker.

CASELLA, Louis, & Co. — **LONDON**
23 Hatton Garden — 1848 – 1860

CASELLA, Louis P.
23 Hatton Garden — 1865 – 1870
147 Holborn Bars — 1875 – 1901

CASELLA, C. F., & Co.
11–15 Rochester Row — 1907
Gurney Street Works — 1913
49 & 50 Parliament Street — 1920
Southampton Street — 1931

Mathematical instrument makers and opticians. Prior to setting up on his own, from 1838 to 1846, Louis Cassella worked in a partnership with ***Caesar Tagliabue***, his father-in-law, at 23 Hatton Garden.

CATTERALL, James — **LIVERPOOL**
1 Threlfall Street, Toxteth Park — 1841

Rule maker.

CAUSTON, Sir Joseph & Sons — **LONDON**
9 Eastcheap — < 1906–1924 >
Sardinia Street — < 1938 >

Manufacturing and law stationers, printers, framers etc. The name has been recorded recorded on a mount cutters' rule but it is unlikely that they made the rules they sold. The business was a continuation of Jones & Causton, in business in Eastcheap from at least 1850 but that name has not been recorded on any rules.

CHADBURN, William — **SHEFFIELD**
81 The Wicker — 1816 – 1830

CHADBURN & Co.
Albion Works, Nursery Street 1830 – 1876
and Stanley Street 1871
CHADBURN Bros.
Nursery Street 1842
CHADBURN, Charles Henry **LIVERPOOL**
71 Lord Street 1842 – 1864
CHADBURN, C. H., & Son
71 Lord Street 1864 – 1878
71 Lord Street and 11 Waterloo Road 1878 – 1890
17 James Street and 11 Waterloo Road 1890 – 1900
CHADBURN TELEGRAPH Co. Ltd.
Cyprus Road Bootle
and 11 Waterloo Road 1905 – 1935

Opticians and mathematical instrument makers. Charles Henry Chadburn was a branch of the Sheffield business operating in Liverpool.

CHADBURN & Son Ltd. **LONDON**

Unidentified but possibly a retailer or alternatively a branch of ***C. H. Chadburn & Son***. The name has been recorded on a Farmar's spirit rule.

CHADWICK, J. **MANCHESTER**
post 1900

Engineer and designer of an engineers' slide rule.

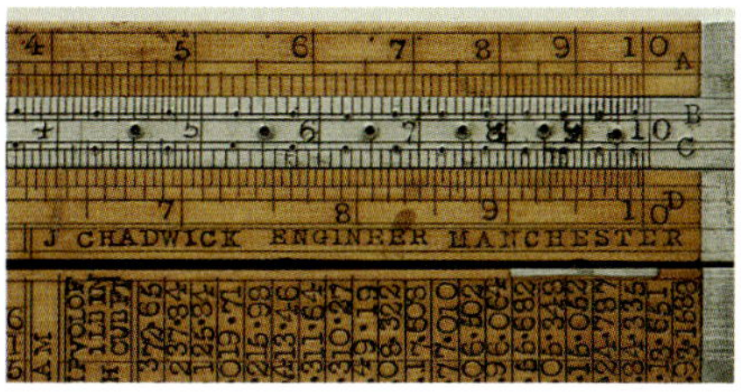

CHADWICK & DIXON **LONDON**
23A Cornwall Road, Lambeth 1921

Rule makers.

CHAPMAN, A., & Co. **LONDON**
128 Farringdon Road 1924
22 Camden Street, Islington 1928 – 1931

Rule makers.

CHARLES & Co. **BIRMINGHAM**

This firm variously described themselves as plane makers and turners inthe second half of the 19th century. See *British Planemakers from 1700,* 3rd ed.

CHESTERMAN, James **SHEFFIELD**
Bow Works, Nursery Street 1821
39 Division Street
CHESTERMAN & BOTTOM
39 Division Street 1845
CHESTERMAN, James, & Co.
Bow Works, Plum Street 1859 – 1963

1841: Patentee and manufacturer of the self-acting spring tape, blind and map rollers, roasting jacks, etc. 1845: model steam engines added to their products. 1868: manufacturers of patent measures; engineers' tools; land chains; legging springs; rules; scissors and pen, pocket and spring knives. In 1841 ***John Preston Cutts*** worked from the same premises and there were

July 23rd 1880

JAMES CHESTERMAN & Co.

BOW WORKS,

Ecclesall Road, SHEFFIELD,

SOLE MANUFACTURERS OF

CHESTERMAN'S PATENT

MEASURING TAPES,

LAND CHAINS, BAND CHAINS,

Steel Rules, Steel Straight Edges, T Squares, Scales,

ENGINEERS' TOOLS, &c.

ALL GOODS MARKED WITH THE NAME AND

TRADE J C MARK.

ENTERED AT STATIONERS' HALL.

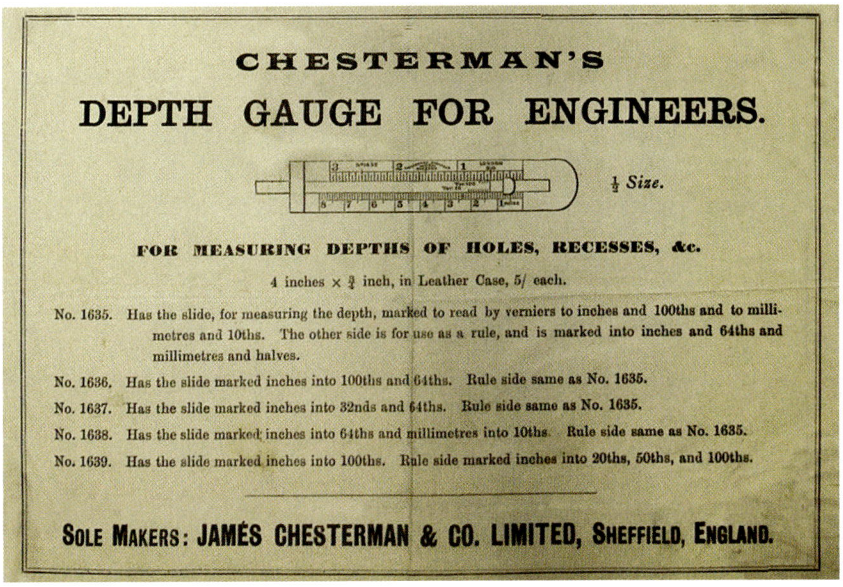

CHESTERMAN'S

DEPTH GAUGE FOR ENGINEERS.

½ Size.

FOR MEASURING DEPTHS OF HOLES, RECESSES, &c.

4 inches × ⅜ inch, in Leather Case, 5/ each.

No. 1635. Has the slide, for measuring the depth, marked to read by verniers to inches and 100ths and to millimetres and 10ths. The other side is for use as a rule, and is marked into inches and 64ths and millimetres and halves.

No. 1636. Has the slide marked inches into 100ths and 64ths. Rule side same as No. 1635.

No. 1637. Has the slide marked inches into 32nds and 64ths. Rule side same as No. 1635.

No. 1638. Has the slide marked inches into 64ths and millimetres into 10ths. Rule side same as No. 1635.

No. 1639. Has the slide marked inches into 100ths. Rule side marked inches into 20ths, 50ths, and 100ths.

SOLE MAKERS: JAMES CHESTERMAN & CO. LIMITED, SHEFFIELD, ENGLAND.

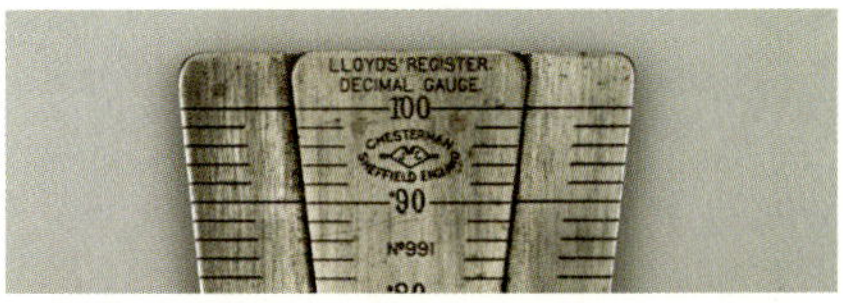

James Chesterman trade marks

connections with ***Cutts, Chesterman, & Beddington*** and ***James Beddington*** of Birmingham. In the 20th century, the company also had an outlet at 5 Eyre Street Hill, Hatton Garden in London. In 1963 Chesterman merged with ***John Rabone & Sons*** to form ***Rabone Chesterman Ltd.*** The cover of James Chesterman & Co.'s 1880 catalogue is shown opposite. For further information, see Hallam, Douglas J., *The First 200 Years, A Short History of Rabone Chesterman Ltd.*, Birmingham, 1984

CHRISTIE, James, & Son — **SHEFFIELD**

Photographic and artists' material dealer operating in the late 19th and early 20th centuries.

CHRISTY — **LONDON**

Hat rules and hat measures marked Christy's are frequently reported. They were a firm of hat and cap manufacturers operating from the mid-19th century to ca. 1906. They did not make the rules on which their name is marked.

CLARK, Charles — **BIRMINGHAM**
8½ Louisa Street — 1884 – 1902

Rule maker.

CLARK, Samuel — **BIRMINGHAM**
15 Great Hampton Street — 1839 – 1842

Yard measure maker. He took over the business of ***T. Lawrence***.

CLARK(E) — **LONDON**
The Minories — 1672

Clark is recorded as having sold rules but it is not known if he made the rules he sold.

CLARK(E), Edward — **WOLVERHAMPTON**
40 Dudley Road — 1792
CLARK, William

Dudley Road 1841 – 1852

Rule makers.

CLARK (E), William **BIRMINGHAM**
(7) (9) (122) Hospital Street 1829 – 1842
No. 7 Court, Henrietta Street 1850 – 1852

Box and ivory rule maker.

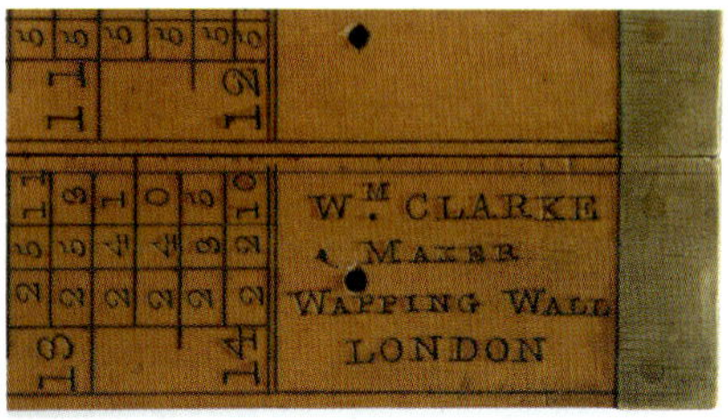

CLARKE, William (I) **LONDON**
Wapping Wall, Shadwell 1722 – 1743
CLARKE, William(II)
13 Wapping Wall, Shadwell 1786 – 1820
CLARKE, Thomas
13 Wapping Wall 1805
CLARKE & Son
13 Wapping Wall 1821 – 1834

1722: Optical instrument makers. 1786: Mathematical and optical instrument makers. The relationship between William Clarke (I) and William Clarke (II) is not known. The relationship between Thomas and William (II) is also not known, but William (II) took over Thomas' business.

CLEMENTS NEWLINS **LONDON**
95 & 96 Wood Street < 1906–1938>
and 28-32 Chiswell Street

Drapers' suppliers. They did not make the rules marked with their name and are listed as customers of ***John Rabone & Sons***.

CLEWITT, Francis **LONDON**
Blue Bell, The Strand 1672

His name may have been Hewitt. He is recorded as having sold rules but it is not known whether he made the rules he sold.

CLOTHIER, John **LONDON**
High Street, Deptford 1839
CLOTHIER, John Thomas
Lower Road, Rotherhithe 1851 – 1855

Rule makers.

COCK & HUGHES **LONDON**
ca. 1790

COCK, S.
Broad Street 1830

Cock & Hughes have been recorded as selling a dipping measure. S. Cock described himself as a gauging instrument maker.

COGGS, John **LONDON**
Fleet Street 1718 – 1733

At the "Globe & Sun". His trade card states "Servant (for many years) to Mr Rowley". This was ***John Rowley***. Coggs made "Hill's new sliding rule" and is

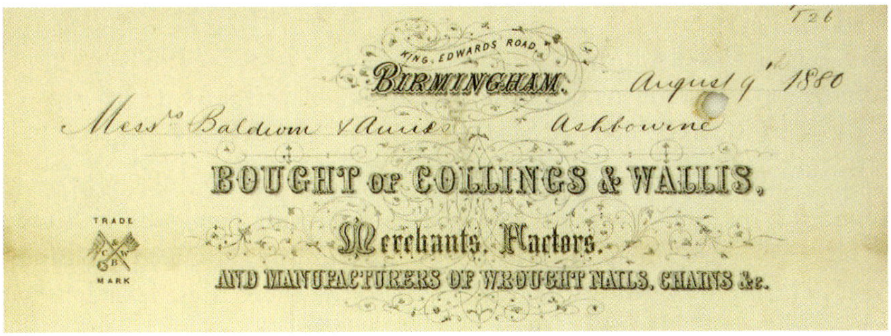

also known to have sold slide rules made by ***Isaac Carver***. A broad sheet advertisement in the Heal Collection of trade cards in the British Library states that "He also sells all sorts of Gauging Rods, and Sliding Rules for Mensuration, made by Isaac Carver (from the Globe on Horley Down), the original maker of all Sliding Rules now in Use Viz. Partridge's, Everard's, Hunt's, Coggeshall's etc..." He was probably succeeded by ***Nathaniel Hill***.

COLE, Humfrey — **LONDON**
Nr. North Door of Street Paul's — 1568 – 1590

Goldsmith, mathematical instrument maker and engraver.

COLES, Christopher
ca. 1664

Unidentified but probably a maker.

COLLINGS & WALLIS — **BIRMINGHAM**
King Edwards Road — 1865 – 1898

Merchants, factors and nail makers. They would not have made the rules that bear their name and are listed as customers by ***John Rabone & Sons***.

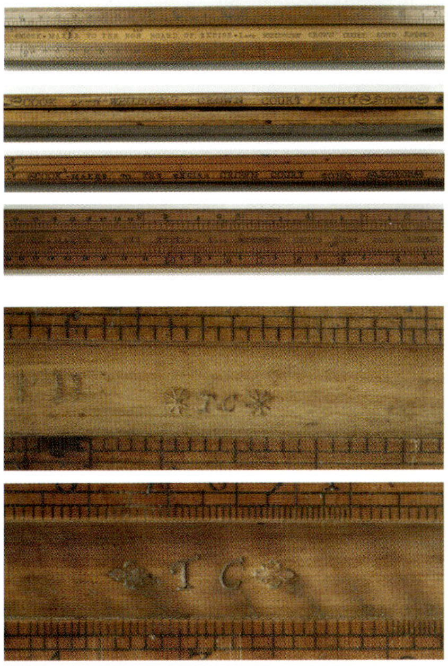

COOK(E), Laban — **LONDON**
21 Crown Court, Soho — 1826 – 1834

Rule maker. He took over the businesses of both ***John Jones*** and ***Alexander Wellington***. Rules are often marked "Maker to the Honble. Board of Excise" or "Maker to the Excise".

COOK(E), Thomas — **LONDON**
Old Jewry — 1704 – 1744

1704: He also listed an address in Threadneedle Street. 1712: "By the Excise Office in Old Jewry". He advertised in the 7th edition of Thomas Everard's *Stereometry, or the Art of Gauging made easy...* published in London in 1712. He took as apprentices ***William Mole*** in 1721, ***John Farmer*** in 1722, ***Edward Roberts*** in 1733, and ***William Rix*** amongst others. He advertised together with a John Worgan in 1695, but the business relationship is not known. The business was almost certainly continued by ***Edward Roberts***.

COOKE, Thomas — **YORK & LONDON**
50 Stonegate, York — 1837 – 1841
12 Coney Street, York — 1844 – 1854
Buckingham Works, Bishopshill, York — 1855 – 1868
31 Southampton Street, London — 1863 – 1868

COOKE, T. & Sons
Buckingham Works, Bishopshill, York — 1868 – 1960s
3 Broadway, Westminster, London — < 1913 – 1960s
COOKE TROUGHTON & SIMMS — **LONDON**
3 Broadway, Westminster, London — 1922 – 1980s.

Opticians. 1922: T. Cooke & Sons amalgamated with ***Troughton & Simms*** to become Cooke Troughton & Simms. This company continued in business until at least the 1980s. T. Cooke & Sons also continued to trade under their own name until at least the 1960s.

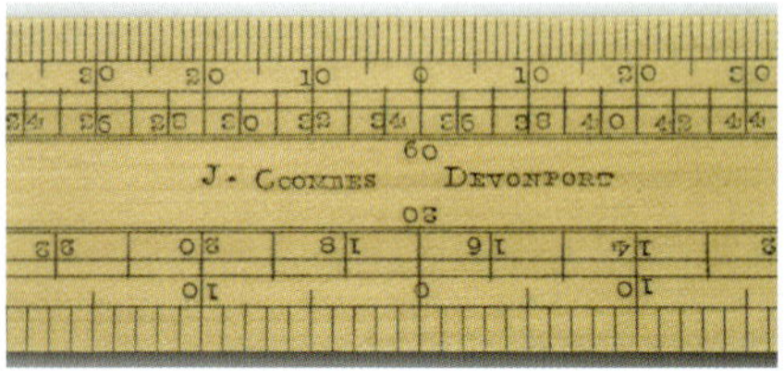

COOMBES, J. — **DEVONPORT**
87 Fore Street — < 1893 >

Opticians.

COOPER, Charles T. — **LONDON**
61 Bland Street, Great Dover Street — 1870
86 Great Dover Street — 1880 – 1882
Russell Place, 9 Old Kent Road — 1890 – 1899
COOPER, Charles T. & Sons (Ltd.)
Russell Place, 9 Old Kent Road — 1905 – 1947 >
Haddenhall Street, 9 Old Kent Road — 1913 – 1928 >

Mathematical and rule makers. 1919: Became a limited company.

COOPER, George — **BIRMINGHAM**
80 Barford Street South — 1863 – 1905

Rule maker.

COOPER, George Charles — **MANCHESTER**
(105)(101)(109) Chapel Street, Salford — 1852 – 1871

Rule maker. In 1871 he also described himself as a "plane maker".

COOPER, George Henry — **LONDON**
25 London Road, Forest Hill — 1930 – 1938

Rule maker.

COOPER, Stephen (Norris) — **LONDON**
119 Great Saffron Hill — 1809 – 1811
MANCHESTER
(25) (16) Miller Street — 1811 – 1841
and 16 Swan Street — 1817
COOPER, Sarah
16 Miller Street — 1847 – 1854
COOPER, Josiah
16 Miller Street — 1858 – 1871

1817: Ivory and box-rule makers. In 1852 Kaziah Cooper was listed in the directory at 18 Miller Street for one year only. This may have been a misspelling of Josiah.

COPE, Thomas Henry — **BIRMINGHAM**
9 A.B. Row — 1852 – 1854

Rule maker.

CORBETT, Thomas — **BIRMINGHAM**
14 Edmond Street — 1849 – 1850
74 Spencer Street — 1852 – 1870
CORBETT, Thomas, & Son
74 Spencer Street — 1875 – 1888

Rule, mathematical instrument and fishing reel makers.

CORBISHLEY, Jephthah Stephen — **LONDON**
62 Alberta Street, Kennington — 1917 – 1964

Rule makers.

CORFIELD & RUSSELL — **BIRMINGHAM**
17 Newton Street — 1845
CORFIELD, Richard
17 Newton Street — 1850 – 1862

Measuring tape and coffin furniture makers.

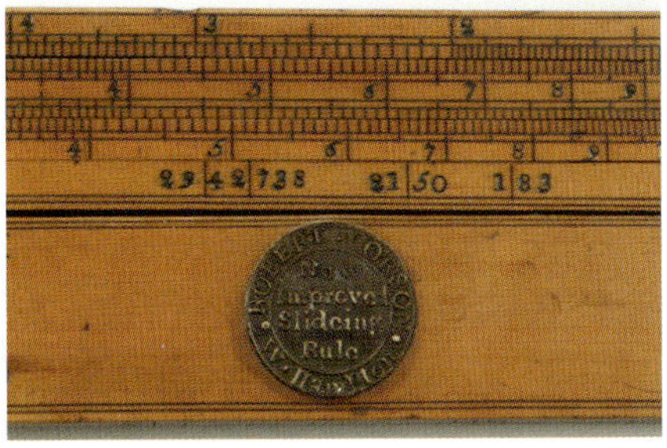

CORSON, Robert — **WOLVERHAMPTON**
Berry Street — 1767 – 1770
CORSON, Thomas
Queen Street — 1770
CORSON, James
Worcester Street — 1770 – 1781
Queen Street — 1780 – 1781

Rule makers. James was succeeded by ***Thomas Smith***.

COTTS, M. — **LONDON**
4 Queen's Row, Bethnal Green — 1809

Rule maker.

COTTER, A. — **LONDON**
13 Hatton Garden — 1880

Rule maker (wholesale). This address as also that of ***Carmouche Coyen***.

COULING, Henry — **LONDON**
6 Hayward's Place, Street John Street — 1845 – 1855
4 Upper Rosamon Street — 1865
COULING, Henry & Sons
98 Rosamon Street — 1890

Measuring tape makers.

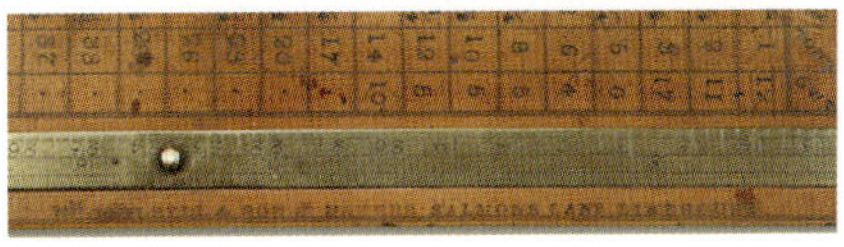

COULSELL, Thomas — **LONDON**
Loman's Pond, Borough — 1794
29 Queen Street, Borough — 1799 – 1814
41 (153) Union Street, Borough — 1815 – 1836
COULSELL, Elizabeth & Matilda
153 Union Street, Borough — 1836
COULSELL, Elizabeth
153 Union Street, Borough — 1838
41 Union Street, Borough — 1839 – 1842

COULSELL, William (I)

	1809
9 Castle Street, Borough	1816 – 1840
41 Union Street, Borough	1845
9 Castle Street, Borough	1849 – 1851

COULSELL, William (II) (& Son)

6 Grange Road, Kent Road	1839
6 Surrey Place, Grange Road	1840
9 Castle Street, Borough	1845 – 1855
163 Union Street, Borough	1850
156 Salmons Lane, Limehouse	1870
102 Grove Road	1880 – 1934>

The family was variously described as rule makers, mathematical and philosophical instrument makers and opticians. 1919: Trades directory entry reads "E.D. scales, protractors, T-set squares & curves of all kinds. Contractors to LCC & War Office." Some rules are marked "& Son" but this never appeared in the directory entries. The firm ceased trading by 1947.

COWLING, Alfred — **SHEFFIELD**

Standard Works, Cavendish Street — 1871 – 1884

Manufacturers of English and foreign rules and measures. The business was taken over by ***Tyzack & Holmes***.

COX — **LONDON**

417 Oxford Street

Unidentified but probably a retailer.

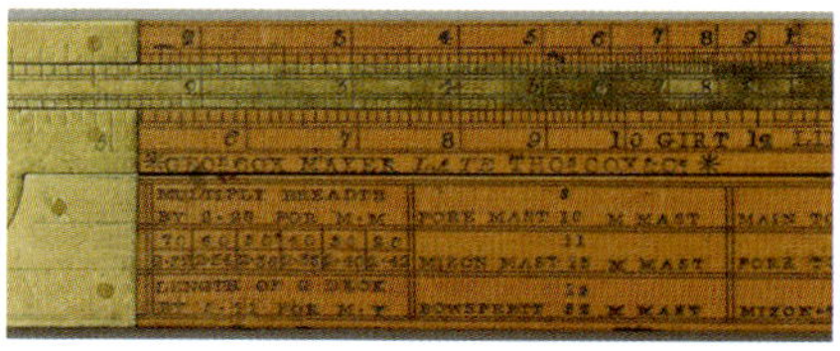

COX, Thomas — **BIRMINGHAM**

28 Bell Street	1777 – 1781
22 Newhall Street	1784 – 1798
Hill Street	1797

COX, Ann

70 Hill Street	1800 – 1801
Suffolk Street	1803

COX, Ann & George
Suffolk Street 1808 – 1811

COX, George
Navigation Street 1815 – 1818
Regent's Place 1821 – 1826
1 Vittoria Street 1828 – 1830

COX, Francis Blakemore
(50) (95) Camden Street 1828 – 1882

Box and ivory rule makers. Between 1800 and 1825, the firm was also known as Thomas Cox & Co. One rule has been recorded marked T. & G. Cox and one marked both Thomas Cox and George Cox. These are likely to have been marks used between 1800 and 1825. 1876: Manufacturers of box and ivory rules, Carrett's, Hawthorn's and Routledge's improved engineering rules, Slater's improved rules for cotton spinning and foreign rules with their correct measures. The company went into liquidation in 1882. In a letter to the Daily Post in 1878, F. B. Cox claimed that the firm were the first to introduce arch joints around 1800. In 1835, 1 Vittoria Street was occupied by ***William Brown***.

COX, William **PLYMOUTH**
<1806 >

COX, William Charles
(86) (89) (83) Fore Street 1822 – 1857

COX & COOMBES

Mathematical and nautical instrument makers. Some rules are marked Devonport, which is the location of the Naval Dockyard in Plymouth. It is likely that Cox & Coombes was a partnership with ***J. Coombes*** but this has not been established.

COYEN Carmouche **LONDON**
13 Hatton Garden 1880

Rule maker (wholesale). This is the same address as ***A. Cotter*** but no connection between the two has been established.

CRAWCOUR, Morris **LONDON**
11 Le Blond's Building, William Street 1870

Rule maker.

CRESWELL, William **LONDON**
239 High Holborn 1816 – 1828

Rule maker.

CRESWICK, William **WOLVERHAMPTON**
3 Church Street 1874 – 1880

Rule maker.

CRICHTON, John **LONDON**
32 Fore Street Limehouse 1831
112 Leadenhall Street 1834 – 1865

Optical, philosophical, mathematical and nautical instrument maker.

CRICK, Charles — **LONDON**
East Smithfield — 1722

Rule maker. His father, Charles, and his two brothers, James and Thomas, were mathematical instrument makers.

CRITCHLEY, F. H. — **LIVERPOOL**

Unidentified. The name has been recorded on a steel rule.

CRONMIRE, John M. & Henry — **LONDON**
5 Seabright Street, Bethnal Green — 1851
10 Cottage Lane, Commercial Road — 1851 – 1860
10 Bromehead Street, Commercial Road — 1865 – 1869
CRONMIRE, John Martin & Sons
10 Bromehead Street, Commercial Road — 1870 – 1880

Rule makers.

CROOKE, John — **LONDON**
Christ's Hospital — 1713
Without Newgate — 1735

Mathematical instrument maker.

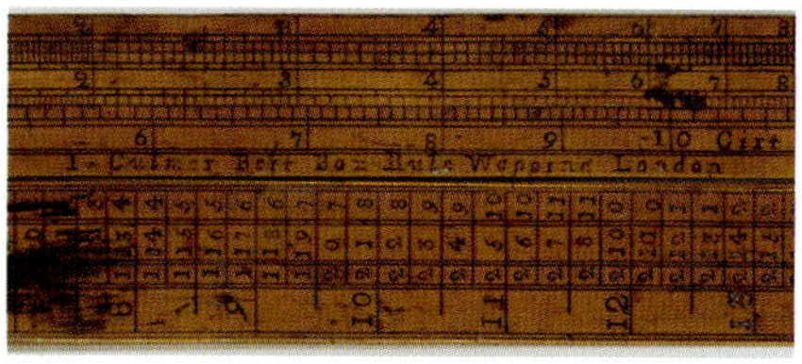

CULMER, Josiah — **LONDON**
(126) (130) Wapping New Stairs — 1783 – 1811

Mathematical instrument maker.

CULPEPER, Edmund (I) — **LONDON**
Cross Daggers, Moorfield — 1697 – 1731
Nr. Royal Exchange — 1737 >
CULPEPER, Edmund (II)
Mile End, Stepney — < 1758
Mare Street, Hackney — 1759 >

Mathematical instrument makers. Edmund (I) was apprenticed to ***Walter Hayes*** and took over his business. he took Edmund (II), his son, as an apprentice.

CUMMINS, Alexander — **LONDON**
148 Leadenhall Street — 1846 – 1851
34 Leadenhall Street

Chronometer maker.

CUSSONS, George (& Son) (Ltd.) — **MANCHESTER**
104 Great Clowes Street, Lower Broughton — 1876 – present

Makers of scientific apparatus.

CUTHBERT, James — **LONDON**
40 Great Windmill Street — 1805
CUTHBERT, John
40 Great Windmill Street — 1809
113 Street Martin's Lane — 1816 – 1817
5 Bridge Road, Lambeth — 1822

445 Strand	1822
22 Bishop's Walk, Lambeth	1826
58 Brook Street, West Square	1833

Rule, mathematical and philosophical instrument makers, opticians. James described himself only as a rule maker.

CUTTS, John Preston	**SHEFFIELD**
39 Division Street	1841
CUTTS, J. P., & Co.	
39 Division Street	1845
CUTTS, CHESTERMAN & BEDINGTON	
43 Division Street	1852
CUTTS, CHESTERMAN & Co.	
45 Division Street	1856
CUTTS, I. P. (CUTTS, SUTTON & Son)	
43 Division Street	1859 – 1871

1841: Mathematical instrument makers. Sole manufacturer of Chesterman's patent self-acting window blind and map rollers, tape measures, etc. ***James Beddington***, Birmingham, was a partner in Cutts Chesterman and Beddington. See also ***James Chesterman***.

D

DANCER, John Benjamin — **LIVERPOOL**
(11) (21) Pleasant Street — 1835 – 1841
MANCHESTER
(13) (43) Cross Street — 1841 – 1878

Mathematical instrument maker and optician. J. B. Dancer's father, Josiah, was the son of Michael Dancer, mathematical instrument maker in London from 1776 to 1817. Josiah moved to Liverpool in 1817 and set up in business, though there is no evidence that he made rules. John Benjamin took over his father's business in 1835 and in 1841 moved to Manchester. Between 1841 and 1844 he was in partnership with ***Abraham Abraham*** of Liverpool.

DARKER, F. & C. — **LONDON**
9 Paradise Street — 1846 – 1851

Opticians.

DARWIN & MILNER — **SHEFFIELD**
Soho Ponds — 1859 – 1871

Cast steel roller, press screw and silversmiths' and braziers' tool manufacturers.

DAVENPORT, Robert — **LONDON**
1635
EDINBURGH
1647

Mathematical instrument maker. He was apprenticed to ***Elias Allen***.

DAVIE, Thomas — **LONDON**
29 Gravel Lane, Southwark — 1851

Rule maker.

DAVIES, James — **LONDON**
17 Amelia Street, Walworth — 1851 – 1855
DAVIES, Henry
17 Amelia Street, Walworth — 1860
DAVIES, James
1 Paradise Street, Lambeth — 1870 – 1890
2 Old Paradise Street, Lambeth — 1891 – 1915 >

Rule and mathematical instrument makers.

DAVIES, John Richard — **BIRMINGHAM**
95 Long Acre — 1888 – 1898

Rule maker.

DAVIS, Gabriel — **LEEDS**
24 Boar Lane — < 1779 – 1847>

DAVIS, Edward

(22) (74) Albion Street	< 1870 –1923

Optical, philosphical and mathematical instrument makers. Edward and ***John Davis*** of Derby were the nephews of Gabriel.

DAVIS, George John	**BIRMINGHAM**
2 Key Hill	1850
29 Ashtead Row	1852 – 1854
15 Upper Windsor Street	1858 – 1865
Queens Road, Aston	1875 – 1876
171 Scholefield Road	1880 – 1884

Rule maker.

DAVIS, John	**DERBY & LONDON**
Rotten Row, Derby	1830
21 Irongate, Derby	1843 – 1865
99 Friargate, Derby	1865 – 1873
DAVIS, John, & Son (Ltd.)	
All Saints Works, Derby	1875 – 1962
118 Newgate Street, London	1894
and 36 Camomile Street, London	1907
(115) (39) Victoria Street, London	1924 – 1941

Mathematical, optical and philosophical instrument makers. John and ***Edward Davis*** of Leeds were the nephews of ***Gabriel Davis***. John was apprenticed to ***Jacob Abraham*** and worked at his Cheltenham premises before setting up on his own in Derby in 1830. He died in 1873 and his son, Henry, took over the running of the business. In 1894 the firm advertised celluloid slide rules on the back cover of Pickworth, C. N., *The Slide Rule*. In 1914 a new slide rule plant and a separate slide rule department were opened. Previusly the slides rules sold by the firm had been imported from France and Germany. The firm is still in business today as Davis Derby Ltd.

DAVIS, Samuel	**WOLVERHAMPTON**
Burry Street	1770
(19) Piper's Row	1780 – 1792
DAVIS, John	
	1798
Queen Street	1805 – 1811
DAVIS, Sarah	
17 Queen Street	1802

Rule makers.

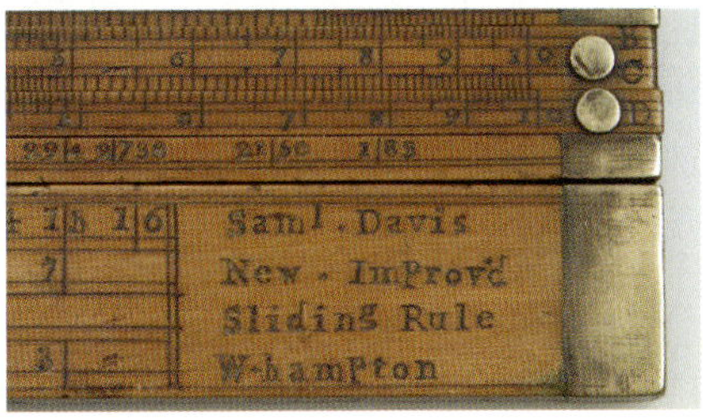

DAVIS, William	**LONDON**
1 Brighton Place, Brixton Road	1839 – 1840

Rule maker.

DAVIS	**LIVERPOOL**
65 Bold Street	ca. 1830

Mathematical instrument maker.

DAWSON, A.G. — **LONDON**
< 1892 >

Unidentified but probably a rule maker as noted on the "Townley Scale Rule" Registered Design No: 186586.

DAWSON, William — **BIRMINGHAM**
70 Hill Street — 1818 – 1821

Rule maker and bed screw maker.

DEAN, Edward — **LONDON**
(30) (7) Frederick Place, Newington — 1845 – 1849
8 Richmond Terrace, East Street — 1855 – 1865
99 East Street, Walworth — 1869 – 1870
DEAN, LERRO & Co.
99 East Street, Walworth — 1890

Measuring tape makers.

DE GRAVE & Son — **LONDON**
59 St. Martin's le Grand — 1817 – 1844
DE GRAVE, SHORT & Co.
59 St. Martin's le Grand — 1845 – 1871
DE GRAVE, SHORT, FANNER & Co.
59 St. Martin's le Grand — 1871 – 1890
13 Farringdon Road — 1890 – 1900
57 Hatton Garden — 1911 >

Scale makers, the continuation of a family firm started in 1780. The name is found on various standard measures. In the early 1900s, the firm became part of the ***Avery*** company, trading under the name of ***Avery De Grave***.

DENCHFIELD, James — **BIRMINGHAM**
12½ Hope Street — 1876

Rule maker.

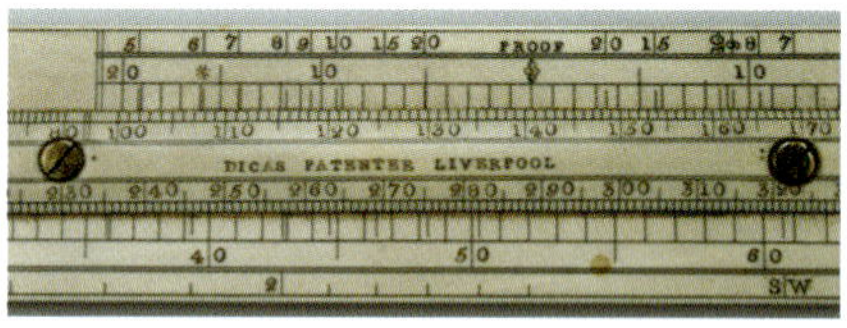

DICAS, John — **LIVERPOOL**
73 Hanover Street — 1774
25 Cleveland Square — 1777
71 Duke Street — 1781 – 1787
(27) (29) Pool Lane — 1790 – 1798
DICAS, Mary
27 Pool Lane — 1800
7 North Side, Old Dock, Strand Street — 1803 – 1805
DICAS & ARSTALL
8 North Side, Old Dock — 1807
DICAS, Ann
20 Trowbridge Street and Bronte Street — 1818
83 Brownlow Street — 1821
DICAS PATENT HYDROMETER MANUFACTORY
(11) (17) Clarence Street — 1829 – 1834
133 Brownlow Street — 1837

Hydrometer and mathematical instrument makers. Before1800 they were also listed in the directories as a navigation warehouse and brandy merchants. In 1780 John Dicas patented a hydrometer containing a notably different slide rule for correcting temperatures. The Dicas Patent Hydrometer Manufactory was run by B. Gammage, the husband of Ann Dicas.

DICK, Alexander — **GLASGOW**
101 Bridgegate — 1827 – 1832
105 Stockwell — 1833 – 1874
DICK, Alexander, & Co.

Surveying chain maker, wire workers.

DIXEY, Edward — **LONDON**
Princes Street — 1805
335 Oxford Street — 1810 – 1843
DIXEY, Charles Wastell
3 New Bond Street — 1839 – 1862
DIXEY & Sons
New Bond Street and Oxford Street — < 1862 – 1883 >

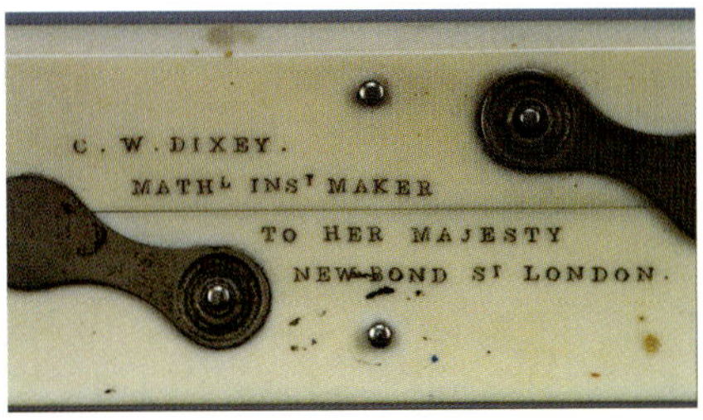

A dynasty of opticians and mathematical instrument makers, with several different branches of the family working from a variety of addresses. The firm is still in existence today as opticians in Wanstead, London. In 1817, Edward Dixey took over Hawks Grice who had previously taken over ***Fraser & Son***.

DIXON, Benjamin — **BIRMINGHAM**
Summer Lane — 1855 – 1867

Rule maker. 1855: Advertised "Patent solid back joint rule".

DOBIE, Alexander — **GLASGOW**
20 Clyde Place — 1844 – 1856
24 Clyde Place — 1857 – 1872
24 & 25 Clyde Place — 1873 – 1885
DOBIE, Alexander, & Son
24 & 25 Clyde Place — 1886 – 1888
DOBIE, John
18 & 19 Clyde Place — 1889 – 1891
44 & 45 Clyde Place — 1893 – 1910 >
DOBIE McINNES

Mathematical, nautical and optical instrument makers. John Dobie was Alexander's son and carried on the business after his father's death in 1887. Dobbie McInnes have not been identified.

DOBSON, JOHN — **LONDON**
268 High Holborn — 1850 – 1852

Opticians and drawing and mathematical instrument makers.

DOLLOND, Peter — **LONDON**
Vine Street — 1750 – 1752
DOLLOND, John, & Son
Nr. Exeter Exchange, Strand — 1752 – 1773
DOLLOND, Peter & John

59 St. Paul's Churchyard — 1766 – 1804

DOLLOND, Peter & George

59 St. Paul's Churchyard — 1805 – 1820

DOLLOND, George

59 St. Paul's Churchyard — 1820 – 1866 >

DOLLOND & Co.

(1) (25) Ludgate Hill — < 1884 – 1915 >

DOLLOND & AITCHISON

1927 – present

Opticians and mathematical instrument makers. Started by Peter Dollond in 1750 as opticians, Peter was joined by his father, John, in 1752 to form John Dollond & Son. The firm was continued by Peter and his brother John and in due course by Peter and his nephew, George Huggins who changed his name to Dollond. ***Jesse Ramsden*** was employed by Peter Dollond. In 1927, a partnership was formed with Irvine Aitchison to create the company Dollond & Aitchison, still in business today as the well-known High Street opticians. ***Henry Barrow*** worked as an outworker for the company.

DOUBLET, T. & H. — **LONDON**

4 City Road and 74 Paul Street — < 1850

39 Moorgate — 1906

118 London Wall — 1924

109 Finsbury Pavement — 1938 >

Opticians, philosophical, mathematical and scientific instrument makers.

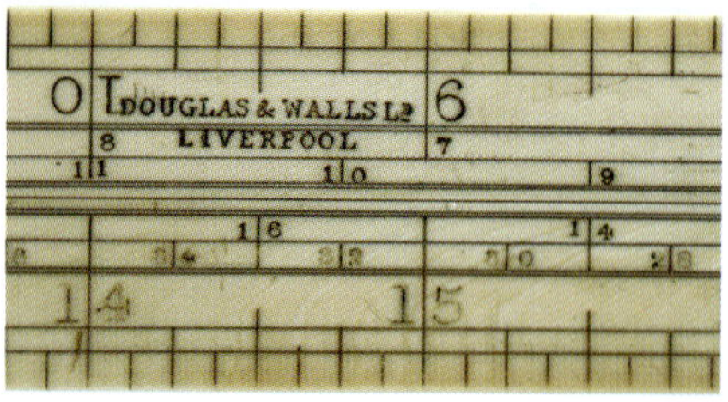

DOUGLAS & WALLS Ltd. — **LIVERPOOL**

Unidentified but probably a retailer.

DOWELL, William — **BIRMINGHAM**

Osler Street — 1865

19 King Edwards Place — 1868 – 1870

Barford Street — 1875

Rule maker.

DRAWING OFFICE SUPPLIES Ltd. — **LONDON**

51 Cheapside — < 1920 – 1936

15-17 St. Ann's Street — 1943 – 1948 >

Also known as DOS Ltd. They also had other addresses in London and branch offices in Manchester and Leeds. In 1943, they were taken over by ***B. J. Hall & Co. Ltd.*** who continued to operate the firm under its own name.

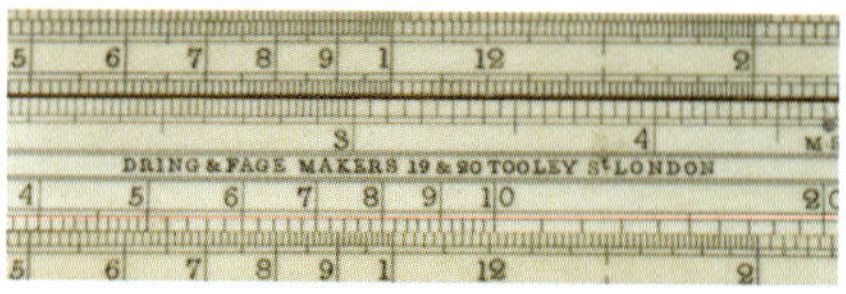

DRING & FAGE — **LONDON**

21 Gracechurch Street — 1790 – 1792

6 Tooley Street — 1792 – 1796

248 Tooley Street — 1796 – 1804

(19) (20) Tooley Street — 1804 – 1882

145 Strand — 1883 – 1902

56 Stamford Street — 1903 – 1946

105 & 152 Tooley Street — 1947 – 1977

Hydrometer, patent cock and mathematical instrument makers, opticians,

Mess. Hind Fellowes & Fellowes LONDON May 21. 1890
King Walden Brewery Hitchin
Bought of Dring & Fage
BY APPOINTMENT,
Hydrometer, Saccharometer & Gauging Instrument Makers,
To Her Majesty's Hon.ble Commissioners of Customs
FOR THE REVENUE OF THE UNITED KINGDOM
AND COLONIES.
To the Hon.ble Victualling Board for Her Majesty's Navy
AND
Dock Companies,
Rules Charts, Navigation Books
Mathematical Philosophical & Optical Instruments of all descriptions.
145, STRAND, W.C.
(A few Doors West of Somerset House)
Not accountable for damages by Carriage
Repairing 1. 14" Thermometer 7 :
Packing Case 1 :
8 :

Dring & Fage trade mark

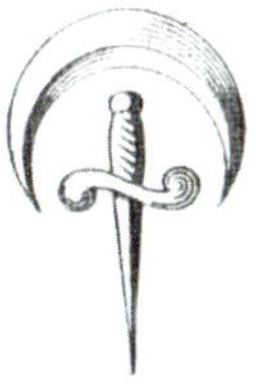

brass workers and barometer, thermometer and scale makers. The business was started by John Dring, who had taken over his brother-in-law's business as a turner and engine maker. Around 1790, he entered into partnership with ***William Fage***, a mathematical instrument maker. The company became Ltd. in 1941.

DRIVER, Thomas — **MANCHESTER**
38 Lower Mosley Street — 1861 – 1871
DRIVER, Edward
98 Lombard Street — 1875
(62) (68) (69) Hardman Street — 1880 – 1890
and Gt. Jackson Street — 1886 – 1899
15 Wood Street — 1899 – 1911 >

Rule makers. Their various premises were called Driver's Rule Works.

DUGMORE, John — **BIRMINGHAM**
17 Newton Street — 1841

Mathematical instrument and measuring tape maker.

DUNSCOMBE, Matthew W. — **BRISTOL**
(10) (20) St Augustine's Parade — 1874 – 1897>
DUNSCOMBE & HUSBANDS

Optician. Matthew was apprenticed to ***John Braham*** in 1855 and took over business of ***L. Braham & Co.*** in 1874. Dunscombe & Husbands was a partnership between M. W. Dunscombe and ***Henry Husbands*** operating sometime before 1876 but no further details have been established.

E

EBSWORTH, Richard — **LONDON**
(41) (54) (68) Fleet Street — 1820 – 1835
EBSWORTH, George Richard
54 Fleet Street — 1829 – 1835
EBSWORTH, Thomas
1824

Opticians, mathematical instrument makers.

EDELSTON, E. & S., & Son

Unidentified but possibly a maker. Name recorded on a wire gauge.

EDGELL, Albert, & Co. — **LONDON**
32 Eglinton Road, Bow — 1895
115 Tredegar Street — 1900 – 1915 >

Mathematical instrument makers.

EDWARDS, James — **LONDON**
25 John Street West, Edgware Road — 1836
201 Kent Road, Borough — 1839
10 Ray Street, Clerkenwell — 1859
(191) (195) (215)King's Cross Road — 1865 – 1885
EDWARDS, Joseph George
Rheidol Mews and 30 Liverpool Road — 1895

Rule makers.

EDWARDS, Joseph — **BIRMINGHAM**
Bull Street — 1767
46 Worcester Street — 1770 – 1777
46 Needless Alley — 1780 – 1781

Rule maker.

EDWARDS, Thomas — **LONDON**
5 Kemp's Court, Berwick Street — 1809

Rule maker.

EDWARDS, William — **BIRMINGHAM**
18 Court, Staniforth Street — 1880

Rule maker.

EGGINGTON, Thomas — **WOLVERHAMPTON**
7 Halletts Buildings — 1792

Rule maker.

ELLIOTT, William **LONDON**
26 Wilderness Row, Goswell Street 1817
21 Great Newport Street, St. Martin's Lane 1817 – 1827
(227) (268) High Holborn 1830 – 1849
ELLIOTT, William, & Sons
56 Strand 1850 – 1853
ELLIOTT Bros.
(56) (449) Strand 1854 – 1886
101 Street Martin's Lane 1887 – 1894
Century Works, Lewisham 1907 – 1913 >
and 36 Leicester Square

Variously described as opticians, drawing instrument and mathematical and philosophical instruments makers. In 1850, William's sons, Frederick Henry and Charles Alfred, joined him in the business continuing together after his death when the firm was renamed Elliott Bros. In 1856, they took over ***Watkins & Hill***. Eventually the firm became part of the General Electric Co. 1894: Advertised in Pickworth, C.N., *The Slide Rule*.

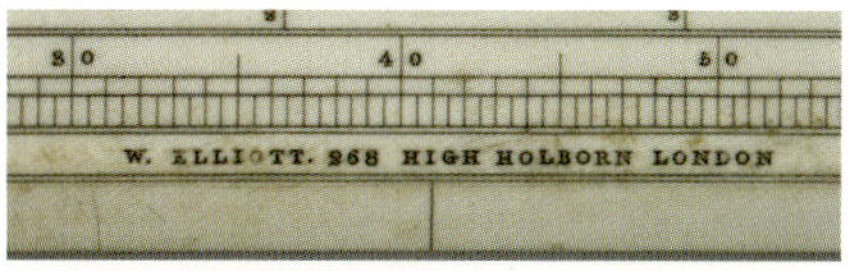

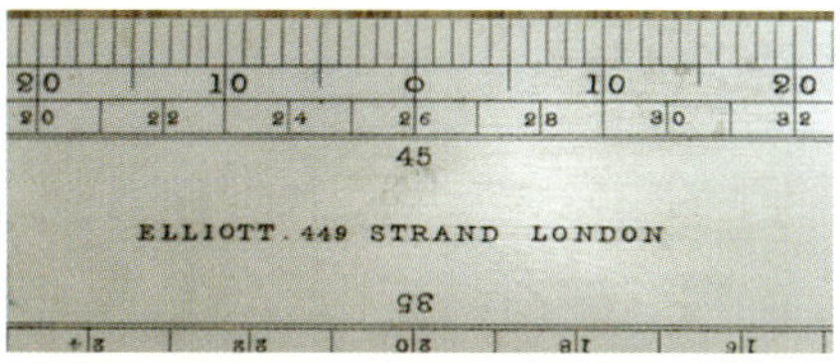

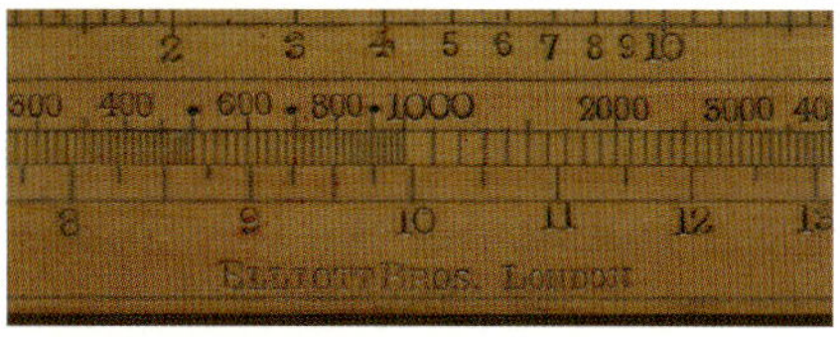

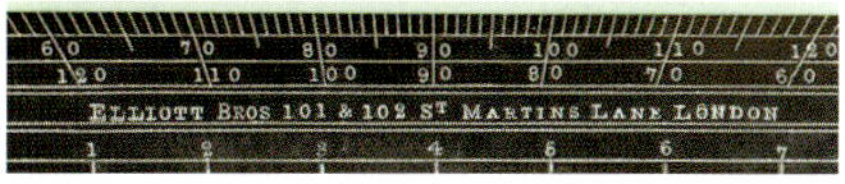

ELLIS, Rufus **SHEFFIELD**
95 Napier Street 1879

Advertised his trade as "Manufacturer of every description of brass and German silver rules, marked by special machinery which ensures the greatest accuracy."

ELLIS, C., & Co. **MAIDSTONE**

The name appears on a rule by ***T. Tustian***. It is likely that they were retailers.

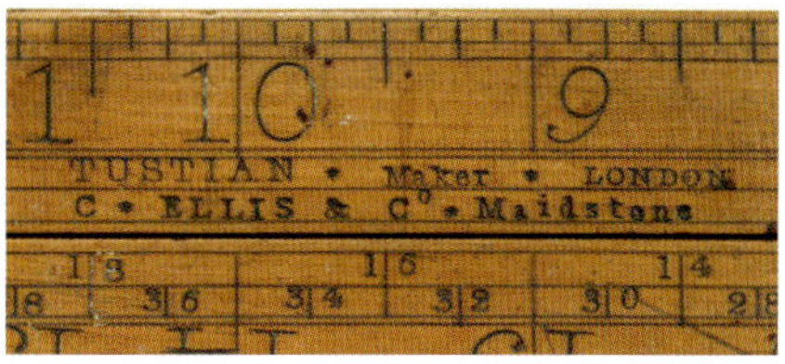

ELMES, William **LONDON**
Moorfields 1668 – 1675

Rule maker.

ENDERSBEE, W. (& Son) **LONDON**
1 Little Thames Street 1809 – 1827
335 High Street Wapping 1828 – 1840

Mathematical instrument makers.

ENDICOTT, John **LONDON**
10 Norwich Court, Fetter Lane 1832 – 1834
23 Little Saffron Hill, Hatton Wall 1845 – 1850

Barometer, thermometer and rule maker.

ESA **LONDON**
High Holborn 1875 – 1966

Stationary suppliers, specialising in equipment for schools. They have various other addresses in London.

EVANS, Frederick **LONDON**
2 Albany Cottages, Harling Street 1913

Rule maker.

EVANS, William **LONDON**
10 Dean Street, Fetter Lane 1771 – 1811
138 Wapping 1793
EVANS, John
Sherrard Street, Golden Square 1775
Paris Street, Southwark 1801
8 Dean Street, Fetter Lane 1809

William: Rule and mathematical instrument maker, printer. John: Rule maker.

EVANS & WINDRED **LONDON**
32 Kennington Road 1880 – 1892
EVANS, George, & Son
71 Lower Kennington Lane 1895 – 1899
EVANS, O.
9 Bell Gardens Road 1901

Rule makers.

EVANS & SONS **LONDON**

A firm of medical suppliers, their name is found on items such as catheter gauges and other medical measuring devices. They probably did not make the items on which their name appears.

EVERLEIGH, Saml. & Sons **MANCHESTER**

A firm of furriers and hatters, from 1820 to 1858, whose name can be found on hatters' rules. They would not have made the rules on which their name appears.

EVERS, Henry **LONDON**
3 Ely Court, Holborn Hill 1840

Rule maker.

EWART, John **NEWCASTLE ON TYNE**
< 1844 - 1847 >

Land surveyor and designer of a cattle gauge for determining the weight of carcases. All examples recorded to date have been made by ***J. Tree*** who exhibited the gauge at the Great Exhibition in 1851.

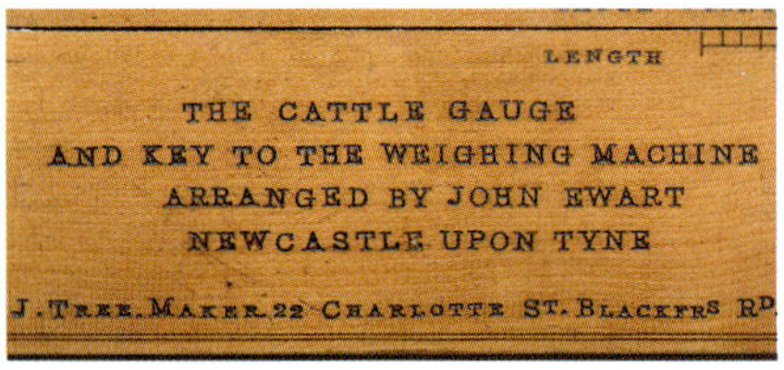

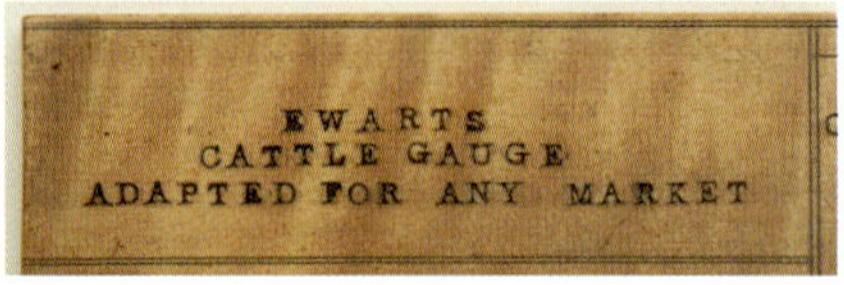

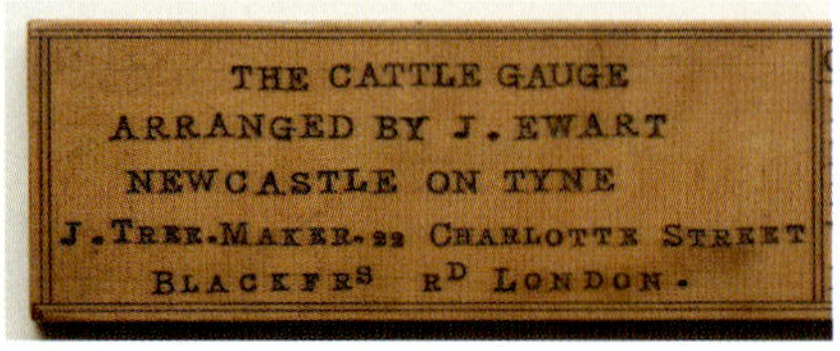

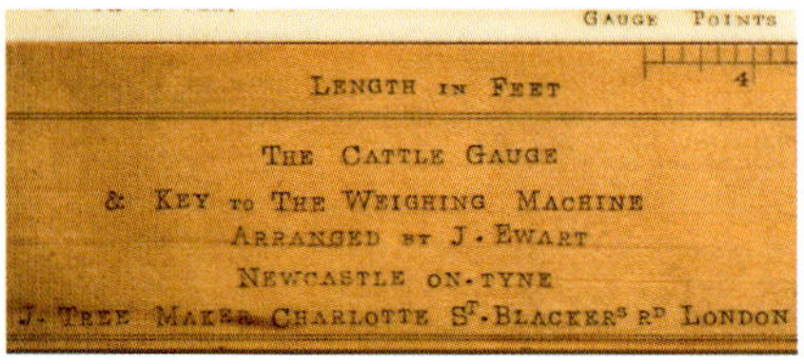

EYRE & SPOTTISWOODE (Publishers) Ltd. **LONDON**

Printers and stationers operating during the 19th and 20th centuries. Some boxwood scales and several sets of cardboard scales have been found bearing this name but it is unlikely that they made the products that they sold.

F

FAGE, William **LONDON**
109 Upper East Smithfield 1799 – 1805
Tooley Street 1805

Mathematical instrument maker. In 1790 he entered a partnership with John Dring to form the company ***Dring & Fage***.

FAIRCLOUGH, R., & Co. **LIVERPOOL**

Plane and edge tool makers, operating from 1816 to 1883. Although they are entered in the list of rule makers in Liverpool from 1870 until 1885, it is unlikely that the firm made any rules. For full details see *British Planemakers from 1700,* 3rd ed.

FARMAR, F. C. **LIVERPOOL**
38 Thorndale Road, Waterloo 1897 – 1900
FARMAR & SON

In 1899 F. C. Farmar, a Revenue Officer, took out two patents, one for an improved slide rule and one for an improved dip rod. He took out further patents in 1901 and 1908 for improvements to slide rules. All these rules were connected with the wine and spirit trade. He did not make any of the rules that are marked with his name. One rule and an advertisement hasbeen noted with the address 10 Gloucester Terrace, Thorpe Bay and an advertisement, shown below, with the address Elderslie, Thorpe Bay, Essex.

FARMER, John **LONDON**
Dove Court, Old Jewry 1731
Whitefriars 1737
St. Ann's Lane, Aldersgate 1739
Cateaton Street 1743
Coach & Horses Yard, Wood Street 1750
Hart Street, Newgate 1752
Pye Corner, West Smithfield 1754 – 1777

Mathematical instrument and rule maker. He was apprenticed to ***Thomas Cooke*** (Old Jewry, London) in 1722. He took as an apprentice ***William***

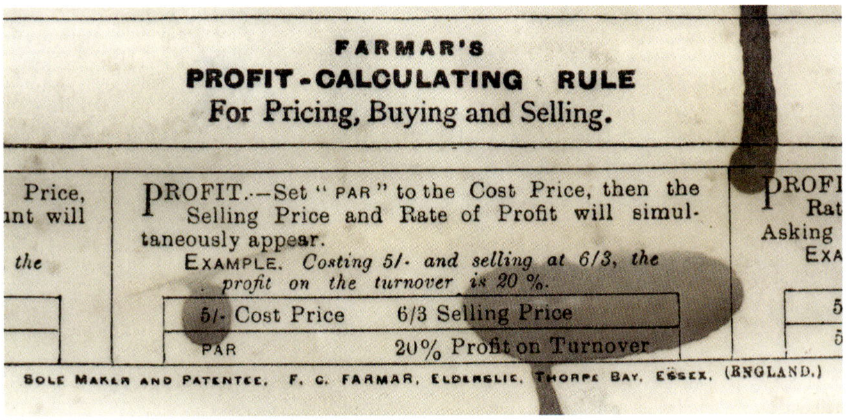

Hussey in 1739. From around 1757 the business was run by ***Richard Gearing***, his nephew.

FARNOLL, John — **BIRMINGHAM**
Constitution Hill 1818 – 1821
(1) (11) Northwood Street 1823 – 1825
11 Constitution Hill 1828 – 1830
2 Court, Snowhill 1835 – 1839

Rule maker.

FARROW, Charles — **LONDON**
18 Great Tower Street 1850 – 1859
FARROW & JACKSON (Ltd.)
18 (16) Great Tower Street 1860 –1930
24 & 25 Eastminster Street, Royal Mint 1940 >

Charles Farrow traded as an ironmonger and manufacturer of machines, tools and utensils for the wine and spirit trade. In 1860 the firm became Farrow & Jackson, the name remaining the same into the 20th century.

FASSBENDER, J. — **LONDON**
93 Upper Thames Street 1938

Rule maker.

FAULKINGHAM — **LONDON**
S.E. corner of Royal Exchange 1672

Rule seller. Rules obtained from *John Nash.*

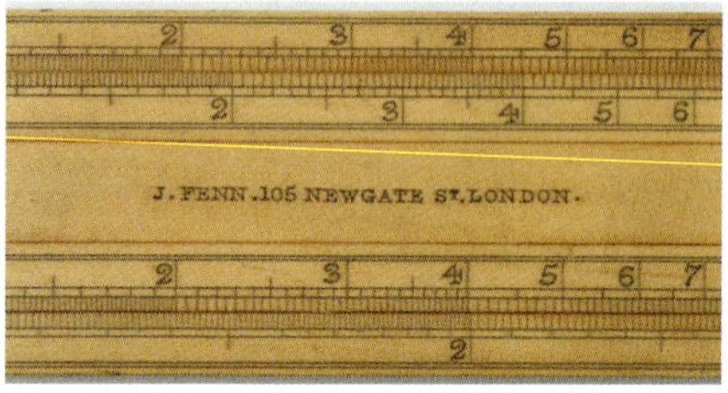

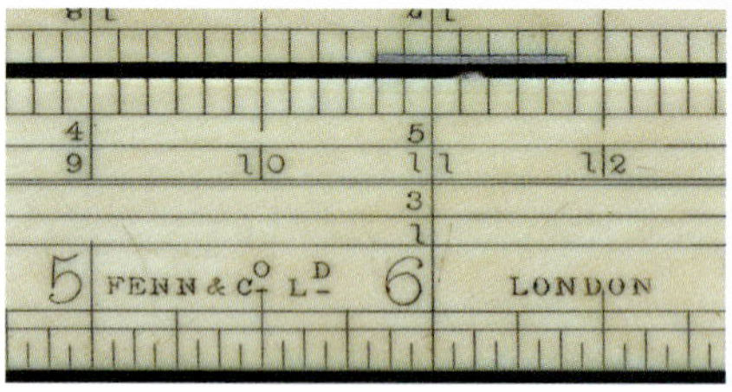

FENN, Daniel — **LONDON**
Newgate Street 1753
FENN, Daniel & Sons
105 Newgate Street 1771
FENN, Daniel & Samuel
105 Newgate Street 1780 – 1800
FENN, Samuel
105 Newgate Street 1805 – 1824
FENN, Joseph
105 Newgate Street 1821 – 1857
105 & 106 Newgate Street 1857 – 1873
FENN & Co.
10 Railway Approach 1874 – 1903
FENN & Co. Ltd.
136 Borough High Road 1905 – 1909 >

Watch tool makers. The firm was founded in 1753 in Newgate Street by Daniel Fenn. It continued until at least 1909. In 1874 it was styled Fenn & Co. and in 1905 became a limited company. The name is found in specialist watch measuring tools and also on some drawing instruments as well as a variety of rules and also gentlemen's tools. For full details see *British Planemakers from 1700,* 3rd ed.

FIELD, R., & Son (& Co.) — **BIRMINGHAM**
33 Navigation Street 1830

111 New Street 1835 – 1870
99 Suffolk Street 1875 – 1880

Opticians and philosophical instrument makers. In 1870 they stated that they were "the sole makers of the Society of Arts microscopes."

FIGGINS, V. & J. **LONDON**
Farringdon Road < 1835 >
Ray Street 1865 – 1910

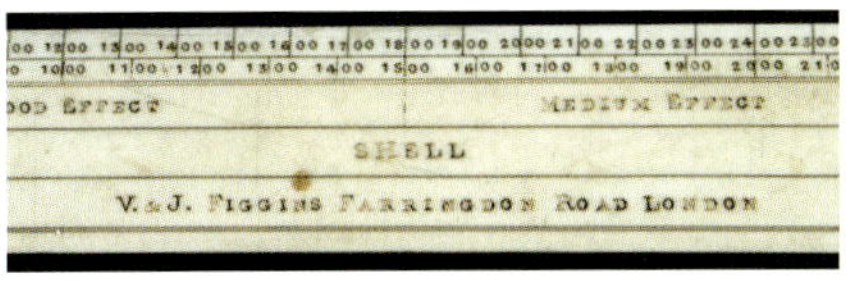

Type founders and manufacturers of printing materials. Their name is mostly found on rules relating to the printing trade, although other types of rules have been found. They would not have made the rules bearing their name.

FINCHERS **BIRMINGHAM**

Unidentified. Unlikely to have been a rule maker.

FINNESEY, R. O. **LIVERPOOL**
11 Williamson Street

Unidentified. Unlikely to have been a rule maker.

FISHER, Wm. **MANCHESTER**
65 Shudehill < 1850 >

Rule maker.

FITCH & JONES **LONDON**
20 Crown Street, Princes Street, Soho 1818 – 1819

Mathematical instrument makers. The address is also given as Crown Court, Wardour Street which is the same premises. It was a partnership between ***John Jones*** and James Fitch, mathematical instrument maker. James Fitch's father, another James, was apprenticed to ***John Bennett*** who had previously worked in Crown Court. James Fitch was apprenticed to ***John Search.*** However, he only appeared in the trades directories in partnership with John Jones. ***Mary Wellington & Son*** also worked at these premises between 1813 and 1827. ***John Jones*** was also listed in the trades directories at these premises under his own name, continuing after the last entry for Fitch & Jones.

FLOWER, Thomas **LONDON**
The Dyall, by the Tower 1642
FLOWER, Christopher
In the Bulwark by the Tower 1650

Thomas is known to have sold a rule and Christopher a slide rule but it is not known if they were makers.

FLOYD, William **BIRMINGHAM**
32 Milk Street 1839
Barford Street 1852 – 1858
FLOYD, John
33 Nelson Street 1863 – 1876
FLOYD, William Henry
(63) (213) Barford Street 1876 – 1903

Box and ivory rule makers.

FOWLER, John **LONDON**
The Globe by the Royal Exchange 1721 – 1750

Mathematical instrument maker. He is known to have made gaugers slide rules. He was apprenticed to ***Thomas Cooke*** (Old Jewry, London) and took ***Henry Gregory*** as an apprentice.

FOWLER, William **BRISTOL**
The Bell, Hillgrove Street 1825

Rule and gauging instrument maker. He took over the business of ***John Wood*** (Bristol).

FRANCE, Archer **LONDON**
6 White Street
and 72 Bethnal Green Road 1836 – 1838

Rule maker.

FRANKLIN, Edmund John **BIRMINGHAM**
35½ Constitution Hill 1855
34½ Unett Street 1858

Manufacturer of measuring tapes, mathematical instruments, mariners' and miners' compasses, rules.

FRANKLIN, J. A. **MANCHESTER**
1 St Ann's Place 1834 – 1838

Optician, mathematical and philosophical instrument maker.

FRANKS, Abraham **MANCHESTER**
114 Deansgate (Dean Street) 1838 – 1865
FRANKS, Abraham & Joseph
44 Market Street 1851 – 1858
FRANKS, Joseph
44 Market Street 1861 – 1865

Mathematical instrument makers and opticians. J. Franks was previously at other addresses in Manchester from 1825. They firm also had a branch in Newcastle under Lyme.

FRASER, William **LONDON**
3 New Bond Street < 1775 – 1799
FRASER & Son
New Bond Street 1799- 1815

Mathematical instrument makers. The firm was taken over by Hawks Grice in 1815 who do not appear to have made rules. This business was subsequently taken over by ***Edward Dixey*** in 1817.

FREEMAN, Edward **LONDON**
87 Long Lane 1817 – 1818

Rule maker.

FREWEN **LONDON**
Sign of the Naked Boy, Cheapside 1672

He is known to have sold rules but it is not known if he made the rules he sold.

FROST & WITHNOLL **BIRMINGHAM**
14 Litchfield Street 1767 – 1776
FROST
20 Great Charles Street 1777 – 1781

Rule makers. The name is also spelt Frost & Withnal. They were the continuation of the business of ***Wood & Lort***. By 1775, ***Wood & Withnol*** were working in Great Charles Street, although Frost is also still listed in the trades directories under his own name.

FRY, John **LONDON**
Little Old Bailey < 1747 – 1752 >

Mathematical instrument maker.

FULLFORD, John **BIRMINGHAM**
Swallow Street 1800 – 1801

Rule maker.

G

GABRIEL, William — **BIRMINGHAM**
78 Barr Street — 1845 – 1852

Parallel rule maker.

GALE & POLDEN Ltd. — **ALDERSHOT**
Grove Road — 1892 – 1974 >

Military stationers and printers.

GALLIE & Son — **GLASGOW**

Unidentified but possibly successors to George Gallie, bookseller and stationer, who operated from various addresses between 1818 and 1830. The name has been recorded on an ivory triangular scale.

GALLOW & Son — **LONDON**
Park Lane — < 1890 >

Unidentified but probably a retailer. The name has been recorded on a horse measure (walking stick type), hallmarked 1890.

GALLOWAY & Co. — **EDINBURGH**

Unidentified but probably a retailer. The name has been recorded on boxwood rules. Rules have also been found location of Leith.

GARBETT, Henry J. — **BIRMINGHAM**
31 Masshouse Lane — 1870 – 1875

Rule maker.

GARCIA, S. Ltd.

Unidentified but possibly a maker. The name has been recorded on a standard yard marked "Borough of Scunthorpe".

GARDNER, J. — **GLASGOW**
Bell's Wynd — 1773 – 1790
GARDNER, J. & J.
(43) Bell Street — 1799 – 1818
GARDNER, Jamieson, & Co.
43 Bell Street — 1819 – 1823
GARDNER & Co.
(43) (93) Bell Street — 1823 – 1831
44 Glassford Street — 1832 – 1838
(21) (53) Buchanan Street — 1839 – 1882
53 Vincent Street — 1883

Mathematical, optical and philosophical instrument makers.

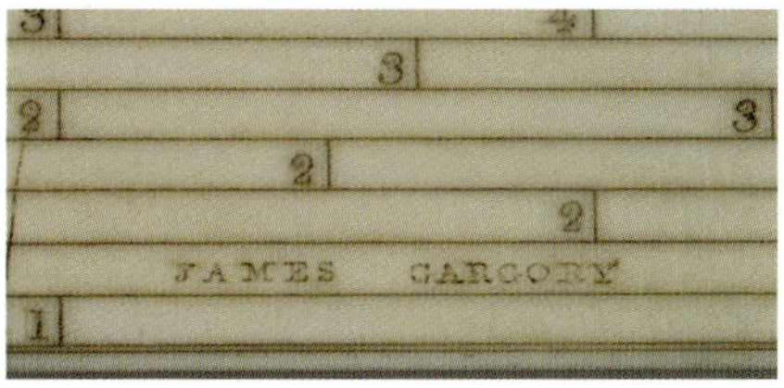

GARGORY, James — **BIRMINGHAM**
(4) (5) (41) Bull Street — 1835 – 1862

Mathematical, optical and philosophical instrument maker.

GARTON & KING — **EXETER**

Unidentified but probably a retailer. The name has been recorded on a 1 ft. 4-fold ivory rule.

GASKELL & CHAMBERS Ltd. — **BIRMINGHAM**
< 1850s – present

Suppliers of equipment for the public house trade. They also had offices in London, Cardiff and Leeds.

GEAR, Henry (& Co) — **BIRMINGHAM**
9 Mott Street — 1861 – 1866
54 Buckingham Street — 1867 – 1887
85½ Constitution Hill — 1888 – 1908

1861: Rule makers. 1867: Mathematical instrument makers.

GEARING, Richard Bates — **LONDON**
2 Horse Shoe Court, Cock Lane, West Smithfield — 1757 – 1776

Rule maker. From around 1757 Gearing took over his uncle ***John Farmer's*** business in Pye Court, West Smithfield, possibly running both businesses. His trade card states "rules mended in the best manner".

GEORGE — **LONDON**
Battersea

Ironmonger.

GEORGE, W. & J. Ltd — **BIRMINGHAM**
33-36 Cambridge Street — 1900 – 1913

Chemical and physical apparatus makers. They also had an address in Hatton Wall, London.

GERREDD, James — **BIRMINGHAM**
85 Wearman Street — 1835

Rule maker.

GERRETT, James (& Co.) — **BIRMINGHAM**
147 Bath Row — 1850
146½ Latimer Street — 1854 – 1856

Rule makers.

GIBBONS, William — **WOLVERHAMPTON**
(41) Stafford Street — 1811 – 1835
GIBBONS, T. — **BIRMINGHAM**
St Pauls Square — 1839
GIBBONS, Mrs Elizabeth
Stafford Street — 1839 – 1851 >

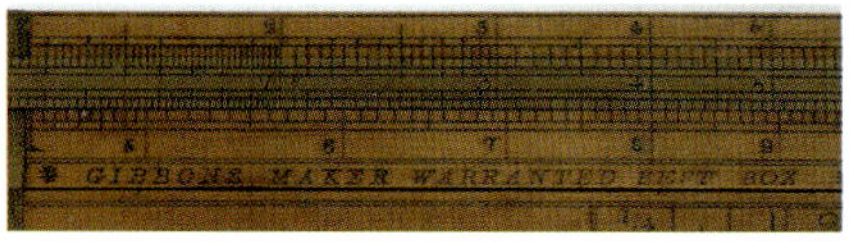

GIBBONS, William
38 Little Hampton Street 1875

Rule makers.

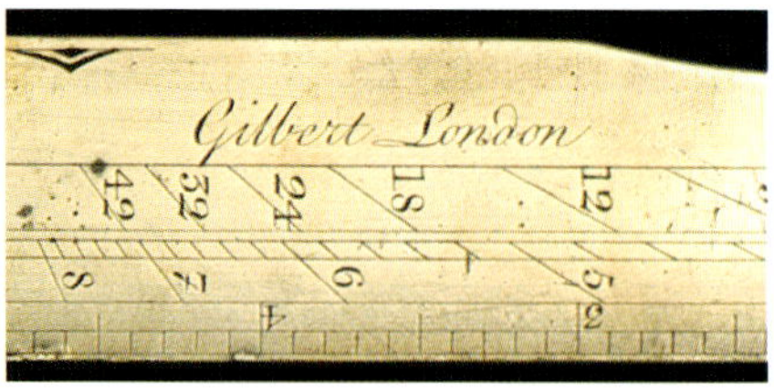

GILBERT, John (I) **LONDON**
Postern Row, Tower Hill 1719 – 1750
GILBERT, John (II)
Postern Row 1751 – 1791
GILBERT & WRIGHT
148 Leadenhall Street 1792 – 1805
GILBERT, William (I) & GILKERSON
8 Postern Row 1793 – 1809
GILBERT, William (I)
148 Leadenhall Street 1795 – 1813
GILBERT & Sons
148 Leadenhall Street 1806 – 1819
GILBERT William (II) & Thomas
148 Leadenhall Street 1819 – 1831
GILBERT William (II)
138 Fenchurch Street 1831 – 1845

Mathematical, nautical, optical and philosophical instrument makers. Around 1790, William Gilbert (I) formed a partnership with ***Gregory & Wright*** in Leadenhall Street, trading as ***Gregory, Gilbert & Wright.*** In 1794 this partnership was styled Gilbert, Wright & Hooke. Between 1819 and 1831 it was styled Gilbert, William & Thomas. After Gilbert moved to Leadenhall Street, the business in Postern Row was continued by ***Gilkerson***.

GILKERSON & Co. **LONDON**
8 Postern Row, Tower Hill 1809 –1825
GILKERSON & McCALL
Postern Row, Tower Hill 1826 – 1830

Mathematical instrument makers. See ***Gilbert, William (I) & Gilkerson***.

GILL, A. & Sons **SHEFFIELD**
St. Mary's Road 1879

Steel and ivory rule makers.

GILL, D., & Son **ABERDEEN**

Unidentified but probably a retailer. The name has been recorded on a lead gauge.

GILL, R. **LEEDS**

Unidentified but probably a specialist supplier of tailoring equipment. The name has been recorded on a tailors' square.

GILL, T. (& Co.) **LONDON**
42 Greek Street 1814 – 1817
20 Little Tower Street 1818 – 1820

Hydrometer makers. Connected with the Liverpool ***Dicas*** family, Gill's name is found on some Dicas patent proof rules.

GILL, T. G. **STROUD**

Unidentified but probably a retailer. The name has been recorded on a 2 ft. 2-fold rule marked "One Shilling Rule".

GLAHOLM & ROBSON **SUNDERLAND**
< 1910 –1916 >

Sail and rope manufacturers and steel hawser makers. The name is found on various measuring tools relating to rope. It is unlikely that they were manufactured by this firm. They also had premises in London

GLASER, William H. **MANCHESTER**
10 Clowes Street, Withington < 1903 >

Engineer. He designed a slide rule for the cloth industry, marked "Pat. 487/1903 No. 1006", although the patent has not been identified.

GLEAVE & Son **MANCHESTER**

Joseph Gleave & Son are a well-known tool wholesalers and trade supplier. The firm started in 1832 as plane makers and continues today. See *British Planemakers from 1700,* 3rd ed.

GODDARD, J. & J. **LONDON**

Piano manufacturers and suppliers of tools for musical instrument making, situated in Tottenham Court Road. Their name is found on tools associated with musical instrument making and tuning, such as string gauges. It is unlikely that they made any of the tools that bear their name.

GOGERTY, Robert **LONDON**

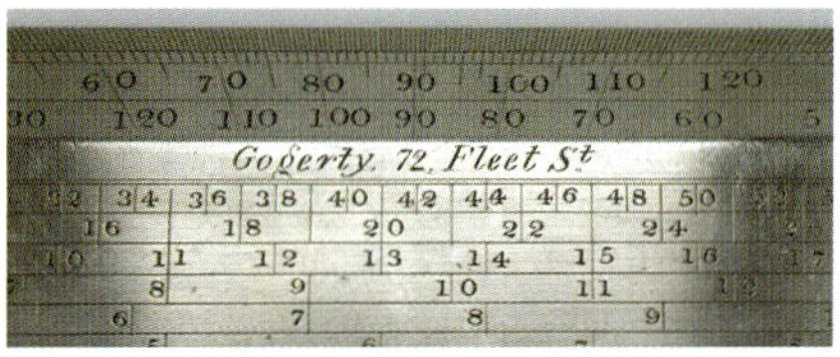

25 St. John Street	1834 – 1835
19 Great Sutton Street	1839
32 King Street	1845
72 Fleet Street	1849 –1855

Opticians and mathematical instrument makers.

GOODWIN **LONDON**
42 Red Lion Street < 1880 >

Clock case maker. The name has been recorded on a lens sizer.

GOUGH, Walter **LONDON**
21 Middle Row Holborn 1784 – 1811

Mathematical instrument maker.

GOUGH & BOWEN **LONDON**
1840 – 1855

Gunsmiths

GRANGER, James **BIRMINGHAM**

Granville Street	1875
Wood Street, Bath Row	1876

1875: Brass rule maker. 1876: Rule maker.

GRAVES, J. G. — **SHEFFIELD**

J. G. Graves operated a mail order firm and was one of the first "bob-a-week" merchants in the UK. See *British Planemakers from 1700,* 3rd ed.

GREATOREX, Thomas — **BIRMINGHAM**
45 Hatchet Street — 1863 – 1882

Rule maker.

GREEN & PICKSLAY — **SHEFFIELD**
High Street and George Street — < 1807 – 1828
Cutlers, ironmongers, etc. After 1828, the business was continued by Pickslay, Appleby and Bertram.

GREEN, John — **BIRMINGHAM**
23 Lower Windsor Street — 1852
GREEN, Thomas
29 Lister Street — 1865 – 1870
31 Burbury Street — 1875 – 1882

Rule makers.

GREENUP, Charles — **LONDON**
Great Saffron Hill — 1790 – 1794

Rule maker.

GREENWAY, Thomas — **BIRMINGHAM**
23 Ward Street — 1839

Rule maker.

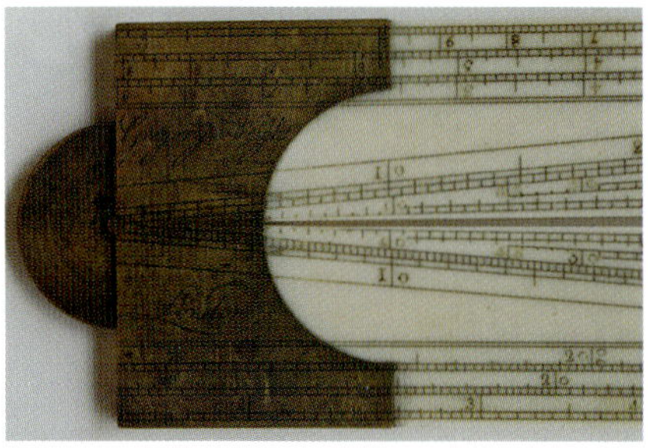

GREGORY, Henry (& Son) — **LONDON**
(8) (148) Leadenhall Street — 1748 – 1782
GREGORY & WRIGHT, Henry & Gabriel
148 Leadenhall Street — 1782 – 1790
GREGORY, GILBERT & WRIGHT
148 Leadenhall Street — 1790 – 1792
GREGORY, GILBERT & Co.
148 Leadenhall Street — 1794

Mathematical instrument makers. Henry Gregory was apprenticed to ***John Fowler***. Gabriel Wright was possibly a descendant of ***George Wright***.

GREGORY, William — **LONDON**
8 Berry Street, Clerkenwell — 1834
9 Hooper Street, Clerkenwell — 1835
2 Francis Court, Berkeley Street — 1836

Rule, optical instrument, philosophical instrument and mathematical instrument maker.

GREW, T. — **PLAISTOW**

Mathematical instrument maker, probably mid- to late 19th century.

GREW, William — **BIRMINGHAM**
Bradford Street — 1785 – 1788

Rule maker.

GRIFFIN & Co. — **PORTSEA**
The Hard

Unidentified but likely to have been a retailer.

GRIFFITHS, Thomas — **BIRMINGHAM**
32 (33) Kenion Street — 1841 – 1867
32½ Gt. Hampton Street — 1860 – 1868
32½ Kennion Street — 1870
11 Court, Sun Street — 1875
GRIFFITHS, Francis Henry
11 Court, Sun St — 1875 – 1876

Ivory rule makers.

GROOME, Henry — **MANCHESTER**
Gleden Street, Bradford Road — 1875

Rule maker.

GROSVENOR, Richard — **BIRMINGHAM**
73 Barford Street — 1882

Rule maker.

GROVE, R.H. — **LONDON**

Unidentified. The name has been recorded on an ivory watchmaker's gauge and a gauging rule.

GROVE, W. — **LONDON**

Unidentified but the name is marked as "maker" on a proof rule.

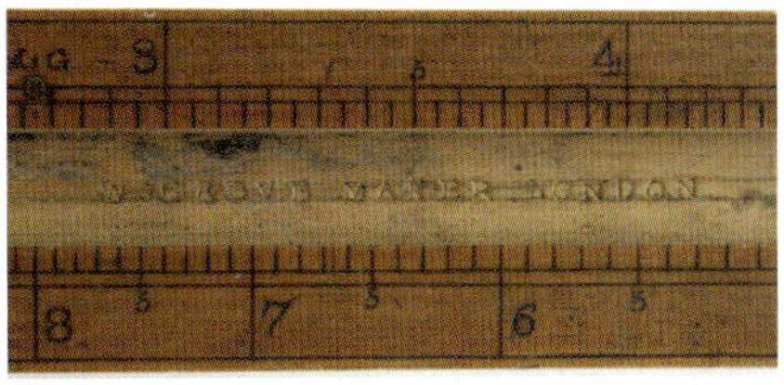

GROVES & BARKER — **LONDON**
16 Market Street, Clerkenwell — 1851 – 1865 >

Wholesale makers of ships', pocket and prismatic compasses; sextants; telescopes; marine and pocket aneroids; drawing and surveying instruments; Barker's patent altitude instruments; sun dials; etc. The firm continued as ***F. Barker & Sons***.

GUEST, Richard — **WOLVERHAMPTON**
67 Brick Kiln Street — 1812 –1818

Rule maker.

GURNEY — **MAIDSTONE**
GURNEY & HAYNES

Unidentified. The name Gurney & Haynes is recorded on a rule by ***T. Tustian***. It is likely that the firms were connected and were retailers.

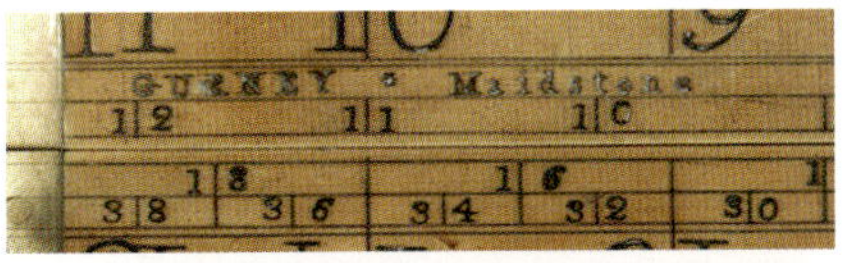

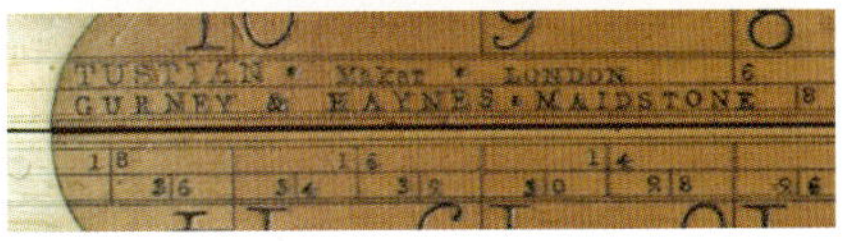

GUTTERIDGE **LONDON**
GUTTERIDGE DOWNING & Son

< 1827 >

Unidentified but possibly makers. "Gutteridge patent" has been recorded on a sector; "Gutteridge Downing" has been recorded on gaugers' and proof slide rules.

GWILLYM, Joshua **BRISTOL**
3 Earl Street 1832 – 1865
GWILLYM, Mrs Ann
Earl Street 1856 – 1859

Rule and gauging instrument makers. Joshua took over the business of ***Thomas Webb***. No rules bearing their names have been recorded.

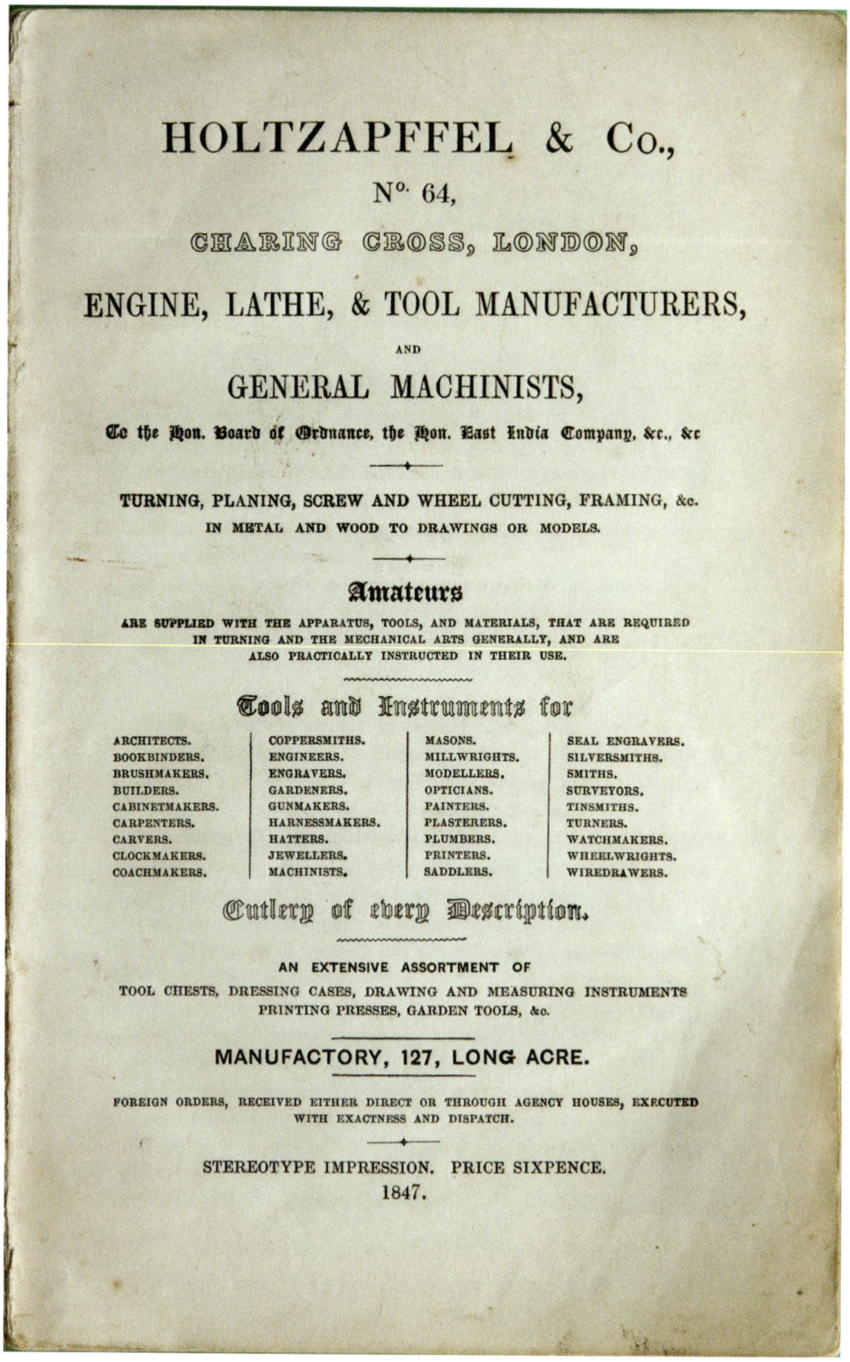

HOLTZAPFFEL & Co.,
No. 64,
CHARING CROSS, LONDON,
ENGINE, LATHE, & TOOL MANUFACTURERS,
AND
GENERAL MACHINISTS,
To the Hon. Board of Ordnance, the Hon. East India Company, &c., &c

TURNING, PLANING, SCREW AND WHEEL CUTTING, FRAMING, &c.
IN METAL AND WOOD TO DRAWINGS OR MODELS.

Amateurs
ARE SUPPLIED WITH THE APPARATUS, TOOLS, AND MATERIALS, THAT ARE REQUIRED IN TURNING AND THE MECHANICAL ARTS GENERALLY, AND ARE ALSO PRACTICALLY INSTRUCTED IN THEIR USE.

Tools and Instruments for

ARCHITECTS.	COPPERSMITHS.	MASONS.	SEAL ENGRAVERS.
BOOKBINDERS.	ENGINEERS.	MILLWRIGHTS.	SILVERSMITHS.
BRUSHMAKERS.	ENGRAVERS.	MODELLERS.	SMITHS.
BUILDERS.	GARDENERS.	OPTICIANS.	SURVEYORS.
CABINETMAKERS.	GUNMAKERS.	PAINTERS.	TINSMITHS.
CARPENTERS.	HARNESSMAKERS.	PLASTERERS.	TURNERS.
CARVERS.	HATTERS.	PLUMBERS.	WATCHMAKERS.
CLOCKMAKERS.	JEWELLERS.	PRINTERS.	WHEELWRIGHTS.
COACHMAKERS.	MACHINISTS.	SADDLERS.	WIREDRAWERS.

Cutlery of every Description.

AN EXTENSIVE ASSORTMENT OF
TOOL CHESTS, DRESSING CASES, DRAWING AND MEASURING INSTRUMENTS PRINTING PRESSES, GARDEN TOOLS, &c.

MANUFACTORY, 127, LONG ACRE.

FOREIGN ORDERS, RECEIVED EITHER DIRECT OR THROUGH AGENCY HOUSES, EXECUTED WITH EXACTNESS AND DISPATCH.

STEREOTYPE IMPRESSION. PRICE SIXPENCE.
1847.

Title page from Holtzapffel & Co. 1847 catalogue.

H

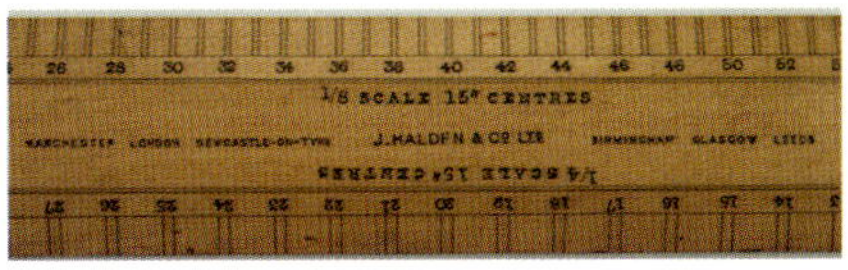

HALDEN, J., & Co. (Ltd.) **MANCHESTER**
8 Albert Square 1877
East Street 1889
Lund Street, Cornbrook 1900
and Altringham Street 1903
and Brazenose Street 1904
Reddish, Stockport 1908 – 1969

Manufacturers of drawing instruments and slide rules. Joseph Halden originally set up in business with ***A. G. Thornton*** but this partnership lasted only two years. By the 1920s Halden also had branches in London, Birmingham, Glasgow, Leeds and, later, Berlin and Paris. In 1969 the business was taken over by the Ozalid Group.

HALL & HAWKES **LONDON**
Topping's Wharf, London Bridge 1839
10 Duke Street, Borough 1839 – 1842

Mathematical instrument makers and opticians.

HALL, B. J. (& Co. Ltd.) **LONDON**
39 Victoria Street < 1900 – 1924
Stourton House, Dacre Street 1938 – 1943
HALL HARDING Ltd.
Camberwell 1943 – 1966

Drawing instrument makers. By 1906 there were several branches around west London. B. J. Hall & Co. Ltd. also had offices in Birmingham, Glasgow and Newcastle. In 1943, B. J. Hall & Co. Ltd. took over ***Drawing Office Supplies Ltd.*** In 1953 Hall Harding Ltd. took over ***J. A. Reynolds & Co.***

HAMLETT, William **WOLVERHAMPTON**
50 Salop Street 1792

Rule maker.

HAND, George **BIRMINGHAM**
19 Henrietta Street 1860 – 1876

Rule maker.

HARDING, T. R. **LEEDS**

Unidentified.

HARDMAN, James **WOLVERHAMPTON**
74 Goat Street 1802

Rule maker. ***Joseph Marryman*** was working at the same address in 1792. The connection between the two has not been established.

HARLEY, J. J. — **LIVERPOOL**

A firm of tool dealers, planemakers and edge-tool makers, operating from 1875 to 1922. Although the firm was entered in the list of rule makers from 1873 to 1905, it is highly unlikely that they ever made rules. From 1880 to 1890 Mary Harley, cutler, was also entered in the list of rule makers. See *British Planemakers from 1700,* 3rd ed.

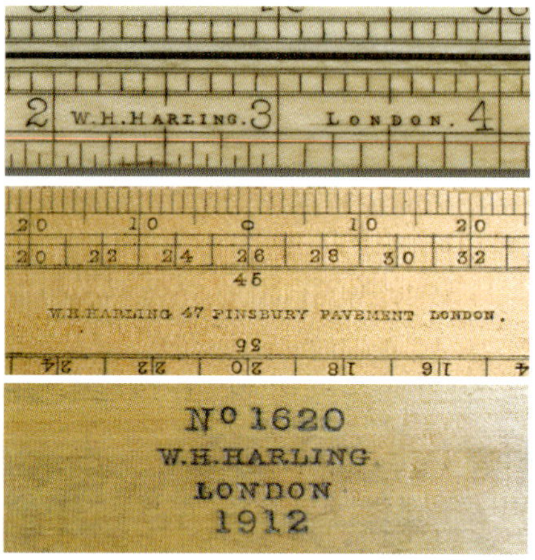

HARLING, William Henry — **LONDON**

47 Finsbury Pavement	1851 – 1961
117 Moorgate	1930s

Drawing instrument and rule maker. By 1880 Harling also had premises in London Place, London Fields, Hackney and 40 Hatton Garden. In 1961 the firm became part of ***Blundell Harling***, Weymouth.

HARMER, James — **BIRMINGHAM**

352 Farm Street	1880
Mount Place, Hockley Hill	1895
Whitmore Street	1906 – 1910

Rule maker.

HARPER, Nathan — **BIRMINGHAM**

Brighton Place	1849 – 1850

HARPER, Alfred (& Sons)

42 Westone Lane	1900 – 1902
Ellis Street	1925

Rule makers. On 16 June 1925, the premises of Messrs. Alfred Harper & Sons were destroyed by fire.

HARRILD & Sons — **LONDON**

75 Farringdon Street	< 1861 >

Printers' brokers and roller manufacturers. They would not have made the rules marked with their name.

HARRIMAN, Thomas — **BIRMINGHAM**

Aston Road	1800 – 1801
Ashtead Row	1805
Loveday Street	1809 – 1818
24 (40) Loveday Street	1821 – 1852

HARRRIMAN, John

New John Street	1835
(58) (60) Church Street	1839 – 1855
100 Pritchet Street	1858 – 1865

Rule makers. In the 1823 trades directory, the name was spelt Harrisman. In 1825 there is also an entry in the list of rule makers in the trades directory for T. Harris, but this name does not appear in the main listing and therefore is probably a misprint for this maker.

HARRIS, Charles — **BIRMINGHAM**

24 Hall Street	1865

Rule maker.

HARRIS, Philip (& Co. Ltd.) — **BIRMINGHAM**
Bull Ring — < 1875 – 1889 >
144 Edmund Street — < 1893 – 1914 >

Makers and suppliers of scientific apparatus. They were unlikely to have made the rules that bear their name.

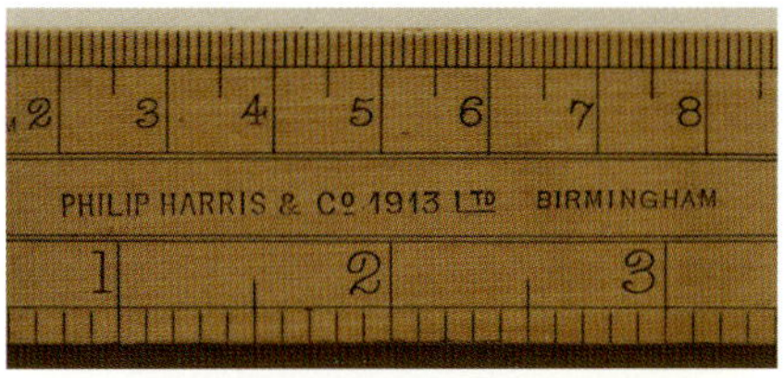

HARRIS, William — **LONDON**
(47) (50) High Holborn — 1800 – 1815
HARRIS & Co.
50 High Holborn — 1816 – 1840
HARRIS & Son
50 High Holborn — 1840 – 1855

Optician and mathematical and drawing instrument makers. William's father, Richard, was an optician, working from 1788 – 1810. Between 1816 and 1839 the firm also had a branch in Liverpool. William Harris noted in his catalogue "late ***Nairne & Blunt***".

HARRISON, Joshua — **BIRMINGHAM**
Gopsall Street — 1839
HARRISON, William
394 Summer Lane — 1839 – 1850

Rule makers.

HARRYMAN, Joseph — **WOLVERHAMPTON**
74 Goat Street — 1792

Rule maker. ***James Hardman*** was working at the same address in 1802. The connection between the two has not been established.

HART, Emmanuel — **LIVERPOOL**
33 Portland Street — 1865

Rule maker.

HART, Saul — **LIVERPOOL**
114 Dale Street — 1841

Rule maker.

HASKINS, Daniel — **BIRMINGHAM**
Henley Street — 1863 – 1882

Rule maker.

HASSALL, Abraham — **BIRMINGHAM**
< 1773 – 1780

HASSALL & RABONE
12 Bull Street — 1784 – 1791
HASSALL & Co.
12 Bull Street — < 1797 >
HASSALL, Joseph
12 Bull Street — < 1800

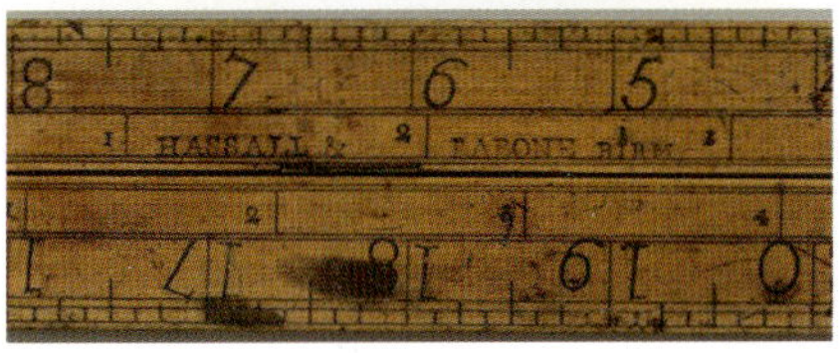

Rule makers. In 1784, a partnership was formed between Joseph Hassall

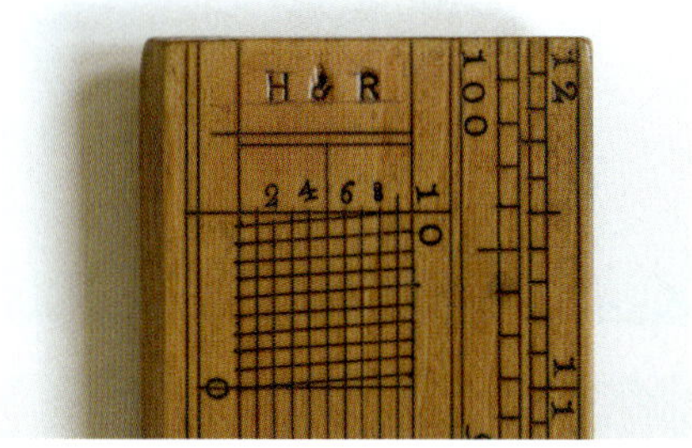

and ***Michael Rabone***. The relationship between Abraham and Joseph Hassall has not been discovered, though it is known that Abraham was not Joseph's father. In 1791 Joseph formed a second partnership with ***Zachariah Belcher***, trading as ***Belcher & Hassall***. In 1794, Belcher moved to Sheffield and this partnership was dissolved. Hassall & Co. is an alternative name for Hassall & Rabone. Joseph Hassall and ***Michael Rabone*** continued in 12 Bull Street until Joseph's death in 1800, although neither the name Hassall & Rabone nor Hassall & Co. are listed in the trades directories. In his will, Joseph left the business to Michael for a nominal sum of 1 shilling.

HATTON, Thomas — **PRESTON** 1740 – 1769; **LONDON** 1769 – 1774

Mathematical instrument maker and clock maker. The name has been recorded on a joint rule dated 1772.

HAWKES, J. & W. — **BIRMINGHAM**

Unidentified but unlikely to have been a maker. The name has been recorded on a 2 ft. 2-fold boxwood architects' rule, brass bound and with brass slide.

HAWKESFORD, Edward — **WOLVERHAMPTON**
57 Canal Street — 1792 – 1802

Rule maker.

HAWTHORNE, Robert — **NEWCASTLE UPON TYNE**

Civil engineer and designer of an engineers' slide rule.

HAY, William — **DUMBARTON**
< 1867 >

Designer of an engineers' rule, which was patented in America in 1867.

HAYCOCK, Samuel (I) — **BIRMINGHAM**
Litchfield Street — 1816
Price Street — 1816 – 1817
Bagot Street — 1818 – 1821
HAYCOCK, Samuel (II)
5 Cock Street — 1821
42 Loveday Street — 1828 – 1830
40 Woodcock Street — 1835 – 1845
HAYCOCK Edward Samuel (& Co.)
40 Woodcock Street — 1850 – 1858

Manufacturers of mathematical drawing instruments and land surveyors' measuring tapes, brass compasses and plated and brass dog collars. In 1850 they described themselves as rule makers.

HAYES, Walter — **LONDON**
Burchin Lane, Cornhill — 1648
Cross Daggers, Moorfields — 1653 – 1687

HAYES, Jane
Cross Daggers, Moorfields 1696 – 1697

Mathematical instrument maker. he took as apprentices ***John Nash*** and ***Edmund Culpeper*** amonst others. Jane was Walter's wife. The business was taken over by ***Edmund Culpeper***.

HAYNES, J.P. (Ltd.) **BIRMINGHAM**
Park Lane, Tipton < 1896 >

Iron merchant. It is unlikely that they made the rules on which their name appears.

HEATH, James **BIRMINGHAM**
11 Court, Dartmouth Street 1852

Rule maker.

HEATH, Thomas **LONDON**
Hercules & Globe, Strand 1720 –1753
HEATH & WING
Nr. Exeter Exchange, Strand 1751 – 1777

Mathematical instrument makers.

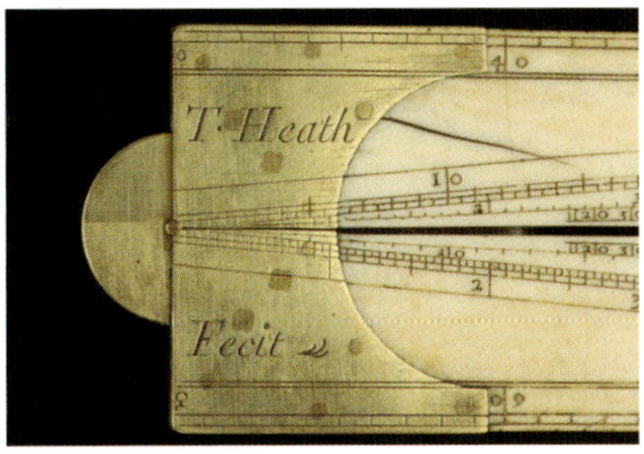

HEATH & Co. **LONDON**
43 Piccadilly < 1899
2 Tower Royal, Cannon Street 1913 – c. 1935
Eltham 1950 – 1960 >

Wholesale nautical instrument and barometer manufacturers and manufacturing opticians. Continuation of ***Murray & Heath***. Taken over in 1926 by ***W. F. Stanley***.

HEATHER, William **LONDON**
157 Leadenhall Street 1793 – 1812

Navigation warehouse, engraver and mathematical instrument maker. The business was taken over by ***John Norie***.

HELLYER, H. W. **LONDON**
98 Newgate Street

Unidentified but unlikely to be a maker. The name has been recorded on a drapers' rule.

HEMMING, Edwin **BIRMINGHAM**
76 Cardigan Street 1847
(10) (49) Vauxhall Road 1849 – 1855
HEMMING, Henry
23 Court, High Street, Bordesly 1854

Rule makers.

HEMMINGS, E. **LONDON**

Unidentified but maybe an owner. The name has been recorded on a Routledge rule.

HENDERSON, D. M., & Co. **LONDON**
141 Moorgate 1924
66A Farringdon Street 1928
45 Hatton Garden 1931

Rule makers. The firms advertised "Folding rules, tapes, levels, calipers, etc."

HENSHAW, Walter **LONDON**
The Globe & Marriner, Wapping 1659 – 1711

Mathematical instrument maker. Apprenticed to ***Thomas Brown(e)***. he advertised in *Timber Measure by a Line...* by Henry Coggeshall, London, 1677.

HEULETT, Ernest **BIRMINGHAM**
78 Talbot Street 1910

Rule maker.

HEWITT, Samuel **LIVERPOOL**
123 Dale Street 1822
HEWITT, Sarah
19 Christian Street 1828

Rule makers.

HEYER, Henry **LONDON**
101 Leather Lane 1891

Rule maker.

HIBBS Bros. **NOTTINGHAM**

Unidentified but probably a retailer. The name has been recorded on 2 ft. 4-fold ivory rule.

HICKS, James J. **LONDON**
8, 9, 10 Hatton Garden 1862 – 1926

Mathematical and scientific instrument maker. The firm was taken over by ***W. F. Stanley*** in 1911.

HIDSON, Frank **BIRMINGHAM**
6 Court, Darwin Street, Lombard Street 1860 – 1868
73 Mole Street 1888 – 1895

Rule maker.

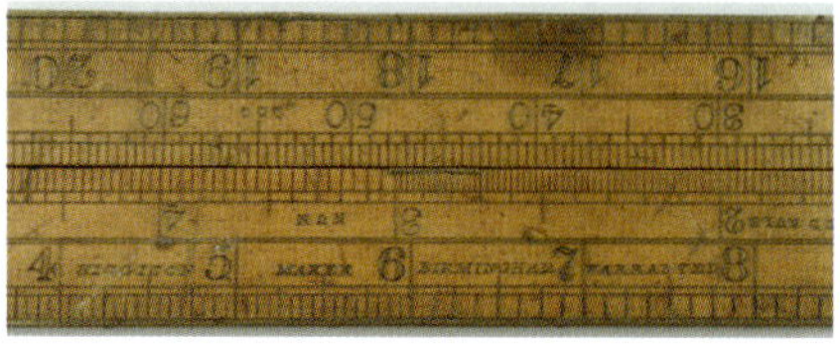

HIGGISON, John **BIRMINGHAM**
23 Court, Bromsgrove Street 1833 – 1839
14 Court, Bristol Street 1839 – 1854
(7) (46) Gooch Street 1855 – 1884
Bishop Street 1886 – 1894

Rule maker. The name is also spelt Higginson. In 1886 he advertised that he was a manufacturer of all kinds of English and foreign rules, mathematical scales, etc., in boxwood and ivory. He claimed to have been established in 1831. In 1894 he combined with ***Charles Bees*** to became ***Bees, Higgison & Co.***

HIGGS, John **WOLVERHAMPTON**
Queen Street 1780 – 1796

Rule maker and japanner.

HILL, Abraham **BIRMINGHAM**
13 Smallbrook Street 1829 – 1831

Rule maker and victualler.

HILL, Isaac **LONDON**
13 Little Compton Street 1841

Rule maker. He operated from the same premises as ***Frederick Mant***.

HILL, John **BIRMINGHAM**
354½ Farm Street 1875

Rule maker.

HILL, John **LONDON**
69 Noble Street 1833 -1836

Rule maker.

HILL, J. V. **LONDON**

Saw maker and tool seller, in business from 1834 –1905. He would not have made rules. See *British Planemakers from 1700,* 3rd ed.

HILL, Nathaniel **LONDON**
Globe & Sun, Chancery Lane, Fleet Street 1746 - 1764

Mathematical instrument maker, globe maker and engraver. He is thought to have taken over the business of ***John Coggs.*** He was the successor to ***John Newton***, his father-in-law.

HILL, S. **LIVERPOOL**
26 Duke Street 1842

Yard measure seller. He was the agent for ***Samuel Clark***.

HIMMEN, Charles **BIRMINGHAM**
Love Street 1854

Rule maker.

HINTON, Henry **BIRMINGHAM**
52 New John Street West. 1868

Rule maker.

HITCHEN, Thomas **BIRMINGHAM**
14 Horsefair 1849 – 1851
(51) (52) New John Street 1852 – 1861

Rule maker.

HOARE, Charles **SOUTHAMPTON & LONDON**

7 York Street Avenue, Southampton < 1855 – 1885 >

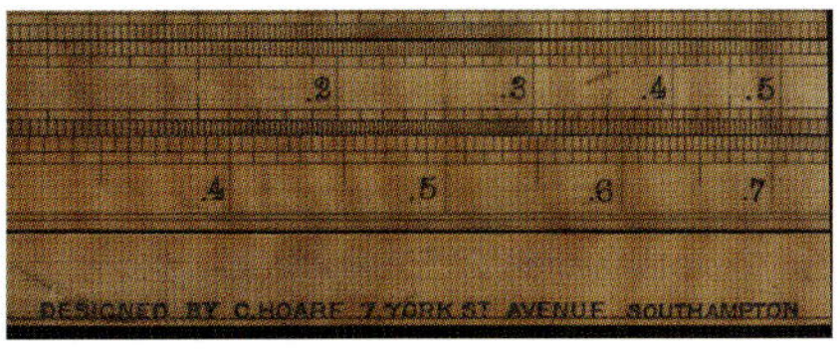

12 Billiter Square, London

Author. He wrote a number of books between 1855 and 1885 concerning slide rules, mathematics and a number of other related subjects. He also designed a range of "Improved" slide rules; some of these are marked "Hoare's Improved Slide Rule" and some "Designed by Charles Hoare". It would appear that his early rules were made by ***Joseph Stebbing***. After Stebbing died in 1874, Hoare formed an association with ***Aston & Mander*** who both published his books and made his slide rules.

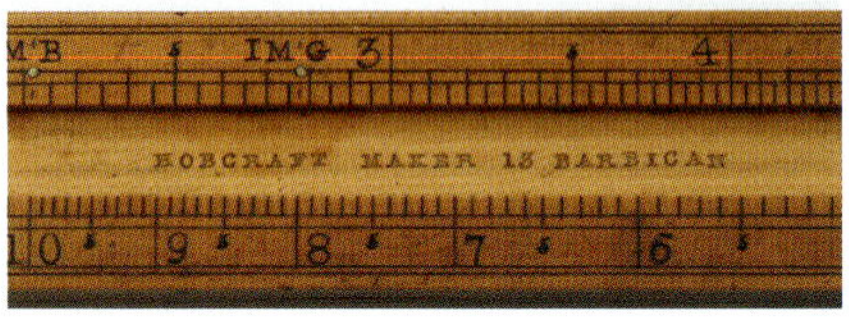

HOBCRAFT, William	**LONDON**
56 Whitecross Street, Fore Street	1817
12 Beech Street, Barbican	1822
13 (18) (14) Barbican	1825 – 1852

Rule, measure and mathematical instrument maker. He was apprenticed to ***Alexander Wellington***. The name was also spelt Hopcroft.

HODGETTS, Thomas	**BIRMINGHAM**
23 Little Hampton Street	1829 – 1830
92 Upper Tower Street	1835

Rule maker.

HODSON, G.	**LONDON**
1 Farringdon Street	
2 Skinner Street	

Unidentified but probably a retailer. The name has been recorded on tailors' squares.

HOILE & NEWBURY **MANCHESTER**

Unidentified but probably a retailer. The name has been recorded on a hatters' rule.

HOLDING, T. H.	**LONDON**
3 Adelaide Street	
7 Madox Street	1892 – 1906

Tailor and fashion publisher. The name is found on tailors' rules, squares, etc. However, it is known that the rules were bought in from ***John Rabone & Sons*** and probably also other makers.

HOLLOWAY, John	**BIRMINGHAM**
28 (west) Grosvenor Street	1852 – 1855
7 Regent Place	1858
98 Mott Street	1860
Chequers Walk	1862
(90) (21) Mott Street	1870 – 1895

Rule maker.

HOLMES, James **BURNLEY**

Designer of a slide rule for the textile industry.

HOLMES, Thomas — **BIRMINGHAM**
41 Cecil Street — 1855
Rodway Street — 1858

Rule maker.

HOLT, James — **MANCHESTER**
15 Red Lion Street, Church Street — 1848

Rule maker.

HOLTZAPFFEL, John — **LONDON**
118 Long Acre — 1794 – 1803
HOLTZAPFFEL & DEYERLAIN
118 Long Acre — 1804 – 1811
10 Cockspur Street — 1812 – 1820
64 Charing Cross Road — 1821 – 1828
HOLTZAPFFEL & Co.
64 Charing Cross Road — 1828 – 1901
13 & 14 New Bond Street — 1902 – 1907
53 Haymarket — 1907 – 1922
HOLTZAPFFEL & Co. Ltd.
53 Haymarket — 1923 – 1930
79 Wigmore Street — 1931

Tool makers, dealers in ivory and wood, etc. See *British Planemakers from 1700,* 3rd ed.

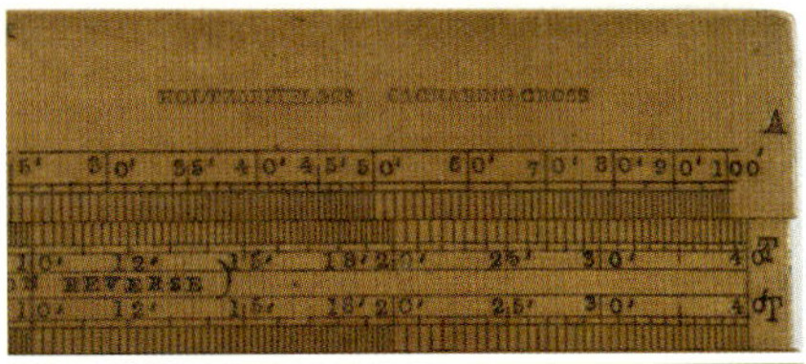

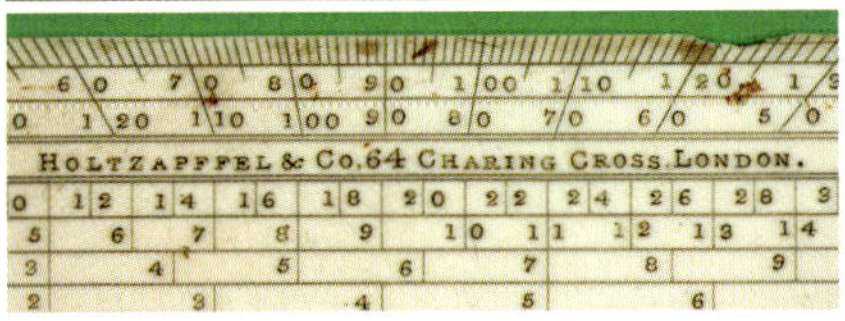

HOOD & HAGGIE & Sons — **NEWCASTLE-UPON-TYNE**

Rope manufacturers, late 19th to 20th centuries. In 1901, they were listed as customers of ***John Rabone & Sons***.

HORSEMAN, Stephen — **LONDON**
King's Arms, Exchange Alley — 1709 – 1724

He is known to have sold rules but it is not known if he also made rules.

HORTON, Edward — **WOLVERHAMPTON**
3 Bilston Street — 1792
5 Salop Street — 1802

Rule maker.

HORTON, Edward — **BIRMINGHAM**
18 Court, Staniforth Street — 1880
HORTON, J. T. & L.

Rule makers. The name was also spelt Hortin.

HORWICK, Isaac — **BIRMINGHAM**
9 Court, Warstone Road — 1858

Rule maker.

HOWELL, William — **BIRMINGHAM**
19 King Edward Place — 1867

Rule maker.

HOUNSFIELD, B. — **SHEFFIELD**
Little Pond Street — 1833

Rule maker.

HUDDY, Francis — **LONDON**
37 Duke Street, Smithfield — 1843 – 1844
(24) (34) Red Lion Street — 1845 – 1865

Rule and mathematical instrument maker.

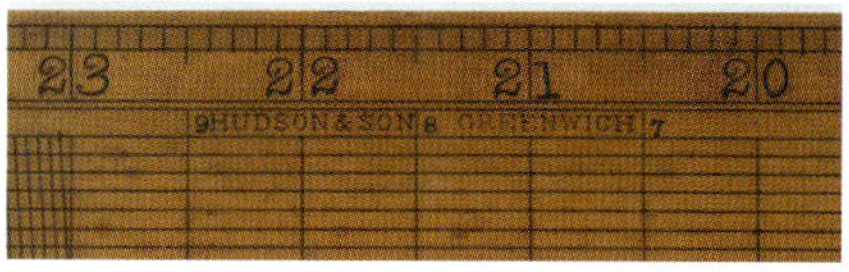

HUDSON, Charles — **SHEFFIELD**
Upper Heely — 1852 – 1855
81 Arundel Street — 1871
Cadman Lane — 1876
Exchange Gateway, Fargate — 1883

Manufacturer of mathematical scales and rules and ruler-knives.

HUDSON, C. — **BIRMINGHAM**
38½ Hanley Street — 1861

Rule maker.

HUDSON, Frederick T. — **LONDON**
Greenwich Road — 1830
Stockwell Street — 1845
HUDSON & KEARNS
HUDSON & Son

Opticians. The trade card of Hudson & Sons states "Manufacturers of Mathematical, optical and nautical instruments to the Lords Commissioners of the Admiralty, Royal Naval College, etc."

HUGHES, Joseph (I) — **LONDON**
16 Queen Street, Ratcliffe — 1818 – 1845
HUGHES, Henry
Commercial Road — 1835 – 1839
(120) (59) Fenchurch Street — 1840 – 1843
HUGHES, Joseph (II)
37 & 38 (40) Queen Street, Ratcliffe — 1843 – 1875
HUGHES, H., & Son (Ltd.)
59 Fenchurch Street — 1844 – 1940

Optical, nautical and mathematical instrument makers. Henry and Joseph (II) were the sons of Joseph (I), both of whom were apprenticed to their father.

HUGHES, Thomas — **BIRMINGHAM**
26 Duke Street — 1785 – 1788

Aston Road	1792 – 1793
Colmore Row	1797

1785: Rule, saw handle and case maker. 1787: Rule maker. 1797: Rule and paper case maker.

HUGHES, Thomas	**LONDON**
52 Gee Street, St. Lukes	1802 – 1835
156 Whitecross Street	
68 Noble Street St. Lukes	
HUGHES, George	
28 (46) Gee Street	1831 – 1839
35 Northampton Street	1845 – 1855

Thomas: Rule maker, gauge maker. George: Rule maker.

HUGHES, William John	**BIRMINGHAM**
75 Coventry Road	1875 – 1876

Manufacturer of ivory, box and brass rules. Advertised that he made rules of every description for home and export with all measurements of the known world. In 1872 he took over the business of ***S. A. Smith***.

HUME, Jane(t)	**LONDON**
10 High Street, Wapping	< 1836 >

Mathematical and philosophical instrument maker.

HUNT & HOLDITCH	**LONDON**
52 Southwark Bridge Road	1924 – 1938 >

Hat trimming manufacturers. Name recorded on hatters' rule. It is unlikely that they made the rules on which their name appears.

HUNTER, William Thomas	**NORWICH**
St. Lawrence Steps	1830

Rule and toy maker.

HUSBANDS & CLARKE	**BRISTOL**
1 Denmark Street	1857 – 1863
(8) (20) St. Augustine's Parade	1864 – 1872
HUSBANDS, Henry	
8 St. Augustine's Parade	1873 – 1876 >

Opticians. In 1857, they took over the business of ***T. D. King***.

HUSSEY, William	**LONDON**
Butler's Alley, Grub Street	1764
Moor Lane, Moorfields	1777
Sweetings Passage, Moor Lane	1779 – 1797
HUSSEY, John	
1 Tenter Alley, Little Moorfields	1806

Rule makers. William was apprenticed in 1739 to ***John Farmer;*** John was apprenticed in 1785 to William, his father.

HYDSON, Robert **LEEDS**
< 1686 >

Unidentified. The name has been recorded on a dated boxwood sector.

HYRONS, G. **LONDON**
82 Queen Street, Southwark 1761 – 1798

Mathematical instrument maker.

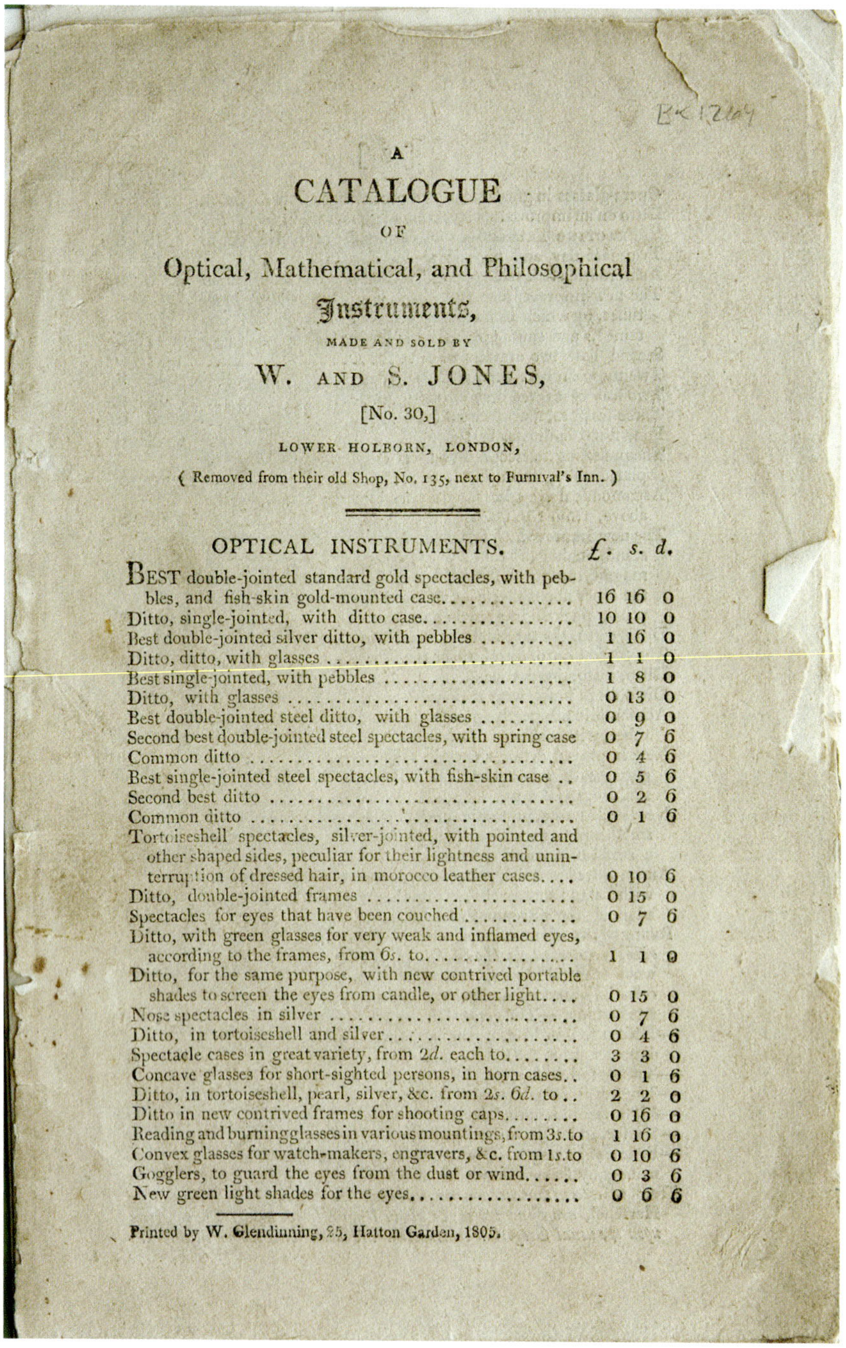

A

CATALOGUE

OF

Optical, Mathematical, and Philosophical

Instruments,

MADE AND SOLD BY

W. AND S. JONES,

[No. 30,]

LOWER HOLBORN, LONDON,

(Removed from their old Shop, No. 135, next to Furnival's Inn.)

OPTICAL INSTRUMENTS.	£.	s.	d.
Best double-jointed standard gold spectacles, with pebbles, and fish-skin gold-mounted case	16	16	0
Ditto, single-jointed, with ditto case	10	10	0
Best double-jointed silver ditto, with pebbles	1	16	0
Ditto, ditto, with glasses	1	1	0
Best single-jointed, with pebbles	1	8	0
Ditto, with glasses	0	13	0
Best double-jointed steel ditto, with glasses	0	9	0
Second best double-jointed steel spectacles, with spring case	0	7	6
Common ditto	0	4	6
Best single-jointed steel spectacles, with fish-skin case	0	5	6
Second best ditto	0	2	6
Common ditto	0	1	6
Tortoiseshell spectacles, silver-jointed, with pointed and other shaped sides, peculiar for their lightness and uninterruption of dressed hair, in morocco leather cases	0	10	6
Ditto, double-jointed frames	0	15	0
Spectacles for eyes that have been couched	0	7	6
Ditto, with green glasses for very weak and inflamed eyes, according to the frames, from 6*s*. to	1	1	0
Ditto, for the same purpose, with new contrived portable shades to screen the eyes from candle, or other light	0	15	0
Nose spectacles in silver	0	7	6
Ditto, in tortoiseshell and silver	0	4	6
Spectacle cases in great variety, from 2*d*. each to	3	3	0
Concave glasses for short-sighted persons, in horn cases	0	1	6
Ditto, in tortoiseshell, pearl, silver, &c. from 2*s*. 6*d*. to	2	2	0
Ditto in new contrived frames for shooting caps	0	16	0
Reading and burning glasses in various mountings, from 3*s*. to	1	16	0
Convex glasses for watch-makers, engravers, &c. from 1*s*. to	0	10	6
Gogglers, to guard the eyes from the dust or wind	0	3	6
New green light shades for the eyes	0	6	6

Printed by W. Glendinning, 25, Hatton Garden, 1805.

Title page from W. & S. Jones 1805 catalogue.

I & J

IBBOTSON, Thomas, & Co.	**SHEFFIELD**

A Sheffield tool maker and seller who operated from 1823 to 1909. See *British Planemakers from 1700,* 3rd ed.

IGOD, George T.	**LONDON**
8 Turnagain Lane, City	1841

Brass rule cutter.

INMAN, William	**BIRMINGHAM**
Market Street	1835
INMAN, William, & Son	
13 Court, Coleshill Street	1839
36 Ashtead Row	1842
INMAN, Charles	
13 Great Hampton Street	1849
40 Lowe Street	1852
66 Hope Street	1858 – 1865
INMAN, Charles William (& Son)	
Edward Street	1867 – 1875

Rule makers.

ISAACS, Henry	**LEEDS**
32 Trinity Street	1880 – present

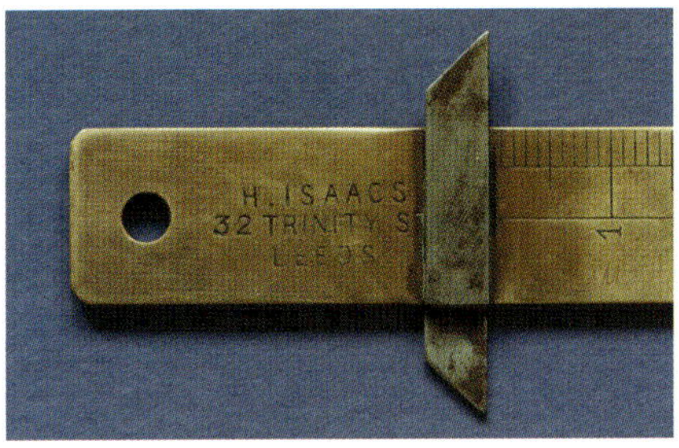

Supplier of wholesale jewellers' sundries and watchmAkers' sundries, etc.

JACKSON, John	**LIVERPOOL**
9 Canning Pl.	1883 – 1887

Rule maker.

JACOB, Christopher	**BIRMINGHAM & WOLVERHAMPTON**
Snow Hill	1761 – 1763
and Bilstone Street, Wolverhampton	1763
Smallbrook Street	1767
Chapel Row	1770
Gosty Green	1774 – 1776
JACOB, William	
29 Smallbrook Street	1770

Rule makers.

JACOB & HALSE	**LONDON**
	1809 – 1810

Mathematical instrument makers.

JESSON — **LONDON**
Nr. the Bars without Aldgate — 1672

Ironmonger. Jesson is known to have sold rules but it is not known if he made the rules he sold.

JOB Bros. & Co. — **LONDON**
37 Bread Street — < 1850 >
JOB, Charles Horatio, & Co.
42 Upper Thames Street — 1860 – 1870

Wholesale stationers, tracing paper manufacturer, etc. Their name is recorded on scale rules but they are unlikely to have made the rules that bear their name.

JOEL, Jacob — **MANCHESTER**
Swan Street — 1852

Birmingham and Sheffield warehouse.

JOHNSON, Joseph — **PETWORTH**
Market Pl. — 1890

Rule maker.

JOHNSON, Joseph S. — **LONDON**
72 Baxter Road, Essex Road — 1880 – 1882
442 White Horse Road — 1896

Rule maker. An address in Horsham is also listed, but this is almost certainly his residence.

JOHNSON, George — **BIRMINGHAM**
9½ Canal Street — 1882
JOHNSON, William
Rowlandson Mill, Cecil Street — 1882

Rule makers.

JOHNSON, William Henry — **BIRMINGHAM**
7 Court, Warstone Road — 1852

Rule maker.

JOHNSON, Samuel — **LONDON**
(13) (23) New Street, Cloth Fair — 1839 – 1846

Rule maker.

JOLE, Robert — **LONDON**
Durham Yard, Strand — 1664
Fleet Street — 1672
St. Johns, Clerkenwell — 1701

He is known to have made gauging rods and other rules.

JONES, Cooper, W., & Co.

Unidentified but probably a retailer. The name has been recorded on various horse measures.

JONES, Edward **LONDON**
14 Somerset Place, New Road < 1768 – 1780 >

Optical instrument maker.

JONES, John **LONDON**
74 Wardour Street, Soho 1817 – 1822
20 Crown Court, Wardour Street 1823 – 1827

Rule maker. A partnership between John Jones and James Fitch, trading as ***Fitch & Jones***, is also listed in 1818 and 1819. ***Mary Wellington & Son*** also worked at these premises between 1813 and 1827. The business was continued by ***Laban Cook***.

JONES, Thomas **LONDON**
120 Mount Street 1806 – 1808
21 Oxenden Street 1811 – 1814
62 Charing Cross 1816 – 1850
4 Rupert Street 1851 – 1859

Mathematical instrument makers. Between 1831 and 1835 the firm traded as Jones & Son.

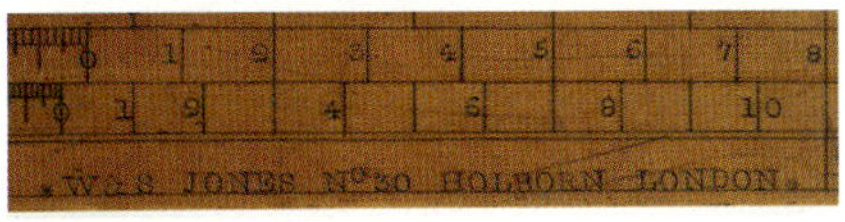

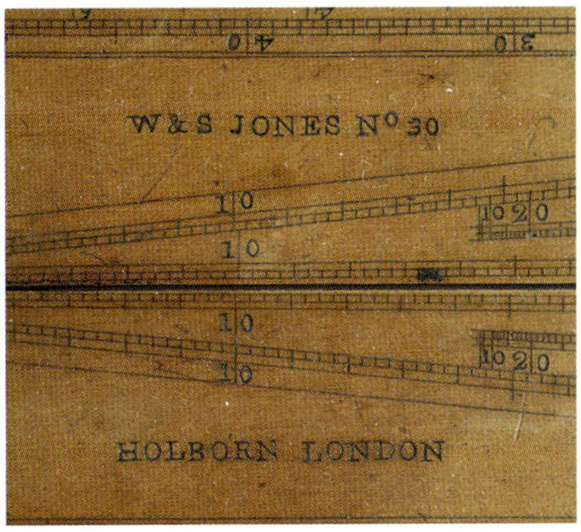

JONES, John **LONDON**
1759
135 Holborn Bars 1776 – 1784
JONES & Son
135 Holborn Bars 1784 – 1790
JONES William & Samuel
135 Holborn 1792 – 1795
30 Holborn 1795 – 1860

Mathematical instrument makers. The business was established by William and Samuel's father, John. William was apprenticed to his father in 1774. Samuel Jones was employed by ***George Adams*** before going into business with his brother. In 1820 W. & S. Jones took over the stock of ***Dudley Adams***.

K & L

KEAT, E. — **LONDON**
69 Banner Street — <1831 – 1852 >

Watch and clock maker.

KELVIN, BOTTOMLEY & BAIRD — **NEWCASTLE-UPON-TYNE**
61 Westgate Road — < 1914 >
GLASGOW
Cambridge Street — < 1922 >

Nautical instrument makers.

KENNING, G. — **LONDON**
Little Britain — 1880 – 1938 >

1880: Gold and silver lacemen. 1906: Army accoutrement makers and printers and stationers, etc.

KENT, Robert Geo. — **LONDON**
20 Old Compton Street — 1860 – 1900 >

Tool and metal warehouse.

KENT & Co. — **LONDON**
293 Euston Road — 1880 – 1899
and 385 (389) Oxford Street — 1888 – 1893

Ironmongers.

KERBY, Joseph — **BIRMINGHAM**
18 Court, Great Hampton Row — 1852

Rule maker.

KERR, Robert — **LONDON**
(81) Chancery Lane — < 1850 – 1880

Law stationer. The name has been recorded on a 36 in. mount cutters' rule. It is unlikely that he made the rules that bear his name.

KEYZOR & BENDON — **LONDON**
50 High Holborn — 1861
KEYZOR & BENDON & Co.
Hatton Garden — 1880

Wholesale opticians. They are listed as customers of ***John Rabone & Sons***.

KIMBERLEY, Samuel — **BIRMINGHAM**
7 Derley Street — 1852

Rule maker.

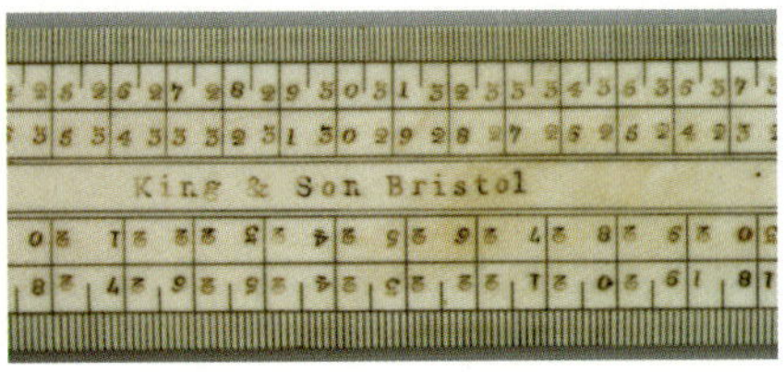

KING, John (& Son) — **BRISTOL**
2 Clare Street — 1821 – 1863
KING, Thomas Davis
2 Clare Street — 1847 – 1851
1 Denmark Street — 1854 – 1856
KING & COOMBES
2 Clare Street — 1851 – 1853

Optical and mathematical instrument makers. There were three generations of the King family in the business. They took over the business of ***Charles Beilby***. In 1857 ***Husbands & Clarke*** took over the business of T.D. King.

KING, Thomas — **LONDON**
Little Britain (?) — 1667 – 1677

Gauging instrument maker.

KIRKHAM, James — **BIRMINGHAM**
199 New John Street — 1895

Rule maker.

KNIGHT, William — **LONDON**
Foster Lane — 1787 – 1795
KNIGHT, Richard & George
Foster Lane — 1794 – 1838
KNIGHT, George & Sons
Foster Lane — 1839 – 1877

Ironmongers and scientific instrument makers.

KOSINE Ltd. — **LONDON**
1 Blenheim Green — < 1940 – 1947 >

Slide rule makers.

KUTZ, Andrew — **LONDON**
4 Cannon Street Road, St. George's East — 1816 – 1817

Rule and drawing instrument maker.

LADD & STREATFIELD — **LONDON**
1 Old Jewry — 1851 – 1859
LADD & OERTLING
192 Bishopsgate Street Without — 1861 – 1896

Hydrometer and gauging instrument makers. They described themselves as "Successors to R. B. Bate. Makers to the Hon. Board of Inland Revenue and standard weight and measure makers." Edward Ladd and John Streatfield were two of ***R. B. Bate's*** workmen. When Ladd & Streatfield folded in 1859, Edward Ladd formed a partnership with ***Ludwig Oertling*** who had also been employed by Bate.

LAKIN, Francis — **WOLVERHAMPTON**
11Berry Street — 1792

Rule maker.

LANDELL, W. — **GLASGOW**

A Glasgow saw, edge-tool and plane maker, working from 1846 to ca. 1931. See *British Planemakers from 1700,* 3rd ed.

LANGFORD, Thomas — **BIRMINGHAM**

137 Great Hampton Row — 1839

Rule maker.

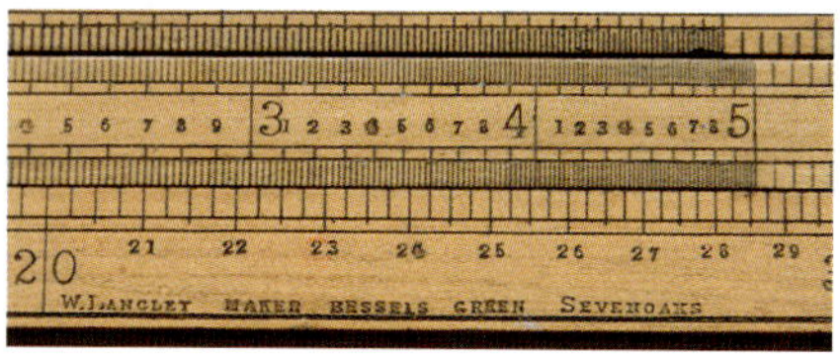

LANGLEY, George N. (& Son) — **LONDON & SEVENOAKS**

(46) (48) Tooley Street — < 1850 – 1880

LANGLEY, William

35 Friday Street — 1895

Hop & Malt Exchange, Tooley Street

(150) (104) Tooley Street — < 1900–1938 >

Bessels Green, Sevenoaks — < 1903 >

1850: Ironmongers. 1906: Ironmongers, hardware merchants, coopers' and carpenters' tool seller. Although they did not describe themselves as rule makers in the trades directories, it appears that they did make rules as William described himself as "Maker to H. M. Custom Bill of Entry in the rule also marked "Bessels Green, Sevenoaks".

LAPWORTH, William — **BIRMINGHAM**

15 Barr Street — 1852

LAPWORTH, Charles

15 West Bar Street — 1852

Upper Tower Street — 1854

Rule maker.

LAURENCE, John — **LIVERPOOL**

10 Bridgwater Street — 1851– 1865

Measuring tape maker.

LAW, Charles — **WOLVERHAMPTON**

New Street — < 1850

121 Salop Street — 1850 – 1865 >

LAW, John

8 Court, Upper Tower Street — 1852

Rule makers.

LAWLEY, George — **BIRMINGHAM**

15 Summer Lane — 1835

22 Court, Lower Tower Street — 1839

44 Newtown Row — 1842

Box and ivory rule maker.

LAWRENCE, T. — **BIRMINGHAM**

37 Upper Temple St — 1829 – 1833

Ivory box, case and caddy maker. The business was taken over by ***Samuel Clark***.

LEAH, Jeremiah — **MANCHESTER**
420 Oldham Road — < 1876
Lincow (?) Street, Salford

Rule maker.

LEALAND, P. — **LONDON**
24 Great Clarendon Street — 1826 – 1841

Mathematical instrument makers. The firm continued into the 20th century trading as of Power & Lealand.

LEALE, John — **MANCHESTER**
77 Hanover Square — 1848 –1859

Rule maker.

LECHERTIER BARBE — **LONDON**
60 Regent Street — 1827 – 1880
95 Jermyn Street — 1912 >

Artists' colour men and retailers of artists' requisites. They claimed to be established 1827. There were also other addresses in London.

LEE, George — **PORTSEA**
33 The Hard — c. 1830 >
LEE, George, & Son
Ordnance Row, The Hard — 1847 – 1880
3 Palmerston Road, Southsea

Mathematical, nautical and optical instrument makers. They described themselves as "Optical and Nautical Instrument Makers to the Hon. Corporation of Trinity House and the Admiralty".

LEE & HAYWARD — **BIRMINGHAM**
96 Lionel Street — 1861
LEE, Joseph
18 Court, New Summer Street — 1888

Manufacturers of every description of measuring tapes, mathematical instruments, plotting scales, box and ivory rules, spirit levels, etc. They also had premises at 61 Ludgate Hill. This could be Birmingham, London or Manchester but is probably Birmingham.

LEJEUNE & PERKIN — **LONDON**
24 Hatton Garden — 1880 – 1882

Rule makers.

LENNIE, James — **EDINBURGH**
14 Leith Street — 1840 – 1854
LENNIE, Eliza
14 Leith Street — 1854 >

Opticians and jewellers.

LEWIS & BRIGGS — **LONDON**
52 Bow Lane, Cheapside — 1795 – 1799

Mathematical and gauging instrument makers. The firm sometimes styled themselves Briggs & Lewis. A rule has been recorded marked "William Lewis, Maker, 52 Bow Lane" but this name is not recorded in the trades directories.

LEWTY, George — **BIRMINGHAM**
51 Summer Lane — 1835

Rule maker.

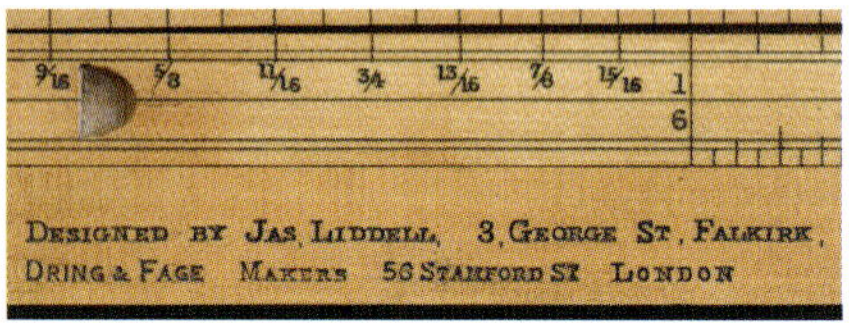

LIDELL, James — **FALKIRK**
3 George Street

Unidentified. He designed a number of slide rules related to the building trades. One is marked "Dring & Fage, 56 Stamford Street", which puts his dates between 1903 and 1946 but nothing further has been established.

LILLEY, John — **LONDON**
Hatton Garden — 1811
Globe Street — 1818 – 1821
7 Jamaica Terrace — 1823 – 1845
LILLEY, John, & Son
7 Jamaica Terrace — 1846 – 1865
Fenchurch Street — 1870 – 1922 >

Nautical and mathematical instrument makers.

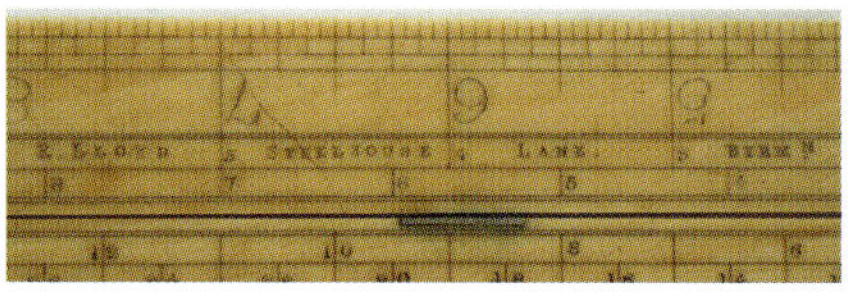

LLOYD, R. — **BIRMINGHAM**
Steelhouse Lane

Unidentified but possibly a maker.

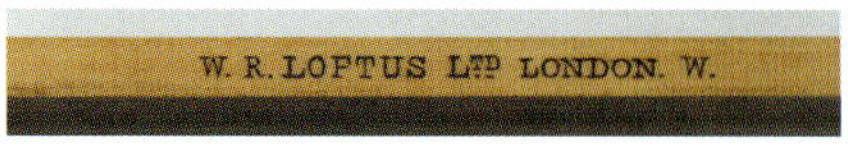

LOFTUS, William R. — **LONDON**
29 Pulford Street — 1855
6 Beaufoy Terrace, Edgware Road — 1856 – 1868
(146) (321) Oxford Street — 1870 – 1910
(18) (24)Tottenham Court Rd — 1911 – 1947 >

W. R. Loftus commenced business as a bookseller and publisher but soon also became a manufacturer of instruments for the Inland Revenue and Her Majesty's Stationary Office. By 1899 he described his business as hydrometer, saccharometer and gauging instrument makers to the Government and manufacturer of bottling and corking machines and all utensils for the wine and spirit trade.

LOGIN, John — **BIRMINGHAM**
97 Well Street — 1876

Rule maker.

LONG, Joseph — **LONDON**
20 Little Tower Street — 1821 – 1884
43 East Cheap — 1885 – 1936

Hydrometer, mathematical instrument and optical instrument maker.

LONGFIELD & CHAPLIN — **BIRMINGHAM**
Park Lane, Aston — 1870

Rule makers.

LONGMANS GREEN & Co. — **LONDON**
Paternoster Row

Booksellers and publishers from at least 1850 until the early 20th century, they published sets of cardboard scale rules. They also had offices in New York and Bombay.

LORENZ, John — **LEEDS**
6 Cookridge Street — 1870 >

Manufacturers of mathematical, drawing and surveyors' instruments and artists' colourmen.

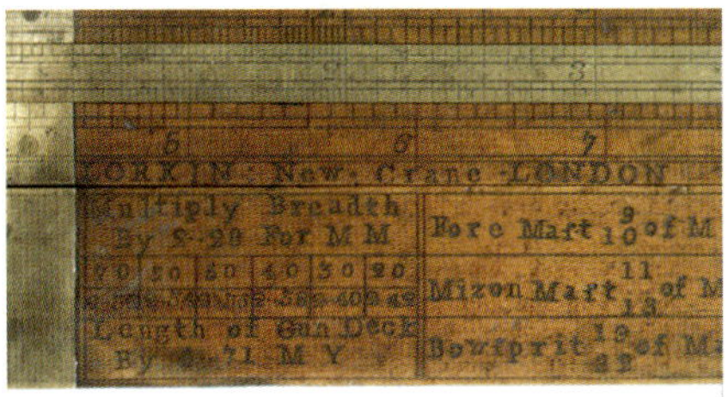

LORKIN, T. — **LONDON**
89 New Crane, Wapping — 1789 – 1830
89 New Gravel Lane

Mathematical instrument maker. These two addresses may be the same premises.

LOWE, Henry — **BIRMINGHAM**
21 Thorp Street — 1852

Rule maker.

LUCAS, James — **LONDON**
11 Wilderness Row, Clerkenwell — 1851
LUCAS, John
58 Langdale Road, Peckham — 1896
37 Parkhouse Street, Clerkenwell — 1914

Rule makers.

LUCAS, William — **BIRMINGHAM**
Constitution Hill — 1841 – 1842
312½ Summer Lane — 1870

Rule maker.

LUCKMAN, Peter — **BIRMINGHAM**
Bradford Street — 1821 – 1828

1821: Ivory and bone turner. 1828: Rule and pen maker.

LUDLOW, T. — **BIRMINGHAM**
81 Allison Street — 1823 – 1826
32 Park Street — 1829 – 1842
and 1 Bordesley Street — 1839
2 Bartholomew Street — 1843
LUDLOW, William
2 Bartholomew Street — 1847 – 1849
11 George Street — 1852

76 Great Colemore Street 1854

Ivory and box rule makers.

LUFKIN RULE Co.

Founded in 1869 in Cleveland Ohio, U.S.A., the company moved to Saginaw, Michigan in 1892. One of the major rule makers in the U.S.A., they had sufficient UK business to mark some of the rules sold in this country as "Lufkin Rule Co., London".

LUMLEY, L., & Co. **LONDON**
(118) (120) The Minories 1868 – 1930s
and 1 America Square

Commencing business as cork merchants, by the 1900s Lumley & Co. described their business as cork merchants, bottling wax, bottling wire, metallic capsule, tinfoil and engineers to the wine, spirit and mineral water trade, general bottlers and cider and wine-making machines and sundriesmen.

LUNT, Joseph **BIRMINGHAM**
(20) (22) Camden Street 1835 –1861

Rule maker.

M

MACHORD & ARCH **LONDON**

Unidentified but likely to have been a retailer. The name has been recorded on a Customs & Excise rule.

MACKENZIE, Alexander **LONDON**
15 Cheapside 1815 – 1830
38 De Beaur'r Square 1856 – 1858

Mathematical, optical and philosophical instrument maker. He also traded as Mackenzie & Co.

MACKIE, A. **EDINBURGH**

Edge-tool maker working from 1875 to 1960. See *British Planemakers from 1700,* 3rd ed.

MACQUEEN & Co. **LONDON**

Hat manufacturers working from 1906 to 1938. They would not have made the rules that bear their name.

MALLANDIAN, A. E. **LONDON**
51 Cheapside < 1906 –1938 >

Manufacturing stationers and paper makers. They would not have made the rules that bear their name.

MALLET, Louis **LONDON**

See ***Société des Lunetiers***.

MALLIN & Co. **BIRMINGHAM**
12 Hampton Court 1841 – 1842
MALLIN, Abraham
5 Chapel Street 1845 – 1855
88 Vauxhall Road 1858 – 1882
MALLIN, William
50 New Town Row 1850 – 1854

Rule makers.

MALLIN, William **MANCHESTER**
118 London Road 1861 – 1875
149 London Road 1880
25 Trans Street, London Road 1886

Rule maker.

MAN, Thomas Harry **BIRMINGHAM**
24 Loveday Street 1849 – 1850

Rule maker.

MANDER, George **LONDON**
21 (Great) Crown Court, Soho 1830 – 1851
25 Old Compton Street 1845 – 1870
64 Charlotte Street, New Road 1880

Rule and gaugers' instrument makers. Succeeded by ***Aston & Mander***.

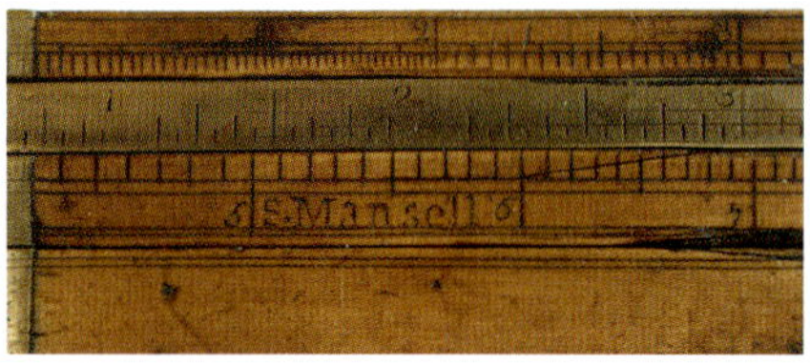

MANSELL, John **WOLVERHAMPTON**
17 Queen Street 1770 – 1792
MANSELL, Samuel
Salop Street 1793
16 Brick Kiln Street 1802
19 Salop Street 1818 – 1827

Rule makers.

MANSELL, William **LONDON**
28 Freeschool Street, Horsleydown 1826 – 1834

Rule maker.

MANSER, Thomas **LONDON**
(49) (74) Rhodeswell Road 1880 – 1890
2 SE Row, Upper North Street 1895
9 Rich Street, Limehouse 1899 – 1907
12 Wellington Row 1910
1 Priscilla Road, Bow 1913 – 1919

Rule maker.

MANT, Frederick **LONDON**
13 Little Compton Street 1837 – 1841
2 Ebenezer Place, Somers Town 1842
4 Claremont Place, New Road 1846 – 1854

Rule and mathematical instrument maker. ***Isaac Hill*** also operated from 13 Little Compton Street in 1841.

MARPLES, Wm. & Sons **SHEFFIELD**

William Marples was one of the largest tool makers and sellers in Sheffield, operating from 1856 to 1983. See *British Planemakers from 1700,* 3rd ed.

MARRATT, J. S. **LONDON**
New Street 1829 – 1830
Shoe Lane 1833
Great Winchester Street 1839 – 1844
King William Street 1845 – 1875
MARRATT & SHORT
King William Street 1859 – 1865

Mathematical and optical instrument makers. Marratt & Short: Opticians and spectacle makers.

MARSHALL, A. **GLASGOW**

Tool makers and sellers operating from 1879 to at least 1934. See *British Planemakers from 1700,* 3rd ed.

THE MARSHFIELD Co. Ltd. **BIRMINGHAM**

Unidentified but may have been manufacturers. The name has been recorded on a wire gauge.

MATHESON, Robert **LIVERPOOL**
22 Renshaw Street 1873 – 1878

Rule maker.

MATHIAS, John **BRISTOL**
6 Stoney Hill 1826 – 1838

Rule and gauging instrument maker.

MATHIESON, A., & Co. **GLASGOW & LIVERPOOL**

Probably the largest of the Scottish tool makers and sellers, operating from several towns in Scotland and later in England from 1822 to around 1960. See *British Planemakers from 1700,* 3rd ed.

MAUDSLEY, T. **BIRMINGHAM**

Unidentified. A carpenters' rule has been recorded marked "T. Maudsley, Maker, Birmingham Warranted Best Box".

MAVITTA, William **BIRMINGHAM**
10 Fisher Street 1882
171 Unett Street 1902 – 1910
Park Road 1912 – 1918
THE MAVITTA DRAFTING MACHINE Ltd.
Park Road 1921 – 1948 >

Rule and mathematical instrument makers.

McCARTHY, Charles **LONDON**
28 Primrose Street 1870
24 White Lion Street, Norton Folgate 1880 – 1890
15 Blossom Street Norton Folgate 1891 – 1895
McCARTHY, Charles & Frank
15 Blossom Street, Norton Folgate 1899 – 1923
McCARTHY, Frank & Sons
51 Summerley Street, Wimbledon 1925
20 High Street, Wandsworth 1931
27 Malva Road, Wandsworth 1936 – 1946

Rule and mathematical instrument makers.

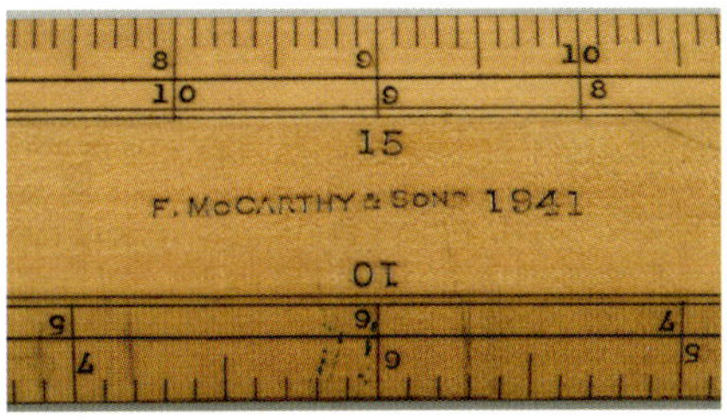

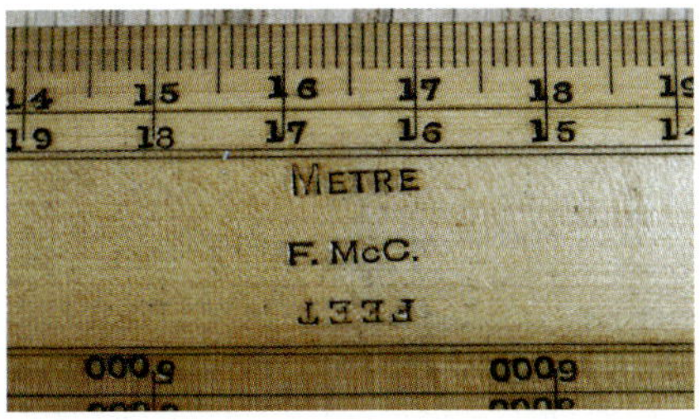

McGLASHAN & Co. **GLASGOW**

Unidentified. The name has been recorded on a dip rod.

McINTOSH, Thomas **LONDON**
Great Queen Street 1776 – 1784

Optician.

McPHERSON Bros. **GLASGOW**

Tool makers and sellers operating from 1883 to 1922. See *British Planemakers from 1700,* 3rd ed.

MDS **LONDON**

Unidentified but unlikely to be a maker. The name has been recorded on a Captain Field's parallel rule.

MELLISH, Robert **LONDON**
Parish of St. Saviour, Southwark 1760

Rule maker.

MERCER **LONDON**
The Feathers, Fleet Street 1672

Instrument seller (i.e. scientific instruments). The rules he sold were obtained from ***John Nash***.

MIDDLETON, William **WOLVERHAMPTON**
65 Bilston Road 1792

Rule maker.

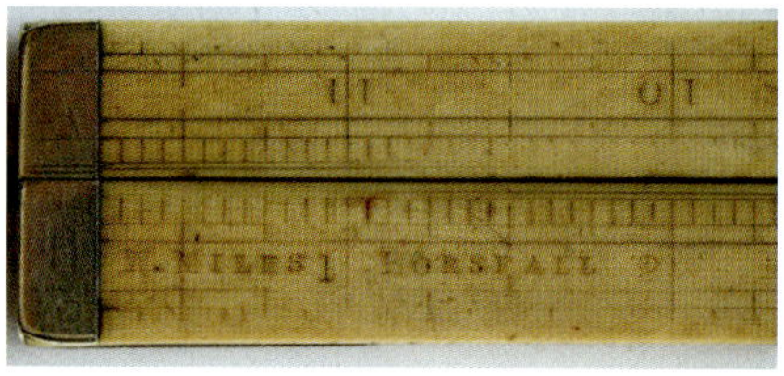

MILES, William **LONDON**
21 Stonecutter Street 1838 – 1839
24 Great New Street 1840 – 1842

Rule maker.

MILLER & RICHARD **EDINBURGH**
Pasternoster Row < 1850
LONDON
14 Water Lane < 1906–1938 >

Type founders. The name has been recorded on printers' rules. They would not have made the rules that bear their name.

MILLWARD, Henry **BIRMINGHAM**
57 Mott Street 1852 – 1854
146½ Moor Street 1855
29½ Bishop Street 1858 – 1867
MILLWARD, Catherine
29½ Bishop Street 1868 – 1870

Rule makers.

MOLE, William (I) **LONDON**
Mariner & Globe, Chiswell Street 1750
MOLE, Widow
Chiswell Street 1770
MOLE, William (II)
3 Star Court, Grub Street 1811

William (I): Gauging instrument maker. William (II): rule maker. William (I) was apprenticed to ***Thomas Cooke***, Old Jewry, London, in 1721.

MOORE, T. — **LONDON**

Unidentified but unlikely to have been a maker. The name has been recorded on a 2 ft. 2-fold rule, which probably dates to the early 19th century.

MOORE & WRIGHT — **SHEFFIELD**
< 1908 to present

Makers of all sorts of tools and equipment including steel rules for the engineering industry.

MORDAN, S., & Co. — **LONDON**
41 City Road — < 1924– 1938 >

Manufacturing gold and silversmiths, including pencils, etc. The firm claimed to have been established in 1815.

MORLEY & PINDER — **LONDON**

Unidentified but probably a maker.

MORRIS, John — **MANCHESTER**
65 Upper Pitt Street — 1824

Measuring tape maker.

MORRIS, John B. — **LONDON**
14 New Walnut Tree Walk, Lambeth — 1833 – 1834
MORRIS, Thomas S.
11 Chester Street, Lambeth — 1839 – 1842
4 Trafalgar Street, Walworth — 1844
89 East Street, Walworth — 1880

Rule, mathematical, optical and philosophical instrument makers.

MORRIS, William — **LONDON**
Strand — 1771
Cooper's Row — 1783
3 St. Catherine's Square — 1794 – 1796
and 54 Upper East Smithfield — 1794 – 1803

Mathematical instrument maker.

MORRIS, William — **BIRMINGHAM**
Great Charles Street — 1800 – 1801
Summer Row — 1805 – 1821
17 Court, Great Charles Street — 1828 – 1830
Northwood Street — 1830
MORRIS, Joshua
360½ Lodge Road — 1876

Rule makers.

MORRIS, KING & GOODING — **LONDON**
35 Ludgate Street — 1846 – 1851
MORRIS, Joseph, & Co.
35 Ludgate Street — 1860

Rule and bookbinders' tool makers and engravers.

MORTIMER, H. H., & Co. **LONDON & BIRMINGHAM**

Great Bush Lane, London
and Birmingham 1860 – 1900>

Wholesale and shipping ironmongers, tinplate and iron merchants. They would not have made the rules on which their name appears.

MOSELEY, John **LONDON**
MOSELEY & SIMPSON

One of the major London tool makers and sellers, in business from 1778 until 1910 after which date the business was transferred to Sheffield, having been purchased by ***Wm. Marples & Sons*** in 1892. See *British Planemakers from 1700,* 3rd ed.

MOSS & WINDRED **LONDON**
5 Crescent, Southwark Bridge Road 1838
5 Winchester Place, Borough 1839
MOSS, Thomas
5 Winchester Place, Borough 1839

Rule, mathematical and philosophical instrument makers. Thomas: Rule maker.

MOTTERAM, R. **BIRMINGHAM**
26 Floodgate Street 1862

Rule maker.

MURRAY, Lewis **SHEFFIELD**
19 Carver Street 1879

Rule and pencil knife maker.

MURRAY, Robert **LONDON**
122 Regent Street 1826 >
MURRAY & HEATH
43 Piccadilly < 1856 – 1866 >

Wholesale nautical instrument and barometer manufacturers and manufacturing opticians. The firm continued as ***Heath & Co.***

MUSGRAVE **LINCOLN**

Musgrave were tool retailers operating in Lincoln from 1851 to the present. See *British Planemakers from 1700,* 3rd ed.

N & O

NAIRNE, Edward — **LONDON**
20 Cornhill — 1749 – 1774
NAIRNE & BLUNT
20 Cornhill — 1774 – 1793

Optical, philosophical and mathematical instrument makers. 20 Cornhill was described as "opposite the Royal Exchange" in the address and before 1772 was known as the Golden Spectacles, Cornhill. ***Thomas Blunt*** was apprenticed to Edward Nairne in 1760 and in 1774 went into partnership with him. After Nairne's death he continued the business.

NARREIN, John — **LONDON**
70 St. James Street — 1805 – 1825

Mathematical, optical and scientific instrument maker.

NASH, John — **LONDON**
The Globe, Aldgate — 1668 – 1707

Mathematical instrument maker. He was apprenticed to ***Walter Hayes***. He supplied rules to ***Faulkingham*** and ***Mercer***.

NEGRETTI & ZAMBRA — **LONDON**
11 Hatton Garden — 1850 – 1859
and 59 and 68 Cornhill — 1857 – 1859
1 Hatton Garden — 1859 – 1867
and 107 Holborn Hill — 1859 – 1860
and 59 Cornhill — 1859 – 1862

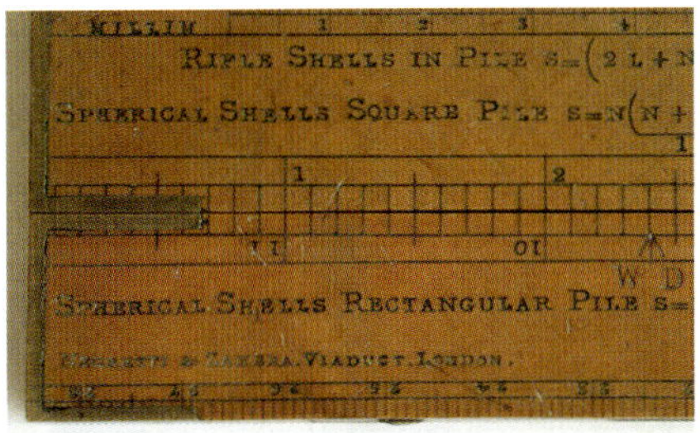

H. Capt. Doneldson — London 31 Decem. 1773

Bo. of Nairne & Blunt,
Optical, Mathematical & Philosophical Instrument Makers,
No. 20 opposite the Royal Exchange.

	£	s	d
A pr. dble joint Silver Spectacles Concaves to pattern in fish Case	1	3	—
A pr. dble joint Steel Do. green Eyes in fish Case		15	—
a Concave in horn box		1	6
	£ 1	19	6

Receiv'd the Contents for Nairne & Blunt
John Poyney

122 Regent Street	1862 – 1901
and 153 Fleet Street	1865 – 1873
and 103 Hatton Garden	1867 – 1870
and Holborn Circus, Holborn Viaduct	1870 – 1890
and 2 Charterhouse Street	1870 – 1882
and 45 Cornhill	1872 – 1901 >
and 38 Holborn Viaduct	1895 – 1920 >

Optician and optical, mathematical, philosophical, photographic, nautical and meteorological instrument makers. Henry Negretti and Joseph Warren Zambra became partners in 1850 and this continued until the death of Negretti in 1879; the firm was carried on by the sons of Negretti and Zambra and still exists under the same name today. Their principal addresses were those in Hatton Garden and later, Holborn Viaduct. The majority of the other addresses were branch shops.

NELSON, Joseph **BIRMINGHAM**
7 Court, Nelson Street 1839 – 1845

Rule maker.

NESBITT, John **MANCHESTER**
42 Market Street

Philosophical instrument maker.

NESTLER, Albert L. **LONDON**
30 Snow Hill 1905 – 1907

Rule maker, specialising in slide rules.

NEWTON, E. T., & Son **CAMBORNE**

Unidentified but probably a retailer. The name has been recorded on a set of drafting scales.

NEWTON, E. **GRANTHAM**
NEWTON, W.
< 1729 >

Unidentified but likely to be makers. "E. Newton" has been recorded on a rule, dated 1729; "W. Newton" has been recorded on a horse measure.

NEWTON, W. **LONDON**
NEWTON, John
NEWTON (Son) & BERRY
NEWTON & Co.
NEWTON, W. E. & F., & Co.
(128)(97) 66 Chancery Lane 1783 – 1883
3 Fleet Street 1851 – 1911 >
NEWTON, Henry William
63 Bancroft Road 1890 – 1913

A family firm of mathematical instrument makers, globe makers and opticians. John Newton was succeeded by ***Nathaniel Hill***, his son-in-law.

NICHOLL, J. A., & Co. — **LONDON**
13 Waterloo Road — 1919 – 1938
and Mathema Works, Cornwall Road

Rule makers.

NICHOLLS, John Merritt — **BIRMINGHAM**
21 St Mary's Row — 1852
NICHOLLS, T. M.
St Mary's Row — 1858
NICHOLLS, John
(6 Court) (73) Warwick Street — 1867 – 1884

Rule makers and dealers in rules.

NICHOLLS, Thomas — **LONDON**
14 Barbican — 1860
7 Old Bailey — 1870
87 Farringdon Street — 1880
78 Fore Street — 1882

Rule maker.

NISBITT, John — **BIRMINGHAM**
209 Lawley Street — 1849 – 1850

Rule maker.

NOBLE, James — **SHEFFIELD**
40 Broomhall Street — 1871
114 Matilda Street — 1876
166 (184) St. Mary's Road — 1879 –1883
NOBLE, James Arthur
26 Welby Place, Meersbrook — 1902 – 1909

1871: Dealer in engineers' rules and journeyman engineer. 1883: Foreman engineer. 1906: Sole maker of the "Eureka" slide rule. The designer of Noble's rule for engineers but whether Noble actually made the rules, even those that state "Manufactured by", has not been established and it is more likely that they were all bought in from rule makers. On the rules, his name is often spelt as Nobles, but the directory entries all use Noble.

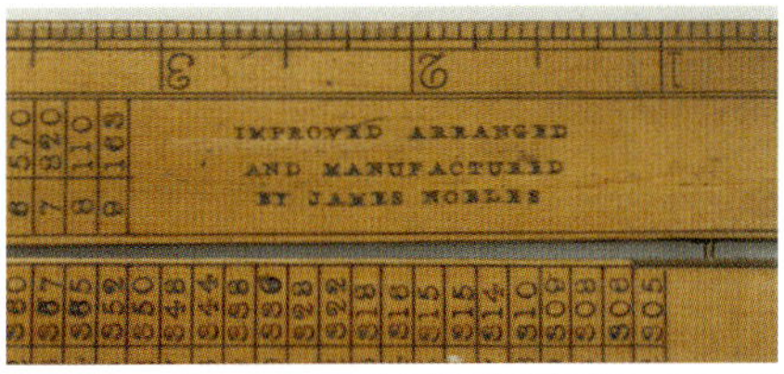

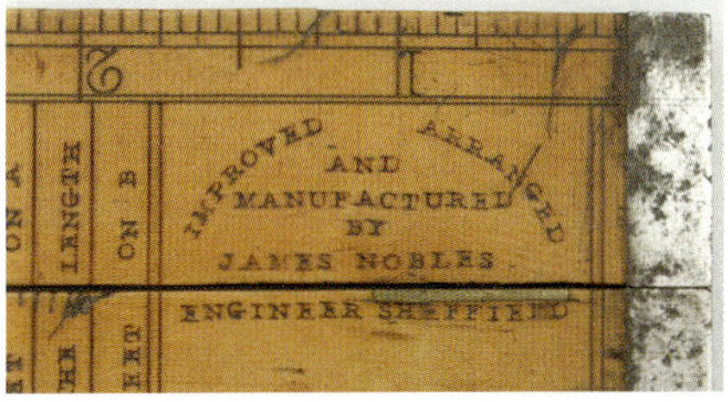

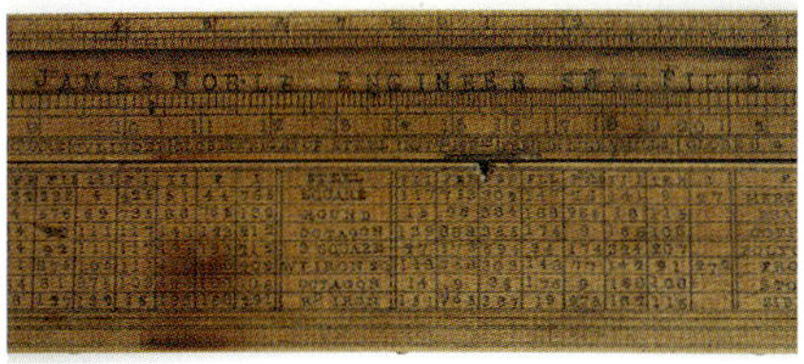

NORIE, John William (& Co.) — **LONDON**
157 Leadenhall Street — 1799 – 1843
NORIE & WILSON
157 Leadenhall Street
and 156 Minories — 1812 – 1820

Naval academy and navigation warehouse. He took over the business of ***William Heather***.

NORRIS & MELLER — **BIRMINGHAM**
1 Court, Ward Street — 1845 – 1850
98 Litchfield Street — 1858 – 1861
NORRIS & Son

98 Lichfield Street 1865 – 1876

NORRIS & FLETCHER

33 Cardigan Street 1880 – 1882

Norris & Mellor: Mathematical instrument and tape measure makers. Norris & Son: Rule makers.

NORTON & GREGORY **LONDON**

3 Victoria Street 1898 – 1900

Castle Lane 1906–1938 >

GLASGOW

34 Robertson Street

71 Queen Street < 1924 >

1900: Tracing office (for reproducing drawings). 1906: Drawing office stationers. The makers of a wide range of mathematical instruments including boxwood scales.

NOTTING, William **LONDON**

Bowling Green Lane 1880

Rule maker.

NURSE, Charles (& Co.) **MAIDSTONE & LONDON**

A leading tool maker and seller, operating in Maidstone from 1841 to 1886 and in London from 1886 to 1937. See *British Planemakers from 1700,* 3rd ed.

OERTLING, Ludwig (Ltd.) **LONDON**

(12) (13) Store Street 1849 – 1861

27 Moorgate 1869 – 1874

Turnmill Street 1874 – 1925

58 Cow Cross Street 1883 – 1888

Scale, hydrometer, and chemical, mathematical and philosophical instrument makers. In 1861, Ludwig formed a partnership with ***Edward Ladd***, although Oertling also continued to be listed in the trades directories on his own account. Both had been employed by ***R. B. Bate***.

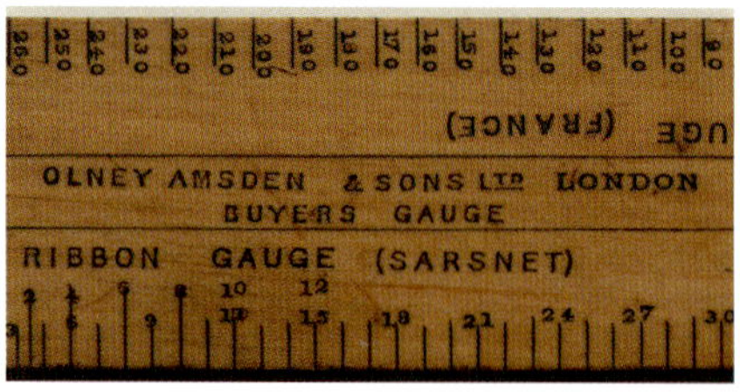

OLNEY AMSDEN & Sons **LONDON**

9–11 Falcon Street < 1880–1938 >

Wholesale haberdashers. The continuation of Thos. Olney & Co. who were in business from the 1860s. They are unlikely to have made any of the rules which bear their name.

OMER, J. **LONDON**

99 Minories 1860 – 1880

Optician and philosophical instrument maker.

ONIONS, Thomas **WOLVERHAMPTON**

Brickhill Lane 1767 – 1781

BIRMINGHAM

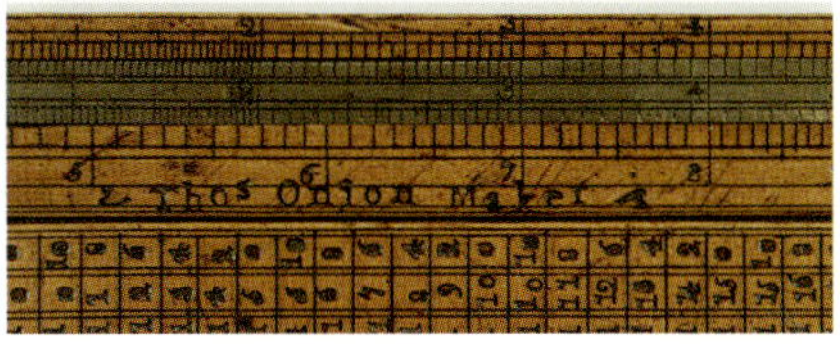

Cherry Street 1784
Walmer Lane 1785
Lancaster Street 1787 – 1798
Great Hampton Street 1797 – 1801

ONIONS, Benjamin
2 Northwood Street 1823 – 1826
(29 Court) (137) Livery Street 1839 – 1852

Rule makers. In 1788, box and ivory rules were included in the trade description.

ORDISH, T., & Co **LONDON**
27 Lamb's Conduit Street < 1850 >
97–101 City Road < 1906

1850: Stationers. 1906: Fancy goods dealers and stationers' sundrymen. They would not have made the rules which bear their name and were listed as customers of ***John Rabone & Sons***.

ORME, Charles **ASHBY DE LA ZOUCH**
1731 – 1747

His trade is unidentified. He is known to have sold rules but it is not known if he made the rules he sold.

ORME, George **BIRMINGHAM**
13 Guest Street 1910

Rule maker.

OSBORN(E), David **LIVERPOOL**
2 Cropper Street 1823 – 1824
Park Lane 1825 –1839
25 Bridgewater Street 1941 – 1845

Rule, mathematical, optical and philosophical instrument maker.

P & Q

PAINE, Christopher — **LONDON**
Hosier Lane, West Smithfield — 1584 – 1590

Mathematical instrument maker.

PALSER, William Hands — **BIRMINGHAM**
180 New John Street West — 1865 – 1876

Rule maker.

PARKER, Henry — **BIRMINGHAM**
Vauxhall Road — 1835

Rule maker.

PARKER, Benjamin — **LONDON**
Isaac Newton's Head, Holborn — 1725 – 1753

Rule maker.

PARKES, Ebenezer — **LONDON**
142 Fetter Lane — 1840 – 1869
PARKES, J.
142 Fetter Lane — < 1843 >

Ebenezer: Rule maker. J. Parkes: Brass rule maker.

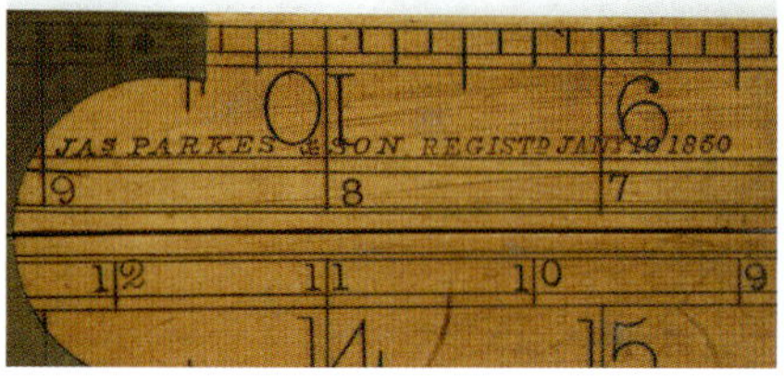

PARKES, James — **BIRMINGHAM**
5 St. Mary's Row — 1839 – 1843
PARKES, James, & Son
5 St. Mary's Row — 1845 – 1882
PARKES, James, & Co.
16 and 18 Vesey Street — 1910 – 1914 >

1839-43: Measuring tape, compass and gilt toy makers. 1845: Mathematical instrument makers. 1882: Opticians.

PARKES, Samuel (I) — **BIRMINGHAM**
8 Sand Street — 1777 – 1780
St. Mary's Row — 1787
Legge Street — 1797 – 1801
PARKES, Samuel, & Son
Legge Street — 1801
(66) Bull Street — 1805 – 1818
PARKES, Samuel (II)
(16) Little Hampton Street — 1821 – 1830
PARKES, Henry
Little Hampton Street — 1830
PARKES, Samuel Edward
296½ Summer Lane — 1871 – 1876

PARKES, Mrs Sarah Elizabeth
84 Summer Lane 1880 – 1882

Box and ivory rule makers, mathematical instrument and compass makers.

PARKINS, William **LONDON**
25 Oxford Street < 1850
PARKINS & GOTTO
54–62 (167) Oxford Street 1860–1938 >

1850: Manufacturing stationers. 1906: Wholesale and retail stationers and fancy goods sellers. 1924: Described themselves as "The Court Stationers". It is unlikely that they made the rules that bear their name.

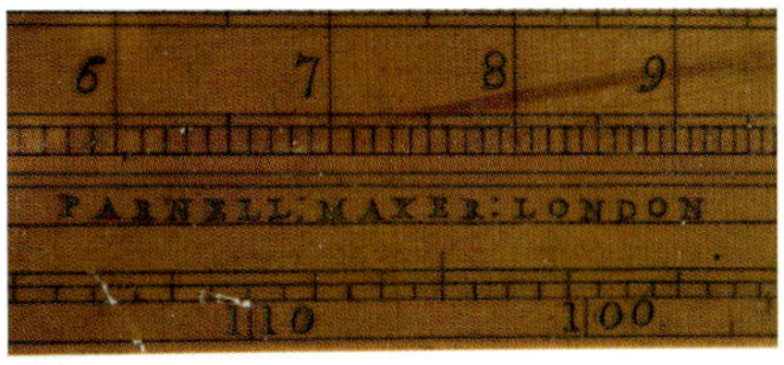

PARNELL, Thomas (I) **LONDON**
94 Hermitage Bridge, Smithfield 1776 – 1811
PARNELL, William
94 Lower East Smithfield 1813 – 1839
PARNELL, W., & Son
2 Lower East Smithfield 1839 – 1840
PARNELL, Thomas (II)
2 Lower East Smithfield 1840 – 1869
PARNELL & Co.

Mathematical, nautical and optical instrument makers. Thomas (I) and William were also ships' chandlers. 94 Hermitage Bridge and 94 Lower East Smithfield may well have been the same premises.

PARTRIDGE, J., & Son **DARLASTON**
New Road < 1850 >
PARTRIDGE, George & Sons
8 Smith Street < 1896 >

Wire gauge makers.

PASLEY, Richard F. **SHEFFIELD**
Rockingham Street 1879 – 1892 >.

Tool merchant and factors.

PASTORELLI, F., & Co. **LONDON**
4 Cross Street, Hatton Garden 1851 – 1869
208 Piccadilly 1859 – 1875
10 New Bond Street 1880
46 Hatton Garden 1890

Mathematical, optical and philosophical instrument makers. The firm was established in 1805 as barometer and thermometer makers in High Holborn and continued through several generations.

PAYNE, William **LONDON & LIVERPOOL**
South Castle Street, Liverpool 1831
163 (and 62) New Bond Street, London 1831 – 1846

In 1831, William Payne patented a pocket pedometer.

PENTON, Edward (& Son) **LONDON**
1–11 Mortimer Street < 1906–1924 >

Leather merchants and factors and boot and shoe manufacturers. It is unlikely that they made the rules that bear their name.

PERCIVAL, William **LIVERPOOL**
5 Maddox Street 1848

Rule maker.

PERCIVAL, William **MANCHESTER**
38 Lower Mosley Street 1850 – 1854

Rule maker.

PERKS, Robert **WOLVERHAMPTON**
8 Berry Street 1802
Temple Street 1805 – 1811
PERKS, Francis **BIRMINGHAM**
Waddam's Hill 1818
3 New Street 1828 – 1829
New Street 1841
PERKS, William
Bilston Street 1833 – 1834
PERKS, William
Court, 6, Little Charles Street 1855 – 1882
and Steward Street
137 Suffolk Street 1882

Rule makers.

PERRY, J. & Sons **LONDON**
349 City Road 1934 – 1938

Printers' rule makers.

PEWTRESS, Josiah Ltd. **LONDON**
Upper Thames Street < 1924 – 1938 >

Export packers and finishers and shipping agents. They would not have made the rules that bear their name.

PHILLIP, HARRIS, & Co. (Ltd) **BIRMINGHAM**
9 Bull Ring 1870
144 & 146 Edmund Street 1875 – 1914 >

Wholesale chemist and manufacturer of physical instruments for experimental science. At other addresses in Birmingham and Dublin. It is unlikely that they made the majority of the rules that bear their name.

PHILLIPS, George **BIRMINGHAM**
14 Court, Brearly Street 1829 – 1830

Rule maker.

PHILLIPS, William — **BIRMINGHAM**
Theodore Street — 1861 – 1869
Vyse Street, Aston — 1870

Rule maker.

PICK, Andrew — **LONDON**
Parish of St. Giles, Cripplegate — 1767

Rule maker.

PICKARD & DANE — **MANCHESTER**

Unidentified but likely to have been a retailer. The name has been found on an ivory rule.

PIGGOTT — **LONDON**
Fleet Street — 1672

Rule seller who sold rules made by ***William Elmes***.

PIKE, Joseph — **BIRMINGHAM**
92½ Carver Street — 1875

Rule maker.

PILLISCHER, Moritz — **LONDON**
398 Oxford Street — 1851 – 1853
88 New Bond Street — 1854 – 1887
PILLISCHER, Jacob
88 New Bond Street — 1887 – 1913 >

Opticians, mathematical and philosophical instrument makers.

PLATT, James, & Co. — **LONDON**
77 & 78 St. Martin's Lane — < 1906 >

Woollen and tailors' trimmings manufacturers. The name has been recorded on tailors' rules. They would not have made the rules that bear their name.

POTTER, C. — **LONDON**
(133) (108) Albany Road — 1847 – 1864
POTTER, J. D.
31 Poultry
and 11 King Street — 1851 – 1882
POTTER, Septimus Charles
31 Poultry
and 11 King Street — 1882 – 1890 >

Mathematical, optical and philosophical instrument makers. J. D. Potter, who was one of ***R. B. Bate's*** workmen. He took over Bates' Chart Agency in 1850 when the Bate busines closed.

POWELL, Robert — **BIRMINGHAM**
Court 4, Lionel Street — 1823
20 King Edwards Place — 1825 – 1826

Box and ivory rule maker and joiners' tool maker. A partner in ***Butler & Powell***.

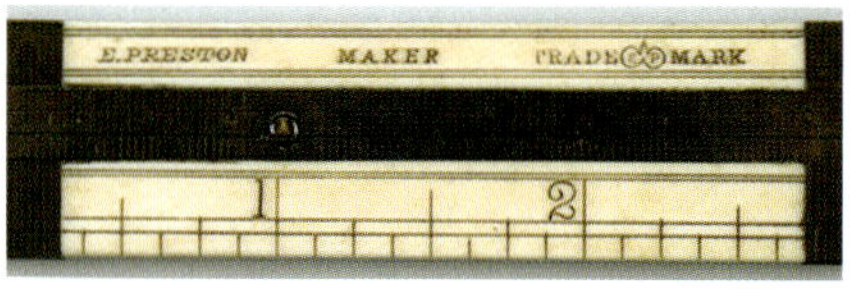

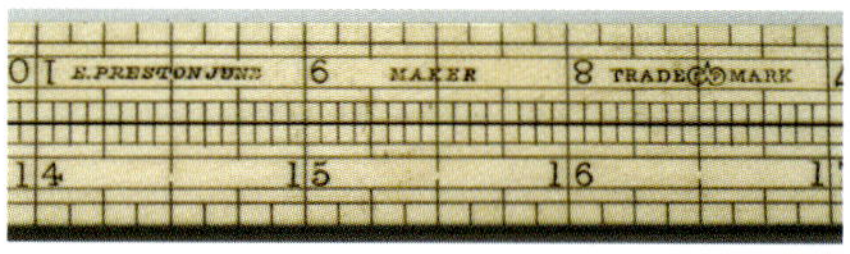

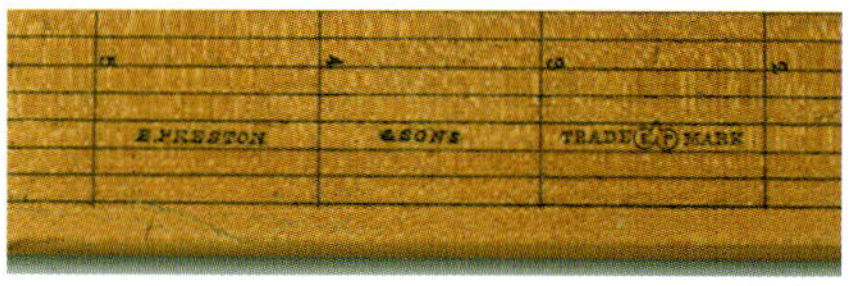

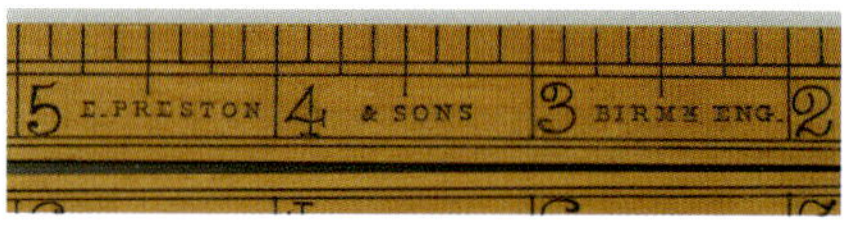

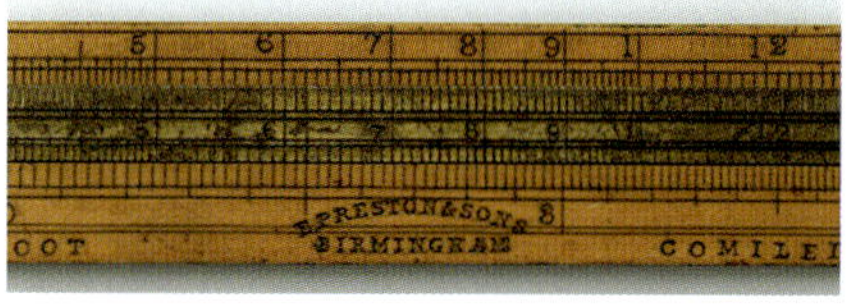

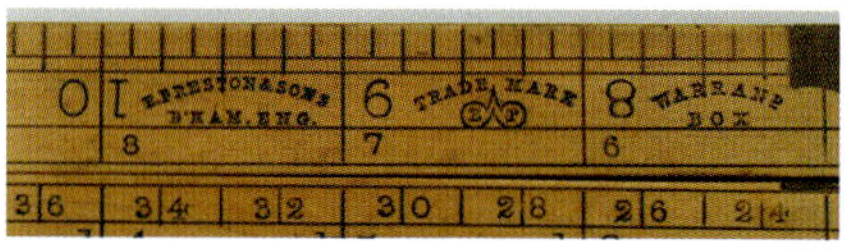

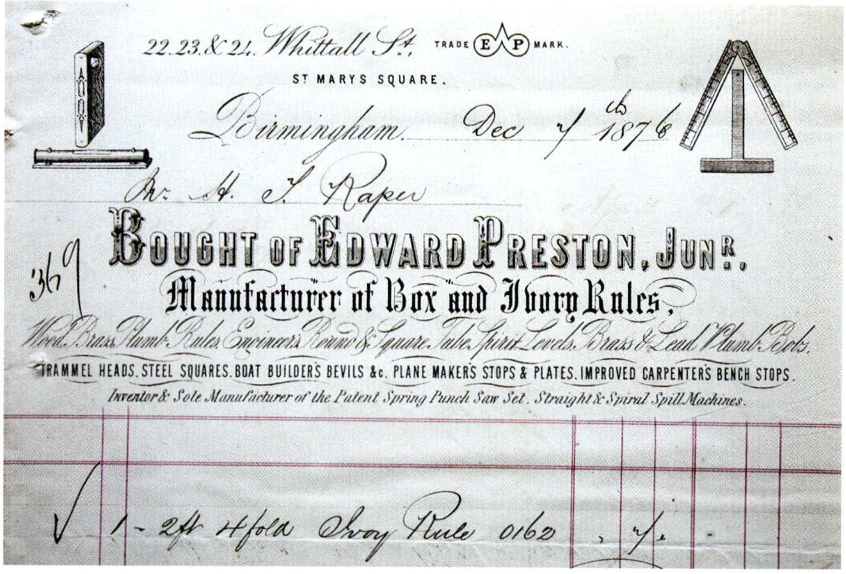

22.23.&24. Whittall St. TRADE EP MARK.
ST MARYS SQUARE.
Birmingham. Dec 7th 1876
Mr. H. J. Raper
369
BOUGHT OF EDWARD PRESTON, JUNR.
Manufacturer of Box and Ivory Rules,
Wood, Brass, Plumb Rules, Engineers Rouna & Square, Tube, Spirit Levels, Brass & Lead Plumb Bobs,
TRAMMEL HEADS, STEEL SQUARES, BOAT BUILDERS BEVILS &c. PLANE MAKERS STOPS & PLATES, IMPROVED CARPENTERS BENCH STOPS.
Inventor & Sole Manufacturer of the Patent Spring Punch Saw Set, Straight & Spiral Spill Machines.
1 - 2ft 4fold Ivory Rule 0162 ,, 7/-

PRATT, Joseph — **BIRMINGHAM**
20 Heneage Street — 1850

Rule maker.

PRESTON, Edward (II) — **BIRMINGHAM**
97 Lichfield Street — 1864 – 1876
22 – 24 Whittall Street — 1876 – 1894
PRESTON, Edward & Sons
22 – 24 Whittall Street — 1894 – 1900
PRESTON, Edward & Sons Ltd.
Whittall Works, Aston — 1901 – 1932

Edward Preston (II) was apprenticed to his father, a Birmingham plane maker in 1850. By 1864 he had started in business on his own and soon expanded the product range to include the rules, levels, etc. for which the firm was to become famous. In 1867 he described himself as a wholesale manufacturer of box and ivory rules. The firm closed in 1932 and the rule making operations were taken over by ***John Rabone & Sons***. A fuller history of the firm can be found in the reprint of the 1909 *Preston Catalogue,* Mendham, NJ, 1991.

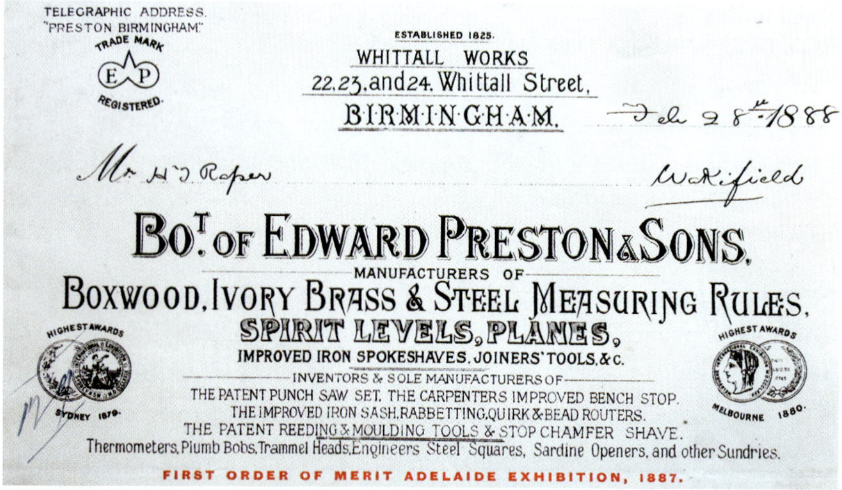

TELEGRAPHIC ADDRESS.
"PRESTON BIRMINGHAM."
TRADE MARK EP REGISTERED.
ESTABLISHED 1825.
WHITTALL WORKS
22.23.and 24. Whittall Street,
BIRMINGHAM. Feb 28th 1888
Mr. H. J. Raper
W.K. field
BOT. OF EDWARD PRESTON & SONS.
MANUFACTURERS OF
BOXWOOD, IVORY BRASS & STEEL MEASURING RULES,
SPIRIT LEVELS, PLANES,
IMPROVED IRON SPOKESHAVES, JOINERS' TOOLS, &C.
HIGHEST AWARDS SYDNEY 1879.
HIGHEST AWARDS MELBOURNE 1880.
INVENTORS & SOLE MANUFACTURERS OF
THE PATENT PUNCH SAW SET. THE CARPENTERS IMPROVED BENCH STOP.
THE IMPROVED IRON SASH, RABBETTING, QUIRK & BEAD ROUTERS.
THE PATENT REEDING & MOULDING TOOLS & STOP CHAMFER SHAVE.
Thermometers, Plumb Bobs, Trammel Heads, Engineers Steel Squares, Sardine Openers, and other Sundries.
FIRST ORDER OF MERIT ADELAIDE EXHIBITION, 1887.

PRICE, Benjamin — **BIRMINGHAM**
41 Loveday Street — 1865 – 1867
PRICE Bros.
41 Loveday Street — 1867 – 1870

Rule makers.

PRICE, John — **LONDON**
(19) (20) (27) Castle Street — 1826 – 1833

Optician, mathematical and philosophical instrument and rule maker.

PRICE, Richard — **BIRMINGHAM**
Mills Court, Slaney Street — 1800 – 1801

Rule maker.

PROBIN, Wm. — **BIRMINGHAM**
62 Summer Street — 1845 – 1850

Rule maker.

PULLEN, William — **LONDON**
28 Cyrus Street — 1905
PULLING & PULLEN
28 Cyrus Street — 1905 – 1907

Rule makers.

PURSER, G. & W. — **BIRMINGHAM**
47 Warwick Street — 1876

Manufacturers of English and foreign measuring tapes, garden engine and syringe maker and horse tail singeing apparatus makers.

PYEFINCH, Henry — **LONDON**
(45) (67) Cornhill — 1763 – 1791

Optician, optical and mathematical instrument maker.

QUINTON & Co. — **BIRMINGHAM**
West Icknield Street — 1852
QUINTON, Jonathan
Ledsam Street and Clark Street — 1855
QUINTON, William
Osler Street — 1856 – 1860
Globe Works, Icknield Port Road — 1862

Rule makers.

R

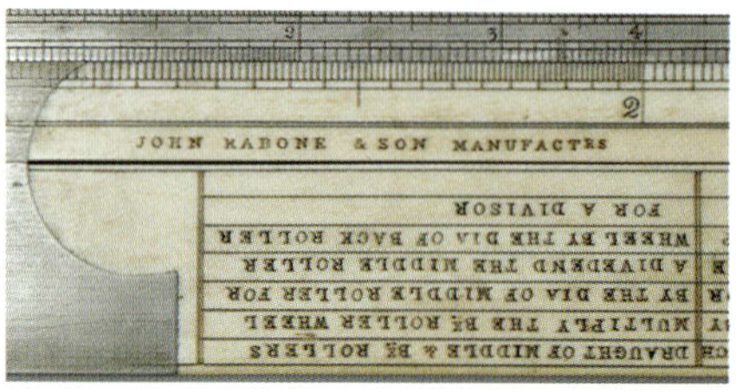

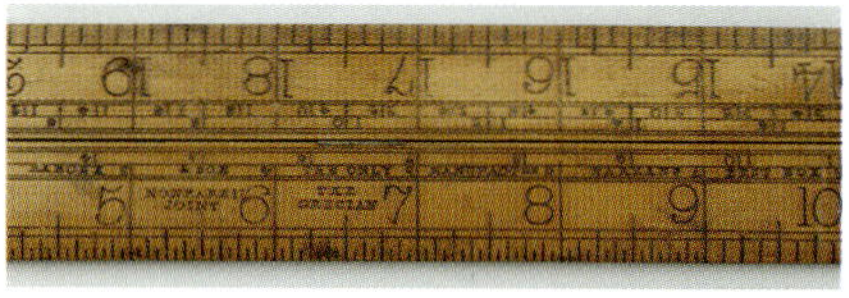

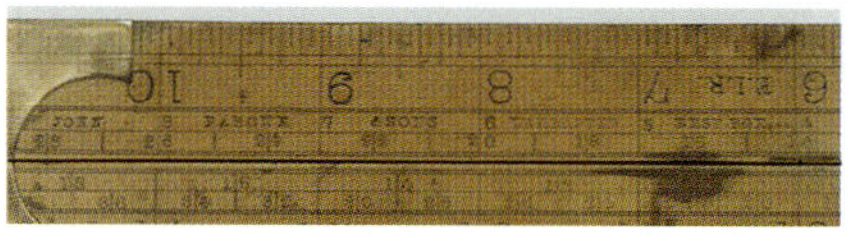
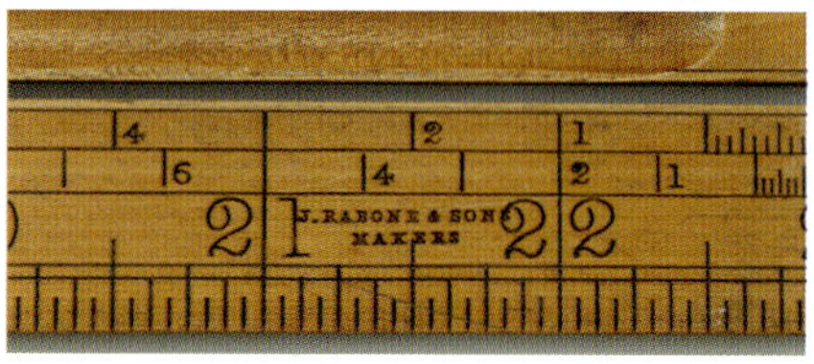

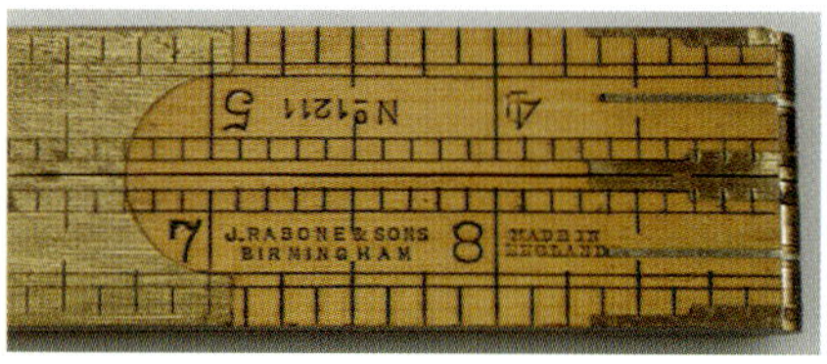

RABONE, Michael	**BIRMINGHAM**
12 Bull Street	1784 – 1800
8 Snow Hill	1801 – 1803
RABONE, Elizabeth (& Son)	
(33) (38) Water Street	1808 – 1830
RABONE, John	
61 St. Paul's Square	1823
33 Ludgate Hill	1825 – 1830
61 St. Paul's Square	1830 – 1860
RABONE, John, & Son	
61 & 62 St. Paul's Square	1845 – 1871
RABONE, John, & Son	
Hockley Abbey Works	1871 – 1877
RABONE, John & Sons	
Hockley Abbey Works	1877– 1963
RABONE CHESTERMAN Ltd.	
Hockley Abbey Works	1963 –1984
Summer Hill Works	1984 – 1990
Sheffield	1990 - present

Rule makers. The first record of Michael Rabone is in 1784 when he married Elizabeth Smith and, the same year, set up in partnership with ***Joseph Hassall***, trading from 12 Bull Street under the name of ***Hassall & Rabone***. In 1800, Joseph Hassall died leaving his rule making and wood turning businesses to Michael Rabone. In 1801 Michael set up on his own in Snow Hill but died shortly after, in 1803. (In later years, the Rabone firm claimed to have been established in 1784, which presumably was from the date of this first partnership.) The business was continued by his widow from premises in Water Street. In 1808 their son, Ephraim, joined the business and the name was changed to Elizabeth & Son. In 1811, ***Ephraim Rabone*** set up his own thermometer and rule making business, operating from another premises in Water Street and in 1817, Michael and Elizabeth's other son, John, took over the management of the original business. Whether this was due to Ephraim wanting to concentrate on his own business, or his ill health (he died in 1819) has not been established. In 1823, the business moved to 61 St. Paul's Square and the name was changed to John Rabone. In the 1841 census, John

English & Foreign Rule Manufactory,
ESTABLISHED 1784,
St. Paul's Square, Birmingham,
Aug. 31: 1857

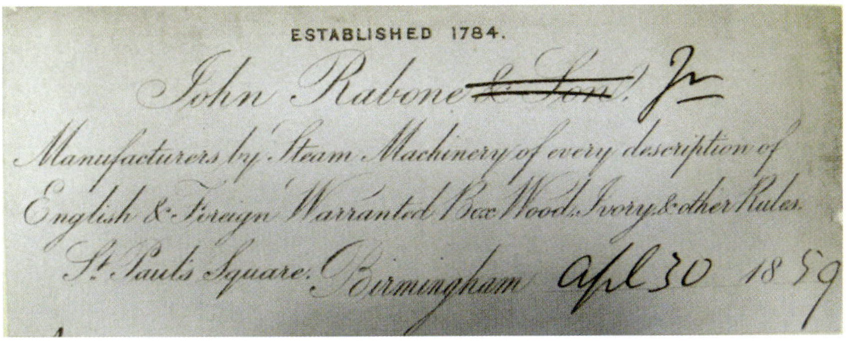
ESTABLISHED 1784.

John Rabone ~~& Son~~ Jr

Manufacturers by Steam Machinery of every description of English & Foreign Warranted Box Wood, Ivory & other Rules.

St. Paul's Square, Birmingham Apl 30 1859

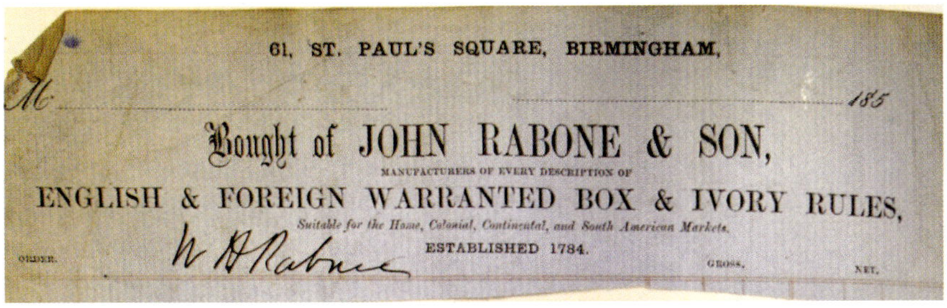
61, ST. PAUL'S SQUARE, BIRMINGHAM,

M ... 185

Bought of JOHN RABONE & SON,

MANUFACTURERS OF EVERY DESCRIPTION OF

ENGLISH & FOREIGN WARRANTED BOX & IVORY RULES,

Suitable for the Home, Colonial, Continental, and South American Markets.

ESTABLISHED 1784.

ORDER. GROSS. NET.

W H Rabone

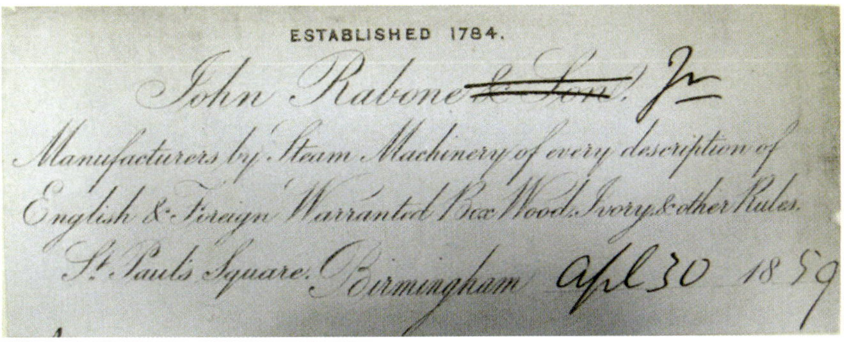
ESTABLISHED 1784.

John Rabone ~~& Son~~ Jr

Manufacturers by Steam Machinery of every description of English & Foreign Warranted Box Wood, Ivory & other Rules.

St. Paul's Square, Birmingham Apl 30 1859

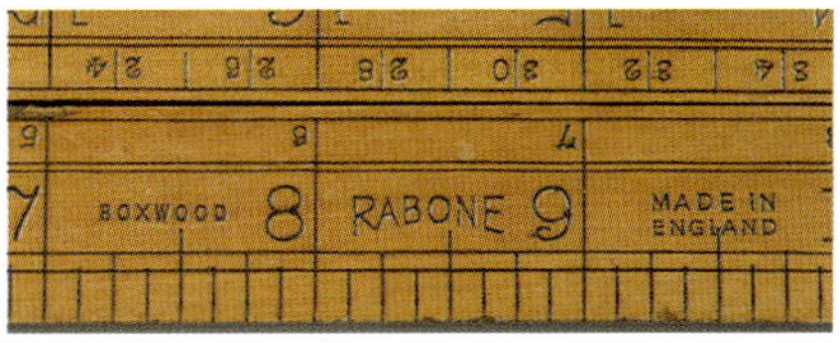

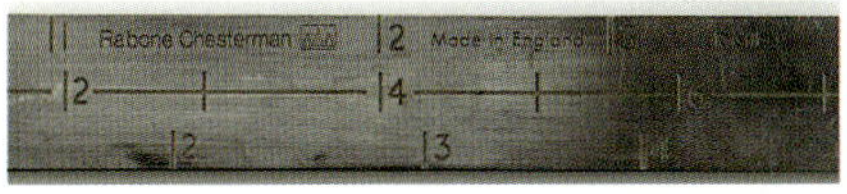

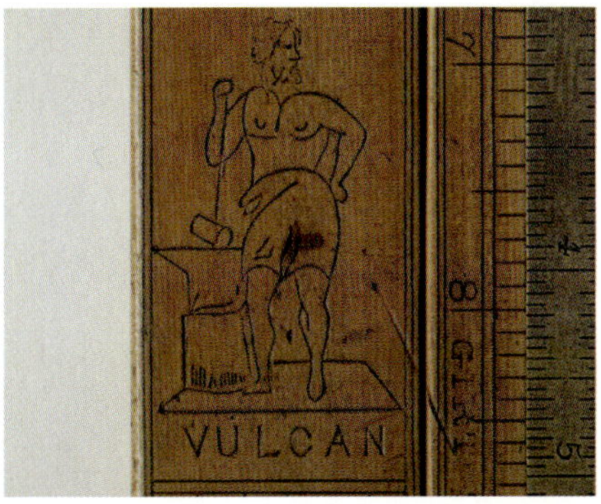

Rabone is listed at 61 St. Paul's Square as the head of the household. Living with him were his wife, Mary, and his sons John, William and Samuel, his daughter Elizabeth and his mother Elizabeth, aged 75. In 1845 his eldest son, also named John, was brought into the family firm and the name was changed to John Rabone & Son. In 1862, Rabone advertised that they were the only manufacturers of the celebrated Nonpareil joint rules. In 1867, the directory description was "Manfrs. by steam machinery of every description of English and foreign warranted box-wood, ivory and metal rules, suitable for the home, colonial, continental Indian and South American markets". 1871: A new factory was built at Hockley Abbey. 1873: They took over the business of ***Sampson Aston***. 1876: The Hockley Abbey and Vulcan trade marks registered (the Vulcan trade mark referred to a design of rule joint). 1877: The name changed to John Rabone & Sons. 1910: They became a limited company. 1932: They took over ***Edward Preston & Sons***. 1963: John Rabone & Sons merged with ***James Chesterman*** to form Rabone Chesterman Ltd. In the early 1990s Rabone Chesterman Ltd. were taken over by the American firm, The Stanley Works, the British division of ***Stanley Tools Ltd.*** and the rule factory was moved to Sheffield, to join Stanley's other British tool businesses.

RABONE, Ephraim — **BIRMINGHAM**
13 Water Street — 1811 – 1818
RABONE, Thomas
8 Court, 31-33 Water Street — 1829

12 Court, Broad Street 1835
(171) (172) Hockley Hill 1839 – 1854
(61) (63) Great Hampton Street 1858 – 1862

RABONE, Mrs Elizabeth
Summer Street Place 1835
171 Hockley Hill 1839 – 1851

Rule makers, thermometer makers and mathematical instrument makers. Ephraim died in 1819 and the business was continued first by his widow, Elizabeth, and then in 1829 by his son Thomas Rutter Rabone, although Elizabeth still continued to be listed in the trades directories as a rule maker.

RABONE BROS. (& Co.) **BIRMINGHAM**
Broad Street 1797 – 1939 >

Merchants. They had no business connection with ***John Rabone & Sons*** but did buy rules from them.

RAMSDEN, Jesse **LONDON**
Strand 1763 – 1766
Haymarket 1767 – 1771
199 Piccadilly 1772 – 1800

Optician and mathematical instrument maker. He worked for ***Jeremiah Sissons***. He took ***William Cary*** (amongst many others) as apprentice. In 1777, Ramsden made the first practical and successful dividing engine. The business was taken over by ***Matthew Berge***.

Opposite is a plate from either the 1808 or 1819 Philadelphia edition of Rees' *Cyclopedia or Universal dictionary of arts, sciences, and literature* showing Ramsden's dividing engine.

RANSLEY, George **LONDON**
10 Barnet Street, Hackney Road 1851 – 1855

RANSLEY, H.

Rule makers.

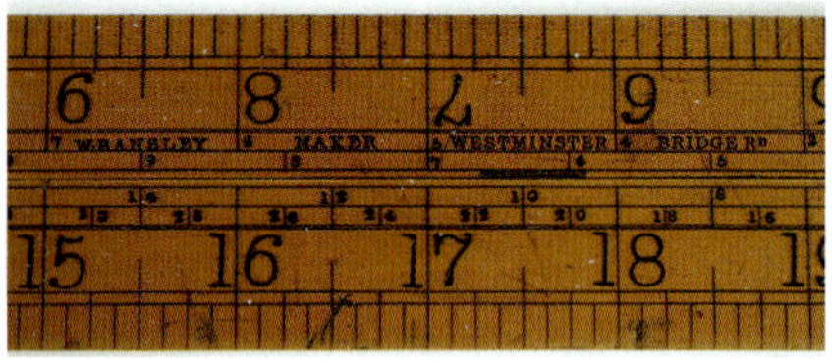

RANSLEY, William **LONDON**
(20) (30) Stangate, Lambeth 1839 – 1851
4 Bridge Rd, Lambeth 1852 – 1860
229 Westminster Bridge Road 1865 – 1885

Rule and tool maker.

RAPHAELS, J., & Co. **LONDON**
3 and 13 Oxford Street 1886
51 Clerkenwell Road 1899
57 b&c Hatton Garden 1911 – 1947 >

Manufacturing opticians and makers of spectacles, lenses cases, opera glasses, field and marine glasses, telescopes, barometers and thermometers. By the 1900s the company had become Raphaels Ltd.

REA, Mark **WOLVERHAMPTON**
2 Court, 6 Dudley Street 1792

Rule maker. See ***William Rea***.

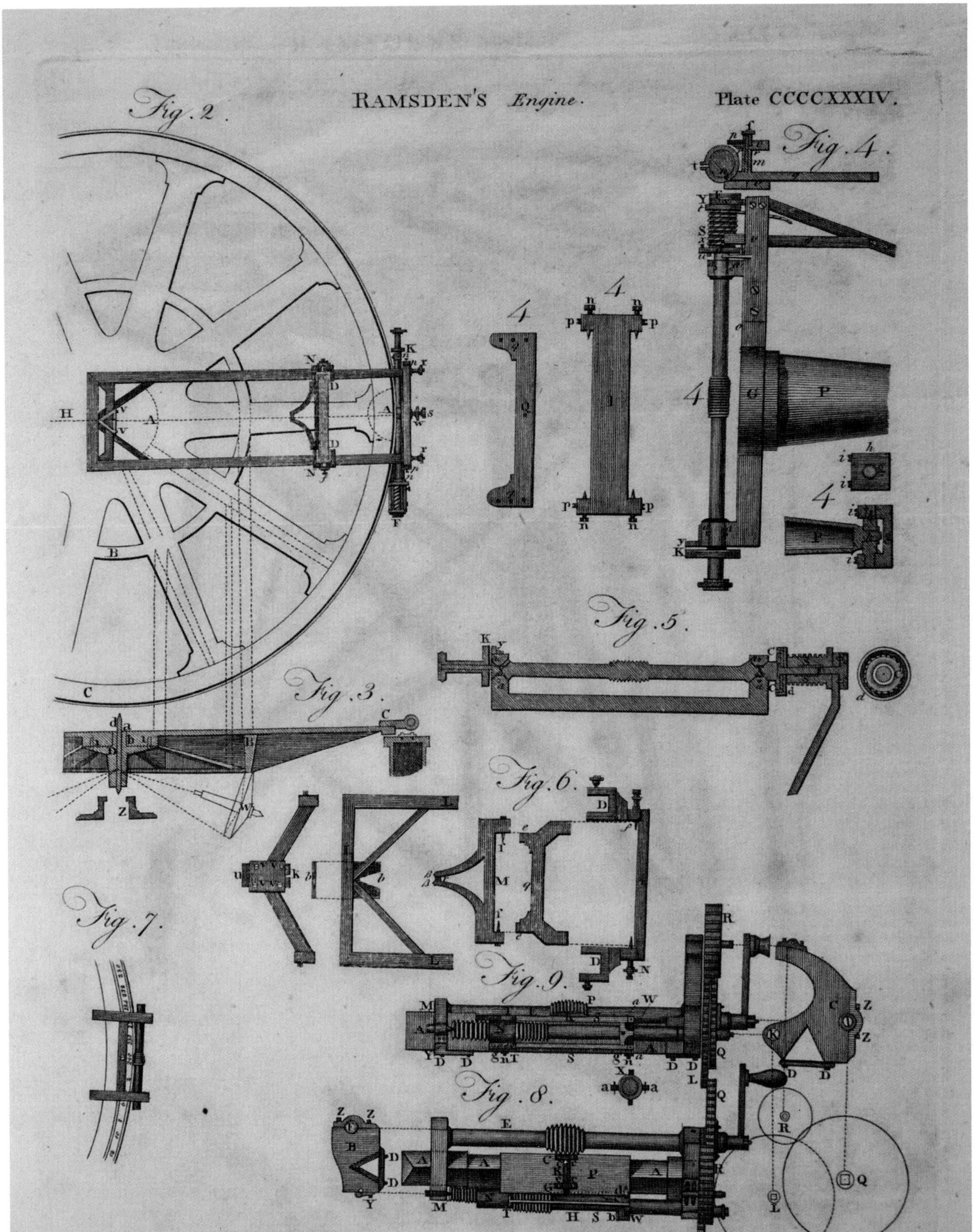
RAMSDEN'S Engine.
Plate CCCCXXXIV.
Fig. 2.
Fig. 3.
Fig. 4.
Fig. 5.
Fig. 6.
Fig. 7.
Fig. 8.
Fig. 9.
Vallance sc.

REA, William — **BIRMINGHAM**
3 Court, Coleshill Street — 1852

Rule maker. Although no evidence has been found, it seems likely that there was a family connection with ***Mark Rea***.

READ, Benjamin, & Co. — **LONDON**
12 Hart Lane, Bloomsbury — 1850

Tailors. It is unlikely that they made any of the tailors' rules that bear their name.

REES, Hugh Ltd. — **LONDON**
124 Pall Mall — < 1903 – 1914
5 (& 7) Regent Street — 1915 – 1934 >

Stationers, publishers and booksellers. The firm specialized in supplying the military.

REEVES, D. — **BIRMINGHAM**
30 Ashtead Row — 1860

Rule maker.

REEVES, W., & Co. — **LONDON**
(26) (146) Minories — < 1914 >
Lombard Road, Morden — 1930s

Instrument and appliance makers to the brewing and distilling trade.

REEVES & Son — **LONDON**

Manufacturers and suppliers of artists' materials. The business was founded in 1781 and continued at least until the 1980s. It is unlikely that they made the rules that bear their name and are listed as customers of ***John Rabone & Sons***.

REVED, Mark — **WOLVERHAMPTON**
1 Court, 2 High Green — 1802

Rule maker.

REYNOLDS, J. A. (& Co.) — **BIRMINGHAM**
43 Colmore Row — 1888 – 1958

Manufacturers of drawing materials and surveying instruments. 1888: The directory entry reads "See Thornton, A.G." and the entry for ***A. G. Thornton*** reads "Proprietor J.A. Reynolds". However, the exact connection between the two firms has not been established. They were taken over by ***Hall Harding Ltd.*** in 1958.

REYNOLDS & BRANSON — **LEEDS**
Commercial Street — < 1893 – 1914 >

Scientific instrument makers.

REYNOLDS, James — **WOLVERHAMPTON**
31 Worcester Street — 1792

Rule maker.

RICHARDS, William — **BIRMINGHAM**
14 Lichfield Street — 1785 – 1798
RICHARDS, Esther
(94) (95) Lichfield Street — 1800 – 1818
RICHARDS, John
(95) (98) Lichfield Street — 1815 – 1826
RICHARDS & LOCKYER
20 Jamaica Row — 1828 – 1831
RICHARDS, Daniel
15 Colmore Row — 1876
RICHARDS, Joseph
101½ Bissell Street — 1875 – 1876

Rule makers.

RICHTER — **NORTHAMPTON**

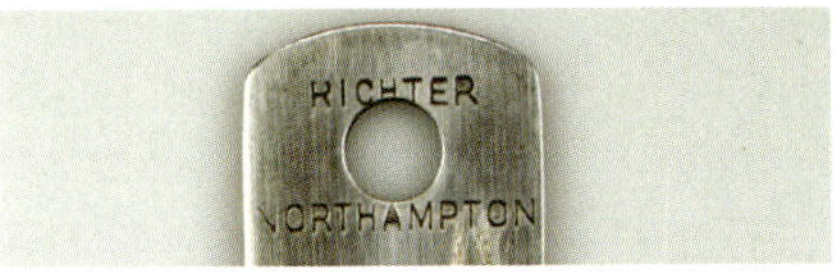

Unidentified. The name has been recorded on a substance gauge for shoe leather so it is likely that they were retailers for tools used in the shoe trade.

RIDER, Charles — **LONDON**
Castle Yd. Holland Street Blackfriars — 1860
RIDER, W., & Son

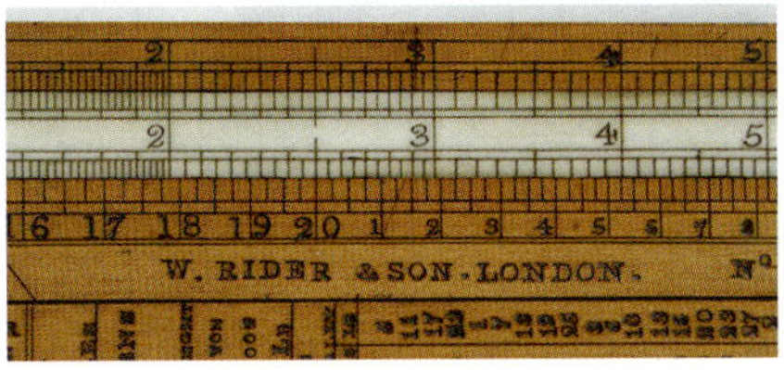

Rule makers. W. Rider & Son were the manufacturers of a rule marked Burt's Provisional Patent No. 3551.

RIX, William — **LONDON**
Mariner & Globe, Old Jewry — <1750
RIX, J.
Shrewsbury Court, White Cross Street — 1750 – 1760 >

Rule makers. William was apprenticed to ***Thomas Cooke***, Old Jewry, London.

ROBERTS, Edward (I) — **LONDON**
Shrewsbury Court, White Cross Street
Grocer's Alley, Old Jewry — 1749 – 1756
3 Dove Court, Old Jewry — 1759 – 1776
ROBERTS, Edward (II)
3 Dove Court — 1788 – 1795

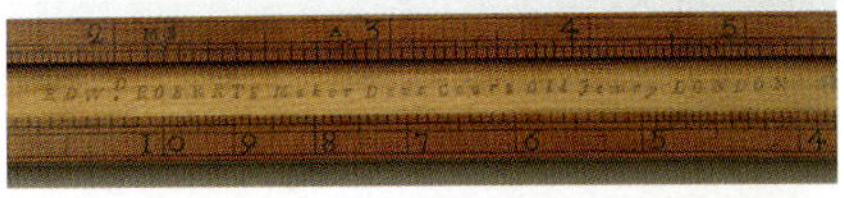

Rule and mathematical instrument makers. Edward (I) was apprenticed to ***Thomas Cooke***, Old Jewry, London in 1733 and it is likely that he continued the business after Cooke's death. He took as apprentices his son Edward and ***Joseph Stutchberry***, amongst others.

ROBERTS, J. — **BIRMINGHAM**
20 Lower Essex Street — 1858
80 Birchall Street — 1861

Rule maker.

ROBERTS, William — **SHEFFIELD**
21 Westbar — 1822
1 Sportsman's Inn Yard, Westbar — 1822 – 1825

(23) (22) Gibraltar Street | 1828 – 1833
14 Steelhouse Lane | 1837 – 1847
and 15 Westbar Green | 1847

ROBERTS, Mrs Fanny
15 Westbar Green | 1852

ROBERTS, Richard
16 Steelhouse Lane | 1852 – 1855
5 Court, Spring Street | 1856
2 Court, Pyebank | 1871
22 North Church Street | 1876

Box and ivory rule makers. 1868: Dealer in carpenters' rules and small wares.

ROBINSON, E. H. (& Sons) — **LONDON**
53 Bishopsgate Street | < 1850–1924 >

Silversmiths, opticians and cutlers. They also had a branch in Shrewsbury.

ROBINSON, Thomas Charles — **LONDON**
38 Devonshire Street | 1823 – 1841

ROBINSON & BARROW
26 Oxenden Street | 1842 – 1869

Mathematical instrument makers. In 1842 the business was taken over by ***Henry Barrow***, who used Robinson & Barrow as one of his trading names.

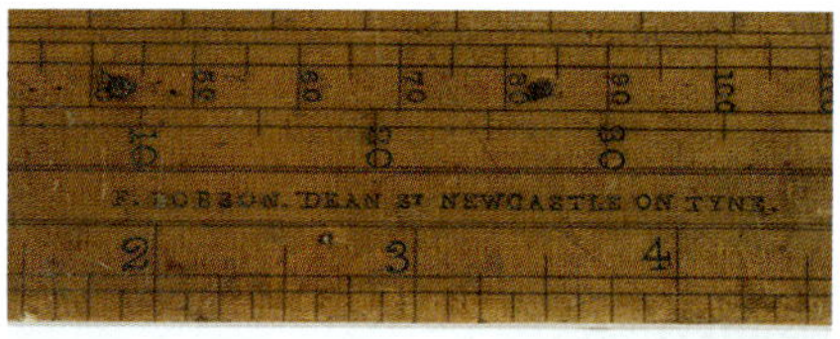

ROBSON, F., & Co. — **NEWCASTLE ON TYNE**
45 Dean Street | 1890 – present

Opticians. Name found on scales, slide rules, calculating rules and other drawing instruments etc. frequently related to the ship building trade. See also ***Glaholm & Robson***.

RO(D)GERS, Walter — **BIRMINGHAM**
Miles Street | 1875
1 Laurel Place | 1888

Rule maker.

ROGERS

Name found on 2 ft. 2-fold timber rule. There are a number of mathematical instrument makers working in London from the late 18th to the mid 19th centuries with the name Rogers, but without an initial it is impossible to identify the maker of this rule.

ROLPH, Thomas — **LONDON**
Ball Yard, Golden Lane | 1750
37 Beach Lane, White Cross Street | 1773

Mathematical instrument, pencil and rule maker. He was apprenticed to ***Richard Bates*** and took as an apprentice ***Thomas Truman*** (London) (amongst others).

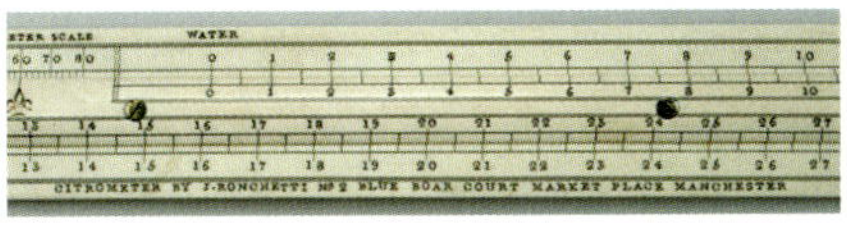

RONCHETTI, Baptist (& Co.) **MANCHESTER**
15 High Street 1785 – 1805
Warwick Street 1800
51 Spear Street 1804
RONCHETTI, Joshua
29 Balloon Street 1817 – 1825
St Ann Place 1828 – 1829
Cateaton Street 1830
43 Market Square 1832 – 1841
RONCHETTI, J. B. & J.
43 Market Street 1842 – 1851
RONCHETTI, J., & Son **LONDON**
(2) (25) Hatton Garden 1836 – 1839

Opticians and mathematical and philosophical instrument makers. The London branch of the firm was taken over by ***J. Solmalvico & Co.*** in 1839 and the Manchester branch by ***J. Casartelli*** in 1851.

ROUTLEDGE, Joshua **BOLTON & LEEDS**
Murray & Woods, Leeds 1808 – 1810
Thomas Swift & Co. Foundry, Bolton 1810 – 1814
Bolton 1814 – 1818
Warsaw 1824 – 1829

Engineer and inventor of an Improved Engineers' Slide Rule, which was invented in 1808. He was born in 1773. In 1800 he became manager for Murray & Wood in Leeds. In 1810 he moved to Bolton to work for Thomas Swift & Co. Foundry and from 1814 to 1818 he was in business there on his own as an ironmonger. In 1824, he moved to Warsaw to work for Thomas Evans & Co., Engineers. He died in Warsaw in 1829.

ROUTLEDGE, R., & Son **BIRMINGHAM**

A leading firm of tool makers and sellers operating from before 1876 to 1944. There was no connection between this firm and ***Joshua Routledge*** of Bolton. See *British Planemakers from 1700,* 3rd ed.

ROSS, Robert, & Son **MANCHESTER**
39 Green Street 1880 – 1890
North Street 1895 –1899

Rule makers.

ROTTON, Ambrose **BIRMINGHAM**
12 Price Street 1829 – 1830

Land surveyors' tape maker.

ROWE, James (& Co.) **BIRMINGHAM**
24 Heneage Street 1852
(45) (39 & 40) Woodcock Street 1858 – 1868
101 Moland Street 1868 – 1880
7 Bath Street 1882 – 1884
32 Carver Street 1895 – 1908

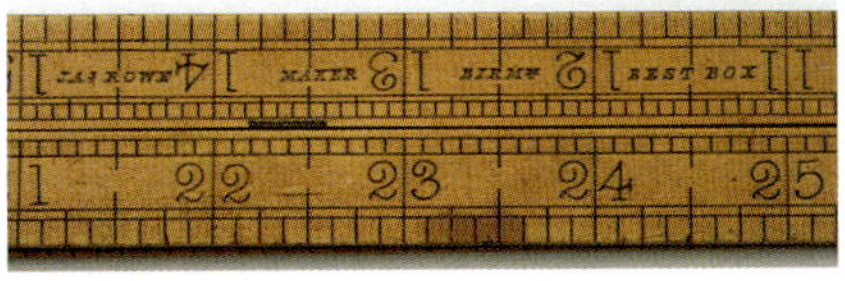

Rule makers. 1878: The business was sold due to James Rowe's ill health for the sum of £1,200. In the sale particulars, it was claimed that the business was established in 1846.

ROWLEY, John — **LONDON**
Threadneedle Street — 1691
Johnson's Court, Fleet Street — 1710
The Globe, Fleet Street — 1712 – 1728

Mathematical instrument maker. Advertised in the 7th edition of Thomas Everard's *Stereometry, or the Art of Gauging made easy...* published in London in 1712. ***John Coggs*** described himself as "servant to John Rowley".

ROWLINSON, John — **BIRMINGHAM**
11 Court, Cheapside — 1839

Rule maker.

ROWNEY G., & Co. — **LONDON**
1849 – 1959

Suppliers of artists' materials and drawing instruments. It is unlikely that they made the rules bearing their name.

RUBERGALL, Thomas — **LONDON**
27 Piccadilly — 1802
27 Coventry Street — 1805 – 1823
24 Coventry Street — 1826 – 1854

Mathematical and optical instrument maker and optician.

RULERS Ltd. — **BIRMINGHAM**
135 Park Road, Aston — 1910 – 1918

The continuation of ***Bees, Higgison & Co.***

RUSSELL, Edward, & Co. — **BIRMINGHAM**
29 Whittal Street — 1849 – 1858

Rule, mathematical instrument and measuring tape makers.

RUST, Richard — **LONDON**
The Minories — 1753 – 1778
Catherine Stairs — 1780 – 1781
RUST, R., & Son
Catherine Stairs — 1785 – 1794

Mathematical instrument makers and inventors.

RYAN, Geo. — **LONDON**

Tool maker and seller in business from 1893 to 1898. He continued after that date as one partner in Buck & Ryan. See also ***The Buck family***. See *British Planemakers from 1700,* 3rd ed.

S

SADLER, William — **CHELTENHAM**

Unidentified but probably a retailer.

SALT, Joseph — **WOLVERHAMPTON**
1793
Cock Street — 1811 – 1827
SALT, Joseph & Abraham — **BIRMINGHAM**
Smallbrook Street — 1797
Windmill Street — 1800 – 1801
SALT, Abraham
Worcester Street — 1805 – 1816
SALT, Abraham, & Son
19 Worcester Street — 1811 – 1826
and Bristol Street — 1823
SALT, Isaac
(18) (19) (20) Worcester Street — 1820 – 1839

Rule makers. After 1801, rule makers, cutlers and ironmongers.

SAMPSON, L. — **LIVERPOOL**
Brownlow Hill — 1823 – 1825
Gildart Street — 1827
Roden Place — 1829 – 1837
Upper Pitt Street — 1839 – 1845

Mathematical instrument maker. Name also spelt Samson.

SANDERS & Sons — **LONDON**
89 New Oxford Street — < 1906 >

Bar fitters and beer engine makers. One rule is known marked Ramsgate.

SANDS, Thomas — **LONDON**
164 Church Street — 1850

Rule maker.

SANFORD, Henry — **LONDON**
180 Bishopsgate Without — 1850

Wholesale ironmonger.

SARJANT, Thomas — **LONDON**
1712 – 1713

Unidentified but possibly a maker. The name has been recorded on a slide rule dated 1713.

SCATTERGOOD, Samuel — **BIRMINGHAM**
23½ Livery Street — 1845

Mathematical instrument and measuring tape maker.

SCHMALCALDER, C. A. **LONDON**
(82) (399) Strand 1806 – 1840
SCHMALCALDER, Joseph
399 Strand 1830 – 1831
SCHMALCALDER, John
400 Strand 1841 – 1845

Opticians and mathematical and philosophical instrument makers.

SEALE, Charles **LONDON**
48 St John's Square, Clerkenwell 1851 – 1859
63 Nightingale Vale, Woolwich 1892
60 Jackson Street, Woolwich 1891 – 1901

Rule and measure maker.

SEARCH, James **LONDON**
Crown Court, Soho 1771 – 1781

Mathematical instrument maker. He was apprenticed to ***John Bennett*** and took as an apprentice ***Alexander Wellington***, who succeeded him.

SETTON & DURWARD Ltd. **BIRMINGHAM**
15 Wood Street 1877
36 & 37 George Street 1892 – 1935
Criterion House, 114 Flat Street 1934
Fleet Street 1938
Benson Road, Hockley 1940 – 1948 >

Rule makers. An advertisement states "Mentor school rules, protractors, set squares etc."

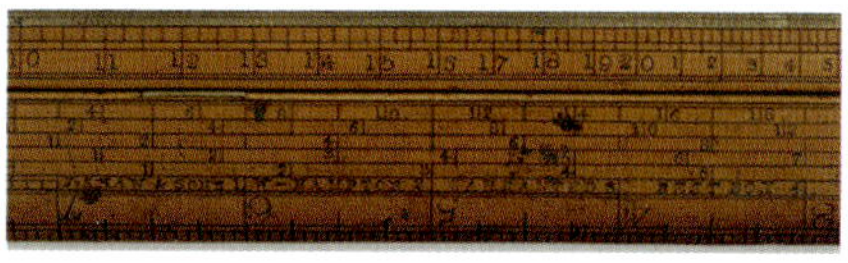

SHAW, J., & Sons **WOLVERHAMPTON**
1795
George Street 1805 – 1852
64 Church Lane 1852 – 1899
Fryer Street 1899 – 1937 >

Wholesale hardware and factoring business. The stock book of John Shaw and Co. ca. 1805 lists 90 different types of wooden and ivory rules as well as numerous other types of tools. They did not make the rules they sold and would have bought them in from other rule makers, including ***John Rabone & Sons***.

SHAW, William **BIRMINGHAM**
18 Smallbrook Street 1823 – 1826
SHAW, Thomas
14 Harford Street 1870

Rule makers.

SHELLARD, Alfred **BRISTOL**
5 Lower Maudlin Street 1854 – 1855
Castle Mill Street 1856
24 Peter Street 1857
SHELLARD, A. & F.

23 Castle Garden 1860
SHELLARD, F.
Lime Kiln Lane 1861 – 1865
St. George's Road 1866 – 1877
SHELLARD, A.
St. George's Road 1878
14 Charles Street 1902

Rule and gauging instrument makers. They took over the business of ***William Blainey***.

SHORT & MASON (Ltd.) **LONDON**
40 Hatton Garden 1876 – 1895
98 & 100 Clerkenwell Road 1900 – 1910
Aneroid Works, McDonalds Road 1913 – 1960s

Aneroid barometer, thermometer, saccharometer, salinometer and pocket compass manufacturers. They were listed as customers of ***John Rabone & Sons***.

SHUKER, Richard **LONDON**
10 Crown Court, Curtain Road 1805

Rule maker.

SHUTTLEWORTH, Joseph **BIRMINGHAM**
(63) (73) Lichfield Street 1828 – 1830
Islington 1829
133 Broad Street 1841
2 Newhall Street 1845
SHUTTLEWORTH, Joseph
(2) (73) Newhall Street 1848 – 1850

Mathematical instrument and measuring tape makers.

SILVESTER, Joseph **LONDON**
The Sign of the Frying Pan, Ludgate 1672

He sold rules supplied by ***Isaac Webb*** but it is not known whether he also made rules himself.

SIMMS, William (I) **BIRMINGHAM**
44 Coleshill Street 1780 – 1781
1793
LONDON
1794
Bowman's Buildings, Aldersgate Street 1808 – 1812
4 Broadway, Blackfriars 1818 – 1822
SIMMS, William (II)
1 Bowman's Buildings, Aldersgate Street 1821 – 1826

1793: Dial, compass and mathematical instrument makers. 1821: Opticians, mathematical instrument and mariners' compass makers. In 1826 William Simms was taken into partnership by ***Edward Troughton***. They traded as ***Troughton & Simms***.

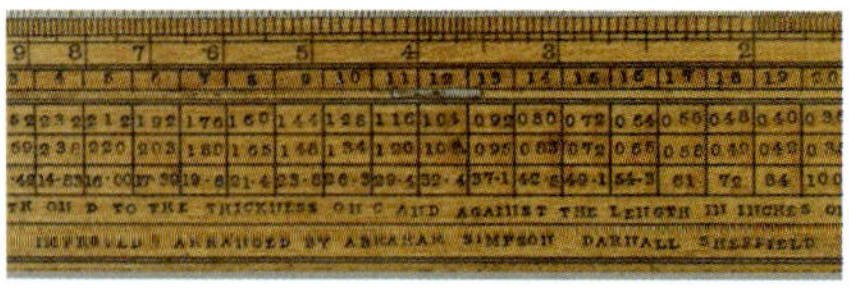

SIMPSON, Abraham — **SHEFFIELD**
Darnall — post 1900

Designer of an engineers' slide rule.

SISSONS, Jonathan — **LONDON**
Strand — 1722 – 1747
SISSONS, Jeremiah
Strand — 1749 – 1780

Mathematical instrument makers. Jeremiah employed ***Jesse Ramsden***.

SKETCHLEY, John — **BIRMINGHAM**
3 Horse Fair — 1854

Rule maker.

SLATER, J.T. — **LONDON**
3 Bartlett Passage

Unidentified, The name has been recorded on a rule also marked "Richardson, 16 Leather Lane, London". Richardson were saw makers working from 1842 to 1872.

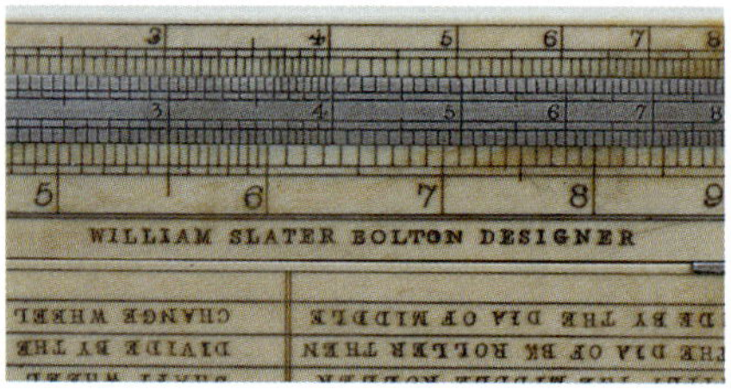

SLATER, William — **BOLTON**

Designer of a slide rule for the textile industry.

SLEEMAN, H. B., & Co. — **SHEFFIELD**
1878

They are not listed in any trades directory, but there is a quarterly price list for January 1878 in the Rabone scrapbooks in the Hawley Collection, Sheffield

SLOPER, Alfred — **LONDON**
14 Wellington Street, Blackfriars Road — 1846
3 Brighton Place, New Kent Road — 1859 – 1859
42 New Kent Road — 1865 – 1875
SLOPER, Alfred & Sons
42 New Kent Road — 1880 – 1882
4A Rockingham Street — 1890 – 1895
14 Occupation Road, Walworth — 1905 – 1907

Rule and mathematical and drawing instrument makers.

SLOPER, Thomas — **LONDON**
28 Free School Street, Horsleydown — 1833 – 1847
15 Shaftesbury Street, New North Road — 1870

Rule maker.

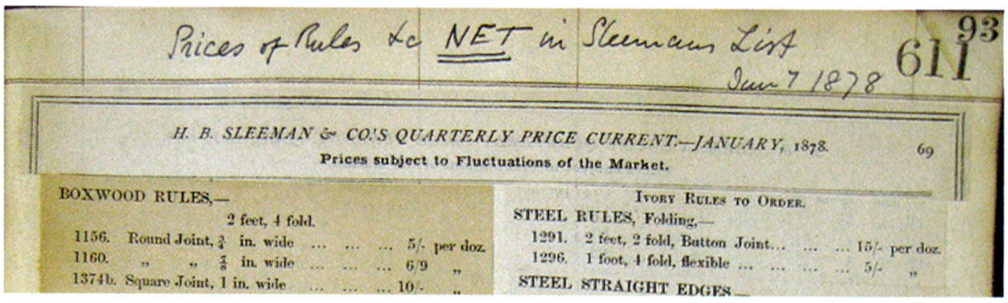
Prices of Rules &c NET in Sleeman List Jan 7 1878 — 611 93

H. B. SLEEMAN & CO.'S QUARTERLY PRICE CURRENT.—JANUARY, 1878. — 69
Prices subject to Fluctuations of the Market.

BOXWOOD RULES,—
2 feet, 4 fold.
1156. Round Joint, ½ in. wide ... ... 5/- per doz.
1160. " " ⅝ in. wide ... ... 6/9 "
1374b. Square Joint, 1 in. wide ... ... 10/- "

Ivory Rules to Order.
STEEL RULES, Folding,—
1291. 2 feet, 2 fold, Button Joint... ... ... 15/- per doz.
1296. 1 foot, 4 fold, flexible ... ... 5/- "
STEEL STRAIGHT EDGES,—

SLOPER, William **LONDON**

38 Great Suffolk Street 1833 – 1855

(15) (30) Lambeth Road 1859 – 1880

Rule and mathematical instrument maker.

SMALLWOOD, William **BIRMINGHAM**

35 Barford Street 1823 – 1830

Moseley Road 1839

SMALLWOOD, Josiah

1863

SMALLWOOD, John & Daniel

66 Moseley Street 1858

Ryland Street 1862 – 1875

Leopold Street 1866 – 1999

John and Daniel were the sons of William and his wife Sarah, Daniel was born in 1824, John in 1828. 1823: Heavy steel toy maker. 1839: Iron square maker, rule makers, also iron and steel makers, spirit levels, squares, etc. In their 1897 catalogue there is a notice announcing the taking over of ***Thomas Aston*** of Birmingham. Although the partners were John and Daniel, the company continued to use the old habit of spelling the "J" as an "I" on their rules.

SMITH, A.J., & Co. **LONDON**

14 and 18A Dartmouth Road 1947

Unidentified but likely to be retailers.

SMITH, Edward **BRADFORD**

Unidentified. The name has been recorded on a 12 in. 4-fold boxwood rule

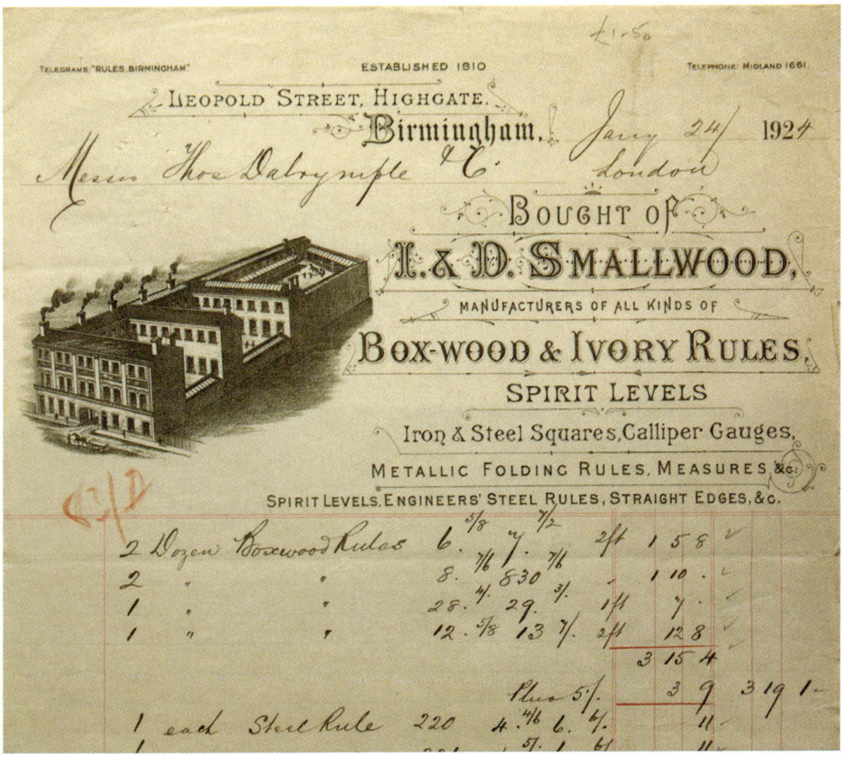
TELEGRAMS "RULES, BIRMINGHAM." ESTABLISHED 1810 TELEPHONE: MIDLAND 1661.

LEOPOLD STREET, HIGHGATE.

Birmingham, Jany 24/ 1924

Messrs Thos Dalrymple & Co London

BOUGHT OF

I. & D. SMALLWOOD,

MANUFACTURERS OF ALL KINDS OF

BOX-WOOD & IVORY RULES,

SPIRIT LEVELS

Iron & Steel Squares, Calliper Gauges,

METALLIC FOLDING RULES, MEASURES &c.

SPIRIT LEVELS, ENGINEERS' STEEL RULES, STRAIGHT EDGES, &c.

2 Dozen Boxwood Rules

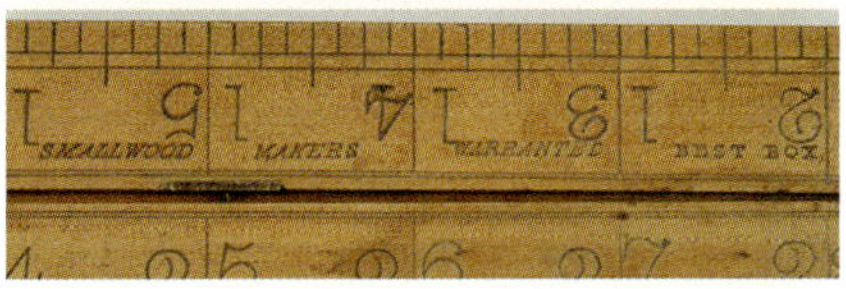

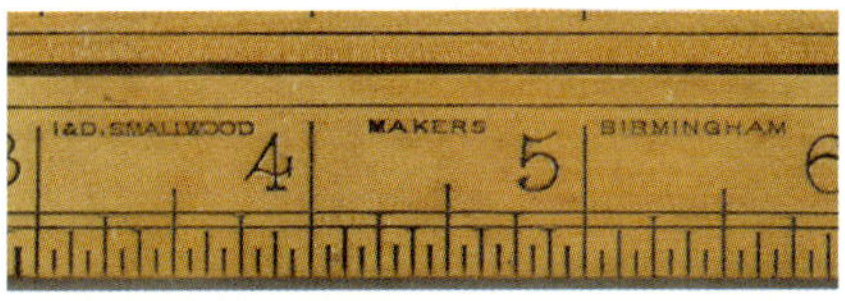

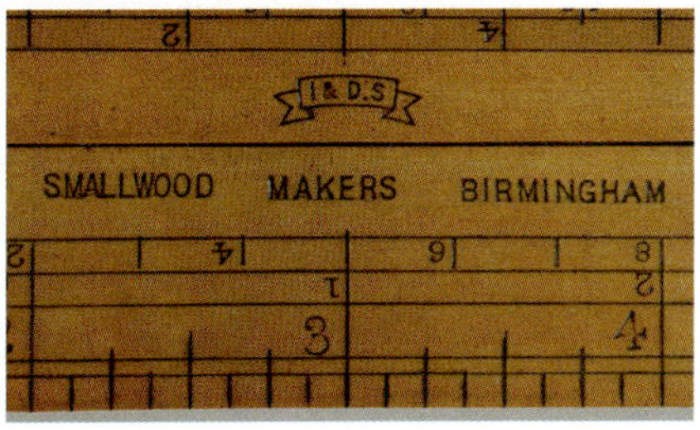

marked "Cloth Dissector Smith's Pat Improved". However, it is unlikely that he made the rule.

SMITH, J. Ciceri — **EDINBURGH** 1890; **LONDON**
61 Hatton Garden — < 1908 – 1914 >

Maker of micrometers.

SMITH, George — **BIRMINGHAM**
18½ Digby Street — 1861

Carpenters' rule maker.

SMITH, Samuel Albert — **BIRMINGHAM**
(68) (75) (77) Coventry Road — 1850 – 1872

Rule maker. His successor, ***W. J. Hughes***, claimed that the business had commenced in 1837 but no evidence has been found of this. The company went into liquidation in 1872.

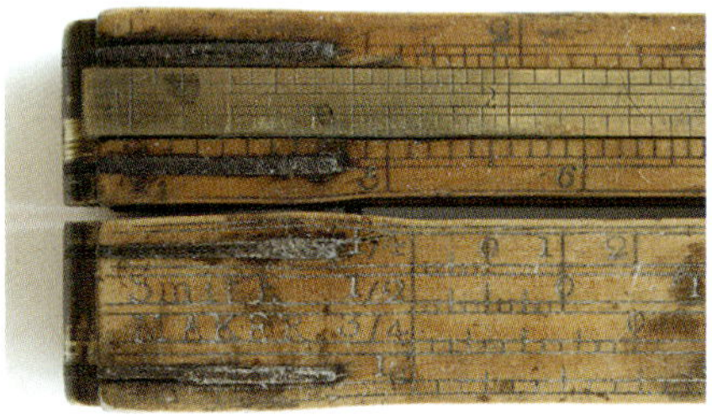

SMITH, Thomas — **WOLVERHAMPTON**

6 St. John Street	1792 – 1798
Bond Street	1800 – 1818
Penn Street	1827
Bloomsbury Street	1828 – 1850
Stafford Street	1865
SMITH, Benjamin	
St. John Street	1828 – 1829
Bloomsbury	1834
SMITH, John	
Bloomsbury	1828 – 1829
SMITH, Edward	
Prospect Place	1850

Rule makers. Thomas Smith succeeded ***James Corson***.

SMITH, William — **BIRMINGHAM**
6 Court, Loveday Street — 1830 – 1837
29 Loveday Street — 1841 – 1850

Mathematical and drawing instrument and measuring tape maker.

SMITH'S, HASLUCK & CAMPBELL

Unidentified but likely to be a maker. The name has been recorded on a number of late 18th century rules.

SOCIÉTÉ DES LUNETIERS — **LONDON**
(13) (36) (56) Hatton Garden — 1880 – 1913

Rule makers. In the 1890s trades directories ***Louis Mallet*** was listed separately and was described as "at the Société des Lunetiers". The firm was founded in France in 1849. The company is still in business today making spectacle lenses, trading as Essilor.

SOLOMONS, S. & B. | **LONDON**
5 New Road | 1838
39 Albermarle Street | 1840 – 1875

Opticians and mathematical instrument makers.

SOMOLVICO, John | **LONDON**
Hatton Garden | 1817 – 1822
SOMOLVICO, Joseph, & Son
Hatton Garden | 1829 – 1839
SOMOLVICO, Joseph, & Co.
Hatton Garden | 1839 – 1865 >
81 Holborn | 1901

Barometer makers, mathematical instrument makers. In 1839 they took over the business of ***J. Ronchetti & Son***.

SNELSON, Joseph | **BIRMINGHAM**
8 Shadwell Street | 1829 – 1830

Rule maker.

SORBY, I. | **SHEFFIELD**

One of the numerous Sheffield tool makers and sellers. See *British Planemakers from 1700,* 3rd ed.

SPRINGER, Joshua | **BRISTOL**
1 St. Stephen's Lane | 1759 – 1760
2 Clare Street | 1775 – 1807

Mathematical instrument maker and electrician. The business was taken over by ***R. & C. Beilby***.

STANDLEY, BELCHER, MASON Ltd | **BIRMINGHAM**
Church Street | <1902 >

Scientific instrument makers.

STANLEY, Wm. Ford (& Co. Ltd.) | **LONDON**

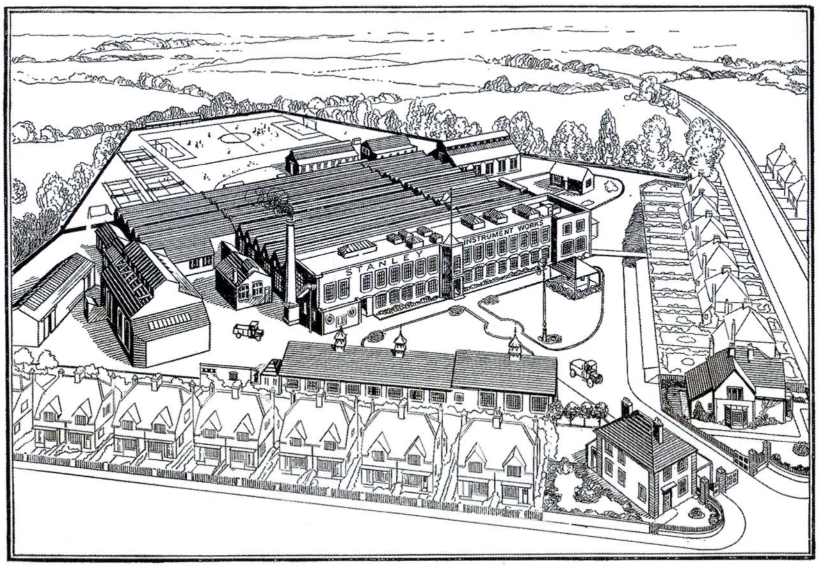

The Stanley Scientific Instrument Works, Avery Hill Road, New Eltham, S.E.9.

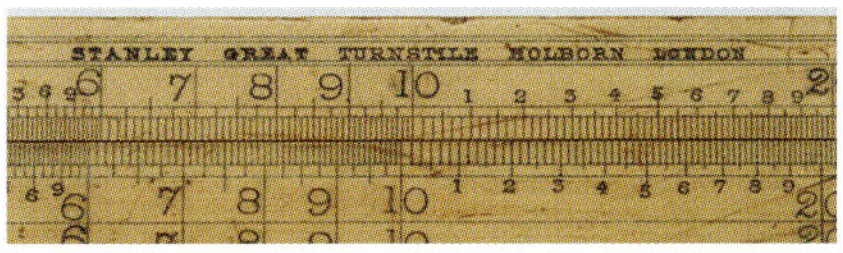

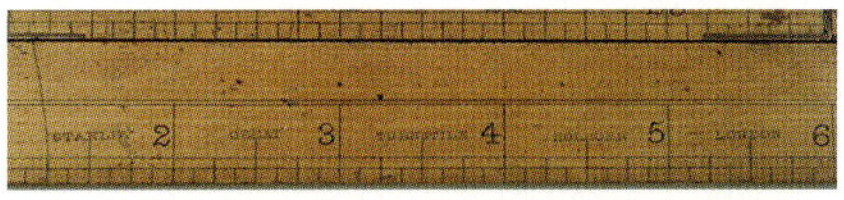

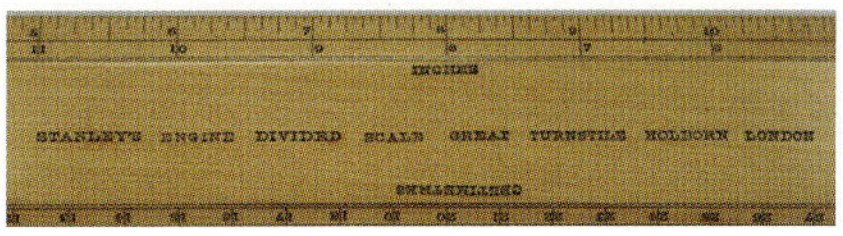

Great Turnstile, Holborn 1853 – 1880
3 & 4 Great Turnstile Road 1880 – 1947
New Eltham, SE1 1951 – 1967

Rule. matjematical and scientific instrument maker. There were a number of other addresses in London, Glasgow and Singapore which were the various retail shops run by the company. In 1894 they advertised in Pickworth, C.N. *The Slide Rule.* In 1911 they took over ***J. J. Hicks*** and in 1926 ***Heath & Co.*** For a full history of the company, see Allen, Cecil J., *A Century of Scientific Instrument Making 1853 – 1953*, London, 1953.

THE STANLEY RULE & LEVEL Co.

The principal American maker of rules and other measuring equipment which was extensively exported to Britain. Boxwood and ivory folding rules were the first products of A. Stanley & Co. which was set up in 1854 in New Britain, Connecticut, U.S.A.. In 1858, after the acquisition of several other rule manufacturers, the company was renamed The Stanley Rule & Level Co.. Over the years they added many other tool lines including planes in 1869, braces in 1870, chisels in 1911, mechanics' tools in 1913. In 1920 the company purchased The Stanley Works and became The Stanley Rule & Level Plant of the Stanley Works. In 1937 this name changed to the Stanley Tool Division of the The Stanley Works. Today the company is known as Stanley Tools Ltd. A wide ranging description of the company and its products can be found in Walter, J., *Antique and Collectible Stanley Tools: A Guide to identity and value*, Marietta OH, 1996, and Stanley, P. E., *Boxwood & Ivory*, Westborough, MA, 1984.

STANTON, John **LONDON**
73 Shoe Lane 1813 – 1840
STANTON, John Alfred
1837 –1874

Optical turners.

STARMER, Thomas **BIRMINGHAM**
103 Broad Street 1868
Ladywood Road 1875
156½ Broad Street 1876 – 1880
10 Grosvenor Place 1882

Rule maker.

THE L.S. STARRETT Co.

Founded in 1880 in Athol, Mass., U.S.A., the firm grew steadily to become one of the major suppliers of machinists' tools in both the U.S.A. and the U.K. Their name is found frequently on steel rules and other machinists' measuring devices.

STATHAM, John, & Son **MANCHESTER**

A firm of joiners and builders from 1852 to 1861 whose name has been found on an ivory scale rule.

STEAD, William — **LONDON**
Old Fish Street — 1763 – 1767

Mathematical instrument and rule maker.

STEBBING, J. R. & H. — **SOUTHAMPTON**
63 High Street — 1833
STEBBING, Joseph Rankin
47 High Street — 1845 – 1851
STEBBING & WOOD
47 High Street — 1851 - 1853
9 Canute Road — 1859 – 1870

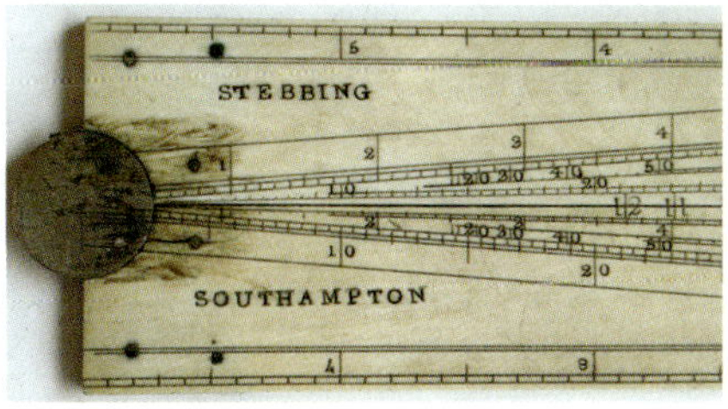

Mathematical and nautical instruments maker, optician and ships' chandlers. J. R. Stebbing was the maker of the earlier rules designed by ***Charles Hoare***.

STEDMAN, Christopher (I) — **LONDON**
The Globe, London Bridge — 1747 – 1750
Leadenhall Street — 1758 –1774
STEDMAN Christopher (II)
Leadenhall Street — 1785

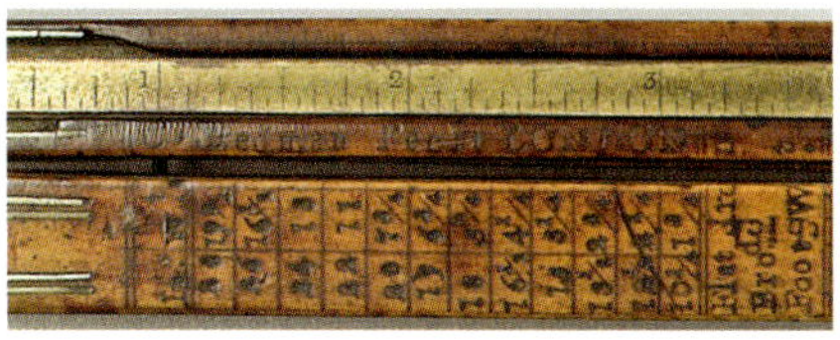

Mathematical instrument makers.

STEPHENSON, BLAKE & Co. — **LONDON**
33 Aldersgate Street — < 1906–1938 >

1906: Letter founders. 1924: Type founders, brass rule makers for printers, joiners, etc.

STEVENS, William. F. — **LONDON**
3 Kingsland Road — 1910 – 1914

Drapers' rule maker.

STEVENSON, Peter (Ltd.) — **EDINBURGH**
9 Lothian Street — 1836 – 1861
51 George 4th Bridge — 1862 – 1870
(5) (9) (7) George 4th Bridge — 1874 – 1900 >

Philosophical instrument makers.

STEWARD, J.H. (Ltd.) — **LONDON**
406 (457) Strand — 1867 – c. 1975
Hove, Sussex — c. 1975
Saxmundham, Suffolk — 1977 – 1982

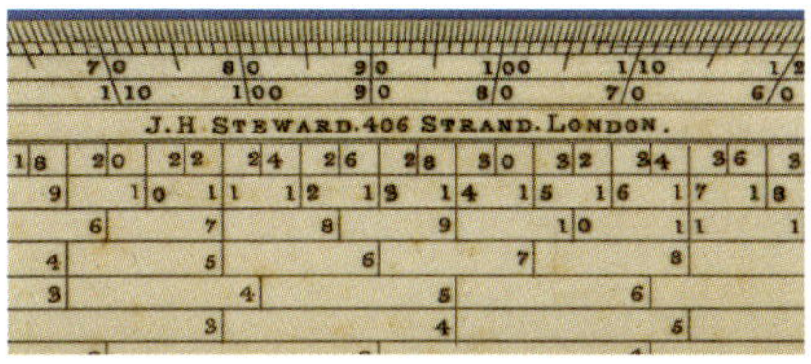

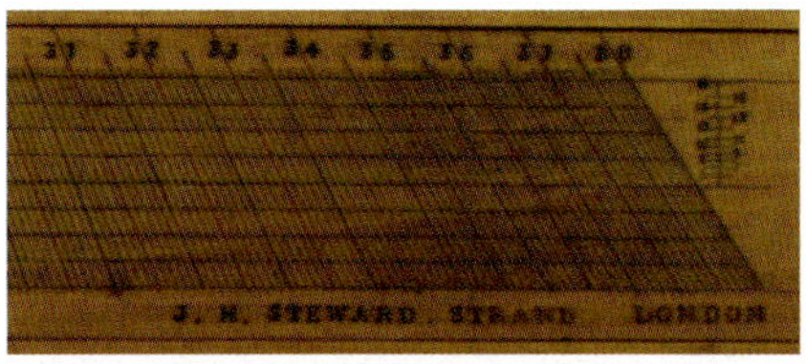

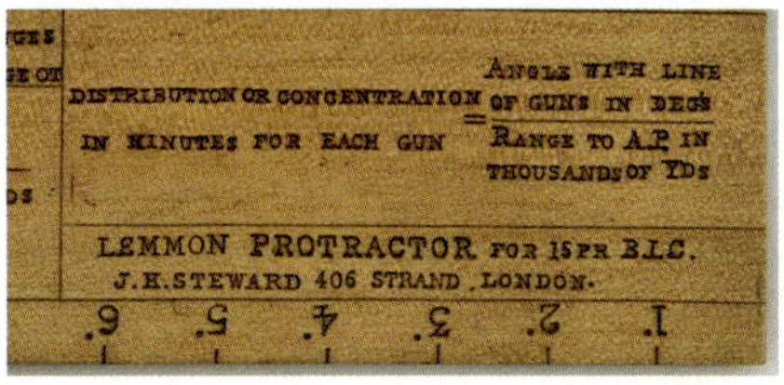

Opticians and makers of field glasses, telescopes, optical lanterns, barometers, meteorological and surveying instruments. The company prided itself as being suppliers to the British and foreign governments and to the National Rifle Association of England, India and Canada.

STINTON, Horatio(I) (& Son) — **BIRMINGHAM**
1 Ladywell Road — 1855
3 Essex Street and 71 Loveday Street — 1858

69 Loveday Street	1860 – 1867
40 Livery Street	1870
71 Moland Street	1875 – 1905
STINTON, Horatio (II)	
35 Lancaster Street	1910 – 1918

Parallel and rolling rule makers.

STRAW, William **LONDON**

Tool maker and seller operating from 1826 to 1879. See *British Planemakers from 1700,* 3rd ed.

STRAWSON, George W. (& Co) **LONDON**

15 Percy Street	1870
9 Pancras Lane	1880
6 & 8 Eastcheap	1900 – 1906

Mathematical instrument makers. 1900: Wholesale stationers.

STUBBS, George **BIRMINGHAM**

18 Tenby Street	1864
73 Northwood Street	1880 – 1882

Rule maker.

STUBS, Peter S. **WARRINGTON**

Bridge Street	1777 – ca. 1800
Scotland Street	ca. 1800 – 1950s

Manufacturer of Lancashire files and tools, particularly for the clock and watch trade. By 1876 he had also established a steelworks at Rotherham. This was sold in 1958, but the name of Peter Stubs continues today as a firm of precision engineering manufacture and metal suppliers.

STUTCHBERRY, Joseph **LONDON**

Bird in Hand Court, Cheapside	1791 – 1796
3 Dove Court, Old Jewry	1797 – 1826
STUTCHBERRY, David	
3 Dove Court, Old Jewry	1807 – 1828

Rule and mathematical and gauging instrument makers. Joseph was apprenticed to ***Edward Roberts***.

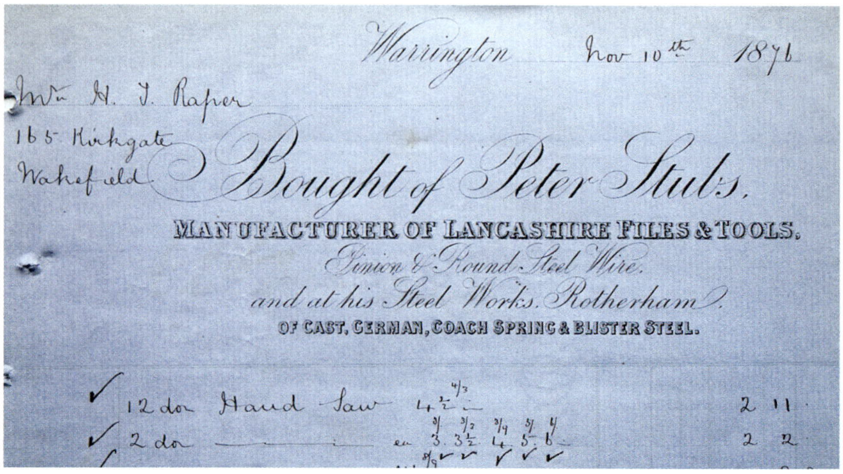
Warrington Nov 10th 1876
Mr H. T. Raper
165 Kirkgate
Wakefield
Bought of Peter Stubs,
MANUFACTURER OF LANCASHIRE FILES & TOOLS,
Pinion & Round Steel Wire,
and at his Steel Works, Rotherham,
OF CAST, GERMAN, COACH SPRING & BLISTER STEEL.
12 doz Hand saw

SUFFELL, C. **LONDON**
(132) (122) London Acre 1839 – 1877

Mathematical instrument maker.

SUTTON, Baptist **LONDON**
Chancery Lane 1636 – 1653
SUTTON, Henry
Threadneedle Street 1649 – 1665

Baptist was either Henry's father or uncle. They both made brass Gunter's scales. Henry was apprenticed to ***Thomas Brown(e)***.

SUMMERS, John **BIRMINGHAM**
120 Lichfield Street 1888 – 1892

Rule maker.

SWAINE & ADENEY **LONDON**
185 Piccadilly 1860 – 1938 >

Whip manufacturers, sponge importers and general warehouse. They are unlikely to have made the rules that bear their name.

SYM, James (I) **GLASGOW**
(2) (16) Bell Street 1801 – 1806
236 High Street 1807 – 1816
SYM, J., & Co.
236 High Street 1817 – 1825
SYM, James (II)
167 High Street 1826 – 1847

Mathematical instrument makers and opticians.

T, U & V

TAGLIABUE, Caesar — **LONDON**
Holborn — 1795 – 1816
Cross Street, Hatton Garden — 1822 – 1838
TAGLIABUE & CASSELLA
23 Hatton Garden — 1838 – 1846

Opticians. An extensive family working at various addresses in the Holborn and Hatton Garden area until 1852. Tagliabue & Cassella was a partnership between Caesar and ***Louis Casella***, his son-in-law

TAPLIN, William M. (& Co.) — **BIRMINGHAM**
Pools Street, Aston Brook — 1854 – 1856

Rule makers. They also had an address at 50 Hatton Garden, London.

TAYLOR, J. — **NOTTINGHAM**

Unidentified. The name has been recorded on a dip rod, but he is unlikely to have been a maker.

TAYLOR, J. & H. — **BIRMINGHAM & LONDON**

Unidentified but probably a retailer.

THE TAYLOR ROLPH Co. — **LONDON**
Mortlake

Manufacturers of lawn bowls equipment, they are still in business today. They would not have made the measures that bear their name.

TERRILL, Edmund Henry — **BIRMINGHAM**
196 New John Street — 1875 – 1882

Rule maker.

THOMAS — **LONDON**
High Street, Deptford

Unidentified but unlikely to be a maker. The name has been recorded on a contraction rule.

THOMAS, John — **BIRMINGHAM**
Howe Street — 1839

Rule maker.

THOMPSON, John — **LONDON**
Hosier Lane — 1609 – 1648

Mathematical instrument maker.

THOMPSON, John, & Co. — **LIVERPOOL**
24 Manchester Street and 65 Lord Street — 1872 – 1887

Opticians and mathematical instrument makers.

THOMPSON, Samuel — **BIRMINGHAM**
2 Jennens Row — 1908 – 1918
441 Victoria Road, Aston — 1919 – 1947

Rule maker.

THORNTON, A. G. — **MANCHESTER**
39 Great Cheetham Street — 1878
100 Deansgate — 1879
South King Street — 1895
St. Mary Street — 1897
(2) (41) King Street West — 1904 – 1960>

A.G. Thornton was apprenticed to ***W. H. Harling***. He orginally set up in business with ***J. Halden*** but this partnership lasted only two years. In 1888 the directory entry states that the proprietor was ***J. A. Reynolds***. In 1967 the firm became British Thornton and later Education and Science Furniture. British Thornton is still in business, now based in Yorkshire.

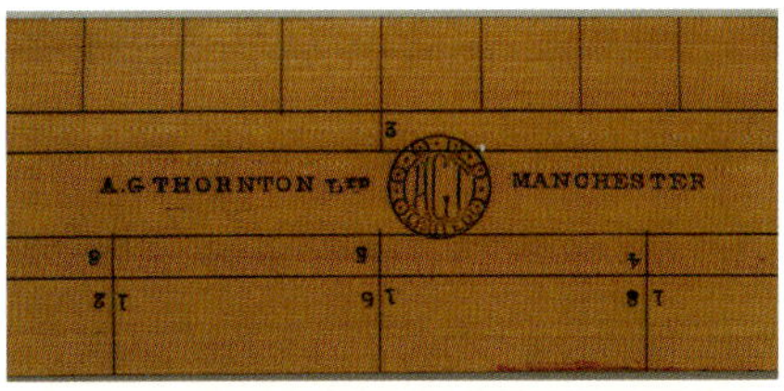

THORNTON, W. C. — **HULL**

Unidentified but possibly a maker. The name has been recorded on a 42 in. 2-fold rule marked "Holdsworth Patent No. 6747-91" and other rules.

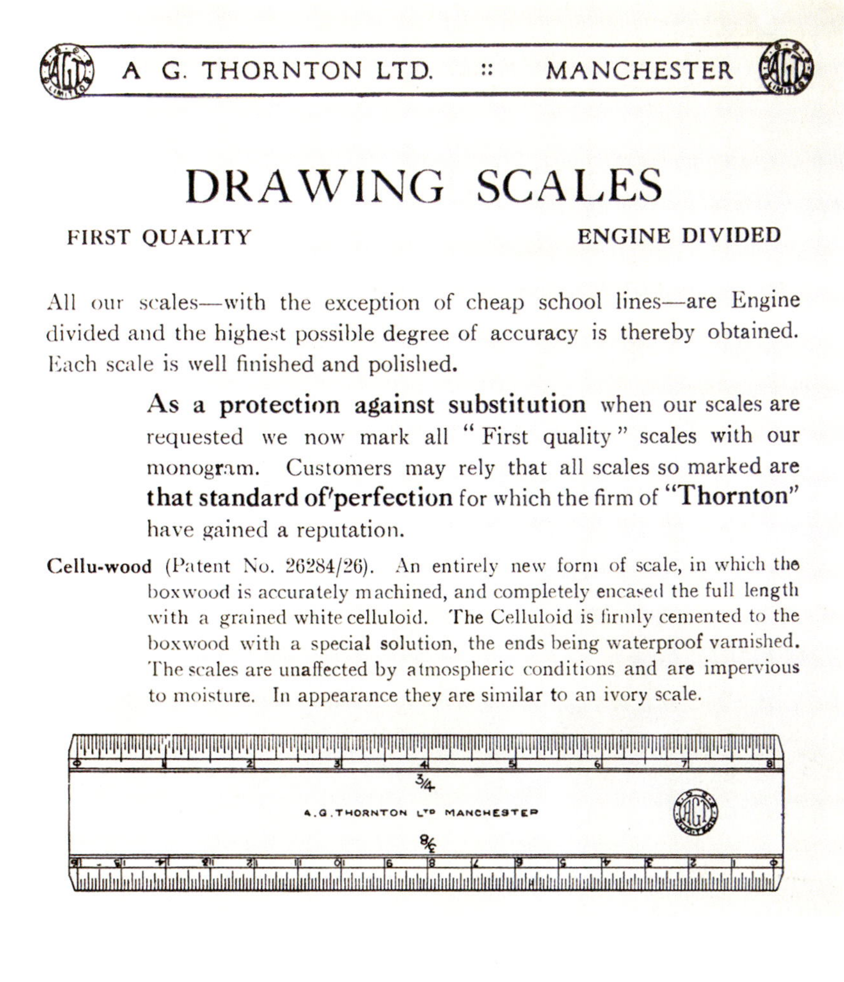
A G. THORNTON LTD. :: MANCHESTER

DRAWING SCALES

FIRST QUALITY — ENGINE DIVIDED

All our scales—with the exception of cheap school lines—are Engine divided and the highest possible degree of accuracy is thereby obtained. Each scale is well finished and polished.

As a protection against substitution when our scales are requested we now mark all " First quality " scales with our monogram. Customers may rely that all scales so marked are **that standard of perfection** for which the firm of **"Thornton"** have gained a reputation.

Cellu-wood (Patent No. 26284/26). An entirely new form of scale, in which the boxwood is accurately machined, and completely encased the full length with a grained white celluloid. The Celluloid is firmly cemented to the boxwood with a special solution, the ends being waterproof varnished. The scales are unaffected by atmospheric conditions and are impervious to moisture. In appearance they are similar to an ivory scale.

THREADWELL, J. J. & Sons — **LONDON**
High Street, Stratford — 1860
127 Dawes Road, Forest Gate — < 1939
Raven Road South, Woodford — 1945 – 1986

Rule makers. According to the reminiscences of one of the instrument makers employed by the company, little of their output was sold under their own name. The majority was stamped with that of other makers who included ***W. F. Stanley***, ***C. Baker*** and Gamages, the London department store.

TICKET, Joseph — **LONDON**
25 Mape Street, Bethnal Green — 1843
2 Albert Street, Stepney — 1847
1888

Rule maker.

TIMMINS & Son — **BIRMINGHAM**

Birmingham tool and "steel toy" makers. The business was established in the 18th century and continued until the mid-20th century. In 1866 combined with ***W. & C. Wynn*** to form ***Wynn Timmins***. They made a variety of steel slot gauges. For a full history see Rees, Jane, *A Pattern Book of Tools and Household Goods,* Early American Industries Assocoation, South Dartmouth, MA, U.S.A., 2006.

TINSLEY, J. — **BIRMINGHAM**
7 Court, Duke Street, Aston — 1866

Rule maker.

TITTERTON, Thomas W. — **BIRMINGHAM**
17 Jennens Row — 1852 – 1866
6 Richard Street — 1875
TITTERTON, George
40 Curzon St — 1868

Parallel and rolling rule makers.

TOMPKINS, Alfred — **LONDON**
86 Farringdon Street — 1880

Rule maker.

TOWNLEY — **BIRMINGHAM**
75/76 Bull Street — < 1910 >

Cutlers.

TRAFFORD, John — **OXFORD**
Windlebury — < 1772 – 1783 >

Mathematical instrument maker.

TREE, James (I) — **LONDON**
1826
20 Great Suffolk Street, Borough

1 Great Guildford Street, Borough
22 Little Charlotte Street, Blackfriars Road 1832 – 1842

TREE, Mrs Emily
22 Little Charlotte Street 1843 – 1849

TREE, James (II)
22 Little Charlotte Street 1850 – 1851

TREE, James, & Co.
22 Little Charlotte Street 1851 – 1895
7 Lawrence Lane 1895

Ivory and box rule makers. J. Tree is the only recorded maker of ***John Ewart's*** Cattle Gauge.

TRESS & Co. **LONDON**
3 Stamford Street < 1906–1924 >

Wholesale hat manufacturers. It is unlikely they made the rules that bear their name.

TROUGHTON, John (I) **LONDON**
Standgate Lane, Lambeth 1752
Surrey Street, Strand 1755 – 1778

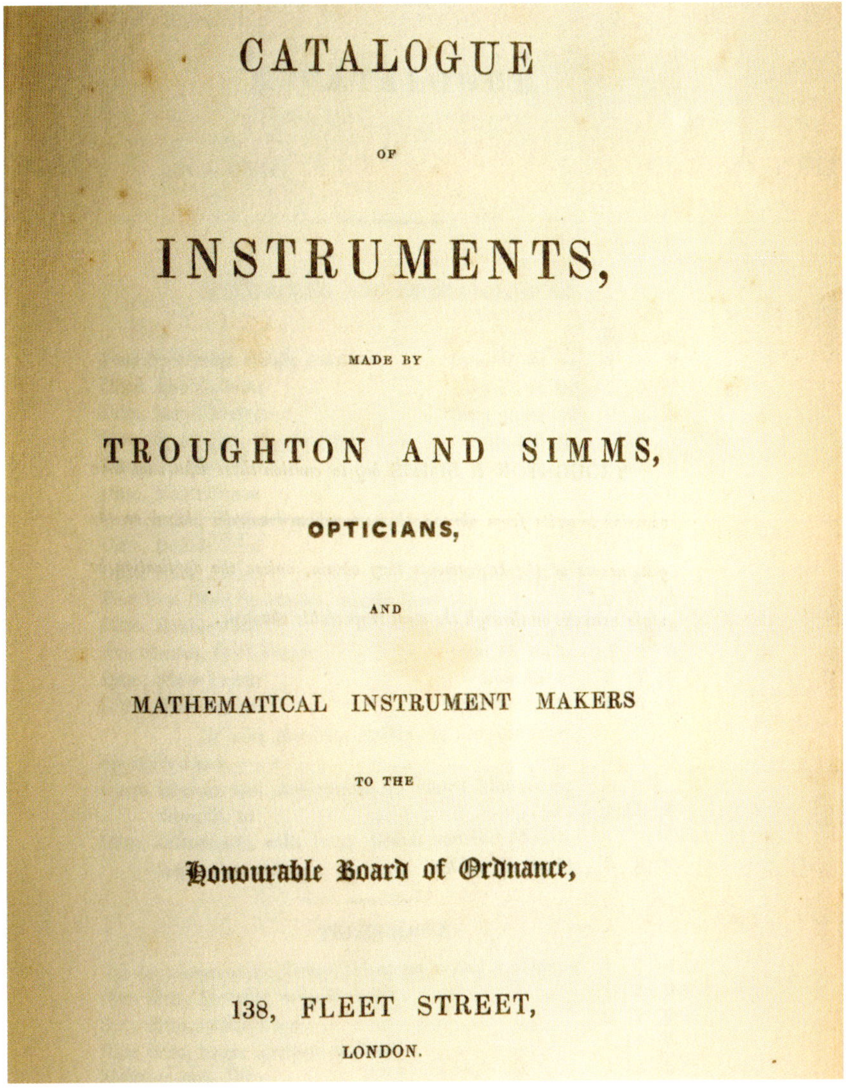

CATALOGUE

OF

INSTRUMENTS,

MADE BY

TROUGHTON AND SIMMS,

OPTICIANS,

AND

MATHEMATICAL INSTRUMENT MAKERS

TO THE

Honourable Board of Ordnance,

138, FLEET STREET,

LONDON.

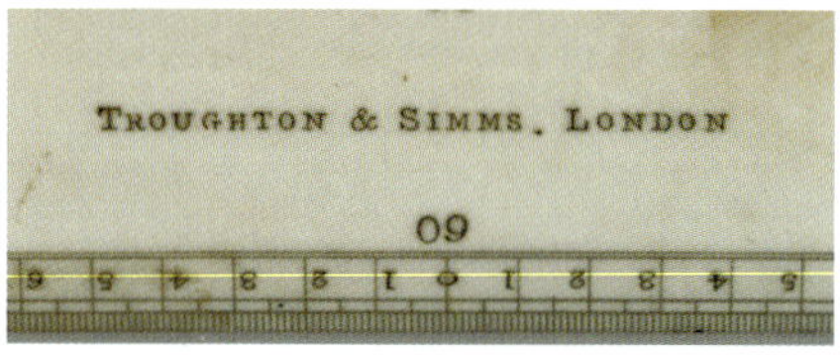

TROUGHTON, John (II)
Crown Court, Fleet Street 1768 – 1771
17 Dean Street, Fetter Lane 1771 – 1778
1 Queen's Square, Bartholomew Close 1778 – 1782
136 Fleet Street 1782 – 1788
TROUGHTON, John & Edward
136 Fleet Street 1788 – 1804
TROUGHTON, Edward
The Orrery, 136 Fleet Street 1804 – 1826
TROUGHTON & SIMMS
136 Fleet Street 1826 – 1844
138 Fleet Street 1844 – 1915
340 Woolwich Road (factory) 1866 – 1915

1756: Mathematical instrument makers. 1826: Opticians, and mathematical and philosophical instrument makers. Edward Troughton formed a partnership with ***William Simms (II)*** to form Troughton & Simms. In 1922 the firm combined with ***Thomas Cooke & Sons*** of York and London to form ***Cooke Troughton & Simms*** although Cooke also continued trading under his own name. ***Henry Barrow*** worked as an outworker for the company.

TROW, Isaac **BIRMINGHAM**
(40) (42) Loveday Street 1829 – 1858

Rule, measuring tape, compass and dog collar maker. He took over the business of ***Richard Bakewell***.

TRULL & Co. **BRISTOL**

Unidentified. The name has been recorded on a boxwood diagram square so they are probably a tailors' equipment supplier.

TRU(E)MAN, Thomas **LONDON**
(7) (89) Chiswell Street, Moorfields 1789 – 1790
87 Long Lane, Smithfield 1793 – 1805
TRUMAN, Elizabeth
87 Long Lane, Smithfield 1817 – 1836
TRUMAN, Edward
87 Long Lane, Smithfield 1832 – 1834

1789: Rule and drawing instrument makers. 1817: Rule makers. Thomas was apprenticed to ***Thomas Rolph***.

TRU(E)MAN, Thomas **BIRMINGHAM**
134 Vauxhall Road 1861 – 1866

Rule maker.

TURNER, John **LONDON**
24 Brook Street, Holborn 1839 – 1840

Rule maker.

TURNER, Joseph **BIRMINGHAM**
Livery St 1839
1 Court, Adam Street 1850

TURNER, John
17 Court, Wharf Street — 1852 – 1865
TURNER, Thomas
11 Caroline Street — 1870
TURNER, Frederick John
105A Vyse Street — 1902

Rule makers.

TURNER, Joseph — **LONDON**
19 Chapel Row, Exmouth Street — 1851 – 1855

Rule maker.

TURNER, ATHERTON & Co. — **MANCHESTER**
Denton

Unidentified but they are unlikely to have been a rule maker. The name has been recorded on an unusual cruciform rule, marked in metric and imperial.

TURTLE, L. H. — **CROYDON**

Ironmongers trading in Croydon from 1885 to present. See *British Planemakers from 1700,* 3rd ed.

TUSTIAN, Thomas — **LONDON**
1 Princes Square, Ratcliff Highway — 1805
18 Princes Square, Ratcliff Highway — 1836 – 1839
TUSTIAN, Thomas, & Co.
18 Princes Square, Ratcliff Highway — 1840 – 1913
and 11 Antclife Street — 1891 – 1899
27 Rectory Square — 1907 – 1915
TUSTIAN & GROVE

Rule makers. The partners of Tustian & Grove have not been identified, nor have their dates of operation.

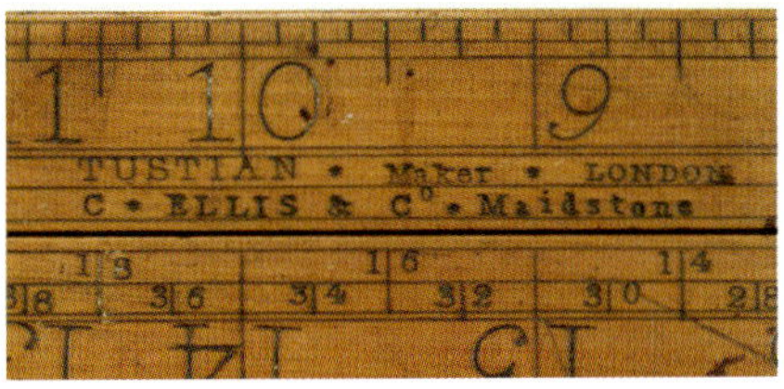

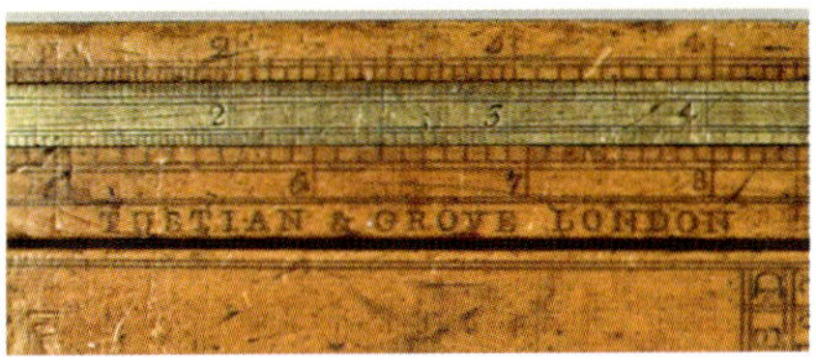

TUTTELL, Thomas — **LONDON**
Charing Cross
and Royal Exchange, Cornhill — < 1695 – 1702

Mathematical instrument maker. Samuel Pepys bought a rule from him. Thomas Tuttell drowned in the Thames in 1701/2 whilst carrying out a survey.

TYE, G.S. — **BIRMINGHAM**
41 Caroline Street — < 1910 >

Dye sinker. He is unlikely to have made the rules that bear his name.

TYZACK & HOLMES — **SHEFFIELD**
Standard Works, Cavendish Street — 1889 –1913 >

Rule makers. 1895: Steel rule makers. 1909: Tape measure makers. Standard Works was previously the address of ***Alfred Cowling*** whose business was taken over by Tyzack & Holmes.

ULLERTHORNE (& Co.)

Suppliers of shoe tools. Their name is recorded on various shoe measures but they are unlikely to have made any rules.

ULLMER, William	**LONDON**
6 West Harding Street	< 1839 – 1841 >

Brass rule cutter.

UNDERHILL, Thomas	**LIVERPOOL**
124 Whitechapel	1824
6 Cooper's Row	1825
32 Mersey Street	1827 – 1829

Mathematical instrument and rule maker, optician.

UNDERHILL, Thomas	**MANCHESTER**
15 Withy Grove	1828
40 Water Street	1834 – 1838
70 Bridge Street, Deansgate	1841
(4) (8) Old Millgate	1847 – 1854
1 St. Mary's Gate	1850 – 1852
(2) (4) Corporation Street	1858 – 1868
53 Princess Street	1873 – 1881

Rule and mathematical and philosophical instrument maker, optician. It would seem likely that there is a connection with ***Thomas Underhill***, Liverpool, but this has not been established.

UNDERWOOD	**BIRMINGHAM**
34 Gopsal Street	

Unidentified but probably a retailer. Name noted on a printers' rule with paper sizes.

UNIVERSAL WOODWORKING Co. Ltd.	**BIRMINGHAM**
51 Newhall Hill	1925 – 1940
17 George Street, Balsall Heath	1948

Rule makers. In 1848 advertised that they were makers of boxwood rules, brass compasses and dividers, mathematical sets, etc.

VARDON, J., & Son	**LONDON**

A substantial London ironmongers in business from the 1790s to at least the 1840s. See *British Planemakers from 1700,* 3rd ed.

VARVILL & Son	**YORK**

The largest and most prominent of the York plane makers in business from 1793 to 1904. See *British Planemakers from 1700,* 3rd ed.

W & Y

WAKERLY, Andrew **LONDON**
Redcliff (Rotherhithe) Wall 1631 – 1665

He was known to have made Gunter's scales and carpenters' rules.

WALKER, Richard **WOLVERHAMPTON**
2 Timber Yard 1792 – 1802

Rule maker.

WALKER, T. F. **BIRMINGHAM**

Designer of an engineers' slide rule.

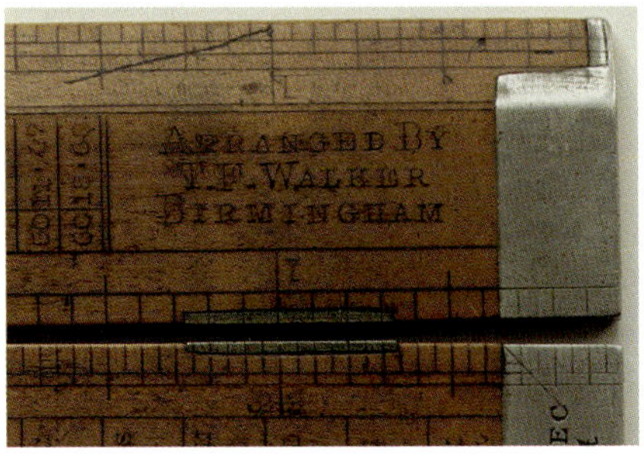

WALKER, Thomas **SHEFFIELD**
9 Westbar Green 1864

Rule maker.

WARD, T. **BIRMINGHAM**
3 Court, Lichfield Street 1852

Rule maker.

WARD & PAYNE **SHEFFIELD**

Edge-tool makers in business from 1803 until 1970. See *British Planemakers from 1700,* 3rd ed.

WARING & DIMES **LONDON**
91 Great Russell Street < 1841 >

Artists' colour-men. They are unlikely to have made the rules which bear their name.

WARNER, John **LONDON**
Lincoln Inn's Field 1684 – 1722

Mathematical instrument maker. He made a modified version of Coggeshall's rule.

WARNER, Samuel **LONDON**
Without Temple Bar 1731

Mathematical instrument maker.

WARREN, Richard **LONDON**
9 Arundel Place, Coventry Street 1850 – 1859

Box and ivory rule maker.

WARREN MANUFACTURING Co. **BRIGHTON**
38 Ship Street 1905

Rule makers.

WARWICK, Henry — **BIRMINGHAM**
22 Clissold Street — 1888 – 1893
21 Great Tower Street — 1908 – 1915

Rule maker.

WATKINS, F. (& SMITH) — **LONDON**
5 Charing Cross Road — 1747 – 1784
WATKINS, J. & W. (& Co.)
5 Charing Cross Road — 1784 – 1798
WATKINS, J.
5 Charing Cross Road — 1798 – 1810
WATKINS & HILL
5 Charing Cross Road — 1819 – 1856

Mathematical, optical and philosophical instrument makers. The business was continued by ***Elliott Bros***.

WATKINS, William — **BRISTOL**
12 St. Michael's Hill — 1803 –1811
13 (21) Clare Street — 1812 – 1813
16 St. Augustine's Back — 1814 – 1832

Mathematical, optical and nautical instrument maker.

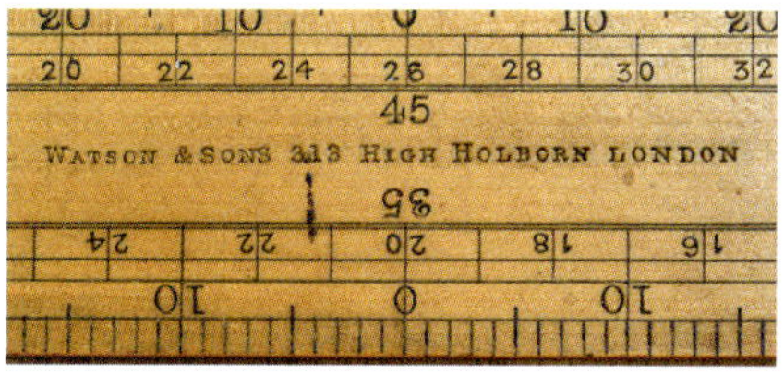

WATSON, W. (& Son) — **LONDON**
313 High Holborn — 1869 – 1914 >

Opticians.

WATSON, Robert — **BLACKBURN**

Unidentified. The name has been recorded on a Routledge rule. He may have been a retailer or an owner.

WATT, James — **GLASGOW**
In the College — 1757 – 1771
Saltmarket — 1759 – 1764
Trongate — 1764 – 1774
BIRMINGHAM
Soho Works — 1774 – 1800

In 1755, James Watt spent a year in London working for the mathematical instrument makers John Morgan and John Neal learning his trade before setting up in Glasgow, where he was instrument maker to the University. In 1774 he went into partnership with Matthew Boulton and moved to Birmingham where they developed their famous steam engine. Whilst in Birmingham Watt designed the slide rule that is now known as the Soho rule.

WATT, James — **GLASGOW**
(407) (179) Argyll Street — 1827 – 1841
7 South Albion Street — 1837 – 1840
3 and 11 St. Enoch Square — 1841 – 1853
16 Croy Place — 1854 – 1861
11 Turner's Court — 1862 – 1863

WATT & BLAIR
1864 >

Surveying chain and measuring line maker.

WATTERS, Thos. — **LEICESTER**

Unidentified.

WATTS, E. R. — **LONDON**
5 Northampton Road — < 1861
123 Camberwell Road — < 1899 – ca. 1945

1861: Mathematical instrument maker. 1913: Makers of surveying, drawing, optical, meteorological, philosophical and scientific instruments and drawing office supplies.

WEBB, Isaac — **LONDON**
1668 – 1704

He supplied rules to ***Joseph Silvester***.

WEBB, Thomas — **BRISTOL**
Earl Street, — 1793 – 1832
WEBB, Ann
Earl Street — 1830
WEBB, G.
3 Earl Street — 1852

Rule and gauging instrument makers. Thomas's business was taken over by ***Joshua Gwillym***.

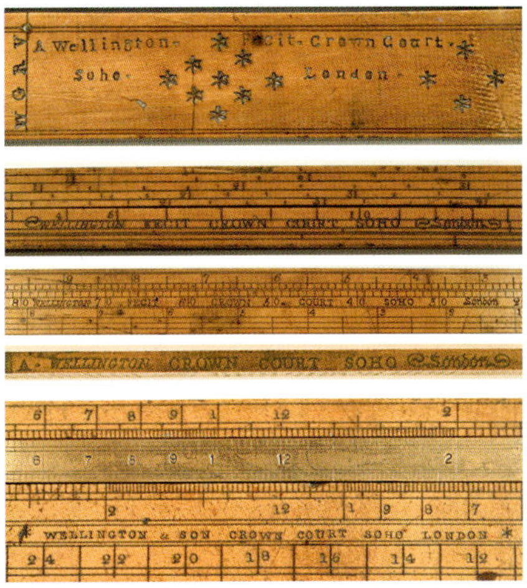

WELLINGTON, Alexander — **LONDON**
20 Crown Court, Wardour Street, Soho — 1784 – 1812
WELLINGTON, Mary, & Son
Crown Court, Soho — 1813 – 1815
WELLINGTON, Mary
20 Crown Court, Soho — 1816 – 1827

Mathematical instrument and rule makers, opticians. Alexander was apprenticed to ***James Search***. Between 1818 and 1827 ***Fitch & Jones***, followed by ***John Jones***, worked from the same premises. The businesses of both Jones and Wellington were taken over by ***Laban Cook***.

WENMAN, William — **LONDON**
10 Snow's Fields, Borough — 1832 – 1834

Rule maker.

WEST, A. & Partners — **LONDON**
(15–17) 91-98 Petty France — 1888 – 1953>
and 36 Broadway

Drawing office suppliers.

WEST, Francis — **LONDON**
Drury Lane — 1822 – 1828

(83) (92 & 93) Fleet Street 1829 – 1852

WEST, F. L.

39 Southampton Street 1848 – 1885

and 31 Cockspur Street 1859 – 1845

Opticians, mathematical and philosophical instrument makers.

WESTON, George **WOLVERHAMPTON**

Worcester Street 1833 – 1850

WESTON, James

30 Temple Street 1853 – 1860

WESTON, William

Worcester Street 1874

7 Gatis Street 1878 – 1888 >

Rule makers.

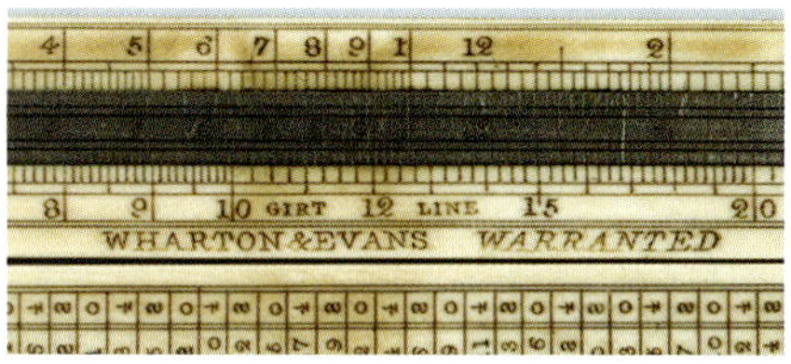

WHARTON & EVANS **BIRMINGHAM**

98 Newhall Street 1858

6 Dean Street 1867 – 1877>

Merchants and factors. They were listed as customers of ***John Rabone & Sons***.

WHIBLIN, John

1669

He is known to have made scales and Gunter's scales.

WHITE, Matthew **MANCHESTER**

3 Baptist Street, Sharp Street 1861

Rule maker.

WHITE, W. B. & Sons

Unidentified but probably a retailer or specialist supplier to the textile industry. The name is recorded on a boxwood caliper rule also marked "Colne Looms, Dobbies, Forks and Temples".

WHITE & BATES **BIRMINGHAM**

< 1858 >

Merchants.

WHITTAM (& Co.) **MANCHESTER**

25 Islington Street 1875

(148) Chapel Street, Salford 1880 – 1890

Dawson Croft Mill, Greengate 1899

Dickinson Street, Salford 1911 >

Steel rule makers.

WHITEHOUSE, William **WOLVERHAMPTON**

Summer Lane 1851

Rule maker.

WHITTINGHAM, William — **WOLVERHAMPTON**
43 Stafford Street — 1793
WHITTINGHAM, Thomas
27 Wardster Street — 1802
Church Lane — 1833 – 1839
Bond Street — 1841 –1842
Snowhill — 1845 – 1850

Rule makers.

WHITWORTH, Sir J., & Co. — **MANCHESTER**
Openshaw — 1833 –1887

Machinists, engineers and tool makers. The inventor of the Whitworth screw thread and the first mass produced taps and dies.

WILDBOARE, Tobias
1683

He is known to have sold rules.

WILKINSON, Thomas — **BRADFORD**
< 1877 >

Designer of a slide rule for engineers and the textile industry.

WILKINSON, ROWLATT & Co. — **LONDON**
62 Barbican — < 1818 >

Merchants. They would not have made the rules that bear their name.

WILLIAMS & HAYDON — **LONDON**
61 Queen Street, Cheapside — < 1841 >

Merchants and factors. They would not have made the rules that bear their name.

WILLIAMSON, John, & Co. Ltd. — **LONDON**
42 Gerard Street
Drury Lane

Unidentified but probably retailers or specialist suppliers to the tailoring trade. The name has been recorded on tailors' rules.

WILSON, Richard — **LONDON**
Within Newgate — 1671 – 1672

Ironmonger. He is known to have sold rules.

WIL(L)SON, William — **BIRMINGHAM**
9 Mott Street — 1845 – 1855
43 (Lower) Brearley Street — 1858 – 1868
331 Summer Lane — 1860 – 1862
WILLSON, Alfred
(31) (32) St. Paul's Square — 1875 –1876
68 Carver Street — 1882

Rule makers.

WINDRED, Henry Barnard — **LONDON**
7 Walnut Tree Walk — 1891 – 1895

Rule maker.

WINSOR & NEWTON — **LONDON**

Manufacturers and suppliers of artists' materials, their name is found on scale rules etc. The business started in 1832 and is still continuing today. They are unlikely to have made the rules that bear their name.

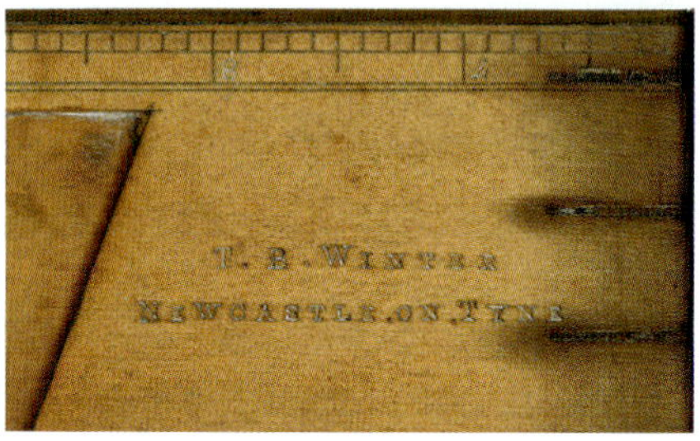

WINTER, T. B. (& Son) — **NEWCASTLE UPON TYNE**
21 Grey Street — <1858 – 1914 >

Opticians.

WITHNOL, William — **BIRMINGHAM**
20 Ann Street — 1777

Rule maker. The successor to ***Frost & Withnol.***

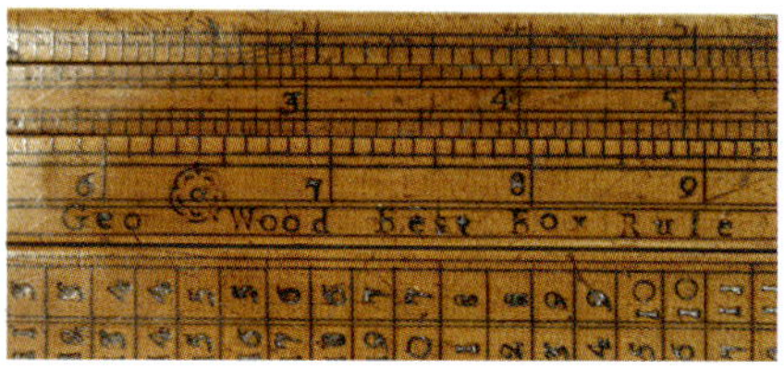

WOOD, George — **WOLVERHAMPTON**
1761

WOOD, Thomas
Brickhill Lane — 1767 – 1770

Rule makers. The address is also given as Brick Kiln Lane.

WOOD, George S. (& Co.) — **LIVERPOOL & MANCHESTER**
Prescot Street, Liverpool — 1847 – 1851
20 Lord Street, Liverpool — 1878 – 1900
24 Market Street, Manchester — 1890 – 1900
WOOD ABRAHAM'S
20 Lord Street, Liverpool — 1900 – 1935
24 Market Street, Manchester — 1900 – 1935

Opticians and mathematical instrument makers. They took over the business of ***A. Abraham & Co.*** in 1878 and were listed in the trades directories as "Successors to A. Abraham & Co."

WOOD, Henry Joseph — **LONDON**
185 Oxford Street — 1890 – 1895
418 Oxford Street

Optician.

WOOD, John — **BRISTOL**
Sign of Bell, Hillgrove Street — 1820 – 1824

Rule and gauge maker. The business was taken over by ***William Fowler.***

WOOD & LORT — **BIRMINGHAM**
1750 – 1760

WOOD, Henry
29 Great Charles Street — 1777 –

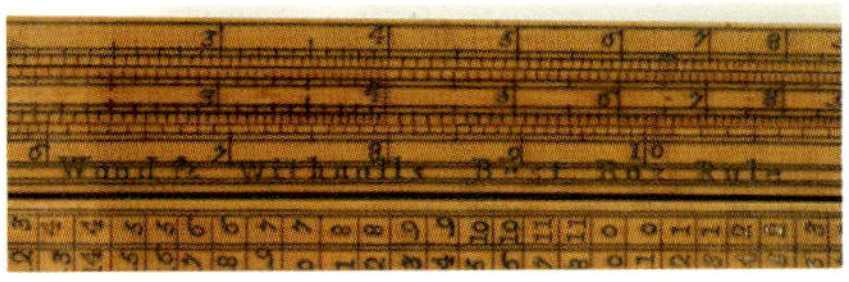

WOOD, John
16 London-'prentice Street — 1777
WOOD & WITHNOL
20 Great Charles Street — 1775 – 1781

Rule makers. Wood & Lort were succeeded by ***Frost & Withnoll*** who, in 1775, became Wood & Withnoll. Both Henry Wood and ***Frost*** are listed in the directories in 1777 in Great Charles Street.

WOOLF Bros. — **LONDON**
8 Stoney Lane, Houndsditch — 1910
135 Kingsland Road — 1913 – 1924 >

Rule makers.

WORBEY, Thomas & George, & Co. — **LONDON**
229A Camberwell Road — 1910

Rule makers.

WORMALL, Henry, & Co. — **SHEFFIELD**
47 Eyre Lane — 1876

English and foreign rule makers.

WORSLEY, S., & Sons — **BIRMINGHAM**

Unidentified but likely to be retailers. The name appears on a cloth measuring rule.

WORTHINGTON & ALLAN — **LONDON**
196 Piccadilly — 1821 – 1834
WORTHINGTON, Nathaniel
196 Piccadilly — 1835 – 1851

Opticians. They took over the business of ***Matthew Berge***.

WRIGHT, Alfred — **SHEFFIELD**
Spring Works, Clough Road — 1883 – 1890

Rule maker.

WRIGHT, George — **LONDON**
Clerkenwell — 1736

Rule maker. He is possibly the predecessor to Gabriel Wright of ***Gregory & Wright***.

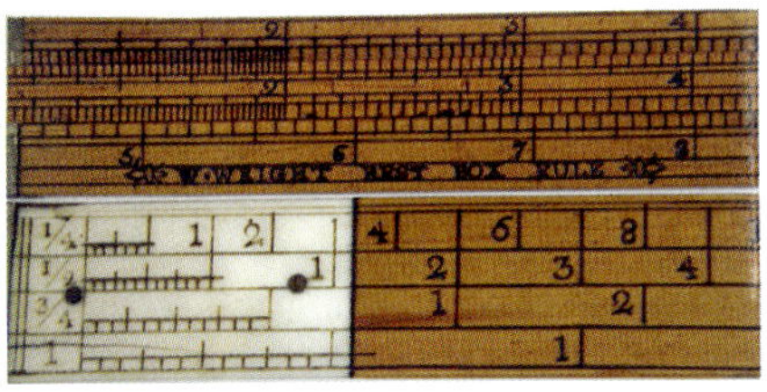

WRIGHT, John — **BIRMINGHAM**
(72) (167) Clifton Road — 1888 – 1904

Rule maker.

WRIGHT, Thomas — **LONDON**
Orrery & Globe, Fleet Street — 1718 – 1747

Mathematical instrument maker and toyman. He is known to have made Coggeshall rules and a waywiser.

WRIGHT, W. **MANCHESTER**
Glossop

Unidentified but unlikely to be a maker. The name has been recorded on a mill gear wheel sector.

WRIGHT, W. H. **LONDON**
40 Brompton Road 1870–1924 >

Saddler. He is unlikely to have made the rules that bear his name.

WRIGHT, William **BIRMINGHAM**
46 Worcester Street 1785 – 1788
14 Philip Street 1791 – 1801
Queen Street 1805

Rule maker in box and ivory.

WYNN, W. & C. **BIRMINGHAM**
Caroline Street 1797 – 1812
Suffolk Street 1813 – 1871
Commercial Street 1872 – 1866
WYNN TIMMINS (Ltd.)
Commercial Street 1866-1969

Birmingham tool and "steel toy" makers. The business was established in 1797 and continued until the mid-20th century. In 1866 W. & C. Wynn combined with ***Richard Timmins*** to form Wynn Timmins. For a history of the firm see Rees, Jane, *A Pattern Book of Tools and Household Goods,* Early American Industries Assocoation, South Dartmouth, MA, U.S.A., 2006.

YATES, Henry **BIRMINGHAM**
Salop Street 1855
129 Cheapside 1860

Rule maker.

YEFF, Robert **BRISTOL**
1693 – 1720

Unidentified but probably a maker. The name has been recorded on a Gunter's scale.

YOUNG, Mrs Elizabeth **BIRMINGHAM**
31 Masshouse Lane 1875

Rule maker.

Index

Index of names

Names in the Checklist of rule makers and dealers and in the Appendices are not included in this index.